COMPUTATIONAL MODELLING OF OBJECTS REPRESENTED IN IMAGES:
FUNDAMENTALS, METHODS AND APPLICATIONS III

PROCEEDINGS OF THE INTERNATIONAL SYMPOSIUM COMPIMAGE 2012, ROME, ITALY, 5–7 SEPTEMBER 2012

Computational Modelling of Objects Represented in Images: Fundamentals, Methods and Applications III

Editors

Paolo Di Giamberardino & Daniela Iacoviello
Sapienza University of Rome, Rome, Italy

R.M. Natal Jorge & João Manuel R.S. Tavares
Faculdade de Engenharia da Universidade do Porto, Porto, Portugal

CRC Press
Taylor & Francis Group
Boca Raton London New York Leiden

CRC Press is an imprint of the
Taylor & Francis Group, an **informa** business

A BALKEMA BOOK

Organized by

DEPARTMENT OF COMPUTER, CONTROL, AND
MANAGEMENT ENGINEERING ANTONIO RUBERTI

SAPIENZA
UNIVERSITÀ DI ROMA

CRC Press/Balkema is an imprint of the Taylor & Francis Group, an informa business

© 2012 Taylor & Francis Group, London, UK

Typeset by MPS Limited, Chennai, India
Printed and bound in Great Britain by CPI Group (UK) Ltd, Croydon, CR0 4YY

Published by: CRC Press/Balkema
                P.O. Box 447, 2300 AK Leiden, The Netherlands
                e-mail: Pub.NL@taylorandfrancis.com
                www.crcpress.com – www.taylorandfrancis.com

ISBN: 978-0-415-62134-2 (Hbk)
ISBN: 978-0-203-07537-1 (eBook)

Computational Modelling of Objects Represented in Images – Di Giamberardino et al. (eds)
© 2012 Taylor & Francis Group, London, ISBN 978-0-415-62134-2

Table of contents

VII

Preface

This book collects the contributions presented at the *International Symposium CompIMAGE 2012: Computational Modelling of Objects Represented in Images: Fundamentals, Methods and Applications*, held in Rome at the Department of Computer, Control and Management Engineering Antonio Ruberti of Sapienza University of Rome, during the period 5–7 September 2012.

This was the 3rd edition of *CompIMAGE*, after the 2006 Edition held in Coimbra (Portugal) and the 2010 Edition held in Buffalo (USA).

As for the previous editions, the purpose of *CompIMAGE 2012* was to bring together researchers in the area of computational modelling of objects represented in images. Due to the intrinsic interdisciplinary aspects of computational vision, different approaches, such as optimization methods, geometry, principal component analysis, stochastic methods, neural networks, fuzzy logic and so on, were presented and discussed by the expertise attendees, with reference to several applications. Contributions in medicine, material science, surveillance, biometric, robotics, defence, satellite data, architecture were presented, along with methodological works on aspects concerning image processing and analysis, image segmentation, 2D and 3D reconstruction, data interpolation, shape modelling, visualization and so on. In this edition, following the cultural and historical background of Italy, a particular session on artistic, architectural and urban heritages was included to put in evidence a wide and important field of application for vision and image analysis.

CompIMAGE 2012 brought together researchers coming from about 25 countries all over the World, representing several fields such as Engineering, Medicine, Mathematics, Physics, Statistic and Architecture. During the event, five Invited Lectures and 85 contributions were presented.

The Editors
Paolo Di Giamberardino & Daniela Iacoviello
(Sapienza University of Rome)
Renato M. Natal Jorge & João Manuel R. S. Tavares
(University of Porto)

Acknowledgements

The Editors wish to acknowledge:

- The Department of Computer, Control and Management Engineering Antonio Ruberti
- Sapienza University of Rome
- The Italian Group of Fracture – IGF
- The Consorzio Interuniversitario Nazionale per l'Informatica – CINI
- Sapienza Innovazione
- Zètema Progetto Cultura S.r.l
- Universidade do Porto – UP
- Faculdade de Engenharia da Universidade do Porto – FEUP
- Fundação para a Ciência e a Tecnologia – FCT
- Instituto de Engenharia Mecânica – IDMEC-Polo FEUP
- Instituto de Engenharia Mecânica e Gestão Industrial – INEGI

for the help and the support given in the organization of this Roman 3rd Edition of the *International Symposium CompIMAGE*.

Computational Modelling of Objects Represented in Images – Di Giamberardino et al. (eds)
© 2012 Taylor & Francis Group, London, ISBN 978-0-415-62134-2

Invited lectures

During *CompIMAGE 2012*, five Invited Lectures were presented by experts from four countries:

Current scenario and challenges in classification of remote sensing images
Lorenzo Bruzzone
University of Trento, Italy

Towards robust deformable shape models
Jorge S. Marques
Instituto Superior Técnico, Portugal

Fast algorithms for Tikhonov and total variation image restoration
Fiorella Sgallari
University of Bologna, Italy

Can make statistical inference on predictive image regions based on multivariate analysis methods?
Bertrand Thirion
INRIA, France

Incorporating global information into active contours
Anthony Yezzi
Georgia Tech, USA

Computational Modelling of Objects Represented in Images – Di Giamberardino et al. (eds)
© 2012 Taylor & Francis Group, London, ISBN 978-0-415-62134-2

Thematic sessions

Under the auspicious of *CompIMAGE* 2012, the following Thematic Sessions were organized:

Functional and structural MRI brain image analysis and processing
Organizers: Elisabetta Binaghi and Valentina Pedoia, Università dell' Insubria Varese, Italy

Materials mechanical behavior and image analysis
Organizer: Italian Group of Fracture – IGF
Vittorio Di Cocco and Francesco Iacoviello, Università di Cassino, Italy

Images for analysis of architectural and urban heritages
Organizer: Michela Cigola, Università di Cassino, Italy

Standard format image analysis and processing for patients diagnostics and surgical planning
Organizer: Mauro Grigioni, Department of Technology and Health–ISS, Italy

Scientific committee

All works submitted to *CompIMAGE 2012* were evaluated by the International Scientific Committee composed by experts from Institutions of more than 20 countries all over the World.

The Organizing Committee wishes to thank the Scientific Committee whose qualification has contributed to ensure a high level of the Symposium and to its success.

Lyuba Alboul	Sheffield Hallam University, UK
Enrique Alegre	Universidad de Leon, Spain
Luís Amaral	Polytechnic Institute of Coimbra, Portugal
Christoph Aubrecht	AIT Austrian Institute of Technology, Austria
Jose M. Garcia Aznar	University of Zaragoza, Spain
Simone Balocco	Universitat de Barcelona, Spain
Jorge M. G. Barbosa	University of PortoUniversity of Porto, Portugal
Reneta Barneva	State University of New York, USA
María A. M. Barrutia	University of Navarra, Spain
George Bebis	University of Nevada, USA
Roberto Bellotti	Istituto Nazionale di Fisica Nucleare, Italy
Bhargab B. Bhattacharya	Advanced Computing & Microelectronics Unit, India
Manuele Bicego	University of Verona, Italy
Elisabetta Binaghi	Università dell' Insubria Varese, Italy
Nguyen D. Binh	Hue University, Vietnam
Valentin Brimkov	Buffalo State College, State University of New York, USA
Lorenzo Bruzzone	University of Trento, Italy
Begoña Calvo	Universidad de Zaragoza, Spain
Marcello Castellano	Politecnico di Bari, Italy
M. Emre Celebi	Louisiana State University in Shreveport, USA
Michela Cigola	University of Cassino, Italy
Laurent Cohen	Universite Paris Dauphine, France
Stefania Colonnese	Sapienza University of Rome, Italy
Christos Costantinou	Stanford University, USA
Miguel V. Correia	University of Porto, Portugal
Durval C. Costa	Champalimaud Fundation, Portugal
Alexandre Cunha	California Institute of Technology, USA
Jérôme Darbon	CNRS – Centre de Matématiques et de Leurs Applic., France
Amr Abdel-Dayem	Laurentian University, Canada
J. C. De Munck	Department of Physics and Medical Technology, The Netherlands
Alberto De Santis	Sapienza University of Rome, Italy
Vittorio Di Cocco	University of Cassino, Italy
Paolo Di Giamberardino	Sapienza University of Rome, Italy
Jorge Dias	University of Coimbra, Portugal
Ahmed El-Rafei	Friedrich-Alexander University Erlangen-Nuremberg, Germany
José A. M. Ferreira	University of Coimbra, Portugal
Mario Figueiredo	Instituto Superior Técnico, Portugal
Paulo Flores	University of Minho, Portugal
Irene Fonseca	Carnegie Mellon University, USA
Mario M. Freire	University of Beira Interior, Portugal
Irene M. Gamba	University of Texas, USA
Antionios Gasteratos	Democritus University of Thrace, Greece
Carlo Gatta	Universitat de Barcelona, Spain
Sidharta Gautama	Ghent University, Belgium
Ivan Gerace	Istituto di Scienza e Tecnologie, Italy
J.F. Silva Gomes	University of Porto, Portugal
Jordi Gonzalez	Universitat Autonoma de Barcelona, Spain

Zeyun Yu   University of Wisconsin-Milwaukee, USA
Yongjie Zhang   Carnegie Mellon University, USA
Jun Zhao   Shanghai Jiao Tong University, China
Huiyu Zhou   Queen's University Belfast, UK
Djemel Ziou   Université de Sherbrooke (Québec), Canada
Alexander Zisman   Central Research Institute of Structural Materials-'Prometey', Russia

Computational Modelling of Objects Represented in Images – Di Giamberardino et al. (eds)
© 2012 Taylor & Francis Group, London, ISBN 978-0-415-62134-2

3D modelling to support dental surgery

D. Avola, A. Petracca & G. Placidi
Department of Health Sciences, University of L'Aquila, L'Aquila, Italy

ABSTRACT: In a previous work, we developed a general purpose framework to support the three-dimensional reconstruction, rendering and processing of biomedical images, **3D Bio-IPF**. In this paper we present a structured component of **3D Bio-IPF**, the plug-in **Implant**, to model customised dental implants on a three-dimensional representation of the oral cavity derived from diagnostic images. The proposed tool was tested on different cases and a result is reported. It has been proven it is very effective for dental surgery planning, implant design and positioning. Moreover, if integrated with a position indicator system and a numerically positionable drilling machine, it could be employed for semi-automatic surgery.

1   INTRODUCTION

The need to have detailed high-resolution visual information of the tissues and organs of the human body has led to a growing research of techniques to highlight their morphological structures and physiological processes. The choice of the appropriate set of techniques depends on the specific diagnostic needs. However, the ability to represent in a suitable way the interleaved set of 2D slices, belonging to a given diagnostic exam, can greatly improve their reading and usage. For this reason, in the last years, there have been many efforts to provide *3D Computer Aided Diagnosis Systems* (3D CAD Systems) to support different kinds of activities such as 3D reconstruction and rendering starting from their 2D representation derived by a descriptive set of slices (Archirapatkave et al. 2011, Wu et al. 2010). Moreover, these systems allow skilled users (physicians, radiologists) to use a set of suitable functionalities (e.g. fly-around, fly-through, multiple-view) to support different analytical processes. Summarizing, 3D CAD Systems enable users to optimize their work maximising the analysis process, while reducing time.

In our previous work (Maurizi et al. 2009), we have developed a *3D Biomedical Image Processing Framework* (**3D Bio-IPF**), a general purpose multi-platform system to manage, analyse, reconstruct, render and process sequences of slices (i.e. a dataset) with different image formats, including DICOM (*Digital Imaging and COmmunications in Medicine*). The developed system has been designed according to the plug-in architecture principles by which different structured components (i.e. plug-ins) extend the core environment inheriting from it functionalities and features, while adding functionalities developed for a specific target. In this way it is possible to improve modularity and extensibility while minimizing support and maintenance costs. In our context, **3D Bio-IPF** makes available to any plug-in a set of *primitives* to support different aspects of the image processing including volumetric three-dimensional reconstruction and rendering of coherent datasets. This last aspect is becoming very important in clinical applications (in particular for surgery planning) where it is having a growing acceptance from medical community. Recently, we focused on improving the technical aspects of the mentioned system allowing the addition of different plug-ins. The further step is to select suitable medical applications where implement, by different plug-ins, specific CAD Systems.

This paper describes the plug-in **Implant**, the first developed and fully tested plug-in for **3D Bio-IPF**. It is designed to model customised dental implants on a 3D reconstruction of the oral cavity from a diagnostic dataset. The used volume rendering algorithm is similar to those used in (Meissner et al. 2000, Reider et al. 2011). It is important to underline the role of the volumetric 3D rendering within the medical imaging field. Until not long ago, 3D reconstruction of anatomical structures from image tomography was of poor quality compared to the original set of two-dimensional images collected from the diagnostic scanner. In particular, the main criticism concerned the reconstruction process where the 3D transformation and rendering steps caused a significant loss of the morphological details.

Recent advancements in 3D methodologies are due to volume rendering algorithms which tend to maintain the informative content of the source images during the rendering process (Smelyanskiy et al. 2009). Obviously, this advantage is achieved at the expense of a high computational complexity that forces the developers to carefully program the CAD Systems and to use a suitable hardware configuration. Fortunately, in recent years, the 3D oriented open source frameworks, packages and libraries allowed a wide spread of customised solutions for the implementation of efficient algorithms for volume rendering.

The paper is structured as follows. Section 2 discusses the related work on volume rendering applications in medical imaging. Both the role and the usefulness of the volumetric rendering are described and justification for our choices are provided. Section 3 presents an overview of the **3D Bio-IPF** architecture. Section 4 describes the proposed plug-in, **Implant**. Section 5 provides experimental results on a specific case study. Section 6 concludes the paper.

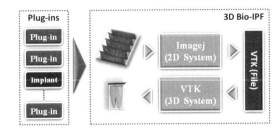

Figure 1.   Framework architecture.

2   RELATED WORK

Due to the huge literature on the volume rendering in medical imaging, we focus only on works which directly interested our approach. A particular attention regarded the study of the OsiriX project (Ratib et al. 2006), an image processing application dedicated to the DICOM images and specifically designed for navigation and visualization of multimodal and multidimensional images. Our purpose is to provide an alternative framework whose core is more oriented to the next challenges of the virtual and augmented reality in medical imaging (Liao et al. 2010). For this reason, we based our image processing engine on the ImageJ application (Burger et al. 2009), a completely programmable environment designed to support the existing image processing algorithms as well as the interaction of different types of acquiring devices and exchanging protocols. We have also considered the CEREC technology (CEREC 2012) to compare our functionalities, including usability, user interfaces and technical features, with ones of a real practical and advanced CAD system.

A remarkable point of view related to the volume visualization on different medical imaging fields is reported in (Zhang et al. 2011), where the authors make a comprehensive overview of the rendering algorithms highlighting the importance of these techniques for surgery. An interesting basic work is (Jani et al. 2000). Here the authors explored the volume rendering approaches applied to different investigation tasks.

Another interesting work is described in (Kuszyk et al. 1996) where the authors show both surface rendering and volume rendering applied to CT images for 3D visualization of the skeletal. Although in the last years both imaging techniques and rendering algorithms reserved other improvements, the clarifications made within their paper have to be considered milestones. In particular, on morphological structures similar to ones that made up the oral cavity, the authors showed that volume rendering is considered a flexible 3D technique that effectively displays a variety of skeletal pathologies, with few artifacts. This last factor is of great importance in dental implantology where anatomical features surrounding each tooth can represent a critical aspect. All the mentioned aspects have been considered both during the improvement of our framework and during the development of the **Implant** plug-in.

3   3D BIO-IPF ARCHITECTURE

This section describes an overview of the **3D Bio-IPF** architecture. As shown in Figure 1, it supports a plug-in strategy by which different developers, in parallel way, can dynamically extend the core environment with ad-hoc components according to specific image processing needs. The basic feature is that every plug-in inherits from the core environment the access to all available functionalities including those implemented by other plug-ins. The working of the **3D Bio-IPF** is based on a simplified *pipeline* methodology in which the *output* of a system provides the *input* for another. Regardless a specific approach, the three-dimensional reconstruction of an object, starting from a set of 2D representations, must always involve the following two logical steps:

− Step 1: a dataset containing a coherent collection of 2D images has to be filtered. The resulting dataset has to be processed to extract the information needed to create a spatial mapping between data coming from the 2D images and the 3D model. The result is a new set of information (i.e. volume information) that will provide the 3D volume rendering of the object;

− Step 2: a rendering techniques has to be adopted on the volume information to provide a related visual 3D interpretation.

In our architecture the first step is performed by a system based on the ImageJ application (ImageJ 2012), while the second one is achieved by another system based on the Visualization ToolKit (VTK) library (VTK 2012). Note that the entire framework has been designed through open source technologies.

The communication process between the two systems is accomplished by a support file in VTK format. The structure of the file is quite complex (VTK-File 2012). It contains, within specific structures (e.g. *dataset structure* and *dataset attributes*), the required geometrical and topological information to build the 3D model reconstruction of an object. The main aspect of the system (shown in the next two sub-sections) regards both the process used by ImageJ to derive and store (within the VTK file) the mentioned information, and the process used by VTK to retrieve and use (from the VTK file) them. For completeness, we point out that the framework uses other VTK files to perform

Figure 3. Steps of the dental implant surgery.

a   b

Figure 2. Usage of a dental implant (a) and its structure (b).

the switch between the different spatial visualization (from 2D to 3D and vice versa), when necessary.

3.1 ImageJ

ImageJ is a public domain multi-platform image analysis and processing application developed using one of the most popular programming languages: Java. ImageJ has been chosen as result of a deep comparison of its features with similar applications. Summarizing, it has been considered compliant to our purposes for the following strength points:

- *user community:* it has a large and knowledge-able worldwide developers community supporting a growing set of research applications;
- *runs everywhere:* it has been written in Java that implies it is multi-platform;
- *toolkit:* it has libraries that allow the development of web-services oriented architecture;
- *image enhancement:* it implements the most relevant and recent image filtering algorithms for different image formats;
- *speed:* it is the world's fastest pure Java image processing program. It can filter a 2048x2048 image in 0.1 seconds (on middle level hardware configurations), corresponding to 40 millions of pixels per second.

Moreover, ImageJ has a virtually infinite growing collection of native filters/operators, macros and plug-ins that allow developers to face research issues using different attractive solutions.

We used and customised the set of 2D management functionalities provided by the native packages of ImageJ to extract the geometrical and topological features useful to create the volume information. The system starts from the information (e.g. sequential number, thickness, interval) contained within the DICOM source images (the header) to build an initial empty space where the informative content of each 2D image is linked. This operation is performed by considering both colour patterns of the correlated portions of the source images and dimension of the 3D empty space. In this way, the system creates the descriptive functions to perform the transposition activity. Each function can be defined as a *primitive* able to identify mathematically the mentioned patterns.

3.2 VTK library

The VTK library is an open source, freely available cross-platform for 3D computer graphics, image processing and visualization. It consists of a C++ class and several interpreted interface layers including the Java language. Like ImageJ also VTK is supported by a wide community that ensures a professional support and a growing availability of heterogeneous solutions.

The library supports a wide variety of visualization algorithms including: scalar, vector, texture and volumetric methods; and advanced modeling techniques such as: implicit modeling, polygon reduction, mesh smoothing, cutting and contouring. In our context, it has been chosen to support the rendering process of the volume information. Usually, in medical image area, this aim is reached by the following techniques: *MultiPlanar rendering* (MPR), *Surface Rendering* (SR) and *Volume Rendering* (VR). We have adopted the last technique, favouring the visualization of the entire volume transparence of the object. In particular, starting from the previously computed volume information, our system processes the correlated set of pixels, according to the three planar spaces, and builds the relative volumetric picture element (voxel). In this way, each voxel represents the set of pixels, suitably arranged, coming from the different orthogonal planes.

4 IMPLANT PLUG-IN

Implantology is a complex field of the dentistry dealing with the replacement of missing teeth with synthetic roots (the implants), anchored to mandibular or to maxillary bone, on which are mounted mobile or fixed prosthesis for the restoration of the masticatory functions. Implant solutions help to maintain healthy teeth intact since their installation does not require the modification of nearest healthy teeth. Figure 2a shows a typical replacing of a molar with a common implant, while Figure 2b shows the basic components of the prosthesis (implant/root, abutment/support and prosthesis/crown).

A dental replacement takes between 45 and 90 minutes under local anesthesia. The procedure requires first the incision of the gum. Then, by a suitable set of surgical drills, the cavity inside the bone is perforated. Subsequently, within the just created implant site, the synthetic root is carefully screwed. Finally, the gum over the implant is sutured and a temporary prosthesis (mobile of fixed) is applied. After a technical time that allows the physiological osseointegration process (about 6-12 weeks) a new little surgery to fix the definitive prosthesis (e.g. single crown, bridge) over the implant is performed. Figure 3 shows the whole process.

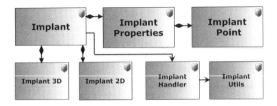

Figure 4. Class diagram of the Implant plug-in.

A synthetic root is an object made in bio-compatible material (usually titanium) having the shape of cylindrical or conical screw. It can have different length and diameter according to several factors such as bone structure and space availability.

The main purpose of the **Implant** plug-in is to support all the aspects of planning of the surgery. This task can be accomplished by using an accurate virtual three-dimensional reconstruction of the oral cavity. In this way, it is possible to determine, in advance and with greater precision, shape, length, diameters, position and orientation of the implants that have to be inserted paying attention to avoid nerve ending and important blood vessels rupture. Moreover, it is possible to reduce pain, bleeding and surgery execution time, while improving the recovery time for the patient after the intervention. It is also possible to perform simulation sessions in which to test different implant solutions. Finally, **Implant** information could be also used to drive a numerically controlled high precision drill to perforate automatically the bone.

4.1 Technical description

Implant has been designed to directly interact with the 3D system of the framework. In this way, the 3D environment (managed from the VTK library) will be enriched with new functionalities specifically conceived to support implantology. For completeness, follows a brief discussion of the basic functions:

– *add*, *update*, *remove*, *lock* and *unlock*: manage a single virtual implant performing the corresponding functionalities within the 3D volumetric representation of the oral cavity;
– *width* and *colour*: set the size and the specific colour of a single implant;
– *point₁* and *point₂*: set respectively upper and lower coordinates (i.e. respect the three-dimensional space (x, y, z)) of a single implant.

To provide an overview of the **Implant** implementation Figure 4 shows a really simplified diagram of the related main classes, where:

– *Implant:* allows creation and management of the main features of an implant;
– *Implant Properties:* allows the management of the physical features of an implant;
– *Implant 2D:* allows the management of the orthogonal projections of the implant on the axial, coronal and sagittal sections;

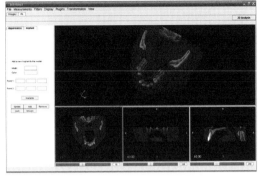

Figure 5. 3D reconstruction and planar sections of the 3D model.

– *Implant 3D:* allows the management of the three-dimensional reconstruction of the implant on the 3D oral cavity model;
– *Implant Handler:* main class of the plug-in, allows its integration with the whole framework;
– *Implant Point:* allows the management of the spatial information related to an implant;
– *Implant Utils:* allows the management of the operations related to the XML file.

The implant is represented, on the 3D model of the oral cavity, as a cylindrical or truncated cone solid object. Its physical features can be changed according to specific needs. The plug-in allows user to specifically set the coordinates of the mounted virtual implant simply managing the spatial parameters (*point₁* and *point₂*) which are automatically referred to the three-dimensional space containing the volume rendering of the patient oral cavity.

Once positioned the implant, the previous operations (e.g. *lock*, *unlock*, *color changing*) can be performed. It should be observed that the chosen functionalities are very important in a 3D real analysis session. For example, the *lock* function anchors the virtual implant on the 3D reconstructed model after it has been correctly positioned, thus preventing for involuntary modification when positioning other implants and, in the same time, saving implant spatial information in a XML file (see below). To make another example, the *colour* distinction between different implants can allow easier distinction and analysis.

As shown in the screenshot presented in Figure 5, during the analysis of the implant positioning, the system allows to display both the 3D model (superior part of the image) and three orthogonal sections (axial, bottom left, coronal, bottom central, sagittal, bottom right). This aspect, as in other medical fields, has a huge importance during surgery planning. All the information related to different implants are stored on a XML file (XML 2012) containing different sets of information both patient-specific (e.g. patient id) and technical (e.g. *implant type*, *implant dimensions*, *point₁*, *point₂*). Technical information, referred to the object coordinate system, are very useful both to

Table 1. Case study - DICOM image features.

Format	Num. Images	Resolution	Num. Bits
DICOM	38	512×512	16

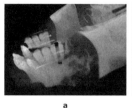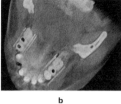

a   b

Figure 6. Different visual prospectives during the implant design for: (a) a molar, (b) multiple teeth. Note that, to simplify implants positioning, the model has been shown upside down.

choose the correct implants and to drive an automatic drilling machine (if used).

5 EXPERIMENTAL RESULTS

We have tested the three-dimensional reconstruction process on different sources images coming from 30 patients. Here, we focus only on a specific case study. Table 1 summarizes the reported dataset characteristics. It represents a Computed Tomography (CT) examination of the superior oral cavity.

During the 3D reconstruction, geometrical and topological features of the DICOM images (arranged within the logical stack of the ImageJ application) are computed and stored into a VTK file (to execute this operation, the process takes few seconds and some supporting files), used as data input for the rendering process. The quality of the reconstruction process was evaluated through comparison measurements between the three sets of planar images and the reconstructed one, where 2D image zones and the related 3D area were analysed by a set of distance measurement tools (provided by ImageJ). The results showed a perfect superposition of the 2D images into the 3D model, where the reliability of the representation is strongly tied to the number and quality of the source images. Besides, we have also analysed the efficiency and the accuracy of the system in positioning the implants in the 3D model. In Figure 6 two different visual prospectives of implants insertion are shown.

Also in this case we have tested the quality of the measurements related to the implants positioning. The empirical results, show that both volume information and rendering process have faced with success the corresponding tasks.

6 CONCLUSION

In this paper we have presented **Implant**, a linked structured component to model customised dental implants on a 3D reconstruction of the oral cavity derived from diagnostic tomography. It has been structured as a plug-in of **3D Bio-IPF**, a general purpose framework to support the 3D reconstruction, rendering and processing of biomedical images. Experimental tests shown that the system can perform the design activity easily, very fast and in a really accurate way.

The developed component presents several advantages for implantology: it provides a set of suitable functionalities to support implant design and positioning. Moreover, it allows skilled user to perform simulation sessions to test different implant solutions. Finally, the high resolution of the 3D reconstructed model allows to support all the aspects of prevention related to the surgery planning (e.g. pain, bleeding and execution time reduction).

We focused different next targets of our plug-in. One of the closest is to use the virtual 3D model of the oral cavity to replace the common dental imprints (in chalk or resin). Another real interesting target is to interface **Implant** with a position indication tool and a numerically positionable drilling machine to perform automatically, in a supervised way, the bone perforation. In this way, it could be possible to reduce errors and, consequently, to improve the recovering time for the patient after surgery.

REFERENCES

Archirapatkave, V. & Sumilo, H. & See, S.C.W. & Achalakul, T. 2011. GPGPU acceleration algorithm for medical image reconstruction. In *Proceedings of the 2011 IEEE 9th International Symposium on Parallel and Distributed Processing with Applications*, ISPA '11. IEEE Computer Society Publisher, 41–46, USA.

Burger, W. & Burge, M.J. (Eds.) 2009. Principles of digital image processing: fundamental techniques. *Book Series: Undergraduate Topics in Computer Science*. Springer Publishing Company, Incorporated, 1(2): 261.

CEREC 2012. http://www.cereconline.com/cerec/software. html.

ImageJ 2012. http://rsbweb.nih.gov/ij/index.html.

Jani, A.B., Pelizzari, C.A., Chen, G.T.Y., Roeske, J.C., Hamilton, R.J., Macdonald, R.L., Bova, F.J., Hoffmann, K.R. & Sweeney, P.A. 2000. Volume rendering quantification algorithm for reconstruction of CT volume-rendered structures: I. cerebral arteriovenous malformations. *IEEE Transactions on Medical Imaging*. IEEE Computer Society Publisher, 19(1): 12–24.

Kuszyk, B.S. & Heath, D.G. & Bliss, D.F. & Fishman, E.K. 1996. Skeletal 3-D CT: advantages of volume rendering over surface rendering. *International Journal on Skeletal Radiology*. Springer Berlin/Heidelberg Publisher, 25(3): 207–214.

Liao, H. & Edwards, E. & Pan, X. & Fan Y. & Yang G.-Z. (Eds.) 2010. Medical imaging and augmented reality. In *Proceedings of the 5th International Workshop on Medical Imaging and Augmented Reality*, MIAR '10. Springer Publisher, Lecture Notes in Computer Science (LNCS), 6326(2010):570, Beijing, China.

Maurizi, A. & Franchi, D. & Placidi G. 2009. An optimized java based software package for biomedical images and volumes processing. In *Proceedings of the 2009 IEEE International Workshop on Medical Measurements and Applications*, MEMEA '09. IEEE Computer Society Publisher, 219–222, USA.

Meissner, M. & Huang, J. & Bartz, D. & Mueller, K. & Crawfis, R. 2000. A practical evaluation of popular volume rendering algorithms. In *Proceedings of the 2000 IEEE International Symposium on Volume Visualization*, VVS '00. ACM Publisher, 81–90, New York, NY, USA.

Ratib, O. & Rosset, A. 2006. Open-source software in medical imaging: development of OsiriX. *International Journal of Computer Assisted Radiology and Surgery*. Springer Berlin/Heidelberg Publisher, 1(4): 187–196.

Rieder, C. & Palmer, S. & Link, F. & Hahn, H.K. 2011. A shader framework for rapid prototyping of GPU-based volume rendering. *International Journal on Computer Graphics Forum*. Blackwell Publishing Ltd, 30(3): 1031–1040.

Smelyanskiy, M. & Holmes, D. & Chhugani, J. & Larson, A. & Carmean, D.M. & Hanson, D. & Dubey, P. & Augustine, K. & Kim, D. & Kyker, A. & Lee, V.W. & Nguyen, A.D. & Seiler, L. & Robb, R. 2009. Mapping high-fidelity volume rendering for medical imaging to CPU, GPU and many-core architectures. *IEEE Transactions on Visualization and Computer Graphics*. IEEE Educational Activities Department Publisher, 15(6): 1563–1570, Piscataway, NJ, USA.

VTK 2012. http://www.vtk.org/.

VTK File Formats 2012. http://vtk.org/VTK/img/file-formats.pdf.

Wu, X. & Luboz, V. & Krissian, K. & Cotin, S. & Dawson, S. 2010. Segmentation and reconstruction of vascular structures for 3D real-time simulation. *International Journal on Medical Image Analysis*. Elsevier B.V. Publisher, 15(1): 22–34.

XML 2012. http://www.w3.org/XML/.

Zhang, Q. & Peters, T.M. & Eagleson, R. 2011. Medical image volumetric visualization: Algorithms, pipelines, and surgical applications. *Book Title: Medical Image Processing*. Springer Publisher, 291–317, New York, USA.

Detection system from the driver's hands based on Address Event Representation (AER)

A. Ríos
Face Recognition and Artificial Vision Group (FRAV), ETSI Informática, Universidad de Rey Juan Carlos

C. Conde, I. Martin de Diego & E. Cabello
Departamento de Arquitectura y Tecnología de Computadores, ETSI Informática, Universidad de Rey Juan Carlos

ABSTRACT: This paper presents a bio-inspired system capable of detecting the hands of a driver in different zones in the front compartment of a vehicle. For this it has made use of a Dynamic Vision Sensor (DVS) that discards the frame concept entirely, encoding the information in the form of Address-Event-Representation (AER) data, allowing transmission and processing at the same time. An algorithm for tracking hands using this information in real time is presented. Detailed experiments showing the improvement of using AER representation are presented.

1 INTRODUCTION

The slightest lapse of concentration, while driving, is one of the reasons that bring about the high percentage of fatal-traffic accidents. According to the DGT, the 36% of those fatal accidents took place on Spanish road in 2010 (DGT 2010). Manipulating any element of the vehicle during a tiny interval of the time may prevent people from having their hands in the proper position and, as a result, the risk of facing an accident is been indirectly assumed. That is why new automobiles are been built with security systems are been increasingly built in order to reduce the number of traffic accidents.

In this paper, it is presented an artificial vision system capable of monitoring and detecting the hands position on the steering wheel in teal time. To do this, it has been used a dynamic vision sensor (DVS) which encodes the information as AER. The Address-Event-Representation protocol allows asynchronous massive interconnection of neuromorphic processing chip. This protocol was first proposed in Caltech, in 1991, to solve the problem of massive interconnectivity between populations of neurons located in different chips (Sivillotti 1991).

AER is based on the transmission of neural information in the brain, by an electrical spike. Similarly, in the case of AER, the pixel activity on chip 1, along the time, is traduced to spikes or events that are sent one by one to the receptor chip using high-speed interchip buses. These pulses are arbitrated and the address of the sending pixel is coded on the fast digital asynchronous bus with handshaking. Chip 2 (or the receiver chip) decodes the arriving address and sends a spike to the corresponding neuron so that the original signal can be reconstructed (see Fig. 1).

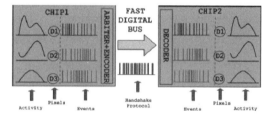

Figure 1. AER communication scheme (Gómez, F et al. 2010).

There are many applications of the AER technology in the area of artificial vision, but all of them are laboratory applications not tested in real environments. In (Camunas-Mesa et al. 2011) presents a 2-D chip convolution totally based on pulses, without using the concept of frames. In (Jimenez-Fernandez et al. 2010) sensor is exposed tilt correction in real time using a layer of high-speed algorithmic mapping. Another application is explained in (Goìmez-Rodriìguez et al. 2010) which proposes a system for multiple objects tracking. On the other hand in (Linares-Barranco et al. 2007) is shown a control system visuo-motor for driving a humanoid robot. Another example visuo-motor control is presented in (Conradt et al. 2009) a Dynamic Vision Sensor for Low-Latency Pole Balancing.

The work presented in this paper is an application of AER technology designed for working in real world conditions, which takes a further step in bringing this technology to more complex environments.

A driver-hand supervising system based on frame artificial vision techniques has been presented in (Crespo et al. 2010) (McAllister et al. 2000). Video systems usually produce a huge amount of data that likely saturate any computational unit responsible for

1 bit	7 bits	7 bits	1 bit
NC	Y coordinate	X coordinate	Polarity

Figure 2. AE Address.

data processing. Thus, real-time object tracking based on video data processing requires large computational effort and is consequently done on high-performance computer platforms. As a consequence, the design of video-tracking systems with embedded real-time applications, where the algorithms are implemented in Digital Signal Processor is a challenging task.

However, vision systems based on AER generate a lower volume of data, focusing only on those which are relevant in avoiding irrelevant data redundancy. This makes it possible to design AER-tracking systems with embedded real-time applications, such as the one presented in this paper, which indeed, it is a task much less expensive, e.g. in (Conradt et al. 2009) (Gómez-Rodríguez et al. 2010).

The paper will be structured as follows. In section II, it is explained the DVS and the labeling events module. In section III, hands-tracking algorithm is described. The experimental results of the presented algorithm on real test are in section IV. The last, section 5 concludes with a summary.

2 SYSTEM HARDWARE COMPONENTS

The data acquisition is implemented by two devices, the sensor and the USBAERmini2 DVS, presenting the information encoded in AER format.

The DVS sensor (silicon retina) contains an array of pixels individually autonomous real-time response to relative changes in light intensity by placing of address (event) in an arbitrary asynchronous bus. Pixels that are not stimulated by any change of lighting are not altered, thus scenes without motion do not generate any output.

The scene information is transmitted event to event through an asynchronous bus. The location of the pixel in the pixel array is encoded in the address of events. This address, called AE (Address Event), contains x,y coordinate of the pixel that generated the event. DVS sensor has been used with an array of 128x128 pixels, so is needed 7 bits to encode each dimension of the array of pixels. It also has a polarity bit indicating the sign of contrast change, whether positive or negative (Lichtsteiner et al 2008).

As shown in (Fig. 2), the direction AE consists of 16 bits, 7 bits corresponding coordinates X, Y coordinate 7 bits, one bit of polarity and a bit NC.

The AE generated is transmitted to USBmini2 on a 16-bits parallel bus, implementing a simple handshake protocol. The device USBAERmini2 is used to label each event that is received from the DVS with a timestamp.

The main elements are FX2 and FPGA modules. The FX2's 8051 microcontroller is the responsible for

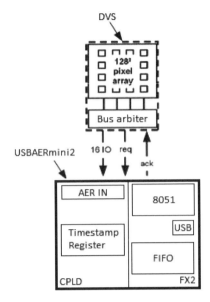

Figure 3. Schematics of the data collection system architecture.

setting the "endpoints" of the USB port, in addition to receiving and interpreting commands from the PC. On the other hand the CPLD is responsible for the "handshaking" with the AER devices, connected to the ports, and for the reading and writing events in the FIFO's FX2 module.

CPLD module has a counter that is used to generate the "timestamp" to label the events. Using only 16 bits is not enough, as only $216 = 65536$ timestamps can be generated, which is the same as 65ms when using a tick of $1\mu s$. Using 32 bit timestamps consumes too much bandwidth, as the higher 16 bits change only rarely. To preserve this bandwidth, a 15 bit counter is used on the device side, another 17 bits (called wrap-add throughout this report) are later added by the host software for monitored events (Berner 2006).

Both devices have been set in a black box, which has three infrared lights, for easy installation of the system in different vehicles. We used this type of lighting to avoid damaging the driver's visibility in dark environments. The box has a visible light filter in front of the sensor to avoid sudden changes in the brightness outside.

3 HANDS TRACKING ALGOTIRHM

The proposed algorithm performs a permanent clustering of events and tries to follow the trajectory of these clusters. The algorithm focuses on certain user-defined regions called regions of interest (ROI), which correspond to certain parts of the front compartment of a vehicle (wheel, gearshift, ...). The events received are processes without data buffering, which can be assigned to a cluster or not, based on a distance criterion. If the event is assigned to a cluster, it will update

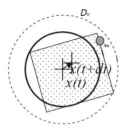

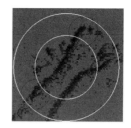

Figure 4.   Continuous clustering of AE.

Figure 5.   Wrong detection situation.

the values of the cluster, such as position, distance, and number of events (Litzenberger et al. 2008).

The performance of the algorithm can be described as follows:

a) When there is a new event belonging to one ROI, a cluster from the list of clusters is located whose distance from this center to the event is less than D_v ,for all clusters (see Fig. 4)

$$D = |x - x_e| < D_v \qquad (1)$$

b) If a cluster is found where the above condition is true, update all features accordingly.

c) If no cluster is found, create a new one (if possible) with the center x_e and initialize all parameters.

For the creation of new clusters should be borne in mind that the maximum number of clusters is two (hands) and in the R.O.I's wheel not to have a special situation detailed below.

The cluster update process is sketched in Fig. 3. Let x_e be the coordinate of an AE produced by the edge of a moving cluster. Let $x(t)$ be the original cluster center, then the new center-coordinate $x(t + dt)$ is calculated as:

$$x(t + dt) = x(t) \cdot \alpha + x_e \cdot (1 - \alpha) \qquad (2)$$

Where $0 < \alpha < 1$ is a parameter of the algorithm and dt is the timestamp difference between the current and the last AE's that were assigned to the cluster. This shifts the clusters center by a certain amount controlled by α, which is usually chosen near 1 to obtain a smooth tracking.

For speed calculation is performed similar to ensure small changes in the velocity vector:

$$x(t + dt) = x(t) \cdot \alpha + x_e \cdot (1 - \alpha) \qquad (3)$$

Previously explained that the algorithm focuses on certain R.O.I. Specifically, in one of them, corresponding to the wheel is sometimes presented a situation of special interest. This region of interest has the form of a circular ring and when the driver uses one hand to manipulate wheel the system detects two hands being wrong (see Fig. 5).

As shown in Fig 5, when the forearm is inserted fully into the ROI, there are two oblique parallel lines. To remedy this situation has been used pattern recognition

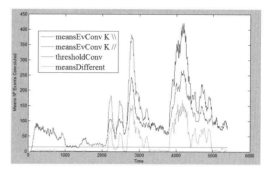

Figure 6.   Graphical values difference between the two convolutions.

by AER convolution. The AER convolution is based on the idea of matrix Y integrators (neurons). Every time an event is received, a convolution kernel is copied to the residents of that event in the array "Y". When a neuron (a cell of Y) reaches a threshold, it generates an event and is reset (Linares-Barranco et al. 2010) (Pérez-Carrasco et al. 2010).

Two convolution kernels have been used simultaneously, one for each arm (left "//", right "\\"). When either arm appears on scene, it will be registered a maximum in the average output of the convolutioned events, where the convolution kernel matches and a minimum (with respect to previous output) where the convolution kernel does not match. However, the output of the two convolution kernels, in a normal situation (both hands on the wheel), is very similar, not registering any maximum output.

Looking at the average of both convolutions, it is possible to identify the situation in which either arm is in scene and know with which arm it is been driving.

Concerning the graph (see Fig. 6), when the driving situation is normal, the mean number of events convolutioned is similar to both kernel, so that their difference is practically zero. When a forearm is detected on scene, the convolution kernel that coincides triggers the number of convolutioned events, while the convolution kernel which does not coincide is not able to reach so high levels. Therefore, when the difference between the means of both convolution kernel events exceeds a threshold, we can detect that there is a forearm on scene and know to which hand it belongs (see Fig. 7).

9

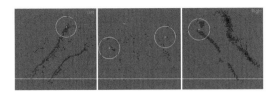

Figure 7. Left: Left forearm on scene. Center: Normal situation (both hands on the wheel) Right: Right forearm on scene

Table 1. Results of different tests.

	Time min:seg	Total n° of events	Means motion evt/seg	Effectiveness
Test 1	5:33	6836386	31194	89.545339%
Test 2	5:30	9557951	16872	81.008254%
Test 3	3:54	4075592	3588	75.460864%
Test 4	5:50	10993454	12525	80.400553%

4 RESULTS

There have been several different tests with different scenarios on a simulator where a subject has made a practice of driving while performing a capture with the system presented in this paper. In each capture the following parameters were obtained:

- Total exercise time (capture)
- Total number of events generated
- Means motion

The last parameter represents the average amount of movement they has occurred in the catch. As explained in section II, the sensor used is a motion sensor so if you over there have been capturing scenes with much movement, the number of events per second will be high. Thus, the greater number of events generated, the more information you have and the more accurate the calculations and operations of the algorithm are presented.

Tests have been carried out in a very realistic simulator. In each test have been used different types of driving scenarios. In the test one has made driving very dynamically, by an urban layout, with lots of curves where the driver has to make a lot of movement. The next test, the number two, In the next test, the number two, has been used a rally stage, which it has driven to a slightly higher velocity. In test three, the scenario used was a highway, where the movements made by the driver are very scarce. The last test has been developed in an urban ride, in which the driver was not found many obstacles and made movement has not been too high.

To calculate the system's effectiveness, it has been compared to the result that the system automatically gives the result that should really give, i.e. to compare the number of hands that detects the system in each region of interest (wheel and gearshift) with the number of hands the real situation in each of these areas at a time.

Section I of the paper indicated Event rate sensor saturation (1 M-Events per second) and Section II details the labeling of timestamp resolution is 1us. Therefore we must take into account the high speed processing system capable of generating large numbers of events in very small movements.

Scenes so with much movement (even very fast), it captures a large number of events per second, increasing the effectiveness of the system (see Tab. 1).

5 CONCLUSION

In this paper a bio-inspired system for driver hands tracking has been presented. The information considered is not based on frames, but Address Event Representation (AER), allowing the processing and transmission at the same time.

The work presented in this paper is an application of AER technology to a real problem with high constrains, which takes a further step in bringing this technology to a real environment. An embedded system could be designed with this application to be used in smart cars. This paper demonstrates the advantage of using AER systems in such scenarios (cabin of a vehicle) because it detects the small and fast movements that high-speed cameras are not able to notice.

Experimental results acquired in a realistic driving simulator are presented, showing the effectiveness of AER-processing applied to an automotive scenario.

As future-work plans, it will be carried out a comparison between both systems, the based AER and the based on frames, to get results, concerning the processing speed, amount of relevant information generate, efficiency...to improve the system performance that is proposed in this paper when little data is generated.

ACKNOWLEDGMENT

Supported by the Minister for Science and Innovation of Spain project VULCANO (TEC2009-10639-C04-04).

REFERENCES

The main figures of road accidents. Dirección General de Tráfico (DGT). Spain 2010. Available: http://www.dgt.es
M. Sivilotti, "Wiring considerations in analog VLSI systems with application to field-programmable networks", Ph.D. dissertation, Comp. Sci. Div., California Inst. Technol., Pasadena, CA, 1991
R. Crespo, I. Martín de Diego, C. Conde, and E. Cabello, "Detection and Tracking of Driver's Hands in Real Time," Progress in Pattern Recognition, Image Analysis, Computer Vision, and Applications, Vol. 6419, pp. 212–219, 2010.
McAllister, G., McKenna, S.J., Ricketta, I.W.: Tracking a driver's hands using computer vision. In: IEEE International Conference on Systems, Man, and Cybernetics (2000)

Conradt, J.; Berner, R.; Cook, M.; Delbruck, T., "An embedded AER dynamic vision sensor for low-latency pole balancing", *Computer Vision Workshops (ICCV Workshops), 2009 IEEE 12th International Conference on,* pp. 780–785.

F. Gómez-Rodríguez, L. Miró-Amarante, F. Diaz-del-Rio, A. Linares-Barranco, G. Jimenez, "Real time multiple objects tracking based on a bio-inspired processing cascade architecture" *Circuits and Systems (ISCAS), Proceedings of 2010 IEEE International Symposium on,* pp.1399–1402.

P. Lichtsteiner, C. Posch, and T. Delbruck, "A 128 × 128 120 dB 30 mW asynchronous vision sensor that responds to relative intensity change," *IEEE J. Solid-State Circuits,* vol. 43, pp. 566–576, Feb. 2008.

R. Berner, "Highspeed USB 2.0 AER Interface", Diploma Thesis, Institute of Neuroinformatics UNI – ETH Zurich, Department of Architecture and Computer Technology, University of Seville, 14th April 2006.

M. Litzenberger, C. Posch, D. Bauer, A.N. Belbachir, P. Schön, B. Kohn, and H. Garn, "Embedded vision system for real-time object tracking using an asynchronous transient vision sensor" *Digital Signal Processing Workshop, 12th – Signal Processing Education Workshop, 4th,* September 2006, pp. 173–178.

A. Linares-Barranco, R. Paz-Vicente, F. Gómez-Rodríguez, A. Jiménez, M. Rivas, G. Jiménez, A. Civit, "On the AER Convolution Processors for FPGA", *Circuits and Systems (ISCAS), Proceedings of 2010 IEEE International Symposium on,* pp. 4237–4240.

J. A. Pérez-Carrasco et al, "Fast vision through frameless event-based sensing and convolutional processing: Application to texture recognition", *Neural Networks, IEEE Transactions on,* Vol. 21, pp. 609–620, April 2010.

Camunas-Mesa, L.; Acosta-Jimenez, A.; Zamarreno-Ramos, C.; Serrano-Gotarredona, T.; Linares-Barranco, B.; , "A 32x32 Pixel Convolution Processor Chip for Address Event Vision Sensors With 155 ns Event Latency and 20 Meps Throughput," *Circuits and Systems I: Regular Papers, IEEE Transactions on ,* vol.58, no.4, pp.777–790, April 2011

Jimenez-Fernandez, A.; Fuentes-del-Bosh, J.L.; Paz-Vicente, R.; Linares-Barranco, A.; Jimeinez, G.; , "Neuro-inspired system for real-time vision sensor tilt correction," *Circuits and Systems (ISCAS), Proceedings of 2010 IEEE International Symposium on ,* vol. , no. , pp.1394–1397, May 30 2010-June 2 2010

Linares-Barranco, A.; Gomez-Rodriguez, F.; Jimenez-Fernandez, A.; Delbruck, T.; Lichtensteiner, P.; , "Using FPGA for visuo-motor control with a silicon retina and a humanoid robot," *Circuits and Systems, 2007. ISCAS 2007. IEEE International Symposium on,* vol., no., pp.1192–1195, 27–30 May 2007

Computational Modelling of Objects Represented in Images – Di Giamberardino et al. (eds)
© 2012 Taylor & Francis Group, London, ISBN 978-0-415-62134-2

Quantitative analysis on PCA-based statistical 3D face shape modeling

A.Y.A. Maghari, I. Yi Liao & B. Belaton
School of Computer Sciences, Universiti Sains Malaysia, Pulau Pinang, Malaysia

ABSTRACT: Principle Component Analysis (PCA)-based statistical 3D face modeling using example faces is a popular technique for modeling 3D faces and has been widely used for 3D face reconstruction and face recognition. The capability of the model to depict a new 3D face depends on the exemplar faces in the training set. Although a few 3D face databases are available to the research community and they have been used for 3D face modeling, there is little work done on rigorous statistical analysis of the models built from these databases. The common factors that are generally concerned are the size of the training set and the different choice of the examples in the training set. In this paper, a case study on USF Human ID 3D database, one of the most popular databases in the field, has been used to study the effect of these factors on the representational power. We found that: 1) the size of the training set increase, the more accurate the model can represent a new face; 2) the increase of the representational power tends to slow down in an exponential manner and achieves saturity when the number of faces is greater than 250. These findings are under assumptions that the 3D faces in the database are randomly chosen and can represent different races and gender with neutral expressions. This analysis can be applied to the database which includes expressions too. A regularized 3D face reconstruction algorithm has also been tested to find out how feature points selection affects the accuracy of the 3D face reconstruction based on the PCA-model.

1 INTRODUCTION

The reconstruction of 3D faces is a very important issue in the fields of computer vision and graphics. The need for 3D face reconstruction has grown in applications like virtual reality simulations, plastic surgery simulations and face recognition (Elyan & Ugail 2007), (Fanany et al. 2002) It has recently received great attention among scholars and researchers (Widanagamaachchi & Dharmaratne 2008). 3D facial reconstruction systems are to recover the three dimensional shape of individuals from their 2D pictures or video sequences. There are many approaches for reconstruction 3D faces from images such as Shape-from-Shading (SFS) (Smith & Hancock 2006), shape from silhouettes (Lee et al. 2003) and shape from motion (Amin & Gillies 2007). There are also learning-based methods, such as neural network (non-statistical). (Nandy & Ben-Arie 2000) and 3D Morphable Model (3DMM) (statistical based) (Volker & Thomas 1999). 3DMM is an analysis-by-synthesis based approach to fit the 3D statistical model to the 2D face image (Widanagamaachchi & Dharmaratne 2008). The presence of 3D scanning technology lead to create a more accurate 3D face model examples (Luximon et al. 2012). Examples based modeling allows more realistically face reconstruction than other methods (Widanagamaachchi & Dharmaratne 2008), (Martin & Yingfeng 2009). However, the quality of face reconstruction using examples is affected by

the chosen examples. For example (Kemelmacher-Shlizerman & Basri 2011), and (Jose et al. 2010) urged that learning a generic 3D face model requires large amounts of 3D faces. However, this issue has not been statistically analyzed in terms of representational power of the model, to the best of our knowledge. Although (Iain et al. 2007) has studied the representational power of two example based models, i.e. Active Appearance Model (AAM) and 3DMM, they did not conduct any statistical analysis or testing on the two models.

A PCA-based model with relatively small sample size (100 faces) was used for face recognition and has obtained reasonable results .(Volker & Thomas 2003). The generation of synthetic views from 2D input images was not needed. Instead, the recognition was based on the model coefficients which represent intrinsic shape and texture of faces. In case of 3D face reconstruction, a more diverse set of 3D faces would build a more powerful PCA-based model to generate an accurate 3D representation of the 2D face. On the other hand, PCA-based models have not been quantitatively studied in terms of the effect of 3D reconstruction accuracy on face recognition.

Although in some statistical modeling methods both shape and texture are modeled separately using PCA (e.g. 3DMM), it is suggested that shapes are more amenable to PCA based modeling than texture, as texture varies dramatically than the shape. In many situations, the 2D texture can be warped to the 3D geometry to generate the face texture (Dalong et al. 2005). Therefore, in this paper we focus on shape modeling. When shapes are considered, the reconstruction of 3D face from 2D images using shape models is relatively simple. A popular method is a regularization based reconstruction where a few feature points are selected as the observations for reconstruction (Dalong et al. 2005). The results based on this method will not go beyond the representational power of the model. Even if a more powerful model trained with more examples would generate a better representation of the true face, the generated face remains within the boundaries of the model.

In this study we propose an empirical study to test the representational power of 3D PCA-based face models using USF Human ID 3D database (Volker & Thomas 1999). We define the representational power as the Euclidian Distance between the reconstructed shape surface vector and the true surface shape vector divided by the number of points in the shape vector. A series of experiments are designed to answer the questions:

1. What is the relationship between the size of training set and the representational power of the model?
2. How many examples will be enough to build a satisfactory model?
3. What is the effect on the representational power if the model is trained with a different sample for the same number of faces?

Finally, the regularization based 3D face reconstruction algorithm was analyzed to find the relationship between the number of feature points and the accuracy of reconstruction.

This paper is organized as follows: section 2 provides a theoretical background. The Experimental design and statistical analysis are explained in Section 3, whereas in Section 4 the findings and discussion are reported. In the last section, the paper is summarized.

2 THEORITICAL BACKGROUND

2.1 Modeling shape using PCA

The 3D face shape model is a linear combination of eigenvectors obtained by applying PCA to training set of 3D shape faces. The linear combination is controlled by shape parameters called α, where shape vectors are given as follows:

$$s_i = (x_{i1}, y_{i1}, z_{i1}, ..., x_{in}, y_{in}, z_{in})^T$$

where s_i has the dimension $n \times 3$, n is the number of vertices and $i = 1, ..., m$ (number of face shapes).

All vertices must be fully corresponded using their semantic position before applying PCA to get a more compact shape representation of 3D shape face. Let

$$s_0 = \frac{1}{m} \sum_{i=1}^{m} s_i$$

where s_0 be the average face shape of m exemplar face shapes and $S = [s_1, s_2, ..., s_m] R^{(3*n)*m}$.

The covariance matrix of the face shapes is defined as

$$C = \sum_{i=1}^{m} (s_i - s_0)(s_i - s_0)^T$$

The eigenvectors e_i and the corresponding eigenvalues λ_i of the covariance matrix C are such that $Ce_i = \lambda_i e_i$ where $i = 1, ..., m$.

After PCA modeling, every shape of the m face shapes can be decomposed into the form

$$s_j = s_0 + \sum_{i=1}^{m} \alpha_i e_i \qquad (1)$$

where e_i represent the i-th eigenvector of the covariance matrix C and α_i is the coefficient of the shape eigenvector e_i.

The coefficient of a face shape s_i can be calculated using the following equation

$$\alpha = E^T(s_i - s_0) \qquad (2)$$

where $E = [e_1, e_2, ..., e_m]$ are the eigenvectors of the covariance matrix C. The projected new face shape can be represented by applying Equation 1.

2.2 Representational power of the model

In this study, we define the representational power (RP) as the Euclidean Distance between the reconstructed shape vector and the true shape vector divided by the number of points in the shape vector.

In Cartesian coordinates, if $p = (p_1, p_2, ..., p_n)$ and $q = (q_1, q_2, ..., q_n)$ are two points in Euclidean n-space, then the Euclidean Distance from p to q is given by:

$$d = \sum_{i=1}^{n} \sqrt{(p_i - q_i)^2}$$

Let s be a shape face belongs to the testing data set and s_r be the face shape that is represented by PCA-model using Equations 2 and 1 then:

a. Calculate the coefficient of the testing face shape s using Equation 2.
b. Apply Equation 1 to represent s_r using the PCA-model.
c. Determine the Euclidean Distance between the true shape face s and the reconstructed shape face s_r.

RP is the Euclidean Distance divided by shape dimension $n \times 3$.

2.3 Reconstruction based on regularization

Since all points of different shape faces in the database are fully corresponded, PCA is used to obtain a more compact shape representation of face with the primary components (Dalong et al. 2005).

14

Let t be the number of points that can be selected from the 3D face shape in the testing set, $s_f = (p_1, p_2, \ldots, p_t) \in \mathbb{R}^t$ be the set of selected points on the 3D face shape (p_i can be x, y or z of any vertex on the 3D face shape, whereas every vertex has 3 axis x, y and z), $s_{f0} \in \mathbb{R}^t$ is the t corresponding points on s_0(the average face shape) and $E_f \in \mathbb{R}^{t \times m}$ is the t corresponding columns on $E \in \mathbb{R}^{3n \times m}$ (the matrix of row eigenvectors). Then the coefficient α of a new 3D face shape can be derived as

$$\alpha = (E_f^T E_f + \lambda \Lambda^{-1})^{-1} E_f^T (s_f - s_{f0}) \tag{3}$$

where Λ is a diagonal $m \times m$ matrix with diagonal elements being the eigenvalues and λ is the weighting factor. Then we apply α to Equation 1 to obtain the whole 3D face shape.

3 EXPERIMENTAL DESIGN & STATISTICAL ANALYSIS

3.1 USF human ID database

A case study is conducted using USF Raw 3D Face Data Set. This database includes shape and texture information of 100 3D faces obtained by using Cyberware head and face 3D color scanner. The 3D faces are aligned with each other as explained by (Volker & Thomas 1999). They developed the 3D morphable model (3DMM) which has been widely used in many facial reconstruction systems. A basic assumption is that any 3D facial surface can be practically represented by a convex combination of shape and texture vectors of 100 3D face examples, where the shape and texture vectors are given as follows:

$$S_i = (x_{i1}, y_{i1}, z_{i1}, \ldots, x_{i75972}, y_{i75972}, z_{i75972})^T$$

$$T_i = (R_{i1}, G_{i1}, B_{i1}, \ldots, R_{i75972}, G_{i75972}, B_{i75972})^T$$

Information from each of these exemplar heads was saved as shape and texture vectors (S_i, T_i) where $i = 1, \ldots, 100$ (Martin & Yingfeng 2009). For each face shape there are 75972 vertices saved in a text file, one line per vertex and 3 points each line.

This study uses only the shape vectors for training and testing the models.

3.2 Representational power analysis

RP of 3D PCA-based models is analyzed using USF Human ID 3D database. A series of experiments are designed to find the relationship between the size of the training data set and RP of the trained model. On the other hand, the effect of different training set for the same number of faces has been analyzed.

3.2.1 Size of training set

We divided the current 100 shape faces into different sets in term of the number of samples. For example PCA15 is a shape model that has 15 shape faces as training data, PCA18 is a model with 18 training face

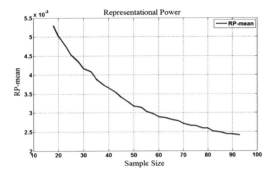

Figure 1. The relation between the sample size and the RP.

shapes and so on until PCA93 with 93 training face shapes. The models with training data between 15 and 70 face shapes were tested with 30 testing shape faces. The other models from 73 to 93 were tested with the remaining faces out of 100 faces. For example the PCA80 is tested with 20 shape faces.

The testing face shape s is projected on the model by calculating the shape coefficient vector α using Equation 2. The projected new face shape s_r can be obtained by applying α to Equation 1. Figure 5 in the appendix shows a 3D face shape represented by the model PCA80. RP can then be calculated for all testing face shapes according to the following equation

$$RP = \frac{\sum_{i=1}^{3 \times 75972} \sqrt{(s_i - s_{ri})^2}}{3 \times 75972}$$

Then the mean of the RP "RP-mean" for all test faces is used to represent the RP of the model. Figure 1 shows the relationship between the sample size and RP-mean. It shows that there is exponential relationship between the sample size and the RP-mean.

An exponential regression model is applied to fit the curve in Figure 1, as follows

$$y = b_0 \times e^{b_1 x}$$

where y is the RP-mean, x is the sample size and b_0 and b_1 are the regression factors. Thus,

$$\ln(y) = \ln(b_0) + b_1 x$$

The linear relationship between the sample size x and the natural logarithm $\ln(y)$ of the RP-mean is shown in Figure 2.

Two variable linear regression was run using MS Excel to find the two factors $\ln(b_0)$ and b_1. The regression result is shown in Table 1.

$b_0 = \exp(\ln(b_0)) = \exp(-5.1425) = 0.0057849$

The exponential relation is presented as

$$y = 0.0057849 \times e^{-0.01046x}$$

where y is the representational power (RP-mean) and x is the sample size. Figure 3 shows the functional relationship.

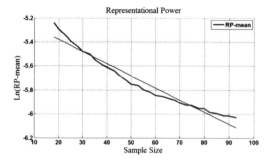

Figure 2. The relation between the sample size and the natural logarithm of representational power.

Table 1. Regression results.

Regression statistics	
Ln(b0)	−5.1525
b1	−0.01046
Multiple R	0.974219
R Square	0.9491019
Standard Error	0.141481178
Observations	32
SSE	0.0998032
SST	1.960840125
SSR	1.861036941

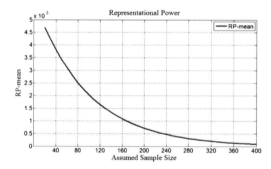

Figure 3. The functional relation between the assumed sample size and the RP.

3.2.2 Different sets of training data

Four pairs of different models with the same sample size but different training data were trained to see if the variations of the data set influence the representational power of the model.

Table 2 shows the comparative results between each pairs of models with significant $\alpha = 0.05$.

We made 4 comparisons and applied t-Test. Two cases showed significant differences in the RP-mean of the models. For example the two learning models PCA40 and PCA40R have been trained with 40 faces from different sets. As illustrated in Table 2, the t-Test value corresponding to the two learning models is 0.01322 which is less than $\alpha = 0.05$. This means that

Table 2. Comparative result of model-pairs with different samples.

Learning model	RP-mean %	Std%	p-value of t-Test
PCA 40	0.371	0.075	0.01322
PCA 40R	0.343	0.06	
PCA 50	0.32	0.048	0.26031
PCA 50R	0.313	0.053	
PCA 60	0.289	0.04	0.105
PCA 60R	0.278	0.037	
PCA 70	0.272	0.039	0.00857
PCA 70R	0.251	0.026	

there is a statistically significant difference between the RPs of the two models.

3.3 Regularization based reconstruction

The regularized algorithm was categorized as one of the main existing four methods for 3D facial reconstruction (Martin & Yingfeng 2009). The algorithm was used by Dalong (Dalong et al. 2005) to reconstruct 3D face for face recognition purposes. The selected feature points are used to compute the 3D shape coefficients of the eigenvectors using Equation 3. Then, the coefficients are used to reconstruct the 3D face shape using Equation 1. Figure 5 in the appendix shows some examples of reconstructed face shapes using different number of feature points.

3.3.1 Number of feature points

Three models with different dataset sizes 60, 70, 80 face shapes were used to analyze the algorithm. The selected points are between 10 and 500 randomly selected from face shape inside and outside the training set. If the test face is from the training set, any number of selected points greater than or equal the number of training face shapes can reconstruct exact 3D face. If the test face is not from the training set, it hardly reconstructs the exact 3D face. Figure 4 shows the relation between the number of feature points (from faces outside the training data) and the accuracy of the reconstructed face shapes using PCA80 (PCA-based model with 80 training shapes).

The algorithm was analyzed for different values of the weighting factor $\lambda = 0.005, 0.05, 0.1, 1, 10, 20, 50, 100$, and 200 as shown in Figure 4. 20 faces outside the training data set are used to test the regularized algorithm. The following steps demonstrate the experiment procedures for each different value of the weighting factor λ

a) For each shape we do the following 30 times:

    I. Select randomly a set of feature points.
    II. Apply the reconstruction algorithm and calculate the RP between the original face shape and reconstructed shape.

b) Determine the mean of 50 RP, termed as RP-mean.

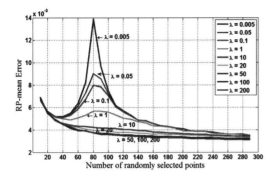

Figure 4. The relationship between the number of feature points and the accuracy of the reconstructed face shapes on PCA80 with different values of λ.

Table 3. ANOVA results with λ = 1.

Source of variation	Between groups	Within groups	total
SS	5.82E-05	0.000968	0.001027
Df	49	0.000968	999
MS	1.19E-06	1.02E-06	
F	1.165887		
P-value	0.206254		
F crit	1.367567		

c) Determine the mean of the 20 RP-means (RP-mean Error)
d) RP-mean Error is used to measure the accuracy of the reconstructed shape faces from the set of selected points.

3.3.2 Location of feature points

For each of the 20 face shapes chosen from outside the training set, a number of points were randomly selected 50 times to reconstruct face shapes. The 50 selections of feature points were repeated with each of the weighting factors 0.05, 1, 10, and 100. The reconstruction results of all repeated 50-time selections showed the similarity of RP.

Single factor ANOVA was run using MS Excel to compare the 50 alternatives for each of the selected 30, 40, 50, 60, 80 and 100 feature points. In all cases, the ANOVA results show that there is no significant difference among the 50 feature selections. Table 3 shows sample results using 60 feature points with λ=1.

The ANOVA results in Table 3 show that there is no statistically significant difference among the 50 alternatives of 60 points selection, as the F-value is smaller than the tabulated critical value (F crit).

4 FINDINGS AND DISCUSSION

4.1 Representational Power

The experimental results in Figure 1 show that there is an exponential relationship between the sample size and the RP mean. To determine this relation we applied linear regression on the sample size and the natural logarithm of the RP mean. Based on this relationship, a minimum number of 250 shape faces is expected to build a satisfactory 3D face model, see Figure 3. This suggests that we may need a large amount of 3D faces to build a generic 3D faces model and this is consistent with the comments made in (Kemelmacher-Shlizerman & Basri 2011), (Jose et al. 2010).

Note that there may be other methods to compensate for this need if large data is not available. However, this issue is beyond the topic of this study.

4.2 The dataset variations & PCA-based Models

The results show that as far as the small sample size training set is concerned, in many cases the representational power may vary if we train the model with a different sample. It was inferred that the representational power of the model may be different if the training data set is changed. Table 2 show the results of t-test conducted on four cases. Two cases (PCA40, PCA40R) and (PCA70, PCA70R) show significant differences in the RP, whereas the p-value of t-test is less than $\alpha = 0.05$ in the two cases.

4.3 Feature points selection for reconstruction

For the selection of feature points, we found that, regardless of the value of the weighting factor λ, the accuracy of reconstruction by a large number of randomly selected points (greater than 200 points) is relatively the same in all cases even with different locations of points on the face. However, if the number of feature points is equal to the number of training faces, the produced face will have the most inaccurate shape particularly when $\lambda \leq 1$. This is because the algorithm became unstable when the number of training faces is almost the same as the number of feature points. When $\lambda > 20$ and the number of selected points is greater than the number of training faces, the accuracy of reconstruction is relatively the same in all cases and is further associated with slight improvement related to the larger number of selected points as shown in Figure 4.

Furthermore, we found that different locations of a consistent number of feature points have no significant quantitative effect on the reconstruction results. Whether the feature points are dependent or independent of each other may be an issue of interest that needs to be further experimented, the matter which is beyond the scope of this study.

5 SUMMARY AND FUTURE WORK

The representational power of the 3D shape model which is based on PCA was evaluated by analyzing the 3D face database we obtained from University South Florida (USF) with a series of experiments and statistical analysis. The current USF Human ID 3D database has 100 faces. All 100 faces were used for training and testing purposes. The functional relationship between the number of independent faces in the training data set and the representational power of the model were estimated. Based on this relationship, a minimum number of 250 shape faces is suggested to build a satisfactory 3D face model, especially if the faces are neither too identical nor too different in their shapes. The findings of this relationship show that we need to have a way to increase the representational power of the model by involving more face or predicting the increasing behavior of the model based on increasing the number of faces.

On the other hand, one type of regularization-based 3D face reconstruction algorithm was analyzed to find the relationship between the number of the feature points and the accuracy of the reconstructed 3D face shape. The extensive experimental results showed that if the test face is from the training set, then any set of any number greater than or equal to the number of training faces -1 can reconstruct exact 3D face. If the test face does not belong to the training set, it will hardly reconstruct the exact 3D face using 3D PAC-based models. However, it could reconstruct an approximate face depending on the number of feature points and the weighting factor. This is consistent with the finding on the representational power of the current database. This indicates that we may choose any set of feature points from the most reliable part of the face as long as they are independent of each other, rather than the ones that are from the parts undergoing the facial expressions. Further studies are suggested to test this idea.

ACKNOWLEDGMENT

The work presented in this paper is sponsored by RU grant 1001/PKCOMP/817055, Universiti Sains Malaysia.

APPENDIX

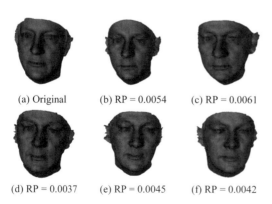

(a) Original   (b) RP = 0.0054   (c) RP = 0.0061

(d) RP = 0.0037   (e) RP = 0.0045   (f) RP = 0.0042

Figure 5.   (a) test face. (b) and (c) are reconstructions from 55 randomly selected points. (e) and (f) are reconstructions from 250 points. (d) PCA representation of (a).

REFERENCES

Amin, S. H. and Gillies, D. (2007). Analysis of 3d face reconstruction. In Image Analysis and Processing, 2007. ICIAP 2007. *14th International Conference on, pp. 413–418.*

Dalong, J., Yuxiao, H., Shuicheng, Y., Lei, Z., Hongjiang, Z. and Wen, G. (2005) Efficient 3D reconstruction for face recognition. *Pattern Recogn.,* 38(6): 787–798.

Elyan, E. and Ugail, H. (2007) Reconstruction of 3D Human Facial Images Using Partial Differential Equations. JCP, 1–8.

Fanany, M. I., Ohno, M. and Kumazawa, I. (2002) Face Reconstruction from Shading Using Smooth Projected Polygon Representation NN.

Iain, M., Jing, X. and Simon, B. (2007) 2D vs. 3D Deformable Face Models: Representational Power, Construction, and Real-Time Fitting. *Int. J. Comput. Vision,* 75(1): 93–113.

Jose, G.-M., Fernando De la, T., Nicolas, G. and Emilio, L. Z. (2010) Learning a generic 3D face model from 2D image databases using incremental Structure-from-Motion. *Image Vision Comput.,* 28(7): 1117–1129.

Kemelmacher-Shlizerman, I. and Basri, R. (2011) 3D Face Reconstruction from a Single Image Using a Single Reference Face Shape. *Pattern Analysis and Machine Intelligence, IEEE Transactions on,* 33(2): 394–405.

Lee, J., Moghaddam, B., Pfister, H. and Machiraju, R. (2003) Silhouette-Based 3D Face Shape Recovery. *In Proceedings of Graphics Interface:* 21–30.

Luximon, Y., Ball, R. & Justice, L (2012)The 3D Chinese head and face modeling. *Computer-Aided Design,* 44(1): 40–47.

Martin, D. L. and Yingfeng, Y. (2009) State-of-the-art of 3D facial reconstruction methods for face recognition based on a single 2D training image per person. *Pattern Recogn. Lett.,* 30(10): 908–913.

Nandy, D. and Ben-Arie, J. (2000) Shape from recognition: a novel approach for 3d face shape recovery. *In Proceedings of the International Conference on Pattern Recognition – Volume 1IEEE Computer Society.*

Smith, W. A. P. and Hancock, E. R. (2006) Recovering Facial Shape Using a Statistical Model of Surface Normal Direction. *Pattern Analysis and Machine Intelligence,* IEEE Transactions on, 28(12): 1914–1930.

Volker, B. and Thomas, V. (1999) A morphable model for the synthesis of 3d faces. *In Proceedings of the 26th annual conference on Computer graphics and interactive techniquesACM Press/Addison-Wesley Publishing Co.*

Volker, B. & Thomas, V. (2003) Face recognition based on fitting a 3D morphable model. *Pattern Analysis and Machine Intelligence,* IEEE Transactions on, 25(9): 1063–1074.

Widanagamaachchi, W. N. and Dharmaratne, A. T. (2008) 3d face reconstruction from 2d images. *In Computing: Techniques and Applications,* 2008. DICTA '08.Digital Image, pp. 365–371.

Computational Modelling of Objects Represented in Images – Di Giamberardino et al. (eds)
© 2012 Taylor & Francis Group, London, ISBN 978-0-415-62134-2

Ordinary video events detection

Md. Haris Uddin Sharif
Computer Engineering Department, Gediz University, Izmir, Turkey

Sahin Uyaver
Vocational School, Istanbul Commerce University, Istanbul, Turkey

Md. Haidar Sharif
Computer Engineering Department, Gediz University, Izmir, Turkey

ABSTRACT: In this paper, we have addressed mainly a detection based method for video events detection. Harris points of interest are tracked by optical flow techniques. Tracked interest points are grouped into several clusters by dint of a clustering algorithm. Geometric means of locations and circular means of directions as well as displacements of the feature points of each cluster are estimated to use them as the principle detecting components of each cluster rather than the individual feature points. Based on these components each cluster is defined either lower-bound or horizon-bound or upper-bound clusters. Lower-bound and upper-bound clusters have been used to detect potential ordinary video events. To show the interest of the proposed framework, the detection results of ObjectDrop, ObjectPut, and ObjectGet at TRECVid 2008 in real videos have been demonstrated. Several results substantiate its effectiveness, while the residuals gives the degree of the difficulty of the problem at hand.

1 INTRODUCTION

Event detection in surveillance video streams is an essential task for both private and public places. As large amount of video surveillance data makes it a backbreaking task for people to keep watching and finding interesting events, an automatic surveillance system is firmly requisite for the security management service. Video event can be defined to be an observable action or change of state in a video stream that would be important for the security management team. Events would vary greatly in duration and would start from two frames to longer duration events that can exceed the bounds of the excerpt. Some events occur ofttimes e.g., in the airport people are putting objects, getting objects, meeting and discussing, splitting up, and etc. Conversely, some events go off suddenly or unexpectedly e.g., in the airport a person is running, falling on the escalator, going to the forbidden area, and etc. Withal both type of events detection in video surveillance is an appreciable task for public security and safety in areas e.g., airports, banks, city centers, concerts, hospitals, hotels, malls, metros, mass meetings, pedestrian subways, parking places, political events, sporting events, and stations.

A lot of research efforts have been made for events detection from videos captured by surveillance cameras. Many single frame detection algorithms based on transfer cascades (Viola and Jones 2001; Lienhart and Maydt 2002) or recognition (Serre, Wolf, and Poggio 2005; Dalal and Triggs 2005; Bileschi and Wolf 2007) have demonstrated some high degree of promise for pedestrian detection in real world busy scenes with occlusion. However, most of those pedestrian detection algorithms are significantly slow for real time applications. An approach to trackingbased event detection focuses on multi-agent activities, where each actor is tracked as a blob and activities are classified based on observed locations and spatial interactions between blobs (Hamid et al. 2005; Hongeng and Nevatia 2001). These models are well-suited for expressing activities such as loitering, meeting, arrival and departure, etc. To detect events in the TRECVid 2008 (TRECVID 2008) many algorithms have been proposed for detecting miscellaneous video events (Xue et al. 2008; Guo et al. 2008; Hauptmann et al. 2008; Yokoi, Nakai, and Sato 2008; Kawai et al. 2008; Hao et al. 2008; Orhan et al. 2008; Wilkins et al. 2008; Chmelar et al. 2008; Dikmen et al. 2008). Common event types are PersonRuns, ObjectPut, OpposingFlow, PeopleMeet, Embrace, PeopleSplitUp, Pointing, TakePicture, CellToEar, and ElevatorNoEntry (TRECVID 2008). Available detectors are based on trajectory and domain knowledge (Guo et al. 2008), motion information (Xue et al. 2008), spatiotemporal video cubes (Hauptmann et al. 2008), change detection (Yokoi, Nakai, and Sato 2008), optical flow concepts (Kawai et al. 2008; Hao et al. 2008; Orhan et al. 2008), etc. Orhan (Orhan et al. 2008) proposed detectors to detect several video events. For example, ObjectPut detector relies heavily on the optical flow concept to track feature points throughout the scene. The tracked feature points are grouped into clusters. Location, direction, and speed of the feature points are used to create clusters; the main tracking component then becomes the clusters rather than the individual feature points. The final step is determining whether or not the tracked cluster is moving or not, achieved by using trained motion maps of the scene. The decision is then made by analyzing cluster flow over the scene map. A number of low-level detectors including foreground region and person detectors can be seen in (Wilkins et al. 2008). Low-level information within a framework to track pedestrians temporally through a scene. Several events are detected from the movements and interactions between people and/or their interactions with specific areas which have been manually annotated by the user. Based on trajectories analysis some video event detectors have been proposed by (Chmelar et al. 2008). Each trajectory is given as a set of the blob size and position in several adjacent time steps. Pseudo Euclidian distances obtained from the trigonometrically treatments of motion history blobs are used to detect video events (Sharif and Djeraba 2009). Though many promising video event detectors have been identified which can be directly or indirectly employed for events detection, their performances have been limited by the existing challenges.

We have proposed mainly a detection based method for ordinary video event detection. Our proposed detector is relied on optical flow concept and related to the video event detector of Orhan (Orhan et al. 2008). Feature points or points of interest or corners are detected by Harris corner detector (Harris and Stephens 1988) after performing a background subtraction. Those points are tracked over frames by means of optical flow techniques (Lucas and Kanade 1981; Shi and Tomasi 1994). The tracked optical flow feature points are grouped into several clusters using *k-means* algorithm (Lloyd 1982). Geometric means of locations and circular means of directions and displacements of the feature points of each cluster are estimated. And then they are used to demonstrate the principal components for each cluster rather than the individual feature points. Based on estimated direction components each cluster is possessed any bounds among lower-bound, upper-bound, and horizon-bound. Lower-bound clusters have been taken into consideration for ObjectDrop and Object-Put events, while upper-bound clusters have been fixed for ObjectGet events. The key difference between our method and Orhan (Orhan et al. 2008) is that our method involves mainly detection. However, we are tracking low level features like optical flow but clusters are not being tracked. Conversely, Orhan's method concerns detection as well as tracking both optical flow and clusters. So our method is easier to implement and also faster than that of Orhan.

The rest of the paper is ordered as follows: Section 2 illustrates the implementation steps of the framework; Section 3 demos detection abilities of the proposed detector; and Section 4 makes a conclusion.

2 IMPLEMENTATION STEPS

In this section, extended treatment of implementation steps of the proposed approach has been carried out.

2.1 *Feature selection*

Moravec's corner detector (Moravec 1980) is a relatively simple algorithm, but is now commonly considered out-of-date. Harris corner detector (Harris and Stephens 1988) is computationally demanding, but directly addresses many of the limitations of the Moravec corner detector. We have considered Harris corner detector. It is a famous point of interest detector due to its strong invariance to rotation, scale, illumination variation, and image noise (Schmid, Mohr, and Bauckhage 2000). It is based on the local auto-correlation function of a signal, where the local auto-correlation function measures the local changes of the signal with patches shifted by a small amount in different directions. We have deemed that in video surveillance scenes, camera positions and lighting conditions admit to get a large number of corner features that can be easily captured and tracked. Images of Fig. 1 from 1st to 3rd demonstrate an example of Harris corner and optical flow estimation. For each point of interest which is found in the next frame, we can get its position, direction, and moving distance easily. *How much distance has it moved? Which direction has it moved?* Suppose that any point of interest i with its position in the current frame $p(x_i, y_i)$ is found in the next frame with its new position $q(x_i, y_i)$; where x_i and y_i be the x and y coordinates, respectively. Now, suppose that the traveled distance d_i and its direction β_i. It is easy to calculate the displacement or Euclidean distance of $p(x_i, y_i)$ and $q(x_i, y_i)$ using Euclidean metric as $d_i = \sqrt{(q_{x_i} - p_{x_i})^2 + (q_{y_i} - p_{y_i})^2}$. Moving direction of the feature i can be calculated using $tan^{-1}(q_{y_i} - p_{y_i}/q_{x_i} - p_{x_i})$. But it incurs a few potential problems [Sharif, Ihaddadene, and Djeraba 2010]. As a remedy, we have used *atan2* function to calculate the accurate direction of motion β_i of any point of interest i by dint of: $\beta_i = atan2(q_{y_i} - p_{y_i}, q_{x_i} - p_{x_i})$.

2.2 *Analysis of cluster information*

The geometric clustering method, *k-means*, is a simple and fast method for partitioning data points into k clusters, based on the work done by (Lloyd 1982) (so-called Voronoi iteration). The rightmost image in Fig.1 illustrates clustering performed by k-means. On clustering we have estimated the *geometric means* of all x and y coordinates for each cluster to get single origin. If any cluster c will contain m number of x coordinates, then the number of y coordinates, displacements, and directions will be also m. If x_{g_c} and y_{g_c} symbolize geometric means of all x and y coordinates of a cluster, respectively, then they can be formulated as: $x_{g_c} = \left[\prod_{j=1}^{m} x_j \right]^{1/m}$ and $y_{g_c} = \left[\prod_{j=1}^{m} y_j \right]^{1/m}$.

Figure 1. Green points in first image depict Harris corners. Second and third images respectively, represent optical flow before and after suppression of the static corners. Fourth image shows the clustering using k-means.

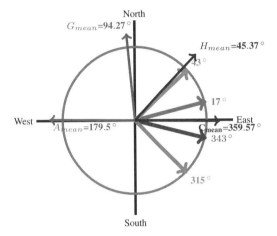

Figure 2. Arithmetic, geometric, and harmonic means of $43\,°$, $17\,°$, $343\,°$, and $315\,°$ provide inaccurate $179.5\,°$, $94.27\,°$, and $45.37\,°$ estimation, respectively. Solely $C_{mean} = 359.57\,°$ confers correct result.

2.3 Circular mean

Consider four persons starting walking from a fixed location to four different directions, i.e., north-east $43\,°$, north-east $17\,°$, south-east $343\,°$, and south-east $315\,°$. Arithmetic, geometric, and harmonic means of these directions are $179.5\,°$, $94.27\,°$, and $45.37\,°$, respectively. All estimations are clearly in error as depicted in Fig. 2. As arithmetic, geometric, and harmonic means are ineffective for angles, it is important to find a good method to estimate accurate mean value of the angles. Merely *circular mean* C_{mean} is the solution of this problem. Deeming that in a cluster we have m number of corners of single origin (x_{g_c}, y_{g_c}) with displacement vectors $V_1, V_2, V_3, \ldots, V_m$ and their respective directions $\beta_1, \beta_2, \beta_3, \ldots, \beta_m$. Then their *circular mean*, denoted by β_{g_c}, can be formulated by dint of:

$$\beta_{g_c} = \begin{cases} tan^{-1}\frac{\sum_{i=1}^{m} sin\beta_i}{\sum_{i=1}^{m} cos\beta_i} & if \sum_{i=1}^{m} sin\beta_i > 0, \sum_{i=1}^{m} cos\beta_i > 0 \\ tan^{-1}\frac{\sum_{i=1}^{m} sin\beta_i}{\sum_{i=1}^{m} cos\beta_i} + 180\,° & if \sum_{i=1}^{m} cos\beta_i < 0 \\ tan^{-1}\frac{\sum_{i=1}^{m} sin\beta_i}{\sum_{i=1}^{m} cos\beta_i} + 360\,° & if \sum_{i=1}^{m} sin\beta_i < 0, \sum_{i=1}^{m} cos\beta_i > 0 \end{cases} \quad (1)$$

and their circular resultant displacement vector length L_{g_c} can be expressed by Pythagorean theorem as: $L_{g_c} = \sqrt{\left(\sum_{i=1}^{m} sin\beta_i\right)^2 + \left(\sum_{i=1}^{m} cos\beta_i\right)^2}$. Using Eq. 1,

we can express the accurate mean of angles in Fig. 2 which is south-east $359.57\,°$.

2.4 Events detection

For each cluster (x_{g_c}, y_{g_c}), β_{g_c}, and L_{g_c} demo the principal component rather than the individual feature point. Based on estimated β_{g_c} each cluster is named by lower or upper or horizon bounds. A cluster will be called a *lower-bound* if β_{g_c} ranges from $224.98\,°$ to $314.98\,°$. A cluster will be called an *upper-bound* if β_{g_c} ranges from $134.98\,°$ to $44.98\,°$. A cluster will be called a *horizon-bound* if β_{g_c} limits either from $134.99\,°$ to $224.99\,°$ or from $44.99\,°$ to $314.99\,°$. An ObjectDrop or ObjectPut event can be detected if any cluster on the video frame is proclaimed by lower-bound. An Object-Get event can be detected if any cluster on the video frame is proclaimed by upper-bound. Horizon-bound cluster would play a vital role to detect video event e.g., somebody is running. We draw a circle with center (x_{g_c}, y_{g_c}), direction β_{g_c}, and displacement vector L_{g_c} for each lower-bound and/or upper-bound cluster on the camera view image to show the detector's delectability.

3 EXPERIMENTAL RESULTS

A large variety in the appearance of the event types makes the events detection task in the surveillance video streams selected for the TRECVid 2008 extremely difficult. The source data of TRECVid 2008 comprise about 100 hours (*10 days* × *2 hours per day* × *5 cameras*) of video streams obtained from Gatwich Airport surveillance video data (TRECVid 2008). A number of events for this task were defined. Since all the videos are taken from surveillance cameras which means the position of the cameras is still and cannot be changed. However, it was not practical for us to analyze 100 hours of video except few hours. The value of k has been considered as 7. If there is a video event on the real video and the algorithm can detect that event, then this case is defined as a *true positive* or correct detection. If there is no video event on the real video but the algorithm can detect a video event, then this case is defined as a false positive or *false alarm* detection. If there is a video event on the real video but the algorithm cannot detect, then this case is defined as a *false negative* or miss detection.

Figure 3.   True positive ObjectDrop: Something is dropping from the hand of a baby which has been detected.

Figure 4.   True positive ObjectPut: A person is putting a bag from one place to other which has been detected.

Figure 5.   True positive ObjectPut: A person is putting a cell phone on the self which has been detected.

Figure 6.   False positive ObjectPut: A person is going to sit on a bench which has been detected.

Fig. 3 shows four sample frames of an ObjectDrop event detected by our proposed detector. Some stuff from the hands of a baby suddenly fall off on the floor. The in-flight stuff has been detected as true positive ObjectDrop event. Object centered at green circle and falling direction with red arrow have been drawn by the algorithm. Fig. 4 demos four sample frames of an ObjectPut event detected by the detector. A person is putting a hand bag from one location to another. That event has been detected as true positive ObjectPut. Fig. 5 depicts four sample frames of another ObjectPut event detected by our proposed detector. A person is putting a hand-phone from one place to another. The event has been detected as true positive ObjectPut. Fig. 6 describes four sample frames of an ObjectPut event detected by the detector. Someone is going to sit on the bench in the waiting area. But this event has been

detected as false positive. Fig. 7 traces four sample frames of an ObjectGet event detected by the detector. A person is getting some stuff from the floor and placing some upper location. The event has been detected as true positive ObjectGet.

Nonetheless, our proposed event detector cannot detect accurately event e.g., Fig. 8. These type of events occur with partial occlusion normally. However, up to to this point, we can conclude that some results represent the effectiveness of the proposed event detector, while the rests show the degree of the difficulty of the problem at hand. TRECVid 2008 surveillance event detection task is a big challenge to test the applicability of such detector in a real world setting. Yet, we have obtained much practical experiences which will help to propose better algorithms in future. Challenges which make circumscribe the performance of our event

Figure 7. True positive ObjectGet: A person is getting some stuff from the floor which has been detected.

Figure 8. False negative: One person is putting and another is getting some stuffs but both cannot be detected.

detector include but not limited to: (i) a wide variety in the appearance of event types with different view angles; (ii) divergent degrees of imperfect occlusion; (iii) complicated interactions among people; and (iv) shadow and fluctuation. Besides these, video events have taken place significantly far distance from the camera and hence the considerable amount of motion components were insufficient to analyze over the obtained optical flow information. These extremely challenging reasons also directly reflected on the detectors proposed by (Hauptmann et al. 2008; Orhan et al. 2008; Chmelar et al. 2008), and (Dikmen et al. 2008). Among those detectors, the best ObjectGet results obtained by the detector of Orhan (Orhan and al. 2008). Their proposed detector used 1944 Object- Put events and detected 573 events. This record limited the performance of their detector by about 29.5% only, which is still far behind for real applications.

To detect events from the TRECVid 2008 are extremely difficult tasks. To solve the existing challenges of TRECVid 2008, it would take some decades for computer vision research community. Future direction would be proposing new efficient algorithms citing the solution and/or minimization of the exiting limitations. Even so our short term targets cover detection of events like OpposingFlow along with the performance improvement of the proposed event detector. Our current event detector cannot detect OpposingFlow video event. But we can detect that event by using the unused horizon-bound cluster. On the other hand, our long term targets would be detecting various events e.g., PeopleMeet, PeopleSplit, Embrace, CellToEar, Pointing, etc. by proposing better algorithms concerning some degree of minimization of the exiting limitations. In this aspect, the comprehended practical experiences will assist substantially.

4 CONCLUSION

We presented mainly a detection based approach to detect several video events. Optical flow techniques track low level information such as corners, which are then grouped into several clusters. Geometric means of locations as well as circular means of direction and displacement of the corners of each cluster are estimated. Each cluster is interpreted either lower or upper bounds to detect potential video events. The detection results of object drop, put, and get at TRECVid 2008 have been exhibited. Some results substantiate the competence of the detector, while the rests represent the degree of the difficulty of the problem at hand. The achieved practical experiences will help to put forth imperious algorithms and consequently future investigation would provoke superior results.

REFERENCES

Bileschi, S. and L. Wolf (2007). Image representations beyond histograms of gradients: The role of gestalt descriptors. In *Computer Vision and Pattern Recognition (CVPR)*, pp. 1–8.

Chmelar, P. et al. (2008). Brno University of Technology at TRECVid 2008. In *Surveillance Event Detection Pilot. http://www-nlpir.nist.gov/projects/trecvid/.*

Dalal, N. and B. Triggs (2005). Histograms of oriented gradients for human detection. In *Computer Vision and Pattern Recognition (CVPR)*, pp. 886–893.

Dikmen, M. et al. (2008). IFP-UIUC-NEC at TRECVid 2008. In *Surveillance Event Detection Pilot. http://www-nlpir.nist.gov/projects/trecvid/.*

Guo, J. et al. (2008). Trecvid 2008 event detection by MCG-ICT-CAS. In *Surveillance Event Detection Pilot. http://www-nlpir.nist.gov/projects/trecvid/.*

Hamid, R. et al. (2005). Unsupervised activity discovery and characterization from event-streams. In *Conference in Uncertainty in Artificial Intelligence (UAI)*, pp. 251–258.

Hao, S. et al. (2008). Tokyo Tech at TRECVid 2008. In *Surveillance Event Detection Pilot. http://www-nlpir.nist.gov/projects/trecvid/*.

Harris, C. and M. Stephens (1988). A combined corner and edge detector. In *Alvey Vision Conference*, pp. 147–152.

Hauptmann, A. et al. (2008). Informedia TRECVid 2008: Exploring new frontiers. In *Surveillance Event Detection Pilot. http://www-nlpir.nist.gov/projects/trecvid/*.

Hongeng, S. and R. Nevatia (2001). Multi-agent event recognition. In *International Conference on Computer Vision (ICCV)*, pp. 84–93.

Kawai, Y. et al. (2008). NHK STRL at TRECVid 2008: High-level feature extraction and surveillance event detection. In *Surveillance Event Detection Pilot. http://www-nlpir.nist.gov/projects/trecvid/*.

Lienhart, R. and J. Maydt (2002). An extended set of haar-like features for rapid object detection. In *International Conference on Image Processing (ICIP)*, pp. 900–903.

Least squares quantization in pcm. *IEEE Transactions on Information Theory 28*(2), 129–136.

Lucas, B. and T. Kanade (1981). An iterative image registration technique with an application to stereo vision. In *International Joint Conference on Artificial Intelligence (IJCAI)*, pp. 674–679.

Obstacle avoidance and navigation in the real world by a seeing robot rover. In *Technical Report CMU-RI-TR-80-03, Robotics Institute, Carnegie Mellon University & doct. diss., Stanf. University*.

Orhan, O. B. et al. (2008). University of Central Florida at TRECVid 2008: Content based copy detection and surveillance event detection. In *Surveillance Event Detection Pilot. http://www-nlpir.nist.gov/projects/trecvid/*.

Schmid, C., R. Mohr, and C. Bauckhage (2000). Evaluation of interest point detectors. *International Journal of Computer Vision 37*(2), 151–172.

Serre, T., L. Wolf, and T. Poggio (2005). Object recognition with features inspired by visual cortex. In *Computer Vision and Pattern Recognition (CVPR)*, pp. 994–1000.

Sharif, M. H. and C. Djeraba (2009). PedVed: Pseudo euclidian distances for video events detection. In *International Symposium on Visual Computing (ISVC)*, Volume LNCS 5876, pp. 674–685.

Sharif, M. H., N. Ihaddadene, and C. Djeraba (2010). Finding and indexing of eccentric events in video emanates. *Journal of Multimedia 5*(1), 22–35.

Shi, J. and C. Tomasi (1994). Good features to track. In *Computer Vision and Pattern Recognition (CVPR)*, pp. 593–600.

TRECVid, B. (2008). *Surveillance Event Detection Pilot.* The benchmark data of TRECVid 2008 available from http://www-nlpir.nist.gov/projects/trecvid/.

Viola, P. and M. Jones (2001). Rapid object detection using a boosted cascade of simple features. In *Computer Vision and Pattern Recognition (CVPR)*, pp. 511–518.

Wilkins, P. et al. (2008). Dublin City University at TRECVid 2008. In *Surveillance Event Detection Pilot. http://www-nlpir.nist.gov/projects/trecvid/*.

Xue, X. et al. (2008). Fudan University at TRECVid 2008. In *Surveillance Event Detection Pilot. http://www-nlpir.nist.gov/projects/trecvid/*.

Yokoi, K., H. Nakai, and T. Sato (2008). Toshiba at TRECVid 2008: Surveillance event detection task. In *Surveillance Event Detection Pilot. http://www-nlpir.nist.gov/projects/trecvid/*.

Computational Modelling of Objects Represented in Images – Di Giamberardino et al. (eds)
© 2012 Taylor & Francis Group, London, ISBN 978-0-415-62134-2

Sclera segmentation for gaze estimation and iris localization in unconstrained images

Marco Marcon, Eliana Frigerio & Stefano Tubaro
Politecnico di Milano, Dipartimento di Elettronica e Informazione,
Milano, Italy

ABSTRACT: Accurate localization of different eye's parts from videos or still images is a crucial step in many image processing applications that range from iris recognition in Biometrics to gaze estimation for Human Computer Interaction (HCI), impaired people aid or, even, marketing analysis for products attractiveness. Notwithstanding this, actually, most of available implementations for eye's parts segmentation are quite invasive, imposing a set of constraints both on the environment and on the user itself limiting their applicability to high security Biometrics or to cumbersome interfaces. In this paper we propose a novel approach to segment the *sclera*, the white part of the eye. We concentrated on this area since, thanks to the dissimilarity with other eye's parts, its identification can be performed in a robust way against light variations, reflections and glasses lens flare. An accurate sclera segmentation is a fundamental step in iris and pupil localization with respect to the eyeball center and to its relative rotation with respect to the head orientation. Once the sclera is correctly defined, iris, pupil and relative eyeball rotation can be found with high accuracy even in non-frontal noisy images. Furthermore its particular geometry, resembling in most of cases a triangle with bent sides, both on the left and on the right of the iris, can be fruitfully used for accurate eyeball rotation estimation. The proposed technique is based on a statistical approach (supported by some heuristic assumptions) to extract discriminating descriptors for sclera and non-sclera pixels. A Support Vector Machine (SVM) is then used as a final supervised classifier.

1 INTRODUCTION

Thanks to the increasing resolution of low cost security cameras and to widespread diffusion of Wireless Sensor Networks, eye parts tracking and segmentation is getting a growing interest The high stability of the iris pattern during the whole life and the uniqueness of its highly structured texture have imposed iris recognition as one of the most reliable and effective biometric feature for high-secure identity recognition or verification(Daugman and Downing 2007). At the same time the small size of iris details, with respect to the whole face area, and the high mobility of the eyeballs require the introduction of several constraints for the final user in order to obtain good performance from gaze tracking and iris-based recognition systems (Chou, Shih, Chen, Cheng, and Chen 2010; Duchowski 2007; Zhu and Ji 2007). Among these constraints the most common ones are: Uniform and known light conditions, head position almost frontal with respect to the camera and/or system calibration for each user(Shih and Liu 2004). Currently a significant research effort is oriented toward relaxation of these constraints maintaining a high level of performance. The small size of iris details, with respect to the whole face for iris recognition biometrics,(Chou, Shih, Chen, Cheng, and Chen 2010) or the high displacement of the observed point due to

the leverage of a small rotation of the eyeball in gaze tracking (Duchowski 2007), required the introduction of many constraints for the final user to make such systems effective and robust (Zhu and Ji 2007). Further requests concern uniform and known light conditions, head positioning almost frontal with respect to the camera and/or system calibration for each user (Shih and Liu 2004). Anyway, despite the aforementioned drawbacks, the high stability of the iris pattern during the whole live and the uniqueness of its highly structured texture made iris recognition one of the most reliable and effective biometrics for high-security applications (Daugman and Downing 2007). Many efforts are actually oriented in making this technique reliable even in almost unconstrained environments. Due to the high variability of possible acquisition contexts iris recognition or gaze tracking techniques have to be integrated with other techniques like face localizers (Viola and Jones 2004), facial features trackers (Zhu and Ji 2006), pose estimators, eye localizers and accurate eyes' parts detectors. In this paper we focus on the last topic, concerning eye's parts segmentation and in particular we tackle Sclera localization. The Sclera is the white part of the eye and its accurate segmentation can offer considerable advantages in localizing the iris, possible eyelids occlusions, and accurate estimation of eyeball rotation with respect to the facial pose.

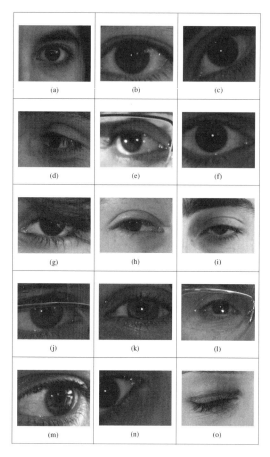

Figure 1. Noise types, see text above for a description.

2 DATABASE DEFINITION

The database we used for algorithm tuning and test evaluation is the UBIRIS v.2 database (Proença and Alexandre 2004), this database differs from other iris databases, like CASIA (Tan 2010) or UPOL (Dobes and Machala 2004), since in UBIRIS the acquired iris images are from those relative to perfect acquisition conditions and are very similar to those that could be acquired by a non-invasive and passive system. In particular UBIRIS is composed of 1877 images collected from 241 persons in two different sessions. The images present the following types of "imperfections" with respect to optimal acquisitions (shown in fig. 1):

- *Images acquired at different distances* from $3\,m$ to $7\,m$ with different eye's size in pixel (e.g. *a* and *b*).
- *Rotated images* when the subject's head is not upright (e.g. *c*).
- *Iris images off-axis* when the iris is not frontal to the acquisition system (e.g. *d* and *e*).
- *Fuzzy and blurred images* due to subject motion during acquisition, eyelashes motion or out-of-focus images (e.g. *e* and *f*). item *Eyes clogged by hair* Hair can hide portions of the iris. (e.g. *g*)

- *Iris and sclera parts obstructed by eyelashes or eyelids* (e.g. *h* and *i*)
- *Eyes images clogged and distorted by glasses or contact lenses* (e.g. *j* and *k* and *l*)
- *Images with specular or diffuse reflections* Specular reflections give rise to bright points that could be confused with sclera regions. (e.g. *l* and *m*)
- *Images with closed eyes or truncated parts* (e.g. *n* and *o*)

3 COARSE SCLERA SEGMENTATION

The term sclera refers to the white part of the eye that is about 5/6 of the outer casing of the eyeball. It is an opaque and fibrous membrane that has a thickness between 0.3 and 1 mm, with both structural and protective function. Its accurate segmentation is particularly important for gaze tracking to estimate eyeball rotation with respect to head pose and camera position but it is also relevant in Iris Recognition systems: since the Cornea and the Anterior Chamber (*Aqueous humor*), the transparent parts in front of the Iris and the Pupil, present an uneven thickness; Due to the variation of their refraction indices with respect to the air, if the framed Iris is not strictly aligned with the camera, its pattern is distorted accordingly to the refraction law. Eyeball orientation with respect to the camera should then be estimated to provide the correct optical undistortion.

The first step in the described Sclera segmentation approach is based on a dynamic enhancement of the R,G,B channel histograms. Being the sclera a white area, this will encourage the emergence of the region of interest. Calling x_{min} and x_{max} the lower and the upper limit of the histogram of the considered channel, assuming that the intensities range from 0 to 1, we apply, independently on each channel, a non-linear transform based on a sigmoid curve where the output intensity y is given by:

$$y = \frac{1}{1 + exp\left(-\alpha\frac{x - \bar{x}}{\sigma}\right)}$$

where $\bar{x}$ is the mean value of x, σ is the standard deviation and we assumed $\alpha = 10$; this value was chosen making various tests trying to obtain a good contrast between sclera and non-sclera pixels.

3.1 Training dataset

We manually segmented 250 randomly-chosen images from the whole database dividing pixels into sclera (ω_S) and non-sclera (ω_N) classes; each pixel can then be considered as vector in the three dimensions (Red, Green and Blue)color space $\mathfrak{R}^3$. 100 of these vectors are then used in a Linear Discriminant Analysis classifier (Fukunaga 1990) with the two aforementioned classes. Using this simple linear classifier we can obtain a coarse mask for Sclera and Non-Sclera pixels using the Mahalanobis distances, $D_S(y)$ and $D_N(y)$

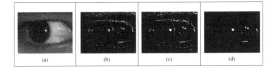

Figure 2. (*a*) is the original, normalized image, (*b*)is a binary mask where '1' represents pixels whose gradient modulus is above 0.1, (*c*)is the mask with filled regions and (*d*) is the mask only with regions of high intensity pixels.

respectively from ω_S and ω_N; we recall (Fukunaga 1990) that $D_i(\mathbf{x}) = \sqrt{(\mathbf{x} - \mu_i)^T \Sigma_i^{-1} (\mathbf{x} - \mu_i)}$ where μ_i is the average vector and Σ_i is the Covariance Matrix. Accordingly to a minimum Mahalanobis distance criterium we define a Mask pixel as:

$$M_S = \begin{cases} 1 & \text{if } D_S(y) \leq D_N(y) \\ 0 & \text{if } D_S(y) > D_N(y) \end{cases}$$

4 DEALING WITH REFLECTIONS

Specular reflections are always a noise factor in images with non-metallic reflective surfaces such as cornea or sclera. Since sclera is not a reflective metallic surface, the presence of glare spots is due to incoherent reflection of light incident on the Cornea. Typically, the pixels representing reflections have a very high intensity, close to pure white. Near to their boundaries, there are sharp discontinuities due to strong variations in brightness. The presented algorithm for the reflexes identification consists mainly of two steps:

- Identification of possible reflection areas by the identification of their edges through the use of a sobel operator;
- Intensity analysis within the selected regions.

The first step, using a gray-scaled version of the image, is based on an approximation of the modulus of the image gradient, $\nabla I(x, y)$, by the Sobel 3×3 operator: $G_x = \begin{bmatrix} -1 & 0 & 1 \\ -2 & 0 & 2 \\ -1 & 0 & 1 \end{bmatrix}$.

So $|\nabla I(x, y)| = \sqrt{G_x^2 + G_y^2}$ where $G_y = G_x^T$. Due to the sharp edges present on reflexes we adopted a threshold of 0.1 to define reflex edges; the threshold value was chose as the best choice in the first 100 used images (the ones used for training). For each candidate region we then check, through a morphological filling operation for 8-connected curves (Gonzalez and Woods 1992), if it is closed, and, if this is the case, we assume it as a reflex if all pixels inside have an intensity above the 95% of the maximum intensity present in the whole image. These steps are shown in fig. 2 where reflexes are insulated. Subtracting reflex regions from the previously defined candidates for the Sclera regions provides us with the first rough estimation of Sclera Regions (Fig. 3).

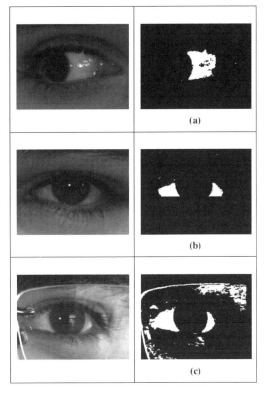

Figure 3. Preliminary results with reflexes removal, in (*c*) classification errors with glasses are presented.

4.1 *Results refinement*

A first improvement with respect to the aforementioned problems is based on morphological operators, they allow small noise removal, holes filling and image regions clustering: we applied the following sequence:

- Opening with a circular element of radius 3. It allows to separate elements weakly connected and to remove small regions.
- Erosion with a circular element of radius 4. It can help to eliminate small noise still present after the opening and insulate objects connected by thin links.
- Removal of elements with area of less than 4 pixels.
- Dilation with a circular element of radius 7 to fill holes or other imperfections and tends to joint neighbor regions.

Radii of structuring elements to perform aforementioned tasks were heuristically found. They performed well on UBIRIS images but may be that different acquisition set-ups could require a different tuning.

Intensity analysis within the selected regions fits well with the UBIRIS database, but, in case of different resolution images should scale accordingly to the eye size. The results are presented in fig. 4

The center of the different regions are then used as seeds for a watershed analysis that will allow us to obtain accurate edges for each region in the

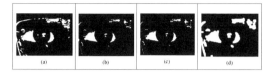

Figure 4. Results for the application of the 4 morphological operators as described in section 4.1.

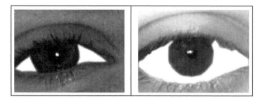

Figure 5. Two examples of segmented sclera regions

image. We performed the watershed algorithm accordingly to the description of F. Meyer and reported in (Vincent and Soille 1991). Different regions, accordingly to the gradient formulation, are then defined whenever a basin area is not significantly changing for consistent intensities variation (water filling approach; for further details see (Zhao, Liu, Li, and Li 2008; Colletto, Marcon, Sarti, and Tubaro 2006)).

4.2 Final classification step

The aforementioned algorithm provided no false negative regions but many false positives are present in images where ambiguous regions, due to glasses or contact lenses reflexes, are confused with sclera. Analyzing all the images included in the considered database we observed that the true sclera region have only two possible configurations: (1) like a triangle with two extruded sides and one intruded (fig. 5 left) or, when the subject is looking up, (2) like two of the previous triangles, facing each other, and connected with a thin line on the bottom (fig. 5 right).

On the basis of these considerations we decided to define a set of features that can be used to separate, among all the selected regions, those that represent real scleras. The vector is composed by the following elements (invariant to translations and rotations):

- Region area in pixels,
- Ratio between region area and its perimeter,
- 7 Hu invariant moments (Hu 1962).

Note: concerning the seventh moment I_7 (Hu 1962) we used its module since we do not want to distinguish mirror images.

We decided to use this shape-based feature vector since sclera regions are, for all the considered database cases, of two types: like a triangle with two extruded sides and one intruded (fig. 5 left) or, when the subject is looking up, like two of the previous triangles, facing each other, and connected with a thin line on the bottom (fig. 5 right). The final region classifier

is then based on a Support Vector Machine accordingly to the Cortes-Vapnik formulation:(Cortes and Vapnik 1995). where sclera and non-sclera regions are classified accordingly to: $\hat{y} = sign\left(f(\mathbf{x})\right)$, where

$$f(\mathbf{x}) = \sum_{i \in S} \alpha_i y_i \Phi\left(\mathbf{x}_i\right) \cdot \Phi\left(\mathbf{x}\right) + b = \sum_{i \in S} \alpha_i y_i \mathcal{K}\left(\mathbf{x}_i, \mathbf{x}\right) + b$$

α_i are Lagrange multipliers, b is a constant, S are elements for which the relative multiplier α_i is non-null after optimization process and $\mathcal{K}$ is the non-linear kernel, in our case it is: $\mathcal{K}(\mathbf{a}, \mathbf{b}) = (\mathbf{a} \cdot \mathbf{b} + 1)^2$; the problem, then becomes a Quadratic Programming Problem and optimization algorithms can be used to solve it (Hiller and Lieberman 1995).

5 RESULTS

Analyzing results over all the 250 chosen images (100 of them were also used for training) of the Ubiris database we obtained a 100% recognition of closed eyes (no sclera pixels at all), 2% false negative (with respect to the handmade segmented region) in the rare cases of very different brightness from left to right eye parts (due to a side illumination), and we also report a 1% of false positive (again with respect to the handmade segmented region) for particular glass-frames that locally resemble sclera regions.

REFERENCES

Chou, C., S. Shih, W. Chen, V. Cheng, and D. Chen (2010). Non-orthogonal view iris recognition system. *IEEE Transactions on Circuits and Systems for Video Technology* 20(3), 417–430.

Colletto, F., M. Marcon, A. Sarti, and S. Tubaro (2006, October). A robust method for the estimation of reliable wide baseline correspondences. In *Proceedings of International Conference on Image Processing, 2006*, pp. 1041– 1044.

Cortes, C. and V. Vapnik (1995). Support-vector networks. *Machine Learning 20*, 273–297. Daugman, j. and C. Downing (2007). Effect of severe image compression on iris recognition performance.

Dobes, M. and L. Machala (2004). Upol iris image database, link: http://phoenix.inf.upol.cz/iris/.

Duchowski, A. (2007). *Eye Tracking Methodology: Theory and Practice, 2nd ed.* Springer.

Fukunaga, K. (1990). *Introduction to Statistical Pattern Recognition* (2nd edn. ed.).

Gonzalez, R. and R. Woods (1992). *Digital Image Processing.* Addison-Wesley Publishing Company.

Hiller, F. and G. J. Lieberman (1995). *Introduction to Mathematical Programming.*

Hu, M. K. (1962). Visual pattern recognition by moment invariants. *IRE Trans. Info. Theory 8*, 179–187.

Proença, H. and L. A. Alexandre (2004). Ubiris iris image database, link: http://iris.di.ubi.pt. Shih, S. W. and J. Liu (2004). A novel approach to 3-d gaze tracking using stereo cameras. *IEEE Trans. Syst. Man Cybern. B 34*, 234.

Tan, T. (2010). Chinese accademy of science institute of automation casia iris database, link: http://biometrics. idealtest.org/.

Vincent, L. and P. Soille (1991). Watersheds in digital spaces: an efficient algorithm based on immersion simulations. *IEEE Transactions on Pattern Analysis and Machine Intelligence* 13(6), 583–598.

Viola, P. and M. J. Jones (2004). Robust real-time face detection. *International Journal of Computer Vision* 57(2), 137–154.

Zhao, Y., J. Liu, H. Li, and G. Li (2008). Improved watershed algorithm for dowels image segmentation. In *Intelligent Control and Automation, 2008. WCICA 2008. 7th World Congress on*, pp. 7644.

Zhu, W. and Q. Ji (2007, Dec). Novel eye gaze tracking techniques under natural head movement. *Biomedical Engineering, IEEE Transactions on* 54(12), 2246.

Zhu, Z. and Q. Ji (2006). Robust pose invariant facial feature detection and tracking in real-time. In *Proceedings of the 18th International Conference on Pattern Recognition (ICPR'06)*.

Computational Modelling of Objects Represented in Images – Di Giamberardino et al. (eds)
© 2012 Taylor & Francis Group, London, ISBN 978-0-415-62134-2

Morphing billboards for accurate reproduction of shape and shading of articulated objects with an application to real-time hand tracking[1]

Nils Petersen & Didier Stricker
German Research Center for Artificial Intelligence, DFKI

ABSTRACT: We present a novel approach to image based rendering of articulated objects that gives reliable estimates of the object's shape and shading in new, previously unseen poses. The main idea is to combine a kinematic 3D model with a novel extension to billboard rendering in combination with a computationally lightweight pixel-wise morphing technique.

The result is both visually appealing and accurate with respect to resolving texture, shading, and elastic deformation of the articulated object. Additionally, we can show experimentally that using our morphing technique allows the formulation of bias-free pixel-wise objective functions with significantly less local optima than using a static texture. Our method is extremely lightweight which makes it very suitable for real-time tracking using a generative model. We demonstrate the applicability of our approach in a hand tracking context on synthetic, ground-truth labeled data and show several examples constructed from synthetic and real hand images.

1 INTRODUCTION

Fast and accurate rendering of objects plays an important role in analysis-by-synthesis frameworks, particularly in the hand tracking context using a generative model. Therefore a variety of methods has been proposed for accurately predicting the appearance of a human hand in certain poses. Most of these methods rely on hand silhouette and edge features only. Especially in a monocular setting recovering an accurate hand posture using these features is a severely ill-posed problem as many postures create the same or very similar silhouette and edge information. The authors of (de La Gorce, Paragios, and Fleet 2008) have proposed the use of texture and shading information to alleviate the observation ambiguities using a textured, deformable 3D mesh model. A complementary approach is the database indexing approach that relies on similarity search in large database of object descriptors labeled with the resp. generating parameters, for example (Gionis et al. 1999) or (Shakhnarovich et al. 2003).

We present a method to synthesize a low-textured, articulated object from other views through image-based rendering that could be used to explore the parameter space between a set of nearest neighbors.

Since the human hand is kinematically complex with 22–30 degrees of freedom with prevalent self-occlusion we resort to a strong geometric proxy to gain robustness. Our approach faithfully approximates both shape and shading of a hand in an unseen target pose even with large unobservable hand parts in the prototype images used for synthesizing. This allows the use of less generalizing, pixel-wise objective functions that incorporate shading information.

The two main contributions of this work are

- an extension to billboard-rendering that we call 2.5D billboards that well describes ellipsoid 3D objects.
- an efficient morphing technique to minimize ghosting and preserve shape in presence of elastic deformation and model alignment errors.

Compared to existing methods our approach has the benefit of being computationally extremely lightweight and requiring only a coarse model fit and object segmentation.

Although the prediction quality improves with shorter distance of the prototypes to the target pose, we will show that our method can handle unconstrained, arbitrary input views. The method is not limited to hands but is in principle applicable to all articulated low-textured objects.

In the remainder of this paper we proceed with a review of related work followed by a detailed description of the method. Afterwards we demonstrate the applicability for tracking using a pixel-wise distance function to incorporate shading information in the pose recovery.

---

[1]The work presented in this paper was performed in the context of the Software-Cluster project EMERGENT (www.software-cluster.org). It was funded by the German Federal Ministry of Education and Research (BMBF) under grant no. "01IC10S01". The authors assume responsibility for the content.

2 RELATED WORK

Relevant related work is found in the hand tracking and image-based rendering (IBR) literature.

A general overview of hand tracking techniques can be taken from (Erol and Bebis 2005).

The authors (de La Gorce, Paragios, and Fleet 2008) use a deformable 3D mesh model to incorporate texture and shading information for frame-by-frame monocular hand tracking. Their method minimizes a pixel-wise distance between the current observation and a deformable mesh model with the texture taken from the last frame. The use of a deformable skinning model is computationally comparatively expensive and therefore achieving real-time performance is difficult. Also the shading information is captured from one frame alone and then approximated for the subsequent frame using Gouraud-shading. However, there is no straight forward way to extending this approach to greater pose difference.

The authors (Stenger et al. 2001) use truncated quadrics to approximate the shape of the human hand. As quadrics can be efficiently projected to the image plane the resulting silhouette/edge model is computationally lightweight. Though incorporation of accurate, measured shading information is not directly possible.

Also a combination of the two aforementioned approaches with our texturing approach is also not feasible. Both models require well defined object boundaries and would therefore require a very precise fit and non-blurred image material to produce decent results. These prerequisites are mostly prohibitive in realistic applications.

In the work of (Lin et al. 2002) the authors propose a "cardboard model". They use model-aligned rectangular shapes to approximate the hand's contour edges. To prevent collapsing of their rectangular segments they strictly limit the possible viewpoints of their model – a restriction that does not apply with our method.

An Image based rendering (IBR) method that is related to ours is the work of (Germann and Hornung 2010). The method generates new views of football players from wide-baseline stadium cameras. The players are modeled as articulated bodies with billboard-fans aligned to each bone. Thus, the authors call their approach "articulated billboards". Instead of performing a pixel-wise morph as proposed in our work they use a separate billboard per segment and camera, arranged in a fan around the kinematic bone. Thus, they have less control over elastic deformation. However, just as in the cardboard model ((Lin et al. 2002)) their billboards are not stable at all relative viewpoints (that do not occur in their application).

The idea of morphing between prototype views images was pursued by (Jones and Poggio 1998) and (Cootes et al. 1998). Their approach is based on establishing point correspondences which is not easily possible in our application due to the high articulatory complexity of the hand. Also since we are decomposing the necessary warp into a segment-wise rigid

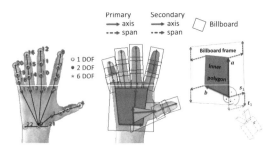

Figure 1. Schematic view of the kinematic hand model (**left**). The 2.5D billboards associated with this model (**middle**). Illustration of a single 2.5D billboard (**right**).

(predicted) transformation and an axis-aligned efficient pixel-wise warp we expect our method to be faster by orders of magnitude.

3 METHOD DESCRIPTION

Our proposed method relies on a kinematic skeleton with adjacent billboards whose distinct textures are being morphed according to the view change. Figure 1 illustrates the kinematic model used to pose the hand (left) and the aligned billboards (right).

The method consists of two main parts. The first part consists of capturing the appearance information from labeled views of the object, the so-called prototypes $\mathcal{P}$. This has to be conducted only once per prototype $p \in \mathcal{P}$ and could very well be stored in a database as this extraction step is only dependent on the prototype and not on the desired target view of the object.

The second part is the synthesis at run-time. We distinguish between three main effects: segment-wise rigid transformation, elastic deformation, and shading change due to changed relative lighting. The rigid transformation which has the strongest impact on the appearance is conducted through positioning and deforming the billboards along the articulated model to match the target pose. To account for elastic effects and to compensate for model alignment errors we propose an efficient axis-aligned pixel-wise warping approach. As we assume relatively low texture, simple blending is sufficient to interpolate the shading changes between two not too distant prototypes after these two formation steps. Compare (Kemelmacher-Shlizerman et al. 2011) for a comment on why this is a sufficient interpolation in this case. In Figure 2 one can see a comparison between rendering with and without the axis-aligned morphing.

In the following subsection we start to explain how we model the hand using our proposed extension to billboards. After that we describe the two steps of our method, capturing and synthesis in detail.

3.1 Billboard appearance model

As mentioned above, the hand is modeled as a set of billboards connected through a kinematic model.

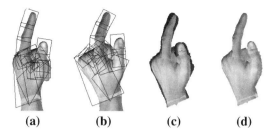

| (a) | (b) | (c) | (d) |

Figure 2. Using prototype views **(a)** and **(b)** cross-fading the re-articulated models **(c)** leads to visible artifacts due to model-alignment errors and elastic deformation. This does not occur using our morphing technique **(d)**.

Billboards can faithfully simulate rotation-symmetric objects[2] such as the finger segments, as long as the viewpoint is not changing substantially other than around the symmetry axis.

Like (Germann and Hornung 2010) we are aligning each billboard with a bone **a** of the kinematic model. The billboard is then spanned using a vector $\mathbf{s}_i = \alpha_i(\mathbf{a} \times [\mathbf{x}_i - \mathbf{c}])$ perpendicular to **a** and parallel to the image plane, i.e. perpendicular to the vector from camera **c** to joint $\mathbf{x}_i$. The scalar α_i is a normalization resp. scaling factor.

This billboard definition suffers from the problem that if the camera view direction is aligned with **a**, the billboard collapses to a single line. Additionally, for instance the palm of the hand exhibits a strong change in shape with change of viewpoint. Therefore we propose an extension to billboards solving both issues which is capable of reproducing non rotation-symmetric objects more faithfully from arbitrary viewpoints. Compare right illustration of Figure 1.

Our approach can be interpreted as combination a 3D planar polygonal patch and a surrounding billboard "frame" always aligned with the image plane. Therefore we call this extension 2.5D billboards. It allows viewpoint dependent minimum and maximum shapes that can be used to describe rotational asymmetry and thus a larger class of convex 3D objects.

Formally, we achieve this by introducing a secondary axis **b** and a secondary perpendicular span vector $\mathbf{t}_i = \beta_i(\mathbf{b} \times [\mathbf{x}_i - \mathbf{c}])$ with the normalization factor β_i. The two axes span a plane in 3D space. Let $\{\mathbf{x}_i\}, |\{\mathbf{x}_i\}| \geq 3$ be the set of vertices of a convex polygon in this plane. In fact the convexity-constraint is too strict but simplifies the construction process with respect to avoiding self-intersection. We build a "billboard frame" around this by adding axis aligned span vectors to obtain the billboard vertices

$$\mathbf{y}_i = \mathbf{x}_i + \mathbf{s}_i + \mathbf{t}_i \qquad (1)$$

Although the orientation of the billboard vectors is entirely determined, their lengths given by α_i and β_i are free parameters for each vertex $\mathbf{x}_i$. To avoid

---

[2] Also objects, where rotational appearance change is not directly apparent, e.g. trees.

self-intersection of the resulting billboard polygon we do enforce the following three constraints when determining α_i and β_i:

- $\{\mathbf{y}_i\}$ represents a convex polygon.
- The billboard frame on the outside of $\{\mathbf{x}_i\}$.
- Let **m** be the centroid of $\{\mathbf{x}_i\}$, $(\mathbf{y}_i - \mathbf{m})(\mathbf{x}_i - \mathbf{m}) > 1$.
- The order is maintained, i.e. if $\mathbf{x}_i$ is the clockwise neighbor of $\mathbf{x}_{i+1}$ the same holds for $\mathbf{y}_i$ and $\mathbf{y}_{i+1}$.

Since inner and outer polygons are both convex, enforcing these constraints for an arbitrary viewpoint is sufficient for the constraints to hold for all viewpoints. The right side of Figure 1 shows the billboards the hand model consists of. The points with the numbers 5, 9, 17, 22, 21 were used as vertices for the palm polygon using two linear combinations for the primary and secondary axis. The root of the thumb was modeled using the points 0, 2, 2, and a 4th point determined empirically, approximately in the center of the palm. Point 2 is used twice but each time with different span vector weights in the billboard construction.

Applied to finger segments there is no obvious choice for a secondary axis. To solve this we use the segment's bone as primary axis and reuse the primary span vector as secondary axis. If the bone aligns with the camera's viewing direction this span vector would become zero and the billboard would collapse to single point. Therefore we substitute the secondary axis with the primary span vector of the preceding segment in the kinematic chain until we have found a non-zero span-vector.

3.2 *Capturing prototype appearance*

Each prototype consists of the set of generating parameters (position and joint angles) and the corresponding appearance information. To sample this appearance information from an image with given generating parameters we assume that a rough segmentation is available to distinguish the hand from background. Typically, this is achieved through skin color segmentation.

To identify the association between pixels and billboards we first articulate our kinematic model according to the given parameters. We then project the resulting model onto the image and label the segments accordingly. Pixels that are located within two billboard areas, e.g. within overlaps along each finger, get copied into both billboard textures.

To avoid sampling neighboring fingers into the same billboard texture we prune each texture line-wise by only keeping the biggest connected segment. If there is at least a pixel gap between neighboring fingers this solves the issue. We have experimented with grab-cut to solve unintentional co-sampling more generally for the case of touching occluding and occluded parts. The results however did not improve upon always co-sampling as the deteriorative effects are alleviated by the texture morphing phase.

Also, the prototype views often contain unobservable parts, e.g. fingers occluded by the palm. We

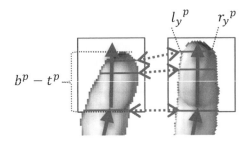

$l_y{}^p \quad r_y{}^p$

$b^p - t^p$

Figure 3. Illustration of the axis-aligned morphing scheme.

nevertheless sample the area where the occluded segment would be located. Although this leads to rectangular artifacts (since our billboards are mostly rectangular), this assures a smooth transition away from this prototype which is important for derivability of the objective function. Our evaluation shows that this method leads to a dominantly monotonic pixel-wise objective function despite the sampling errors.

3.3 Axis-aligned morphing between prototypes

The biggest influence, the segment-wise rigid transformation, is handled through the billboard articulation and accounts for rotational alignment of prototype and target billboard axes. As we show in the following evaluation only relying on the 2.5D billboard transformation is already providing useful results.

However, to allow exploitation of shading information, a more precise approximation is required. Morphing the billboard texture, i.e. blending and simultaneous warping could improve the results, though the incorporated warping step is generally too costly to be used as an objective function. Due to the preceding rigid transformation we can formulate an axis-aligned warping technique that produces satisfying results while being computationally comparable to a non-uniform scale.

We are exploiting two observations about the problem: Firstly, the fact that the visual effects due to warping within the object's silhouette is negligible compared to warping that affects the boundary because of the relatively low texture. Secondly, we use that the "bones" of the kinematic chain are always roughly aligned with the object boundary.

Figure 3 illustrates the method. When capturing the prototypes we make sure that we store the billboard texture always aligned on the principal axis. We additionally compute the left and right contour boundaries of every billboard texture which we do implicitly when cropping each line on the largest connected segment. Left and right denotes left and right of the kinematic bone, i.e. the principal axis of the billboard. Additionally we calculate the top and bottom boundary of the texture t^p and b^p, $\forall p \in \mathcal{P}$. In practice, we only calculate top and bottom on billboard textures representing the finger tips and assume zero resp. the texture's height elsewhere.

For the resulting texture we take the weighted average for top $t^{res} = \sum_{p \in \mathcal{P}} \gamma^p t^p$ and bottom $b^{res} = \sum_{p \in \mathcal{P}} \gamma^p b^p$ and row-wise left boundary $l_y^{res} = \sum_{p \in \mathcal{P}} \gamma^p l_y^p$ and right boundary $r_y^{res} = \sum_{p \in \mathcal{P}} \gamma^p r_y^p$.

The computation of the weights γ^p will be described later. The morphing can then be formulated as

$$\mathbf{T}^{res}(x,y) = \sum_{p \in \mathcal{P}} \gamma^p \mathbf{T}^p(x^p(y^p), y^p) \tag{2}$$

$\forall y \in [t^{res}, b^{res}]$ with

$\forall y \in [t^{res}, b^{res}]$ with

$$y^p = (y - t^{res}) \frac{b^p - t^p}{b^{res} - t^{res}} + t^p \tag{3}$$

$$x^p(y^p) = (x - l_{y^p}^{res}) \frac{r_{y^p}^p - l_{y^p}^p}{r_{y^p}^{res} - l_{y^p}^{res}} + l_{y^p}^p \tag{4}$$

where $\mathbf{T}(x,y)$ is a placeholder for billboard textures at pixel coordinate x, y with $\mathbf{T}^p(x,y)$ denoting the texture of prototype p and $\mathbf{T}^{res}(x,y)$ denoting the resulting texture.

The posed and warped prototypes are similar enough that the subsequent cross-fade is well approximating the fine-grained changes from relative movement of the light source and fine-grained deformation as (Kemelmacher-Shlizerman et al. 2011) pointed out. Only small areas of the hand violate this assumption and exhibit a certain degree of ghosting, most importantly the shirt-sleeve and the finger nail area.

Without warping there would be substantial ghosting effects at the entire boundary, see Figure 2. Particularly since we do not dedicatedly treat unobservable image content these boundary effects would be very evident. The respective billboard texture is then likely to fill the whole (rectangular) billboard area.

As we do not perform any deformation to account for bending at joints we do have visible boundaries between the single segments. A deformation method like (Jacobson et al. 2011) would allow a smooth transition but at the cost of a disproportionally increased computational cost. However, since pixel-wise distance functions are not very sensitive to slight texture discontinuities this does not result in an accuracy loss.

3.4 Determining blending weights

The definition of the blending weight γ^p has a crucial influence. Simply taking γ^p reciprocal to the distance of the target parameter $\mathbf{t}$ to $\mathbf{p}^p$ does not give appropriate results which is illustrated in the following examples given the three prototypes $(0,0)^T$, $(10,0)^T$, and $(20,0)^T$. The neighborhood problem: trying to synthesize $\mathbf{t} = (5,0)^T$ would still incorporate the prototype $(20,0)^T$ though only using $(0,0)^T$ and $(10,0)^T$ produces a "cleaner" result. The second, is called lateral-position problem: trying to synthesize $\mathbf{t} = (0,10)^T$ should only take the prototype $(0,0)^T$

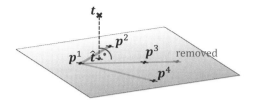

Figure 4. Illustration of the prototype subspace used to calculate blending weights.

Figure 5. Difference images between (synthetic) observation and our rendering $|\mathbf{I}^{obs} - \mathbf{I}^{render}_{\alpha}|$ for several values of α. Center image is $\alpha = 0.5$.

as neither prototype contains information about the second component.

To solve this, we are using the following procedure to define the blending weights. First we are sorting the prototypes in $\mathcal{P}$ such that $\mathbf{p}^1$ is the nearest neighbor to $\mathbf{t}$ in parameter space, $\mathbf{p}^2$ the second nearest and so on. We chose the closest prototype $\mathbf{p}^1$ as support vector for a subspace, see Figure 4. Whenever a span vector $\mathbf{p}^i - \mathbf{p}^1$ is representable using a linear combination of any $\langle \mathbf{p}^j - \mathbf{p}^1 \rangle, 2 \le j < i$ it is removed from the set of prototypes until we have the k linearly independent prototypes $\mathbf{p}^1$ to $\mathbf{p}^k$ in $\mathcal{P}_k$.

We then collect span vectors for the prototype subspace in $\mathbf{P} = (\mathbf{p}^2 - \mathbf{p}^1 \mid \ldots \mid \mathbf{p}^k - \mathbf{p}^1)$.

The orthogonal projection $\hat{\mathbf{t}}$ of $\mathbf{t}$ in the subspace $\mathbf{P}$ then satisfies

$$\mathbf{P}^T(\mathbf{t} - \mathbf{p}^1 - \hat{\mathbf{p}}) = \mathbf{0} \qquad (5)$$

Since we want $\hat{\mathbf{t}}$ to be included in the subspace we substitute $\mathbf{P}\bar{\mathbf{t}} = \hat{\mathbf{t}}$ and write

$$\mathbf{P}^T(\mathbf{t} - \mathbf{p}^1) = \mathbf{P}^T\mathbf{P}\bar{\mathbf{t}} \qquad (6)$$

As the inverse of $\mathbf{P}^T\mathbf{P}$ exists we can solve this as

$$\bar{\mathbf{t}} = (\mathbf{P}^T\mathbf{P})^{-1}\mathbf{P}^T(\mathbf{t} - \mathbf{p}^1) \qquad (7)$$

The factor γ^p is the blending weight of the p-th prototype is then defined to be

$$\gamma^p \propto 1/|\mathbf{p}^p - \mathbf{P}\bar{\mathbf{t}} + \mathbf{p}^1|_2 \qquad (8)$$

and is normalized to satisfy $\sum_{p \in \mathcal{P}} \gamma^p = 1$.

4 EVALUATION

As already stated, we propose our method as appearance model for tracking with a pixel-wise objective function. To investigate its feasibility in this regard and to obtain quantitative measures we use renderings generated with Poser with ground-truth available. For increased comparability with real scenes, the texture of the Poser model was augmented with a green stripe around the wrist to represent the shirt-sleeve and the model was always render on top of of cluttered background.

For our experiments we have generated 2000 random hand poses including random finger articulation

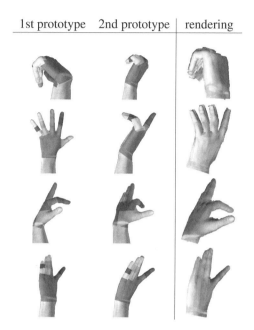

| 1st prototype | 2nd prototype | rendering |

Figure 6. Examples using synthetic prototypes. Upper two rows show results for partly unobservable areas, lower two rows for fully observable prototypes.

and random viewpoints except from impossible angles (e.g. from within the arm joint and intersecting fingers) used as target views. Each target view was then rendered from viewpoints differing by $\pm 20°$, $\pm 40°$, and $\pm 60°$ degrees out-of-plane rotation. These views were used as prototype views $\mathbf{p}^-, \mathbf{p}^+$ to sample our model from. To measure the improvement due to morphing we perform each experiment once with both prototypes and once only using the prototype $\mathbf{p}^-$ at $-20°$, $-40°$, resp. $-60°$. We did not enforce any further visibility constraints other than rejecting impossible poses. Particularly with higher viewpoint distance the prototype views exhibit substantial amounts of unobservable areas with respect to the object parts visible in the target view.

To simulate the performance of a pixel-wise objective function, we have rendered the target view onto a cluttered background, further denoted as $\mathbf{I}^{obs}$. Using our method we render the model onto the same background for all parameters $\mathbf{p}_{\alpha} = (1 - \alpha)\mathbf{p}^- + \alpha\mathbf{p}^+$ with $\alpha \in [0, 1]$, denoted as $\mathbf{I}^{render}_{\alpha}$. So ideally $\mathbf{I}^{obs} = \mathbf{I}^{render}_{0.5}$ should hold.

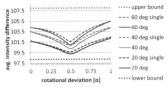 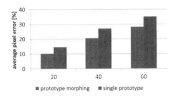 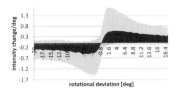

Figure 7. Average pixel error in percent between observation and our rendering given the true parameters (**left**) and objective function between prototype parameters (**middle**) for $\pm 20°$, $\pm 40°$, and $\pm 60°$ degrees prototype distance using one or two prototypes. Derivative of the objective function (**right**) using two prototypes at $\pm 20°$. The gray area contains 90% of all derivatives.

Figure 8. Example using real data. The leftmost and rightmost images show the frames used to extract the prototypes. The images in between show interpolated frames using our method.

To measure the image difference we use the average pixel-wise difference $f_\alpha = (1/pixel\ count) \sum_{\forall x,y} |\mathbf{I}^{obs}(x,y) - \mathbf{I}_\alpha^{render}(x,y)|$, see Figure 5.

The resulting graphs are seen in Figure 7 for all viewpoint distances, resp. two prototypes or only the $\mathbf{p}^-$ one. For easier understanding we have calculated an upper bound which is the average pixel value of $\mathbf{I}^{res}$ and a lower bound by subtracting the true target view from the image and then determining the average pixel value.

Averaged across our 2000 repetitions all experiments show clearly defined optima around $\alpha = 0.5$. As expected, closer prototype viewpoints lead to better results and morphing between two prototypes increases the accuracy of the prediction significantly. One thing to notice though is that when only using the $\mathbf{p}^-$ prototype, the optimum is slightly ($\leq 1.2°$) biased to the negative side while using both, the optimum is at the true optimum.

To better rate the improvements we have compared the function at the true optimum. We show these results in the left graph of Figure 7 normalized so that 0% corresponds to the lower and 100% to the upper bound. The improvement between using one and two prototypes is always between 20% and 30%. With two prototypes at $\pm 20°$ viewpoint deviance our prediction and the ground truth image differentiate by only 10%.

The improvement when using the morphing scheme is substantial. With 28% error using 2 prototypes at $\pm 60°$ our method almost achieves the same score as using one prototype at $\pm 40°$ (27% error). The closer the prototype views are to the target, the steeper becomes the objective function around the true optimum.

More important than the value of the absolute optimum is the existence of local optima in the objective function. The right plot of Figure 7 shows the derivative

of the objective function using two prototypes at $\pm 20°$. The gray area contains all derivatives between 5% and 95% α-quantiles, i.e. contains 90% of all occurring derivatives. The colored plot represents the mean. As the numbers indicate, the majority of trials exhibit a good monotonicity. This is also due to the pixel-wise distance functions since edge-based objective functions typically exhibit a vast amount of local optima. For $\pm 20°$ using two prototypes about 17% of the randomly picked target views do not have a single local optimum which is 52% more than when using only a single prototype.

Figure 6 shows further examples using synthetic images both with and without unobservable areas. Eventually, Figure 8 shows example results of our method using real-world prototypes to show its applicability in this context. The prototype views shown left- and rightmost were labeled manually with their generating parameters extracted through inverse kinematics. The transition views were performed by linearly interpolating between the two parameter sets. As no real ground-truth is available for this we do not perform an quantitative evaluation.

Run-time performance is an important aspect, particularly for real-time tracking. Our current implementation runs entirely on the CPU at 60 frames per second or approx. 16 ms per frame. For each frame 6 ms are spent computing the morphed texture and the remaining 10 ms to render the billboards. Since the billboards can be rendered using only 33 textured triangles and the morphing step also well suitable for a GPU implementation we expect our frame rate to be much higher when ported to the GPU.

Since the texture-morphing step is not necessary in every iteration the performance can be further increased. The morphed billboard texture can be reused for a certain distance as the rigid transformation due to billboard movement is sufficiently large compared to elastic deformation.

5 CONCLUSION AND FUTURE WORK

We have presented an image-based rendering approach to realistic and accurate rendering of low-textured articulated objects such as hands. Our method is extremely lightweight and well suited as hand appearance model for real-time hand tracking, in particular for refining results in presence of nearest neighbor hypotheses. The proposed morphing technique using two prototypes improves pixel-wise error by up to 30% compared to using only one prototype. More importantly it improves the monotonicity of a pixel-wise objective function (measured as the number of repetitions without a single non-global optimum) by up to 50%. Future work comprises an optimized implementation on the GPU and a combination with a hand tracking approach providing nearest neighbor estimates.

REFERENCES

Cootes, T., G. Edwards, and C. Taylor (1998). Active appearance models. In *ECCV*, pp. 484–498. Springer.

de La Gorce, M., N. Paragios, and D. J. Fleet (2008). Model-based hand tracking with texture, shading and self-occlusions. In *CVPR*. IEEE.

Erol, A. and G. Bebis (2005). A review on vision-based full dof hand motion estimation. In *CVPRWS*. IEEE.

Germann, M. and A. Hornung (2010). Articulated Billboards for Video-based Rendering. In *Eurographics*.

Gionis, A., P. Indyk, and R. Motwani (1999). Similarity search in high dimensions via hashing. In *VLDB*, pp. 518–529. Springer.

Jacobson, A., I. Baran, J. Popovic, and O. Sorkine (2011). Bounded biharmonic weights for real-time deformation. In *SIGGRAPH*, pp. 78. ACM.

Jones, M. and T. Poggio (1998). Multidimensional morphable models: A framework for representing and matching object classes. *IJCV 29*(2), 107–131.

Kemelmacher-Shlizerman, I., E. Shechtman, R. Garg, and S. Seitz (2011). Exploring Photobios. In *SIGGRAPH*. ACM.

Lin, J., Y. Wu, and T. Huang (2002). Capturing human hand motion in image sequences. In *WMVC*, pp. 99–104. IEEE.

Shakhnarovich, G., P. Viola, and T. Darrell (2003). Fast pose estimation with parameter-sensitive hashing. In *ICCV*, pp. 750–757. IEEE.

Stenger, B., P. Mendonça, and R. Cipolla (2001). Model-based 3D tracking of an articulated hand. In *CVPR*, Volume 2, pp. II–310. IEEE.

Computational Modelling of Objects Represented in Images – Di Giamberardino et al. (eds)
© 2012 Taylor & Francis Group, London, ISBN 978-0-415-62134-2

Independent multimodal background subtraction

Domenico Bloisi and Luca Iocchi
Department of Computer, Control, and Management Engineering – Sapienza University of Rome, Italy

ABSTRACT: Background subtraction is a common method for detecting moving objects from static cameras able to achieve real-time performance. However, it is highly dependent on a good background model particularly to deal with dynamic scenes. In this paper a novel real-time algorithm for creating a robust and multimodal background model is presented. The proposed approach is based on an on-line clustering algorithm to create the model and on a novel conditional update mechanism that allows for obtaining an accurate foreground mask. A quantitative comparison of the algorithm with several state-of-the-art methods on a well-known benchmark dataset is provided demonstrating the effectiveness of the approach.

1 INTRODUCTION

Background subtraction (BS) is a popular method for detecting moving objects from static cameras able to achieve real-time performance. BS aims to identify moving regions in image sequences comparing the current frame to a model of the scene background (BG). The creation of such a model is a challenging task due to illumination changes (gradual and sudden), shadows, camera jitter, movement of background elements (e.g., trees swaying in the breeze, waves in water), and changes in the background geometry (e.g., parked cars).

Different classifications of BS methods have been proposed in literature. In (Cristani and Murino 2008), BS algorithms are organized in: 1) per pixel, 2) per region, 3) per frame and 4) hybrid. Per-pixel approaches (e.g., (Cucchiara et al. 2003; Stauffer and Grimson 1999)) consider each pixel signal as an independent process. Per-region algorithms (e.g., (Heikkila and Pietikainen 2006)) usually divide the frames into blocks and calculate block-specific features in order to obtain the foreground. Frame-level methods look for global changes in the scene (e.g., (Oliver et al. 2000)). Hybrid methods (e.g., (Wang and Suter 2006; Toyama et al. 1999)) combine the previous approaches in a multi-stage process.

In (Cheung and Kamath 2004) two classes of BS methods, namely recursive and non-recursive, are identified. Recursive algorithms (e.g., (Stauffer and Grimson 1999)) maintain a single background model that is updated with each new input frame. Nonrecursive approaches (e.g., (Cucchiara et al. 2003; Oliver et al. 2000)) maintain a buffer of previous video frames and estimate the background model based on a statistical analysis of these frames.

A third classification (e.g., (Mittal and Paragios 2004)) divides existing BS methods in predictive and non-predictive. Predictive algorithms (e.g., (Doretto et al. 2003)) model the scene as a time series and develop a dynamical model to recover the current input based on past observations. Non-predictive techniques (e.g., (Stauffer and Grimson 1999; Elgammal et al. 2000)) neglect the order of the input observations and build a probabilistic representation of the observations at a particular pixel.

Although all the above mentioned approaches can deal with dynamic background, a real-time, complete, and effective solution does not yet exist. In particular, water background is more difficult than other kinds of dynamic background since waves in water do not belong to the foreground even though they involve motion. Per-pixel approaches (e.g., (Stauffer and Grimson 1999)) typically fail because these dynamic textures cause large changes at an individual pixel level (see Fig. 1) (Dalley et al. 2008). A non-parametric approach (e.g., (Elgammal et al. 2000)) is not able to learn all the changes, since in the water surface the changes do not present any regular patterns (Tavakkoli and Bebis 2006). More complex approaches (e.g., (Sheikh and Shah 2005; Zhong and Sclaroff 2003; Zhong et al. 2008)), can obtain better results at the cost of increasing the computational load of the process.

In this paper, a per-pixel, non-recursive, non-predictive BS approach is described. It has been designed especially for dealing with water background, but can be successfully applied to every scenario. The algorithm is currently in use within a real 24/7 video surveillance system for the control of naval traffic. The main novelties are 1) an on-line clustering algorithm to capture the multimodal nature of the background without maintaining a buffer with the previous frames, 2) a model update mechanism that can detect changes in the background geometry. Quantitative experiments show the advantages of the proposed

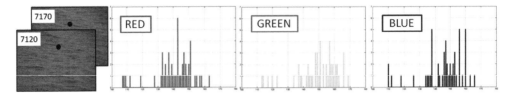

Figure 1. RGB values of a pixel (black dot) from frame 7120 to frame 7170 of *Jug* sequence.

method over several state-of-the-art algorithms and its real-time performance.

The reminder of the paper is organized as follows. In Section 2 the method is presented and in Section 3 a shadow suppression module is described. The model update process is detailed in Section 4. Experiments demonstrating the effectiveness of the approach are reported in Section 5 and Section 6 provides the conclusions and future work.

2 THE IMBS METHOD

The first step of the proposed method is called Independent Multimodal Background Subtraction (IMBS) algorithm and has been designed in order to perform a fast and effective BS. The background model is computed through a per-pixel on-line statistical analysis of a set L of N frames in order to achieve a high computational speed. According to a sampling period P, the current frame I is added to L, thus becoming a background sample S_n, $1 \leq n \leq N$.

Let $I(t)$ be the $W \times H$ input frame at time t, and $F(t)$ the corresponding foreground mask. The background model $\mathfrak{B}$ is a matrix of H rows and W columns. Each element $\mathfrak{B}(i,j)$ of the matrix is a set of tuples $\langle r, g, b, d \rangle$, where r, g, b are RGB values and $d \in [1, N]$ is the number of pixels $S_n(i,j)$ associated with those r, g, b values. Modelling each pixel as a tuple has the advantage of capturing the statistical dependences between RGB channels, instead of considering each channel independently.

The method is detailed in Algorithm 1, where t_s is the time-stamp of the last processed background sample. IMBS takes as input the sampling period P, the number N of background samples to analyse, the minimal number D of occurrences to consider a tuple $\langle r, g, b, d \geq D \rangle$ as a significant background value, and the association threshold A for assigning a pixel to an existing tuple. The procedure *RegisterBackground* (see Algorithm 2) creates the background model, while *GetForeground* (see Algorithm 3) computes the binary foreground image F representing the output of the process.

Each pixel in a background sample S_n is associated with a tuple according to A. Once the last sample S_N has been processed, if a tuple has a number d of associated samples greater than D, then its r, g, b values become a significant background value. Up to $\lfloor N/D \rfloor$ tuples for each element of $\mathfrak{B}$ are considered at the same time allowing for approximating a multimodal probability distribution, that can address the problem

Algotithm 1. IMBS.

Input: P, N, D, A
Initialize: $n = 0$, $t_s = 0$, $\mathfrak{B}(i,j) = \oslash \; \forall \; i, j$
foreach $I(t)$ **do**
    if $t - t_s > P$ **then**
        $\mathfrak{B} \leftarrow RegisterBackground(I(t), D, A, n)$;
        $t_s = t$;
        $n = n + 1$;
        if $n = N$ **then**
            $n = 0$;
    $F(t) \leftarrow GetForeground(I(t), \mathfrak{B}, A)$;

of waves, gradual illumination changes, noise in sensor data, and movement of small background elements. Indeed, the adaptive number of tuples for each pixel can model non-regular patterns (Fig. 1) since IMBS do not model the BG with a predefined distribution (e.g., Gaussian), but produces a "discretization" of an unknown distribution.

F is computed according to a thresholding mechanism shown in Algorithm 3, where $|\mathfrak{B}(i,j)|$ denotes the number of sets in $\mathfrak{B}(i,j)$ with $|\mathfrak{B}(i,j) = \oslash | = 0$. The use of a set of tuples instead of a single one makes IMBS robust with respect to the choice of the parameter A, since a pixel that presents a variation in the RGB values larger than A will be modelled by a set of contiguous tuples.

An example of the 4D background model space for a single pixel is depicted in Fig. 2. The background point is modelled as a set of tuples $\langle r, g, b, d \rangle$ in the RGB color space. Every tuple can be represented graphically by a bin of width $2A + 1$ and height d. Bins having height lower than D are not considered as valid background values.

IMBS requires a time $R = NP$ for creating the first background model; then a new background model, independent from the previous one, is built continuously according to the same refresh time R. The independence of each BG model is a key aspect of the algorithm, since it allows to adapt to fast changing environments avoiding error propagation and do not affect the accuracy for slow changing ones.

The on-line model creation mechanism allows for avoiding to store the images belonging to L, that is the main drawback of the non-recursive BS techniques (Piccardi 2004). In order to manage illumination changes, N and P values are reduced if the percentage of foreground pixels in F is above a certain threshold

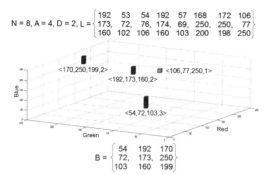

$$N = 8, A = 4, D = 2, L = \left\{ \begin{matrix} 192 & 53 & 54 & 192 & 57 & 168 & 172 & 106 \\ 173 & 72 & 76 & 174 & 69 & 250 & 250 & 77 \\ 160 & 102 & 106 & 160 & 103 & 200 & 198 & 250 \end{matrix} \right\}$$

$$B = \left\{ \begin{matrix} 54 & 192 & 170 \\ 72 & 173 & 250 \\ 103 & 160 & 199 \end{matrix} \right\}$$

Figure 2. An example of the background model space for a single pixel.

Algotithm 2. Register background.

Input: I, D, A, n

foreach $pixel\ p(i,j) \in I$ **do**
  if $n = 0$ **then**
    | $add\ tuple\ \langle p^R, p^G, p^B, 1 \rangle\ to\ \mathfrak{B}(i,j)$;
  else if $n = N - 1$ **then**
    foreach $tuple\ T := \langle r, g, b, d \rangle \in \mathfrak{B}(i,j)$ **do**
      if $d < D$ **then**
         | $delete\ T$;

  else
    foreach $tuple\ T := \langle r, g, b, d \rangle \in \mathfrak{B}(i,j)$ **do**
      if $max\{|p^R - r|, |p^G - g|, |p^B - b|\} \leq A$
      then

          $r' \leftarrow \left\lfloor \frac{r \cdot d + p^R}{d+1} \right\rfloor$;

          $g' \leftarrow \left\lfloor \frac{g \cdot d + p^G}{d+1} \right\rfloor$;

          $b' \leftarrow \left\lfloor \frac{b \cdot d + p^B}{d+1} \right\rfloor$;

          $T \leftarrow \langle r', g', b', d+1 \rangle$;
          break;
      else
        | $add\ tuple\ \langle p^R, p^G, p^B, 1 \rangle\ to\ \mathfrak{B}(i,j)$;

(e.g., 50 %). Moreover, the computational load can be spread over the time interval P until the arrival of the next background sample, thus further increasing the speed of the algorithm.

3 SHADOW SUPPRESSION

When BS is adopted to compute the foreground mask, the moving pixels representing both objects and shadows are detected as foreground. In order to deal with the erroneously classified foreground pixels that can deform the detected object shape, a shadow suppression module is required. IMBS adopts a strategy that is a slight modification of the HSV based method proposed by Cucchiara et al. in (Cucchiara et al. 2003).

Algotithm 3. Get foreground.

Input: $I, \mathfrak{B}, A$

Initialize: $F(i,j) = 1\ \forall\ i, j$

foreach $pixel\ p(i,j) \in I$ **do**
  if $|\mathfrak{B}(i,j)| \neq \varnothing$ **then**
    foreach $tuple\ T := \langle r, g, b, d \rangle \in \mathfrak{B}(i,j)$ **do**
      if $max\{|p^R - r|, |p^G - g|, |p^B - b|\} < A$
      then
        | $F(i,j) \leftarrow 0$;
          break;

Figure 3. Shadow suppression example. Original frame (left), foreground extraction without shadow removal (center), and with shadow suppression (right).

Let $I^c(i,j)$, $c = \{H, S, V\}$ be the HSV color values for the pixel (i,j) of the input frame and $B_T^c(i,j)$ the HSV values for the tuple $T \in \mathfrak{B}(i,j)$. The shadow mask M value for each foreground point is:

$$M(i,j) = \begin{cases} 1 & if\ \exists\ T : \alpha \leq \frac{I^V(i,j)}{B_T^V(i,j)} \leq \beta\ \wedge \\ & |I^S(i,j) - B_T^S(i,j)| \leq \tau_S\ \wedge \\ & |I^H(i,j) - B_T^H(i,j)| \leq \tau_H \\ 0 & otherwise \end{cases}$$

The parameters $\alpha, \beta, \tau_S, \tau_H$ are user defined and can be found experimentally. We analysed the sequences from ATON database (ATON) in order to find the combination of parameters minimizing the error caused by shadows (see Fig. 3). In our OpenCV (OpenCV) based implementation[1] we set $\alpha = 0.75$, $\beta = 1.15$, $\tau_S = 30$, and $\tau_H = 40$. We set β to a value slightly greater than 1 to filter out light reflections.

The shadow suppression module is essential for the success of the algorithm. The HSV analysis can effectively remove the errors introduced by a dynamic background, since it is a more stable color space with respect to RGB (Zhao et al. 2002).

After the removal of the shadow pixels, F can be further refined by exploiting the opening and closing morphological operators. The former is particularly useful for filtering out the noise left by the shadow suppression process, the latter is used to fill internal holes and small gaps.

4 MODEL UPDATE

Elgammal et al. in (Elgammal et al. 2000) proposed two alternative strategies to update the background. In

---

[1] http://www.dis.uniroma1.it/ bloisi/software/imbs.html

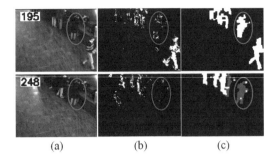

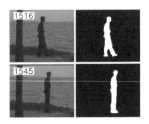

(a)	(b)	(c)

Figure 5. Water surface sequence.

Figure 4. IMBS model update. (a) The target remains in the same position over several frames. (b) A blind update (obtained by OpenCV function CreateGaussianBGModel) includes it in the background model. (c) IMBS model update is able to identify the target as a potential foreground region (grey pixels).

Frame #36   Zhong & Sclaroff   Dalley et al.   IMBS

Figure 6. Jug sequence.

selective update, only pixels classified as belonging to the background are updated, while in *blind update* every pixel in the background model is updated. The selective (or *conditional*) update improves the detection of the targets since foreground information are not added to the background model, thus solving the problem of ghost observations. However, when using selective updating any incorrect pixel classification produces a persistent error, since the background model will never adapt to it. Blind update does not suffer from this problem since no update decisions are taken, but it has the disadvantage that values not belonging to the background are added to the model.

We propose a different solution that aims to solve the problems of both the selective and blind update. Given the background sample S_n and the current foreground mask F, if $F(i, j) = 1$ and $S_n(i, j)$ is associated to a tuple T in the background model under development, then T is labelled as a "foreground tuple". When computing the foreground, if $I(i, j)$ is associated with a foreground tuple, then it is classified as a potential foreground point.

Such a solution allows for identifying regions of the scene representing not moving foreground objects, as in Fig. 4 where a target that remains in the same position over several frames is detected as a potential foreground. The decision about including or not the potential foreground points as part of the background is taken on the basis of a *persistence map*. If a pixel is classified as potential foreground consecutively for a period of time longer than a predefined value (e.g., $R/3$), then it becomes part of the background model. Furthermore, the labelling process provides additional information to higher level modules (e.g., a visual tracking module) helping in reducing ghost observations.

Figure 7. IMBS output with real data.

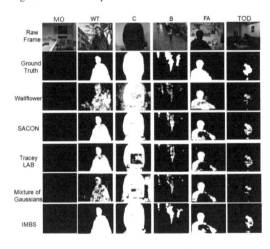

Figure 8. IMBS results (last row) for Wallflower sequences.

surface[2]: A person walks in front of a water surface with moving waves (see Fig. 5) and 2) *Jug*[3]: A foreground jug floats through the background rippling water (see Fig. 6). IMBS is able to correctly model the background in both the situations, extracting the foreground with great accuracy.

5 RESULTS

To test IMBS on water background, we selected two different publicly available sequences: 1) *Water*

---

[2] http://perception.i2r.a-star.edu.sg/bk_model/bk_index.html
[3] http://www.cs.bu.edu/groups/ivc/data.php

Table 1. A comparison of IBMS with state-of-the-art methods on Wallflower dataset.

Method	Err.	MO	WT	C	B	FA	TOD	Tot.E.
Wallflower (Toyama et al. 1999)	f. neg.	0	877	229	2025	**320**	961	10156
	f. pos.	0	1999	2706	365	649	25	
	tot. e.	0	2876	2935	2390	**969**	986	
SACON (Wang and Suter 2006)	f. neg.	0	41	**47**	1150	1508	**236**	4467
	f. pos.	0	230	462	125	521	147	
	tot. e.	0	271	509	1275	2029	**383**	
Tracey LAB	f. neg.	0	191	1998	1974	2403	772	8046
LP (Kottow et al. 2004)	f. pos.	1	**136**	**69**	**92**	**356**	54	
	tot. e.	1	327	2067	2066	2759	826	
Bayesian	f. neg.	0	629	1538	2143	2511	1018	15603
Decision (Nakai 1995)	f. pos.	0	334	2130	2764	1974	562	
	tot. e.	0	963	3688	4907	4485	1580	
Eigen-background	f. neg.	0	1027	350	**304**	2441	879	16353
(Oliver et al. 2000)	f. pos.	1065	2057	1548	6129	537	**16**	
	tot. e.	1065	3084	1898	6433	2978	895	
Mixture of Gaussians	f. neg.	0	1323	398	1874	2442	1008	11251
(Stauffer and Grimson 1999)	f. pos.	0	341	3098	217	530	20	
	tot. e.	0	1664	3496	2091	2972	1028	
	f. neg.	**0**	**23**	83	388	1438	294	
	f. pos.	**0**	152	345	301	564	117	
IMBS	tot. e.	**0**	**175**	**428**	**689**	2002	411	**3696**

We use also real data coming from a video surveillance system for the control of naval traffic (see Fig. 7). Both people and boats are present in the same scene, allowing the test of the algorithm with different targets. The IMBS conditional update mechanism allows for detecting parked boats (grey pixels in the second row of Fig. 7) before considering them as background. The real data will be made available as part of a public dataset.

To quantitatively compare IMBS with other BS algorithms, we selected the Wallflower sequences (Toyama et al. 1999), a widely used benchmark. In particular, we analysed the following sequences: 1) Moved object (MO), a person enters into a room, makes a phone call and leaves. The phone and the chair are left in a different position; 2) Waving trees (WT), a tree is swaying and a person walks in front of it; 3) Camouflage (C), a person walks in front of a monitor, having rolling interference bars on the screen of the same color of the person's clothing; 4) Bootstrapping (B), a busy cafeteria where each frame contains foreground objects; 5) Foreground aperture (FA), a person with uniformly coloured shirt wakes up and begins to move; 6) Time of day (TOD), the light in a room is gradually changes from dark to bright. Then, a person enters the room and sits down.

The remaining Wallflower sequence "Light switch" presents a sudden global change in the illumination conditions. In this case, our method fails since it does not include a per-frame module. Anyway, the problem can be faced using the previous computed models and choosing the most appropriate (Ohta 2001).

Fig. 8 shows the IMBS results and the ground truth images for Wallflower dataset as well as the results obtained by applying others methods (Cristani and Murino 2008; Wang and Suter 2006).

Table 2. IMBS computational speed.

Data	Acquisition	Frame Dim.	FPS
Wallflower	off-line	160×120	115
ATON	off-line	320×240	70
Live 1	on-line	320×240	25
Live 2	on-line	640×480	14
Live 3	on-line	768×576	10

In order to make a quantitative comparison with state-of-the-art BS methods, in Table 1 the results are provided in terms of false negatives (the number of foreground points detected as background) and false positives (the number of background points detected as foreground). The results show the effectiveness of the IMBS approach, that performs better than the other methods in terms of total error. Moreover, the computational speed for IMBS on the Wallflower dataset using an Intel Xeon Quad Core 2.26 GHz 8 GB RAM CPU is 115 fps (see Table 2), while for the most accurate of the other methods, SACON (Wang and Suter 2006), the authors claim a speed of 6 fps.

The computational speed of the algorithm has been tested also with on-line data coming from a camera. The results (see Table 2) show the real-time performance of the proposed approach.

For all the test sequences, IMBS parameters have been set as: $A = 5$, $N = 30$, $D = 2$, and $P = 1000\ ms$. We investigate the influence of the parameter N on the results (see Fig. 9): Given the same refresh time $R = 30$ sec., various N and P values are used. For all the sequences, except for the sequence B, there are similar total error values even with a limited number

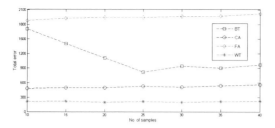

Figure 9.   Plot of total error vs. different background sample number.

of background samples. Since B is the sequence with the most dynamical background, there is a significant increase in the total error if N is small. However, starting from $N = 25$, the total error value becomes stable, as for the other sequences.

6   CONCLUSIONS

In this paper, a fast and robust background subtraction algorithm has been proposed and quantitatively compared to several state-of-the-art methods using a well-known benchmark dataset. Thanks to an on-line clustering algorithm to create the model and a novel conditional update mechanism, the method is able to achieve good accuracy while maintaining real-time performances. As future work, we intend to investigate the use of the HSV color space in building the background model, and to add per-region and per-frame modules to the algorithm.

REFERENCES

ATON. http://cvrr.ucsd.edu/aton/testbed.

Cheung, S. and C. Kamath (2004). Robust techniques for background subtraction in urban traffic video. In *Visual Comm. and Image Proc.*, Volume 5308, pp. 881–892.

Cristani, M. and V. Murino (2008). Background subtraction with adaptive spatio-temporal neighbourhood analysis. In *VISAPP*, Volume 2, pp. 484–489.

Cucchiara, R., C. Grana, M. Piccardi, and A. Prati (2003). Detecting moving objects, ghosts, and shadows in video streams. *PAMI 25*(10), 1337–1342.

Dalley, G., J. Migdal, and W. Grimson (2008). Background subtraction for temporally irregular dynamic textures.

In *IEEE Workshop on Applications of Computer Vision*, pp. 1–7.

Doretto, G., A. Chiuso, Y. N. Wu, and S. Soatto (2003). Dynamic textures. *IJCV 51*(2), 91–109.

Elgammal, A. M., D. Harwood, and L. S. Davis (2000). Non-parametric model for background subtraction. In *ECCV*, pp. 751–767.

Heikkila, M. and M. Pietikainen (2006). A texture-based method for modeling the background and detecting moving objects. *PAMI 28*(4), 657–662.

Kottow, D., M. Koppen, and J. Ruiz del Solar (2004). A background maintenance model in the spatial-range domain. In *SMVP*, pp. 141–152.

Mittal, A. and N. Paragios (2004). Motion-based background subtraction using adaptive kernel density estimation. In *CVPR*, pp. 302–309.

Nakai, H. (1995). Non-parameterized bayes decision method for moving object detection. In *Asian Conf. Comp. Vis.*, pp. 447–451.

Ohta, N. (2001). A statistical approach to background subtraction for surveillance systems. *IEEE Int. Conf. on Computer Vision 2*, 481–486.

Oliver, N. M., B. Rosario, and A. P. Pentland (2000). A bayesian computer vision system for modeling human interactions. *PAMI 22*(8), 831–843.

OpenCV. http://opencv.willowgarage.com.

Piccardi, M. (2004). Background subtraction techniques: a review. In *Proc. of the IEEE Int. Conf. on Systems, Man & Cybernetics*, pp. 3099–3104.

Sheikh, Y. and M. Shah (2005). Bayesian object detection in dynamic scenes. In *CVPR*, pp. 74–79.

Stauffer, C. and W. Grimson (1999). Adaptive background mixture models for real-time tracking. *CVPR 2*, 246–252.

Tavakkoli, A. Nicolescu, M. and G. Bebis (2006). Robust recursive learning for foreground region detection in videos with quasi-stationary backgrounds. In *ICPR*, pp. 315–318.

Toyama, K., J. Krumm, B. Brumitt, and B. Meyers (1999). Wallflower: principles and practice of background maintenance. In *ICCV*, Volume 1, pp. 255–261.

Wang, H. and D. Suter (2006). Background subtraction based on a robust consensus method. In *ICPR*, pp. 223–226.

Zhao, M., J. Bu, and C. Chen (2002). Robust background subtraction in hsv color space. In *SPIE: Multimedia Systems and Applications*, pp. 325–332.

Zhong, B., H. Yao, S. Shan, X. Chen, and W. Gao (2008). Hierarchical background subtraction using local pixel clustering. In *ICPR*, pp. 1–4.

Zhong, J. and S. Sclaroff (2003). Segmenting foreground objects from a dynamic textured background via a robust kalman filter. In *ICCV*, pp. 44–50.

Computational Modelling of Objects Represented in Images – Di Giamberardino et al. (eds)
© 2012 Taylor & Francis Group, London, ISBN 978-0-415-62134-2

Perception and decision systems for autonomous UAV flight

P. Fallavollita, S. Esposito & M. Balsi
Dipartimento di Ingegneria dell'Informazione, Elettronica e Telecomunicazioni,
"La Sapienza" University, Rome, Italy
Humanitarian Demining Laboratory, Via delle Province snc, Cisterna di Latina (LT), Italy

ABSTRACT: This work consists of the first feasibility study, simulation, design and testing of a new system for autonomous flight of UAVs for civil applications.

After identifying an appropriate set of sensors, computer vision, data fusion and automatic decision algorithms will be defined and implemented on specific embedded platforms.

1 INTRODUCTION

1.1 Sense and avoid

In recent years Unmanned Aerial Vehicles (UAV) have had an unprecedented growth in both military as well as civilian applications, (Fig 1). During normal manned flight procedures, current regulations state that the pilot, during any phase of flight, should be in charge of obstacle separation using "see and avoid" rules, independently from the assigned flight plan or the Flight Control Centre instructions. The main request and constraints for the futuristic scenario, where UAVs could fly together with manned vehicles, is to mount some kind of device on-board such vehicles, that replaces the pilot and works well as if he/she is seated on board.

Such kind of multi-disciplinary researches and technologies is generally referred as the "sense and avoid" activity. This scenario has generated an exponentially growing interest of the scientific community and commercial world to invest in this research, and the challenge is obtaining as soon as possible a working and approved system ready to operate on board

Figure 1. An airship UAV (Photo courtesy from: www.oben.it).

UAVs. For this reason, international civil flight agencies are working over a complete redesign of existing flight and sky rules, planning in the near future to permit UAVs flying in shared civil flight space. Certainly, sense-and-avoid functionalities are being developed for military application, but under secrecy, so that research and development for the civil sector must proceed in parallel. As of today, such a device, destined to civil operations, is still not available or not capable of being certified. Such considerations mean that autonomous flight is currently an open problem rather far from being solved.

1.2 State of art

The first activity of our research group has consisted in assessing requirements for autonomous UAV flight and determining which sensors and algorithms are most suitable to be integrated in a light UAV, with strict constraints of weight and power. Some words need to be spent explaining the meaning of automatic flight vs. autonomous flight for UAVs. Both terms describe the way to fly, without any human pilot manually controlling the plane, but with a fundamental difference. In the first one the UAV has only a standard autopilot system (AP) permitting to cruise following waypoints and in some case also automatic takeoff and landing. In the second one, the UAVs has also, in addition to the above mentioned AP, an obstacles detection system able to sense fixed or moving objects, cooperative or not, estimate a time to contact, and a processing unit capable of re-planning the route, and in particular of devising an escape strategy. So, referring from this point only to autonomous flight, at the moment, state-of-art techniques for UAV can be divided into two different areas: the first one includes technologies based on data fusion from visible, infrared and heavy radar systems flying on aerial vehicles allowing a considerable payload (more than 100 kg), permitting high processing capabilities thanks to large power source available onboard and a large space allocated in the avionic bay. Typically these systems are used or derived from military applications. The second area refers to flying systems allowing small payload (some tens of kg down to few grams), under strict computing power limitations. This is the field where this research project moves and where most efficient techniques have been tested. Optical flow for motion detection, and more generally bio-inspired algorithms (Zufferey 2008) appear particularly promising for this purpose.

2 A NEW APPROACH

2.1 *Mosaicking vs. full frame processing*

As the work focuses on determining a computer vision methodology for obstacle avoidance, ready to be embedded and integrated with other sensors on quite light UAVs (typically in the range from 10kg until 150kg of mass), the main constrains of this work require:

– Efficient and robust obstacle detection.
– Time-to-contact estimation between UAV and approaching obstacles (cooperative or not).
– Good response in possibly complex scenarios.
– Fast reaction.
– Use of limited vehicle resources in terms of power consumption, weight and dimensions.
– Efficient and potential integration with heterogeneous sensors (i.e. light laser or radar).

In this work, preliminary feasibility study and first results are presented, using and tuning algorithms developed in Matlab environment. While this paper is being written, the hardware prototype system is being implemented on a PandaBoard device, a light, yet powerful, embedded system, (PandaBoard 2012). The adopted choice of using a C/OpenCV environment guarantees efficient computation and portability on different embedded platforms, of actual or future conception. Optical flow (OF) was chosen as the main cue for motion detection, applied in such a way that makes computation manageable. Computation of OF (Horn 1985) relies mainly on the brightness intensity changing of pixel between two frames sequentially captured from the video sequence, (Fig 2). Assuming the brightness of a moving object should remain unchanged as it moves, using the related equation:

$$\nabla E \cdot (u, v) = -\frac{\partial E}{\partial t} \qquad (1)$$

where $E(x, y)$ is the brightness intensity calculated in each pixel, (x, y) and u, v are the horizontal and vertical component of the optical flow vector, it assumes that the time variation of brightness intensity comes as dot product between the space gradient of brightness and the optical flow vector (based on a conservation

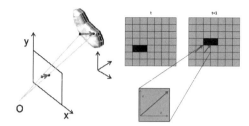

Figure 2. Optical Flow (OF) computation procedure.

principle). The computational cost C of OF limits its applicability to embedded computation applied on the full frame extracted from video streaming. In fact, considering that:

$$C \propto V^2 S \qquad (2)$$

(where S is the pixel resolution of individual video frames and V is the maximum speed detectable in the scene), since V is not modifiable, other authors, in order to speed up the computation, try to decrease S, but the main risk is aliasing problems and/or loss of accuracy and significance, in terms of minimum size of recognized targets, (Liu 1998). An alternative approach relies on the temporal down-sampling by reducing the processed frames frequency, but this produces a strong impact against the accuracy of the time-to-contact estimation based on the "number of remaining frames to contact". Therefore, we decided to take inspiration from bio-mimicry of insect vision, since they have very simple perception structures for movement detection, but very efficient, (Beyeler 2009, Floreano 2009). The strategy adopted in this work divides the whole frame in several smaller sub-frames, using such attention mechanism to compute OF only on relevant Regions of Interest (RoI), at high resolution, keeping S relatively low throughout most of the image, thereby avoiding loss of accuracy where necessary, and at the same time without reducing the field of view (FoV). The above discussed procedure doesn't yield any improvement in term of computational cost with respect to the analysis of OF for the whole frame unless a strategy is chosen to limit the number of RoIs where the OF is calculated.

2.2 *The processing chain*

Each frame of the captured video flow is divided into an array (e.g. 3×4) of sub-images (like a mosaic), possibly with overlapping, (Fig. 3).

In this work we have used gray scale images, but color information could be used to enhance performance. Within every frame, a rough motion signal has been estimated, to trigger motion alert, using a very simple and efficient algorithm based on change detection (attention mechanism). In order to make the motion detection (attention) process more robust to high-frequency noise, denoising pre-processing has

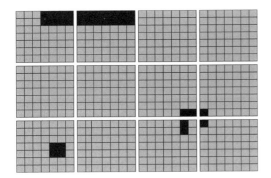

Figure 3. A 3 × 4 single frame mosaicking scheme.

(a)   (b)

Figure 4. Orphan pixel removal and minimum area zones re-moval (a), Closing operation (b).

been applied using a two-dimensional Gaussian low-pass filter. The following step consists of a background removal process which takes place by considering as background all those areas where no remarkable variation has been detected in at least two consecutive frames. Several algorithms have been investigated: frames difference, weighted mean, rounded mean and Gaussian mixture. The resulting most useful solution has been the adoption of a simplified version of the rounded mean algorithm, (Tian & Hampapur 2005). After background initialization (using the first frame), for each pixel belonging to the actual sub-frame, the absolute difference with respect to the corresponding background pixel from the previous iteration was computed. Each pixel is classified as object if:

$$\left| I^{k+1}(x,y) - B^k(x,y) \right| > \vartheta \qquad (3)$$

noindent where I and B are the frame and background brightness arrays and θ is a chosen threshold. At each step the background is updated using the rule:

$$B^{k+1}(x,y) = \begin{cases} B^k(x,y)+1 & if\ I^k(x,y) > B^k(x,y) \\ B^k(x,y)-1 & if\ I^k(x,y) < B^k(x,y) \\ B^k(x,y) & if\ I^k(x,y) = B^k(x,y) \end{cases} \qquad (4)$$

As the previous step ensures extraction of the scene from the background, the following phase, still operated using only the inferred scene in each sub-frame, is composed of a chain of classic morphological operations, useful to lower computational cost:

– Binarization
– Orphan pixels and minimum area zones removal
– Closing (Dilation, Erosion)

Starting from the gray-scaled sub-frame, a first binarization operation has applied by thresholding:

$$I_{scene}^{BW}(x,y) = \begin{cases} 1 & if\ I_{scene}(x,y) > \xi \\ 0 & elsewhere \end{cases} \qquad (5)$$

The result is a black and white array, where ξ is obtained from the gray scale value histogram of the scene's pixels, chosen to maximize the gray tone value variance. The next step is a two-phase process: an orphan pixel and minimum area zone removal processing, (Fig 4a). In the former each pixel is marked as isolated then deleted from the scene, if the cardinal eight neighboring pixels don't belong to the scene.

The latter is the deletion of pixels belonging to areas whose numerical area value is lower than a threshold. In order to avoid that connected areas of the scene could result split in several parts, the sub-frame's last core processing step performs a standard Closing operation applying a Erosion after a Dilation treatment, (Fig. 4b). Finally, for each sub-frame, a simple alert criterion is used to easy trigger and detect only the sub-image where a real movement is happening, (Fig. 5). This is permitted by a simple Boolean operation between pixels of two following sub-images (as just processed), represented by binary arrays, using:

$$I_{alert}^{k+1}(x,y) = AND\,(I_{scene}^{k+1}(x,y), \overline{I_{scene}^k(x,y)}) \qquad (6)$$

This approach permits to easy detect translating or approaching objects and reject those that are moving out and could be considered as not significant. When and where motion is detected, more accurate estimation is made using optical flow and, after evaluating either Horn-Schunck or Lucas-Kanade (Barron et al. 1994) methods for its computation, the latter algorithm has been implemented, having provided more strength against brightness variability.

3 A SIMULATION CASE

A real captured video has been adopted to test the algorithms efficiency, displaying an approach to a tethered balloon used as safe and practical target. The original video frame resolution (1280x720) has been reduced to 240x320 and the initial 30 fps frame rate lowered to 10 fps. After a gray scale conversion (8 bits for pixel), 35.7 seconds of video streaming have been processed using a mosaicking scheme of 3 rows and 4 columns with no overlapping.

As shown in Figure 6, the five areas where motion was detected have been alerted and only here the OF is calculated by a 20×20 pixels window. To quantify the computation cost saving with respect to the whole frame processing, indicating with n_z the number of alerted zones, Z_{tot} the whole zones number, τ_{of} the OF computation time on the whole frame and

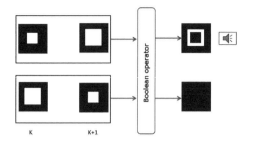

Figure 5. The alert Boolean operation.

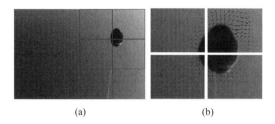

(a)   (b)

Figure 6. Real video processing: mosaicking and alerted zones (a); optical flow vectors displayed for target areas (b).

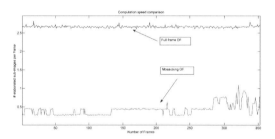

Figure 7. Computation speed comparison: average OF time calculation per frame in seconds, whole frame (upper) and only alerted frames (lower).

τ_{extra} the time spent for auxiliary operations related to mosaicking, the total computation time τ_{tot} is:

$$\tau_{tot} = \frac{n_z}{Z_{tot}} \tau_{of} + \tau_{extra} \qquad (7)$$

As $\tau_{extra} \ll \tau_{of}$, the τ_{extra} term can be eliminated and the speed-up factor of this processing respect to the whole frame computation is:

$$SU_{factor} = \frac{Z_{tot}}{n_z} \geq 1 \qquad (8)$$

It has demonstrated that such processing strategy permits a speed-up of computational cost (with respect to whole-frame OF computation) which is proportional to the rate between the total number of windows and the number of those that triggered an alarm. Results of the experiment are quantitatively demonstrated in Figure 7.

4 CONCLUSIONS

By only relying on the mosaicking procedure, the proposed system yields good efficiency in scenarios where there is a quite small target showing significant relative motion (like a near fixed obstacle or an approaching aerial vehicle) over a relatively stable background. The mosaicking procedure, contrary to other different OF-enhanced approaches, doesn't decrease the accuracy and is easy to integrate for parallel architectures. The future step can be the easy conversion into a hardware system, using a moving mirror mount to scan, like an array, the whole actual field of view. In this way, very wide angle of view can be obtained without falling into nonlinear distortion typical of wide angle lenses, reaching a good resolution on a single sub frame (e.g. equivalent to that of a VGA covering a wide FoV). In case of more complex scenarios, for example inside confined environments, the expected number of alerted windows increases, not all including dangerous or interesting obstacles For this reason the continuation of this research will consider adoption of a higher-level intelligent system that include machine learning and automatic reasoning for path planning and reaction only to unforeseen obstacles.

REFERENCES

Barron, J.L. et al. 1994. Performance of Optical Flow Techniques. *International Journal of Computer Vision* 12(1): 43–77.
Beyeler, A. et al. 2009. Vision-Based Control of Near-Obstacle Flight. *Autonomous Robots* 27(3): 201–219.
Floreano, D. et al. 2009. *Flying Insects and Robots*. Berlin: Springer.
Horn, B. & Schunck, B. 1981. Determing Optical Flow. *Artificial Intelligence* 17: 185–203.
Liu, H. et al. 1998. Accuracy vs. Efficiency Trade-offs in Optical Flow Algorithms. *Computer Vision and Image Understanding* 65(1): 271–286.
PandaBoard website 2012. *http://www.pandaboard.org*. Last access April 2012.
Tian, Y. L. & Hampapur, A. 2005. Robust Salient Motion Detection with Complex Background for Real-time Video Surveillance. *Application of Computer Vision. WACV/MOTIONS '05*.
Zufferey, J.C. 2008. *Bio-inspired Flying Robots Experimental Synthesis of Autonomous Indoor Flyers*. EPFL Press.

Computational Modelling of Objects Represented in Images – Di Giamberardino et al. (eds)
© 2012 Taylor & Francis Group, London, ISBN 978-0-415-62134-2

Flow measurement in open channels based in digital image processing to debris flow study

C. Alberto Ospina Caicedo & L. Pencue Fierro
GOL, Universidad del Cauca, Popayán, Cauca, Colombia

N. Oliveras
Servicio Geológico Colombiano, Popayán, Cauca, Colombia

ABSTRACT: This project presents a progress in developing a system that will support the scientific study of debris flows and the staff of the Observatorio Vulcanológico y Sismológico de Popayán (OVSPop). This progress is achieved through a system to measure flow in an open channel using a camera and digital image processing. The measurements require analysis of the super elevation of a fluid when there is a bend in its path, further prior knowledge of the channel dimensions, as a cross section, followed the system uses motion detection algorithms, accumulation of bottom, and snakes over the images, getting through them the value of the mean flow at a given time, which are corroborated with a system of flow reference measurements. The system has proven reliable in obtaining the measurements and allows for a positive vision for the future system implementation on the banks of a river, like the Páez River.

1 INTRODUCTION

The volcano Nevado del Huila (VNH) in the South of Colombia's Cordillera Central has erupted three times in the last 6 years, February 19 and April 18 of 2007 and November 20 of 2008 (Pulgarín et al, 2009), which had associated different effects as debris flows (lahars), which damage could be limited to infrastructure thanks to timely warning. However, the order of magnitude of these lahars and the prevailing potential for similar or even larger events poses significant hazards to local people, and make an appropriate lahar modeling a challenge which affected the population that live near to the Páez River banks (Worni et al, 2009).

In order to support prevention strategies to the population that could be affected by the activity of the VNH an particularly by debris flows, the Colombian Geological Service (before INGEOMINAS), and specifically, the Volcanological and Seismologic Observatory of Popayán, had developed studies about lahars, some of them was based on scenario-defined lahar modeling obtained from LAHARZ and FLO-2D (Cardona et al, 2008; Worni et al, 2009), from geophones and seismometers detectors of debris flows (Oliveras, 2008), furthermore some studies based on super elevation measurements on deposits left by the lahar (Pulgarín et al, 2008), those studies was developed on lahars in the River Páez basin of 2007 and 2008 eruptions, as well as debris flow generated in 1994 due an event of 6.4 in Richter scale in a rainy time (Pulgarín et al, 2009).

Must be clarified that design, build and fine tuning a system or tool for make studies of debris flows may be constrained by two fundamental problems: the order of magnitude and the unpredictable nature of debris flows

Also in other places have been developed studies about debris flows using different kinds of systems (Massimo & Marchi, 2008), such surveillance cameras which trough machine-vision detection had identified the average velocity and superficial velocity field of debris flows occurred in Japan and Italy (Inama & Itakura 1999; Arattano & Marchi, 1999).

This paper show a prototype system to debris flows studies and particularly to estimate the volume flow rate of a flow like this, based on digital image processing (DIP) of a cloudy fluid achieved in a bend cross section of a small open channel, regarding the Páez River basin, without illumination control and it use a super elevation model.

2 ESTIMATION OF FLOW RATE IN OPEN CHANNELS – SUPERELEVATION MODEL

The volumetric flow rate in a cross section of bend part of an open channel, can achieved from dot product between the wet area and the average fluid speed in this specific cross section.

The Equation 1 below shows the volumetric flow rate obtained from a mathematical model which predict that the average fluid speed depends on superelevation obtained by the fluid when it arrive into a bend

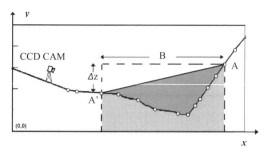

Figure 1. Location of the camera CCD, wet area, Δz correspond to the difference in high between the right (A') left margins (A). And B corresponds to the distance in plain between these right and left margins.

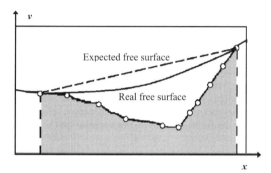

Figure 2. Overestimation of volumetric flow rate Q, when the real free surface of the fluid differs from the free surface expected to the superelevation model.

(Vide, 2003; Dingman, 2009) furthermore, it assume that free surface of the fluid acquires a linear shape as is possible to see in Figure 1, reason which is plausible to overestimate the wet area and by this way the volumetric flow rate when the turbulent behavior of the fluid can't allow a free surface linear of the fluid and instead, acquires a free surface curved (Fig. 2).

$$Q = A_m \cdot \sqrt{\frac{g \cdot r \cdot \Delta z}{B}} \qquad (1)$$

Where Q = volumetric flow rate in across section in a bend; A_m = wet area; g = gravity; r = bend radius; Δz = superelevation; B = width of the fluid in a cross section (free surface).

Among the parameters of Equation 1, g and r are known, A_m, Δz and B can be obtained whether a map of the cross section and the right and left margins (A and A' respectively, Fig. 1) are known.

By other side is possible to find the total volume V in a time interval T_0 a T_n, whether is known the volumetric flow rate Q in several consecutive intervals ΔT_i, thought the Equation 2:

$$V_{(Tn-To)} = \sum_{To}^{Tn} Q_i \cdot \Delta T_i \qquad (2)$$

Where V = total volume, $T_n - T_0$ = time interval for the total volume V, ΔT_i = time interval i

SIDE VIEW

IMAGE FROM THE CCD CAM

PLAIN VIEW

CONVENTIONS:
- ° Reference points
- — Cross section
- ⇒ Flow direction
- · Calibration point

Figure 3. Calibration method, point by point. Line of calibration (320) on the image and its correspond cross section on the open channel.

where $\Delta T_i < T_n - T_0$, Q = volumetric flow rate in the time interval ΔT_i.

Was designed a methodology to calibrate point by point the cross section with the coordinates of the image obtained from the CCD camera and an algorithm of digital image processing was developed to detect the margins right and left of the fluid on the images obtained and finally it identifies their corresponds positions in the map of the cross section.

3 CAMERA CALIBRATION AND CROSS SECTION

The calibration method used gets points with its spatial coordinates in a cross section and its corresponding positions in a column of the images (line 320 in the image from CCD camera, Fig. 3). This method requires at first, to identify a cross section in a bend of the channel, and this cross section has to fit with the column 320 of the image.

The calibration consist in identify 14 points in a vertical line of the image and measure their corresponding positions in the real space based on a horizontal squad (side and plan view, Fig. 3). From the 14 calibration points are interpolated in the real space, so that it obtain a position in the real space by each pixel of the line 320 of the acquire image from the CCD camera.

4 FLOW ESTIMATION ALGORITHM

The algorithm of volumetric flow rate estimation gets the margins of the fluid in the open channel on the image and then it determinates their corresponding positions on the map of the cross section, so that is possible to get the parameters A_m, Δz y B to apply the Equation 1.

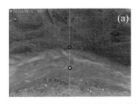
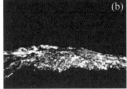
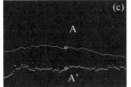

Figure 4. View from CCD camera. (a) Image of the cloudy fluid in the bend of the open channel. (b) Detection of moving fluid based on motion templates. (c) Left (A') and right (A) detected over the image.

Table 1. Summary of results of different tests, with variations in fps and detected changes in light intensity of the scene.

Test No.	Lux*	fps	Total Actual Vol (L)**	Total Measured Vol (L)***	%Error Total Measured Vol
1	620	15	304.9	373.6	23
2	550	15	192.2	182.4	5
3	200	15	224.1	227.3	1
4	17	15	272.6	91.8	66
5	850	1	327.8	288.2	12
6	950	0.5	284.1	269.6	5
7	1200	0.33	315.3	305.3	3

* Light intensity detected with the LX-102 lux meter at the test time.
** These values are associated with a 7% error.
*** These values are associated with a 8% error

To detect and threshold the fluid in movement the algorithm was based on the technique of motion templates and subtraction of consecutive images, that allows to obtain a history of moving objects of the scene (Bradski & Kaehler, 2008), also it take account the *Thalweg* of the channel on the images to eliminate objects in movement which are not part of the fluid, obtaining results as is shown in the Figure 4b.

Once is detected he fluid in movement, it identifies the left (A') and right (A) margins that correspond to the higher and lower edges on the images, then the edges are smoothed (noise elimination) and finally are identify the margins or positions A' and A (Fig. 4c) associated to the used cross section in the calibration, with the objective of determinate the parameters A_m, Δz y B and with them estimate the volumetric flow rate Q.

5 RESULTS AND DISCUSSION

Several tests were performed by flowing a turbid fluid from a 500 L tank through a 3'' valve, achieving different reference flow rates (measured at the output of the tank valve) in different camera frames per second. Tests in which, from the estimation algorithm is flow-based model of super elevation and using the cross section of the calibration process made from space camera, yielded the results summarized in Table 1, Figure 5 also shows the measured flow and flow rate reference in test No. 1, Figure 6 shows the absolute value of the difference between the margins A' and A detected in the images and margins A' and A real (identified by hand) that were obtained in test No. 1.

Table 1 shows good results in most tests (No. 2, 3, 5, 6 and 7), in which the total volume measured has a relative error of up to 12%, tests at 15, 1, 0.5 and 0.33 fps and 200 to 1200 lux, while tests No. 1 and 4 had errors of 23 and 66% respectively in the total volume measured. In test No. 4 the error was caused by the

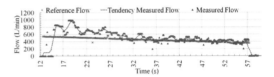

Figure 5. Rated flow from CCD camera images (Measured Flow) and Reference Flow during a test No. 1 (at approx. 600 lux, 15 fps and 640x480 pixels image resolution).

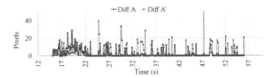

Figure 6. Difference between the detected edges from DIP algorithm implemented and the actual margins.

misidentification of the left and right margins due to the low lighting of the scene (test performed at 17 lux).

It was noted that prevailed during testing a turbulent regime of flow in the curve, resulting overestimation of the wet area (Fig. 2) and therefore the estimated flow, becoming more noticeable the effect of overestimation in the first moments of the measurements. In Figure 5 Measured Flow became 1.9 times Reference Flow at 18 s, but over the free surface of test fluid was shown to be more and more linear and consequently Flow Measured showed a trend to Reference Flow, compensating for the overestimation of the first moments of the measurements.

Results of tests showed that the estimates achieved by this system may have a wrong value if taken a isolated data, but as seen in Figure 5, is the set Flow Measured Data which is close to Reference Flow value during test No. 1, giving reliable results.

It should be clarified regarding the results of the test No. 1, presented in Figure 5, which ranges from 12 to 16 s and 57 to 60 s to test No. 1 (Fig. 5), during which a sample Measured Flow increase from 0 L/min and decreased to 0 L/min respectively represent the instants during which the fluid appears in the image and the disappearance time of the same camera vision field.

On the other hand of Figure 6 is appreciable from 23 seconds to 56 seconds was detected with improved accuracy over the range points 15 to 23 s, which is reflected in differences close to 0 between the margins detected and actual margins in Figure 6. Achieving effectiveness in detecting these margins from 81% in test No. 1, which may vary depending on lighting conditions, the amount of fps, the effects on self-compensation can offer the CCD camera on the images and its resolution.

6 CONCLUSIONS

The study of debris flows and in particular the estimation of volumetric flow rate is important because it helps support prevention strategies for populations that may be affected by flows of this type. This article was proposed and presented the results of a new prototype system for estimating volume flow of turbid flow in a curved section of a small open channel, on the Páez River, who based on a super elevation model and digital processing of images from a CCD camera achieved in a scenario without lighting control, achieved good results, which were confronted with a wealth of reference and presented in this article.

Furthermore, although the presented system provides good results, they are valid mainly for the channel on which testing was developed. However, the guidelines and requirements identified for the system to work, to allow an important first step in building a proposal for a system that considers flow of a debris flow naturally in a real setting and offers a real support to the prevention strategies for populations that may be affected by debris flows.

ACKNOWLEDGMENTS

The staff of the Volcanological and Seismological Observatory of Popayan, who supported at all times the development of this work. MSc. Bernard who contributed their knowledge and readiness. In Grupo de Óptica y Láser (GOL) at the Universidad del Cauca with whom we discussed the gradual progress of this work.

REFERENCES

Arattano, M. Marchi, L. 1999. Video-Derived Velocity Distribution Along a Debris Flow Surge. *Physics and Chemistry of the Earth* 25:781–784.

Bradski, G. & Kaehler, A. 2008. *Learning OpenCV Computer. Vision with the OpenCV library*. Sebastopol: O'Reilly.

Cardona, C., Pulgarín, B., Agudelo, A., Calvache, M., Ordoñez, M., Manzo, O. 2008. Ajuste del Método Lahar-Z a la cuenca del Rio Páez, con base en los flujos de lodo ocurridos en los años de 1994 y 2007 en el sector del Volcán Nevado del Huila. *VII ENCUENTRO INTERNACIONAL DE INVESTIGADORES DEL GRUPO DE TRABAJO DE NIEVES Y HIELOS DE AMERICA LATINA DEL PHI-UNESCO.*

Chow, V.T. 2004. *Hidráulica de canales abiertos*. Santafé de Bogotá: Nomos S.A.

Dingman, L. 2009. *Fluvial Hydraulics*. Oxford: Oxford University Press. Inc.

Inama, H., Itakura, Y. & Kasahara., M. 1999. Surface Velocity Computation of debris Flows by Vector Field Measurements. *Physics and Chemistry of the Earth* 25(9): 741–744.

Massimo, A. & Marchi, L. 2008. Systems and Sensors for Debris-flow Monitoring and Warning. *Sensors* 8(4): 2436–2452.

Oliveras. N. 2008. Informe de velocidades estimadas a partir de datos de las estaciones de flujos de lodo en los ríos Símbola y Páez en la erupción del 20 de Noviembre del 2008. Popayán. *Informe interno INGEOMINAS.*

Pulgarin, B., Hugo, M., Agudelo, A., Calvache, M., Manzo, O., Narváez, A. 2008. Características del lahar del 20 de noviembre de 2008 en el Río Páez, causado por la Erupción del Complejo Volcánico Nevado del Huila – Colombia. *Informe interno INGEOMINAS.*

Vide, J. 2003. *Ingeniería Fluvial*. Barcelona: Centro de publicaciones Nacional de Colombia sede Medellín.

Worni, R., Huggel, C., Stoffel, M., Pulgarín, B. Challenges of modeling current very large lahars at Nevado del Huila Volcano, Colombia. *Bulletin of Volcanology. Official Journal of the International Association of Volcanology and Chemistry of the Earth's Interior (IAVCEI)* 74(2): 309–324.

Computational Modelling of Objects Represented in Images – Di Giamberardino et al. (eds)
© 2012 Taylor & Francis Group, London, ISBN 978-0-415-62134-2

Incorporating global information in active contour models

Appia Vikram & Anthony Yezzi

Georgia Institute of Technology, Atlanta, US

ABSTRACT: In this paper, we present some traditional models which incorporate global information in active contour models in the form of edge-, region-statistic- or shape-based constraints. We then develop an *active geodesic* contour model, which imposes a shape constraint on the evolving active contour to be a geodesic with respect to a weighted edge-based energy through its entire region-based evolution, rather than just at its final state (as in the traditional *geodesic active contour* models). We therefore device purely region-based energy minimization under the shape- and edge-based constraints. We show that this novel approach of combining edge information as the geodesic shape constraint in optimizing a purely region-based energy yields a new class of active contours which exhibit both local and global behaviors. By imposing edge- and shape-constraints on the region-based active contour model we incorporate global information into active contour models to make them robust against local minima.

1   INTRODUCTION

Curve evolution based approaches pose image segmentation as an energy optimization problem. In these models, an image feature-based energy functional is defined on the curve. Minimizing the energy drives the curve towards near-by image features generating desired segmentation results. Such energy optimization based curve evolution models have been used extensively for image segmentation since their introduction in (Kass et al. 1988).

The basic idea behind the active contour model developed in (Kass et al. 1988) is to minimize an energy functional of the form

$$E_{snake} = \int_0^1 \{E_{int}(C(s)) + E_{ext}(C(s))\}ds, \qquad (1)$$

where $C(s) = (x(s), y(s))$ is the deformable curve with parameter $s \in [0,1]$. E_{int} represents internal energy associated with the curve that governs the smoothness of the curve, and E_{ext} represents external energy that depends on image features or constraints added by the user. The energy in (1) can be written as

$$E(C) = \int_0^1 \{a_1 \|C''(s)\|^2 + a_2 \|C'(s)\|^2 + P(C(s))\}ds, \qquad (2)$$

where C' and C'' are derivatives of the curve with respect to the parameter s, $P \geq 0$ is a cost function associated with image features, and $a_1 \geq 0$, $a_2 \geq 0$ are scalar constants. The first two terms in the integral in Equation 2 are internal forces associated with the curve, and the third term is the external image dependent force.

The active contour model in (Kass et al. 1988) uses a local gradient based edge detector to stop the evolving curve on object boundaries. A major drawback associated with this model is that the curve has a tendency to get stuck at subtle (undesired) image features/edges. These subtle features create undesired "*local minima*" in energy minimization. This drawback makes the model sensitive to initialization and noise. So active contour models are typically initialized by the user with a contour near the desired object of interest to avoid local minima.

The local minima problem has received considerable attention in the past two decades. One of the earliest approaches proposed to overcome these local minima and drive the curve towards desired image features was proposed in (Cohen 1991). Cohen proposed a simple solution to deal with the problem. He proposed including an additional balloon force to either shrink or expand the contour to drive the contour towards desired edges. But, one requires prior knowledge of whether the object is inside or outside the initial contour. Besides, the final contour will be biased towards smaller and larger segmentation while using the shrinking and expanding balloon forces, respectively. Since the introduction of active contours and the subsequent introduction of the balloon force model, several variations and modifications have been proposed to alleviate the local minima problem (Appleton and Talbot 2005, Chan and Vese 2001, Cohen 1991, Paragios and Deriche 2000, Ronfard 1994, Samson et al. 1999, Tsai et al. 2001, Yezzi et al. 2002).

In the following sections, we discuss some existing active contour models with edge-based constraints, region-based constraints and shape-based constraints, which are robust to local minima problem in certain class of images. These models add constraints with global considerations to avoid extraneous local minima in energy minimization.

1.1 Active contour models with edge-based constraints

If a monotonically decreasing, positive function of the gradient of the image ($g(\nabla I)$) is used as the local potential (cost function) in (2), the minimum of the energy drives the curve towards edges in the image. The final segmentation varies with initialization and parameterization of the curve. (Caselles et al. 1997, Kichenassamy et al. 1995) modify the energy functional in (2) by defining the parameter s as the arc length of the contour. This modification yields a geometric interpretation to the classical energy minimization problem in (2). Since s denotes the arc length of the curve, we have $\|C'(s)\| = 1$. Neglecting the term associated with the second derivative $C''(s)$, we modify (2) as

$$E(C) = \int_\Omega \{a_1 + P(C(s))\}ds = \int_\Omega \{\tau(C(s))\}ds,$$
(3)

where $\tau = P + a_1(\Rightarrow \tau \geq 0)$. For the given length of the curve, L, the domain becomes $\Omega = [0, L]$. The internal forces are now included in the cost function τ. Thus, the energy minimization problem is reduced to finding the minimum path length of a contour in a Riemannian space.intensity. Cohen and Kimmel (Cohen and Kimmel 1997) converted the energy minimization problem in (3) into a global cost accumulation problem. By treating τ as the traveling cost associated with each point on the image plane, one can find the minimum of the energy by finding the accumulated cost (u) in traveling from a given point on the image plane to every other point on the image plane. According to (Cohen and Kimmel 1997), the accumulated cost u can be found by solving the Eikonal equation,

$$\|\nabla u\| = \tau.$$
(4)

Authors in (Adalsteinsson and Sethian 1994) developed a single-pass algorithm to find the accumulated cost u over the entire image domain, and they named the algorithm, "Fast Marching" technique. Thus, transforming the problem of generating open geodesics into a minimal path problem. (Cohen and Kimmel 1997) used the minimal path technique to develop an interactive edge-based segmentation algorithm, which can be initialized by a single point on the object boundary, to generate closed geodesics. Later, (Appleton and Talbot 2005) used minimal paths to develop a segmentation algorithm, which can be initialized by a single user-given point inside the object of interest.

1.2 Active contour moderls with region-based constraints

In contrast to edge-based models, the region-based curve evolution techniques, which were developed

later, define energy functionals based on region statistics rather than local image gradients (Chan and Vese 2001, Mumford and Shah 1989, Ronfard 1994, Samson et al. 1999, Tsai et al. 2001, Yezzi et al. 2002). In general, these region-based segmentation models are less sensitive to noise and initialization when compared to the edge-based models. Since these region-based models make strong assumptions about homogeneity of the image, they fail to capture the relevant edges in certain cases.

While region-based segmentation algorithms are less susceptible to local minima, edge-based segmentation algorithms have a better chance of detecting edges along the object boundary. (Paragios and Deriche 2000) used a linear combination of a probability based active region model with the classical edge-based model to exploit the benefits of both approaches. The final segmentation curve had certain desirable properties of both models. But a fixed weight for the linear combination may not be suitable for all kinds of images. (Chakraborty and Duncan 1999) introduced a game-theory based approach to combine region- and edge-based models. In addition to sensitivity to the choice of linear weighting factors, such energy-based schemes that employ combination of edge- and region-based terms may yield new classes of local minima that represent unsatisfactory *compromises* of these two criteria.

1.3 Active contour models with shape-based constraints

Segmentation purely based on image information becomes difficult when the image has inherent noise, low contrast, and missing/diffused edges. Such difficult scenarios are very common in medical images. In such cases, it is desirable to integrate prior shape knowledge in the active contour model. The concept of using prior shape knowledge for image segmentation was introduced in (Cootes et al. 1995). They used an explicit parametrized curve representation framework for training, as well as curve fitting. Subsequently, research has been conducted actively in the last two decades to incorporate prior shape knowledge in image segmentation. Similar shape prior-based segmentation models with explicit parameterization (Chen et al. 2001, Davatzikos et al. 2003) and implicit level set approaches (Rousson and Paragios 2002, Tsai et al. 2003, Rousson and Paragios 2008, Leventon et al. 2002, Sundaramoorthi et al. 2007) were also developed.

While global minimal path approaches introduce global information in a local cost accumulation framework, the region-based and shape prior based model enforce strong global image statistics and shape based constraints on the contour. In this paper, we develop an *active geodesic* model that minimizes region-based energy with an in-built edge- and shape-based constraint, to exploit the benefits of these three approaches without imposing rigid constraints on the contour. Rather than imposing shape constraint on the evolving contour based on prior training data, we impose a geodesic contraint based on image features. This geodesic contraint is less stringent than the shape prioir constraint giving the model flexibiltiy to capture relevant edges.

2 COUPLING REGION- AND EDGE-BASED SEGMENTATION

In this section, we describe the coupling of region- and edge-based segmentation using the minimal path approach (Cohen and Kimmel 1997). We first discuss an extension of the minimal path approach to detect closed geodesics and then incorporate region-based energy to find a globally accurate segmentation.

2.1 Edge-based segmentation using minimal paths

For edge-based minimal paths, we use the Eikonal equation from (4) with an edge-based local potential of the form

$$\tau(x) = g\left(\frac{1}{1 + \|\nabla I\|}\right) + \epsilon, \qquad (5)$$

where $g(\cdot)$ is a monotonically increasing function, $\epsilon > 0$ is a regularizer and ∇I denotes the gradient of the image at a given location. In all the examples presented here we use Interpolated Fast Marching scheme (Appia and Yezzi 2010) with a monotonic function of the form $g(x) = x^m$, where $m \geq 1$. To extract the shortest path between two points, we calculate u by propagating wavefronts from one of the two points (source point) to the other (end point). Then, by following the gradient descent in the vector field $\vec{\nabla} u$, we trace our path back to the source point. The path obtained is the globally optimal open geodesic (shortest path) between the two points.

2.2 Detecting closed curves using minimal paths

If two different global minimal paths exist between two points on the image, the two open geodesics will complete a closed contour. Now, consider a single source point given in the image domain. To detect closed curves, we have to find points on the domain from which two global minimal paths (back to the single source) exist, i.e, the two paths have the same accumulated cost. These special points are called saddles of u, and they can be interpreted as the points where the propagating fronts collide. We use the saddle point detection technique described in (Cohen and Kimmel 1995).

In the cardiac image shown in Figure 1(a), we place the source point (marked 'X') on the object boundary. Figure 1(b) shows the level set representation of the wavefronts propagating from the source. Figure 1(c) and Figure 1(d) show the various detected saddle points and closed curves associated witheach detected saddle point.

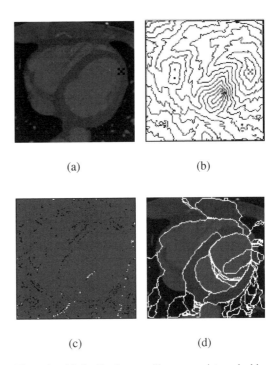

(a)   (b)

(c)   (d)

Figure 1. (a) Cardiac image with source point marked by an 'X'. (b) Level set representation of the propagating wavefronts. (c) White pixels in the image indicate saddle points. (d) Closed curves associated with the saddle points.

2.3 Shock curves

Figure 2(a) shows an illustration of the fronts emanating out of the source in different directions. Each gray level indicates a different neighbor of the source point from which the front at any given location propagated from. We observe that the fronts arriving from two different directions will form shock curves when they meet. By definition, the locations on these shock curves where the fronts arrive from two exactly opposite directions (collide) are the saddle points. The minimal paths from saddle points lying on the shock curves arrive at the source from two different directions, forming a closed geodesic. The minimal paths from all other points along the shock curve also arrive at the source from two different directions, but the two paths will not have the same accumulated cost. Saddle points are isolated points on a shock curve which form curves that are closed geodesic at the shock curve.

2.4 Incorporating region-based energy

The next step is to choose the appropriate saddle point to segment the object. For this, we assume that the object surrounded by an edge also exhibits certain region-based (statistical) properties. We can now compare the region-based energy for each closed geodesic associated with the saddle points. Let us consider the Chan-Vese (Chan and Vese 2001) energy function of the form

$$E = (I - \mu)^2 + (I - \nu)^2, \qquad (6)$$

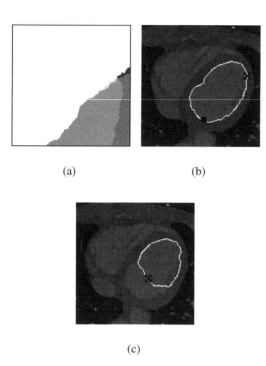

(a)   (b)

(c)

Figure 2. (a) Each gray level indicates a different neighbor of the source point from which the front at the given location propagated. The intersection of these labeled regions forms the shock curves. (b) Closed curve with minimum region-based energy. The source point is marked by an 'X' and the saddle point by a 'dot'. (c) Segmentation with source point lying inside the object of interest.

where μ is the mean of pixel intensities inside the curve and ν is the mean of pixel intensities outside the curve. Among all detected closed curves, the curve with the minimum region-based energy segments the object. From the various geodesics shown in Figure 1(d), the closed geodesic with the minimum region-based energy segments the left ventricle (Figure 2(b)).

Until this point, we assumed that our source point was on the boundary of the object of interest. This meant that the saddle point associated with the closed geodesic exhibiting least region-based energy also fell on the object boundary. Consider the source point shown in Figure 2(c), which is placed away from the object boundary. The segmentation obtained using the approach described previously does not segment the object (left ventricle). Thus, for a given source point, the minimal path approach guarantees the global minimum for edge-based energy, but it does not guarantee that the closed curve will also correspond to the minimum of the region-based energy.

3 ACTIVE GEODESICS: INCORPORATING SHAPE CONSTRAINTS IN A REGION-BASED ACTIVE CONTOURS WITH GLOBAL EDGE-BASED CONSTRAINTS

We use the region-based energy of the closed geodesic to perturb the saddle point along the shock curve,

which indirectly evolves the closed geodesic. Thus, the curve evolution has only two degrees of freedom, the two co-ordinates of the saddle point. We continue the evolution until we reach a minimum for the region-based energy. The minimal path approach imposes a shape constraint on the curve, which ensures that the evolving contour stays a closed geodesic throughout the entire evolution process.

Let S denote the saddle point on the given curve C. The shock curve corresponding to S will form a boundary of the vector field $\vec{\nabla} u$. We parametrize the boundary (shock curve) with a linear function $\psi(x) = p$. By solving

$$\nabla \psi \cdot \nabla u = 0, \tag{7}$$

we propagate ψ in the direction of the characteristics of $\vec{\nabla} u$ to form a level set function in the image domain. This level set function ψ forms an implicit representation of the curve C. If we choose a function such that $\psi(S) = 0$, then the closed geodesic will be embedded as the zero level set of ψ.

Now, consider a general class of region-based energies,

$$E(C) = \int_{\Omega} f(x) dA. \tag{8}$$

We can represent the gradient of $E(C)$ w.r.t the parameter p as the line integral

$$\nabla_p E = \int_C f \cdot \nabla_p C \cdot \vec{N} ds, \tag{9}$$

where s is the arc length parameter of the curve and $\vec{N}$ is the outward normal to the curve C. Since the level set function ψ is a function of the parameter p as well as the curve C, we have

$$\psi(C, p) = p. \tag{10}$$

Taking the gradient of 10, we get

$$\nabla \psi \cdot \nabla_p C = 1 \Rightarrow \frac{\nabla \psi}{\|\nabla \psi\|} \cdot \nabla_p C = \frac{1}{\|\nabla \psi\|}$$

$$\tag{11}$$

$$\Rightarrow \nabla_p C \cdot \vec{N} = \frac{1}{\|\nabla \psi\|}.$$

Since C is embedded in level set function ψ, both C and ψ have the same outward normal, $\vec{N} = \frac{\nabla \psi}{\|\nabla \psi\|}$. Thus, substituting 11 in 9 we get

$$\nabla_p E = \int_C \frac{f}{\|\nabla \psi\|} ds. \tag{12}$$

The factor of $\|\nabla \psi\|$ in the denominator of the line integral varies the contribution of each point on the curve.

The points on C closer to the saddle point will have a higher contribution to the integral than the points further away.

For the Chan-Vese (Chan and Vese 2001) energy model described in (6), f takes the form

$$f = 2(\mu - \nu)\left\{I - \frac{\mu + \nu}{2}\right\}. \qquad (13)$$

We now perturb the saddle point S along the shock curve, against the gradient $\nabla_p E$. The value of the line integral in (12) governs how far along the shock curve we perturb the saddle point. Once we perturb the saddle point, the two open geodesics back to the source from the new location cease to form a closed geodesic. Thus, we make the perturbed saddle point (at the new location) our new source. We now recompute u from this new source point and pick a saddle point satisfying the following two conditions:

1. The associated closed geodesic has lower region-based energy when compared to the energy of the closed geodesic obtained prior to perturbing the saddle point.
2. It lies closest to the previous source point.

The accumulated cost (u) from the previous iteration is used as the metric to measure the distance from the previous source point and not the Euclidean distance. Since we move the saddle point against the gradient of E, the region-based energy for the closed geodesic will decrease with each evolution of the curve. We follow this procedure until we converge to a minimum for the region-based energy.

4 INTERACTIVE SEGMENTATION ALGORITHM

Using the *active geodesic* contour evolution, we present an interactive segmentation algorithm. By placing poles and zeros the user can attract/repel the *active geodesic* towards a desired segmentation.

4.1 *Attractors and repellers*

We refer to the poles and zeros placed by the user as *repellers* and *attractors*, respectively. For each repeller (P) and attractor (Z) the user places, we update the local potential (traveling cost) as

$$\tau'(x) = \tau(x) \circ h_1\left(\frac{1}{distance(x,P)}\right), \qquad (14)$$

$$\tau'(x) = \tau(x) \circ h_2\left(distance(x,Z)\right), \qquad (15)$$

where h_1 and h_2 are monotonically increasing functions, the 'o' operator represents Hadamard product[1]

---

[1] Hadamard product is the entry-wise product of two matrices. For two given matrices A_{mxn} and B_{mxn}, $(A \circ B)_{i,j} = (A)_{i,j} \cdot (B)_{i,j}$.

and *distance($\cdot$, $\cdot$)* is the Euclidean distance. By placing these *attractors* and *repellers* the user is locally modifying the weighted edge function. This local variation has a global effect on the *active geodesic* evolution as we see in the examples in Section 5.

4.2 *Algorithm details*

We initialize the algorithm by asking the user to place a single point within the desired object of interest. We then place a *repeller* at this location. This artificially placed *repeller* serves the following two purposes:

1. It identifies the object of interest.
2. It ensures that the propagating wavefronts wrap around the pole to guarantee that at least one closed geodesic exists.

Now, we randomly pick a point in the vicinity of the *repeller* (different from the *repeller*), as the source point, and follow the procedure described in Section 3. In the very first iteration we do not have a reference for source from previous iteration, hence we choose the saddle point closest to the *repeller* placed by the user. We move the saddle point against the gradient $\nabla_p E$ to minimize the region-based energy of the *active geodesic*. This saddle point becomes the source for the second iteration. We continue the evolution described in Section 3 until we converge to a minimum. Figure 3 shows the evolving *active geodesic* and the final segmentation of the left ventricle for the cardiac image.

Once we converge to a minimum, we present the user with the resultant closed geodesic. If the user is not satisfied with the segmentation result he can add an *attractor* or a *repeller* to drive the *active geodesic* towards the desired edges. Consider the segmentation result shown in Figure 4(a). Placing another *repeller* inside the closed contour further evolves the *active geodesic* away from the new *repeller* as shown in Figure 4(b).

The *repellers* placed by the user are classified as interior or exterior *repellers* based on their location with respect to the current state of the *active geodesic*. We choose only those saddle points that form closed geodesics that separate all the interior *repellers* from the exterior *repellers*. This ensures that a *repeller* placed inside the closed curve lies inside the final segmentation and a *repeller* placed outside the closed curve stays outside the final segmentation. No such constraint is placed on the *active geodesic* based on the location of *attractors* added by the user.

Placing a few more *attractors* and *repellers*, the user can converge to the final desired segmentation of the right ventricle as shown in Figure 4(b). A brief outline of the algorithm is given in Table 1.

5 EXPERIMENTAL RESULTS

The *active geodesic* model combines a global edge-based constraint with a region-based energy minimization model in a framework that enables intuitive

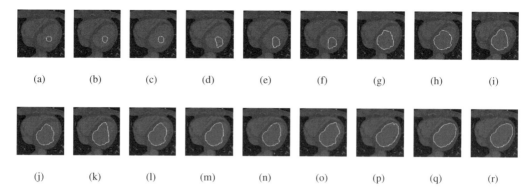

(a)   (b)   (c)   (d)   (e)   (f)   (g)   (h)   (i)

(j)   (k)   (l)   (m)   (n)   (o)   (p)   (q)   (r)

Figure 3. Left Ventricle segmentation: (a) Segmentation after the first iteration. (b-r) Evolution of the closed curve to minimize the region-based energy. (s) Final converged segmentation after 19 iterations.

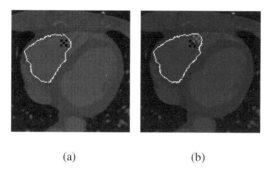

(a)   (b)

Figure 4. The initial *repeller* is marked by a black 'X'. (a) Converged segmentation of the right ventricle with a single *repeller* inside the ventricle. (b) Converged segmentation after the user places a second *repeller*.

interactions to generate non-arbitrary segmentation results with a sense of global optimality. Since our model minimizes region-based energy with a global edge constraint in an interactive segmentation algorithm, we compare the results of our method with the following existing segmentation algorithms:

5.1 Active contours without edges (chan-vese segmentation) model

As opposed to the proposed global edge-constrained *active geodesic* model, the Chan-Vese model (Chan and Vese 2001) is an unconstrained region-based energy minimization model. The algorithm is initialized with a user specified initial contour. By minimizing the energy given in (6), we converge to the final segmentation result. Figure 5 shows segmentation results for the left and right ventricles in the cardiac image shown in Figure 1 (a). We can see that this purely region-based model fails to segment the ventricles because it does not take any edge information into consideration.

5.2 Globally Optimal Geodesic Active Contour (GOGAC) model

GOGAC (Appleton and Talbot 2005) is a purely edge-based segmentation model, which generates globally

optimal edge-based segmentation. The algorithm is initialized with a single user-given point inside the object of interest. The edge-based potential in 5 is then modified to

$$\tau'(x) = \frac{1}{r}\left(g\left(\frac{1}{1+\|\nabla I\|}\right) + \epsilon\right), \tag{16}$$

where r is the Euclidean distance from the user-specified point. A globally optimal edge-based segmentation is obtained using a minimal path approach by inducing an artificial cut in the image domain. The procedure followed by the algorithm does not allow further user interactions to improve segmentation results.

In Figure 6, the GOGAC-model-based left and right ventricle segmentations are initialized by the user with points marked by black 'X's. We obtain the globally optimal closed geodesic with the vertical cut induced in the image plane as shown in the Figures 6 (a) and (b), respectively. Although, the final segmentation curves are global minima with respect to the edge-based potential in (16), they fail to segment the ventricles due to the presence of strong, misleading edges inside the ventricles.

5.3 Linear combination of edge- and region-based energies

In this model (Paragios and Deriche 2000), the active contour minimizes a linear combination of edge- and region-based energies,

$$E = \alpha \cdot E_{reg} + (1-\alpha) \cdot E_{edge}, \tag{17}$$

where $\alpha \in (0, 1)$. We again use the Chan-Vese energy (6) as our region-based energy in this linear combination. The edge-based energy is of the form

$$E_{edge} = \int_C \phi \cdot ds, \tag{18}$$

58

Table 1. Pseudo-code for the interactive segmentation algorithm.

(1)	**do**
(2)	**if** First iteration = *TRUE*
(3)	Initialize algorithm with a *repeller* inside the object of interest.
(4)	Update the local potential.
(5)	Choose a random point other than the *repeller* as the initial source point.
(6)	**else**
(7)	Add a *repeller* or an *attractor* and update the local potential.
(8)	**end**
(9)	**do**
(10)	**if** First iteration = *FALSE*
(11)	Perturb the saddle point based on region-based energy and make it the new source point.
(12)	**end**
(13)	Propagate wavefronts from the source point.
(14)	Detect shock curves and the associated saddle points.
(15)	**if** First iteration = *TRUE*
(16)	Find the saddle point closest to the *repeller*.
(17)	**else**
(18)	Find the saddle point lying closest to the source in the previous iteration, which also minimizes region-based energy and separates the interior and exterior *repellers*.
(19)	**end**
(20)	**while** Convergence = *FALSE*
(21)	Complete the closed contour using the current source and saddle point to obtain the converged segmentation.
(22)	**while** Desired Segmentation = *FALSE*

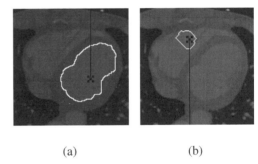

(a)   (b)

Figure 6. (a,b) Left and Right Ventricle segmentation using the purely-edge-based-GOGAC model initialized with the points marked by 'X' within the ventricles.

Figure 7 shows the left and right ventricle segmentation initialized with contours shown in Figures 5 (a) and (b), respectively. As the contour converges to the edges of the ventricles, it gets stuck in certain local minima within the ventricles. Further, it may become necessary to heuristically change the weight of the linear combination for different scenarios (Results presented here were generated with $\alpha = 0.75$). In addition to sensitivity to the choice of α, such linear combination yield new classes of local minima that represent unsatisfactory *compromises* of edge- and region-based models.

Figure 8(a) shows the final segmentation of the left ventricle in the cardiac image shown in Figure 1(a) with the proposed *active geodesic* model. A single *repeller* placed by the user is sufficient to

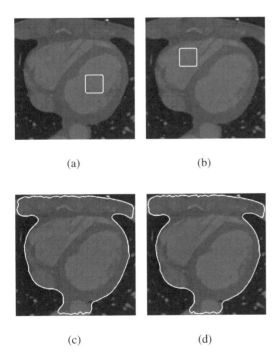

(c)   (d)

Figure 5. Segmentation of ventricles with the purely region-based Chan-Vese model: (a,b) Initialization for the left and right ventricle segmentation. (c,d) Final region-based, Chan-Vese segmentation.

where,

$$\phi = \frac{1}{(1 + ||\nabla I||^2)}. \tag{19}$$

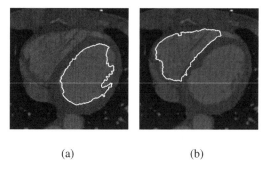

(a)　　　　　　　　　　(b)

Figure 7. (a,b) Final segmentation of left and right ventricles obtained by minimizing the energy in 17, when initialized with curve given in Figures 5 (a) and (b), respectively.

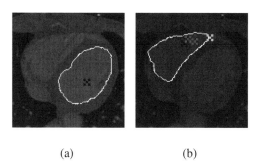

(a)　　　　　　　　　　(b)

Figure 8. The initial *repeller* is marked by a black 'X'. (a) Desired Left Ventricle segmentation achieved in 19 iterations (b) Right Ventricle segmentation after the user has placed a few *attractors* (marked by green 'X's) and *repellers* (marked by red 'X's). Desired segmentation achieved in 27 iterations.

segment out the left ventricle. Since, the final segmentation tries to optimize a region-based energy with an edge-based constraint, it overcomes the minima that hampers the Chan-Vese segmentation model, GOGAC model and the linear combination model. Further, the global edge-constraint ensures that our model generates meaningful segmentation results rather than arbitrary intermediate results. Thus, we can achieve desired segmentation with fewer user interactions (with a single *repeller* in the case with left ventricle). Similarly, Figure 8(b) shows the final segmentation of the right ventricle after modifying the metric by placing a few *attractors* and *repellers*.

In Figure 9, we present cell segmentation results. We can see that the edge-based GOGAC approach fails due to the presence of several strong edges within the cell. Using our approach, the user can accurately segment the cell by placing a few additional *attractors* and *repellers* (Figures 9(a,c,e)). Figures 9(a,c,e), also illustrate how we can segment the cell with three different initializations.

In Figure 10, we compare the segmentation of the two nuclei in a cell image with multiple nuclei. We see that using the *active geodesic* model segments both nuclei with a few user interactions, where as the other techniques fail to segment the nuclei. Since, it is

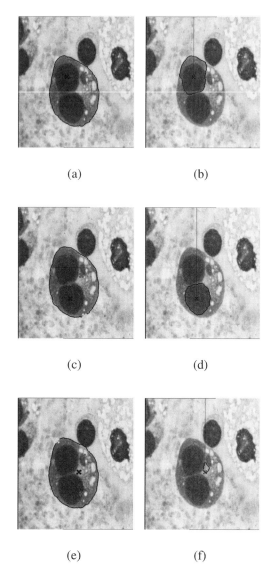

(a)　　　　　　　　　　(b)

(c)　　　　　　　　　　(d)

(e)　　　　　　　　　　(f)

Figure 9. (a,c,e) *Active geodesic* based segmentation: The initial *repeller* is marked by a black 'X'. Subsequent *attractors* are marked by green 'X's and *repellers* are marked by red 'X's. Desired segmentation achieved in 22, 34 and 40 iterations, respectively. (b,d,f) Cell segmentation results with GOGAC.

not possible to interact with the algorithm to improve segmentation, we fail to segment the nuclei accurately.

6 CONCLUSION

We presented the *active geodesic* model, which constrains an evolving active contour to continually be a geodesic with respect to an edge-based metric throughout the evolution process. The geodesic shape-constraint reduced the infinite dimension region-based optimization problem into a finite dimension problem

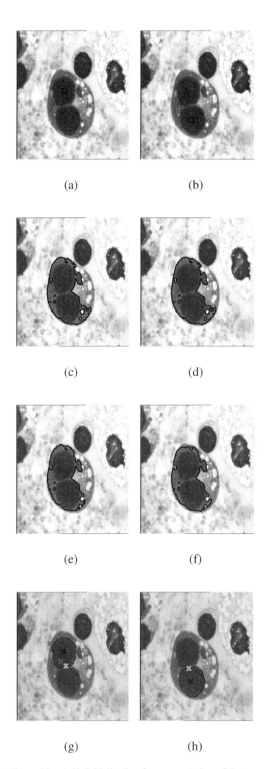

Figure 10. (a,b) Initialization for segmentation of the two nuclei. (c,d) Final region-based Chan-Vese segmentation. (e,f) Final segmentation optimizing the combination of region- and edge-based energies. (g,h) *Active geodesic* based segmentation: *Attractors* – Green 'X's and *Repellers* – Red 'X's. Desired segmentation achieved in 49 and 21 iterations, respectively.

by reducing the search space. Further, using minimal path technique to generate geodesics during the "active" evolution of the contour ensured that the edges captured by the curve corresponds to edges of a global minimizer rather than the unwanted local minima. Minimizing a region-based energy subject to shape- and edge-based constraint yields closed geodesics that exhibit both local and global behaviors rather than being compromises achieved by weighted combination of region- and edge-based energies. We showed the relationship of this new class of globally constrained active contours with traditional minimal path methods, which seek global minimizers of purely edge-based energies without incorporating region-based criteria.

REFERENCES

Adalsteinsson, D. and J. A. Sethian (1994). A fast level set method for propagating interfaces. *Journal of computational physics 118.*

Appia, V. and A. Yezzi (2010). Fully isotropic fast marching methods on cartesian grids. ECCV'10, Berlin, Heidelberg, pp. 71–83. Springer-Verlag.

Appleton, B. and H. Talbot (2005). Globally optimal geodesic active contours. *Jorunal of mathematical imaging and vision 23*, 67–86.

Caselles, V., R. Kimmel, and G. Sapiro (1997). Geodesic active contours. *International Journal of Computer Vision 22*, 61–79.

Chakraborty, A. and J. Duncan (1999, jan). Game-theoretic integration for image segmentation. *Pattern Analysis and Machine Intelligence, IEEE Transactions on 21*(1), 12–30.

Chan, T. and L. Vese (2001). A level set algorithm for minimizing the mumford-shah functional in image processing. In *Variational and Level Set Methods in Computer Vision, 2001. Proceedings. IEEE Workshop on*, pp. 161 –168.

Chen, Y., S. Thiruvenkadam, H. Tagare, F. Huang, D. Wilson, and E. Geiser (2001). On the incorporation of shape priors into geometric active contours. In *Variational and Level Set Methods in Computer Vision, 2001. Proceedings. IEEE Workshop on*, pp. 145 –152.

Cohen, L. and R. Kimmel (1995). Edge integration using minimal geodesics. *Technical report*.

Cohen, L. and R. Kimmel (1997). Global minimum for active contour models: A minimal path approach. *International Journal of Computer Vision 24*(1), 57–78.

Cohen, L. D. (1991). On active contour models and balloons. *Computer vision, graphics, and image processing 53*, 211–218.

Cootes, T. F., C. J. Taylor, D. H. Cooper, and J. Graham (1995). Active shape models-their training and application. *Computer Vision and Image Understanding 61*(1), 38–59.

Davatzikos, C., X. Tao, and D. Shen (2003, march). Hierarchical active shape models, using the wavelet transform. *Medical Imaging, IEEE Transactions on 22*(3), 414 –423.

Kass, M., A. Witkin, and D. Terzopoulos (1988). Snakes: Active contour models. *International Journal of Computer Vision 1*(4), 321–331.

Kichenassamy, S., A. Kumar, P. J. Olver, A. Tannenbaum, and A. J. Yezzi (1995). Gradient flows and geometric active contour models. In *Computer Vision (ICCV), 1995 IEEE International Conference on*, Washington, DC, USA, pp. 810–. IEEE Computer Society.

Leventon, M., W. Grimson, and O. Faugeras (2002, june). Statistical shape influence in geodesic active contours. In *Biomedical Imaging, 2002. 5th IEEE EMBS International Summer School on*, pp. 8 pp.

Mumford, D. and J. Shah (1989). Optimal approximations by piecewise smooth functions and associated variational problems. *Communications on pure and applied mathematics 42*(5), 577–685.

Paragios, N. and R. Deriche (2000). Coupled geodesic active regions for image segmentation: A level set approach. In *ECCV*, pp. 224–240.

Ronfard, R. (1994). Region-based strategies for active contour models. *International Journal of Computer Vision 13*(2), 229–251.

Rousson, M. and N. Paragios (2002). Shape priors for level set representations. In *ECCV (2)*, pp. 78–92.

Rousson, M. and N. Paragios (2008). Prior knowledge, level set representations & visual grouping. *International Journal of Computer Vision 76*(3), 231–243.

Samson, C., L. Blanc-Féraud, G. Aubert, and J. Zerubia (1999). A level set model for image classification. In *Scale-Space Theories in Computer Vision*, pp. 306–317.

Sundaramoorthi, G., A. J. Yezzi, and A. Mennucci (2007). Sobolev active contours. *International Journal of Computer Vision 73*(3), 345–366.

Tsai, A., J. Yezzi, A., W. Wells, C. Tempany, D. Tucker, A. Fan, W. Grimson, and A. Willsky (2003, feb.). A shape-based approach to the segmentation of medical imagery using level sets. *Medical Imaging, IEEE Transactions on 22*(2), 137–154.

Tsai, A., J. Yezzi, A., and A. Willsky (2001, aug). Curve evolution implementation of the mumford-shah functional for image segmentation, denoising, interpolation, and magnification. *Image Processing, IEEE Transactions on 10*(8), 1169–1186.

Yezzi, A. J., A. Tsai, and A. Willsky (2002). A fully global approach to image segmentation via coupled curve evolution equations. *Journal of Visual Communication and Image Representation 13*, 195–216.

Computational Modelling of Objects Represented in Images – Di Giamberardino et al. (eds)
© 2012 Taylor & Francis Group, London, ISBN 978-0-415-62134-2

Hexagonal parallel thinning algorithms based on sufficient conditions for topology preservation

Péter Kardos & Kálmán Palágyi
Department of Image Processing and Computer Graphics, University of Szeged, Szeged, Hungary

ABSTRACT: Thinning is a well-known technique for producing skeleton-like shape features from digital binary objects in a topology preserving way. Most of the existing thinning algorithms presuppose that the input images are sampled on orthogonal grids. This paper presents new sufficient conditions for topology preserving reductions working on hexagonal grids (or triangular lattices) and eight new 2D hexagonal parallel thinning algorithms that are based on our conditions. The proposed algorithms are capable of producing both medial lines and topological kernels as well.

1 INTRODUCTION

Various applications of image processing and pattern recognition are based on the concept of skeletons (Siddiqi and Pizer 2008). Thinning is an iterative object reduction until only the skeletons of the binary objects are left (Lam et al. 1992; Suen and Wang 1994). Thinning algorithms in 2D serve for extracting medial lines and topological kernels (Hall et al. 1996). A topological kernel is a minimal set of points that is topologically equivalent to the original object (Hall et al. 1996; Kong and Rosenfeld 1989; Kong 1995; Ronse 1988). Some thinning algorithms working on hexagonal grids have been proposed (Deutsch 1970; Deutsch 1972; Staunton 1996; Staunton 1999; Wiederhold and Morales 2008; Kardos and Palágyi 2011)

Parallel thinning algorithms are composed of reduction operators (i.e., some object points having value of "1" in a binary picture that satisfy certain topological and geometric constrains are changed to "0" ones simultaneously) (Hall 1996).

Digital pictures on non–orthogonal grids have been studied by a number of authors (Kong and Rosenfeld 1989; Marchand-Maillet and Sharaiha 2000). A hexagonal grid, which is formed by a tessellation of regular hexagons, corresponds, by duality, to the triangular lattice, where the points are the centers of that hexagons, see Figure 1. The advantage of hexagonal grids over the orthogonal ones lies in the fact that in hexagonal sampling scheme, each pixel is surrounded by six equidistant nearest neighbors, which results in a less ambiguous connectivity structure and in a better angular resolution compared to the rectangular case (Lee and Jayanthi 2005; Marchand-Maillet and Sharaiha 2000).

Topology preservation is an essential requirement for thinning algorithms (Kong and Rosenfeld 1989). In order to verify that a reduction preserves topology, Ronse and Kong gave some sufficient conditions for

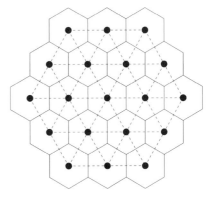

Figure 1. A hexagonal grid and the corresponding triangular lattice.

reduction operators working on the orthogonal grid (Kong 1995; Ronse 1988), then later, Kardos and Palágyi proposed similar conditions for the hexagonal case, that can be used to verify the topological correctness of the thinning process (Kardos and Palágyi 2011).

In this paper we present some new alternative sufficient conditions for topology preservation on hexagonal grids that make possible to generate deletion conditions for various thinning algorithms and we also introduce such algorithms based on these conditions.

The rest of this paper is organized as follows. Section 2 reviews the basic notions of 2D digital hexagonal topology and some sufficient conditions for reduction operators to preserve topology. Section 3 discusses the proposed hexagonal parallel thinning algorithms that are based on the three parallel thinning schemes. Section 4 presents some examples of the produced skeleton-like shape features. Finally, we round off the paper with some concluding remarks.

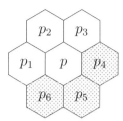

Figure 2. Indexing scheme for the elements of $N_6(p)$ on hexagonal grid. Pixels p_i ($i = 4, 5, 6$), for which p precedes p_i, are gray dotted.

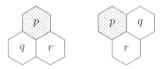

Figure 3. The two possible kinds of unit triangles. Since p precedes q and q precedes r, pixel p is the first element of the unit triangle.

2 BASIC NOTIONS AND RESULTS

Let us consider a hexagonal grid denoted by H, and let p be a pixel in H. Let us denote $N_6(p)$ the set of pixels being 6-*adjacent* to pixel p and let $N_6^*(p) = N_6(p) \backslash \{p\}$. Figure 2 shows the 6–neighbors of a pixel p denoted by $N_6(p)$. The pixel denoted by p_i is called as the *i-th neighbor* of the central pixel p. We say that p *precedes* the pixels p_4, p_5, and p_6. (It is easy to see that the relation "precedes" is irreflexive, antisymmetric, and transitive, therefore, it is a partial order on the set $N_6(p)$.)

The sequence S of distinct pixels $\langle x_0, x_1, \ldots, x_n \rangle$ is called a 6-*path* of length n from pixel x_0 to pixel x_n in a non-empty set of pixels X if each pixel of the sequence is in X and x_i is 6-adjacent to x_{i-1} ($i = 1, \ldots, n$). Note that a single pixel is a 6-path of length 0. Two pixels are said to be 6-*connected* in set X if there is a 6-path in X between them.

Based on the concept of digital pictures as reviewed in (Kong and Rosenfeld 1989) we define the *2D binary* (6, 6) *digital picture* as a quadruple $\mathcal{P} = (H, 6, 6, B)$. The elements of H are called the *pixels* of $\mathcal{P}$. Each element in $B \subseteq H$ is called a *black pixel* and has a value of 1. Each member of $H \backslash B$ is called a *white pixel* and the value of 0 is assigned to it. 6-adjacency is associated with both black and white pixels. An *object* is a maximal 6-connected set of black pixels, while a *white component* is a maximal 6-connected set of white pixels. A set composed of three mutually 6-adjacent black pixels p, q, and r is a *unit triangle* (see Fig. 3). Pixel p is called the *first element* of a unit triangle (see Fig. 3).

A black pixel is called a *border pixel* in a (6, 6) picture if it is 6-adjacent to at least one white pixel. A black pixel p is called an *d-border pixel* in a (6, 6) picture if its d-th neighbor (denoted by p_d in Fig. 2) is a white pixel ($d = 1, \ldots, 6$).

A *reduction operator* transforms a binary picture only by changing some black pixels to white ones (which is referred to as the deletion of 1's). A 2D reduction operator does *not* preserve topology (Kong 1995) if any object is split or is completely deleted, any white component is merged with another white component, or a new white component is created.

A *simple pixel* is a black pixel whose deletion is a topology preserving reduction (Kong and Rosenfeld 1989). A useful characterization of simple pixels on (6,6) pictures is stated as follows:

Theorem 1. (Kardos and Palágyi 2011) Black pixel p in picture $(H, 6, 6, B)$ is simple if and only if both of the following conditions are satisfied:

1. p is a border pixel.
2. Picture $(H, 6, 6, N_6^*(p) \cap B)$ contains exactly one object.

Note that the simplicity of pixel p in a (6, 6) picture is a local property; it can be decided in view of $N_6^*(p)$.

Reduction operators delete a set of black pixels and not only a single simple pixel. Kardos and Palágyi gave the following sufficient conditions for topology preserving reduction operators on hexagonal grids (Kardos and Palágyi 2011):

Theorem 2. A reduction operator $\mathcal{O}$ is topology preserving in picture $(H, 6, 6, B)$, if all of the following conditions hold:

1. Only simple pixels are deleted by $\mathcal{O}$.
2. If $\mathcal{O}$ deletes two 6-adjacent pixels p, q, then p is simple in $(H, 6, 6, B \backslash \{q\})$, or q is simple in $(H, 6, 6, B \backslash \{p\})$.
3. $\mathcal{O}$ does not delete completely any object contained in a unit triangle.

While the above result states conditions for pixel-configurations, we can derive from Theorem 2 some new criteria that examine if an individual pixel is deletable or not:

Theorem 3. A reduction operator $\mathcal{O}$ is topology preserving in picture $(H, 6, 6, B)$, if each pixel p deleted by $\mathcal{O}$ satisfies the following conditions:

1. p is a simple pixel in $(H, 6, 6, B)$.
2. For any simple pixel $q \in N_6^*(p)$ preceded by p, p is simple in $(H, 6, 6, B \backslash \{q\})$, or q is simple in $(H, 6, 6, B \backslash \{p\})$.
3. p is not the first element of any object that forms a unit triangle.

Proof Condition 1 of Theorem 3 corresponds to Condition 1 of Theorem 2. Furthermore, it is obvious that if p fulfills Condition 2 of Theorem 3 for a given $q \in N_6^*(p)$ but the set $\{p, q\}$ does not satisfy Condition 2 of Theorem 2, then q must precede p, and as the relation "precedes" is a partial order, this implies that q is not deleted by $\mathcal{O}$. We show that Condition 3 of Theorem 2 also holds. $\mathcal{O}$ does not delete a single pixel object by Condition 1. Objects composed by two 6-adjacent black pixels may not be completely deleted by Condition 2. Finally, from Condition 3 of Theorem 3 follows that exactly one element of any object composed by 3 mutually 6-adjacent pixels is retained by $\mathcal{O}$.

Therefore, $\mathcal{O}$ satisfies all conditions of Theorem 3.

Besides the topological correctness, another key requirement of thinning is shape preservation. For this aim, thinning algorithms usually apply reduction operators that do not delete so-called end pixels that provide important geometrical information related to the shape of objects. We say that none of the black pixels are *end pixels of type $E0$* in any picture, while a black pixel p is called an *end pixel of type $E1$* in a $(6, 6)$ picture if it is 6-adjacent to exactly one black pixel. Using end pixel characterization $E0$ leads to algorithms that extract topological kernels of objects, while criterion $E1$ can be applied for producing medial lines.

3 HEXAGONAL THINNING ALGORITHMS

In this section, eight thinning algorithms on hexagonal grids composed of reduction operations satisfying Theorem 3 are reported.

3.1 *Fully parallel algorithms*

In fully parallel algorithms, the same reduction operation is applied in each iteration step (Hall 1996).

Algorithm 1 introduces the general scheme of our two fully parallel algorithms H-FP-$E0$ and H-FP-$E1$.

Algorithm 1.   Algorithm H-FP-ε

1: *Input*: picture (H, 6, 6, X)
2: *Output*: picture (H, 6, 6, Y)
3: Y = X
4: repeat
5:    $D = \{p \mid p$ is H-FP-ε-deletable in Y$\}$
6:    Y = Y \ D
7: until $D = \emptyset$

H-FP-ε-deletable pixels ($\varepsilon \in \{E0, E1\}$) are defined as follows:

Definition 1. Black pixel p is H-FP-ε-deletable ($\varepsilon \in \{E0, E1\}$) if it is not an ε-end pixel and all the conditions of Theorem 3 hold.

The topological correctness of the above algorithm can be easily shown.

Theorem 4. Both algorithms H-FP-$E0$ and H-FP-$E1$ are topology preserving.

Proof. It can readily be seen that deletable pixels of the proposed two fully parallel algorithms (see Definition 1) are derived directly from conditions of Theorem 3. Hence, both algorithms preserve the topology.

3.2 *Subiteration-based algorithms*

The general idea of the subiteration-based approach (often referred to as directional strategy) is that an iteration step is divided into some successive reduction operations according to the major deletion directions. The deletion rules of the given reductions are determined by the actual direction (Hall 1996).

For the hexagonal case, here we propose our two 6-subiteration algorithms H-SI-ε ($\varepsilon \in \{E0, E1\}$) sketched in Algorithm 2. In each of its subiterations only d-border pixels ($d = 1, \ldots, 6$, see Definition 2) are deleted.

Algorithm 2.   Algorithm H-SI-ε

1: *Input*: picture (H, 6, 6, X)
2: *Output*: picture (H, 6, 6, Y)
3: Y = X
4: **repeat** 5: $D = \emptyset$
6:    **for** $d = 1$ **to** 6
7:      $D_d = \{p \mid p$ is H-SI-d-ε-deletable in Y $\}$
8:      Y = Y \ D_d
9:      $D = D \cup D_d$
10:   **endfor**
11: **until** $D = \emptyset$

Here we give the following definition for deletable pixels:

Definition 2. Black pixel p is H-SI-d-ε-deletable ($\varepsilon \in \{E0, E1\}, d = 1, \ldots, 6$) if all of the following conditions hold:

1. p is a simple but not an ε-end pixel and it is an d-border pixel in picture (H, 6, 6, B).
2. If $\varepsilon = E0$, then p is not the first element of any object $\{p, q\}$, where q is a d-border pixel.

Again, we can use Theorem 3 to prove the following result:

Theorem 5. Algorithms H-SI-$E0$ and H-SI-$E1$ are topology preserving.

Proof. If the conditions of Theorem 3 hold for every pixel deleted by our subiteration-based algorithms, then they are topology preserving by Theorem 3. Let p be a deleted d-border pixel ($d = 1, \ldots, 6$) that does not satisfy the mentioned conditions. Condition 1 of Definition 2 corresponds to Condition 1 of Theorem 3. Furthermore, if Condition 2 of Definition 2 holds, then so does Condition 2 of Theorem 3. Consequently, p must be the first element of an object $\{p, q, r\}$ that forms a unit triangle. However, it can be easily seen that in this case, q or r is not a d-border pixel, which means that the object $\{p, q, r\}$ may not completely removed by the algorithm. This shows that, even if Condition 3 of Theorem 3 is not fulfilled, algorithms H-SI-$E0$ and H-SI-$E1$ are topology preserving.

3.3 *Subfield-based algorithms*

In subfield-based parallel thinning the digital space is decomposed into several subfields. During an iteration step, the subfields are alternatively activated, and only pixels in the active subfield may be deleted (Hall

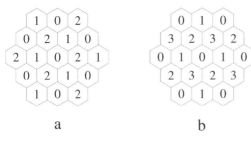

a   b

Figure 4. Partitions of H into three (a) and four (b) sub-fields. For the k-subfield case the pixels marked i are in $SF_k(i)$ ($k = 3, 4$; $i = 0, 1, \ldots, k - 1$)

1996). To reduce the noise sensitivity and the number of unwanted side branches in the produced medial lines, a modified subfield-based thinning scheme with iteration-level endpoint checking was proposed (Németh et al. 2010; Németh and Palágyi 2011). It takes the endpoints into consideration at the beginning of iteration steps, instead of preserving them in each parallel reduction as it is accustomed in the conventional subfield-based thinning algorithms.

We propose two possible partitionings of the hexagonal grid H into three and four subfields $SF_k(i)$ ($k = 3, 4$; $i = 0, 1, \ldots, k - 1$) (see Fig. 4).

We would like to emphasize a useful straightforward property of these partitionings:

Proposition 1. If $p \in SF_k(i)$, then $N_6^*(p) \cap SF_k(i) = \emptyset$ ($k = 3, 4$; $i = 0, \ldots, k - 1$).

Using these partitionings, we can formulate our four subfield-based algorithms H-SF-k-ε ($\varepsilon \in \{E0, E1\}$, $k \in \{3, 4\}$) (see Algorithm 3).

Algorithm 3. Algorithm H-SF-k-ε.

1: *Input*: picture (H, 6, 6, X)
2: *Output*: picture (H, 6, 6, Y)
3: Y = X
4: **repeat** 5: $D = \emptyset$
6:      $E = \{p \mid p$ is a border pixel but not an end pixel of type ε in Y$\}$
7:    **for** $i = 0$ **to** $k - 1$
8:      $D_i = \{p \mid p \in E$ and p is H-SF-k-i-deletable in Y$\}$
9:      Y = Y $\setminus D_i$
10:      $D = D \cup D_i$
11:    **endfor**
12: $D = \emptyset$

SF-k-i-deletable pixels are defined as follows:

Definition 3. A black pixel p is H-SF-k-i-deletable in picture $(H, 6, 6, Y)$ if p is a simple pixel, and $p \in SF_k(i)$ ($k \in \{3, 4\}, i = 0, \ldots, k - 1$).

Now, let us discuss the topological correctness of algorithm H-SF-k-ε ($\varepsilon \in \{E0, E1\}, k \in \{3, 4\}$).

Theorem 6. Algorithm H-SF-k-ε is topology preserving ($\varepsilon \in \{E0, E1\}, k \in \{3, 4\}$).

Proof. By Definition 3, the removal of SF-k-i-deletable pixels satisfies Condition 1 of Theorem 3. Proposition

(a) H-FP-$E0$

(b) H-FP-$E1$

Figure 5. Topological kernels (a) and medial lines (b) produced by the new fully parallel hexagonal thinning algorithms.

(a) H-SI-$E0$

(b) H-SI-$E1$

Figure 6. Topological kernels (a) and medial lines (b) produced by the new subiteration-based hexagonal thinning algorithms.

1 implies that if $p \in D_i$, then $N_6^*(p) \cap D_i = \emptyset$. Consequently, the antecedent of Condition 2 of Theorem 3 never holds, while Condition 3 is always satisfied in the case of algorithms H-SF-k-ε. Therefore, all the four subfield-based algorithms are topology preserving by Theorem 3.

(a) H-SF-3-$E0$

(b) H-SF-3-$E1$

Figure 7. Topological kernels (a) and medial lines (b) produced by the new 3-subfield hexagonal thinning algorithms.

(a) H-SF-4-$E0$

(b) H-SF-4-$E1$

Figure 8. Topological kernels (a) and medial lines (b) produced by the new 4-subfield hexagonal thinning algorithms.

4 RESULTS

In experiments the proposed algorithms were tested on objects of various images. Due to the lack of space, here we can only present the results for one picture containing three characters, see Figures 5–8, where the extracted skeleton-like shape features are superimposed on the original objects.

To summarize the properties of the presented algorithms, we state the followings:

- All the eight algorithms are different from each other (see Figs. 4–7).
- The four algorithms H-FP-$E0$, H-SI-$E0$, H-SF-3-$E0$, and H-SF-4-$E0$ produce topological kernels (i.e., there is no simple point in their results).
- Medial lines produced by the four algorithms H-FP-$E1$, H-SI-$E1$, H-SF-3-$E1$, and H-SF-4-$E1$ are minimal (i.e., they do not contain any simple point except the endpoints of type $E1$).

5 CONCLUSIONS

In this work some new sufficient conditions for topology preserving parallel reduction operations working on (6,6) pictures have been proposed. Based on this result, eight variations of hexagonal parallel thinning algorithms have been reported for producing topological kernels and medial lines. All of the proposed algorithms are proved to be topologically correct.

ACKNOWLEDGEMENTS

This research was supported by the European Union and the European Regional Development Fund under the grant agreements TÁMOP-4.2.1/B-09/1/KONV-2010-0005 and TÁMOP-4.2.2/B-10/1-201-0012, and the grant CNK80370 of the National Office for Research and Technology (NKTH) & the Hungarian Scientific Research Fund (OTKA).

REFERENCES

Deutsch, E. S. (1970). On parallel operations on hexagonal arrays. *IEEE Transactions on Computers C- 19*(10), 982–983.

Deutsch, E. S. (1972). Thinning algorithms on rectangular, hexagonal, and triangular arrays. *Communications of the ACM 15*(9), 827–837.

Hall, R. W. (1996). Parallel connectivity-preserving thinning algorithms. In T. Y. Kong and A. Rosenfeld (Eds.), *Topological Algorithms for Digital Image Processing*, pp. 145–179. New York, NY, USA: Elsevier Science Inc.

Hall, R. W., T. Y. Kong, and A. Rosenfeld (1996). Shrinking binary images. In T. Y. Kong and A. Rosenfeld (Eds.), *Topological Algorithms for Digital Image Processing*, pp. 31–98. New York, NY, USA: Elsevier Science Inc.

Kardos, P. and K. Pal´agyi (2011). On topology preservation for hexagonal parallel thinning algorithms. In *Proceedings of the International Workshop on Combinatorial Image Analysis*, Volume 6636 of *Lecture Notes in Computer Science*, pp. 31–42. Springer Verlag.

Kong, T. Y. (1995). On topology preservation in 2-d and 3-d thinning. *IJPRAI 9*(5), 813–844. Kong, T. Y. and A. Rosenfeld (1989). Digital topology: introduction and survey. *Comput. Vision Graph. Image Process.* 48, 357–393.

Lam, L., S. Lee, and C. Suen (1992). Thinning methodologies-a comprehensive survey. *IEEE Transactions on Pattern Analysis and Machine Intelligence 14*, 869–885.

Lee, M. and S. Jayanthi (2005). *Hexagonal Image Processing: A Practical Approach (Advances in Pattern Recognition)*. Secaucus, NJ, USA: Springer-Verlag New York, Inc.

Marchand-Maillet, S. and Y. M. Sharaiha (2000). *Binary digital image processing – a discrete approach*. Academic Press.

N´emeth, G., P. Kardos, and K. Pal´agyi (2010). Topology preserving 3d thinning algorithms using four and eight subfields. In *Proceedings of International Conference on Image Analysis and Recognition, Volume 6111 of Lecture Notes in Computer Science*, pp. 316–325. Springer Verlag.

Németh, G. and K. Palágyi (2011). Topology preserving parallel thinning algorithms. *International Journal of Imaging Systems and Technology 21*, 37–44.

Ronse, C. (1988). Minimal test patterns for connectivity preservation in parallel thinning algorithms for binary digital images. *Discrete Applied Mathematics 21*(1), 67–79.

Siddiqi, K. and S. Pizer (2008). *Medial Representations: Mathematics, Algorithms and Applications* (1st ed.). Springer Publishing Company, Incorporated. Staunton, R. (1996). An analysis of hexagonal thinning algorithms and skeletal shape representation. *PR 29*, 1131–1146.

Staunton, R. (1999). A one pass parallel hexagonal thinning algorithm. *In Image Processing and Its Applications, 1999. Seventh International Conference on (Conf. Publ. No. 465)*, Volume 2, pp. 841–845 vol.2.

Suen, C. Y. and P. Wang (1994). *Thinning Methodologies for Pattern Recognition*. River Edge, NJ, USA: World Scientific Publishing Co., Inc. Wiederhold, P. and S. Morales (2008). Thinning on quadratic, triangular, and hexagonal cell complexes. In *IWCIA'08*, pp. 13–25.

Computational Modelling of Objects Represented in Images – Di Giamberardino et al. (eds)
© 2012 Taylor & Francis Group, London, ISBN 978-0-415-62134-2

Evolving concepts using gene duplication

Marc Ebner
Ernst-Moritz-Arndt-Universität Greifswald, Institut für Mathematik und Informatik,
Walther-Rathenau-Straße, Greifswald, Germany

ABSTRACT: We have developed an on-line evolutionary vision system which is able to evolve detectors for a variety of different objects. The system processes a video stream and uses motion as a cue to extract moving objects from this video sequence. The center of gravity of the largest object, currently in the image, is taken as the teaching input, i.e. the object which should be detected by the evolved object detector. Each object detector only takes a single image as input. It uses image processing operations to transform the input image into an output image which has a maximum response as close as possible to the teaching input. Since the detectors only work on single images, they are basically appearance detectors. However, the appearance of the object may change as the object moves or changes its orientation relative to the position of the camera. Hence multiple different detectors will need to be combined to represent the concept of an object. We use gene duplication to evolve sub-detectors for the different appearances of an object. The difficulty in evolving these sub-detectors is that only one type of appearance is visible for any given image. Hence, the genetic operators could disrupt the genetic material, i.e. the sub-detectors which are currently not in use.

1 MOTIVATION

The human visual system is a product of natural evolution. It is highly advanced. Indeed, to date, no artificial systems have been created which match the abilities of the human visual system. Obviously, it is of considerable interest to learn how the visual system works and how exactly it was created during the course of natural evolution.

Our long term goal is to get a deeper understanding of the human visual system and to reproduce at least parts of human perceptual behavior. Ebner (2009) has created an online evolutionary vision system which takes an image stream as input and evolves object detectors which locate interesting objects (as determined by the user). In an early prototype system, interesting objects had to be pointed out using the mouse pointer. The user would follow the object using the mouse as the object moves across the screen. The user created the teaching input (position of mouse pointer) by pressing the mouse button. Ebner (2010b) then went on to remove the user input from the system. Motion was found to be a valuable cue to provide the teaching input. Hence, independently moving objects were extracted from the image stream and the center of gravity of the detected motion was taken as the teaching input. Evolved object detectors worked on single images, i.e. were able to detect the desired object without the motion cue. After all, humans are also able to correctly name objects which are shown in images.

In Ebner's work, it became apparent that detectors for objects without unique colors are more difficult to

Figure 1. The spatial distribution of colors changes as on object (toy train) moves along a track.

evolve than detectors which can be based on unique colors by which the object can be identified. If an object cannot be uniquely identified by a single color, one has to take the spatial relationship between colors or the shape of the object in general into account to create a successful detector for such an object. However, the shape of the object as well as the spatial relationship of the colors changes as the orientation of the object relative to the position of the camera changes. This is illustrated in Figure 1. In theory, an overall object detector can be based on several sub-detectors each of which is tuned to a single appearance of an object on the screen. In the human visual system the detectors are independent of the object's size on the retina due to the special type of mapping (complex-logarithmic) from the retinal receptors to visual area V1 (Schwartz 1977).

With this contribution, we have made a move towards creating an artificial system, shaped by evolution, which is able to evolve detectors responding to an overall concept, i.e. detectors which consist

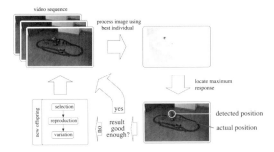

Figure 2.   System overview.

different aspects shape, motion and color of a stimulus are processed via separate areas of the brain.

So called matched filters (Schrater, Knill, and Simoncelli 2000; Simpson and Manahilov 2001) could be used to detect objects of known shape. An overall detector which responds in a generalized manner, i.e. whenever a random view of an object appears, can be created by simply merging (adding) the output of several matched filters. Such a detector will respond whenever one of the sub-detectors, i.e. matched filters responds. We investigate how such general detectors can be generated through artificial evolution.

of sub-detectors each of which tuned to a particular appearance of an object. The system is illustrated in Figure 2. The system takes a video stream as input. It maintains a population of image processing algorithms. The best individual of the population is used to detect the desired object. If the detected position is close enough to the actual position of the object then the next image is processed. Otherwise, evolutionary operators are used to create a new generation of offspring. Evolution continues through additional images as long as it is required, i.e. until the detection accuracy is good enough. We show that gene duplication is an important evolutionary operator which allows us to incrementally evolve concept detectors. Without gene duplication the problem is more difficult to solve.

The paper is structured as follows. In section 2, we provide a brief introduction into the human vision system. Section 3 puts our work into the context of research in the field of evolutionary computer vision. Section 4 describes our approach to concept learning using gene duplication. Section 5 describes the experiments that we have performed. Conclusions are given in Section 6.

2   THE HUMAN VISION SYSTEM

The visual system is highly structured (Zeki 1993). Processing of visual information of course starts with the retinal receptors (Dowling 1987). Three types of receptors can be distinguished for color perception which measure the light in the red, green and blue parts of the spectrum (Dartnall, Bowmaker & Mollon 1983). This information is then sent to the primary visual cortex, which is located in the back of the brain. The primary visual cortex, also called visual area 1 or V1 is also highly structured (Livingstone and Hubel 1984). Inside the so called ocular dominance segments, neurons respond primarily to stimuli presented to either the left or the right eye. Within these segments, neurons can be found which respond to light of certain wavelengths or to lines with different orientations. In short, the visual information is analyzed locally with respect to color and lines. Higher visual areas receive their input from V1 and further analyze it to recover shape, motion and the color of objects. It appears that the

3   EVOLUTIONARY COMPUTER VISION

Evolutionary computer vision is an active research field since the early 1990s with the pioneering work of Lohmann (1991). An recent overview is given by Cagnoni (2008). Evolutionary computer vision can be used to optimize a given algorithm for a particular task. In this case, the algorithm is either well known from the literature or developed by a human and an evolutionary algorithm is used to optimize parameters of the algorithm. However, evolutionary algorithms can also be used to evolve an entire computer vision algorithm from scratch using Genetic Programming (Koza 1992). This is of course a highly interesting approach as it would eventually allow the construction of systems which automatically create vision algorithms based on some type of fitness function. Ebner (2008) has been working towards the creation of an adaptive vision system. Creating such adaptive systems is especially important as many computer vision algorithms are very fragile. They work well in the lab but when taken to a different environment the algorithms often no longer work because the ambient light has changed, e.g. from artificial light to direct sunlight. The human visual apparatus adapts easily to such changing conditions. Some artificial systems compute so called intrinsic images (see Matsushita et al. (2004)). Even though it is possible to apply Color Constancy algorithms to arrive at a descriptor which is independent of the illuminant, current computer vision algorithms are far from being fully adaptive. Using evolutionary algorithms it could be possible to create an artificial vision system which also adapts to environmental conditions.

Such an adaptive vision system needs to adapt to an image stream on an image by image basis. That's why our system works with a population of algorithms which are all applied to an incoming image. The best algorithms are allowed to reproduce and create offspring which are then tested on the next incoming image. Of course this requires extensive computational resources. Computer vision algorithms usually require a lot of processing power. However, we have to apply multiple different algorithms to each input image. The time required to evaluate all of the algorithms would be extensive using single CPU processing. We have used GPU accelerated image processing to speed up the evaluation. GPU accelerated image processing is

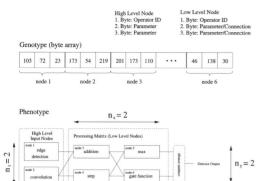

High Level Node
1. Byte: Operator ID
2. Byte: Parameter
3. Byte: Parameter

Low Level Node
1. Byte: Operator ID
2. Byte: Parameter/Connection
3. Byte: Parameter/Connection

Genotype (byte array)

| 103 | 72 | 23 | 173 | 54 | 219 | 201 | 173 | 110 | $\cdots$ | 46 | 138 | 30 |

node 1   node 2   node 3   node 6

Phenotype $n_x = 2$

High Level Input Nodes   Processing Matrix (Low Level Nodes)

node 1 edge detection   node 3 addition   node 5 max

node 2 convolution   node 4 step   node 6 gate function

Detector Output $n_y = 2$

$n_1 = 2$

Figure 3. Each individual (byte array) is decoded into a computer vision algorithm which consists of n_1 high level operators and a processing matrix of $n_x \times n_y$ low level operators. The output of the n_y sub-detectors is averaged to compute the output of the entire detector.

Set of operators	
High level input node operators (used in column 1)	`Image, DX, DY, Lap, Grad,` `ImageGray, ImageChrom,` `ImageLogDX, ImageConv1,` `ImageConv4, ImageConv16,` `ImageConvd, ImageSeg,` `0.0, 0.5, 1.0`
Low level node operators (used in the processing matrix)	`id, abs, dot, sqrt, norm,` `clamp(0,1), step(0),` `step(0.5), smstep(0,1), red,` `green, blue, avg, minChannel,` `maxChannel, equalMin,` `equalMax, gateR, gateG,` `gateB, gateRc, gateGc,` `gateBc, step, +, -, *, /, min, max,` `clamp0, clamp1, mix, step,` `lessThan, greaterThan,` `dot, cross, reflect, refract`

Figure 4. Set of image processing operators. See Ebner (2009) for a detailed description of these operators.

known to provide a speedup of over 40 if ten or more high level image operators are applied Ebner (2010a).

4 GENE DUPLICATION

Ebner has used a variant of the Cartesian Genetic Programming approach (Miller 1999) for his experiments as shown in Figure 3. Each evolved individual consists of a set of n_1 high level image processing operators and a processing matrix of low level image operators of size $n_x \times n_y$ which are applied to the input image. Each high level processing operator transforms the input image in some way to produce a modified image. Operators include edge detection, Laplacian, convolution or segmentation. In contrast to the high level operators which also take the surrounding of an image pixel into account, the operators used in the processing matrix use point operations to combine the output obtained so far on a pixel by pixel basis. The list of both types of operators is shown in Figure 4. This set of operators is fully described in (Ebner 2009). Due to the lack of space it is not possible to explain the functions of these operators in detail. The interested reader is referred to (Ebner 2009).

Ebner has called this the $n_1 - n_x \times n_y$ representation. Since there are n_y output operators on the right hand side of this representation, there are basically n_y sub-detectors in this representation. The output of these n_y sub-detectors is averaged to obtain a single output image for this detector. The output RGB pixel is interpreted as a 24 bit value and the maximum response over all image pixels denotes the position of the detected object. If multiple image pixels have the same maximum value, then we compute the center of gravity for all these image pixels to obtain a single position inside the image. This position denotes the position inside the image at which the detector has located the object.

Unfortunately, the evolved detectors did not generalize to detect different views of the same object. We address this problem using gene duplication and gene deletion. Gene duplication and gene deletion have also been used by Koza (1995a, 1995b) in the context of tree-based genetic programming to evolve solutions to the parity problem. Haynes (1996) experimented with duplication of code segments in genetic programming which resulted in a speedup of the learning process. Hoai et al. (2005) have used gene deletion and duplication for tree-adjoining grammar guided genetic programming.

We add both genetic operators, gene duplication and gene deletion, to the set of operators in addition to the standard evolutionary operators crossover and mutation. In our approach, each individual may consist of several segments or genes. Each gene is basically a description of nodes as in the representation shown in Figure 3. Hence, each individual with multiple genes can be decoded into multiple image processing algorithms as shown in Figure 5. We call them sub-algorithms. The output of these sub-algorithms is averaged to obtain the overall output of the entire individual. When a gene duplication happens, then one of the genes is selected at random. This gene is most likely useful in one way or another because it has been shaped by evolution. The selected gene is duplicated and appended to the individual as shown in Figure 6. When a gene deletion occurs, then a randomly selected gene, i.e. sub-algorithm, is deleted.

The output of an individual with a single gene which has just been duplicated is almost identical to the original individual. However, mutation is now free to change the contents of one the sub-algorithms without compromising on the function of the original gene. Hence, it is now possible to evolve concept detectors. Each concept detector consists of several sub-algorithms whose output is averaged. As long as

Genotype (byte array)

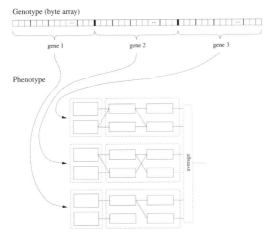

Figure 5. Each individual (byte array) consists of several genes, i.e. sub-algorithms. Each gene is decoded into a $n_1 + n_x \times n_y$ detector. The output of all detectors is averaged to obtain the overall response of the individual.

Figure 6. A randomly selected gene is duplicated and appended at the end of the individual.

one particular view is present in the current image, then one of the sub-algorithms will respond.

5 EXPERIMENTS

We have used the toy train video sequence that has already been used by Ebner (2010b, 2010a) as shown in Figure 7. It is not possible to detect the toy train through color alone. The color of the toy train is mostly yellow and red. However, other objects in the video sequence feature the same colors, e.g. the wagon in the center. The spatial arrangement of the different colors must also be taken into account to successfully detect the train on its track. A successful algorithm could detect the object showing a large yellow area on the top left and a large red area on the bottom right. Of course this spatial arrangement of colors changes as the train takes turns on the track. During previous runs of this experiment, it was not possible to find an overall detector which would detect the train as it completes a full circle on the track. Therefore, we increased the length of the sequence to ten times its size by playing it forward, backward, forward and so on. However, the problem still remains difficult.

Fitness is computed as the Euclidean distance between the position of the moving object (automatically detected as described in (Ebner 2010b) and the object position as detected by the individual. A perfect

Figure 7. Two images from the toy train image sequence. The sequence was taken with a stationary camera.

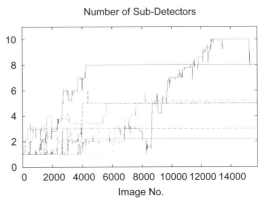

Figure 8. Evolution of the number of sub-detectors (shown independently for all 10 runs of experiment b).

individual would have a fitness of zero as it would correctly locate the moving object in the image sequence. Initially, evolution is used to find an individual which is able to correctly locate the moving object in the image sequence. Once the moving object is correctly located, i.e. with an accuracy of 10 pixels or less for five consecutive images, then evolution is turned off. The best individual found so far is then used to detect the moving object for successive images. Of course it may happen that the appearance of the object changes and that the currently used individual is no longer able to correctly detect the moving object. If the accuracy deteriorates beyond 25 pixels difference between the actual and the detected position, then evolution is turned on again. This process continues throughout the entire image sequence. Evolution is turned on whenever re-learning is needed.

Our experiments are carried out using a population of 5 parent individuals which generate 10 offspring for each image whenever evolution has been turned on. In addition to these offspring, 10 offspring are generated completely at random. This ensures an influx of new genetic material at all times. We use four different genetic operators: crossover, mutation, gene duplication and gene deletion. These operators are applied with probabilities p_c, p_m, p_{dup} and p_{del} respectively. We use 2-point crossover with randomly selected crossover points. The mutation operator either increments or decrements a randomly chosen parameter by one or it mutates all of the bits with a probability of $2/l$ per bit where l is the length of the genotype in bits. In other words, on average, two bits are mutated.

Table 1. Experimental results for experiments a), b), and c). The standard deviation is shown in brackets. Best results are printed in bold face.

	generations (all images)		generations (last iteration)		restarts (all images)		restarts (last iteration)	
a) no gene duplication	1407.1	(833.6)	32.6	(97.8)	73.4	(42.0)	1.7	(5.1)
b) with gene duplication	**1014.8**	(915.8)	**11.6**	(34.8)	**50.8**	(40.8)	**0.4**	(1.2)
c) control experiment	1316.4	(1098.1)	21.8	(65.4)	55.9	(47.1)	0.7	(2.1)

The reason for the factor of two is that the genetic representation is redundant and hence a slightly larger mutation rate than $1/l$ is used.

For each new incoming image, fitness is computed for all parents as well as all offspring. Parent and offspring are then sorted according to fitness. Whenever two or more individuals achieve the same fitness they are considered to be identical even though their genetic material may be different. Hence, only one of these individuals is kept. The remaining duplicates are discarded. The best 5 individuals are selected to become parents of the next generation. Usually, approximative solutions are found within 24 generations. The evolved detectors will then be further refined by evolution.

We have carried out three experiments:

a) without gene duplication using a $2 - 2 \times 2$ representation with probabilities $p_c = 0.1$, $p_m = 0.9$, $p_{\text{dup}} = 0$ and $p_{\text{del}} = 0$
b) with gene duplication, using a $2 - 2 \times 2$ representation for the first generation with probabilities $p_c = 0.1$, $p_m = 0.89$, $p_{\text{dup}} = 0.005$ and $p_{\text{del}} = 0.005$
c) without gene duplication using a $10 - 2 \times 10$ representation with probabilities $p_c = 0.1$, $p_m = 0.9$, $p_{\text{dup}} = 0$ and $p_{\text{del}} = 0$ as a control experiment.

Figure 8 shows how the number of genes changes during the course of evolution for experiment b). Performing 10 runs for each of these experiments took almost two days on a Linux system (Intel Core 2 CPU running at 2.13 GHz) equipped with a GeForce 9600GT/PCI/SEE2.

Table 1 summarizes the results. The first two columns show the number of generations that evolution had to be turned on. The last two columns show the number of restarts which were necessary. The data is shown for the entire run (all 15800 images), as well as for the last iteration of the image sequence, i.e. only for the last 1580 images. The number of restarts, i.e. the number of times evolution had to be turned on again for re-learning, is taken as the relevant indicator. A successful individual would not need any restarts because it is able to correctly detect the moving object at all times. The differences are not statistically significant using a t-Test. However, the problem got easier on average by using gene duplication. The total number of generations that evolution was turned on, could be decreased from 1407.1 to 1014.8 generations on average. The number of restarts could be decreased from 73.4 to 50.8 on average.

6 CONCLUSION

We have shown that gene duplication is a tool to incrementally evolve object detectors. If the representation used to evolve an object detector is too small, then the evolved detector may not be able to generalize to other views. If the representation is too complex, then evolution may fail to find a good solution. However with gene duplication, evolution can start off using a small representation. Once an approximative solution has been found, it will eventually be duplicated. Evolution is then free to modify one of the genes while still maintaining the functionality of the original gene. This makes its possible to incrementally evolve good solutions.

REFERENCES

Cagnoni, S. 2008. Evolutionary computer vision: a taxonomic tutorial. In *8th Int. Conference on Hybrid Intelligent Systems*, Los Alamitos, CA, pp. 1–6. IEEE Computer Society.

Dartnall, H. J. A., Bowmaker, J. K. & Mollon, J. D. 1983. Human visual pigments: microspectrophotometric results from the eyes of seven persons. *Proc. R. Soc. Lond. B 220*, 115–130.

Dowling, J. E. 1987. *The retina: an approachable part of the brain*. Cambridge, MA: The Belknap Press of Harvard University Press.

Ebner, M. 2008. An adaptive on-line evolutionary visual system. In E. Hart, B. Paechter & J. Willies (Eds.), *Workshop on Pervasive Adaptation, Venice, Italy*, pp. 84–89. IEEE.

Ebner, M. 2009. A real-time evolutionary object recognition system. In L. Vanneschi, S. Gustafson, A. Moraglio, I. De Falco & M. Ebner (Eds.), *Genetic Programming: Proc. of the 12th Europ. Conf., Tübingen, Germany*, Berlin, pp. 268–279. Springer.

Ebner, M. 2010a. Evolving object detectors with a GPU accelerated vision system. In *Proc. of the 9th Int. Conf. on Evolvable Systems – From Biology to Hardware, York, UK*, Berlin, pp. 109–120. Springer.

Ebner, M. 2010b. Towards automated learning of object detectors. In *Applications of Evolutionary Computation, Proceedings, Istanbul, Turkey*, Berlin, pp. 231–240. Springer.

Haynes, T. 1996. Duplication of coding segments in genetic programming. In *Proc. of the 13th National Conf. on Artificial Intelligenc. Volume 1*, pp. 344–349. AAAI Press/MIT Press.

Hoai, N. X., McKay, R. I. B., Essam, D. & Hao, H. T. 2005. Genetic transposition in tree-adjoining grammar guided genetic programming: The duplication operator. In M. Keijzer, A. Tettamanzi, P. Collet, J. I. van Hemert & M. Tomassini (Eds.), *Proc. of the 8th European Conference on Genetic Programming*, Berlin, pp. 108–119. Springer-Verlag.

Koza, J. R. 1992. *Genetic Programming. On the Programming of Computers by Means of Natural Selection*. Cambridge, MA: The MIT Press.

Koza, J. R. 1995a. Evolving the architecture of a multi-part program in genetic programming using architecture-altering operations. In A. V. Sebald & L. J. Fogel (Eds.), *Proc. of the 4th Annual Conf. on Evolutionary Programming*, Cambridge, MA, pp. 695–717. The MIT Press.

Koza, J. R. 1995b. Gene duplication to enable genetic programming to concurrently evolve both the architecture and work-performing steps of a computer program. In *Proc. of the 14th International Joint Conference on Artificial Intelligence*, San Francisco, CA, pp. 734–740. Morgan Kaufmann.

Livingstone, M. S. & Hubel, D. H. 1984. Anatomy and physiology of a color system in the primate visual cortex. *The Journal of Neuroscience 4*(1), 309–356.

Lohmann, R. 1991. Selforganization by evolution strategy in visual systems. In J. D. Becker, I. Isele & F. W. Mündemann (Eds.), *Parallelism, Learning, Evolution*, pp. 500–508. Springer.

Matsushita, Y., Nishino, K., Ikeuchi, K. & Sakauchi, M. 2004. Illumination normalization with time-dependent intrinsic images for video surveillance. *IEEE Transactions on Pattern Analysis and Machine Intelligence 26*(10), 1336–1347.

Miller, J. F. 1999. An empirical study of the efficiency of learning boolean functions using a cartesian genetic programming approach. In W. Banzhaf, J. Daida, A. E. Eiben, M. H. Garzon, V. Honavar, M. Jakiela & R. E. Smith (Eds.), *Proc. of the Genetic and Evolutionary Computation Conference*, San Francisco, California, pp. 1135–1142. Morgan Kaufmann.

Schrater, P. R., Knill, D. C. & Simoncelli, E. P. 2000. Mechanisms of visual motion detection. *Nature Neuroscience 3*(1), 64–68.

Schwartz, E. L. 1977. Spatial mapping in the primate sensory projection: Analytic structure and relevance to perception. *Biological Cybernetics 25*, 181–194.

Simpson, W. A. & Manahilov, V. 2001. Matched filtering in motion detection and discrimination. *Proc. R. Soc. Lond. B 268*, 1–7.

Zeki, S. 1993. *A Vision of the Brain*. Oxford: Blackwell Science.

FPGA implementation of OpenCV compatible background identification circuit

M. Genovese & E. Napoli
DIBET – University of Napoli Federico II, Napoli, Italy

ABSTRACT: The paper proposes the hardware implementation of the Gaussian Mixture Model (GMM) algorithm included in the OpenCV library. The OpenCV GMM algorithm is adapted to allow the FPGA implementation while providing a minimal impact on the quality of the processed videos.

The circuit performs 30 frame per second (fps) background (*Bg*) identification on High Definition (HD) video sequences when implemented on commercial FPGA and outperforms previously proposed implementations. When implemented on Virtex5 lx50 FPGA using one level of pipeline, runs at 95.3 MHz, uses 5.3% of FPGA resources with a power dissipation of 1.47 mW/MHz.

1 INTRODUCTION

The real time detection of moving objects in a video sequence has many important applications. Various algorithms have been developed during the years. The algorithms based on a *Bg* statistical model are preferred due to good performances and adaptability.

Statistical *Bg* models are proposed in (Chiu et al. 2010; Wren et al. 1997; Stauffer and Grimson 1999). In (Stauffer and Grimson 1999) each pixel is modeled with a mixture of Gaussian distributions. The algorithm, known as Gaussian Mixture Model (GMM), provides good performances in both presence of illumination changes and multimodal *Bg*. A multimodal *Bg* is characterized by objects showing repetitive motions (e.g. waves, moving leaves or flickering light). Due to the good performances, the GMM algorithm has been selected as the *Bg* detection algorithm in the Open Source Computer Vision library, (OpenCV 1999), that provides a common base of computer vision instruments. The OpenCV GMM algorithm is an optimized version of the algorithm of (Stauffer and Grimson 1999). It improves the initial learning phase for the *Bg* model and calculates the learning rate in a simplified yet efficient way.

Main drawback of the GMM algorithm is the computational complexity. In fact, in order to generate the updated *Bg* model, the GMM repeats many non linear computations for each pixel of the image. This paper proposes an FPGA implementation of the luminance based OpenCV GMM algorithm that allows 30 fps processing of HD videos (1920 × 1080), a target not possible with a software implementation.

2 GMM – OPENCV VERSION

The GMM algorithm has been partially modified in the OpenCV libraries. The differences between GMM and the OpenCV GMM are indicated in the text.

2.1 Statistical model

The statistical model for each pixel of the video sequence is composed by a mixture of *K* Gaussian distributions, each one represented by four parameters: weight (*w*), mean (*μ*), variance (*σ²*), and *matchsum*, a counter introduced in the OpenCV algorithm.

Gaussian parameters differ for each Gaussian of each pixel and change for every frame of the video sequence. They are therefore defined by three indexes (p,k,t), where 'p' is the index for the pixel, 'k' is the index for the Gaussian distribution, and 't' is for the frame. In the following the pixel index is omitted since the same operations are repeated for every pixel.

2.2 Parameters update

When a frame is acquired, for each pixel, the *K* Gaussian distributions are sorted in decreasing order of a parameter named Fitness (*F*):

$$F_{k,t} = w_{k,t}/\sigma_{k,t} \qquad (1)$$

A match condition is checked with the *K* Gaussian distributions that model the pixel. Match condition is:

$$M_k = 1 \quad if \mid (pixel - \mu_{k,t}) \mid < 2.5\sigma_{k,t} \qquad (2)$$

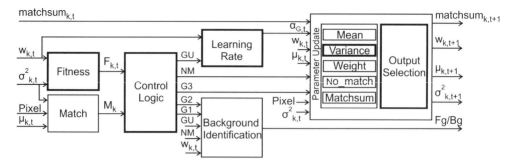

Figure 1. Block diagram of the proposed circuit. The critical path blocks are shown with thick boundary lines. The signals with subscript 'k' refer to a bus of three signals; $w_{k,t} = \{w_{1,t}, w_{2,t}, w_{3,t}\}$.

Eq. (2) establishes if the pixel can be considered part of the *Bg*. A pixel can verify (2) for more than one Gaussian. The Gaussian that matches with the pixel ($M_k = 1$) and has the highest *F* value is considered as the 'matched distribution' and its parameters are updated as follows:

$$\mu_{G,t+1} = \mu_{G,t} + \alpha_{G,t} \cdot (pixel - \mu_{G,t})$$

$$\sigma^2_{G,t+1} = \sigma^2_{G,t} + \alpha_{G,t} \cdot [(pixel - \mu_{G,t})^2 - \sigma^2_{G,t}]$$

$$w_{G,t+1} = w_{G,t} - \alpha_w \cdot w_{G,t} + \alpha_w \qquad (3)$$

$$matchsum_{G,t+1} = matchsum_{G,t} + 1$$

where $\mu_{G,t}$, $\sigma^2_{G,t}$, $w_{G,t}$ and *matchsum*$_{G,t}$ are mean, variance, weight and matchsum of the matched distribution. The parameter α_w is the learning rate for the weight while $\alpha_{G,t}$ is the learning rate for mean and variance. It is derived from α_w as:

$$\alpha_{G,t} = \alpha_w / w_{G,t} \qquad (4)$$

Eq. (4) is employed in the OpenCV algorithm and differs from what is proposed in (Stauffer and Grimson 1999) where $\alpha_{G,t}$ is calculated, being η the Gaussian probability density function, as:

$$\alpha_{G,t} = \alpha_w * \eta(pixel, \mu_{G,t}, \sigma_{G,t}) \qquad (5)$$

The approach of (Stauffer and Grimson 1999) results in a slow convergence if compared with (4).

For the unmatched Gaussian distributions mean and variance are unchanged while the weights are updated as:

$$w_{k,t+1} = w_{k,t} - \alpha_w \cdot w_{k,t} \qquad (6)$$

When the pixel does not match any Gaussians, a specific 'no_match' procedure is executed and the Gaussian distribution with smallest *F* is updated as:

$$\mu_{k,t+1} = pixel \qquad matchsum_{k,t+1} = 1$$

$$\sigma^2_{k,t+1} = vinit \qquad w_{k,t+1} = 1/msumtot \qquad (7)$$

where *vinit* is a fixed initialization value and *msumtot* is the sum of the values of the *matchsum* of the *K-1* Gaussians with highest *F*. The weights of the *K-1* Gaussians with highest *F* are decremented as in (6) while their means and variances are unchanged.

2.3 Background identification

The *Bg* identification is performed using the following algorithm:

$$B = \arg \min_b (\Sigma_{k=1}^b w_{k,t} > T) \qquad (8)$$

Eq. (8) adds in succession the weights of the first *b* Gaussian distributions, sorted beginning from the one with highest *F* value, until their sum is greater than *T*, a fixed threshold belonging to the (0,1) interval. The set of Gaussian distributions that verify (8) represents the *Bg* and a pixel that matches one of these Gaussians is classified as a *Bg* pixel. The algorithm entails that if the 'no_match' condition occurs, the pixel is classified as *Fg*.

3 BANDWIDTH OPTIMIZATION ALGORITHM

The block diagram of the proposed circuit is shown in Fig. 1 and its functionality is detailed in Section 4.

The input data are the 8 bit luminance of the input pixel (*Pixel*), and the statistical model of the pixel for the given frame ($\mu_{k,t}$, $\sigma^2_{k,t}$, $w_{k,t}$, *matchsum*$_{k,t}$). The output data are the updated statistical model ($\mu_{k,t+1}$, $\sigma^2_{k,t+1}$, $w_{k,t+1}$, *matchsum*$_{k,t+1}$) and the *Fg/Bg* tag.

The circuit implements the OpenCV GMM algorithm using, as suggested in Stauffer and Grimson 1999, 3 Gaussian distributions for each pixel ($k = 1,2,3$) providing good quality of the processed images while limiting circuit complexity and memory requirements. As a consequence the processing of each pixel requires 12 parameters and pixel luminance.

The software algorithm of the OpenCV library represents the Gaussians parameters as double precision (64 bit) floating point numbers. A double precision hardware implementation would be impractically large and slow and would require a very high bandwidth towards the memory. As example, if the memory throughput is 128 bit, 12 or 13 clock cycles are needed to load the parameters for each pixel and to store the updated parameters. The target of processing 62.3 Mps implies that the clock frequency towards the memory is around 825 MHz with a required bandwidth

Table 1. AvPSNR values varying the word lengths of the input signals of the circuit of Figure 1.

Video sequence	AvPSNR (average value for 300 frames)								
	52 bit	23 bit	18 bit	14 bit	12 bit	11 bit	10 bit	9 bit	8 bit
Video 1	46.22	44.85	34.31	19.54	18.25	17.09	14.95	8.06	0.96
Video 2	44.81	43.46	34.32	18.13	16.99	15.69	14.84	7.80	0.83
Video 3	37.45	36.97	30.17	15.79	13.81	12.12	10.73	6.18	1.41
Video 4	37.96	37.82	32.36	18.49	16.37	14.94	13.11	7.54	0.82
Video 5	39.33	38.37	30.73	15.25	13.78	12.71	10.02	6.19	1.24

of 13 GBs, not feasible for a low power lightweight electronic system.

In order to reduce the logic and the bandwidth, it is necessary to reduce the number of bits that represent the pixel statistic. The examination of the GMM algorithm reveals that the signals have a limited dynamics. Mean, variance, and weight range in [0–255], [0–127], and [0–1], respectively. When dealing with limited dynamics signals, using a fixed point representation provides improved performances while reducing hardware complexity. This requires that the number of bits of the fixed point representation is based on both the range of the signals and the required accuracy.

In this paper the Gaussian parameters are represented as fixed point numbers and their wordlengths are limited with the target of accommodating all the data for the processing of a pixel in 128 bits. In this way, if 62.3 Mps are processed, the clock frequency towards the memory is reduced to 125 MHz.

3.1 Word lengths optimization algorithm

For a fair evaluation of the performances of the circuit, five test bench videos have been selected and processed varying the number of bits that represent the signals. The effect of the word length reduction on the Bg identification has been evaluated calculating the Peak Signal to Noise Ratio (PSNR) for each frame of the test bench videos followed by the calculation of the average PSNR (AvPSNR), considered as the accuracy performance values for the processed video streams. The PSNR is calculated as:

$$PSNR = 20 \log_{10} \frac{\max(I)}{\sqrt{MSE}} \quad (9)$$

$$MSE = \frac{1}{M \cdot N} \sum_{i=1}^{M} \sum_{j=1}^{N} (I(i,j) - Iref(i,j))^2 \quad (10)$$

where $max(I)$ is the maximum intensity value of a pixel in the input image (I) (equal to 1 for a one bit image). Iref is the reference image. Both I and $Iref$ have size equal to M×N (320 × 240). $I(i,j)$ and $Iref(i,j)$ represent the value of the intensity of the pixel. For a binary image, the difference between $I(i,j)$ and $Iref(i,j)$ is '1' if the pixel has been differently classified in I and $Iref$; '0' otherwise. Eq. (10) is the ratio between the number of pixels differently classified in I with respect to $Iref$

and the number of pixels in a frame. For each video, $Iref$ has been calculated using the double precision floating point OpenCV GMM algorithm.

The word lengths optimization has been carried out in two phases. In the first phase the number of bits has been simultaneously reduced for all Gaussians parameters. Table 1 reports, for the five test bench videos, the AvPSNR values obtained representing the parameters of each Gaussian as fixed point numbers.

Table 1 shows that the AvPSNR value is not acceptable when the signals are represented on less than 10 bits. We chose the 11 bits case as the output of this first optimization phase. Note this allows, on average, to correctly identify 94% of the pixels.

In the second phase of the optimization, the word length of one signal between variance, mean, weight, and matchsum has been further reduced while keeping the calculated AvPSNR values higher than the AvPSNR values obtained representing all the signals with 10 bits (Table 1) that is defined as the performance threshold for the identification circuit.

The word length reduction has been firstly carried out on the signals for which the reduction of the number of bits provides the highest reduction of resource utilization (variance, mean, weight and matchsum).

The AvPSNR values obtained for the second phase of the optimization are in Table 2. Column A shows that when the variance is represented on 10 bits the AvPSRN values are lower than the performance threshold. As consequence, it is not possible to reduce the variance to 10 bits.

Column B and C refer to the mean signal. The AvPSNR values are acceptable when the mean is represented with 10 bits (column B). Column C show the the mean signal cannot be represented with 9 bits.

Column D and E refer to the reduction of the word length of the weight signal while in column F the wordlength of the matchsum signal is reduced. The proposed circuit has been finally implemented using the word length reported in column F. The same column also reports representations of the signals given with $U_{m,n}$ notation, where 2^m is the weight of the MSB and 2^{-n} is the weight of the LSB.

It can be seen that, using the representations of Table 2 (column F), the percentage of differently classified pixels for the five test bench videos is 2.7%, 3.7%, 8.4%, 4.7%, and 8.6%. The analysis shows that the proposed fixed point adaptation of the OpenCV

Table 2. AvPSNR values for the second phase of the word length optimization algorithm.

	AvPSNR (average value for 300 frames)					
Video sequence	A 11 bit σ^2 10 bit	B 11 bit μ 10 bit	C 11 bit μ 9 bit	D 11 bit μ 10 bit w 8 bit	E 11 bit μ 10 bit w 7 bit	F σ^2 11 bit ($U_{13,-3}$) μ 10 bit ($U_{7,2}$) w 8 bit ($U_{-1,8}$) $matchsum$ 4 bit ($U_{3,0}$)
Video 1	14.90	17.09	14.13	16.44	16.35	16.44
Video 2	14.70	15.69	14.43	15.05	14.78	15.05
Video 3	10.70	12.06	11.27	11.53	11.41	11.53
Video 4	12.49	14.87	12.53	14.06	13.25	14.06
Video 5	10.12	12.71	9.06	11.46	10.53	11.46

GMM algorithm provides little impact on the video quality and allows to use 107 bits for each pixel, resulting in a required memory bandwidth for HD 30 fps video processing of 1.5 GBs.

4 PROPOSED CIRCUITS

Target devices for the proposed implementation are FPGA devices, a technology largely used and of great interest for the considered applications. The circuit implements the equations of the algorithm detailed in Section 2 as described in the following.

Fitness: implements eq. (1) with three identical units. Each unit is implemented with a ROM, in which the inverse of the square root of the variance input are stored. The ROM output is then multiplied by the weight of the Gaussian. Since the variance and the inverse of the standard deviation are on 11 and 8 bits, respectively, ROM size is $2^{11} \times 8$ bits. The logic utilization is 216 LUTs and 1 DSP block for Virtex5 xc5vlx50 FPGA. Maximum combinatorial delay and power dissipation are 6.84 ns and 0.5 mW/MHz.

Match: is composed by three units that implement:

$$M_k = 1 \quad if(pixel - \mu_{k,t})^2 < 6,25 \cdot \sigma_{k,t}^2 \quad (11)$$

that is the square of (2). Equation (2) requires the standard deviation as input while (11) requires the variance as input and does not require the square root circuit. When implemented on xc5vlx50 Virtex5 FPGA, it uses 90 LUTs and 6 DSP blocks.

Control Logic: sorts the Gaussians in decreasing Fitness (F) order using three comparators and few logic gates, and establishes which Gaussian has to be updated as in (3), (6), and (7). This block is on the critical path of the circuit. It has a delay of 2.39 ns.

Learning Rate: computes the learning rate $\alpha_{G,t}$ as in (4). In the proposed implementation the α_w and $\alpha_{G,t}$ values are quantized as power of two ($\alpha_w = 2^{nw}$ and $\alpha_{G,t} = 2^{ng}$), allowing the replacement of the multipliers with shifters (dashed lines in Figure 2). The nw value is hardwired while the ng values are stored in a ROM. The resulting 'Learning Rate' block is composed by a single ROM that uses 11 LUT of the Virtex5

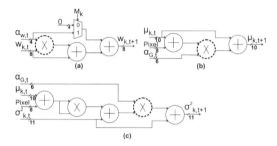

Figure 2. Implementation of the updating equations for (a) weight, (b) mean, (c) variance. The multipliers drawn with dashed lines are replaced by shifters.

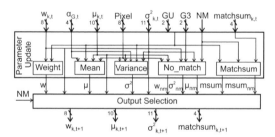

Figure 3. 'Parameter Update' and 'Output Selection' units of Figure 1.

FPGA. The 'Learning Rate' block is on the critical path of the circuit and has a delay of 2.7 ns.

Parameter Update: is shown in Figure 3. If the match condition is verified, 'Weight', 'Mean', and 'Variance' blocks update the parameters according to (3) using the circuits shown in Figure 2. The 'Variance' block is on the critical path of the circuit (Figure 1) and uses 69 LUTs and 1 DSP block with a maximum combinatorial delay of 6.9 ns.

If no Gaussians match the pixel, the 'No_match' block updates the parameters of the Gaussian with smallest F value according to (6).

The 'Matchsum' block updates the matchsum signal according to (3) and (7). The matchsum signal is a counter, introduced in the OpenCV GMM algorithm, to count the number of times that a given Gaussian is matched according to (2). The introduction of the matchsum entails the synthesis of three counters that are not on the critical path. The only drawback is an increase of circuit area and power dissipation.

The initial learning phase of the traditional GMM algorithm of (Stauffer and Grimson 1999) is very slow, (KaewTraKulPong and Bowden 2001; Lee 2005). The use of the matchsum counters improves this phase as shown in Figure 4. Figure 4(a) shows a video sequence extracted from the Wallflower database of (Wallflower). The same database is also used in (Toyama et al. 1999). Figure 4(b) and Figure 4(c) show the output obtained with the proposed OpenCV GMM algorithm and with the algorithm of (Stauffer and Grimson 1999), respectively. Using the OpenCV GMM, the Bg is almost disappeared at the 2nd frame and it is well identified at the 56th frame. On the contrary, the conventional GMM results in a wrong identification at the 2nd frame because all pixels of the scene are classified as Fg and some Fg pixels are still visible to the 56th frame.

Output Selection: establishes the values of the updated parameters depending on the match condition. The 'Output Selection' block is on the critical path and has a maximum delay of 0.81 ns.

Background Identification: Verifies the Bg identification condition shown in (7) determining if the input pixel belongs to the Bg or not. The logic occupation for Virtex5 implementation is 47 LUTs.

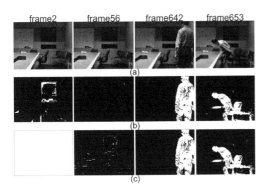

Figure 4. Comparison of the learning phase for the initial Bg model between the conventional GMM and the OpenCV GMM algorithm. (a): original frames; (b): output of the OpenCV version of the GMM algorithm; (c): output of the conventional GMM algorithm.

5 RESULTS AND PERFORMANCES

The proposed implementation of the OpenCV GMM algorithm is synthesized and implemented on Virtex5, VirtexII, VirtexII Pro, and VirtexE (Xilinx) and on StratixII (Altera).

The synthesis for Virtex5, and VirtexII Pro is conducted using Synplify while XST has been used for VirtexII and VirtexE synthesis. Place&Route has been carried out using ISE for Xilinx FPGA devices. Synthesis and Place&Route for Altera FPGA are carried out using QuartusII software.

Circuit simulations use ModelSim XE (Xilinx FPGA) or ModelSim Altera (Altera FPGA) that provide the 'vcd' files for the determination of the power dissipation. Power dissipation is, afterwards, computed using XPower (Xilinx FPGA) or Power-Play Analyzer (Altera FPGA). In the paper, only the dynamic power dissipation, without including the power dissipation due to the I/O pads, is reported.

Table 3 shows the performances of the proposed circuit as a function of the target FPGA and of the number of pipeline levels on Virtex5 and StratixII FPGA. The target speed of 62.3 Mps is reached on both FPGA using one level of pipeline. As example, the circuit implemented on Virtex5 uses 385 Slice (5% of the total number of Slice) and 7 DSP. It runs at 95.32 MHz with a power dissipation of 1.47 mW/MHz.

Table 4 compares the proposed circuit with previous art. Papers (Kristensen et al. 2008; Jiang et al. 2005; Jiang et al. 2009; Shi et al. 2006) propose FPGA implementations of the GMM that do not comply with the OpenCV GMM algorithm. They provide no information regarding pipeline levels or power dissipation of the systems. As example, in (Shi et al. 2006) a GMM based classifier based on distributed arithmetic is implemented on XCV2000E. The circuit, when compared with the proposed architecture with two levels of pipeline, occupies 5.4% more slices while being 35% slower. No information is provided in (Shi et al. 2006) regarding power dissipation and pipeline levels.

Reference (Genovese et al. 2010) proposes a OpenCV GMM algorithm implementation on Virtex5 xc5vlx50. The circuit of (Genovese et al. 2010) is improved in (Genovese and Napoli 2011) in which 43 HD fps are processed with a dynamic power dissipation of 2.2 mW/MHz and using 476 Slice with one pipeline level. The circuit proposed in this paper,

Table 3. Performances of the circuit of Figure 1 implemented on Xilinx and Altera FPGA.

Target FPGA	Pipeline levels	LUT/ALUT	Flip Flop	Slice/LAB	DSP	Frequency (MHz)	Frame rate (1920 × 1080)	Power (mW/MHz)
Virtex5	0	705/28800	0/28800	313/7200	10/48	51.30	24	4.30
xc5vlx50-3	1	832/28800	211/28800	385/7200	7/48	95.32	45	1.47
StratixII	0	918/12480	0/12480	99/780	11/96	44.50	21	2.76
EP2S15F484C3	1	1004/12480	165/12480	116/780	11/96	76.28	36	1.30

Table 4. Performances of the proposed circuit compared with previously proposed FPGA implementations.

Target FPGA	Circuit	Pipeline levels	LUT	DSP-MULT	Flip Flop	BRAM	Slice	Frequency (MHz)	Power (mW/MHz)
VirtexII xc2pro30	(Kristensen) proposed	N.A. 2	3397/27932 1832/27392	7/136 7/136	N.A. 566/27392	13/136 0	N.A. 1114/13696	83 83.91	N.A. 11.89
VirtexII xc2v1000	(Jiang,2005) proposed	N.A. 0	N.A. 1421/10240	N.A. 12/40	N.A. 0/10240	N.A. 0	N.A. 800/5120	19 51.17	N.A. 65.18
VirtexII xc2pro30	(Jiang,2009) proposed	N.A. 1	N.A. 1724/27392	N.A. 7/136	4273/27392 354/27392	84/136 0	6107/13696 993/13696	7.7 55.79	N.A. 29.72
VirtexE xcv2000E	(Shi,2006) proposed	N.A. 2	1845/4704 2340/4704	– –	N.A. 470/4704	N.A. 0	1456/2352 1381/2352	27.0 42.6	N.A. 126.84
Virtex5 xc5vlx50	(Genovese, 2011) proposed	1 1	1182/28800 832/28800	10/48 7/48	282/28800 211/28800	0 0	476/7200 385/7200	43.0 95.32	2.20 1.47

implemented using one pipeline level, on the same target FPGA and using the same number of Gaussians per pixel of (Genovese and Napoli 2011), processes 45 fps (+5%), reduces the power dissipation to 1.47 mW/MHz (−33%) and uses 385 Slice (−19%).

6 CONCLUSIONS

An OpenCV compatible FPGA implementation of the GMM algorithm for *Bg* identification is presented in the paper. The OpenCV GMM algorithm has been implemented using a fixed point ROM based arithmetic, replacing some multipliers with shifters and reducing the bandwidth towards the memory without using compression techniques. The circuit is implemented on both Altera and Xilinx FPGA.

When compared with previously proposed FPGA based *Bg* identification circuits, the proposed circuit provides increased speed, lower power dissipation and lower logic utilization.

REFERENCES

Chiu, C.-C., M.-Y. Ku, and L.-W. Liang (2010). A robust object segmentation system. *TCSVT 20*(4), 518–528.

Genovese, M. and E. Napoli (2011). Fpga-based architecture for real time segmentation. *Journal of Real-Time Image Processing*.

Genovese, M., E. Napoli, and N. Petra (2010). OpenCV compatible real time processor. In *ICM*, pp. 467–470.

Jiang, H., H. Ardo, and V. Owall (2005). Hardware accelerator design for video segmentation. In *ISCAS*, pp. 1142–1145.

Jiang, H., H. Ardo, and V. Owall (2009). A hardware architecture for real-time video segmentation. *TCSVT 19*(2), 226–236.

KaewTraKulPong, P. and R. Bowden (2001). An improved adaptive background mixture model. In *In Proc. of AVBS*.

Kristensen, F., H. Hedberg, H. Jiang, P. Nilsson, and V. Öwall (2008). An embedded real-time surveillance system: Implementation and evaluation. *J. Signal Process. Syst. 52*, 75–94.

Lee, D.-S. (2005). Effective gaussian mixture learning for video background subtraction. *IEEE Trans. on Pattern Analysis and Machine Intelligence 27*, 827–832.

OpenCV (1999). http://sourceforge.net/projects/opencv library/.

Shi, M., A. Bermak, S. Chandrasekaran, and A. Amira (2006). An efficient fpga implementation of gaussian mixture models. In *IEEE Int. Conf. on Electronics, Circuits and Systems*, pp. 1276–1279.

Stauffer, C. and W. Grimson (1999). Adaptive background mixture models for real-time tracking. In *CVPR*, Volume 2, pp. 246–252.

Toyama, K., J. Krumm, B. Brumitt, and B. Meyers (1999). Wallflower: principles and practice of background maintenance. In *Proc. of Int. Conf. on Computer Vision*, Volume 1, pp. 255–261.

Wallflower. http://research.microsoft.com/enus/um/people/ jckrumm/WallFlower/TestImages.htm.

Wren, C., A. Azarbayejani, T. Darrell, and A. Pentland (1997). Pfinder: real-time tracking of the human body. *IEEE Trans. on Pattern Analysis and Machine Intelligence 19*(7), 780–785.

Computational Modelling of Objects Represented in Images – Di Giamberardino et al. (eds)
© *2012 Taylor & Francis Group, London, ISBN 978-0-415-62134-2*

Fourier optics approach in evaluation of the diffraction and defocus aberration in three-dimensional integral imaging

Z.E. Ashari Esfahani, Z. Kavehvash & K. Mehrany
Department of Electrical Engineering, Sharif University of Technology, Tehran, Iran

ABSTRACT: The unwanted effects of diffraction and defocus aberration in three-dimensional integral imaging are taken into account by using the Fourier optics approach. The concepts of point spread function and optical transfer function widely in use for conventional two-dimensional imaging are generalized and applied to three-dimensional integral imaging systems. The effects of diffraction and defocus aberration are then studied and the performance of the conventional single lens imaging system is compared against that of the integral imaging.

1 INTRODUCTION

Integral imaging (InI) is among the promising three-dimensional (3D) imaging techniques and thus has been the subject of many researches [1, 3-8]. While it is capable of providing autostreoscopic images from any desired viewpoint, it does not require a coherent optical source, nor does it mandate wearing a set of special glasses. Nevertheless, it suffers from a limited depth-of-focus (DOF). This point of weakness cannot be remedied unless it is full studied. Given that diffraction and aberration are among the main factors limiting the DOF in a direct reconstruction InI system, it is necessary to study the unwanted image distortion incurred by the inevitable diffraction and aberration in typical InI systems. Unfortunately, a thorough study of such unwanted effects by using the vectorial diffraction theory and Maxwell's equations is quite cumbersome. That is why geometrical optics is normally employed to study and simulate InI [1,3,4,7,8]. Here, we employ Fourier optics as a more accurate yet relatively simple tool to consider the wave nature of light and thus to some extent take into account the unwanted effects of diffraction and aberration.

To this end, the point spread function (PSF) of the InI system is extracted to quantify the amount of distortion added to the image because of the unwanted yet inevitable diffraction and aberration. It should be however noted that even though the PSF is a well known function in a single lens two-dimensional (2D) imaging system [2], it is not a well-defined function in an InI system. The complication of defining the PSF in InI is due to the fact that InI has two different stages: recording and reconstruction. In the recording stage, a set of convex lenslets in the pickup microlens array record several 2D images of a 3D object from different directions. The recorded 2D images are referred to as elemental images (EI). Part of the image distortion

is brought about at this stage, where each EI shows a distorted perspective of the 3D object. This part of the distortion is reflected in the recording stage PSF already given by Martinez et al. in [5]. In the reconstruction stage, a set of similar lenlets in the display microlens array puts EIs on view and thus forms the 3D image. The rest of the image distortion is brought about at this stage, where the limited resolution of the display device and the unwanted effects of diffraction and aberration in the display microlens array are both at work to deteriorate the quality of 3D imaging. Neglecting the limited resolution of the display device, we focus on the diffraction and aberration of the pickup and display microlens arrays and present as generalized PSF for a typical InI system. This is to the best of our knowledge the first attempt to at least partly consider the unwanted effects of diffraction and aberration by using Fourier optics in InI.

The rest of the paper is organized as follows. Fourier optics is utilized to extract the PSF in 2D and 3D imaging in section II. Since the overall PSF in InI is to be obtained by using the convolution operator, the optical transfer function (OTF) of the system is also extracted in this section. Simulation results are then provided in section III, where a typical InI system is considered and images are reconstructed by using the optical transfer function.

2 FOURIER OPTICS IN INTEGRAL IMAGING

2.1 *Single lens imaging point-spread-function*

It is a well-known fact that in conventional single lens imaging system, there exists a plane of focus whereat the best lateral resolution can be achieved as the defocus error is minimum. In reference to Fig. 1 where the structure of a single lens imaging system is schematically shown, the plane of focus satisfies the famous

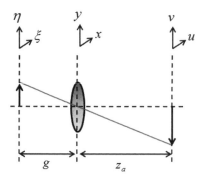

Figure 1. The schematic structure of a single lens imaging system.

lens law. If the object is placed at distance g in front of the lens and the focal length of the lens is f, the plane of focus lies at $z = z_i$ where:

$$\frac{1}{z_i} + \frac{1}{g} = \frac{1}{f} \tag{1}$$

Taking the Fresnel diffraction into account, the amount of image degradation at an arbitrary defocused plane lying at distance z_a behind the lens can be studied by the following PSF:

$$
h(u,v;\xi,\eta) = \Big| \frac{1}{\lambda^2 z_1 z_2} \exp\left[\frac{jk}{2z_a}(u^2 + v^2)\right]
$$
$$
\times \exp\left[\frac{jk}{2g}(\xi^2 + \eta^2)\right] \times \int\int_{-\infty}^{\infty} p(x,y)
$$
$$
\times \exp\left[\frac{jk}{2}(\frac{1}{z_a} + \frac{1}{g} - \frac{1}{f})(x^2 + y^2)\right]
$$
$$
\times \exp\left\{-jk\left[(\frac{\xi}{g} + \frac{u}{z_a})x + (\frac{\eta}{g} + \frac{v}{z_a})y\right]\right\} dxdy \Big|^2 \tag{2}
$$

where (ξ, η) is the coordinate of the object plane, (x,y) is the coordinate of the lens plane and (u, v) is the coordinate of image plane. The parameter λ is the wavelength of the light source, and $P(x, y)$ is the pupil function of the lens.

To better measure the resolution, the OTF of the typical system having a square aperture lens is given below [2]:

$$
H(f_x, f_y) = \Lambda\left(\frac{f_x}{2f_0}\right) \Lambda\left(\frac{f_y}{2f_0}\right) \text{sinc}\left[\frac{8w_m}{\lambda}\left(\frac{f_x}{2f_0}\right)\right.
$$
$$
\left(1 - \left(\frac{|f_x|}{2f_0}\right)\right)\right] \times \text{sinc}\left[\frac{8w_m}{\lambda}\left(\frac{f_y}{2f_0}\right)\left(1 - \left(\frac{|f_y|}{2f_0}\right)\right)\right] \tag{3}
$$

where w_m is a measure of the defocus aberration:

$$
w_m = -\frac{1}{2}\left(\frac{1}{z_a} - \frac{1}{z_i}\right)w^2 \tag{4}
$$

w is the half-width of the square aperture of the lens, f_x, and f_y are the frequency variables in x and y directions,

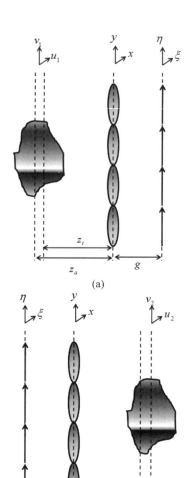

(a)

(b)

Figure 2. Schematic diagram of the (a) recording and (b) reconstruction stage of an InI system.

and:

$$
f_0 = \frac{w}{\lambda z_i} \tag{5}
$$

2.2 Integral imaging point-spread-function

In an integral imaging system, the pickup lens array maps the object to an array of EIs and then the display lens array transforms the array of EIs into the 3D image. The interrelation between the EI array and the 3D object in the recording stage; schematically shown in Fig. 2.a, is already described by using the following impulse response [5]:

$$
h_1(\xi, \eta; u_1, v_1, z_a) = \Big| \sum_{m,n} \frac{j\pi}{\lambda z_a} |mp - u_1|^2 |np - v_1|^2 \int\int p(x, y)
$$
$$
\times \exp\left\{-2j\pi x \frac{\xi + [M(mp - u_1)]}{\lambda g}\right\} \times \exp\left\{\frac{j\pi}{\lambda}\left(\frac{1}{z_a} - \frac{1}{a}\right)(x^2 + y^2)\right\} \tag{6}
$$
$$
\times \exp\left\{-2j\pi y \frac{\eta + [M(np - v_1)]}{\lambda g}\right\} dxdy \Big|^2
$$

82

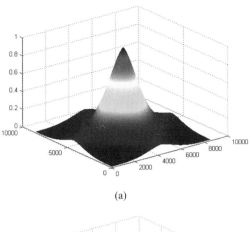

(a)

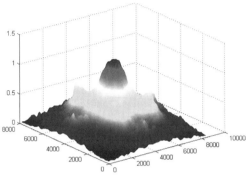

(b)

Figure 3. (a) The OTF of a 2D single lens imaging system and (b) The OTF of a 3D InI system.

where g is the distance between the EI array and the lens array, z_i is the focused image axial distance satisfying the famous lens law in accordance with (1), and z_a is the axial distance of the image plane, p is the lens pitch, $M = z_a/g$, and m, n represent the contribution of the m,nth lens.

Along the same line, the interrelation between the EI array and the 3D image in the reconstruction stage; schematically shown in Fig. 2.b, can be described by using a similar impulse response:

$$h_2(u_2,v_2;\xi,\eta,z_a) = \left| \sum_{m,n} \int\int \exp\left\{ \frac{j\pi}{\lambda g}(|\xi - mp - x|^2 + |\eta - np - y|^2) \right\} \right.$$
$$\times p(x,y)\exp\left\{ \frac{-j\pi(x^2 + y^2)}{\lambda f} \right\} \qquad (7)$$
$$\left. \times \exp\left\{ j\pi \frac{|mp - u_2 + x|^2}{\lambda z_a} + j\pi \frac{|np - v_2 + y|^2}{\lambda z_a} \right\} dxdy \right|^2$$

The overall PSF of the InI system, $h(u_2,v_2;u_1,v_1,z_a)$, can then be straightforwardly obtained by convolving the impulse response of the recording and reconstruction stages and therefore we have:

$$h(u_2,v_2;u_1,v_1,z_a) = h_1(\xi,\eta;u_1,v_1,z_a) * h_2(u_2,v_2;\xi,\eta,z_a) \quad (8)$$

(a)

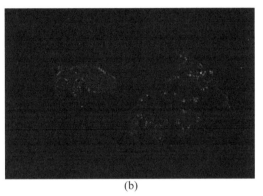

(b)

Figure 4. The 10,7th EI reconstructed by using the 10,7th lenslet as if in a single lens 2D imaging system (a) using geometrical optics (b) using Fourier optics (including the effects of diffraction and defocus aberration).

The OTF of the InI system can then be easily extracted by using the Fourier transform and we have:

$$H(f_x,f_y,z_a) = FT\{h_1(u_2,v_2;u_1,v_1,z_a)\} \qquad (9)$$

To compare the unwanted effects incurred by diffraction and defocus aberration in single lens 2D imaging and in 3D InI, the OTF of the former should be measured against that of the latter. Consider a single lens 2D imaging system made of a square aperture lens with $w = 5\,\text{mm}$, $g = 50\,\text{mm}$, $z_i = 360\,\text{mm}$, and $z_a = 377\,\text{mm}$. The OTF of this simple system is plotted in Fig. 3.a. Now consider a typical InI system with an array of 16×16 lenslets of square aperture with $w = 5\,\text{mm}$, $g = 50\,\text{mm}$, $z_i = 360\,\text{mm}$, and $z_a = 377\,\text{mm}$. The OTF of this simple system is plotted in Fig. 3.b.

Since the lenslets in the lens array of the InI system is similar to the lens used in the conventional single lens imaging system, the comparison between the two is fair enough. This comparison stands witness for the fact that the InI system permits wider range of spatial frequencies to pass through the imaging system. It is therefore no surprise that the image degradation caused by the defocus aberration in InI system is to some extent diminished. This point is shown in the next section.

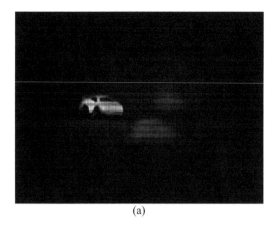

(a)

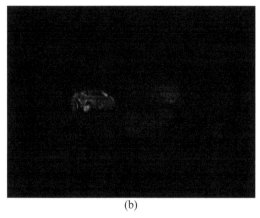

(b)

Figure 5. The 3D image reconstructed in InI system by using (a) geometrical optics and (b) Fourier optics (including the effects of diffraction and defocus aberration).

3 SIMULATION RESULTS

A 3D integral imaging system with a 16×16 array of square aperture lenses is considered in this section. The distance between the EI array and the lens array is $g = 50$ mm, the length of the EIs in the x and y directions are $s_x = 22.4$ mm and $s_y = 14.8$ mm, respectively. The wavelength is $\lambda = 1 \mu m$, the half-width of the square apertures is $w=5$mm, and $z_i = 360$ mm, $z_a = 377$ mm. The 3D scene is composed of three toy cars and 16×16 EIs are optically recorded.

As the first example, the 10,7th EI is considered and its corresponding image obtained by using the 10,7th lenslet in the array is reconstructed at $z_a = 377$ mm by two different methods. First, the geometrical optics is used to reconstruct the EI numerically. The obtained result is shown in Fig. 4(a). The thus reconstructed image has a very good quality because the unwanted effects of diffraction and aberration are all neglected. Second, the Fourier optics is employed and the EI

is once again reconstructed at $z_a = 377$ mm. The obtained result is shown in Fig. 4(b). The quality of the thus reconstructed image is considerably deteriorated. This is due to the unwanted effects incurred by the diffraction and the defocus aberration. It is worth noting that the comparison between Figs. 4(a) and (b) reveals the detrimental effects of diffraction and defocus aberration.

As the final example, the 3D image of the InI system is reconstructed at $z_a = 377$ mm by considering all EIs and all lenslets in the array. The results obtained by using the geometrical optics and the Foureir optics are shown in Figs. 5(a) and (b), respectively. Although the former has a better quality than the latter, the effects of diffraction and defocus aberration in 3D imaging by InI are not as detrimental as are the effects of diffraction and defocus aberration in 2D single lens imaging.

4 CONCLUSIONS

In this manuscript, the effect of diffraction and defocus aberration in InI system is evaluated by generalizing the concept of PSF and OTF. It is shown that the image reconstruction by using lens array is more tolerant to defocus aberration and diffraction effects than is the conventional single lens imaging system.

REFERENCES

Arai, J., Okui, M., Yamashita, T. & Okano, F. 2006. Integral three-dimensional television using a 2000-scanning-line video system. *Applied Optics.* Vol.45, NO.8.

Goodman, J.W. 1996. *Introduction to Fourier Optics.* McGraw-Hill.

Hong, S., Jang, J. & Javidi, B. 2004. Three-dimensional volumetric object reconstruction using computational integral imaging. *Optics Express.*Vol.12, NO.3.

Manolache, S., Aggoun, A., McCormick, M. & Davies, N. 2001. Analytical model of a three-dimensional integral Image recording system that uses circular and hexagonal-based spherical surface microlenses. *Journal of Optical Society of America A.* Vol.18, NO.8.

Martinez-Corral, M., Javidi, B., Martinez-Cuenca, R. & Saavedra, G. 2004. Integral imaging with improved depth of field by use of amplitude-modulated microlens arrays imaging. *Applied Optics.* Vol.43, NO.31.

Kavehvash, Z., Martinez-Corral, M., Mehrany, K., Bagheri S., Saavedra, G. & Navarro H. 2012. Three-dimensional resolvability in an integral imaging system. *J. Opt. Soc. Am. A.*Vol.29, NO.4.

Okano, F., Hoshino, H., Arai, J. & Yuyama, I. 1997. Real-time pickup method for a three-dimensional image based on integral photography. *Applied Optics.* Vol.36, NO.7.

Stern, A. & Javidi, B. 2003. 3-D computational synthetic aperture integral imaging (COMPSAII). *Optics Express.* Vol.11, No.19.

Computational Modelling of Objects Represented in Images – Di Giamberardino et al. (eds)
© 2012 Taylor & Francis Group, London, ISBN 978-0-415-62134-2

Intensity and affine invariant statistic-based image matching

Michela Lecca & Mauro Dalla Mura
Fondazione Bruno Kessler, Povo, Trento, Italy

ABSTRACT: In this paper, we propose a technique to perform image matching, which is invariant to changes in geometry and intensity. In detail, this approach relies on a representation of an image portion as a 3D function simultaneously coding both the 2D spatial coordinates and the intensity value of each pixel. In this equivalent image representation, a geometric and intensity variation between the visual content of two images depicting the same objects leads to an isometric relationship between the image 3D functions. The comparison between the two images is effectively obtained by estimating the 3D linear isometry that connects them. The effectiveness of the proposed technique was assessed on two synthetic, and one real, image data sets.

1 INTRODUCTION

Comparing images to establish whether they depict the same object is an easy task for humans, who can perform it at a glance. On the contrary, although this problem has been investigated for many years, the results provided by the current computer vision systems are still poor (Pinto, Cox, and DiCarlo 2008). In Computer Vision, to compare image regions means matching features describing their visual appearance, like for instance color, shape, and texture. This is the core task of the object recognition challenge, i.e. posing the question of whether an image portion is an instance of a known object. In facing this challenge, the main difficulty is to model the wide variability under which an object can appear in a scene. For example, in different images, the same object could be rescaled, partially occluded, differently oriented and/or illuminated. Thus, the features used to describe the object should be invariant to the largest possible number of circumstances, so that a stable and efficient recognition becomes feasible.

In this work, we focus on the *geometric* and *intensity invariances*. Geometric distortions like changes of size or perspective, rotations and/or translations are often due to the changing of a camera's location. Intensity variations are generally produced by changes of the brightness or of the position of the light source illuminating the scene, or even by the use of another camera for taking the pictures.

In the past decades, several methods have been developed for obtaining a representation of the region shape insensitive to changes of scale, in-plane rotation and translations, e.g. (Hu 1962), (Yajun and Li 1992), (Zahn and Roskies 1972). A few works (Pei and Lin 1995), (Rothganger, Lazebnik, Schmid, and Ponce 2006) also handle skew transforms, which occur, for instance, when the camera is not vertically positioned

with respect to the image plane. Intensity changes are difficult to model, because of their complex dependency on the physical cues of the light and on the spectral sensitivity of the device employed for image acquisition. An illuminant intrinsic image is proposed in (Finlayson, Drew, and Lu 2004), but its computation requires information about the camera responsivity, that is often unavailable. Intensity invariance is thus often achieved by normalizing the image, or a portion of it, by a Gray-World algorithm (Rizzi, Gatta, and Marini 2002), or the problem is solved using features insensitive to light changes, like the oriented gradient of SIFT (Lowe 2004).

Here, we present a method for obtaining simultaneously geometric and intensity invariance without a feature-based image representation. The proposed approach extends the work in (Pei and Lin 1995), which is used to achieve geometric invariance to rescaling, rotation, translation and skew. In particular, the technique of (Pei and Lin 1995) transforms 2D image regions through a set of linear maps, that diagonalize the covariance matrices of the regions. These operations transform two affine regions into isometric regions and the isometry is a 2D rotation. Tensor theory is then used to find a rotation angle making each pattern invariant to orientation changes.

To also cope with intensity variations, we reformulate the technique of (Pei and Lin 1995) in three-dimensional space. Hence, we consider a pair of image regions, that possibly differ by a linear change of their intensity and by an affine map that modifies their spatial coordinates as in (Pei and Lin 1995). We regard any image region as a 2D surface, in which each point corresponds to a pixel and it is identified by its pixel location in the image plane along with its intensity. We model the geometric and intensity change using a 3D affine mapping between the surfaces representing the two image regions. As in 2D case, we linearly transform these surfaces by diagonalizing their covariance matrix to obtain two isometric sets. Differently from the 2D case, applying the tensor theory of (Pei and Lin 1995) in 3D is more complex, because it requires the computation of the 3th and 4th moments of the surfaces, which are very sensitive to noise and digitalization effects. Therefore, in order to match our surfaces accurately, we estimate the isometry (here a rotation or a reflection composed with a rotation) and then compare the isometrically transformed sets. We efficiently perform this estimation using the approach in (Kostelec and Rockmore 2003),, which convolves the Fourier transforms of our isometric sets over the rotation group $SO(3)$.

There are three main advantages to use our approach for matching image regions: (1) we do not employ any representation of image appearance using features, but instead perform the analysis directly on image data; (2) we reduce the computational burden since linear transforms can be efficiently computed; (3) we establish not only if two regions depict the same object upon a geometric and intensity change, but also retrieve the geometric and intensity transform relating them.

The main disadvantage is that the method is based on the computation of the covariance matrix, that is generally sensitive to noise and outliers. In general, some noise is due to intensity saturation, that occurs when the incident light at a pixel causes the maximum response of a color channel. For dealing with this aspect, we exclude saturated pixels, in order to increase the robustness of the proposed technique. The tests we carried out on synthetic and real-world databases showed good performances, also in the presence of saturation.

Outlines – Section 2 recalls some mathematical concepts about affine maps and covariance matrices; Section 3 describes our method; Section 4 illustrates the results we obtained; finally, Section 5 reports some conclusions and our future work.

2   BACKGROUND

In this Section we recall two well known properties (Theorems 1 and 2) of the covariance matrix and we explain how two affine sets of points can be turned into isometric sets.

Let R and Q be two subsets of $\mathbf{R}^n$ ($n \geq 1$) related by an affine map such that $y = Ax + b$, with A being a non singular $n \times n$ matrix, b a vector of $\mathbf{R}^n$, $x \in R$, and $y \in Q$.

Theorem 1 Let C_R and C_Q be the covariance matrices of R and Q respectively, Then $C_Q = AC_RA^t$, where A^t denotes the transpose of A.

Theorem 2 The covariance matrices C_R and C_Q can be simultaneously diagonalized, i.e. there exists a $n \times n$ matrix B such that BC_RB^t is the identity matrix and BC_QB^t is diagonal. Moreover, denoted by D the diagonal matrix with $D_{ii} = 1/\sqrt{[BB^t]_{ii}}$ ($i = 1, ..., n$), DBC_QB^tD is the identity matrix.

Without loss of generality, we can assume that $b = \mathbf{0}$. Otherwise, we can translate the sets R and Q so that their mean points coincide with the origin of $\mathbf{R}^n$. We claim that the sets $R_B = \{x_B = Bx | x \in R\}$ and $Q_B = \{y_B = DBy | y \in Q\}$ are isometric. In fact, for each y_B in Q_B, there exist $y \in Q, x \in R$, and $x_B \in R_B$ such that $y_B = DBy = DBAx = DBAB^{-1}x_B$. Hence, from Theorem 2,

$$C_{Q_B} = (DBAB^{-1})^t C_{R_B}(DBAB^{-1})$$
$$= (DBAB^{-1})^t(DBAB^{-1}).$$

Since C_{Q_B} is the identity matrix, the linear transform associated to the matrix $DBAB^{-1}$ is an isometry F, i.e. R_B and Q_B are related by a rotation or by the composition of a reflection with a rotation.

Figure 1 shows the diagram of the transforms considered here for two 2D surfaces R and Q ($n = 3$).

3   PROPOSED METHOD

In this Section, we explain how the approach in (Pei and Lin 1995) can be extended from 2D to 3D case to cope with both geometric and intensity changes.

Affine Transform Model. We regard any image region R as a 2D surface of $\mathbf{R}^3$, so that each pixel x of R is a triplet $(r, c, f(r, c))$ where r and c are the coordinates of x in the image plane and $f(r, c)$ is its intensity value. The intensity in x is computed as the mean value of the three channel responses at x. Let Q be another image region possibly related to R by a 3D affine map of the form

$$\begin{bmatrix} r \\ c \\ f(r,c) \end{bmatrix} = \begin{bmatrix} a_{00} & a_{01} & 0 \\ a_{10} & a_{11} & 0 \\ 0 & 0 & \alpha \end{bmatrix} \begin{bmatrix} r' \\ c' \\ f(r',c') \end{bmatrix} \qquad (1)$$

with $(r, c, f(r, c))$ and $(r', c', f(r', c'))$ being points of R and Q respectively.

The 2D matrix $(a_{ij})_{i,j=0,1}$ models a geometric distortion of the pixel coordinates (r, c). As in (Pei and Lin 1995), we consider re-scaling, rotation and skew changes. Translation terms are not considered, because we suppose that the 3D points of R and Q have been translated so that their mean points coincide with the origin of $\mathbf{R}^3$. The parameter α is a scaling factor of the intensity coordinate. Thus, the 3×3 matrix in Equation 1 models a linear geometric and intensity change.

In Figure 1, two image regions depicting a glove are shown on top. The second region is rescaled and rotated with respect to the first one, and its brightness is darkered. The red plots show the corresponding surfaces in 3D space.

Isometry Estimation. If R and Q are actually related by the transform (1), the sets R_B and Q_B obtained from R and Q by diagonalizing their covariance matrices as in Section 2, are isometric (see the green plots in Figure 1).

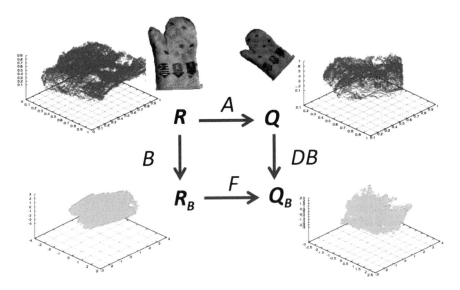

Figure 1. Schema of the transforms applied to two image regions depicting an object upon a change of scale, in-plane orientation and intensity. Pixel coordinates are displayed on the xy plane, while intensity is shown along the z axis.

According to the Cartan-Dieudonné Theorem (Gallier 2001), any linear isometry F on $\mathbf{R}^n$ is the composition of $m \leq n$ reflections. In particular, for $n = 3$, (a) if m is even, F is a rotation; (b) if m is odd, F is a reflection ($m = 1$) or a reflection composed with a rotation ($m = 3$). Since we do not known m in advance, we consider separately the two cases.

Case (a). F is a 3D rotation. The main problem in estimating efficiently F is due to the large number of points to be matched and to the lack of information about their correspondences.

We cope with these issues by considering the *spherical histograms* of the sets R_B and Q_B. A spherical histogram H of a set X is a 2D discrete function computed as follows:

1. the points of X are expressed in spherical coordinates (ρ, ϕ, ψ);
2. the variability ranges $[-\pi, \pi]$ and $[-\pi/2, \pi/2]$ of ϕ and ψ respectively, are partitioned into N equispaced intervals;
3. the ijth entry h_{ij} of H is the mean distance ρ of the points with ϕ and ψ belonging to the ith and jth set of the partitions of $[-\pi, \pi]$ and $[-\pi/2, \pi/2]$ respectively.

Spherical histograms of 3D rotated data are also interrelated by a rotation. Thus, finding a rotation between R and Q is equivalent to find a rotation between H_R and H_Q. The advantage here is that, thanks to the quantization of ϕ and ψ, the histograms provide a more compact representation of the point sets, so that the number of points to be matched is N^2.

We efficiently compute the rotation between spherical histograms H_R and H_Q of R_B and Q_B by convolving the Fourier transforms of H_R and H_Q over the rotation group $SO(3)$ (Kostelec and Rockmore 2003).

The maximum value of the convolution function corresponds to the Euler angles defining the desired rotation F.

Case (b). F is a reflection or the composition of a reflection with a rotation. This case can be reworked to appear like case (a). In fact, the reflection F can be expressed as the composition of a reflection ω with a rotation R, where R is the identity when F is just a reflection. Therefore we can map the set R_B to a set R'_B by an arbitrary reflection ω, then we estimate the rotation R between R'_B and Q_B as in case (a). Function F is given thus by $R \circ \omega$. Since ω is arbitrary, in our experiments we choose ω to be the reflection mapping the Cartesian z coordinates into $-z$.

Given the pair of isometric sets R_B and Q_B, we estimate two possible isometries F_a and F_b, one from Case (a) and the other from Case (b). This lead us to two estimates of the matrix A mapping R onto Q:

$$A_a := B^{-1}D^{-1}F_aB \text{ and } A_b := B^{-1}D^{-1}F_bB.$$

We choose the isometry such that the estimate of A performs the best alignment of R onto Q. The quality of the alignment is defined in terms of the Euclidean distance between the sets R and Q, as explained later.

Similarity Computation. If R and Q are related by a 3D affine map, and F is the isometry aligning R_B and Q_B, then the sets Q and $R_e = \{y = B^{-1}DFBx | x \in R\}$ should coincide. Obviously, since the data we manage can be noisy, an exact overlap between R_e and Q is rarely reached.

In our work, we measure the accuracy of our algorithm as the overlap between the sets R_e and Q. This overlap is expressed in terms of the Euclidean distance between the points of R_e and Q as following:

$$d(R_e, Q) = \frac{1}{2}\left[\frac{1}{|R_e|}\sum_{x\in R_e}\min_{y\in Q}d(x,y) + \right.$$

$$\left. \frac{1}{|Q|}\sum_{y\in Q}\min_{x\in R_e}d(x,y)\right] \tag{2}$$

Here $d(x,y)$ indicates the Euclidean distance between two points x and y of $\mathbf{R}^3$ and $|\cdot|$ is the cardinality of the set between the pipes.

The first term in Equation (3) measures *how much* the set R_e is *contained* in set Q, while the second term measures the converse, i.e. how much the set R_e *contains* the set Q. The lower the distance is, the higher is the similarity between R and Q upon a transform like (1).

Dealing with Saturation. Since intensity values range over [0, 255], any transformation (1) with $\alpha \geq 1.0$ casts the intensities with value greater than $255/\alpha$ to 255. This saturation phenomenon can adversely affect the estimate of the covariance matrices and hence the computation of the isometry. To overcome this problem, we consider the cumulative histograms K_R and K_Q of the intensity of R and Q and we compute the minimum values b_R and b_Q of the intensity such that $K_R(b_R) = K_Q(b_Q) = 1.0$. We then discard from the covariance computation all the pixels of R and Q with intensity greater than b_R and b_Q respectively.

Figure 2 shows the spherical histograms (a) and (b) of the surfaces R_B and Q_B of Figure 1 (green plots). Figure 2(c) shows the spherical histogram of R_B corrected by the isometry F (in this case a rotation) we estimate. In Figure 2 (d) we also see the set R mapped onto Q by the linear transform with matrix $B^{-1}D^{-1}FB$ which approximates matrix A. The two sets overlap very well.

4 EXPERIMENTS

We evaluate the performance of our method on two synthetic databases (SYN1 and SYN2) and on one real-world dataset (RWD).

The synthetic databases SYN1 and SYN2 have been generated from the public dataset PP-GADGETS[1], consisting of 48 images of different objects (see Figure 3 for some examples). In our experiments, we just consider the image portion effectively belonging to objects, assuming the rest to be transparent (Figure 4). By this way, we show how our algorithm works on arbitrarily shaped image regions.

Database SYN1 has been created by modifying each image region of PP-GADGETS by 100 synthetic transforms like in (1), where the sub-matrix $(a_{ij})_{i,j=0,1}$ represents a change of scale and of orientation of

[1] http://www.vision.caltech.edu/html-files/archive.html

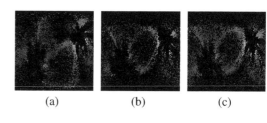

(a)   (b)   (c)

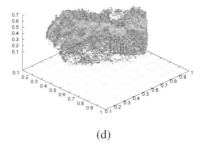

(d)

Figure 2. (a, b) Spherical Histograms related to the image regions (glove) of Figure 1. (c) Spherical Histogram of the first region aligned onto the seocond one. The horizontal and vertical axes refers to ϕ and ψ, respectively. The brightness is proportional to the value of ρ. (d) The set R aligned on Q by the estimated transform: it can be noticed that the resulting sets overlap each other.

Figure 3. Some objects from the dataset PP-GADGETS.

the 2D coordinates (r, c). More precisely, the matrix considered in Equation (1) is

$$\begin{bmatrix} \sigma\cos\theta & -\sigma\sin\theta & 0 \\ \sigma\sin\theta & \sigma\cos\theta & 0 \\ 0 & 0 & \alpha \end{bmatrix} \tag{3}$$

where the scale factor σ and the intensity parameter α range in the set $\{0.5, 0.75, 1.0, 1.25, 1.5\}$ and the rotation angle θ in $\{0°, 90°, 180°, 270°\}$.

Database SYN2 considers also skew: in this case, the sub-matrix $(a_{ij})_{i,j=0,1}$ represents a skew transform with parameters u and v composed with a 2D rotation of the coordinates (r, c). The matrix is given by

$$\begin{bmatrix} u\cos\theta & -u\sin\theta & 0 \\ v\sin\theta & v\cos\theta & 0 \\ 0 & 0 & \alpha \end{bmatrix} \tag{4}$$

where the values of u and v have been randomly chosen in the range [0.5, 1.5], while θ has been randomly selected over $[0, 2\pi]$.

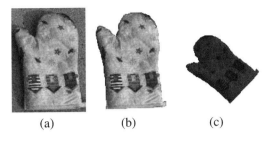

(a)   (b)   (c)

Figure 4. (a) Sample from the PP-GADGETS dataset. (b) Region of (a) considered in the experiments. (c) image in (b) after a synthetic transform of SYN1.

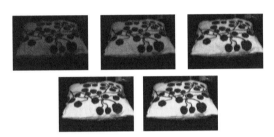

Figure 5. An object from RWD database captured under a light with increasing intensity.

Finally, database RWD consists of 80 images, representing 16 different objects, each captured under 5 different light settings with increasing intensity (see Figure 5).

For SYN1 and SYN2 we compared the original images of PP-GADGETS with the modified versions produced by our synthetic transforms. For RWD, we took as reference images the pictures with darker intensity and we matched them against the pictures captured in a brighter environment.

The spherical histograms we use in these experiments quantize the angles ϕ and ψ by $N = 128$ intervals. Coarser quantizations provide generally worse results, while for finer quantization, the increment of the accuracy is negligible, while the computational time increases. In our implementation, on a PC with IntelProcessor, 2.9 GHz, the algorithm requires 13 seconds to process an image region of 18750 pixels, when $N = 256$, while the time is about 3.80 seconds when $N = 128$. On average, in our experiments, we needed about 2.58 seconds for $N = 128$ and 12 seconds when $N = 256$.

For all the databases, we measured the accuracy of our method by comparing the Euclidean distances (3) d_{before} and d_{after} between the image regions R and Q before and after our correction. As shown in Table 1, in all the cases, we observed that d_{after} is smaller than d_{before}.

Figure 6 displays the distributions of the values of d_{before} (in red) and d_{after} (in green) for database SYN1. The distances after the region alignment we propose are remarkably smaller than those obtained

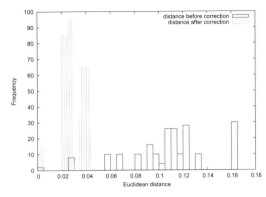

Figure 6. Database SYN1 – Distributions of the Euclidean distance (3) before (red line) and after the region alignment we propose.

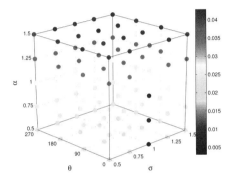

Figure 7. Euclidean distance (3) of the image regions aligned by our method increases by increasing the intensity value, i.e. by increasing the number of saturated points. The data displayed here refers to SYN1.

Table 1. Distance before and after applying the proposed method.

Database	d_{before}	d_{after}
SYN1	0.1085	0.0294
SYN2	0.0719	0.0295
RWD	0.0937	0.0210

before applying our method. The same trend has been observed for SYN2 and RWD.

Although the technique we implement to discard saturated pixels gives good results also for high value of the intensity parameter α, the distance generally increases with α. This fact is shown in Figure 7 which refers to SYN1. The three Cartesian axes represent the parameters θ, σ α of the transform (3) and the colored points are the value of the distance d_{after} for different values of θ, σ α. We note that the higher the intensity is, the higher the distance is. In this case-study, the maximum percentage of saturated pixels is 20% of the total number of image pixels.

5 CONCLUSIONS

In this paper, we presented a method for matching two image regions that represent the same scene under linear changes in the image plane geometry (i.e., rescaling, rotation, translation and skew) and in the image intensity. We proposed to represent the spatial coordinates and the intensity values of a 2D image as a 3D function. The estimation of the linear transformation that relates two images in such 3D representation is obtained in two steps. Firstly, a normalization of the two images is obtained by simultaneously diagonalizing their covariance matrices. Subsequently, the normalized 3D functions are aligned by estimating the linear isometry between them with a technique based on the Fourier transform computed in spherical coordinates. The experiments reported good performances both on synthetic and real-world data sets. Furthermore, this technique proved to be robust to the presence of saturated pixels. According to the preliminary results obtained, we believe that the proposed technique can be of interest for applications such as the extraction of 3D information from an image and for quality control.

Our future work will extend the approach to a higher number of dimensions by embedding additional features extracted from the scene such as color information, edges, regions, etc. Furthermore, we aim to test this technique for image retrieval tasks since it can get hints on the similarity of the two image regions and retrieve the affine map that relates them.

REFERENCES

Finlayson, G. D., M. S. Drew, and C. Lu (2004). Intrinsic images by entropy minimization. In *Proc. 8th European Conf. on Computer Vision, Praque*, pp. 582–595.

Gallier, J. H. (2001). *Geometric methods and applications: for computer science and engineering*, Volume Vol. 38 of Texts in applied mathematics. Springer, ISBN 0387950443.

Hu, M.-K. (1962, february). Visual pattern recognition by moment invariants. *IRE Transactions on Information Theory 8*(2), 179–187.

Kostelec, P. J. and D. N. Rockmore (2003). Ffts on the rotation group. Technical report, Santa Fe Institute's Working Paper Series.

Lowe, D. G. (2004, November). Distinctive image features from scale-invariant keypoints. *International Journal of Computer Vision 60*, 91–110.

Pei, S.-C. and C.-N. Lin (1995). Image normalization for pattern recognition. *Image and Vision Computing 13*(10), 711–723.

Pinto, N., D. D. Cox, and J. J. DiCarlo (2008, January). Why is Real-World Visual Object Recognition Hard? *PLoS Computational Biology 4*(1), e27+.

Rizzi, R., C. Gatta, and D. Marini (2002). Color correction between gray world and white patch. In *in IS&T/SPIE Electronic Imaging 2002. The human Vision and Electronic Imaging VII Conference., 4662*, pp. 367–375.

Rothganger, F., S. Lazebnik, C. Schmid, and J. Ponce (2006). 3d object modeling and recognition using local affine-invariant image descriptors and multi-view spatial constraints. *International Journal of Computer Vision 66*, 2006.

Yajun and Li (1992). Reforming the theory of invariant moments for pattern recognition. *Pattern Recognition 25*(7), 723–730.

Zahn, C. T. and R. Z. Roskies (1972, March). Fourier descriptors for plane closed curves. *IEEE Transactions on Computers 21*, 269–281.

Computational Modelling of Objects Represented in Images – Di Giamberardino et al. (eds)
© 2012 Taylor & Francis Group, London, ISBN 978-0-415-62134-2

Recognition of shock-wave patterns from shock-capturing solutions

Renato Paciorri
Dip. di Ingegneria Meccanica e Aerospaziale, Università di Roma "La Sapienza" A. Roma, Italy

Aldo Bonfiglioli
Dip. to Ingegneria e Fisica dell'Ambiente – Università degli Studi della Basilicata, Potenza, Italy

ABSTRACT: A new technique for detecting shock waves and recognising the shock patterns in compressible flow solutions computed by means of shock-capturing solvers is proposed and discussed. This newly developed technique uses the Hough transform and a least squares fit to further post-process the data provided by a classical shock detection technique based on local criteria. This two-steps analysis significantly enhances the quality of the shock lines extracted from numerical solutions; moreover, it allows to correctly recognise the complex shock-wave patterns that arise when the flow solution is characterised by the presence of multiple interacting shocks.

1 INTRODUCTION

Starting in 2006, the authors have been developing a new shock-fitting technique for unstructured meshes (Paciorri and Bonfiglioli 2009; Bonfiglioli et al. 2010; Paciorri and Bonfiglioli 2011). In this newly developed technique, the discontinuities (either shock waves or contact-discontinuities) are moving boundaries that are free to float inside a computational domain discretised using triangles, in two space dimensions, or tetrahedra in three. Steady calculations are performed in a transient fashion, whereby the motion of the fitted discontinuities is governed by the Rankine-Hugoniot jump relations and a shock-capturing solver is used to discretise the governing PDEs within the smooth regions of the flow-field. At each time-step, the computational mesh is locally re-generated in the neighbourhood of the fitted discontinuities while ensuring that the shock lines (or the shock surfaces in three dimensions) are always part of the mesh during their motion. Once steady-state is reached, the shock speed vanishes and the computational mesh does not change any longer.

In order to start a calculation, the aforementioned shock-fitting technique requires, beside a reasonable initial solution, the initial, even though approximate, position of the shocks and, whenever shock interactions occur, it also requires an a-priori knowledge of the flow topology. This information is provided by a preliminary shock-capturing calculation of the same flow-field. In a shock-capturing calculation, however, the information concerning the position and shape of the shocks and their mutual connections is not directly available as it is in the shock-fitting solution of the same flow-field; however, these data can be obtained by post-processing the computed shock-capturing solution.

Several techniques for detecting shock waves from a shock-capturing solution have been documented in the literature. Ma et al. (Ma et al. 1996) and Pagendarm and Seitz (Pagendarm and Seitz 1993) proposed a technique based on a criterion which relies upon directional derivatives of the density field. Lovely and Haimes (Lovely and Haimes 1999) suggested a different criterion based on the scalar product between the local velocity vector and pressure gradient. More recently, Kanamory and Suzuki (Kanamori and Suzuki 2011) used the theory of characteristics to identify the presence of shock waves. A common feature of all the aforementioned techniques is that they are based on local criteria. In other words, the presence of a shock at a specific grid-point within the flow-field is recognised by analysing the flow quantities at the grid-point itself and, at most, its immediate neighbours.

The local nature of these techniques is the origin of a number of limitations and drawbacks. Indeed, all these techniques represent the detected shock lines as a sequence of shock points, eventually connected to each other; when looking at the extracted shock lines, these often feature the presence of artificial oscillations, spurious branches and gaps. Moreover, all these techniques fail in the interaction points and, last but not least, they are unable to recognise the topology of shock interactions. For all these reasons, the aforementioned local detection techniques turned out to be of little use in order to provide the information required to initialise our shock fitting calculations. Therefore, in our previous work, a non-negligible amount of user's intervention was required in order to set-up the initial shock location and shape and the position of the interaction points.

A slightly different framework in which automatic shock detection may find a useful application is gridadaptation using overlapping, multi-block, structured grids (Kao et al. 1994; Bonfiglioli et al. 2012). In this context, a good approximation of the shock front extracted from a preliminary, un-adapted, shockcapturing calculation may be used to construct overlapping blocks that are tailored on the shock fronts.

Prompted by the need to develop an automatic procedure to detect the shock fronts from an available shock-capturing calculation, the present work will show how it is possible to post-process the data provided by a local shock detection technique in order to improve its quality and also recognise the shock wave pattern. This newly developed "global" shock detection technique relies upon the Hough transform (Hough 1962; Duda and Hart 1972) and a least squares fit.

2 SHOCK DETECTION AND PATTERN RECOGNITION

A brief description of the shock-detection technique based on the Hough transform is here illustrated by means of a simple flow configuration. Let us consider the shock-capturing solution of a regular shock reflection; the computed density iso-contour lines in the neighbourhood of the reflection point are shown in Fig. 1a, along with the underlying triangular mesh. The use of the local shock detection technique described in (Ma et al. 1996) to post-process the solution shown in Fig. 1a gives the set of shock points shown in Fig. 1b. This is the input data for our newly developed "global" shock detection algorithm; the various steps involved in the algorithm will be described in the following paragraphs.

2.1 *Search of predominant alignments and grouping of shock points*

Figure 1b clearly shows that there are two preferential directions along which most of the shock points identified by the local criterion are aligned. It is also evident, however, that a non-negligible number of candidate shock points falls outside the two straight lines corresponding to the incident and reflected shocks (respectively shown in red and blue in Fig. 2); their identification as shock points is due to the inherent limitations of the local shock detection criterion and has nothing to do with the flow physics.

The Hough transform (Hough 1962; Duda and Hart 1972) allows to distinguish the predominant alignments from the others. Once the two straight lines along which most of the shock points are aligned have been identified, the shock points are grouped by calculating the distance of each shock point from the nearest straight line. Specifically, a shock point belongs to the nearest straight line if its distance from the line is less than a threshold. If not, the shock point is rejected and marked as a spurious point; this is the case of the points marked by open squares in Fig. 2.

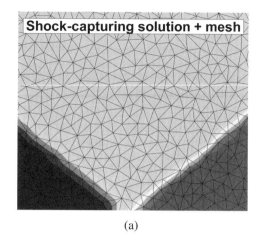

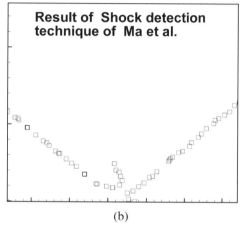

Figure 1. Regular shock reflection off a straight wall: shock-capturing solution and corresponding cloud of shock-points.

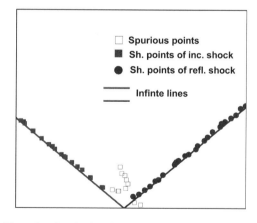

Figure 2. Search of predominant alignments and grouping of shock points.

2.2 *Line refinement and segment definition*

The application of the previous step provides n infinite straight lines along with the corresponding set of

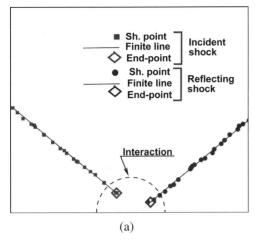

(a)

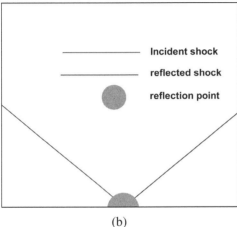

(b)

Figure 3. Shock reflection off a straight wall: identification of the segments corresponding to the incident and reflected shocks and the interaction point.

shock points. The equations defining each of these n lines are modified by applying a linear or quadratic least squares fit to each set of shock points. The use of quadratic functions allows to better fit those sets of shock points that belong to a curved shock. Each of these n straight or curved infinite lines is converted into a segment by identifying two end-points. These end-points bound the portion of the infinite line inside of which the shock points of the corresponding infinite line are found.

2.3 *Shock-pattern recognition*

The straight or curved segments obtained from the previous step are not joined to each other. This is clearly shown in Fig. 3: even though step 2.2 of the algorithm has correctly identified the incident and reflected shocks, the point where these interact with the wall is still missing. This is due to the fact that the cloud of points provided by the local detection technique is particularly noisy in regions where different shocks

mutually interact (see for instance Fig. 2). In order to complete the topological description of the shock pattern, it is therefore necessary to connect the end-points of the various segments and to identify those points where the shocks may eventually cross the far-field boundaries or interact with the solid boundaries of the computational domain.

Two different kinds of connections among neighbouring shock segments can be identified: junction and interaction points.

A junction connects one of the end-points of two distinct segments so that they are joined to form a unique shock front. This typically occurs when the present detection technique is applied to curved shocks.

An interaction point occurs when a shock merges with another shocks and/or impinges on a wall, as in the case of a Mach reflection. In this case, the end-points of two or more different shocks are joined with each other at the interaction point.

In addition to the junction/interaction points, the topological description of the flow-field is completed by those points where the shocks cross the boundaries of the computational domain.

The pattern recognition step is accomplished by trying to identify different, specific shock-patterns in the neighbourhood of the end-points of all the shock segments that have been picked-up in step 2.2 of the algorithm.

Within a two-dimensional flow-field, the fundamental shock-patterns that may occur around the end-point of a shock segment can be grouped into two categories: those that are found far from the boundaries of the computational domain and those that occur on the domain boundaries. The former set is made of three distinct, fundamental patterns: shock junctions, triple points and quadruple points, as shown in frames a, b and c of Fig. 4; the remaining four fundamental patterns (incoming shock, outgoing shock, regular reflection and shock normal to a wall) belong to the latter set (see frames d, e, f and g of Fig. 4) as they all involve the interaction of the shock with the boundaries of the computational domain.

Each of the different shock-patterns described in Fig. 4 may be recognised by testing each end-point against a set of rules. The various rules take into account the following information: the distance (δ in Fig. 4a) between the end-points of neighbouring shock segments, the difference in slope ($\Delta\theta$ in Fig. 4a) of neighbouring segments, the distance of an end-point from the nearest boundary (δ_b in Fig. 4d), the kind of boundary condition (inlet, outlet or solid wall, see frames d), e) and g) of Fig. 4) applied to the nearest boundary and the angle (σ in Fig. 4f) formed with the tangent to the nearest boundary.

For example, a regular reflection pattern, which is sketched in Fig. 4f), is identified when an end-point complies to all of the following rules:

- the end-point is close "enough" to the end-point of a unique, different shock segment;
- the end-point is close "enough" to a solid wall;

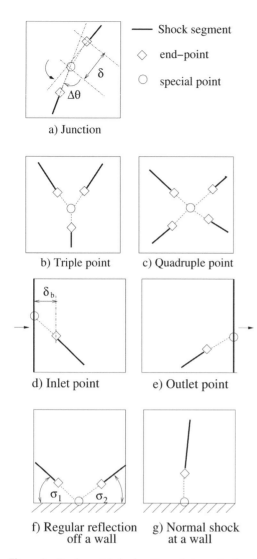

— Shock segment

◇ end–point

○ special point

a) Junction

b) Triple point   c) Quadruple point

d) Inlet point   e) Outlet point

f) Regular reflection   g) Normal shock
off a wall   at a wall

Figure 4. Fundamental shock patterns that may be found around the end-points of a shock segment.

• the angles σ formed by either of the two segments and the tangent at the wall in the nearest boundary point are "approximately" equal.

A junction, sketched in Fig. 4a), is recognised as such if:

• the end-point is close "enough" to the end-point of a unique, different segment;
• the difference ($\Delta\theta$) between the slopes of the two segments at the two neighbouring end-points is "sufficiently" small.

At present, the various sets of rules are implemented using the classical Boolean logic (true-false logic) and the vagueness of the aforementioned rules (see the use of adverbs such as: "enough", "approximately", "sufficiently") is handled by using large threshold values in the actual coding. It is our opinion that the use of the

Fuzzy logic may provide a more suitable programming approach to implement these sets of rules and, therefore, the future research activity will be addressed in this direction.

Once the specific shock topology has been identified, the actual position of the junction or interaction points is computed using different rules. For instance: the position of a junction point is obtained by averaging the original position of the two end-points involved in the junction, whereas in the case of the regular reflection shown in Fig. 3, the interaction point is obtained by the intersection between the straight line of the incident shock and the wall.

Once the coordinates of the interaction or junction point have been computed, the coordinates of the two (or more, in the case of an interaction) end-points involved are correspondingly reset. This, in turn, implies (small) changes in the coefficients of the linear or quadratic equations defining the various shock segments involved.

3 APPLICATIONS

During its development, the present shock detection and pattern recognition algorithm has been applied to different kinds of flows with shocks to assess its performances both in terms of computational costs and quality of the results.

The computational cost of the present technique depends on the number of shock points provided as input, but, in any case, it is really negligible compared to the computational cost required to obtain the numerical solution from which the shock points have been extracted.

The capabilities and the quality of the "global" shock detection technique described in the previous section will now be demonstrated using two different test-cases: the first one features the interaction between straight, oblique shocks and the second one is characterised by a curved bow shock.

3.1 *Detection of a shock-shock interaction*

We consider first the interaction between shocks of the same family, which is schematically shown in Fig. 5a. A uniform, supersonic ($M_\infty = 2$) stream is first deflected through an angle $\theta_1 = 10°$ by an oblique shock and further deflected through an angle $\theta_2 = 20°$ by a second oblique shock. These two oblique shocks interact in a point P, giving rise to a stronger shock, also shown in Fig. 5a. A weak wave (either a weak shock or an expansion fan) and a slip line also form in point P, but the detection of these weaker waves has been deliberately ignored. This flow configuration has been computed by means of a shock-capturing solver on an unstructured, triangular mesh. Figure 5b shows the computed density field and the triangular grid in the neighbourhood of the interaction point. The shock points identified by means of the local shock-detection technique described in (Ma et al. 1996) are

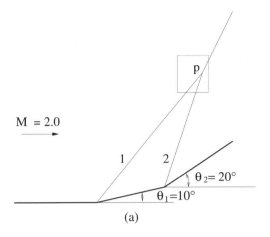

M = 2.0

$\theta_2 = 20°$

$\theta_1 = 10°$

(a)

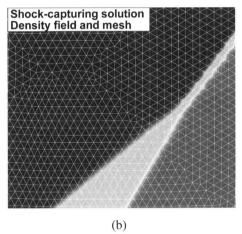

Shock-capturing solution
Density field and mesh

(b)

Figure 5. Interaction of two shocks of the same family: sketch of the flow configuration and detail of the shock-capturing solution.

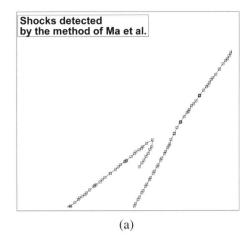

Shocks detected
by the method of Ma et al.

(a)

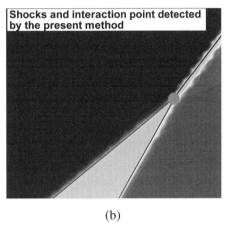

Shocks and interaction point detected
by the present method

(b)

Figure 6. Shock detection results: local (a) vs. global (b) technique.

shown in Figure 6a. The presence of a gap and a spurious branch in the neighbourhood of the interaction point are clearly visible. Moreover, the local technique identifies the second incident shock and the strongest shock as a single shock, rather than two different ones, as they should be. The entire set of shock points detected by the local technique, which is displayed in Fig. 6a, has been successively processed by the present shock-detection technique and the result of the global analysis is shown in Fig. 6b. The three shocks are correctly identified as three distinct entities without any gaps, spurious oscillations or artificial branches. Moreover, the point where the three shocks interact is recognised and correctly located.

3.2 Detection of a curved shock

In order to test the capability of the present technique to correctly deal with curved shocks, we have considered the hypersonic $M_\infty = 10$ flow past the fore-body of a circular cylinder. Due to the bluntness of the body, a bow shock forms which stretches to the far-field boundary of the computational domain. Figure 7a shows the shock-capturing solution computed on an unstructured, triangular mesh. This solution is characterised by the presence of wiggles that are particularly evident within the subsonic shock layer. These anomalies, which are not uncommon when shock-capturing schemes are used to simulate strong shocks (Lee and Zhong 1999), severely challenge the capability of any shock-detection technique to correctly identify the shock front. Indeed, the cloud of shock points obtained using the local detection technique (Ma et al. 1996) is characterised by the presence of spurious branches in the stagnation region and small oscillations along the shock front, see Fig. 7b. These drawbacks are completely removed using the new shock-detection technique. Figure 7c shows the results obtained by applying the global technique to the cloud of shock points shown in Fig. 7b. The bow shock is described using a sequence of curved finite segments that are mutually joined in the junction points. The detected shock line, shown in Fig. 7d, matches very well the captured shock and it does not feature neither oscillations nor artificial branches.

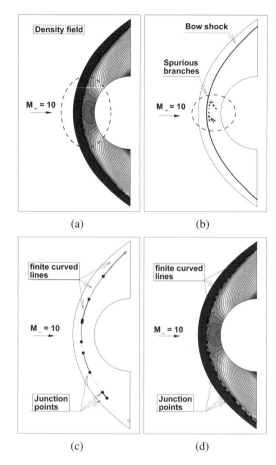

(a)   (b)

(c)   (d)

Figure 7. Hypersonic flow past the fore-body of a circular cylinder.

4 CONCLUSIONS

A new technique that is capable of recognising the shock wave pattern in two-dimensional shock-capturing solutions is proposed. This newly developed technique employs a global analysis, based mainly on the Hough transform, to process the cloud of shock points that have been preliminary identified using a local detection criterion. The shocks are represented as a sequence of mutually connected straight or curved segments. Thanks to the use of a global analysis, it is possible to individually identify different interacting shocks as well as the points where they interact. The results provided by this original detection technique are free from the numerical artefacts, such as spurious branches, gaps and oscillations that plague the classical local detection techniques.

REFERENCES

Bonfiglioli, A., M. Grottadaurea, R. Paciorri, and F. Sabetta (2010). An unstructured, threedimensional, shock-fitting solver for hypersonic flows. *40th AIAA Fluid Dynamics Conference.*

Bonfiglioli, A., R. Paciorri, and A. Di Mascio (2012). The role of mesh generation, adaptation and refinement on the computation of flows featuring strong shocks. *Modelling and Simulation in Engineering.*

Duda, R. and P. Hart (1972). Use of the hough transformation to detect lines and curves in pictures. *Communications of the ACM 15*(1), 11– 15.

Hough, P. (1962, December). Method and means for recognizing complex patterns. *U.S. Patent.*

Kanamori, M. and K. Suzuki (2011). Shock wave detection in two-dimensional flow based on the theory of characteristics from cfd data. *Journal of Computational Physics 230*(8), 3085–3092.

Kao, K.-H., M.-S. Liou, and C.-Y. Chow (1994). Grid adaptation using chimera composite overlapping meshes. *AIAA journal 32*(5), 942–949.

Lee, T. and X. Zhong (1999). Spurious numerical oscillations in simulation of supersonic flows using shock-capturing schemes. *AIAA Journal 37*(2–3), 313–319.

Lovely, D. and R. Haimes (1999). Shock detection from computational fluid dynamics results. *AIAA Computational Fluid Dynamics Conference.*

Ma, K.-L., J. Van Rosendale, and W. Vermeer (1996). 3d shock wave visualization on unstructured grids. *Proceedings of the Symposium on Volume Visualization*, 87–94.

Paciorri, R. and A. Bonfiglioli (2009). A shockfitting technique for 2d unstructured grids. *Computers and Fluids 38*(3), 715–726.

Paciorri, R. and A. Bonfiglioli (2011). Shock interaction computations on unstructured, two-dimensional grids using a shockfitting technique. *Journal of Computational Physics 230*(8), 3155–3177.

Pagendarm, H. G. and B. Seitz (1993). An algorithm for detection and visualization of discontinuities in scientific data fields applied to flow data with shock waves. In *Scientific visualization – advanced software techniques*, pp. 161–177. Ellis Hortwood.

Computational Modelling of Objects Represented in Images – Di Giamberardino et al. (eds)
© 2012 Taylor & Francis Group, London, ISBN 978-0-415-62134-2

Fast orientation invariant template matching using centre-of-gravity information

Andreas Maier & Andreas Uhl
University of Salzburg, Department of Computer Sciences, Salzburg, Austria

ABSTRACT: Template matching is an old and well known technique which is still very prominent in all sorts of object identification tasks. By rotating the template to a (possibly high) number of different angles the template matching process can be made rotation invariant at the price of a high performance penalty. In this paper it is shown that using centre-of-gravity information for an initial rotation estimation can compensate for the additional computational costs of the rotation invariance for certain categories of templates. Preconditions for employing this technique are discussed and results from an image segmentation application are presented.

1 INTRODUCTION

Template matching is used in many applications for object identification and image segmentation because of its simplicity and stability. It frequently occurs as a normalised cross correlation between the template and patches of the image to process(Brunelli 2009). The normalised correlation coefficient $m(x,y)$ of the template T and the patch located at coordinates (x,y) in image I is

$$m(x,y) = \frac{\sum_{j=0,k=0}^{W-1,H-1}(I'(x+j,y+k) * T'(j,k))}{\sigma_I(x,y) * \sigma_T} \quad (1)$$

where

$$I'(x+j,y+k) = I(x+j,y+k) - \overline{I(x,y)} \quad (2)$$

$$\overline{I(x,y)} = \frac{1}{W*H} \sum_{j=0,k=0}^{W-1,H-1} I(x+j,y+k) \quad (3)$$

$$T'(j,k) = T(j,k) - \overline{T} \quad (4)$$

$$\overline{T} = \frac{1}{W*H} \sum_{j=0,k=0}^{W-1,H-1} T(j,k) \quad (5)$$

$$\sigma_I(x,y) = \sqrt{\sum_{j=0,k=0}^{W-1,H-1} I'(x+j,y+k)^2} \quad (6)$$

$$\sigma_T = \sqrt{\sum_{j=0,k=0}^{W-1,H-1} T'(j,k)^2}. \quad (7)$$

$I(x,y)$ and $T(x,y)$ denote the brightness intensity of the pixels at coordinates (x,y) of the image and the template, respectively, while the width W and the hight H are the dimensions of the template and the image patch. Rotation invariance of this technique can be achieved by rotating the template to different angles, matching with all rotated templates and selecting the highest correlation coefficient.

Other methods are known for rotation invariant template matching like the log-polar transform(Brunelli 2009) which reduces rotation to translation, the Hu moment(Hu 1962) which are rotation invariant combinations of centralised image moments or the SURF algorithm(Bay et al. 2008) which relies on the detection of dominant features, amongst others. None of these methods is a universal substitution for the classical cross correlation algorithm because of individual shortcomings like numerical instabilities, high computation effort or hard to determine parameters and thresholds. Therefore, cross correlation based template matching is widely used but achieving rotation invariance through exhaustive matching with rotated templates leads to a high number of matching operations which is computational demanding and thus time consuming. For the special case of non-normalized cross correlation the Discrete Fourier Transform may be used to speed up the calculation, thereby the achieved gain depends on the size of image and template(Brunelli 2009).

In this paper we present a fast orientation invariant template matching algorithm which can be used to speed up normalized and non-normalized cross correlation. The speed-up factor is independent of the image or template size and can be estimated from the number or rotated template. Section 2 gives a detailed description of the proposed algorithm. Section 3 shows the results of the evaluation of the performance and accuracy of the algorithm compared to the classical cross correlation approach and Section 4 concludes the work.

2 FAST ORIENTATION INVARIANT TEMPLATE MATCHING ALGORITHMUS

A fast alternative to the exhaustive search in orientation invariant template matching is based on the idea of intelligently selecting a good template from the set or rotated templates. The following algorithm achieves this by incorporating image centre-of-gravity information into the template matching process to reduce the processing effort and at selecting a reduced set of candidate patches for the exact match value calculation. The coordinates of the image centre-of-gravity (x_{cg}, y_{cg}) are calculated from the first spacial image moments (Teh and Chin 1988) M and are

$$x_{cg} = \frac{M_{1,0}}{M_{0,0}} \quad and \quad y_{cg} = \frac{M_{0,1}}{M_{0,0}} \quad (8)$$

$$M_{q,r} = \sum_{x=0,y=0}^{W-1,H-1} I(x,y) * x^q * y^r \quad (9)$$

We are only interested in the orientation of the centre-of-gravity relative to the image centre point, so the calculation can be reduced for images with odd horizontal and vertical sizes to

$$x_{rel} = \sum_{x=-(W-1)/2, y=-(H-1)/2}^{(W-1)/2, (H-1)/2} I(x,y) * x \quad (10)$$

$$y_{rel} = \sum_{x=-(W-1)/2, y=-(H-1)/2}^{(W-1)/2, (H-1)/2} I(x,y) * y \quad (11)$$

$$\alpha = arctan \frac{y_{rel}}{x_{rel}} \quad (12)$$

where x_{rel} and y_{rel} denote the orientation of the "gravity vector" from the image centre point to the centre-of-gravity of the image and α denotes its angle.

In the case of rotation invariant template matching the highest correlation coefficient is expected to be achieved if the gravity vector of the template and the image patch do have the same orientation. This is true if some constraints are considered. First of all, templates have to have a distinct gravity vector. Otherwise, if the centre-of-gravity is at the image centre, the gravity vector vanishes and its discriminative power is lost. Unfortunately, a fraction of useful templates do have a zero gravity vector. This is especially the case for (N-fold) rotation symmetric templates, where a N-fold rotation symmetric template is one that repeats itself for every $2 * \pi/N$ angle rotation around its image centre. Templates with a zero gravity vector may have subsections for which a distinct gravity vector exists. Limiting the centre-of-gravity calculation

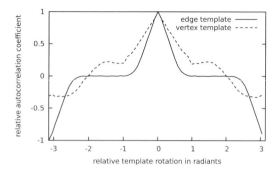

Figure 1. Autocorrelation trend of an example edge and vertex template with its rotated variants.

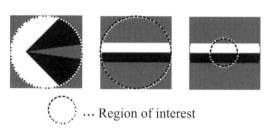

$\vdots$ $\cdots$ Region of interest

Figure 2. Example vertex and edge templates with different regions of interest for centre-of-gravity computation.

and comparison to such a subsection re-establishes the discriminative nature but alters at the same time the view on the template. Thus the practicability of this approach depends on the specific template or application and is most promising for templates which are not rotation symmetric.

The second constraint is that the cross correlation of the template with its rotated variants (rotation autocorrelation) must have a unique maximum at the point where the rotation is zero and must be monotonically decreasing with increasing rotation angles. See Figure 1 for the progression of the cross correlation of example edge and vertex templates with their rotated variants.

A third constraint is that the template should not have areas that are ignored in the matching process (i.e. the coefficients of these pixels are all zero). While ignored areas are not considered in the matching operation they are well part of the centre-of-gravity calculation and can therefore dominate the result and give an incorrect gravity vector. Figure 2b shows an example of an edge template that has large areas that are ignored in the matching process (grey shaded areas). A possible work around for such situations is to narrow the region of interest for the centre-of-gravity calculation (see Figure 2c) so that the computed values become meaningful again.

The accelerated rotation invariant template matching algorithm relies on the significance of the gravity vector and operates in two steps. In the first step every patch of the image is processed in the following way. For each patch the gravity vector is calculated and used to choose a template with a similar gravity vector from the set of rotated templates. The correlation coefficient of the selected template with the image patch is calculated and stored as the match value for this patch. Although this match value is not necessarily the highest match value achievable due to a possibly imperfect template selection, it is a good enough approximation to take its value as the selection criteria for choosing candidate patches for the next step. The amount of candidate patches depends on the threshold level chosen for the match value but is usually small for typical applications and reasonable threshold levels. See Section 3 for examples and results.

In the second step a candidate patch can either be correlated with all rotated templates to get the best match value or a progressive approach can be used where the templates with angles adjacent to the used template are selected iteratively as long as the correlation coefficient increases. The progressive method is possible due to the specific characteristic of the template autocorrelation function and can save a reasonable amount of matching operations. In any case, because of the sharply reduced number of candidate patches the run-time of the second step is negligible compared to the run-time of the first step or, even more, to the run-time of exhaustive template matching.

For exhaustive template matching the run-time $r_{exhaust}$ depends on the number of different angles n_t the template is rotated to to achieve rotation invariant template matching and can be calculated as

$$r_{exhaust} = n_r * r_{one} \qquad (13)$$

where r_{one} denotes the run-time of calculating the correlation coefficients of all image patches for one template. For the accelerated algorithm this run-time r_{acc} is reduced to

$$r_{acc} = r_{one} + r_{cg} + r_{step2} \qquad (14)$$

$$r_{step2} = k * (n_r - 1) * r_{one} \qquad (15)$$

where r_{cg} denotes the run-time of calculating the centre-of-gravity for all patches and r_{step2} denotes the run-time of step two of the accelerated algorithm in which k is the ratio of candidate patches from the overall number of patches. A simple implementation for the centre-of-gravity calculation could be a correlation of two specific modelled templates (one for vertical and one for horizontal direction) with an image patch and then selecting the correspondent rotation angle from a lookup table. Although faster algorithms for the centre-of-gravity calculation exist, this estimate already delimits the run-time to $r_{cg} \leq 2 * r_{one}$. For an average configuration the run-time for step two of the accelerated algorithm is well bound to be $r_{step2} \leq r_{one}$ because the number of candidate patches is usually small and thus $k * (n_r - 1) \leq 1$. This reduces the upper bound of the run-time estimate for the accelerated algorithm to

$$r_{acc} \leq 4 * r_{one} \qquad (16)$$

Figure 3. Example image from Vickers indentation images database.

and the achieved speed-up in processing time is therefore at least

$$speedup \geq \frac{r_{exhaust}}{r_{acc}} = \frac{n_r}{4}. \qquad (17)$$

which qualifies the proposed algorithm for templates that are rotated to more than four different angles.

3 EXPERIMENTS

The fast rotation invariant template matching algorithm has been evaluated for its accuracy and speed compared to exhaustive cross correlation template matching. A database(Gadermayr et al. 2011) of 150 grey scale Vickers hardness indentation images of 1280×1024 pixels have been processed with an edge template and an vertex template to identify edges and vertexes within the image. See Figure 2 for an illustration of the templates. The diameter of the region of interest circle for the centre-of-gravity calculation has been set to half the template size for the edge template and to full template size for the vertex template. The templates used are of 41×41 pixels each and are rotated to 64 different equally distributed angles, so that each rotation has an angle of $\alpha = 2 * \pi/64 = 0.098 \hat{=} 5.625°$.

An example image (see Figure 3) has been selected from the database to illustrate the progress of the speed-up and error values for different match value threshold levels for the algorithm. Figure 4 shows that for a threshold level of 80% of the maximum match value the number of candidate patches is already below 2% of all patches and it further decreases to below 0.1% (which corresponds to less than thousand patches) at a threshold level of 90%. This gives a speed-up factor of 24-26 for both templates.

Figure 5 presents average error values for the example image. The relative average match value error

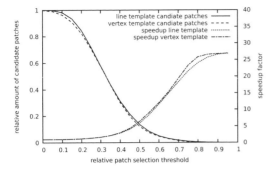

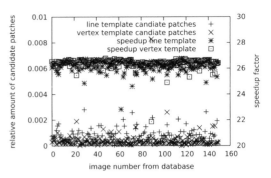

Figure 4. Amount of candidate patches and achieved speed-up related to match value threshold level for the example image.

Figure 6. Amount of candidate patches and achieved speed-up for images from the database (90% match value threshold).

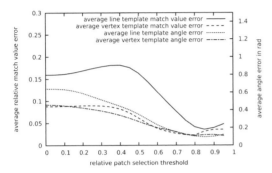

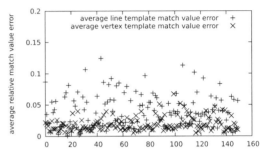

Figure 5. Match value error and angle error related to match value threshold level for the example image.

Figure 7. Average match value error for images from the database (90% match value threshold).

err_{match} of all candidate patches (*cand*) is calculated as

$$err_{match} = \frac{1}{cand} * \sum_{cand} \frac{|m_{cg} - m_{exhaust}|}{m_{max}} \qquad (18)$$

where m_{cg} and $m_{exhaust}$ denote the match value for a patch from the algorithm and from exhaustive cross correlation while m_{max} denotes the highest match value of all patches. The average angel error is the average of the sum of individual angel errors between the template selected from the algorithm and the best template for a patch. One can see that for a match value threshold level of 80% or above the average match value error is around or below 5%. The average angle error is at the same time around 0.1 rad which corresponds to circa one rotation step at 64 template rotations.

For the whole set of images from the database the results are similar. Figure 6 show the number of candidate patches and the achieved speed-up values while Figure 7 and Figure 8 show the average match value and angle errors. This time the match value threshold level has been fixed to 90%. For all figures the ranges have been spread for better visibility.

The distribution in Figure 6 shows a very small number of candidate patches and thus a high speed-up factor of well above 26 for nearly all images of the database. The few exceptions still have a high speed-up factor of at least 22. The average relative match value

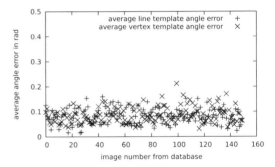

Figure 8. Average angle error for images from the database (90% match value threshold).

error is nearly always below 5% for the vertex template while the line template shows a more spread distribution but which is still below 10% for all except three images (see Figure 7). The average angle error does not show differences between the two templates (see Figure 8). While most results group around 0.1 rad angle error some go up to 0.2 rad which corresponds to more or less two steps of template rotation. The results from this example application show that in accordance with the predictions the number of candidate patches for the second step of the algorithm is small and that the candidate patches are valid because of the low error figures. Therefore, the proposed algorithm qualifies for accelerating cross correlation template matching.

4 CONCLUSION

An algorithm for fast rotation invariant template matching has been presented which is based on the centre-of-gravity calculation of image patches. The run-time gain has been calculated and the break-even point for applying this algorithms has been given. Experiments point out that templates that do follow the constraints of the algorithm achieve accurate results and that the method shows the predicted run-time benefit.

ACKNOWLEDGMENTS

This work has been partially supported by the Austrian Federal Ministry for Transport, Innovation and Technology (FFG Bridge 2 project no. 822682).

REFERENCES

Bay, H., A. Ess, T. Tuytelaars, and L. Van Gool (2008, June). Speeded-up robust features (surf). *Comput. Vis. Image Underst. 110*, 346–359.

Brunelli, R. (2009). *Template matching techniques in computer vision: theory and practice*. Wiley.

Gadermayr, M., A. Maier, and A. Uhl (2011, June). Algorithms for microindentation measurement in automated Vickers hardness testing. In J.-C. Pinoli, J. Debayle, Y. Gavet, F. Cruy, and C. Lambert (Eds.), *Tenth International Conference on Quality Control for Artificial Vision (QCAV'11)*, Number 8000 in Proceedings of SPIE, St. Etienne, France, pp. 80000M–1 – 80000M–10. SPIE.

Hu, M.-K. (1962, February). Visual pattern recognition by moment invariants. *Information Theory, IRE Transactions on 8*(2), 179 –187.

Teh, C.-H. and R. Chin (1988, July). On image analysis by the methods of moments. *Pattern Analysis and Machine Intelligence, IEEE Transactions on 10*(4), 496 –513.

Computational Modelling of Objects Represented in Images – Di Giamberardino et al. (eds)
© 2012 Taylor & Francis Group, London, ISBN 978-0-415-62134-2

Evaluation of the Menzies method potential for automatic dermoscopic image analysis

André R.S. Marcal & Teresa Mendonca
Faculdade de Ciências, Universidade do Porto, Portugal

Cátia S.P. Silva
Faculdade de Engenharia, Universidade do Porto, Portugal

Marta A. Pereira & Jorge Rozeira
Hospital Pedro Hispano, Matosinhos, Portugal

ABSTRACT: There is a considerable interest in the development of automatic image analysis systems for dermoscopic images. The standard approach usually consists of three stages: (i) image segmentation, (ii) feature extraction and selection, and (iii) lesion classification. This paper evaluates the potential of an alternative approach, based on the Menzies method. It consists on the identification of the presence of 1 or more of 6 possible color classes, indicating that the lesion should be considered a potential melanoma. The Jeffries-Matusita (JM) and Transformed Divergence (TD) separability measures were used for an experimental evaluation with 28 dermoscopic images. In the most challenging case tested, with training identified in multiple images, 8 out of 15 class pairs were found to be well separable, or $13 + 2$ out of 21 considering the skin as an additional class.

1 INTRODUCTION

Dermoscopy (dermatoscopy or skin surface microscopy) is a non-invasive diagnostic technique for the in vivo observation of pigmented skin lesions used in dermatology. The automatic analysis of dermoscopic images is of great interest, both to provide quantitative information about a lesion, for example to support the follow up procedures by the clinician, and also as a potential stand-alone early warning tool. In the last few years a number of screening tests have been proposed for dermoscopic images, suitable for health care personal with minimum clinical training. These screening tests, such as the ABCD Rule algorithm (Marghoob & Braun 2004) and the 7-point check-list algorithm (Argenziano et al. 2011) are used to reduce the number of cases that need to be evaluated by a dermatologist.

Various attempts have been made to implement computer based systems inspired on the human based screening tests (Mendonça et al. 2007). The standard approach of these automatic dermoscopic image analysis systems usually consists of three stages: (i) image segmentation, (ii) feature extraction and selection, and (iii) lesion classification. The segmentation procedure alone is a challenging task as the various algorithms produce different segmentation results (Silveira et al. 2009). Furthermore, even when the lesion is segmented, there is still considerable work to be done in order to establish a link between the human based

criteria and the features extracted by automatic computer based algorithms.

An alternative approach for the evaluation of dermoscopic images is the Menzies method, where presence of 1 or more out of 6 color classes indicates that the lesion should be considered a potential melanoma (Menzies 2001). The implementation of an image processing system based on the Menzies method does not require the segmentation stage, nor the subsequent extraction of geometric features, avoiding some of the potential errors in those stages. The purpose of this work is to investigate the applicability of the Menzies method for the development of an image processing tool for dermoscopic image analysis.

2 METHODS

2.1 *Separability measurements*

Dermoscopic images are usually RGB color images, having 3 independent channels (the Red, Green and Blue color channels). An image pixel, or observation, is characterized by a vector $x = [x_1, x_2, ..., x_n]^T$, where x_i is the grey level intensity of image channel i. Thus, for a typical dermoscopic RGB color image (n=3), there are only 3 features. Each color class corresponds to a specific location and volume in the 3D feature space. The regions of the feature space associated with each color class would ideally be non-overlapping and well spread. However, this is rarely the case in real life problems, as the data from the various channels tend to be highly correlated, which results in the color class regions being mostly located along the feature space diagonal. It is therefore important to quantify the separability between the various classes in order to evaluate the potential applicability of the Menzies method for an automatic image processing system.

Divergence is a separability measure for a pair of probability distributions that has its basis in their degree of overlap. It is defined in terms of the likelihood ratio $L_{ij}(x) = p(x|\omega_i)/p(x|\omega_j)$, where $p(x|\omega_i)$ and $p(x|\omega_j)$ are the values of the probability distributions of the spectral classes i and j, for a feature vector x (Richards & Jia, 2006). For perfectly separable classes, $L_{ij}(x) = 0$ or $L_{ij}(x) = \infty$ for all values of x.

It is worth choosing the logarithm of the likelihood ratio (L') by means of which the divergence of the pair of class distribution is defined as $d_{ij} = E\{L'_{ij}|\omega_i\} + E\{L'_{ij}|\omega_j\}$, where E is the expectation operator (Richards & Jia 2006). For spectral classes modeled by multi-dimensional normal distributions, with means μ_i, and μ_j, and covariances Σ_i and Σ_j, it can be shown that:

$$d_{i,j} = \frac{1}{2} tr\left\{(\Sigma_i - \Sigma_j)(\Sigma_i^{-1} - \Sigma_j^{-1})\right\} + tr\left\{(\Sigma_i^{-1} - \Sigma_j^{-1})(m_i - m_j)(m_i - m_j)^T\right\} (1)$$

where $tr\{\}$ is the trace of the subject matrix (Richards & Jia 2006).

The Bhattacharyya distance $D_B(i,j)$ between two classes can be calculated from the variance and mean of each class (Reyes-Aldasoro & Bhalerao 2006), in the following way:

$$D_B(i,j) = \frac{1}{8}(\mu_i - \mu_j)^T\left(\frac{\Sigma_i^{-1} - \Sigma_j^{-1}}{2}\right)^{-1}(\mu_i - \mu_j) + \frac{1}{2}\ln\left\{\frac{\left|\frac{\Sigma_i + \Sigma_j}{2}\right|}{2|\Sigma_i|^{1/2}|\Sigma_j|^{1/2}}\right\}. (2)$$

The Jeffries-Matusita (JM) distance J_{ij} is used to assess how well two classes are separated. The JM distance between a pair of probability distributions (spectral classes) is defined as:

$$J_{i,j} = \int_x \left\{\sqrt{p(x|\omega i)} - \sqrt{p(x|\omega j)}\right\}^2 dx \qquad (3)$$

which is seen to be a measure of the average distance between density functions (Richards & Jia 2006). For normally distributed classes it becomes:

$$J_{i,j} = 2(1 - e^{-D_B}) \qquad (4)$$

where D_B is the Bhattacharyya distance (Equation 2). The presence of the exponential factor in Equation 4 gives an exponentially decreasing weight to increasing separations between spectral classes. If plotted as a function of distance between class means it shows a saturating behavior, not unlike that expected for the probability of correct classification. It is asymptotic to 2, so that for a JM distance of 2, the signatures can be totally separable (with 100% accuracy). Generally, the classes can be considered separable for JM

values above 1.8 or, preferably, above 1.9. JM values bellow 1.8 indicate the possible confusion between the class pair in the classification process (Richards & Jia 2006).

An useful modification of divergence (Eq.1) becomes apparent by noting the algebraic similarity of d_{ij} to the parameter D_B, used in the JM distance. Since both involve terms which are functions of the covariance alone, and terms which appear as normalised distances between class means, it is possible to make use of a heuristic Transformed Divergence (TD) measure of the form (Swain & Davis 1978):

$$d_{i,j}^T = 2(1 - e^{-\frac{d_{i,j}}{8}}) \qquad (5)$$

Because of its exponential character, TD will have saturating behavior with increasing class separation, as does the JM distance, and yet it is computationally more economical. It is worth noting that for the computational implementation of d_{ij} and $D_B(i,j)$, and consequently for the JM and TD distances, the data must be Gaussian.

2.2 Image calibration

Dermoscopic images are acquired under controlled illumination and should thus produce accurate color images. However, this is not always true in practice and a dermatologist is expected to make the same diagnosis on a particular case even if the image is acquired in different imaging conditions (Iyatomi et al. 2011). Furthermore, the patient skin (background) varies, which results in a different color perception of the lesion by the clinician. Although color calibration in dermoscopy is recognized as important, it is still an open issue (Iyatomi et al. 2011).

Two alternative calibration procedures were considered here. The assumption is that the background (skin) should be normalized to a reference. Sample areas for skin ($\mathbf{m}^{skin}$) and lesion ($\mathbf{m}^{lesion}$) are considered, with $\mathbf{m}$ being a 3-dimensional vector with the average RGB values for the sample. Reference vectors ($\mathbf{r}^{skin}$ and $\mathbf{r}^{lesion}$) are also considered.

Calibration method I uses only the skin intensity. The original RGB values are multiplied by a factor α, computed as $\alpha = ||\mathbf{r}^{skin}||/||\mathbf{m}^{skin}||$.

Calibration method II applies different multiplication factors (β_j) for the 3 color components of the RGB image. The coefficients (β_j) are computed as $\beta_j = (\mathbf{r}_j^{skin} - \mathbf{r}_j^{lesion})/(\mathbf{m}_j^{skin} - \mathbf{m}_j^{lesion})$, with j = 1,2,3.

3 RESULTS

3.1 Experimental data description

An experimental procedure was devised to test the applicability of the Menzies method – presence of at least 1 of 6 possible colors – for an automatic image processing system. A total of 28 dermoscopic images

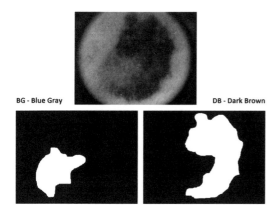

BG - Blue Gray   **DB - Dark Brown**

Figure 1. Example of a dermoscopic image (IMD19) and the medical evaluation (segmentation) for color classes BG – Blue Gray (left) and DB – Dark Brown (right).

Table 1. Menzies colors present in the test image dataset.

Colors	Wh	Re	LB	DB	BG	Bk	Images (IMDxx)
			x				05, 08, 10, 14,
			x				23, 24, 25
1				x			02, 07, 15, 16, 21
						x	18
			x	x			01, 03, 04, 06
			x	x			09, 11, 22
2				x	x		12, 19, 26
		x	x				27
			x			x	13
	x			x	x		17
3	x				x	x	20
			x	x	x		28

(IMD01 to IMD28) were selected and evaluated by a dermatologist, identifying the presence and location of the six Menzies color classes – Black (Bk), Blue Gray (BG), Dark Brown (DB), Light Brown (LB), Red (Rd) and White (Wh). The identification was done using database and segmentation tools developed in the Automatic computer-based Diagnosis system for Dermoscopy Image (ADDI) project (Amorim et al. 2011, Ferreira et al. 2011). As an example, Figure 1 shows the original dermoscopic image (IMD19) and the segmented regions associated with the color classes Dark Brown (DB) and Blue Gray (BG), both identified as present in this image. Most test images have a single color (13) or two colors (12), with only 3 out of 28 images having three Menzies colors present in a single lesion. A summary of the color classes present in the various test images is available in Table 1.

3.2 *Single image training*

Initially, training areas for each class were identified in a single image. The images used were IMD17 (for white), IMD27 (red), IMD05 (light brown), IMD01

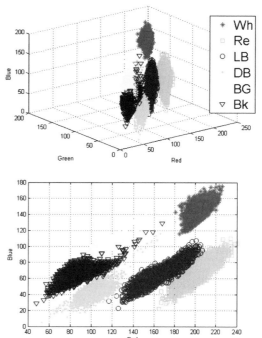

Figure 2. Scatterplots for the single image training test: on the RGB feature space (top) and a projection on the 2D Red, Blue plane (bottom).

(dark brown), IMD12 (blue gray) and IMD20 (black). The 3D scatterplot and the projection in the 2D plane (Red/Blue) are presented in Figure 2. As expected, the feature space is scarcely used and there is a certain amount of overlap between some classes.

The Kolmogorov-Smirnov (KS) test was used to determine if the samples can be considered as originating from a normal (Gaussian) distribution (Kolmogorov 1993). The KS test provides the mean (μ_i) and standard deviation (σ_i) parameters for each sample i, as well as the probability of the data to have a normal distribution (p-value). The color classes follow Gaussian distributions, as demonstrated by the p-values results for significance level α = 5% presented in Table 2 (all above 0.90). Table 2 also presents the minimum, maximum, average (μ) and standard deviation (σ) values for each color class.

Considering that the color classes have a normal distribution, it is thus possible to use the JM (Eq.4) and TD (Eq.5) distances to evaluate the class separabilities. Table 3 presents the two triangular matrices (TD and JM) together, the JM distance as a top right triangular matrix and the TD as a bottom left triangular matrix. For most class pairs, both TD and JM have either a perfect separability (2.00), or very high values (above 1.9). The only major difficulty is to distinguish between color classes BG and Bk, which have low values of both TD and JM distances. According to this test, considering the RGB color features and training data used, these two classes are not distinguishable, but all other class pairs are well separable.

Table 2. Characteristics summary for the six classes in the RGB color channels for the single image training case.

	Class	Min	Max	μ	σ	p
	Wh	181	226	205	6.1	0.99
R	Re	165	240	205	10.4	0.95
E	LB	117	207	164	15.3	0.91
D	DB	53	181	102	9.4	0.96
	BG	59	161	101	9.5	0.95
	Bk	48	168	83	8.4	0.96
	Wh	133	181	156	5.8	0.99
G	Re	70	152	114	11.1	0.94
R	LB	63	145	104	14.4	0.90
E	DB	14	142	59.2	7.5	0.98
E	BG	51	147	86.0	10.6	0.95
N	Bk	38	133	68.1	6.0	0.98
	Wh	116	171	148	8.5	0.96
B	Re	22	110	70.4	10.7	0.93
L	LB	23	107	64.5	12.5	0.94
U	DB	5	131	42.9	7.9	0.95
E	BG	44	153	84.4	12.1	0.92
	Bk	29	129	64.8	6.4	0.99

Table 3. Jeffries-Matusita - JM (top right triangular matrix) and Transformed Divergence - TD (bottom left) distance values, for the single image training test.

	Wh	Re	LB	DB	BG	Bk
Wh	–	2.00	2.00	2.00	2.00	2.00
Re	2.00	–	2.00	2.00	2.00	2.00
LB	2.00	2.00	–	1.98	2.00	2.00
DB	2.00	2.00	2.00	–	1.99	1.99
BG	2.00	2.00	2.00	2.00	–	1.02
Bk	2.00	2.00	2.00	1.99	1.01	–

3.3 Multiple image training

A second test, more challenging, was performed, by identifying the training areas for each color class using all 28 test images available. As it can be seen in Table 1, the number of images used to train each color class varies considerably, from only a few (1 for Re and 2 for Wh) up to 17 (for both LB and DB). The large number of images used to train these classes (DB and LB) greatly increases the dispersion of their signatures in the RGB feature space, thus increasing the overlap between the various color classes.

The p-values from the KS test are generally lower than those for single image training. For class Bk the p-values are above 0.74 for the 3 color channels (R, G, B), while for classes Wh and Re the p-values are all above 0.92. For classes DB and BG the p-values vary between 0.54 and 0.75 for the three color channels. The worst class is class LB, with p-values of 0.56, 0.57 and 0.50 for the R, G and B color channels. The data was nevertheless still treated as having normal distribution, and the TD and JM distances were thus computed.

Table 4 presents the two triangular matrices for the multiple-image training test (the JM distance at top

Table 4. Jeffries-Matusita - JM (top right triangular matrix) and Transformed Divergence - TD (bottom left) distance values, for the multiple-image training test.

	Wh	Re	LB	DB	BG	Bk
Wh	–	2.00	1.74	1.97	1.69	1.99
Re	2.00	–	1.48	1.97	2.00	2.00
LB	1.98	1.23	–	1.47	1.87	1.99
DB	1.95	2.00	1.21	–	1.03	1.22
BG	1.70	2.00	1.94	0.75	–	0.55
Bk	1.99	2.00	2.00	0.94	0.10	–

Table 5. Jeffries-Matusita – JM (top right triangular matrix) and Transformed Divergence – TD (bottom left) distance values, for the multiple-image training test with calibration I.

	Wh	Re	LB	DB	BG	Bk
Wh	–	2.00	1.86	1.96	1.73	2.00
Re	2.00	–	1.78	1.97	2.00	2.00
LB	1.83	1.77	–	1.48	1.84	2.00
DB	2.00	2.00	1.49	–	0.78	0.96
BG	1.98	2.00	1.78	0.23	–	1.11
Bk	2.00	2.00	(*)	(*)	(*)	–

(*) Not possible to calculate

right and the TD at bottom left). These results indicate that 8 out of 15 classes pairs are clearly distinguishable (for JM) or 9 out of 15 (based on TD values). There is a class pair (Wh-LB) with contradictory indications from JM and TD distances (1.74 and 1.98, respectively). The remaining 6 class pairs have both low JM and TD values (below 1.8) and cannot thus be considered separable. These class pairs are: Wh-LB, Wh-BG, Re-LB, LB-DB, DB-BG, DB-Bk, BG-Bk. More details about the multiple image training evaluation, with a slightly different dataset, are available in Silva et al. (2012).

3.4 Image calibration

Another test was carried out, both for the single and the multiple image training cases, using calibrated versions of the test images. The JM and TD values, computed with the calibrated images, are presented in Table 5 for calibration method I and in Table 6 for method II. There are a few cases where the indications from JM and TD distances are contradictory, and also some cases where it was not possible to calculate TD due to the fact that the data matrix was not symmetrical.

A 3D scatterplot of the multiple-image training data with calibration method I is presented in Figure 3, as well as projections in 2D planes for the Red/Blue and Green/Blue color channels. As in the single-image training case (Figure 2) the feature space is scarcely used and there is a considerable overlap between some classes.

Table 6. Jeffries-Matusita – JM (top right triangular matrix) and Transformed Divergence – TD (bottom left) distance values, for the multiple-image training test with calibration II.

	Wh	Re	LB	DB	BG	Bk
Wh	–	2.00	1.92	1.93	1.65	2.00
Re	2.00	–	2.00	2.00	2.00	2.00
LB	(*)	2.00	–	1.31	1.47	1.79
DB	1.33	2.00	(*)	–	0.83	1.63
BG	1.98	2.00	0.76	(*)	–	1.57
Bk	2.00	2.00	2.00	(*)	(*)	–

(*) Not possible to calculate

Table 7. Summary of JM results for the various scenarios tested (number of class pairs).

Test scenario	JM≥1.9	1.8≤JM<1.9	JM<1.8
Single-image training	14	–	1 [a]
Multiple-image training	7	1	7 [b]
M-image train., Calib. I	7	2	6 [c]
M-image train., Calib. II	8	–	7 [d]

(a) BG-Bk.
(b) Wh-LB, Wh-BG, Re-LB, LB-DB, DB-BG, DB-Bk, BG-Bk.
(c) Wh-BG, Re-LB, LB-DB, DB-BG, DB-Bk, BG-Bk.
(d) Wh-BG, LB-DB, LB-BG, LB-Bk, DB-BG, DB-Bk, BG-Bk.

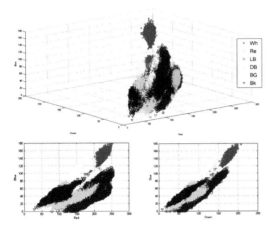

Figure 3. Scatterplots for the multiple image training test with calibration I. RGB feature space (top) and projection on 2D: Green, Blue (bottom left) and Red, Blue (bottom right) planes.

The results presented in Tables 3–6 is summarized in Table 7, where the number of class pairs (15 in total) are grouped in well separable (JM≥1.9), moderately separable (1.8≤JM<1.9) and non-separable or undistinguishable (JM<1.8). As expected, class separability evaluation benefited slightly by performing image calibration to a reference. The best results were obtained by the calibration method I. The results based on TD values are slightly better, but as it was not always possible to compute TD, the evaluation is mostly based on JM values.

3.5 Including the skin

A final test was carried out, including an additional class – the skin. The skin can be seen as the background of the lesion in a dermoscopic image, but also as a class on its own. This additional class increases the difficulty in the classification / discrimination problem. The JM and TD values were computed for the 7 class case (21 pairs). A summary of the JM results for the various scenarios tested is presented in Table 8. The table shows the number of class pairs considered

Table 8. Summary of JM results for the various scenarios tested (number of class pairs).

Test scenario	JM≥1.9	1.8≤JM<1.9	JM<1.8
Single-image training	19	–	2 [a]
Multiple-image training	11	1	9
M-image train., Calib. I	13	2	6 [b]
M-image train., Calib. II	10	–	11

(a) Bk-BG, Wh-skin.
(b) Wh-BG, Re-LB, LB-DB, DB-BG, DB-Bk, BG-Bk.

well separable in the feature space (JM≥1.9), moderately separated (1.8≤JM<1.9) and undistinguishable (JM<1.8). For the single-image training case, most class pairs are distinguishable, even with the additional skin class. However, when the training uses all 28 test images, there are 6 class pairs that cannot be separable using the RGB feature space. When the test images are calibrated to a reference (using method I), the number of distinguishable class pairs increases to 13+2 (out of 21).

4 CONCLUSIONS

An alternative approach to the standard computer based analysis of dermoscopic images is offered by the Menzies diagnosis method. The method consists on the identification of the presence of colors out of 6 possible color classes in a dermoscopic image. This method has some advantages comparing to the standard approaches, as it does not require the lesion segmentation. However, the identification of color classes in dermoscopic images is a subjective task, which poses great challenges for an automatic implementation. The purpose of this work was to evaluate the potential discrimination between the various Menzies color classes in dermoscopic RGB images.

The tests performed using the JM and TD separability metrics indicate that it is possible to identify and distinguish most color classes. In the most challenging case, where the class signature is obtained by multiple images (between 1 and 17 for each class, on a total of 28 test images), 7+1 of the 15 class pairs are distinguishable without image calibration. The result is slightly improved by using a simple calibration procedure, which normalizes the background intensity. Considering the skin as an additional class, a total of 13 class pairs are well separable and another 2 pairs are moderately separable (using calibration method I). Only 6 class pairs remain as non-separable in both cases (with or without the skin class).

Although most class pairs are distinguishable in RGB color images, the Menzies method cannot yet be used as there are considerable confusion between some color classes. Further work is thus required in order to establish more suitable features and calibration procedures. An alternative can be the use of color models that have the color component detached from the intensity (e.g. HSI model). However, the preliminary tests carried out using the HSI color model only produced slightly improved results.

Another line of work should focus on the subjectivity in the human perception of color. In order to evaluate this aspect, a more extensive collection of images and clinical evaluation is needed, preferably with more than one medical evaluation for each dermoscopic image.

ACKNOWLEDGEMENTS

This work was done as part of the ADDI project (http://www2.fc.up.pt/addi/). The authors would like to thank Bárbara Amorim for her support in the acquisition of training data.

REFERENCES

Amorim, B., Marçal, A., Mendonça, T., Marques, J.S. & Rozeira, J. 2011. Database implementation for clinical and computer assisted diagnosis of dermoscopic images. In João Manuel R.S. Tavares & R.M. Natal Jorge (eds), *VipIMAGE; Proc. intern. Symp, Olhão, Portugal, 12–14 October 2011.* Taylor & Francis.

Argenziano, G., Catricala, C., Ardigo, A., Buccini, P., De Simone, P., Eibenschutz, L., Ferrari, A., Mariani, G., Silip, V., Sperduti, I. & Zalaudek, I. 2011. Seven-point checklist of dermoscopy revisited. *British Journal of Dermatology* 164(4): 785–790.

Ferreira, P.M., Mendonça, T., Rocha, P. & Rozeira, J. 2011. A new interface for manual segmentation of dermoscopic images. In João Manuel R.S. Tavares & R.M. Natal Jorge (eds), *VipIMAGE; Proc. intern. Symp, Olhão, Portugal, 12–14 October 2011.* Taylor & Francis.

Iyatomi, H., Celebi, M.E., Schaefer, G. & Tanaka, M. 2011. Automated color calibration method for dermoscopy images. *Computerized Medical Imaging and Graphics* 35 89–98.

Kolmogorov, A.N. 1993. Sulla determinazione empirica di una legge di distribuzione. *Giornale dell'Istituto Italiano degli Attuari* 4: 83–91.

Marghoob, A. & Braun, R. 2004. *Atlas of dermoscopy.* Taylor and Francis.

Mendonça, T., Marçal, A.R.S., Vieira, A., Lacerda, L., Caridade, C. & Rozeira, J. 2007. Automatic analysis of dermoscopy images – A review. In João Manuel R.S. Tavares & R.M. Natal Jorge (eds), *Computational Modelling of Objects Represented in Images*: 281–287. Taylor & Francis.

Menzies, S.W. 2001. A method for the diagnosis of primary cutaneous melanoma using surface microscopy. *Dermatologic Clinics* 19(2): 299–305.

Reyes-Aldasoro, C.C. & Bhalerao, A. 2006. The Bhattacharyya space for feature selection and its application to texture segmentation. *Pattern Recognition* 39(5) 812–826.

Richards, J.A. & Jia, X. 2006. *Remote Sensing Digital Image Analysis: An Introduction*, 4th edition. Springer.

Silva, C.S.P., Marcal, A.R.S., Pereira, M.A., Mendonça, T. & Rozeira, J, 2012. Separability analysis of color classes on dermoscopic images. *Lecture Notes on Computer Science* (in press).

Silveira, M., Nascimento, J.C., Marques, J.S., Marçal, A.R.S., Mendonça, T., Yamauchi, S., Maeda, J. & Rozeira, J, 2009. Comparison of Segmentation Methods for Melanoma Diagnosis in Dermoscopy Images. *IEEE Journal of Selected Topics in Signal Processing* 3(1): 35–45.

Swain, P.H. & Davis, S.M., 1978. Remote Sensing: The Quantitative Approach. New York: McGraw-Hill.

Computational Modelling of Objects Represented in Images – Di Giamberardino et al. (eds)
© 2012 Taylor & Francis Group, London, ISBN 978-0-415-62134-2

Application of UTHSCSA-Image Tool™ to estimate concrete texture

J.S. Camacho
Structural Masonry Study and Research Center (NEPAE/UNESP), Brazil

A.S. Felipe & M.C.F. Albuquerque
São Paulo State University (UNESP), Ilha Solteira – Civil Engineering Department, Brazil

ABSTRACT: The texture of concrete blocks is very important and is often the decisive factor when choosing a product, particularly if the building specifications does not dispense with the high resistance of the blocks, but has the purpose of reducing costs with finishing, therefore preferring exposed blocks with a closer texture. Furthermore, a closer texture, especially for exteriors, may be the vital factor of the building's pathology. However, there is so far no standard to quantify the texture of a structural block. This article proposes to apply the freely available UTHSCSA-Image Tool™ program developed by the University of Texas Health Science Center at San Antonio to evaluate the texture of masonry blocks. One aspect that should never be overlooked when studying masonry blocks is compressive strength. Therefore, this work also gets the compressive strength of the blocks with and without the addition of lime. The addition of small quantities of lime proved beneficial for both texture and compressive strength. However, increasing the amount of lime proved to be feasible only to improve texture.

1 INTRODUCTION

The texture of concrete blocks is very important and often decisive in the choice of a product, particularly if the building requires high-strength blocks allied to low-cost finish, in which case exposed blocks with a closer texture are preferable. However, in order to increase their strength, such blocks require larger quantities of cement than concrete with a rougher and more porous texture.

When buildings require high strength blocks to meet design specifications, especially multi-floor buildings, it is common to use more open-textured blocks containing a higher proportion of coarse aggregates. These blocks are stronger and have a smaller surface area, making it easier for the cement to envelop the aggregates and thus improving their hydration performance. This type of concrete also consumes less water, but may give rise to building pathology problems.

Sham et al. [1] state that cracks in concrete may impair its durability by allowing for the migration of aggressive external agents. In view of the fact that, like cracks, pores will also allow the migration of such agents, a closer texture may be vital to the building's pathology, especially for exteriors.

Sham et al. [1] analyzed the cracks in concrete using flash thermography. This technique consists in exciting the surface of the concrete with a flash and recording the instantaneous reflections of visible light and infrared radiation with an infrared camera.

According to Breul et al. [2], it is vital to check the homogeneity of a concrete to ensure that its desired characteristics are achieved. The author measured the homogeneity of concrete using an endoscopic image processing technique.

Chew [3] analyzed the pathology of concrete in wet areas of the building using visual inspection of photographs as one of his methods of evaluation. In his work, defects were attributed to construction deficiencies (43%), material (37%), design (11%) and maintenance (9%). The author found that material is crucial to the durability of concrete.

Buendía et al. [4] studied the lithological influence of aggregates on the chemical reaction between alkali and carbon. They created a classification for carbon aggregates with a pathological risk for the alkali-aggregate reaction, based on the mechanical behavior of expansion and contraction. The materials were classified as nonreactive, potentially reactive and probably reactive. In this study, the authors clearly show the importance of material in the pathology of concrete.

This study proposes to apply the freely available UTHSCSA-Image Tool™ program developed by the University of Texas Health Science Center at San Antonio to evaluate the texture of masonry concrete blocks.

Limestone improves workability [7]. Therefore, some of the formulations in this study contained hydrated lime added to the concrete in small proportions of aggregates to correct the lack of cohesion found in mixes with low cement content. The influence of lime on the texture and strength was evaluated.

At present, there is no standard to quantify the texture of structural blocks. However, there are studies that quantify it, although they are not standardized. Thus, any quantification of texture should be evaluated carefully by the author.

The UTHSCSA-Image Tool™ program scans images in shades of grey and presents standard deviation values for the frequency histogram to quantify the surface texture of a specific object. This program is widely used in the medical sector.

When using the program for concrete blocks, it is assumed that blocks with a closer texture present lighter shades of grey. In contrast, more open-textured blocks (rougher) present darker shades of grey due to their small surface cavities.

Oliveira [8] recorded images by adapting a common scanner to avoid any loss of light. After recording the image, the author executed the same program used in this study to quantify the standard deviation of shades of grey. The program itself transforms the original colors of the image into shades of grey.

A similar procedure was carried out in this study, but it was adapted to use a 3-megapixel digital camera instead of a scanner. The resulting photographs were analyzed by the program.

In any study of masonry blocks, an aspect that must be kept in mind is compressive strength. The compressive strength of masonry is crucial in design and safety evaluations, since the masonry structure is subjected mainly to compression [5]. Therefore, this work also shows results of compressive strength.

2 RESEARCH MATERIALS AND METHODOLOGY

2.1 Materials used in the research

2.1.1 Cement and lime

The concrete blocks were made with CPII-E 40 SR, a sulfate-resistant composite Portland cement.

Small percentages of hydrated lime with an apparent specific mass of 1.0 g/cm^3 were used in some compositions.

2.1.2 Aggregates

Natural sand (fine and coarse), industrial sand (stone powder) and class 00 crushed stone (gravel) were used in the experiment. Table 1 describes the tests performed to characterize the aggregates.

2.1.3 Proportion of aggregates

The ideal proportion of aggregates was identified [9] and concrete blocks were prepared with and without lime in order to improve cohesion between the aggregates and to evaluate texture and strength. Seven different mixes were prepared (see Table 2) according to the ideal proportion of aggregates, varying only the aggregate/cement (m) ratio and the addition or not of lime (Table 2).

The abbreviations in Table 2 indicate the following: f.s. – fine sand, c.s. – coarse sand, gr. – gravel, powder – stone powder, and C – cement.

Table 1. Characterization of natural and industrial sands, and crushed stone size 00.

Sand classification		Fine	Industrial	Coarse	Crushed stone 00
Max$^{(1)}$ diameter		1.18	4.75	4.75	6.30
Fineness modulus		1.37	2.82	2.54	5.71
Specific	s.s.s.	2.595	2.811	2.607	2.879
mass	dry	2.600	2.930	2.620	2.824
(g/cm^3)	app$^{(2)}$	2.582	2.749	2.599	2.988
Unit	loose	1.466	1.555	1.510	1.410
mass$^{(3)}$	l$^{(4)}$	1.134	1.377	1.262	–
Ab$^{(5)}$ (%)		0.06	2.25	0.30	1.95
PVL$^{(6)}$		4.15	16.94	4.15	0.95
O. m.$^{(7)}$		clear	–	clear	–

$^{(1)}$ Maximum diameter (mm)
$^{(2)}$ apparent
$^{(3)}$ g/cm^3
$^{(4)}$ loose 4% humidity
$^{(5)}$ Absortion
$^{(6)}$ PVL is the percentage of powdery material
$^{(7)}$ Organic matter

Table 2. Mixes for concrete blocks.

Unit dosages

Mix	Dry materials						
	f.s.	c.s.	gr.	po.	C	% Lime*	m
M1	2.13	1.07	5.85	3.95	1	1%	13
M2	2.13	1.07	5.86	3.94	1	0	13
M3	1.47	0.74	4.05	2.73	1	0	9
M4	1.47	0.74	4.05	2.73	1	3%	9
M5	2.13	1.07	5.85	3.95	1	3%	13
M6	0.98	0.49	2.70	1.82	1	3%	6
M7	2.78	1.40	7.65	5.17	1	0	17

*Percentage of hydrated lime in the aggregate mix.

2.1.4 Concrete block volume and area

The blocks were hollow, as illustrated in Figure 1. They were manufactured by vibrocompression at COPEL in Araçatuba, SP, Brazil.

Table 3 lists the average values for volume and net area of the concrete blocks, as well as the net area/gross area ratio. These values were obtained from a total of thirty test specimens, which were produced in seven mixes (Table 2). The duration of vibrocompression in three of these mixes was varied, and three samples of each mix and each vibrocompression time were tested. The net volume of the block was determined based on Pascal's Principle.

2.2 Experimental methodology

2.2.1 Determining the ideal proportion of aggregates

To increase the block's strength at the same cement content and compaction energy, a proportion of fine and coarse aggregates was determined, called the ideal

Figure 1. Concrete block used in the study.

Table 3. Volume of concrete block.

	Average	Standard deviation	Coefficient of variation (%)
Net vol. of block (cm^3)	7303	36	0.50
Net area (cm^2)	384.38	2	0.50
Net A. / Gross A.	0.52	0.003	0.50

proportion, in accordance with the Brazilian NBR NM 45 standard [9].

The ideal proportion is determined by the highest unit mass of the mix composed of two types of aggregates.

2.2.2 Determining compressive strength

The blocks were tested following the Brazilian NBR-7211 standard [10]. The load was applied in the direction perpendicular to that of the cylindrical hole, with the gross area of the block's face in contact with the 19×39 cm^2 load plate. Net tension was calculated based on an average net area of 384.38 cm^2.

2.2.3 Quantifying the surface texture of a concrete block

The surface texture of the blocks was quantified using the UTHSCSA-Image Tool^{TM}, free image processing

and analysis program, developed by the University of Texas Health Science Center at San Antonio.

The UTHSCSA-Image has the property of scanning images in grayscale and present values of standard deviation of the histogram of frequencies capable of quantify the surface texture of a given object.

The premise is that blocks with texture close, have lighter grayscale, in counterpart, more open-textured (rough) have darker grayscale due to the small surface craters.

The program can read and write over several common file formats including BMP, GIF and JPEG. It transforms the original color of image for grayscale. Image analysis functions include gray scale measurements (histogram with statistics – quantify the standard deviations of grayscale).

A picture was taken with a 3-megapixel digital camera, which was evaluated by the program. The use of a camera was one of the particularities of this study.

Another particularity was to paint the block's surface with white latex paint, since darker or lighter stains have been found in different batches of blocks due to the cement or to handling, although they do not cause changes in texture. It was believed that without a painted surface, the program might have interpreted cement stains as small cavities. The procedure was therefore carried out as follows:

– Using a small roller, the block's surface was painted with two layers of latex paint without the addition of water and without using pressure, to cover the larger surface cavities.
– After the paint dried, a representative area on the block where the texture was predominant was marked out with a pen.
– Using a square, a 9×14 cm rectangle was drawn, which would be the area evaluated to determine the standard deviation of the shades of grey. Figure 2 illustrates this procedure.
– Lastly, a photo was taken from a distance of approximately 30 cm perpendicularly to the block, without using the zoom. It is advisable to take pictures always at the same time of day without using a flash. The time chosen was at night in order to have standardized illumination, which was provided by two 40W fluorescent lamps positioned 3 m from the block. The procedure is illustrated in Figure 3.

3 RESULTS AND DISCUSSION

3.1 "Ideal" proportion of aggregates

The ideal proportion for the mix was 30.38% of industrial sand, 16.36% of fine sand, 8.26% of coarse sand and 45% of class 00 crushed stone.

3.2 Compressive strength

Table 4 lists the results of the compression tests. In this table, t.vb indicates the length of time vibro-compression was applied to the blocks during their

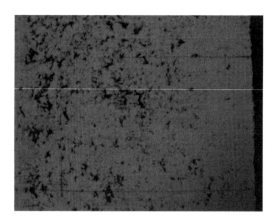

Figure 2. Delimitation of the area in which the surface texture of the block is to be quantified.

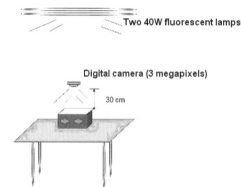

Figure 3. Diagram of the procedure to quantify the surface texture of concrete blocks.

Table 4. Compressive strength of the blocks.

mix	t. vb (s)	Net σ (MPa)	Standard deviation–STDV
M1	7	22.45	2.44
M2	7	20.02	1.41
M3	7	24.69	2.17
M4.1	7	21.51	1.50
M4.2	10	24.58	1.56
M5.1	7	20.17	1.25
M5.2	10	22.28	1.47
M6.1	7	24.26	2.28
M6.2	10	25.03	1.89
M7	7	18.07	1.55

manufacture, and net σ is the stress rupture, taking into account the net area of the block (Table 3).

In Table 4, the number after the letter M (for mix) refers to the number of the mix, as specified in Table 2, and the next number was used for some mixes whose vibrocompression timing varied.

Table 5. Quantification of surface textures of concrete blocks.

Mixes	STDV
M1	20.07
M2	21.9
M3	21.5
M4	19.50
M5	20.06
M6	16.75
M7	24.71

The values of compressive strength for aggregate/cement ratio equal 13 (M1, M2, M5.1 – table 4), i.e., the mix contains the same dosage, the only variable being the addition of 1 and 3% of lime (Table 2), indicates that compressive strength increased in response to a 1% increase in lime. Therefore, the addition of 1% of lime resulted in increased compressive strength, while a slightly larger amount of lime reduced it. These results are consistent with the results of cylindrical test specimens of the same materials reported by Felipe [11].

Also in Table 4 we can observe the effect of reduced compressive strength with other concrete dosages, mixes M3 and M4.1, without lime and with 3% of lime, respectively. This mixes have an aggregate/cement ratio of 9.

The addition of 1% of lime in this study produced results consistent with those reported by Turgut [6], who found increased strength in concrete blocks with limestone powder waste. However, Turgut did not observe a decrease in strength in response to higher limestone content, possibly because the limestone powder waste was added to the concrete together with glass powder. He did not report any variation in compressive strength with the addition of limestone powder alone. He achieved the highest average strength of 33.7 MPa when applying the load over an area of 10,500 mm^2. These results are higher than those found in this study, but Turgut's blocks were solid while the blocks in our study were hollow.

3.3 Surface textures of concrete blocks

Table 5 shows the results of the standard deviation for the shades of grey on the surface of blocks in their respective dosages.

The photographs do not seem to show any significant visual difference in the concrete texture, as can be seen in Figure 4, which shows a photograph of mixed concrete for M2 and M4.

The most significant difference was between mixes 6 and 7, which displayed the densest and the most porous textures, respectively. This finding is confirmed in the histograms of Figures 5 and 6. Therefore, it can be stated that a visual inspection such as that described by Chew [3] would not suffice to determine the different textures of these concretes.

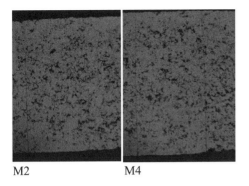

M2   M4

Figure 4.   Photograph of mixed concrete for M2 and M4.

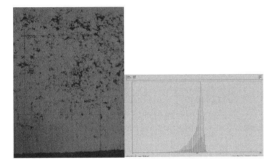

Figure 5.   Photograph of the block representing mix M6 and its histogram produced with the UTHSCSA-Image Tool program.

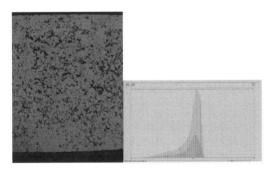

Figure 6.   Photograph of the block representing mix M7 and its histogram produced with the UTHSCSA-Image Tool program.

The histograms indicate that the mix with the lowest standard deviation, and therefore the densest texture, was M6, which contained the highest proportion of fine aggregates among the mixes that were prepared.

The mix with the best texture was M6 and the worst was M7, which were the mixes containing, respectively, the highest and the lowest proportion of fine aggregates (m = 6 and m = 17).

For the same quantity of cement (m = 13 – Table 2), the addition of hydrated lime in mix M1 resulted in a 6% improvement in texture compared to M2, which contained no added lime.

The method employed in this study is simpler than that described by Breul et al. [2] (endoscopic) and does not require sophisticated equipment. The UTHSCSA ImageTool is a free image processing and analysis program and the camera is a much less sophisticated piece of equipment. The same holds true for the method used by Sham et al. [1] (flash thermography).

Compared with Chew's evaluation [3], i.e., visual inspection of photographs, it can be stated that the method employed in this study provides more accurate results.

The advantage of the method used in this work compared to that employed by Oliveira, who used the same program but captured images with a scanner, is that it requires no complex setup to take pictures without loss of light, which is the case when recording images with a scanner.

4   CONCLUSIONS

The addition of 1% of lime in proportion to the aggregates increased the compressive strength and improved the texture of the blocks. However, the addition of 3% lime decreased the strength while improving the texture. The increase in strength with 1% of lime can be explained by the easier compaction, allowing for better grain packing. However, a larger proportion of lime may have had a stronger influence on the chemical reactions of the cement, impairing the strength of the concrete. It is therefore recommended to use small quantities of lime in concrete with low cement content, to obtain the required cohesion and higher strength of these formulations. The improvement in the block's texture in response to the addition of lime was likely due to the larger amount of fine aggregates with the incorporation of lime, indicating that this fine aggregate fills voids in the concrete and improves its texture.

The method used in this study to evaluate texture proved to be simpler and cheaper than other methods reported in the literature. The image processing program used here is available free of charge and the camera is a very accessible piece of equipment.

Text is set in two columns of 9 cm (3.54") width each with 7 mm (0.28") spacing between the columns

ACKNOWLEDGMENTS

The authors thank FUNDUNESP, FAPESP, DEC-UNESP and PPGEC.

REFERENCES

Sham F. C., Chen N., Long L. Surface crack detection by flash thermography on concrete surface. Flash Thermography, 50 (2008) 240–243.

Breul P., Geoffray J. M., Haddani Y. On-site concrete segregation estimation using image analysis. Journal of Advanced Concrete Technology, 6 (2008) 1–10.

Chew M. Y. L. Defect analysis in wet areas of buildings. Construction and Building Materials, 19 (2005) 165–173.

Buendía A. M. L., Climent V., Verdú P. Lithological influence of aggregate in the alkali-carbonate reaction. Cement and Concrete Research, 36 (2006) 1490–2500.

Mohamad G., Lourenço P. B., Roman H. R. Mechanics of hollow concrete block masonry prisms under compression: Review and prospects. Cement & Concrete Composites 29 (2007) 181–192.

Turgut P. Properties of masonry blocks produced with waste limestone sawdust and glass powder. Construction and Building Materials 22 (2008) 1422–1427.

Hendry E. A. W. Masonry walls: materials and construction. Construction and Building Materials, 15 (2001) 323–330.

OLIVEIRA, A. L. Contribuição para a dosagem e produção de peças de concreto para pavimentação. (Contribution to the dosage and manufacture of concrete parts for pavements) PhD Thesis, Federal University of Santa Catarina, Florianópolis, Brazil, 2004.

ASSOCIAÇÃO BRASILEIRA DE NORMAS TÉCNICAS. Agregados – Determinação da massa unitária e do volume de vazios. NBR NM 45, Rio de Janeiro (2006) 8p.

ASSOCIAÇÃO BRASILEIRA DE NORMAS TÉCNICAS. Agregados para concreto – Especificação. NBR-7211, Rio de Janeiro (2009) 9p.

Felipe, A. S. Contribution to optimize concrete dosages used in the production of structural blocks. (in press), MSc dissertation, São Paulo State University, Ilha Solteira, Brazil; 2010.

Computational Modelling of Objects Represented in Images – Di Giamberardino et al. (eds)
© *2012 Taylor & Francis Group, London, ISBN 978-0-415-62134-2*

Texture image segmentation with smooth gradients and local information

Isabel N. Figueiredo & Juan C. Moreno
CMUC, Department of Mathematics, University of Coimbra, Portugal

V.B. Surya Prasath
Department of Computer Science, University of Missouri-Columbia, US

ABSTRACT: We study a region based segmentation scheme for images with textures based on the gradient information weighted by local image intensity histograms. It relies on the Chan and Vese model without any edge-detectors, and incorporates a new input term, defined by the product of the smoothed gradient and a local histogram of pixel intensity measure of the input image. Segmentation of images with texture objects is performed by effectively differentiating regions displaying different textural information via the local histogram features. A fast numerical scheme based on the dual formulation of the energy minimization is considered. The performance of the proposed scheme is tested on different natural images which contain texture objects.

1 INTRODUCTION

Texture image segmentation plays an important role in computer vision based tasks such as object recognition and classification. The main idea in image segmentation is to obtain a partition of the input image into a finite number of disjoint homogeneous objects. Nevertheless, it is not always easy to get a good partition in regions with texture information, since it is difficult to differentiate the boundaries between two textures, due to the lack of sharp differences between them. Therefore, it is necessary to introduce extra features to the usual segmentation algorithm. Different approaches deal with the extraction of homogeneous features from textures. Many of these texture descriptors go from statistical models to filtering methods, and to geometric approaches (Cremers et al. 2007).

Active contour based models are widely used for image segmentation (Caselles et al. 1997) and recently the Chan and Vese model (Chan and Vese 2001) has generated a lot of interest. However, such contour evolution based schemes are mainly guided by the boundary information without considering the local region properties and hence can not handle texture images. Moreover, the minimization of energy can often be trapped in the local solution leading to wrong segmentation results.

We propose in this paper an extension of the Chan and Vese model to manage images with texture. Recall that (Chan and Vese 2001) is based on the piecewise-constant Mumford and Shah model (Mumford and Shah 1989) and the level set method (Osher and Fedkiw 2003), and it is different from the classical geodesic active contour model, whose main idea is based on curve evolution of an edge-detector function

(Caselles et al. 1997). To capture texture objects we proposed to utilize a hybrid input channel for the energy minimization. It makes use of the smoothed image gradient norm, as well as local intensity information in neighbors around image-pixels via local histograms. These two images features are combined by means of the product which makes possible to distinguish different texture information from images. Thus, smoothed gradient image is weighted by the local histograms computed around each pixel of the input image. In this sense, we take advantage of a weighted gradient magnitude information given by objects with texture information together with a variational segmentation model. Further, our method is based on the global minimization version of the Chan and Vese model (Chan et al. 2006) and is implemented using the dual minimization approach of Chambolle (Chambolle 2004). The proposed model is tested and validated in the segmentation of texture images.

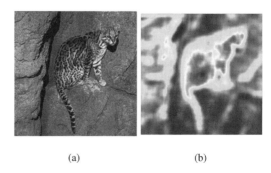

(a)   (b)

Figure 1. (a) Image with texture information. (b) the input feature I_{new} computed from (a) with $\sigma = 10$, see (2).

The rest of this paper is organized as follows. Section 2 describes the proposed model and some extensions. In Section 3, we discuss the numerical approximation using the dual minimization technique, and present some experimental results. Finally, Section 4 concludes the paper.

2 DESCRIPTION OF THE MODEL

Let Ω be an open and bounded set in $\mathbb{R}^2$, and $I: \overline{\Omega} \longrightarrow [0, L]$ a scalar function, representing the observed gray-scale input image. The convex formulation of the Chan and Vese model (Chan and Vese 2001), amounts to solving the minimization problem (Chan et al. 2006):

$$\min_{0 \leq u \leq 1} \left\{ \int_\Omega |\nabla u|\, dx + \lambda \int_\Omega r(x, c_{in}, c_{out})u\, dx, \right\} \tag{1}$$

where $u: \Omega \to [0, 1]$ is a function of bounded variation ($u \in BV(\Omega)$) and $r(x, c_{in}, c_{out}) = (I - c_{in})^2 - (I - c_{out})^2$ which is known as the fitting term. The fixed vector $c = (c_{in}, c_{out})$ represents the averages of I, respectively inside and outside the segmentation curve. Finally, the parameter λ is a positive scalar weighting the fitting term $r(x, c_{in}, c_{out})$. It can be shown that for any λ, there exists a minimizer u of (1), which corresponds to a global minimum, see (Chan et al. 2006) for more details.

2.1 *Proposed weighted gradient based fitting terms*

Given an image with texture information, it is possible to extract texture features using the gradient vector and its norm. A new input image channel is defined by smoothing the norm of its gradient. In short, given a scalar image $I: \Omega \subset \mathbb{R}^2 \to \mathbb{R}$, the new input channel is

$$I_{new} = (|\nabla I|)_\sigma, \tag{2}$$

where the lower sub-script σ in (2) means a smoothed version of the corresponding indexed function. The smoothing is done by a Gaussian function $G_\sigma(x) = (2\pi\sigma)^{-1}e^{-\frac{|x|^2}{2\sigma}}$. This new input channel I_{new} embodies different texture information (Figueiredo et al. 2012), nevertheless, it is not able to differentiate the similar scale textures for natural images, as it can be seen in Figure 1(b). The input *Leopard* image, Figure 1(a) is an example of a natural image with a texture object present. For such an image with comparable texture information both inside and outside the object of interest, its corresponding feature channel I_{new} in (2), can not distinguish well the object itself, and therefore can lead to bad segmentations.

To overcome this difficulty and to get a good input channel for the global minimization scheme (1), we propose to introduce a feature data descriptor of the image using local histograms and the smoothed gradient image (2). This new feature descriptor of the image does not depend on the regions, and moreover the model does not involve histogram differentiation, see (Ni et al. 2009) for more details. For a given gray-scale image $I: \overline{\Omega} \longrightarrow [0, L]$, let $\mathcal{N}_{x,r}$ be the local region centered at a pixel x with radius r. The local histogram of a pixel $x \in \Omega$ and its corresponding cumulative distribution function are then defined by

$$P_x(y) = \frac{|\{z \in \mathcal{N}_{x,r} \cap \Omega \mid I(z) = y\}|}{\mathcal{N}_{x,r} \cap \Omega}$$

$$F_x(y) = \frac{|\{z \in \mathcal{N}_{x,r} \cap \Omega \mid I(z) \leq y\}|}{\mathcal{N}_{x,r} \cap \Omega} \tag{3}$$

for $0 \leq y \leq L$, respectively. Local histograms makes no simplifying assumptions about the statistics of the image intensity values around a pixel image, which allow us to define the following measurable function $\Psi_1: \Omega \to \mathbb{R}$, such that for each $x \in \Omega$

$$\Psi_1(x) = \int_0^L F_x(y)\, dy, \tag{4}$$

which in turn, allowing us to get a weight of how much nonhomogeneous intensity is present in a local region $\mathcal{N}_{x,r}$ of a given pixel x. Then, a new input feature channel is defined by the product

$$J_1 = \Psi_1 I_{new}. \tag{5}$$

The new weighted input channel uses the smoothed gradient to differentiate the object boundaries with the help of local histograms computed at each pixel neighborhood which helps in the identification of the edges between textured object and the background. Implementing the convex formulation of the Chan and Vese model (1), using the input channel J_1 in the definition of the fitting term $r_1(x, c_{in}, c_{out}) = (J_1 - c_{in})^2 - (J_1 - c_{out})^2$, provides good results.

Usually, we encounter two different types of texture images in natural images. The first one is related with a small scale textures corresponding to a lot of oscillations inside the main object present, and the second one consist of different flat regions given by large scale textures within the object of interest in the given image. Figure 2 shows two real images with different texture information. In Figure 2(a) we have a *Cheetah* and a *Tiger* image, together with their corresponding pixel-valued image (scaled gray-scale values) Figure 2(b), and the computed local cumulative distribution function of the interior region of the main object to be segmented, Figure 2(c), with $r = 10$. As we can see, the local cumulative distribution function for the *Cheetah* has one convex shape over all the domain, whereas the *Tiger* image has local cumulative distribution function that oscillates between convex and concave shape. This is due to the fact that, *Tiger* has a lot of different flat regions inside the body which can be considered as large scale textures. In order to apply our measurable function (4), we use the input channel given by

116

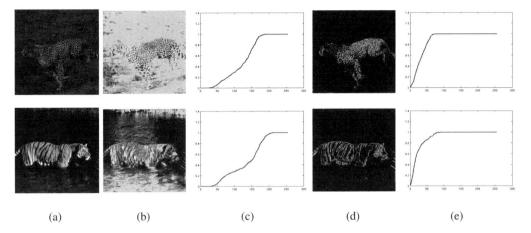

(a)   (b)   (c)   (d)   (e)

Figure 2. Local Cumulative Distribution Functions of two images with different texture information. (a) Input image. (b) The scaled pixel-values of the input images showing the nature of texture present. (c) Corresponding local cumulative distribution function inside the region of interest using $r = 10$. (d) The scaled pixel-values of the gradient image. (e) Corresponding local cumulative distribution function inside the region of interest in (d) using $r = 10$.

the image norm gradient, Figure 2(d), which recovers the oscillations patterns inside the *Tiger* image and therefore its local cumulative distribution function, Figure 2(d), shows an almost single convex shape over all the domain.

2.2 Extensions

- The local histograms computation can also be computed using the gradient image $|\nabla I|$ instead of the input image I in (3). That is, we consider the measurable function $\Psi_2 \colon \Omega \to \mathbb{R}$, such that for each $x \in \Omega$, $\Psi_2(x) = \int_0^L F_x(y)\,dy$, where $F_x(y) = |\{z \in \mathcal{N}_{x,r} \cap \Omega \,|\, |\nabla I(z)| \leq y\}|/(\mathcal{N}_{x,r} \cap \Omega)$. Then, we consider the input feature

$$J_2 = \Psi_2 I_{\text{new}}, \qquad (6)$$

together with the fitting term $r_2(x, d_{in}, d_{out}) = (J_2 - d_{in})^2 - (J_2 - d_{out})^2$.
- By combining (5) and (6), an extension of the fitting term $r(x, c_{in}, c_{out})$ in (1) can be defined as

$$r(x, c_{in}, c_{out}, d_{in}, d_{out}) = \beta r_1 + (1 - \beta) r_2 \qquad (7)$$

where $0 \leq \beta \leq 1$ balances fitting terms given by $r_1 = r_1(x, c_{in}, c_{out}) = [(J_1 - c_{in})^2 - (J_1 - c_{out})^2]$ and $r_2 = r_2(x, d_{in}, d_{out}) = [(J_2 - d_{in})^2 - (J_2 - d_{out})^2]$.

3 NUMERICAL EXPERIMENTS

3.1 Fast dual minimization

A fast numerical scheme is implemented for solving the minimization problem (1). In effect the dual formulation of the total variation (Chambolle 2004), with the global minimization model (1) is considered:

$$\min_{u,v} \left\{ \int_\Omega |\nabla u|\,dx + \frac{1}{2\theta}\|u - v\|^2_{L^2(\Omega)} \right.$$

$$\left. + \int_\Omega \lambda r(x, c_{in}, c_{out}) v + \alpha \nu(v) dx \right\}, \qquad (8)$$

where $r(x, c_{in}, c_{out}) = (I_{\text{new}} - c_{in})^2 - (I_{\text{new}} - c_{out})^2$, θ is chosen to be small, $\nu(\xi) := \max\{0, 2|\xi - \frac{1}{2}| - 1\}$ and $\alpha > \frac{\lambda}{2}\|r\|_{L^\infty(\Omega)}$. Then (8) is split into the following two minimization problems:

1. Minimize for u

$$\left\{ \int_\Omega |\nabla u|\,dx + \frac{1}{2\theta}\|u - v\|^2_{L^2(\Omega)} \right\},$$

for which the solution is given by: $u = v - \theta\,div\,p$. The vector $p = (p_1, p_2)$ is given by $\nabla(\theta div\,p - v) - |\nabla(\theta div\,p - v)|p = 0$ and can be solved by a fixed point method: $p^0 = 0$ and

$$p^{n+1} = \frac{p^n + \delta t\nabla(\theta div(p^n) - v/\theta)}{1 + \delta t|\nabla(\theta div(p^n) - v/\theta)|}.$$

2. Minimize for v

$$\left\{ \frac{1}{2\theta}\|u - v\|^2_{L^2(\Omega)} + \int_\Omega \lambda r(x, c_{in}, c_{out}) v + \alpha \nu(v) dx \right\},$$

for which the solution is given by: $v = \min\{\max(u(x) - \theta\lambda r(x, c_{in}, c_{out}), 0), 1\}$.

3. The constants c_{in} and c_{out} are updated every few iterations of the above algorithm, by

$$c_{in}(u) = \frac{\int_{\Sigma_\mu} u\,dx}{|\Sigma_\mu|} \quad \text{and} \quad c_{out}(u) = \frac{\int_{(\Sigma_\mu)^c} u\,dx}{|(\Sigma_\mu)^c|}, \qquad (9)$$

where $\Sigma_\mu = \{x \in \Omega \colon u(x) \geq \mu\}$, with μ any arbitrary point in $[0, 1]$.

The above scheme helps in finding the global minimizer of the energy functional, see (Bresson et al. 2007) for more details.

117

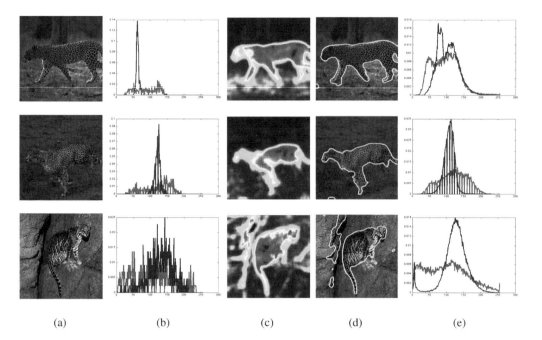

Figure 3. Segmentation results for gray-scale real images using (5). (a) Input image. (b) Neighborhood histograms where the red and blue histograms correspond to the information of the foreground (inside) and background (outside) image respectively, using $r = 10$. (c) Weighted gradient based term $J_1 = \Psi_1 I_{\text{new}}$, where higher values are indicated by red and lower by blue. (d) Segmentation results with 80 iterations. (e) Histograms after segmentations.

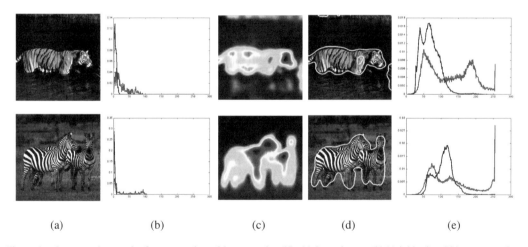

Figure 4. Segmentation results for gray-scale real images using (6). (a) Input image. (b) Neighborhood histograms where the red and blue histograms correspond to the information of the foreground (inside) and background (outside) norm of the gradient image respectively, using $r = 10$. (c) Weighted gradient based term $J_2 = \Psi_2 I_{\text{new}}$, where higher values are indicated by red and lower by blue. (d) Segmentation results with 80 iterations. (e) Histograms after segmentations.

3.2 Experimental results

The proposed scheme is implemented in MATLAB 7 R2009a on a laptop with Intel Core2 Duo, with 2.20GHz. For a 321×321 image, it takes about 3 minutes (with 80 iterations for steps 1 and 2 from section 3.1) to get the final segmentation result. In all the experiments reported here, the parameters were fixed at $\delta t = 1/8, \sigma = 10, r = 10, \lambda = 0.1$ and $\theta = 1$.

Figure 3 present the segmentation results for different gray-scale real images using the input channel (5). The original gray-scale images are presented in the first column Figure 3(a). In the second column Figure 3(b), we show the histograms of the images, where the red and blue histograms correspond to the information of the foreground (inside) and background (outside) images respectively, using $r = 10$. Third,

118

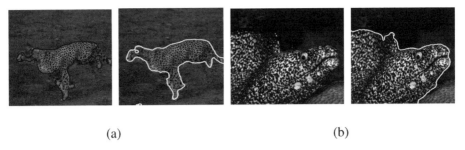

(a)   (b)

Figure 5. Comparison of segmentation results for different texture images. In each sub-figure segmentation results of (left) Ni et al. 2009 and (right) Proposed approach with (7) are shown.

fourth and fifth column, Figure 3(c), Figure 3(d) and Figure 3(e), represent the input channels given by J_1, the segmentation results and the corresponding histograms after segmentations respectively. It can be seen by comparing the neighborhood histograms before and after segmentation, we obtain a smoother histogram separation and the weighted gradient based input image gives good segmentations.

Figure 4 presents similar segmentation results for the extension given in equation (6). The neighborhood histograms of the gradient norm image (without smoothing) is given in Figure 4(b), and the input channels J_2 are given in Figure 4(c). The corresponding segmentation results given in Figure 4(d), and Figure 4(e) shows the histograms after segmentation. As can be seen, overall the proposed extension gives good segmentations.

Finally, we compare our combined extension scheme (7) with $\beta = 0.5$ and the local histogram based scheme from (Ni et al. 2009). The segmentation results given in Figure 5 shows the advantage of using balanced weighted information based on local histograms with smoothed gradients. For example, comparing the segmentation results for the *Cheetah* image in Figure 5(a), we see that better segmentation is obtained using the proposed balanced approach, whereas the scheme from (Ni et al. 2009) gives disjoined segments of the *Cheetah*'s body, see for example the tail section. Similarly, for the *Fish* image given in Figure 5(b) we obtain a single unified segment whereas the scheme from (Ni et al. 2009) gives spurious segments at the left hand side of the image.

4 CONCLUSION

A new framework for texture segmentation using the globally convex formulation of the Chan and Vese model has been proposed. The method takes advantage of local histograms present in different regions with texture information. By taking the multiplication of the smoothed gradient norm image with the local histogram information computed around each pixel, new fitting terms provide better segmentation results. Experimental results on texture images demonstrate the good performance of the method and compared with the previous local histogram based model it gives better segmentation results as well.

ACKNOWLEDGMENT

This work was partially supported by the research project UTAustin/MAT/0009/2008 of the UT Austin | Portugal Program (http://www.utaustinportugal.org/) and by CMUC and FCT (Portugal), through European program COMPETE/FEDER.

REFERENCES

Bresson, X., S. Esedoglu, P. Vandergheynst, J. Thiran, and S. Osher (2007). Fast global minimization of the active contour/snake model. *Journal of Mathematical Imaging and Vision 28*(2), 151–167.

Caselles, V., R. Kimmel, and G. Sapiro (1997). Geodesic active contours. *International Journal of Computer Vision 22*(1), 61–79.

Chambolle, A. (2004). An algorithm for total variation minimization and applications. *Journal of Mathematical Imaging and Vision 20*(1–2), 89–97.

Chan, T. F., S. Esedoglu, and M. Nikolova (2006). Algorithms for finding global minimizers of image segmentation and denoising models. *SIAM Journal on Applied Mathematics 66*(5), 1632–1648.

Chan, T. F. and L. A. Vese (2001). Active contours without edges. *IEEE Transactions on Image Processing 10*(2), 266–277.

Cremers, D., M. Rousson, and R. Deriche (2007). A review of statistical approaches to level set segmentation: integrating color, texture, motion and shape. *International Journal of Computer Vision 72*(2), 195–215.

Figueiredo, I. N., J. C. Moreno, V. B. S. Prasath, and P. N. Figueiredo (2012). A segmentation model and application to endoscopic images. In *International Conference on Image Analysis and Recognition (ICIAR 2012)*, Aveiro, Portugal. Springer LNCS Eds.: A. Campilho and M. Kamel.

Mumford, D. and J. Shah (1989). Optimal approximations by piecewise smooth functions and associated variational problems. *Communications in Pure and Applied Mathematics 42*(5), 577–685.

Ni, K., X. Bresson, T. Chan, and S. Esedoglu (2009). Local histogram based segmentation using the Wasserstein distance. *International Journal of Computer Vision 84*(1), 97–111.

Osher, S. and R. Fedkiw (2003). *Level set methods and dynamic implicit surfaces*. New York, NY, USA: Springer-Verlag.

119

Computational Modelling of Objects Represented in Images – Di Giamberardino et al. (eds)
© 2012 Taylor & Francis Group, London, ISBN 978-0-415-62134-2

Unsupervised self-organizing texture descriptor

Marco Vanetti, Ignazio Gallo & Angelo Nodari
Dipartimento di Scienze Teoriche e Applicate, Universita' degli studi dell'Insubria

ABSTRACT: We propose a local texture descriptor based on a pyramidal composition of Self Organizing Map (SOM). As with the SOM model, our visual descriptor presents two operational steps: a first unsupervised learning phase and a second mapping phase involving a dimensionality reduction of the input data. During the first step a large number of image patches, including different classes of textures, are presented to the model. At the end of the learning process the neural weights on each layer of the SOM pyramid will contain good prototypes of the patches used in training at different level of detail. During the mapping phase a new texture patch is presented to the model and, by using a winner take all principle, a winner neuron is selected and its 2D spatial location is used to describe the input patch. Exploiting the topological order of the SOM, two different texture descriptions can be compared using the common Euclidean distance. In the experimental section we show that a simple clustering algorithm like K-means, applied to the local descriptor responses, is able to segment complex texture mosaics with very good results, even in difficult areas like boundaries which separate two different textures.

1 INTRODUCTION

In order to automatically produce a description of a natural image, a fundamental role is played by texture descriptors. Images representing real objects often do not exhibit regions with uniform intensities but, due to the physical properties of real surfaces, they contain frequent variations of brightness which form certain repeated patterns called visual texture or more simply: texture.

Over the years, many problems involving texture analysis have been proposed, the main ones are listed below. *Texture classification* aims to produce a classification map of an image where each uniform textured region is identified by a particular texture class which belong to. *Texture segmentation* is focused on finding texture boundaries even if it is not possible to classify each region. Figure 1 shows an example of unsupervised texture segmentation obtained applying a K-means clustering to the local descriptor proposed in this paper. *Texture synthesis* is used for image compression applications and in computer graphics, with the aim of rendering object surfaces which need to be as realistic as possible. Finally, with *shape from texture*, we aim to extract the three-dimensional shape of objects in a scene using texture information, distorted by imaging process and the perceptive projection (Tuceryan and Jain 1998). Despite the final purpose is quite different, each of the problems listed above requires a texture descriptor, which becomes an essential tool in many applications.

A common denominator for most successful texture descriptors is that the textured image is submitted to a linear transform, filter or filter bank. Methods using

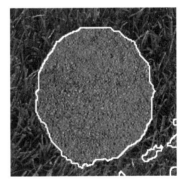

Figure 1. Segmentation between two areas with different textures obtained using the proposed descriptor. Segmentation border is depicted with a white line.

this common scheme are called *filtering approaches*, and received an extensive survey in (Randen and Husy 1999), a comparative study where various filtering approaches have been evaluated within a texture classification framework.

An important issue that characterizes most of the filtering approaches is the selection of an appropriate filter bank. The most commonly are the Gabor filters, inspired by experiments with animal visual systems (Daugman 1980), and signal-processing based filters, designed with desirable band-pass properties in the Fourier domain (Bovik 1991). However, the optimal choice of a filter bank is often influenced by the particular application and may require a lot of experimentation.

A simple and promising strategy to combine multiple filters, resulting in a compact description of the texture, is the spectral histogram, first suggested in psychophysical studies on texture modeling (Bergen and Adelson 1988) and later used for texture analysis and synthesis (Heeger and Bergen 1995) (Zhu, Wu, and Mumford 1997). Spectral histogram is based on the assumption that all of the spatial information characterizing a texture image can be captured in the first order statistics of an appropriately chosen set of linear filter outputs. Spectral histogram can also be used as a local descriptor, using and appropriately sized integration window, in this case the descriptor is often called Local Spectral Histogram (LSH).

LSH is a powerful local texture descriptor, able to seize general aspects of texture as well as non-texture regions. In (Liu and Wang 2006) a LSH based on a filter bank based composed of eight filter (pixel intensity, two gradient filters, two-scales Laplacian of Gaussian and three Gabor filters) has been used for texture segmentation, attaining the state of the art in the field of unsupervised texture segmentation methods based on filter bank.

The main drawback of LSH is that it requires large integration windows to extract meaningful texture features from the image, this results in a poor reliability of the description along texture boundaries. A solution to the aforementioned problem has been proposed in (Liu and Wang 2006) by using asymmetric windows and a refined probability model based on seed points automatically extracted from the segmented regions.

Some work tried to generalize the methods based on multichannel filtering by training: in a supervised fashion, a neural network in order to find a minimal set of specific filters. These methods may delegate to the neural network the dual task of extracting features and classifying textures (Jain and Karu 1996) (Kim, Jung, Park, and Kim 2002), or perform separately the second phase using a most powerful classifiers such as Support Vector Machines (Melendez, Giron'es, and Puig 2011).

In this paper we propose an innovative texture descriptor, based on a pyramidal composition of Self Organizing Map (SOM) (Kohonen 1990), that is capable of extract a powerful local texture feature from an image without requiring any supervision or handcrafted filter bank. The pyramidal nature of the approach perform an image analysis using pixel contexts which become progressively larger. At each layer of the pyramid, only the most relevant feature for the particular context will be extracted by the SOM and the image will be "redrawn", for the upper layer of the pyramid, deprived of redundant information.

Considering the complexity of the non-linear dimensionality reduction introduced by each SOM, the validity of the proposed approach is difficult to prove analytically. However, to evaluate the method, we used a very simple unsupervised texture segmentation strategy, based on a K-means clustering algorithm. In this way we highlight the goodness of the proposed descriptor, excluding contributions

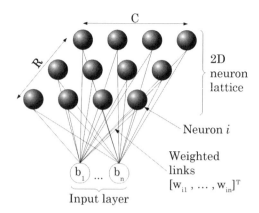

Figure 2. Two dimensional SOM.

attributable to a supervised machine learning method or a post-processing/refinement phase.

This work is organized as follow. In Section 2 is described the proposed texture descriptor, based on the SOM unsupervised learning method. In Section 3 are shown and discussed experimental results, using the K-means clustering algorithm for texture segmentation. Finally, Section 4 gives the conclusions.

2 PROPOSED DESCRIPTOR

As discussed in Section 1, the proposed descriptor is based on a pyramidal composition of SOM. SOM is an artificial neural network first proposed by Teuvo Kohonen in early 1981, able to produce, without supervision, a spatially organized internal representation of various features of input signals (Kohonen 1990). As depicted in Figure 2 we employ a two dimensional SOM, composed of a 2D lattice of neurons, each of which is fully connected to the input layer through a series of weighted links $\mathbf{w_i} = [w_{i1}, w_{i2}, \ldots, w_{in}]^T$ where $0 \leq w_{ij} \leq 1$, i is the index of a single neuron and n is the dimension of the input data.

The proposed method involves an initial training phase, where a large number of training vectors are presented to the network and the neural weights are updated according to a particular rule. Training vectors are extracted from the input image using an overlapping sliding window approach, the window shall henceforth be called *context window*.

No handcrafted feature is extracted from the image: a training vector is composed of the intensity/brightness values of pixels within the context window and denoted by $\mathbf{b} = [b_1, b_2, \ldots, b_n]^T$ where $0 \leq b_j \leq 1$ and n is the total number of pixels. The input image will be properly border-padded[1] so that, in total, $H \cdot W$ training vectors will be extracted, where H is the height and W is the width in pixels of the image.

Let us describe now how the unsupervised learning happens. By presenting a new input vector to the

---

[1] We used a "mirror" border padding strategy, as explained in (Szeliski 2010).

122

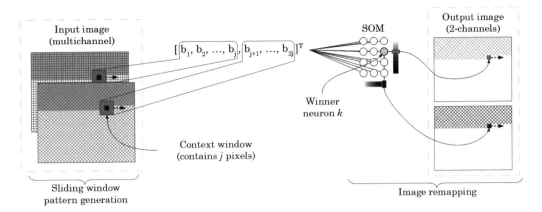

Figure 3. Schema representing the image remapping strategy, performed exploiting the SOM competitive behavior.

SOM, a single neuron k will be activated in a particular location of the network, we call this neuron "winner". The winner selection occurs by satisfying the following identity:

$$\|\mathbf{b} - \mathbf{w_k}\| = \min_i \{\|\mathbf{b} - \mathbf{w_i}\|\} \qquad (1)$$

The step just described is followed by the update of the weights in the neighborhood of the winner neuron. The update is described by the following equation:

$$\mathbf{w_i}(t + 1) = \mathbf{w_i}(t) + \alpha h_{ik}[\mathbf{b}(t) - \mathbf{w_i}(t)] \qquad (2)$$

Referring to the equation ??, α is a scalar constant called *adaptation gain* or *learning rate*, $0 < \alpha < 1$, and the function h_{ik} is a scalar "bell curve" kernel function defined as:

$$h_{ik} = \exp\left(-\frac{\|\mathbf{q_i} - \mathbf{q_k}\|^2}{2\sigma^2}\right) \qquad (3)$$

where the vectors $\mathbf{q_k} = [q_{kr}, q_{kc}]^T$ and $\mathbf{q_i} = [q_{ir}, q_{ic}]^T$ denote the coordinates of the winning neuron k and the neuron to be updated i, and the r in subscript is in reference to the row number inside the neuron lattice and c refers to the column number. To speed up the convergence of the self-organization, σ and α can be chosen as time-variable functions, monotonically decreasing with iterations, however, for simplicity in this work these parameters are chosen constant.

At the end of the training phase, the spatial location, represented by the coordinates of each neuron in the network, corresponds to a particular domain or feature of input signal patterns (Kohonen 1990) and the weights of each neuron contain a good prototype of the input patches (Gersho and Gray 1992). By using a small window of local context around each pixel, the proposed method tries to discover local salient features from the image.

Once the SOM is trained, its neural weights w can be treated as constant values, and employing the same sliding window approach used during the previous training phase, we can map each pixel of the input image in the two dimensional Euclidean space of the activated neurons within the SOM lattice. Using again Equation 1, we thus generate a new image with two channels, the first dependent on the row number of the winner neurons and the second on the column number. The new image, that we call *remapped image*, can be formally calculated from the input image $I_0(x, y)$ using:

$$I_1(x, y) = \left[\frac{q_{kr}}{R}, \frac{q_{kc}}{C}\right]^T \qquad (4)$$

where, for each pattern $\mathbf{b}$ centered on the pixel (x, y) of the input image I_0, the winner neuron k is found using Equation 1. R and C refer to the size, in rows and columns, of the neural lattice. Note that I_1 is a two-channels image, therefore each pixel contains two intensity values.

Since each pattern is extracted by simply concatenating the values of the pixels within the context window, the input image can have an arbitrary number of channels. In this way the learning process and the subsequent remapping can be performed iteratively on more layers, following a pyramidal approach. For each layer of the pyramid, the parameters involved are the context window size, the size of the SOM and the learning parameters σ and α. Figure 3 graphically explains the remapping strategy just described.

In the following section we show a sample configuration based on three layers and applied to some real and synthetic images.

3 EXPERIMENTS

To test the proposed method, we used a configuration based on a three-layers pyramid, with a context window of 2×2 pixels, a SOM with 10×10 neurons, $\sigma = 1$ and $\alpha = 0.1$ for the first layer that operates on the input image. The second layer involves a 4×4 context window and a 15×15 SOM with $\sigma = 3$ and $\alpha = 0.01$. The third layer uses the same SOM parameters of the

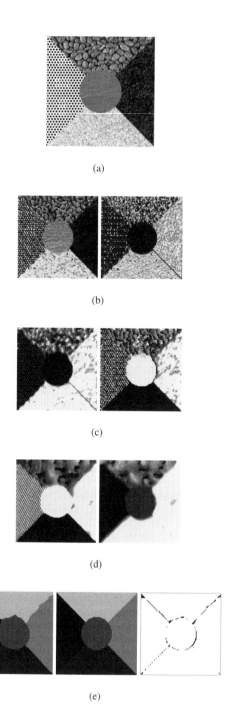

(a)

(b)

(c)

(d)

(e)

Figure 4. (a) Input image, a mosaic composed of 5 textures. (b)(c)(d) Two channels images remapped by the first (b), second (c) and third (d) layers of the SOM pyramid. (e) From the left, the final segmentation, the ground truth segmentation and the segmentation error map. Wrong pixels are shown in black.

second layer but doubles the context window size to 8×8 pixels. The SOM parameters were chosen primarily taking into account the size of the input pattern. Note that a 15×15 sized SOM contains about twice

the neurons of a 10×10 sized SOM. We found that slight changes of the parameters σ and α minimally affect the final segmentation error.

Using the architecture described above, the overall training/mapping pipeline can be schematized as follows: *(1)* the first layer is trained using the input image I_0, *(2)* the first layer performs the remapping, *(3)* the second layer is trained using the remapped image provided by the first layer I_1, *(4)* the second layer performs the remapping, *(5)* the third layer is trained using the remapped image provided by the second layer I_2, *(6)* the third layer performs the final remapping providing the output image I_3.

To evaluate the proposed texture descriptor, we employed a simple segmentation strategy based on the K-means algorithm. Pattern set was created by concatenating pixel intensities from the last layer to their normalized coordinates, in order to create a raw topological constraint. Formally:

$$P = \bigcup_{x=0}^{W} \bigcup_{y=0}^{H} \left(I_3(x,y) \| \left[\frac{x}{W}, \frac{y}{H} \right]^T \right) \quad (5)$$

Figure 4(a) depict a 5-texture mosaic used in (Liu and Wang 2006) to test an unsupervised segmentation method. The authors have obtained a 3.90% error using a LSH texture descriptor with a 19×19 pixels integration window and a filter bank composed of one intensity filter, two gradient filters, two-scales Laplacian of Gaussian and three Gabor filters. By applying a refined probability model to localize the region boundaries, they have reduced the error to 0.95%. The proposed method performs with an error of 1, 83%, Figure 4(e) shows the resulting segmentation, the ground truth segmentation and a map that highlights wrong segmented pixels. The reported errors refer to the percentage of pixels incorrectly segmented.

As can be seen in Figure 4(d), the local texture description is smoothed near the boundary between two different textures, this is due to the context window size. Despite this fact, the local texture description is still reliable, since the K-means clustering, using the common Euclidean distance as a metric of distance, is able to recognize and separate with a good precision the two textures along the boundary. Considering that we do not use any handcrafted feaure/filter and our method does not rely on a specific border localization technique, the result obtained is very challenging.

Figure 5 is another 5-texture mosaic used in (Karoui, Fablet, Boucher, Pieczynski, and Augustin 2008) to test a supervised approach based on empirical marginal distributions of local texture features like co-occurrence distributions, Gabor magnitude distributions, etc. They have obtained a 3.1% error while our error is 5.14%. The two results are comparable, but the problem studied here is essentially more difficult, given the unsupervised nature of the feature extraction process and of the image segmentation.

(Awate, Tasdizen, and Whitaker 2006) proposed the 2-class mosaic in Figure 1 as a challenging image since it show two textures that are both irregular and have

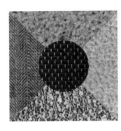

(a)

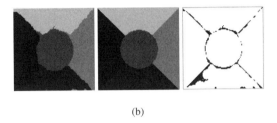

(b)

Figure 5. (a) Texture mosaic composed of 5 textures. (b) From the left, the final segmentation, the ground truth segmentation and the error map. Wrong pixels are shown in black.

Table 1. Segmentation results obtained using different subsets of the proposed 3-layers architecture.

	Figure 4(a) error (%)	Figure 5(a) error (%)
Only Intensity	29.45	52.45
Only Layer 1	17.39	19.98
Only Layer 2	28.97	23.77
Only Layer 3	29.17	35.95
Layer 1 + 2	10.14	9.22
Layer 1 + 3	8.68	4.86
All layers	1.83	5.14

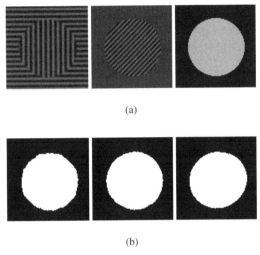

(a)

(b)

Figure 6. (a) Three synthetically created texture-non texture mosaics. (b) Segmentation results obtained with the proposed method.

similar means and gradient-magnitudes. No numerical result is available in their paper, but the results that we obtained is qualitatively comparable with that shown in (Awate, Tasdizen, and Whitaker 2006), obtained using an unsupervised approach that minimizes the entropy-based metric on the probability density functions of image neighborhoods.

All the proposed texture mosaics are composed of textures taken from the Brodatz album (Brodatz 1966) and the *Vision Texture Dataset*[2].

To investigate the contribution of each layer in the overall process, we evaluated the method by excluding different subsets of layers. The worst result is obtained by applying the K-means clustering directly on the intensity levels of the input image, while the best configuration involves all the three levels. Results in Table 1 show that the strength of the descriptor lies primarily in the pyramidal approach and, using a shallow architecture, the segmentation accuracy suddenly decreases.

As a final experiment we tested the method on three synthetically generated images. The first image in Figure 6(a), is composed by two wave-gradient regions with two different orientations. The mean intensity is constant within the two regions and the only discriminant information is the orientation of the wave pattern. The second image shows two regions, one with a wave-gradient texture and one with a solid color. Also in this case both regions have the same mean intensity. The third image contains two non-textured areas with different intensities. As can be seen in Figure 6(b), in all

three cases, the proposed method has been able to distinguish the two regions almost perfectly. This result experimentally prove that the proposed descriptor can handle, at the same time, texture regions as well as non-texture regions.

Before being processed, the input image is scaled to 100×100 pixels. Under these conditions, the computational time required to process an image is about 15 seconds[3], where more than half of the time is spent training the third layer. This is due to the large patches (8×8 pixels) used on the layer, which generate big training patterns.

4 CONCLUSIONS

In this paper we have presented a new texture descriptor that is able to characterize textured as well as non-textured regions with high accuracy. The potential of the method lies in its independence from a feature bank and its ability to automatically extract, without

---

[2] The Vision Texture Dataset is provided by the MIT Vision and Modeling Group, http://vismod.media.mit.edu/

[3] Results were obtained using an unoptimized, single C# thread, on an Intel(R) Core(TM) i5 mobile CPU at 2.30 Ghz.

supervision, salient information using only brightness values of pixels from small image patches. The descriptor exploits the important topological ordering property of the SOM allowing a smoothed and reliable image description even in areas with strong transitions, such as the boundary between two different textures or two different colors.

Comparison with other state of the art methods shows that our solution gives comparable results even without a directly managing of difficult areas, such as texture boundaries. The provided three-layers configuration offers good results on images of different types. Future research is focused in improving the segmentation accuracy and defining a method to automatically find an optimal parameters setting.

REFERENCES

Awate, S. P., T. Tasdizen, and R. T. Whitaker (2006). Unsupervised texture segmentation with nonparametric neighborhood statistics. In *European Conference on Computer Vision*, pp. 494–507.

Bergen, J. R. and E. H. Adelson (1988). Early vision and texture perception. *Nature 333*, 363–364.

Bovik, A. C. (1991). Analysis of multichannel narrow-band filters for image texture segmentation. IEEE *Transactions on Signal Processing 39*, 2025–2043.

Brodatz, P. (1966). Textures: a photographic album for artists and designers.

Daugman, J. (1980). Two-dimensional spectral analysis of cortical receptive field profiles. *Vision Research 20*, 847–856.

Gersho, A. and R. M. Gray (1992). Vector quantization and signal compression.

Heeger, D. J. and J. R. Bergen (1995). Pyramidbased texture analysis/synthesis. In Annual *Conference on Computer Graphics*, pp. 229–238.

Jain, A. K. and K. Karu (1996). Learning texture discrimination masks. *IEEE Transactions on Pattern Analysis and Machine Intelligence 18*, 195–205.

Karoui, I., R. Fablet, J.-M. Boucher, W. Pieczynski, and J.-M. Augustin (2008). Fusion of textural statistics using a similarity measure: application to texture recognition and segmentation. *Pattern Analysis and Applications 11*, 425–434.

Kim, K. I., K. Jung, S. H. Park, and H. J. Kim (2002). Support vector machines for texture classification. *IEEE Transactions on Pattern Analysis and Machine Intelligence 24*, 1542–1550.

Kohonen, T. (1990). The self-organizing map. *Proceedings of the IEEE 78*, 1464–1480.

Liu, X. and D. Wang (2006). Image and texture segmentation using local spectral histograms. IEEE Transactions on Image Processing 15, 3066–3077.

Melendez, J., X. Gironés, and D. Puig (2011). Supervised texture segmentation through a multilevel pixel-based classifier based on specifically designed filters. *In ICIP*, pp. 2869–2872.

Randen, T. and J. H. Husy (1999). Filtering for texture classification: A comparative study. *IEEE Transactions on Pattern Analysis and Machine Intelligence 21*, 291–310.

Szeliski, R. (2010). *Computer Vision: Algorithms and Applications*.

Tuceryan, M. and A. K. Jain (1998). Texture analysis. *The Handbook of Pattern Recognition and Computer Vision*.

Zhu, S. C., Y. N. Wu, and D. Mumford (1997). Minimax entropy principle and its application to texture modeling. *Neural Computation 9*, 1627–1660.

Computational Modelling of Objects Represented in Images – Di Giamberardino et al. (eds)
© 2012 Taylor & Francis Group, London, ISBN 978-0-415-62134-2

Automatic visual attributes extraction from web offers images

Angelo Nodari, Ignazio Gallo & Marco Vanetti
Dipartimento di Scienze Teoriche e Applicate, Universita' degli studi dell'Insubria

ABSTRACT: In this study we propose a method for the automatic extraction of Visual Attributes from images. In particular, our case study concerns the processing of images related to commercial offers in the fashion domain and the results show how the use of the proposed method can be successfully applied in a real context. This method is based on a pre-processing phase in which an object detection algorithm identifies the object of interest, subsequently the visual attributes are extracted using a descriptor based on the Pyramid of Histograms of Orientation Gradients. In order to classify these descriptions, we have trained a discriminative model using a manually annotated dataset of commercial offers, which we released for future comparisons. To increase the performance of the visual attributes extraction, the results provided by the previous step have been refined with an a priori probability which models the occurrence of each visual attribute with a specific product type, opportunely estimated on the dataset.

1   INTRODUCTION

In the recent years we are experiencing a growing interest turned towards visual attributes and their usage in support to many different task: classification, recognition, content-based image retrieval etc...

The concept of visual attribute has been firstly formalized and analyzed by (Ferrari and Zisserman 2007) who propose a generative model for learning simple color and texture attributes from loose annotations and (Farhadi et al. 2009) which learn a richer set of attributes including parts, shape, materials, etc

In the field of face recognition (Kumar et al. 2011) used a set of general and local visual attributes to train a discriminative model which measures the presence, absence, or degree to which an attribute is expressed in images in order to compose a signature of visual attributes. In (Sivic et al. 2006; Anguelov et al. 2007) the main goal consists in finding all the occurrences of a particular person in a sequence of pictures taken over a short period of time. In particular the use of visual attributes to extract information about the hair and clothes of the people has given a consistent contribution in the management of all the cases where the people move around, change their pose and scale, and partially occlude each other.

There are also successful applications of the visual attributes in the field of security and surveillance systems, for example in (Vaquero et al. 2009) the authors used visual attributes instead of the standard face recognition algorithms, which are known to be subject to problems like lighting changes, face pose variation and low-resolution. They search for people by parsing human parts and their attributes, including facial hair, eyewear, clothing color, etc.

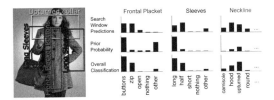

Figure 1.   Example of the attribute extraction phase using the proposed method for the three types of attribute analyzed in this study.

In (Wang and Mori 2010) the authors demonstrate that object naming can benefit from inferring attributes of objects and that, in general, the attributes are not independent each other.

The visual attributes express local or general characteristics of a subject, while color, texture and shape are the global features most commonly used, they are also the less interesting in the particularization of the object of interest (OI). They are also used in the aforementioned work and we have analyzed them in a previous work (Gallo et al. 2011), but in this paper we have focused on the local attributes. These visual attributes are very domain-specific and therefore contain much information that can be used in various fields. In order to be exported to other domains, the extracted attributes must be carefully selected. Moreover, to ensure the applicability of the proposed method in real application contexts, the extraction time of the visual features must not increase disproportionately with the number of attributes that have to be extracted and at the same time we want that these algorithms can be very fast.

Cloth Type	Num	Cloth Type	Num
gown	117	gym suit	57
tunic	108	top	119
tailleur	120	overcoat	118
t-shirt	148	overalls	114
pullover	131	polo	121
padded jacket	130	cloak	111
sweater	120	knitwear	135
raincoat	102	vest	125
short coat	143	jacket	133
sweatshirt	133	turtleneck	102
shrugs	115	cardigan	130
coat	128	camisole	125
shirt	137	blouse	112
blazer	109	short dress	123
dress	202		

Figure 2. Summarization of the types of clothes in the DVA dataset used in this study.

To the best of our knowledge, the use of the visual attributes in the online shopping has not been exploited yet and for this reason we consider this work very innovative in this area. For example the visual attributes can be used by a user as a method to search the products with particular characteristics which may be congenial for him. For example they can be used in a typical faceted navigation, where the user can search for an item in a structure where all the facets of an item are a possible entry point. At the same time these visual attributes can be integrated in Content-Based Image Retrieval engines for the estimation of the similarity between different images, for example in a query by example system.

2 PROPOSED METHOD

The automatic visual attributes extraction method, proposed in this study, involves the use of a search window whose size and position are in function of the type of visual attribute to search and the bounding box of the OI. After have positioned this window, we build a pattern which is then classified by a Support Vector Machine (SVM) (Cortes and Vapnik 1995) in one of the attribute classes which we take into account.

The global features used in the works mentioned in Section 1, such as color and texture, can not be used as visual attributes in this domain because they are not discriminative enough since clothing of the same category may be of different colors and textures. Nevertheless, these attributes have been used successfully in other domains, where the color and texture features are more discriminative such of the case of the dataset of animals (Lampert et al. 2009). For this reason we have chosen more domain specific attributes such as neckline shape, frontal closure and sleeve types.

One problem identified in all the works that extract visual attributes, involves the application of features over the whole image without having first identified, even in a coarse way, the OI. Therefore, to properly extract the visual attributes only from the Object of Interest avoiding to be distracted by the background and introducing noise in the extracted data, we necessarily have to distinguish the subject from the background. To perform this step, we relied on our previous work called MNOD (Gallo et al. 2011; Gallo and Nodari 2011) which consists in an algorithm able to perform a segmentation of the OI in a specific context, in this case applied to the fashion domain.

After have detected the OI, we select all the manually labeled regions of interest and we estimate the relative position in relation with the bounding box that contains the OI. As a result we obtained that to properly look for the neckline attribute we have to position the search window at the upper side of the OI, the sleeves attribute in the lateral side and the frontal placket in the middle. This step can be considered trivial, but its formalization allows to easily transfer the entire visual attributes extraction process from the context addressed in this study to any other one.

Once we have identified the location where to place the search window, we resize it in according to the best parameters estimated in Section 3. After that we read the information within the image and represent it as a PHOG (Bosch et al. 2007) features into a pattern that is classified by an SVM in one of the classes of the visual attributes which we are looking for.

One of the objectives of this study consists in showing how the use of the visual attributes occurrence information, depending on the type of product, can be used in support of the attributes classification phase. This type of use of the attributes has already been addressed in earlier papers such as, for example (Lampert et al. 2009), who have used, however, high-level of attributes manually associated to each image in support of the classification of objects. Instead, in this study, the visual attributes are fully automatically extracted from the image, because in our domain there are not always manual annotations and we want to focus and analyze only the extraction of information from images. The probability distribution of the visual attributes, depending on the types of product, was estimated on the training set and showed according to the representation of the Class Attribute Matrix (Kemp et al. 2006) in Figure 6 and 3.

One of the most difficult task is to place the window for the search of the visual attributes in the correct position. In the experimental phase we have automatically calculated the best position of the window relative to the bounding box of the OI. In spite of this, a variation in the positioning of the window is reflected on the accuracy in the extraction phase of the attributes. To overcome this problem, the window is moved arbitrarily respect to its position in order to obtain k readings. For each of these readings is then performed a prediction using the trained SVM. In this way we obtain a set of predictions which are integrated with the values of the a priori probability in order to obtain a result on the extraction of the visual attribute, which minimizes the

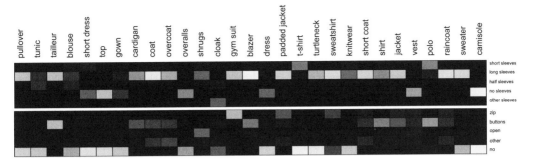

Figure 3. Visual Attribute Signatures for the Sleeves and Frontal Placket Attribute for every product type in the DVA dataset. The gray level represent the probability that a visual attribute may appear associated to a specific product type. To a gray value very high, corresponds a high probability that the pair Visual Attribute and Product Type appears and vice versa.

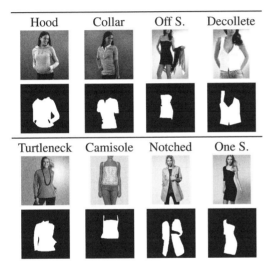

Figure 4. Example of images from the visual attribute Drezzy Dataset grouped by their neckline classes.

risk of committing an error. In the experimental section is shown how this method increases the performance in the extraction of the visual attributes.

Given T, the set of product types, and A, the set of visual attributes, the function $s : T \times A -> \mathbb{N}$ counts the number of occurrences of an attribute $a_i \epsilon A$ given the product type $t \epsilon T$. The function $\widehat{f} : T \times A -> \mathbb{R}$ returns the probability estimated on the training set, that a specific attribute $a_i \epsilon A$ may occurs given a product type $t \epsilon T$. In particular $\widehat{f}$ is computed as

$$\widehat{f}(t, a_k) = \frac{s(t, a_k)}{\sum_{i=0}^{m} s(t, a_i)}$$

where m is the number of types of attribute in A. This information is combined with the predictions of the SVM in order to minimize the classification error.

Given p a pattern from the set of patterns P and an observed attribute a, the prediction function n: $P \times A -> \mathbb{N}$ returns the number of predictions of the class attribute a for the pattern p for all the readings

of the search window. The predicted attribute class c is defined as follows

$$c = \arg \max_i (\max(\epsilon, \widehat{f}(p, a_i)) \cdot n(p, a_i))$$

where ϵ is a constant to avoid the 0 valued case, situation which corresponds to the probability that a particular attribute a doesn't occur associated with a product name p.

3 EXPERIMENTS

In order to evaluate the proposed method we have built a dataset consisting of 3523 images, and for each of them we have manually labeled the visual attributes analyzed in this study. Because our case of study consists in a set of images related to the fashion domain, they were downloaded from an internet web site for the shopping online[1]. The dataset collected was uploaded to our homepage in order to be used by other methods for comparison[2] and it is named *Drezzy Visual Attribute (DVA)* Dataset.

The collected dataset is composed by different types of clothing and, to reduce the variability in this domain, we focused on the woman clothing worn in the upper part of the body, whose images are characterized by a consistency in the appearance of clothes and human poses. The dataset consists of over 3,000 images divided in 29 different product types associated to different sets of visual attributes such as 16 classes of neckline shape, 5 classes of sleeves type and 5 classes of frontal closure type. All the product types are summarized in Figure 2 and a set of example images of the *Neckline* attribute are shown in Figure 4. The same approach followed in this domain can be simply adapted and used in other contexts. These images have a resolution of 200x200 pixels because they came from the online domain where there is a constraint on the size of the image and a consistent

---

[1] http://www.drezzy.com

[2] http://www.dicom.uninsubria.it/arteLab/

129

JPEG compression. These factors significantly complicate the extraction of the visual attributes, because at this resolution is very difficult to extract this kind of information.

3.1 Windows parameters

To select where to place the search window, we estimated on the training set all the manually placed bounding box. Instead to select the best size of the search window, we performed an experiment varying the proportions of width and height of the window and observing the Kappa value (Cohen 1960), estimated on the DVA dataset. The results have shown that the variation of the observation window, with the exclusion of proportions that lead to degenerate dimensions, it is not found significant changes.

3.2 Features configuration

We tried different features applied to the problem of the visual attributes extraction. Features such as color and texture were discarded for the reasons discussed in Section 1. In this context we have tried to apply the state

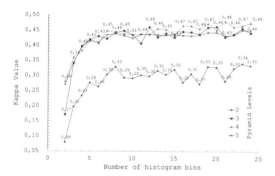

Figure 5. Experimental results of the PHOG parameters tuning using the DVA Dataset. There are plotted 4 series which correspond to the number of pyramid levels, on the horizontal axis are reported the Number of Histogram Bins and on the vertical axis the Kappa value.

of the art in the extraction of local information from the images using the Bag of Visual Words (BOVW) (Yang et al. 2007; Chen et al. 2009; Csurka et al. 2004). We used the SURF algorithm as a feature descriptor and we have performed a trial and error test to select 150 as the best number of words for the BOVW dictionary. The performance in the extraction of visual attributes estimated on the DVA dataset corresponds to a $Kappa = 0, 14$. This result is consistent with the analysis performed on the DVA dataset, which shows how an excessive lossy compression of the images leads to a reduction in the quality and therefore the possibility to extract local information, fundamental to discriminate on the different visual attributes. Therefore the choice of another feature that can capture this type of information, working also with degraded images, fell on the PHOG feature.

In order to select the best PHOG parameters we have performed a tuning experiment on the DVA dataset using a fixed position and a fixed size of the reading window relative to the bounding box of the OI. We have evaluated the number of histogram bins and the number of pyramid levels in according to the Kappa value estimated on the dataset, the result are shown in Figure 5. Considering the computational time and results in figure, a good balance in setting the parameters corresponds to set the number of histogram bins to 12 and the pyramid levels to 2.

3.3 Visual attributes extraction

We have evaluated the extraction of three different visual attributes, with the method proposed in this

Table 1. Comparison on the application of the estimation distribution of the visual attributes on the DVA dataset for each attribute class.

Attribute class	standard method	with estimation
sleeves	0,63	0,66
frontal placket	0,32	0, 43
neckline	0,43	0,59

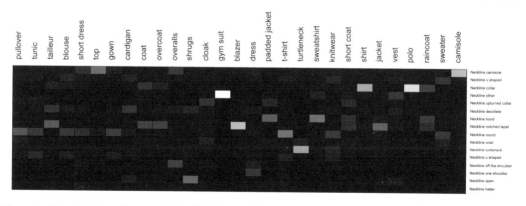

Figure 6. Visual Attribute Signatures for the Neckline Attribute for every product type in the DVA dataset. The gray level represent the probability that a visual attribute may appear associated to a specific product type. To a gray value very high, corresponds a high probability that the pair Attribute Visual and type of product appears and vice versa.

Table 2. Confusion matrix of the classification of the Sleeves visual attribute with the method proposed in this study.

	short	long	half	nothing	other	CO	PA
short sleeves	28	4	2	4	1	39	71,79%
long sleeves	8	250	20	10	5	293	85,32%
half sleeves	3	5	5	3	4	20	25,00%
nothing sleeves	5	12	4	102	4	127	80,31%
other sleeves	0	0	1	2	1	4	25,00%
TO	44	271	32	121	15	483	
UA	63,64%	92,25%	15,63%	84,30%	6,67%		

Overall Accuracy (OA): 79,92%
Kappa-value: 0,66

Table 3. Confusion matrix of the classification of the Frontal Placket visual attribute with the method proposed in this study.

	zip	buttons	open	other	nothing	CO	PA
zip frontal placket	46	11	4	12	13	86	53,49%
button frontal placket	24	172	9	49	45	299	57,53%
open frontal placket	0	3	23	1	0	27	85,19%
other frontal placket	9	10	5	10	6	40	25,00%
nothing	43	101	23	52	503	722	69,67%
TO	122	297	64	124	567	1174	
UA	37,70%	57,91%	35,94%	8,06%	88,71%		

Overall Accuracy (OA): 64,22%
Kappa-value: 0,43

Table 4. Confusion matrix of the Neckline visual attributes classification with the proposed method. The labels correspond respectively to the neckline categories: camisole (Ca.), v shaped (V.S.), collar(Co.), upturned collar (Up.), other (Ot.), decollete (De.), hood (Ho.), notched lapel (N.Lap.), round (Ro.), Cowlneckline (Cowl), turtleneck (Turt.), u shaped (U.S.), off the shoulder (off.), one shoulder (One.), open (Open), halter (Hal.).

	Ca.	V.S.	Co.	Up.	Ot.	De.	Ho.	N.Lap.	Ro.	Cowl	Turt.	U.S.	Off.	One.	Open	Hal.	CO	PA
camisole	32	0	4	2	2	7	1	3	3	0	1	3	0	2	0	3	63	50,79%
v shaped	0	13	1	1	1	3	1	0	2	1	0	2	0	0	0	0	25	52,00%
collar	0	3	35	3	3	1	1	2	2	1	1	0	0	0	0	0	52	67,31%
upturned collar	0	0	2	10	0	0	1	1	1	2	3	0	0	0	0	0	20	50,00%
other	0	1	0	2	4	1	0	2	0	1	0	0	0	0	2	0	13	30,77%
decollete	2	3	0	2	3	20	1	6	0	0	0	1	0	0	5	0	43	46,51%
hood	0	0	0	1	7	0	28	1	1	1	1	0	0	0	0	0	40	70,00%
notched lapel	0	3	0	1	1	2	3	32	0	0	0	0	0	0	0	0	42	76,19%
round	2	5	4	0	4	0	5	0	60	1	1	8	3	4	0	0	97	61,86%
Cowl	0	1	0	1	0	0	1	0	0	3	0	0	0	0	0	0	6	50,00%
turtleneck	0	0	0	5	0	0	0	0	0	0	6	19	0	0	0	0	30	63,33%
u shaped	0	0	0	0	0	1	0	0	2	1	0	9	0	0	0	0	13	69,23%
off the shoulder	0	0	0	0	0	0	0	0	0	0	0	0	15	0	0	0	15	100,00%
one shoulder	0	0	0	0	0	0	0	0	0	0	0	0	0	8	0	0	8	100,00%
open	0	0	0	0	1	0	0	0	0	0	0	0	0	0	11	0	12	91,67%
halter	0	0	0	0	0	0	1	0	0	0	0	0	0	0	0	3	4	75,00%
TO	36	29	46	28	26	36	42	47	71	17	26	23	18	14	18	6	483	
UA	88,89%	44,83%	76,09%	35,71%	15,38%	55,56%	66,67%	68,09%	84,51%	17,65%	73,08%	39,13%	83,33%	57,14%	61,11%	50,00%		

Overall Accuracy (OA): 62,53%
Kappa-value: 0,59

study, using the DVA dataset and for each visual attribute we discuss about the performance results.

The results on the Sleeves Visual Attribute are shown in Table 2 and it is possible to notice how the *Other Sleeves* class is very difficult to predict. This is due to the few number of examples in the dataset and they can be considered as outlayers in the classification task and so this class can be omitted. The result on the *Half Sleeves* class is very low because, as showed in Table 2, it mingles with the class *Long Sleeves*. For this reason a deep analysis on the used feature have to be performed in order to integrate this class in a real application.

The results on the Frontal Placket Attribute are shown in Table 3 and also for this attribute the *Other* class is troublesome for the same aforementioned reasons.

As regards the visual attribute *Neckline* the results are reported in Table 4 where the difficulty of classification of this attribute lies in the high number of classes. Nevertheless, it is possible to note how the proposed method is able to correctly recognize a good part of the classes, despite the problem of classification on the generic images of the DVA dataset is very complex.

Each prediction of the discriminative model is combined with the estimated information on the training set, as explained in Section 2, which associates to each product name the probability that it contains a particular attribute. In order to verify if the introduction of this priori information on the estimated training set has given benefits, we performed an evaluation on all the visual attributes with and without this feature showed in Table 1.

The last experiment concerns the calculation of the computational time for the extraction of a visual attribute using a single C# thread, on a Intel®Core™i5 CPU at 2.30 GHz. The extraction time average of all the elements in the DVA dataset is equal to 304 *ms*.

4 CONCLUSIONS

In this study we have investigated a method for the automatic extraction of visual attributes from images. This technique was applied to the domain of fashion, but as explained in the previous sections, thanks to generic approach that has been adopted, it is possible to extend this method to any other domain assuming to have selected a properly set of attributes on which to work and at the same time it can be applied in a real application to provide new functionalities in CBIR engines. The experimental results show that estimating the occurrence of the visual attributes on the sample data is of fundamental importance in order to significantly improve the accuracy in the extraction of visual attributes. Given the lack of available dataset in the domain of the visual attributes extraction, an important contribution led from this work is the introduction of the DVA dataset which can be used for future comparisons.

REFERENCES

Anguelov, D., K. chih Lee, S. B. Gktrk, B. Sumengen, and R. Inc (2007). Contextual identity recognition in personal photo albums. In *IEEE Conference on In Computer Vision and Pattern Recognition (CVPR)*.

Bosch, A., A. Zisserman, and X. Muoz (2007). Representing shape with a spatial pyramid kernel. In *Conference on Image and Video Retrieval*.

Chen, X., X. Hu, and X. Shen (2009). Spatial weighting for bag-of-visual-words and its application in content-based image retrieval. In *Pacific-Asia Conference on Knowledge Discovery and Data Mining*, pp. 867–874.

Cohen, J. (1960). A coefficient of agreement for nominal scales. *Educational and Psychological Measurement 20*.

Cortes, C. and V. Vapnik (1995). Support-vector networks. *Machine Learning*, 273–297.

Csurka, G., C. R. Dance, L. Fan, J. Willamowski, and C. Bray (2004). Visual categorization with bags of keypoints. In *Workshop on Statistical Learning in Computer Vision, ECCV*.

Farhadi, A., I. Endres, D. Hoiem, and D. Forsyth (2009). Describing objects by their attributes. In *Proceedings of the IEEE Computer Society Conference on CVPR*.

Ferrari, V. and A. Zisserman (2007, December). Learning visual attributes. In *Advances in Neural Information Processing Systems*.

Gallo, I. and A. Nodari (2011). Learning object detection using multiple neural netwoks. In *VISAP 2011*. INSTICC Press.

Gallo, I., A. Nodari, and M. Vanetti (2011). Object segmentation using multiple neural networks for commercial offers visual search. In *EANN2011 Engineering Applications of Neural Networks*. ACM Press.

Kemp, C., J. B. Tenenbaum, T. L. Griffiths, T. Yamada, and N. Ueda (2006). Learning systems of concepts with an infinite relational model. In *AAAI*.

Kumar, N., A. C. Berg, P. N. Belhumeur, and S. K. Nayar (2011, Oct). Describable visual attributes for face verification and image search. In *IEEE Transactions on Pattern Analysis and Machine Intelligence (PAMI)*.

Lampert, C. H., H. Nickisch, and S. Harmeling (2009). Learning to detect unseen object classes by between-class attribute transfer. In *Computer Vision and Pattern Recognition*.

Sivic, J., C. L. Zitnick, and R. Szeliski (2006). Finding people in repeated shots of the same scene. In *Proceedings of the British Machine Vision Conference*.

Vaquero, D. A., R. S. Feris, D. Tran, L. Brown, A. Hampapur, and M. Turk (2009). M.: Attribute-based people search in surveillance environments. In *In: IEEE Workshop on Applications of Computer Vision*.

Wang, Y. and G. Mori (2010). A discriminative latent model of object classes and attributes. In *European Conference on Computer Vision*, pp. 155–168.

Yang, J., Y.-G. Jiang, A. G. Hauptmann, and C.-W. Ngo (2007). Evaluating bag-of-visualwords representations in scene classification. In *Proceedings of the international workshop on Workshop on MIR*. ACM.

Segmenting active regions on solar EUV images

C. Caballero & M.C. Aranda
Department of Languages and Computer Science, Engineering School, University of Malaga

ABSTRACT: Solar catalogs are handmade by a committee of experts in solar physics, which label each of the solar phenomena that appears in the Sun. The first task is to identify which regions of the Sun are regions of interest. The second task of the experts in solar physics is to determine the occurrence of solar phenomena such as solar flares or coronal mass ejections which mainly occur in the regions of interest. The appearance of new tools is very useful because it automates work which is currently performed manually. This paper presents a new segmentation method that identifies candidate regions which will be active regions. The technique presented in this paper is based on a seeded region growing segmentation in order to achieve a robust automatic segmentation of candidate active regions in solar images. The procedure developed has been tested on 3500 full-disk solar images taken from the satellite SOHO. The results of the proposed method were compared with typical segmentation methods with good results showing the robustness of the method described.

1   INTRODUCTION

The automatic processing of information in Solar Physics is becoming increasingly important due to the substantial increase in the size of solar image data archives and also to avoid the subjectivity that may occur when this information is treated manually. The automated detection of solar phenomena such as sunspots, flares, solar filaments, active regions (*AR*), etc, is important for, among other applications, data mining and the reliable forecasting of solar activity and space weather.

In this paper we focus on the automatic detection of AR in solar Extreme Ultraviolet (*EUV*) images obtained from the satellite SOHO. AR are solar regions with intense magnetic activity which can be detected as bright regions in the bands of $H\alpha$ or EUV. AR have been manually detected and numbered for dozens of years by the NOAA organization. Some automated detection methods have been developed in order to avoid the inherent subjectivity of manual detections of solar phenomena (Banda and Angryk 2009; Gill et al. 2010).

An important step in the detection of AR is the segmentation of regions of interest (*ROIs*) in the image. The segmentation of solar phenomena has been addressed in many ways in recent years. There are two main categories: 1. Edge-based methods identifies the boundaries of the ROIs (Veronig et al. 2000; Fuller et al. 2005); 2. Region-based methods use the connectivity and the properties of individual pixels to determine whether the pixels are relevants (McAteer et al. 2005; Wit 2006; Barra et al. 2008; Aranda and Caballero 2010). Although these methods have not been tested on a large set of images.

We have based our method on seeded region growing improving the way seeds and thresholds are chosen. In this paper, different segmentation techniques are analyzed for solar images. The two main problems that arise when segmenting AR on solar images are: 1. The appearance of bright points (*BPs*) which are not related to the emergence of AR. For our aim these BPs are noise; 2. The appearance of solar flares because these cause an increase in the value of the intensity in the pixels of the image where the solar flare appears. This may mean that in one ROI a solar flare cancels the presence of other ROI.

The rest of this paper is organized as follows. Section 2 describes the preprocessing techniques performed on the image. Section 3 describes some classic methods of segmentation to make a comparative study. Section 4 presents our method for segmentation of solar EUV images. Finally, in Section 5 some experimental results are presented.

2   IMAGE ACQUISITION AND IMAGE PREPROCESSING

Image acquisition is provided by another computer system that stores solar images from the satellite SOHO. FITS is the most commonly used digital file format in astronomy. It is designed specifically for scientific data and hence includes many descriptions of photometric and spatial calibration information and image origin metadata.

Prior to the segmentation process, solar images have to be pre-processed in order to correct geometrical or photometric distortions. A charge-coupled device

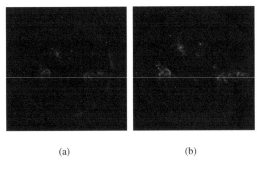

(a)   (b)

Figure 1. (a) Original image in FITS format and (b) preprocessed image without background, halo and contour of the Sun taken on January 15, 2005.

(CCD) is used in all space-borne telescopes operating in EUV wavelengths, such as SOHO. The images acquired by CCD are not always correct, but contain up to 0.1% bad pixels. The calibrations applied to the images from SOHO were dark current subtraction, degridding, filter normalization, exposure time normalization and response correction. After the calibration of the image has been done, the background, the halo, and the contour of the image are completely erased as can be seen in Figure 1.

3 A REVIEW OF REGION-BASED IMAGE SEGMENTATION METHODS

Region-based methods are based on image regions that have similar properties. The main methods are based on thresholding, seeded region growing, and Split & Merge. The following subsections briefly describe each of these three techniques.

3.1 Thresholding-based segmentation

The main objective of thresholding-based methods is to distinguish relevant objects at different levels of intensity. These methods are useful when objects have characteristic and unique levels of intensity. A pixel P of the image $I(x, y)$ belongs to a relevant object if the pixel value is greater than or equal to a threshold σ and is considered irrelevant if the pixel value is not greater than this threshold. Thresholding-based segmentation methods are classified into three categories based on how to select the threshold: global, local, and adaptive. These thresholds are described in Equation (1).

$$
\begin{aligned}
global &: \sigma = f(I(x, y)), \\
local &: \sigma_i = f(N(x_i, y_i), I(x, y)), \\
adaptive &: \sigma_i = f(x_i, y_i, N(x_i, y_i), I(x, y)),
\end{aligned} \tag{1}
$$

where $N(x_i, y_i)$ is a local neighborhood function at point (x_i, y_i) of the image. The purpose of the thresholds is to minimize the number of misclassified pixels.

Segmentation methods based on global thresholding use a single value for the image. A widely accepted value in the literature for the histogram-based segmentation using a global threshold is the optimal value of Otsu (1979).

Thresholding is called local thresholding when a different threshold is used for different regions in the image. In addition, if the thresholding is calculated for each pixel then this threshold is known as adaptive or dynamic thresholding (Wellner 1993).

3.2 Seeded region growing

Thresholding-based methods ensure the homogeneity of the regions but do not guarantee that they are related. This fact leads to situations in which the pixels of relevant objects are not selected. The information from the environment of the pixel is used in the methods of segmentation based on seeded region growing (SRG) (Adams and Bischof 1994). Thus, the pixels of the same environment tend to have similar statistical properties and belong to the same object.

Methods based on SRG obtain satisfactory results without requiring parameters. This method uses a set of n seeds, initially each of these seeds is a set A_i. The method adds a pixel on the edge of one of the sets A_i in each step. The results depend on the seeds because these pixels create the relevant objects in the image. False positives occur if many seeds are used. On the other hand, important objects may not be selected if few seeds are used. SRG is very sensitive to noisy images because if noisy pixels are selected as seeds then the error is propagated. In conclusion, the method is fast and robust if seeds are suitable for the image.

3.3 Split & Merge

The Split & Merge (SM) methods (Horowitz and Pavlidis 1974) divide the image into sub-images, each sub-image is split again if it is not homogeneous. A quaternary tree is generated in which the root node is the image and the child nodes are each of the four sub-images generated. The homogeneity function is applied on each sub-image. The disadvantage of these methods is that adjacent regions with identical properties are split into sub-images. That is, a relevant object is split in two different quadrants. This causes, a merging of sub-images to be performed. It also has a high computational cost.

4 IMAGE SEGMENTATION: THE $GSRG_H^2$ METHOD

This section presents our method called Global Seeded Region Growing based on the Geometry of the Histogram ($GSRG_h^2$) for solar EUV image segmentation, which is a SRG method with a new way to select the seeds. This technique is based on the histogram of the images and consists of three phases: 1. A new procedure of seed selection; 2. SRG is applied using the seeds previously selected; 3. The ROIs are selected using a filter for the specific-domain of solar images.

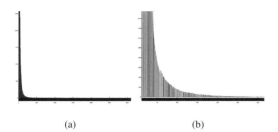

(a)	(b)

Figure 2. (a) Typical histogram of a SOHO solar preprocessed image (b) Histogram of a solar image preprocessed which shows the lines approximated by least-squares in red.

4.1 Seed selection: The GTG_h method

Seed selection is done using a technique based on the histogram, this section presents our method called Global Thresholding based on the Geometry of the Histogram (GTG_h) for seed selection which is not based on a local or global statistical measurement but is based on the geometry of the histogram of the image which is not affected by large changes in values of the histogram. In particular, it uses a pattern that appears in the histogram of solar images after preprocessing. In the preprocessed solar images, most pixels have a value close to zero, so pixels with the highest values of the histogram have few occurrences in the histogram. Figure 2 shows the straight lines that have been calculated using the least-squares approximation for small groups of M points of the histogram values instead of calculating the straight line that represent all values in the histogram. Note that the slope of these lines is negative, and therefore, the trend is decreasing. However, the straight line that represents the last points of the histogram has a slope very close to zero. This inherent characteristic of a pre-processed solar image histogram is used by our method. There are N possible values for the histogram, so there are $\lceil \frac{N}{M} \rceil$ different straigh lines with its corresponding slope.

The threshold that determines the value of seed is calculated using the points of the histogram. If the value of the seeds is determined at the beginning of the fall of the histogram then the intensity level of the pixels that are selected will be too low. This causes that irrelevant pixels (noise) are taken as relevant objects. On the other hand, if the value of the seeds is determined at the end of the fall of the histogram then relevant objects are taken as noise. So, the ideal threshold is close to the beginning of the fall. Algorithm 1 shows the pseudocode of the algorithm GTG_h. So, let f_i be the function that performs the first-degree polynomial fit by least-squares in the points of the histogram $(x_1, y_1), \ldots, (x_M, y_M)$ and $(x_{2i-1}, y_{2i-1}), \ldots, (x_{i \cdot M-1}, y_{i \cdot M-1})$ with $i = 2, \ldots, N-M$ and $M < N$. In addition, the slope of the function f_i is defined as a_i. Finally, let F_M be the family of functions f_i with $i = 1, \ldots, N-M$.

The next step of the method is to select those functions f_i of the set of functions F_M in which there is a big fall in the histogram, i.e., the slope of the function f_i is very small. So, let F_c be the family of functions f_i

Algorithm 1. GTG_h: Seed Selection.

Input: I: Preprocessed solar image.
Output: $seeds$: Set of seeds to be used with the SRG method.
Data: $F_C = \emptyset$, $P = \emptyset$, σ_c = positive small value, σ_p = negative small value, Δ_c = a decreasing factor, Δ_p = an increasing factor.

1 **while** $size(F_C) = 0$ **do**
2   **foreach** f_i **do**
3   **if** $a_i < \sigma_c$ **then**
4        $F_C.add(f_i)$
5    $\sigma_c = \sigma_c \cdot \Delta_c$
6 The functions g_i and its slope c_i are calculated
7 **while** $size(P) = 0$ **do**
8   **foreach** f_i **do**
9   **if** $F_C.contains(f_i)$ and $c_i < \sigma_p$ **then**
10        $P.add(f_i)$
11    $\sigma_p = \sigma_p \cdot \Delta_p$
12 $seeds = I(x, y) < round(P.first().getX())$

---

whose slope satisfies the condition $a_i < \sigma_c$, where the threshold σ_c is defined with a very small value. If any function f_i satisfies the previous condition, then the threshold σ_c is modified by a decreasing factor (Δ_c). This step is repeated until the number of functions f_i selected is greater than 0.

The family of functions g_i is also calculated, where g_i is defined as the function that performs the first-degree polynomial fit by least squares of the points $(x_{(iM)}, y_{(iM)}), \ldots, (x_{(iM+M \cdot NG)}, y_{(iM+M \cdot NG)})$ of the histogram. NG defines the number of lines of size M that are approximated. The points of the functions g_i are related to the points of the functions f_i. Thus, histogram points used to calculate the functions g_i are the following histogram points that have been used to calculate the functions f_i.

The family of functions g_i are used to determine the trend of the points after each of the pre-selected functions f_i. As stated above, the idea is to select the value of the histogram in which the trend is decreasing but the first value or the value where the trend is over are not useful. So, the threshold σ_p is defined as a negative slope and small enough to consider those functions with a small slope and discard those functions with a large fall which are often the first.

Let P be the family of functions f_i of the set F_C which also satisfy the condition $c_i < \sigma_p$. If any function f_i satisfies the previous conditions then the threshold σ_p is modified using an increasing factor (Δ_p). This step is repeated until the number of functions f_i selected is greater than 0. The functions of the set P represent the straight lines between the points $(x_j, y_j), \ldots, (x_k, y_k)$. The value of the seed is determined by the average of the values x_j and x_k from the first function of the set P. The first function of the set P is

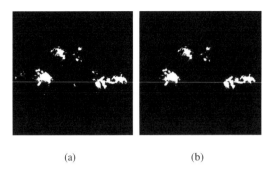

(a)   (b)

Figure 3. (a) Segmented image using the seeds of the method $GSRG_h^2$ (b) image after selecting the candidate regions of the Sun taken on January 15, 2005.

selected because this function has the smallest values of x_j and x_k.

4.2 Selection of candidate regions

After applying the SRG method, there is noise in the images, as well as regions that are not important. Therefore, a more satisfactory result is achieved after applying a selection phase of candidate regions. This phase consists of two steps: 1. Remove all regions that do not exceed a minimum value of area and selecting large regions. This eliminates regions from seeds that were false positives; 2. Once large regions have been selected, the next step is to look for regions close to these, regardless of the size of these surrounding regions. The result of the selection process is positive, as can be seen by comparing Figure 3a and 3b.

4.2.1 Removing noise and selecting large regions
Let R be the set of regions that have been selected after the SRG method and R_i the ith region of the set R. Furthermore, let σ_a be the area threshold that determines which elements of the set R are selected as candidate regions. So, the regions of set R that satisfies the condition $size(R_i) \geq \sigma_a$ are selected, where $size(R_i)$ is the function that calculates the area of the region i^{th}. Similarly, let σ_r be the threshold that determines which elements of the set R are too small and therefore are likely to be noise. The regions that satisfy the condition $size(R_i) < \sigma_r$ are initially discarded as candidate regions.

4.2.2 Adding the surrounding regions
Let S be those elements of the set R that satisfy the condition $R_i \geq \sigma_a$ and let P be the set of elements of R where the area is between the values σ_r and σ_a and have not yet been selected. Finally, P_i and S_j are the regions i^{th} and j^{th} of the sets P and S respectively.

There may be regions belonging to the set P with relevant information. These regions are those that initially have a large area and satisfies the condition $R_i \geq \sigma_a$ but in the segmentation process are divided into several regions and some or all of these regions do not satisfy the condition.

A region $P_i \in P$ becomes an element of S if it satisfies the condition shown in (2). That is, if the number of regions close is significant, then the region is a ROI, although the area of the regions were not initially large. This is achieved using the threshold distance between regions (σ_d) and the threshold of the number of regions (σ_e).

Threshold σ_d is defined as the minimum distance between two elements of the sets S and P and the threshold σ_e determines the minimum number of regions close to one another. That is, those regions of the set P that are near at least one number σ_d of other regions of the set P are selected. Thus, the noise regions, which satisfy the condition $R_i < \sigma_r$, are not selected and the relevant regions, which the segmentation process divided into several smaller regions are selected.

$$\#(d(P_i, P_j) < \sigma_d) \geq \sigma_e, \forall j \in [1 \ldots N_p], \qquad (2)$$

where d is the distance function between two regions which is defined as the smallest Euclidean distance between two points on the two regions to which distance is calculated.

5 EXPERIMENTAL RESULTS

In this study, more than 3500 solar images from the satellite SOHO have been used. The images are from different days to provide a variety of cases to be studied. In some cases there is only one ROI, and in others there are even 6 or 7 ROIs, as well as sequences of images in which ROIs appear and/or disappear.

A committee of experts on solar images has been established to achieve greater objectivity in the results presented. Experts have been asked to mark ROIs that are candidates to become AR. Once images have been marked by the experts and the segmented images have been obtained, a comparison between these images is performed. It leads to the following classification: 1. Satisfactory images; 2. Images with a slight, tolerable noise; 3. Images with considerable, tolerable noise; 4. Useless images.

Figure 4 shows an image labeled by the experts and the images obtained by different segmentation techniques on 28 September, 2009. First, Figure 4(a) shows the initial image that the experts labeled with two ROIs. The edge-based segmentation is performed using the method of Kirsch (1971) which obtains the edges of the ROIs (Figure 4(b)). Although in this particular case the method achieves a satisfactory result, this operator is actually too sensitive to noise, as can be seen in the results of the Table 1, which do not even reach 18% of successful segmentations.

The result of segmentation based on the Otsu threshold (Figure 4(c)) recovers the two regions, but adds several regions of noise that are tolerable. This technique is affected by the occurrence of solar flares because it uses a global threshold. Figure 4(c) shows the method of Wellner which retrieves images useless

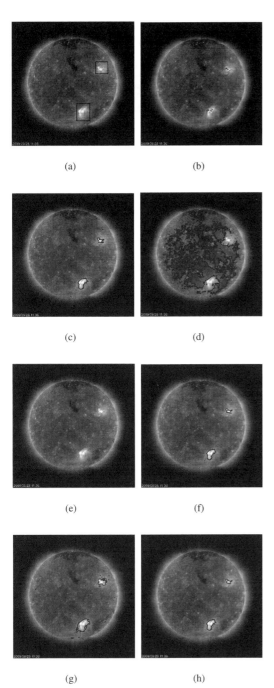

(a)   (b)

(c)   (d)

(e)   (f)

(g)   (h)

Figure 4. (a) Initial image of the day September 28, 2009 with several bright regions (b-f) segmented images using different methods of segmentation: Kirsch, Otsu, Wellner, SM, SRG using seeds obtained by the methods of Otsu and Wellner and $GSRG_h^2$.

Table 2. Percentages of the results obtained for the segmentation methods. *Sat* are satisfactory images, ± 1 are the images with slight noise tolerable, ± 2 are the images with considerable noise tolerable and *Int* are useless images. The last column shows the average running time in seconds.

Methods	Sat.	± 1	± 2	Int.	T_{CPU}
Kirsch	17.84	4.90	7.94	69.31	0.36
Otsu	8.27	24.30	9.71	57.72	0.08
SM	2.59	0.03	3.30	94.08	0.62
Wellner	0.00	0.00	0.00	100.00	0.88
SRG Otsu	7.99	23.05	6.55	62.41	4.45
SRG Wellner	8.97	57.47	27.99	5.56	4.09
$GSRG_h^2*$	46.35	50.49	2.07	1.09	7.20
$GSRG_h^2$	87.99	8.59	3.38	0.02	14.03

as most of the pixels in the image of a dark color (low intensity values) cause sub-images tend to have a value close to 0 (dark). This error is propagated to all sub-images, as a result, the final image does not select any ROI (Figure 4(e)).

The SRG methods perform well for this particular case, the SRG using the Otsu value (Figure 4(f)) produces the same results as the segmentation based on the optimal value of Otsu but wasting more time. The SRG method using the values of Wellner method as seeds has the problem of not relating the pixel information away, causing the BP in a local context, are the brightest pixels in the region. So, Figure 4(g) shows that most of the BP are selected as ROI in the segmentation. Finally, the results of $GSRG_h^2$ method are shown in Figure 4(h), which obtain the two ROI without noise. This is due to the fact that the pixels are considered as seeds which are globally relevant in the image, the appearance of solar flares does not affect other ROI because the selection of these seeds is not a statistical measurement but a geometric measurement on the histogram of values. Finally, a custom filter is applied to eliminate noise. These good results are repeated in most of the images.

Table 1 shows a summarize of the results. Generally, the typical methods of segmentation obtained unacceptable results for a real system, while the $GSRG_h^2$ method obtains results that exceed 87% of satisfactory images. However most images which are not satisfactory in the group of images with slight noise (8.59%), may be corrected at a later stage of interpretation. Futhermore, the table shows the results of the method without applying $GSRG_h^2$ specific filter, which also obtain better results than typical techniques.

6 CONCLUSIONS

This paper has presented a new segmentation method applied to solar physics. The method, called $GSRG_h^2$, is based on a SRG algorithm and consists of three phases: seed selection, SRG and noise removal. This technique is not affected by the two main problems of

in all images, because most pixels are selected as ROI, due to the inherent problems of local thresholding. The SM method divides the image into sub-images and the average value of pixels is assigned to each sub-image,

segmentation of images of solar physics: bright points and solar flares. The $GSRG_h^2$ method has been tested empirically on a database of over 3500 solar images from the satellite SOHO. A committee of experts on solar images has been established to determine what should be the final results of segmentation. The results obtained with the proposed method have been very positive, because the segmentation of ROIs was recovered in 87% of cases. Finally, the results of this new method were compared with results from traditional segmentation methods. The traditional methods of segmentation produce negative results, with about 18% of images being satisfactory, whereas the method without and with specific filter reaches 46% and 87% respectively.

REFERENCES

Adams, R. and L. Bischof (1994). Seeded region growing. *IEEE Transactions on Pattern Analysis and Machine Intelligence 16*, 641–647.

Aranda, M. C. and C. Caballero (2010). Automatic detection of active region on EUV solar images using fuzzy clustering. In *International Conference on Information Processing and Management of Uncertainty in Knowledge-Based System*, pp. 69–78.

Banda, J. and R. Angryk (2009). On the effectiveness of fuzzy clustering as a data discretization technique for large-scale classification of solar images. In *IEEE International Conference on Fuzzy Systems*, pp. 2019–2024.

Barra, V., V. Delouille, and J.-F. Hochedez (2008). Segmentation of extreme ultraviolet solar images via multichannel fuzzy clustering. *Advances in Space Research 42*(5), 917–925.

Fuller, N., J. Aboudarham, and R. Bentley (2005). Filament recognition and image cleaning of meudon halpha spectroheliograms. *Solar Physics*, 61–73.

Gill, C., L. Fletcher, and S. Marshall (2010). Using active contours for semi-automated tracking of uv and euv solar flare ribbons. *Solar Physics 262*, 355–371.

Horowitz, S. and T. Pavlidis (1974). Picture segmentation by a directed split-and-merge procedure. *Proceedings of the Second International Joint Conference on Pattern Recognition*, 424–433.

Kirsch, R. (1971). Computer determination of the constituent structure of biological images. *Computers and biomedical research 4*, 315–328.

McAteer, R., P. Gallagher, J. Ireland, and C. Young (2005). Automated boundary-extraction and region-growing techniques applied to solar magnetograms. *Solar Physics 228*, 55–66.

Otsu, N. (1979). A threshold selection method from grey level histograms. In *IEEE Transactions on System, Man, and Cybernetics*, New York, pp. 62–66. IEEE Press.

Veronig, A., M. Steinegger, W. Otruba, A. Hanslmeier, M. Messerotti, M. Temmer, S. Gonzi, and G. Brunner (2000). Automatic Image Processing in the Frame of a Solar Flare Alerting System. *Hvar Observatory Bulletin 24*, 195–205.

Wellner, P. (1993). Interacting with paper on the digital desk. *Communications of the ACM 36*, 86–96.

Wit, T. D. D. (2006). Fast segmentation of solar extreme ultraviolet images. *Solar Physics 239*, 519–530.

Automated Shape Analysis landmarks detection for medical image processing

Nicola Amoroso & Roberto Bellotti
Università degli Studi di Bari "Aldo Moro", Bari, Italy
Istituto Nazionale di Fisica Nucleare, Bari, Italy

Stefania Bruno
Dipartimento di Neurologia Clinica e di Ricerca
Azienda Ospedaliera "Card. G.Panico", Tricase, Italy
Università degli Studi di Bari "Aldo Moro", Bari, Italy

Andrea Chincarini
Istituto Nazionale di Fisica Nucleare, Genova, Italy

Giancarlo Logroscino
Department of Neurosciences and Sense Organs
Azienda Ospedaliera "Card. G.Panico", Tricase, Italy
Università degli Studi di Bari "Aldo Moro", Bari, Italy

Sabina Tangaro
Istituto Nazionale di Fisica Nucleare, Bari, Italy

Andrea Tateo
Università degli Studi di Bari "Aldo Moro", Bari, Italy

ABSTRACT: A fully automated shape analysis algorithm based on the Point Distribution Model is proposed (APoD). The algorithm identifies automatically the edges of noisy shapes, determining for each shape a fixed number of contour points and the underlying *true* shape. The proposed algorithm has been tested using a database of simulated images with different noise levels. The performance of the model was investigated using 50000 simulated images which differ from a gold standard for approximately 20% of pixels. With this method a Dice index $D = 0.968 \pm 0.004$ is obtained.

1 INTRODUCTION

In medical imaging accurate segmentation of anatomical structures is essential for the detection of change, both within individuals and for group comparisons The main problems faced when dealing with biological shapes are that they are not regular and it is often difficult to obtain real images that can be used as gold standard comparison templates. The first problem can be solved by using learning iterative algorithms, such as the wavelet analysis combined with a principal component analysis (Davatzikos et al. 2003) or deformable representations (Pizer et al. 2003). As to the determination of a gold standard temÂplate task it is worthwhile to note that this problem depends on the type of data and on the analysis models used to describe it. A suitable approach is to determine a constant number of contour points in order to compare different shapes in a point distribution model framework (Cootes et al. 1995). With this aim we have created a database of simulated images with different noise levels allowing us both to test an iterative learning process and to create a sound statistical model.

This study, in particular the design of computer aided detection systems for medical applications (Calvini et al. 2009) and the diagnosis of early stages of Alzheimer's disease (Chincarini et al. 2011), is part of a research project, funded by the Istituto Nazionale di Fisica Nucleare (INFN), aimed to develop suitable techniques in Medical Imaging for Neurodegenerative Diseases (MIND). In recent years the issue of brain segmentation, both in its main components (grey matter, white matter and cerebro-spinal fluid) or in chosen anatomical regions, has been one of the main topics in neuroimaging research. Segmentation of the hippocampus, in particular, represents one of the biggest challenges in medical image processing. The hippocampus is a complex brain structure with a primary role in memory and learning, and involvement in a number of other cognitive functions.

Hippocampal atrophy is perhaps the most important imaging biomarker in Azheimers's Disease (AD) (Frisoni et al. 2009), a progressive neurodegenerative disorder, leading to global cognitive impairment, and affecting more than 30 million people all around the world. The hippocampus loses volume in AD at a rate of 5% per year, thus making hippocampal atrophy the most important imaging biomarker of the condition. The prevalence of AD is expected to quadruplicate by 2050 (World Alzheimer Report 2010), hence the development of processing tools allowing the analysis of large amounts of imaging data is of primary importance. AD is characterised by deposition of abnormal proteins in the brain, leading to abnormal cell function, and resulting in neuronal death. The neuronal loss is reflected in global and regional brain atrophy, both correlated with cognitive decline (Jack Jr. et al. 2009; Schott et al. 2007), and both important outcome measures in AD clinical trials (Fox et al. 2005). The quantification of the volume loss through magnetic resonance is often achieved through accurate detection of the hippocampus contours, performed by the investigator manually or with semi-automated techniques followed by manual editing.

Manual segmentation is time-consuming and it is vulnerable to statistical inhomogeneity, introducing error and increasing the variance of the outcome measure (in turn increasing sample sizes). Intra- and inter-rater reliability, anatomical variability of biological shapes, possible additional pathological changes (*i.e.* vascular disease), absence of a clear distinction between the hippocampus' edges and the neighbouring cerebral structures, are all factors that hinder manual segmentation. In the specific case of the hippocampus the complexity of the shape itself increases the difficulty of the segmentation. For these reasons the development of a general algorithm enabling accurate automatic segmentation the hippocampus is compelling.

We developed a fully Automated Point Distribution (APoD) algorithm which individuates the shape of an object intended as a collection of a fixed number of points. Points are uniformly sampled. The uniform distribution keeps the main aspects of the shapes unchanged and it reduces the contour size. This allows us to describe each shape with the same number of degrees of freedom and therefore to perform a straightforward comparison.

2 THE SIMULATED IMAGE DATABASE

With the aim of studying a complex shape such as the hippocampus, we have built a database of simulated images. First of all we used a standard image I, taken from the Moving Picture Experts Group (MPEG), representing a watch. This is a $(257, 559)$ pixel binary image [Figure 1].

This image has been then processed with iterative algorithms in order to simulate a biological shape variability.

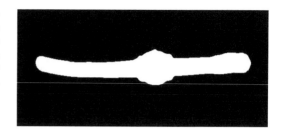

Figure 1.   A watch. This image is from the standard MPEG database.

In order to develop a suitable but controlled software framework we developed a simulated database whose parameters can be easily tuned. This allows both to represent a wider range of shapes and to acquire further insight on the shape model under investigation.

We introduced a gaussian noise to vary the image contours. For this purpose a $(257, 559)$ matrix A of real numbers is extracted from a gaussian distribution. Then the gaussian noise $R = r \times A$ is calculated, where r is a suitable real positive number representing the noise factor amplification. A number N of noisy binary images

$$I_i = I + R_i$$

can be then obtained by iterating the procedure to generate the noise. The gaussian noise was chosen for two different reasons: it is able to take into account the natural variability of biological shapes and at the same time it introduces on the images artifacts which reproduce well real cases. In this regard we have studied the similarity of the simulated images with the reference image I by measuring the sum of the squared pixel intensity differences function SSD as a function of the factor r:

$$SSD(r) = ||I - I_i(r)||^2.$$

The SSD function can be interpreted as a counter of the differences between two images and therefore as a similarity index. As expected the function SSD increases with the value of r. In this study it is important that the simulated images I_i can be related to the reference I therefore we have a chosen the value for r so that all images I_i differ from I of approximately the 20% of the pixels. A range of different values for the variable r was explored; in this way the value of $r = 2.3$ corresponding to $SSD = 0.22$ was chosen. [Figure 2].

The original image undergoes an iterative modification process which gives a simulated database of 50 different binary images. There is no limit to the number of simulated images that could be generated, but the value of 50 was preferred in order to better represent a typical database of medical images. The parameters of the model fix at this point the noise amplitude and the dimensionality of the database itself; an example of the results obtained is shown in the following [Figure 3]. It is worthwhile to note that the study and the characterization of the different parameters' role in the shape

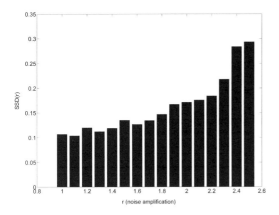

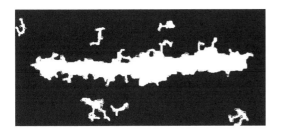

Figure 2. This image shows how the increment of the noise factor yelds an increment for the *SSD* function.

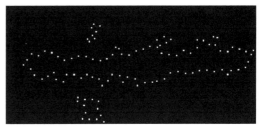

Figure 4. Iterative algorithms extract from each contour a fixed number of points which can therefore used to represent a generic shape. The picture has to be considered a useful example to allow the visualization.

Figure 3. The original shape undergoes a noisy random process which reproduces the biological shape variablity taking into account the possibility of detected artifacts.

model construction is foundamental to obtain significant reliability of the model itself when applied to real images.

3 THE AUTOMATIC POINT DITRIBUTION MODEL

The simulated database is used to study the performance of the APoD model. Every image I_i can be seen as a matrix $I_i(s, t)$ with values 0, 1; the algorithm proposed detects for every column n where $n = 1, ..., 559$ the pixels having value equal to 1. These can be seen as an ordered vector $(I_i(m_\alpha, n), ..., I_i(m_\omega, n))$ whose minimum $m_{i,n} = m_{i,\alpha}$ and maximum $M_{i,n} = m_{i,\omega}$ can be determined. The points $I_i(m_n, n)$ and $I_i(M_n, n)$ are interpreted as the contour points of the n-th column. By iteration of this procedure upon all the different columns the contour can be therefore represented for every image as a vector of coordinates. It is worthwhile to note that each contour $\vec{x}_i$ has a different dimension d_i depending on the particular case under investigation. Once contours have been determined a uniform sampling of $\vec{x}_i$ is performed and a fixed number of contour points k is detected, in literature these points are best known as *landmarks* [Figure 4].

This is a very fast computation so there is no need to explore different values for the number of extracted points. For this study we chose a number of contour points $k = min(d_i)/2$.

These k points are then used to perform a reconstruction of the "true" shape S. First of all the I_i images are summed to obtain the new image S, then a threshold is applied to make S binary. Finally the contour is retrieved through a simple linear interpolation. The choices for the threshold and for the linear interpolation have been tested in cross validation on a different simulated database. No sensible improvements are achievable on this data using more sophisticated algorithms as the cubic or the spline reconstruction for the gold standard I.

4 EXPERIMENTAL RESULTS

To quantify this procedure the Dice index D has been calculated for each reconstruction algorithm; it is usually used as a measure of accuracy for a segmentation process. In fact, by definition:

$$D = \frac{2|S \cap I|}{|S| + |I|}$$

which can be therefore interpreted as a measure of how many pixel of the image unknown I are correctly returned by S.

The "true" shape has to be validated with the reference image I. We have simulated 1000 datasets, each one consisiting of 50 noisy images. These datasets have been simulated according to the procedure described in the section 3. For each dataset the "true" shape has been determined and a Dice index has been calculated using the original I as gold standard.

The distribution obtained for the Dice indexes allowed us to determine a mean Dice $D = 0.968 \pm 0.004$ [Figure 5]. Even from a qualitative point of view it is possible to appreciate how the algorithm is capable to reproduce the unknown "true" shape [Figure 6].

It is worthwhile to note that the only hypothesis of the model is that the dataset consists of similar images. The next task will consist in the validation of the model accuracy on real images, with the aim of developing a gold standard template for the hippocampus. It is worthwhile to note that the described procedure on the simulated images is fully automated.

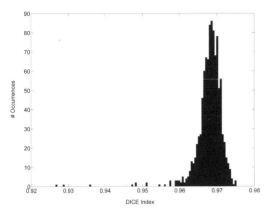

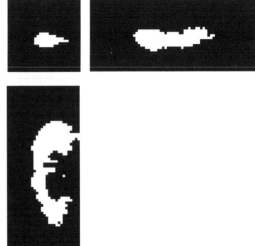

Figure 5. The Dice distribution. It is roughly symmetric so that mean and standard deviation can be used to estimate the Dice and its error: $D = 0.968 \pm 0.004$.

Figure 6. The figure shows the shape S representing the underlying "true" shape of our dataset.

Figure 7. The figure shows the sagittal, the coronal and the axial view of a hippocampus (anticlockwise order). The proposed algorithm will be used to retrieve the "true" hippocampal shape according to a database of manually segmented hippocampal boxes.

is gaining recognition alongside the more established image intensity based techniques. The application of the proposed method to simulated data has yielded a very satisfactory performance $D = 0.968 \pm 0.004$, and further improvement is to be expected.

5 APPLICATION OF APoD TO HIPPOCAMPAL SEGMENTATION

The proposed method will be used for the automatic segmentation of the hippocampus [Figure 7].

Widely used image analysis commercial packages such as SPM (Statistical Parametric Mapping) perform an intensity-based segmentation, which does not take into account shape information. The APoD algorithm can be used to determine a "true" hippocampal shape, which, in turn, can be used to improve the SPM performance by averaging the hippocampus obtained with SPM with the "true" ApoD shape. Further studies are necessary to determine the "best fit" parameters for an optimal hippocampal segmentation, but this approach is a promising development, which can be seen as alternative or complementary to atlas-based segmentation or tissue labeling methods based on image statistics.

6 CONCLUSIONS

This work proposes a new APoD model that can be used as an automated tool for the segmentation of images of biological structures. Its more immediate implementation is related to the segmentation of the hippocampus, an area to which a very substantial amount of work in the field of image processing has been recently devoted, and in which shape analysis

REFERENCES

Calvini et al., P. (2009). Automatic analysis of medial temporal lobe atrophy from structural MRIs for the early assessment of Alzheimer disease. *Medical Physics 36*, 3737–3747.

Chincarini et al., A. (2011). Local MRI analysis approach in the diagnosis of early and prodromal Alzheimer's disease. *NeuroImage 58*, 469–480.

Cootes et al., T. F. (1995). Active ShapeModels- Their Training and Application. *Comput. Vis. Image Underst. 61*, 38–59.

Davatzikos et al., C. (2003). Hierarchical Active Shape Models, Using the Wavelet Transform. *IEEE Trans. on Med. Imaging 22*, 414–423.

Fox et al., N. C. (2005). Effects of Abeta immunization (AN1792) on MRI measures of cerebral volume in Alzheimer's disease. *Neurology 64*, 1563–1572.

Frisoni et al., G. P. (2009). The clinical use of structural MRI in Alzheimer disease. *Nat Rev Neurol 6*, 67–77.

Jack Jr. et al., C. R. (2009). Serial PIB and MRI in normal, mild cognitive impairment and Alzheimers's disease: implications for sequence of pathological events in Alzheimer's disease. *Brain 132*, 1355–1365.

Pizer et al., S. M. (2003). Deformable M-Reps for 3D Medical Image Segmentation. *Int. Journal of Computer Vision 55*, 85–106. Schott et al., J.M. (2007). Neuropsychological correlates of whole brain atrophy in Alzheimer's disease. *Neuropsychologia 34*, 996–1019.

World Alzheimer Report (2010). *Alzheimer's Disease International*.

Computational Modelling of Objects Represented in Images – Di Giamberardino et al. (eds)
© 2012 Taylor & Francis Group, London, ISBN 978-0-415-62134-2

Bayesian depth map interpolation using edge driven Markov Random Fields

Stefania Colonnese, Stefano Rinauro & Gaetano Scarano
DIET, Universitá di Roma "La Sapienza", Italy

ABSTRACT: In this work we present a Bayesian interpolation procedure to perform depth map upsampling. The depth map prior is designed via an edge driven Markov Random Field. The upsampling procedure is computationally efficient and outperforms selected state of the art upsampling procedure; moreover it allows to perform depth map upsampling even without the reference high resolution luminance map.

1 INTRODUCTION AND RELATED WORKS

Acquired images and video are nowadays often transmitted along with a depth map, or range map, displaying information on the distance of each framed object from the camera. Such information enables various advanced video-based functionalities, ranging from free view point television and 3D video, to advanced video surveillance or computer vision, to cite a few. This paper addresses the problem of upsampling a low-resolution depth map to reconstruct the unknown high resolution depth map. Depth maps acquisition and compression are expected to be key technologies for future innovative video services (1). The problem of depth map upsampling emerges primarily from a technological limit in the range acquisition devices; in fact, due to the high cost of the range map acquisition technology, accurate and robust techniques for improving the resolution of measured depth maps are required. While high definition video and photo cameras are commonly available on the market (*e.g.* 1920×1240 or larger), the commercial depth acquisition devices usually provide lower resolution (*e.g.* 176×144) maps. Moreover, in order to reduce the bandwidth occupancy of 3D or multi-view video systems, depth maps are usually downsampled before compression, so that a suitable upsampling stage is needed at receiver side.

Depth maps upsampling techniques recently debated in technical literature involve statistical, possibly Markovian, priors (6) (7), as well as video sequence analysis (2), (5). In (8) an edge adaptive depth map upsampling procedure stemming from an edge map of the original natural image has been proposed. The resulting upsampled depth maps exhibit an increased sharpreness in edge regions with respect to classical upsampling such as bilinear interpolation and linear spatial scalabel filters (9); yet, since no information about the orientation of the edges is considered, some edges still exhibits visual artifacts which, in turn, may severely affect all of the stages where a correct high resolution depth maop is required, such as,

for instance, the rendering of free viewpoint images. Moreover, the work in (8) estimates the edge map from the high resolution luminance map, thus needing the samples of the low resolution depth map to be correctly registered with the sample of the high resolution luminance map. This latter operation is computationally onerous; besides, it may affect the upsampling procedure when not perfectly performed.

In this paper we describe a single pass upsampling procedure based on modeling the unknown depth map as a Markov Random Field (MRF). The herein adopted MRF involves a complex line process recently introduced for natural images resptration (12), based on a fast image edge estimation stage (13). Differently from (6) and (8), the procedure can be applied even without using the high resolution luminance map and therefore it applies even to stand-alone depth estimation system. In the followings, we briefly introduce our MRF model in Sect.2 and the Bayesian interpolator in Sect.3. In Sect.4 we report simulation results and Sect.5 concludes the paper.

2 MRF MODEL USING COMPLEX LINE PROCESS

We model the unknown high resolution depth image as a realization of a bidimensional MRF $\mathbf{d}_\mathcal{L}$ defined on a rectangular lattice ($\mathcal{L}$). Depth maps are made up by homogeneous regions separated by abrupt boundaries. In this respect, MRFs, aiming to assign higher probabilities to configurations in which small differences between the pixel values occur in large regions, are well suited to be used as a statistical prior for depth maps.

Let us then assume that the depth values are observed on a sub-lattice $\mathcal{L}' \in \mathcal{L}$: the observations' set is then given by $\mathbf{d}_{\mathcal{L}'}$. Here, we address the problem of interpolating the HR depth map $\mathbf{d}_{\mathcal{L} \setminus \mathcal{L}'}$ from the observed LR samples, namely $\mathbf{d}_{\mathcal{L}'}$. Specifically we obtain the to-be-interpolated pixels as the Maximum a Posteriori (MAP) estimation of the high-resolution samples $\mathbf{d}_{\mathcal{L}\setminus\mathcal{L}'}$ given the observations $\mathbf{d}_{\mathcal{L}'}$:

$$\hat{\mathbf{d}}_{\mathcal{L}\setminus\mathcal{L}'}^{(\mathrm{MAP})} = \arg\max p\left(\mathbf{d}_{\mathcal{L}\setminus\mathcal{L}'}|\mathbf{d}_{\mathcal{L}'}\right) \qquad (1)$$

where the maximization is conducted over all the possible values of $\mathbf{d}_{\mathcal{L}'}$,.

Let d_{mn} be the value $\mathbf{d}_{\mathcal{L}}$ of the random field $\mathbf{d}_{\mathcal{L}}$ at the pixel $(m,\ n)$ and let $p(d_{mn}|\mathbf{d}_{\mathcal{L}\setminus(m,n)})$ denote the probability density function (pdf) of d_{mn} conditioned to the values $\mathbf{d}_{\mathcal{L}\setminus(m,n)}$. The random field $\mathbf{d}_{\mathcal{L}}$ is said to be a MRF if, for every pixel $(m,\ n)$, a neighborhood η_{mn} is found such that:

$$p\left(d_{mn}|\mathbf{d}_{\mathcal{L}\setminus(m,n)}\right) = p\left(d_{mn}|\mathbf{d}_{\eta_{mn}}\right) \qquad (2)$$

A set of pixels such that all the pixels belonging to the set are neighbors of each other is called a clique. The joint pdf of a random field satisfying (2) takes the form of a Gibbs distribution (10), given by:

$$p\left(\mathbf{d}_{\mathcal{L}}\right) \overset{\mathrm{def}}{=} \frac{1}{Z}\exp\left(-\frac{1}{T}\sum_{c} V_c\left(\mathbf{d}_c\right)\right) \qquad (3)$$

where the functions $V_c(\mathbf{d}_c)$ operate on subsets of pixels $\mathbf{d}_c$ belonging to the same clique (c), and the sum is carried out on all the cliques in the field. In this work we will consider cliques composed by two pixels, i.e. ($\mathbf{d}_c = \{d_{mn} - d_c\}$). The functions $V_c(\mathbf{d}_c)$ are called *potential functions* and the parameter (T) driving the pdf curvature is often referred to as the *temperature of the distribution*.

The MRF is characterized by the neighborhood system defined on ($\mathcal{L}$) and by the form of the potential functions $V_c(\mathbf{d}_c)$, which ultimately determine the energy, and hence the probability, of the configuration $\mathbf{d}_{\mathcal{L}}$.

A key element in the design of the MRF based interpolation procedure is the suitable determination of the neighborhood system η_{mn} and of the potential functions $V_c(\mathbf{d}_c)$ that represent the tightness of the spatial continuity constraints. We employ the homogeneous neighborhood system η_{mn} depicted in Fig.1, formerly introduced in (10), which encompasses two pixels cliques $\mathbf{d} = \{d_{mn}, d_c\}$ oriented along different directions φ_c allowing the clique potential $V_c(\mathbf{d}_c)$ to adapt to up to 8 different edge directions.

We assume the following quadratic form the potential function $V_c(\mathbf{d}_c)$:

$$V_c\left(\mathbf{d}_c\right) \overset{\mathrm{def}}{=} w_c(d_{mn} - d_c)^2 \qquad (4)$$

The parabolic term in (4) measures luminance variations between the pixels d_{mn}, d_c. The factor w_c represents the weight of the clique. From (4), we recognize that large values of w_c amplify the contribution of the cost term $w_c(d_{mn} - d_c)^2$, to the overall configuration energy, and discourage abrupt luminance variations along the direction φ_c of the clique c. On

the contrary, small values of w_c, result in looser luminance constraints. Thereby, the weights w_c definitely associate higher probabilities to pixels configurations in which small differences between the pixel values occur in large regions.

To cope with uniform regions boundaries, where discontinuities are allowed, spatially variant weights are considered. Classical approaches represent the uniform region boundaries by means of suitable binary (10) or real valued (11) line processes. Here, following the guidelines in (12), we generalize these works resorting to a complex valued line process, to formally take into account visually relevant characteristics such as the edge intensity and orientation. Towards this aim, let us introduce a complex edge process e_{mn}, whose magnitude is proportional to the image edge intensity, and whose orientation is parallel to the edge orientation.

To elaborate, let us consider a generic site $(m,\ n)$ and let us associate a clique direction φ_c to every clique $c \in \eta_{mn}$. The weight w_c should vary as a function of the angular distance between the clique direction φ_c and the local edge direction; specifically we define the weighting function so that it reaches its maximum value on the clique oriented along the edge. Besides, since the achieved maximum value represents the contribution of the clique potential function to the overall configuration energy, the weighting function should depend on the edge intensity in $(m,\ n)$. The weight w_c can then be defined as the scalar product between the edge process e_{mn} and the versor $e^{j\varphi_c}$ of the clique direction. In formulas:

$$w_c = <\mathrm{e}_{mn}, e^{j\varphi_c}> = |\mathrm{e}_{mn}| \cdot \cos|\varphi_c - \arg(\mathrm{e}_{mn})| \qquad (5)$$

The edge process e_{mn} generalizes the binary line process $l_{m,n}$ in (10) and the real valued process $\alpha_{m,n}$ in (11) and it allows to elegantly include visually relevant image features in the MRF definition. The feasibility of the herein introduced MRF model strictly depends on the availability of a procedure for estimation of the edge process e_{mn}. The Circular Harmonic Functions (CHFs) provide an efficacious mean for such estimation, as explained in the followings.

The first order CHF $\psi_{mn}^{(1)}$ is a bandpass filter, defined by the following impulse response:

$$\psi_{mn}^{(1)} = \frac{1}{\alpha}\sqrt{m^2 + n^2}e^{-(m^2+n^2)/\alpha^2}\,e^{-j\arctan n/m} \qquad (6)$$

The result of filtering a bidimensional sequence with the first order CHF $\psi_{mn}^{(1)}$ is a complex image $d_{mn}^{(1)}$ (15), whose magnitude is related to the local edge intensity and whose phase is orthogonal to the edge orientation. Thus, each edge represented by a high magnitude value whereas the interior of uniform or textured regions is characterized by low magnitude values. Then, first-order CHF filtering, measures the edge strength and orientation and, apart of a constant π-shift, provides an estimate of the complex edge process e_{mn}.

In Fig.2, we report a detail of a depth map of the video sequence Cones, adopted for performance evaluation in (14) and publicly available at (16), and the estimated complex line process, represented by a field of vectors whose magnitude is proportional to $|e_{mn}|$ and whose phase equal to arg(e_{mn}). Once the edge process is estimated, the MRF is completely characterized.

3 MRFBASED DEPTH MAP INTERPOLATION

Herein we adopt a local Bayesian interpolator in closed form, given the measurements $\mathbf{d}_{\eta mn}$ extracted from a realization of a MRF. Specifically, we estimate d_{mn} given $\mathbf{d}_{\eta mn}$:

$$\hat{d}_{mn}^{(MAP)} = \arg\max_{\xi} p\left(\xi|\mathbf{d}_{\eta mn}\right) \qquad (7)$$

where $\hat{d}_{mn}^{MAP}$ can be proved to exhibit the following form:

$$\hat{d}_{mn}^{(MAP)} = \sum_{c\in\mathcal{C}_{\eta mn}} w_c d_c \left[\sum_{c\in\mathcal{C}_{\eta mn}} w_c\right]^{-1} \qquad (8)$$

The MRF based local Bayesian interpolator (8) is very compact: the interpolator weights the neighboring pixels by taking into account the a priori spatial constraints, represented by the values w_c's. Note that, in the implementation, the effectively available set of neighboring pixels vary for each and every interpolated pixels. Thereby, with reference to the pixel pairs represented in Fig.1-A-C, the weighted average involves differently selected neighbors at each pixel site. For non edge pixels, we refer to the reduced neighboring sets in Fig.1-D, with uniform weights. This choice reflects the fact that, for non edge pixel, no favorite direction need sto be taken in to account in the interpolation. If interpolation is performed by cascading consecutive $2\times$ interpolation stages, at each stage the available pixels may or not contain previously interpolated pixels.

Stemming on the MAP interpolator in (8), the MRF based interpolation procedure reads as follows. The first stage of the interpolation algorithm is the estimation of the complex line process by means of the first order CHF. This task can be accomplished by either filtering the natural image associated to the depth map or by performing a bilinear interpolation of the low-resolution depth map so as to provide a coarse estimate of the high resolution depth map over which the CHF filtering is performed. In this remarkable latter case, no registration is needed between the pixels of the high resolution natural image and those of the to be interpolated depth map. After the CHF filtering has been performed, the clique weights can be correctly evaluated as in (5). Then, for each and every unknown pixel of the high resolution depth map, the edge magnitude is compared to a predefined threshold in order to detect the edge pixels; on the so detected edge pixels, we

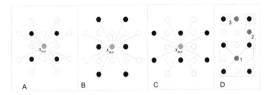

Figure 1. Spatial configuration of known (black), missing (white) and to-be-interpolated (gray) pixels. The neighborhood system η_{mn} along with the clique set is identified for the different geometries of edge pixel interpolation (A, B and C) and for non edge pixels (D1, D2 and D3).

Table 1. PSNR achieved for different magnification factors.

	SQ $\to$ Q ($2\times$)	SQ $\to$ H ($4\times$)	SQ $\to$ F ($8\times$)
PSNR BAY.	30.81	28.31	26.58
PSNR BIL.	30.30	26.22	24.50
PSNR EKME	28.39	27.02	25.75

evaluate the Markovian interpolator (8) on the neighborhood system shown in Fig.1-A-C that can better fit the possible edge directions. On the remaining non edge pixels, we adopt the neighborhood system shown in Fig.1-D.

4 SIMULATION RESULTS

We present here preliminary results obtained on the video sequence Cones, for which different resolutions, namely SQ (225×187)[1], Q (450×375), H (900×750), F (1800×1500), are available (16). In Table I we present the PSNR observed adopting the herein presented MRF-based interpolation technique (BAY), the upsampling procedure presented in (8) (EKME), and the bilinear interpolation (BIL), starting from the SQ resolution to Q ($2\times$ interpolation), H ($4\times$ interpolation) and F ($8\times$ interpolation).

In Fig.3, we show a F depth map as acquired by structured light, and the depth map estimated using the MRF based interpolation from the Q depth map ($4\times$ interpolation); for comparison sake, the Figure also shows the corresponding bilinear estimate and the result obtained by the edge adaptive upsampling technique in (8). The MRF based interpolation favorably compares with the bilinear interpolation; besides, it is viable of different improvements as mentioned in the concluding Section.

5 CONCLUSION

In this paper, we have presented a MRF based interpolation procedure for depth data images. Differently to

---

[1] The SQ map is obtained by decimation of the Q map without anti-aliasing filtering; the other resolutions are available at (16).

state of the art works, our approach doesn't need the reference high resolution luminance map, thus avoiding errors due to erroneous registration between depth map and luminance map. Moreover, no segmentation stage is to be performed over the edge map, this step being implicitly performed by the CHF filtering stage. Finally, the procedure can be modified to take into account different issues, as processing for boundary noise remove, edge estimation refinement. Besides, following the approach in (13), a fast algorithm for depth map upsampling can be envisaged.

REFERENCES

[1] C. Timmerer, K. Müller, "Immersive Future Media Technologies: From 3D Video to Sensory Experiences", *ACM Multimedia 2010* (Tutorial), Florence, Italy, Oct. 25–29, 2010.

[2] H. Wang, C. Huang, J. Yang, "Depth maps interpolation from existing pairs of keyframes and depth maps for 3D video generation", *Proc. of IEEE Int. Symp. on Circuits and Systems (ISCAS 2010)*, Paris, 2010.

[3] V. Garro, C. dal Mutto, P. Zanuttigh, G.M. Cortelazzo, "A novel interpolation scheme for range data with side information" *European Conf. Visual Media Production* CVMP 2009, London, 2009.

[4] G. Zhang, J. Jia, T. T. Wong, and H. Bao, "Consistent depth maps recovery from a video sequence", *IEEE Trans. Patt. Anal. Mach. Intell.*, vol.31, no.6, pp.974–988, Jun. 2009.

[5] Hung-Ming Wang, Chun-Hao Huang, Jar-Ferr Yang, "Block-based depth maps interpolation for efficient multiview content generation", *Trans. on Circ. and Syst. for Video Tech.*, Vol.21, no.12, pp.1847–1585, Dec. 2011

[6] G. Lee, Y. Ho, "Depth map up-sampling using random walk", *Proc. of Fifth Pacific Rim Symposiumv PSIVT 2011*, Gwangju, South Korea, November 20–23, 2011

[7] J. Lu, D. Min, R. Singh Pahwa, M. N. Do, "A revisit to MRF-based depth map super resolution and ehnancement", *Proc. of Int. Conf. on Acoustics, Speech and Signal Processing*, Prague, Czech Rep., May 22–27, 2011.

[8] E. Ekmekcioglu, M. Mrak, S. T. Worrall, A. M. Kondoz, "Edge adaptive upsampling of depth map videos for enhanced free-viewpoint video quality",*IET Electronics Letters*, vol.45, no.7, pp.353–354 March 2009.

[9] Spatial Scalability Filters,ISO/IEC JTC1/SC29/ WG11 and ITU-T SG16 Q.6, July 2005, Doc. JVT-P007, Ponzan, Poland.

[10] S. Geman and D. Geman, "Stochastic relaxation, Gibbs distribution, and the Bayesian restoration of images", *IEEE Trans. on Patt. Anal. and Mach. Intel.*, vol. PAMI-6, no. 6, pp. 721–741, November 1984.

[11] G. K. Chantas, N.P. Galatsanos, A.C. Likas "Bayesian resroration using a new nonstationary edge-preserving image prior", *IEEE Transactions on Image Processing*, vol. 15, no. 10, pp. 2987–2997, October 2006.

[12] S. Colonnese, S. Rinauro, G. Scarano, "Markov Random Fields using complex line process: an application to Bayesian Image Restoration" *Proc. of 3rd Eur. Work. on Vis. Infor. Proc.* Paris, France, July 2–4, 2011

[13] S. Colonnese, S. Rinauro, G. Scarano, "Fast image interpolation using circular harmonic functions", *Proc. of 2nd Eur. Work. on Visual Infor. Proc.*, Paris, France, July 5–7, 2010.

[14] D. Scharstein, R. Szeliski, "A taxonomy and evaluation of dense two-frame stereo correspondence algorithms" *International Journal of Computer Vision*,, vol.47, no.1–3, April–June 2002.

[15] S. Colonnese, P. Campisi, G. Panci, G. Scarano, "Blind image deblurring driven by nonlinear processing in the edge domain", *EURASIP Journal on Applied Signal Processing*, vol. 16, pp. 2462–2475, 2004.

[16] http://vision.middlebury.edu/stereo/

Computational Modelling of Objects Represented in Images – Di Giamberardino et al. (eds)
© 2012 Taylor & Francis Group, London, ISBN 978-0-415-62134-2

Man-made objects delineation on high-resolution satellite images

A. Kourgli
LTIR, Faculté d'Electronique et d'Informatique, U.S.T.H.B., Bab-Ezzouar, Alger

Y. Oukil
Département de Géographie, E.N.S. de Bouzareah, Alger

ABSTRACT: This paper presents a novel framework for man-made objects boundary delineation from high-resolution color images. These later are corrupted with geometric noise and artifacts (trees, shadows, undesirable objects, etc.) that disturb automatic extraction processes. To get robust segmentation results, the process is based on a two steps stage. First, KFCM (Kernel Fuzzy-C-Means) procedure is used to find the highest density regions which correspond to cluster centroids of the modes in the color space. KFCM clustering has been adopted for its robustness to noise and outliers and its ability to cluster complex data. In a second step, we formalized and applied a morphological filtering adapted to the artifacts to be removed or at least reduced. Its formulation is based on a homogeneity measure around the objects to be removed. The proposed approach has been tested on JPEG images extracted from Google Earth Software at different resolutions providing interesting results.

1 INTRODUCTION

In remote sensing, image segmentation is desired to provide meaningful object primitives for further feature recognition and thematic classification (Tian & Chen 2007). Numerous segmentation algorithms specific to high-resolution images have been developed during the past few years. They can be broadly grouped into four categories: point-based, edge-based, region-based, and hybrid/combined techniques. Point-based or pixel-based segmentation is generally performed by applying a global threshold on individual pixels within an image ignoring the spatial context. Thus, they are not efficient on high-resolution images. Moreover, it has been shown that the algorithms belonging to the second group are sensitive to the noise or texture. Region-based algorithms, in particular multi-resolution ones, perform generally better than feature space clustering or thresholding approaches since they take into account both feature space and the spatial relation between pixels simultaneously (Cheng et al. 2001). However, the application of multi-resolution segmentation still poses problems, especially in its operational aspects (Tian & Chen 2007).

Let us note that efficient segmentation approaches, usually, combine more than one technique to achieve a valuable segmentation such as mathematical morphology and neural network (Benidiktsson, 2003), mathematical morphology and wavelet transform (Aksoy & Akcay 2005), local binary pattern texture features and the lossless wavelet transform (Gou et al. 2005), spectral angle mapper, the irregular pyramid and the watershed with markers (Habib et al. 2006), graphs and morphological operations (Omelas 2009), hybrid genetic algorithm and Gaussian Markov random field model (Mridula & Patra 2010), mean-shift segmentation, adjacent graph construction and MRF (Markov Random Fields) modeling (Hong et al. 2011).

In this context, we present a new region-based methodology for the automated segmentation of high spatial resolution images. The proposed method uses the KFCM (Kernel Fuzzy-C-Means) clustering combined to a sequential morphological filtering. KFCM has been adopted because it is robust to noise and outliers and also tolerates unequal sized clusters. We used it to detect the main modes corresponding to the main regions, thus, objects in the image. The proposed filtering may be considered analogous to morphological reconstruction filters. Therefore, the original contribution of the paper is the definition of a morphologic filtering method, which avoids fragmentation or/and over-segmentation. Although the proposed technique may be applied to the study of different types of information, this work focuses on the extraction and segmentation of man-made objects. Indeed, the proposed method is particularly well suited for the segmentation of complex image scenes such as aerial or fine-resolution satellite images. Accordingly, the paper is organized as follows. Section II provides a brief description of the segmentation scheme proposed. Examples of segmentation process applied to satellite images are given in Section III. Finally, conclusions and directions for future work are drawn in Section IV.

2 SEGMENTATION PROCESS

Data clustering is the process of dividing data elements into classes or clusters so that items in the same class are as similar as possible, and items in different classes are as dissimilar as possible. Different measures of similarity may be used to place items into classes, where the similarity measure controls how the clusters are formed (Singla & Mehra 2011). A Kernel function is a generalization of the distance metric that measures the distance between two data points as the data points are mapped into a high dimensional space in which they are more clearly separable (Muller & Mika 2001).

KFCM adopts a new kernel-induced metric in the data space to replace the original Euclidean norm metric in FCM. Use of the kernel function makes it possible to partition data that is linearly non-separable and non hyper-spherical in the original input space, into homogeneous groups in a transformed high-dimensional feature space. Moreover, it permits to identify clusters with arbitrary shapes and has the capability of dealing with noise and outliers. And finally these properties can be utilized to cluster incomplete data (an unobserved part of the data or either corrupt data) such as very high resolution images. In fact, these later and the urban context induce a significant increase in geometric noise and artifacts such as vehicles, road markings, trees, occlusions or shadows that disrupt the process of automatic extraction. The use of KFCM clustering should overcome these weaknesses.

2.1 KFCM clustering

KFCM adopts a new kernel-induced metric in the data space to replace the original Euclidean norm metric in FCM and the clustered prototypes still lie in the data space so that the clustering results can be reformulated and interpreted in the original space (Chen & Zhang 2004). Let us use the notation employed in (Yang & Tsai 2008) to present KFCM alvbgorithm.

Let $X = \{x_1, \ldots, x_n\}$ be a data set in an s-dimensional Euclidean space R^s with its norm $\| \|$. Let c be a positive integer greater than one. A partition of X into c parts can be presented by a mutually disjoint set $X_1, \ldots, X_c$ such that $X_1 \cup \ldots \cup X_c = X$, or by the fuzzy partition $\{\mu_1, \ldots, \mu_c\}$ such that:

$$\sum_{i=1}^{c} \mu_i(x_j) = 1, \text{ for all } x_j \text{ in } X \tag{1}$$

The fuzzy c-partitions of X can be represented in a matrix form as follows:

$$\mu = \{\mu_{ij}\}_{c \times n} \in M_{fcn} \tag{2}$$

Where M_{fcn} is a partition matrix with (Bezdek 1981):

$$M_{fcn} = \{ \mu = [\mu_{ij}]_{cn} | \\ \forall i, \forall j, \mu_{ij} \geq 0, \sum_{i=1}^{c} \mu_{ij} = 1, n > \sum_{j=1}^{n} \mu_{ij} > 0 \} \tag{3}$$

The KFCM objective function is then defined as:

$$J_m(\mu, a) = \sum_{j=1}^{n} \sum_{i=1}^{c} \mu_{ij}^m \left(1 - \exp\left(-\|x_j - a_i\|^2 / \sigma^2\right)\right), \tag{4}$$

$$i = 1, 2, \ldots, c$$

Where $\mu_{ij} \in M_{fcn}$ and $\{a_1, \ldots, a_c\}$ denote the cluster centers of the data set X. The KFCM algorithm is iterated through the necessary conditions for minimizing J_m with its corresponding update equations:

$$a_i = \frac{\sum_{j=1}^{n} \mu_{ij}^m \left(x_j \exp\left(-\|x_j - a_i\|^2 / \sigma^2\right)\right)}{\sum_{j=1}^{n} \mu_{ij}^m \exp\left(-\|x_j - a_i\|^2 / \sigma^2\right)}, \tag{5}$$

$$i = 1, 2, \ldots, c \quad j = 1, 2, \ldots, n$$

and

$$\mu_{ij} = \frac{\left(1 - \exp\left(-\|x_j - a_i\|^2 / \sigma^2\right)\right)^{\frac{1}{m-1}}}{\sum_{k=1}^{c} \left(1 - \exp\left(-\|x_j - a_i\|^2 / \sigma^2\right)\right)^{\frac{1}{m-1}}} \tag{6}$$

As the parameter σ strongly affects the final result, we referred to Yang & Tsai (2008) for its estimation.

The input data of the modified KFCM algorithm are: the data set: $X = \{x_1, x_2, \ldots, x_n\}, x_i \in R_S$, the number of clusters: $c(2 \leq c \leq n)$, and the initial cluster centers: $a^{(0)} = \{a_1^{(0)}, a_2^{(0)}, \ldots, a_c^{(0)}\}$.

The KFCM procedure can, then, be described as follows:

1. Begin with iteration it=1
2. Calculate μ^{it} from a^{it-1} using equation 6
3. Update a^{it} with a^{it-1} and μ^{it} using equation 5
4. If $|a^{it} - a^{it-1}| < \varepsilon$ (stopping criterion) stop else it = it + 1 and go to step 2.

As established, iterative fuzzy clustering methods do not guarantee a unique final partition because different results are obtained with different initializations of a^{it} (or μ^{it}). In particular, it has been shown that iterative fuzzy clustering gives better results when the initials cluster centers are chosen with values sufficiently close to the final centers (LeCapitaine & Frélicot 2011). If this is achieved, this will increase the convergence speed which is essential when dealing with large image sizes. To this purpose, we propose a rather simple way for initializing the c cluster centers based on histogram characteristics (Figure 1).

For each spectral band $b \in S$, the histogram is computed. Then, for all histograms, means M_b and variances σ_b are estimated, and the c vectors of initial cluster centers are taken in the interval $[M_b - 2 * \sigma_b, M_b + 2 * \sigma_b]$ and are equally spaced in this interval. By this way, the computation time has been drastically reduced (less than 30 iterations are required to reach convergence for $\varepsilon < 0.01$), enabling to process very large images in a more reasonable time.

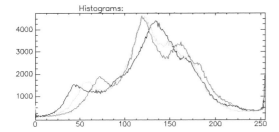

Figure 1. Histograms of image presented at Figure 2a.

2.2 Morphological filtering

Fragmentation is a key problem that exists in HSR image segmentation. Artifacts or small pieces of information can be leftover from segmentation. Fragmentation can also cause undesirable holes in regions. This problem exists more predominately in some solutions than in others and can cause difficulties for successful classification.

To reduce the artifacts and make segmentation results more homogenous, we propose an iterative process of local filtering, which minimizes the average heterogeneity of the generated images. The concept is similar the concepts inherent to morphological profiles. The profiles are obtained by successively applying morphological opening (or closing) or by applying alternately opening and closing operators with an increasing size of structuring element (SE). A progressive simplification of the image is obtained by removing at each step the objects (dark and/or clear) features that are smaller than the SE. Let us remind that our motivation is to remove small objects causing artifacts, these objects are not necessary dark or clear, they are just an amount of undesirable colored pixels inside a quasi homogenous area disturbing its delineation. First, the objects to be removed should be localized. This stage is achieved through the definition of a morphological filter that permits to identify the amount of pixels that are surrounded by pixels possessing the same intensities. Once these pixels are identified, their intensities are replaced by the intensity of the surrounding pixels.

This leads to the construction of the following morphological filter: Let's I be an image defined in the s-dimensional space R^s and a threshold λ^s defining the degree of likeness tolerated. We formulate a morphological operation (MO^s) defined on a s-dimensional window sized ω, that transforms the input image I in a filtered image FI through the filtering window FW_ω^s as:

$$FI = MO^s(I, FW_\omega^s) \qquad (7)$$

Where FW_ω^s is a structuring window (playing the role of a structuring element) whose borders are equal to 1 and the other pixels (X) that correspond to the artifact (to be removed) in the image has different values than

the border supposed quasi-homogenous. An example of such a filter for s = 1 and size $\omega = 5$ is given below:

$$FW_5^1 = 1 \begin{array}{ccccc} 1 & 1 & 1 & 1 & 1 \\ 1 & X & X & X & 1 \\ 1 & X & X & X & 1 \\ 1 & X & X & X & 1 \\ 1 & 1 & 1 & 1 & 1 \end{array} \qquad (8)$$

MO^s operation corresponds to these successive steps:

1. For each dimension s, compute the variance of the border elements (σ_{BE}^s) -of the element whose corresponding elements are equal to 1 in FW_ω^s
2. If $\sigma_{BE}^s < \lambda^s$, replace FI by the mean of the border elements μ_{BE}^s.

This operation is repeated for growing sizes of FW_i^s. The choice of λ^s value is crucial as it determinates the degree of homogeneity supposed. For '0' value, we are looking for bordering pixels having the same intensity values. As the filter is applied on the KFCM segmentation result that is conditioned by the number of clusters c, we calculated image range IR^s for each band and took $\lambda^s = \alpha$. IR^s/c. Where IR^s/c represents the approximate intensity differences between to adjacent clusters. While, $\alpha \in [0, 1]$ to avoid merging more than two regions. These later correspond to the modes detected by KFCM segmentation. Thus, if α equals 0, this means that we are looking for objects (that can potentially be removed) surrounded by pixels sharing the same intensities. For α value close to 1, the intensities of pixels surrounding the objects to be removed could belong to two successive modes. So its tuning is based on the likeness assumed, for a strict similarity, we take an α value equal to 0, but if we tolerate a little dissimilarity (corresponding to two successive modes), we can take an α value near to 1.

3 RESULTS AND DISCUSSION

Examples of the application of the proposed method are now given for the segmentation of satellite high-resolution data with different spatial resolutions. The satellite imagery is taken from a densely built-up area extracted from Google Earth-software.

Figure 2a (up) shows the image extracted. It is easy to note the relatively poor dynamic range of the data recorded (JPEG format). Figure 2b shows the results of the application of KFCM segmentation and the clusters detected for c = 12. We can observe that the man-made objects are not well delineated because of cars, trees and objects on roofs. We applied the morphological filter described at section 2.2 with growing window sizes from 3x3 to 25x25. By looking at Figures 2c and 2d, it can be noted that the filtering method appears to better delineate most of the relevant regions.

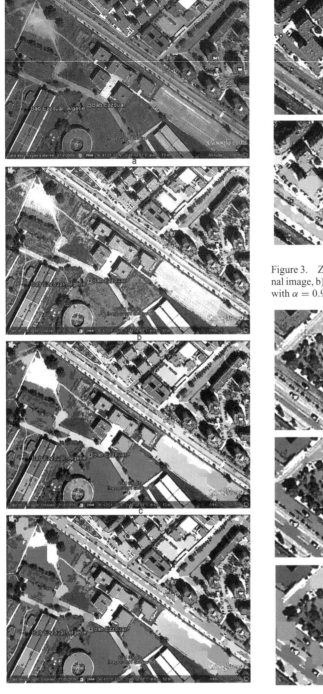

Figure 2. a) original image, b) KFCM modes detection with 12 classes, c) Mophological filtering with $\alpha = 0.9$, d) Mophological filtering with $\alpha = 0.975$.

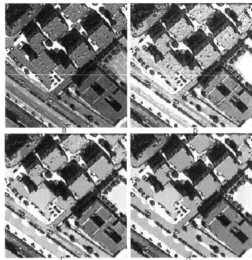

Figure 3. Zoom of an area extracted from Figure 2: a) original image, b) KFCM segmentation, c) Mophological filtering with $\alpha = 0.9$ d) Mophological filtering with $\alpha = 0.975$.

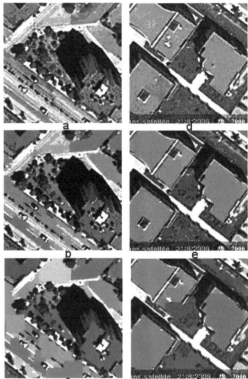

Figure 4. Zoom of two areas extracted from Figure 5: a) & d) KFCM segmentation, b) & e) Mophological filtering with $\alpha = 0.9$ c) & i) Mophological filtering with $\alpha = 1.0$.

The urban structure is well highlighted even though it is very complex.

To a better visualization, we give at Figure 3 a zoom of the upper middle area. We can see from Figure 3d

that most of the objects (satellite receivers) on the roofs of buildings have been removed making the borders of roofs more easily recognized. This can be useful for a successive automatic classification phase. Let us note

150

Figure 5. a) original image, b) KFCM modes detection with 12 classes, c) Mophological filtering with $\alpha = 0.9$ d) Mophological filtering with $\alpha = 0.975$.

that most of edges have been preserved and enhanced including open spaces and roads.

To test the robustness of this process against resolution, another example is given at Figure 5 from which two areas were zoomed (shown at

Figure 4) to a better interpretation. For the alpha parameter, we tested two values: the first worth 0.9 and the second equals 1.0. Recall that this threshold determines the amount of homogeneity allowed.

Even if the objects are better delineated with high value of $\alpha = 1.0$; we note the appearance of block effect and sometimes a false delineation (building on the left of Figure 4i) due to the form of the growing window size.

To overcome this shortcoming, we are currently testing other window forms (diamond, disk) to better fit the structures present in the high resolution images. However, it can be noted that the segmentation method appears to correctly detect and delineate most of the relevant regions including building and open spaces with a correct value of α.

4 CONCLUSION

The classification of objects can only be successful if the segmentation of the image is successful. In this paper, we developed a method and algorithmic framework for automatically segmenting imagery into different meaningful regions. In experiments, the proposed method demonstrated good performance. In particular, the proposed approach gives a better shape description to KFCM segmentation result. The drawback of the proposed filtering concerns the necessity of looking at a range of increasing filtering operations, which may cause a heavy computational burden. However, we gain a more homogenous segmentation and by reducing fragmentation, this filtering leads to a better delineation of man-made objects. For the above reasons, the method presented here is particularly suited for segmentation of complex image scenes such as aerial or satellite images. Currently, our work is focused on the extension of the proposed approach by improving the morphological detection by using other window forms.

REFERENCES

Aksoy, S. & Akcay, H. G. 2005 Multi-resolution Segmentation and Shape Analysis for Remote Sensing Image classification. In Proceedings of *2nd International Conference on Recent Advances in Space Technologies* 2005 (RAST 2005), *Istanbul , Turkey, June 09–11, 2005.*

Benidiktsson, J. A., Pesaresi, M., Arnason, K. 2003. Classification and Feature Extraction for Remote Sensing Images from Urban Areas Based on Morphological Transformations. *IEEE Transactions on Geoscience and Remote Sensing*, 41 (9): 1940–1949.

Bezdek, J. 1981. *Pattern Recognition with Fuzzy Objective Function Algorithms*. Plenum Press, New York.

Cheng, H.D., Jiang, X.H. , Sun, Y. & Wang, J. 2001. Color image segmentation: advances and prospects. *Pattern Recognition*, 34: 259–228.

Chen, S.C. & Zhang, D.Q. 2004. Robust image segmentation using FCM with spatial constrains based on new kernel-induced distance measure. In *IEEE Transactions on Systems, Man and Cybernetics*; Pt. B 34: 1907–1916.

Guo, D., Atluri, V. & Adam, N.R. 2005. Texture-Based Remote-Sensing Image Segmentation. In *Proceedings of IEEE International Conference on Multimedia and Expo,*. ICME'2005, *Amsterdam, Netherlands, July 5–8, 2005:* 1472–1475.

Habib T., Gay M., Chanussot J. & Bertolino P. 2006. Segmentation of high resolution satellite images SPOT applied to lake detection. In: *Proceedings of IEEE International Geoscience and Remote Sensing Symposium* 2006, IGARSS'06, *Denver, Colorado, July 31–August 4, 2006:* 3680–3683.

Hong, L., Pan, X., Gao, Z. & Yang, K. 2011. A novel segmentation method of high resolution remote sensing image based on object-oriented Markov random fields model. in Li, J., ed., *International Symposium on Lidar and Radar Mapping 2011—Technologies and Applications, Nanjing, China, 26-29 May 2011*, Proceedings of SPIE Vol. 8286: Bellingham, Wash., Society of Photo-Optical Instrumentation Engineers (SPIE).

Le Capitaine, H & Frélicot, C.2001. A fast fuzzy c-means algorithm for color image segmentation. Proceedings of *EUSFLAT-LFA, Aix-les-Bains, France, July 2011*: 1074–1081.

Mridula, J. & Patra, D. 2010. Genetic algorithm based segmentation of high resolution multispectral images using GMRF model", *Industrial Electronics, Control & Robotics* (IECR),*Rourkela, India 27–29 December 2010*: 230–235.

Muller, K. R. & Mika, S. et al. 2001. An Introduction to Kernel-based Learning algorithms. *IEEE Transactions on Neural Networks*, 12(2): 181–202.

Ornelas, L. 2009. High Resolution Images: Segmenting, Extracting Information and GIS Integration, *World Academy of Science, Engineering and Technology* 54: 172–177.

Singla, A. & Mehra, R. 2011. Design & Analysis of Fuzzy Clustering Algorithm for Data Partitioning Application. International Journal of VLSI and Signal Processing Applications, 1(2): 52–56.

Tian, J. & Chen, D. 2007. Optimization in multi-scale segmentation of high-resolution images for artificial feature recognition, *International Journal of Remote Sensing*, 28(20): 4625–4644.

Yang, M-S. & Tsai, M-S. 2008. A Gaussian kernel-based fuzzy c-means algorithm with a spatial bias correction. *Pattern Recognition Letters*, 29(12): 1713–172.

Computational Modelling of Objects Represented in Images – Di Giamberardino et al. (eds)
© 2012 Taylor & Francis Group, London, ISBN 978-0-415-62134-2

A hybrid approach to clouds and shadows removal in satellite images

Danilo Sousa, Ana Carolina Siravenha & Evaldo Pelaes
Signal Processing Laboratory, Technology Institute (ITEC), Federal University of Para, Belem, PA, Brazil

ABSTRACT: This work aims to overcome a common problem in many satellite images, which is the presence of undesirable atmospheric components such as clouds and shadows at the time of scene capture. The presence of such elements can hinder the identification of image objects, urban environmental monitoring, and subsequent steps of the digital image processing like segmentation and classification. Thus, this work presents a new way to perform a hybrid approach toward removal and replacing of these elements in satellite images. The authors propose a method of regions decomposition using nonlinear median filter in order to map the regions of structure and texture. In this areas will be applied the methods of inpainting, by smoothing based on DCT, and exemplar-based texture synthesis, respectively. Finally, it was found the effectiveness of this technique through a qualitative evaluation at the same time that a discussion about quantitative analysis is made.

1 INTRODUCTION

It is noticeable that remote sensing images are susceptible to the undesirable presence of the atmospheric interferences, such as clouds, hazes and shadows. These occurrences are largely present in equatorial and tropical warm-to-hot regions and changes the brightness values of pixels at different levels of saturation, which can corrupt the visual representation of the covered land surface. The presence of such elements affect, in many ways, the image processing in an environmental or urban monitoring, and also the segmentation and/or classification methods that are the mainly responsible for the image information extracting.

To remove dense clouds and shadows is commonly related the use of reference images or by the estimative of the covered areas. The first strategy can use a multi-temporal analysis (Zhang, Qin, and Qin 2010) or can use another image of the same scene, captured by a different sensor, as a SAR image (Hoan and Tateishi 2008), both aims to replace the region affected by a non-affected area.

The estimative of covered area, can be done by an interpolation method called inpainting (Liu, Wang, and Bi 2010). Many methods use derivations of this method in order to fill up some damaged regions or to remove large objects.

Another method employed to redefine regions is the texture synthesis (Liu, Wong, and Fieguth 2010). Generally working in blocks, reach better results for areas that contains some textural pattern in heterogeneous regions.

Some studies has exploring hybrid solutions, observing both texture synthesis and inpainting effectiveness by (Bugeau and Bertalmio 2009). In that

investigation, each approach is responsible for redefine the texture and structure regions separately, ensuring images with no interferences and with real textures.

This paper presents a novel hybrid approach using inpainting by smoothing based on multidimensional DCT as proposed by Garcia (2010) and an exemplar-based texture synthesis as proposed in Criminisi, Perez, and Toyama (2004). The structure and texture mapping was based in Vese and Osher (2002) through a novelty approach using nonlinear median filter.

2 METHODOLOGY

2.1 Regions detection algorithm

Before apply the filling algorithm, described further, it is important identify the regions that will be processed. The algorithm used in this algorithm was first presented in Hau, Liu, Chou, and Yang (2008) and lately expanded in Siravenha (2011). In this step, the different image features are identified and separated into four classes: dense cloud, thin cloud, shadow and not affected area. To this, statistical measures of image average value and standard deviation are computed, and the classification is possible.

Besides, Siravenha (2011) added the capability of shadows detection, was also added two constants called cc and sc, increasing the algorithm flexibility. The equation that describes this operation is:

$$m(x,y) = \begin{cases} f(x,y) < sc \times f_{av-sd}, & f(x,y)\epsilon\, 0, \\ f_{av-sd} < f(x,y) < f_{av}, & f(x,y)\epsilon\, 1, \\ f_m < f(x,y) < cc \times f_{av+sd}, & f(x,y)\epsilon\, 2, \\ f(x,y) > cc \times f_{av+sd}, & f(x,y)\epsilon\, 3. \end{cases}$$

(1)

where $f(x, y)$ is the pixel value, f_{av} is the average value of image pixels, f_{av+sd} is the sum of the average value pixels and the standard deviation value of the image and f_{av-sd} is the subtraction of the average value by the standard deviation pixels.

The region labeled as 0 represents a shadow region, labeled as 1 means region not affected by atmospheric interference, while regions labeled as 2 represents thin clouds, and dense cloud are labeled as 3. For images with multiple bands these labels are assigned if and only if the rule is valid for all bands

To complete this process, it is applied a morphological opening operation that aims to remove very small objects that can cause mistakes in following steps.

2.2 Inpainting by smoothing based on multidimensional DCT

This method was proposed by Garcia (2010), and so as in Bertalmio, Sapiro, Caselles, and Ballester (2000), is based on the information propagation by smoothing. The specificity of this approach is related to the use of the Discrete Cosine Transform (DCT) to simplify and to solve linear systems, to an efficient smoothing. In statistics and data analysis, smoothing is used to reduce experimental noise or information and keeping the most important marks of the data set.

Considering the following model for the one-dimensional noisy signal y from the $y = \hat{y} + \varepsilon$, where ε represents a Gaussian noise with zero mean and unknown variance, and $\hat{y}$ is the so-called smoothing, i.e., has continuous derivatives up to some order (usually ≥ 2) throughout the domain. The smoothing of y depends on the best estimate of $\hat{y}$ and this operation is usually performed by a parametric or nonparametric regression.

A classic approach to smooth is the Penalized Least Squares Regression. This technique minimizes a criterion that balances the data fidelity, measured by the Residual Sum-of-Squares (RSS) and by a penalty term (P), which reflects the robustness of the smoothed data. Another simple and straightforward approach to express the robustness is by using a Second-order Divided Difference ($SoDD$), which produces an one-dimensional array of data.

Now, using RSS and the $SoDD$, the minimization of $F(\hat{y})$ results in a linear system, expressed in Eq. 2, which allows the smoothed data determination.

$$(I_n + sD^T D)\hat{y} = y, \tag{2}$$

where I_n is the identity matrix $n \times n$, s is a positive real scalar that controls the grade of smoothing, so that, as it increases, the degree of smoothing of $\hat{y}$ increases too; and D^T represents the transpose of D.

Eq. 2 can be solved using the left division matrix applied to sparse matrices (Garcia 2010). Solving this linear system, however, can be a lot of time expensive for a large amount of data. But, this algorithm can be simplified and accelerated, since the data are evenly spaced, in images where pixels are equally spaced, resulting in the following equation for multidimensional data:

$$\hat{y}_{k+1} = IDCTN(\Gamma^N \circ DCTN(y_k)). \tag{3}$$

Where $DCTN$ and $IDCTN$ refers to the N-dimensional DCT and its inverse, respectively. k is the number of iterations, N is the number of dimensions, $\circ$ is the Schur product (element by element) and Γ^n represents a tensor of rank N defined by

$$\Gamma^n = 1^N \div (1^N + s \wedge^n \circ \wedge^n). \tag{4}$$

Here, the operator $\div$ symbolizes the division element by element, and 1^N is a tensor of rank N composed by 1's. $\wedge$ is the following tensor of rank N (Buckley 1994)

$$\wedge_{i_1, \ldots, s i_N}^N = \sum_{j=1}^{N} \left(-2 + 2 \cos \frac{(i_j - 1)\pi}{n_j} \right), \tag{5}$$

where n_j denotes the size of $\wedge^n$ along the j-th dimension.

Can also be observed that when there are undefined values in the image, smoothing is also responsible for interpolation of data, functioning as an inpainting method. In order to accelerate the convergence, the process starts performing nearest neighbor interpolation on the image to be restored.

2.3 Texture synthesis

The texture synthesis has been an intensive field of study because its purpose variety. It can be applied in objects fill-in tasks, image recovery, video compression, foreground removal and others.

Let us define texture as a visual pattern in a $2D$ infinite plan that, at some scale, has a steady distribution, then, naturally, one can obtain a finite sample of textures present in this plan in order to synthesize other samples from the same texture. This finite sample can be extracted from an uncountable different textures, and it is an ill-posed situation. To contour this, the assumption is that the sample is larger enough to capture the textural stead distribution of the image where the texture elements scale is known (Efros and Leung 1999). Furthermore, the texture synthesis is responsible for merging continuous regions with minimal imperfection and perception of the operation, ensuring the visual quality.

The approach proposed by Criminisi, Perez, and Toyama (2004), aims to remove or redefine larges objects in a digital image with the neighborhood information. This method uses the texture synthesis to fill in regions that contains two-dimensional textural patterns with moderated stochasticity. For this, generates new sampling textures from a source image and make a simple copy to the target areas.

In Figure 1, is presented the Criminisi, Perez, and Toyama (2004), algorithm. Suppose an image where

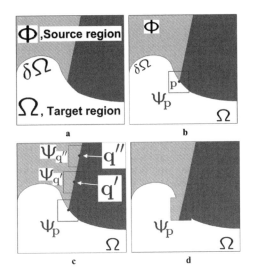

a

b

c

d

Figure 1. Model based texture synthesis: (a) Original image. (b) Ψ_p fragment centralized in $p \in \Phi$. (c) The most probable candidates $\Psi_{q'}$ e Ψ_q''. (d) The most probable candidate is propagated to the target fragment.

there is a target region (Ω) contoured by $\delta\Omega$ and completed by a source region Φ, clearly distinguishable (Figure 1 (a)). The objective is synthesize the area delimited by the fragment named Ψ_p, which is centralized on the $p \in \Phi$ point, illustrated in Figure 1 (b). Then, it is counted the most probable candidates to fill Ψ_p presented in $\delta\Omega$, for example Ψ_q' and Ψ_q'' in Figure 1 (c). Among the candidates, there is one that has the better corresponding to the target fragment, and this candidates is copied to the Ψ_p fragment. This process is repeated until the full fill of Ω. In Figure 1 (d) can be noted that the texture, as well the structure (the line that apart the light and dark gray regions), are propagated to the Ψ_p fragment.

The fill in order is influenced by the linear structures adjacent from the target region. Thus, the model based texture synthesis algorithm with propagation along the isophotes direction of the image, presents efficiency and qualitative performance respecting the restrictions imposed by the linear structures.

2.3.1 Fill in regions algorithm

Taking a source image with a target region Ω to be redefined and a source region Φ, that can be expressed as the subtraction of the image f by the target region ($\Phi = f - \Omega$), one must define the window size to the model called Ψ. It is very common use a window with 9×9 dimension, but it is recommended that the size are lager than the biggest distinguishable textural element in the source region.

In this algorithm, each pixel maintain a color value (or NaN, if is an undefined pixel to be filled) and a confidence value, which reflects the confidence in the color value since the pixel is filled (Criminisi, Perez, and Toyama 2004) During the algorithm execution, the fragments located in $\delta\Omega$ contour receive temporary

priority value, defining the order to be filled. Hence, an interactive process is executed in the following sequence:

1) Computing fragments priority: Because the texture synthesis works with priorities, the strategy called *best-first* fill in the regions according the priority levels and becomes tendentiously to the regions that a) are on strong continuity borders or b) are surrounded by high confidence pixels.

The Figure 2 shows an image to be processed, and given the fragment Ψ_p, n_p is the normal to the contour $\delta\Omega$ of the target region and $\nabla I_p^\perp$ is the isophote at point p. The isophote represents the direction and intensity in that point.

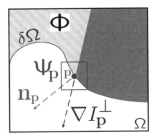

Figure 2. Notation diagram of an image and its components according the synthesis texture.

The priority ($P(p)$) is defined as a product of the confidence term ($C(p)$) and the data terms ($D(p)$), that are defined as:

$$C(p) = \frac{\sum_{q \in \Psi_p \cap (f - \Omega)} C(q)}{|\Psi_p|} \qquad (6)$$

and

$$D(p) = \frac{|\nabla I_p^\perp \cdot n_p|}{\alpha}, \qquad (7)$$

where $|\Psi_p|$ is the total area of Ψ_p, α is the normalization factor (255 in typical applications with gray scale images), n_p is an unit vector orthogonal to the $\delta\Omega$ in p, and $\perp$ is the orthogonal operator. For each border fragment a $P(p)$ is computed, and every distinct pixel represents a fragment on a target region border. During the initialization, taking the $C(p)$ equals to 0 for all point p in the target region and $C(p)$ equals to 1 for all point in source region.

The confidence represents the measure of trusted information surrounding a pixel. Thus, the algorithm aims to fill, firstly, those most reliable fragments, including those with more redefined pixels or fragments whose pixels were never part of the target region.

The $D(p)$ term defines the isophotes strength on $\delta\Omega$ in every iteration and it is responsible by increases the confidence of the linear structures, making a safely filling.

2) Structure and texture spread informations:
Once the priorities are computed, the $\Psi_{\hat{p}}$ high confidence fragment is found and filled by the information extracted from Φ source region more similar to that region. Formally

$$\arg\min_{\Psi_q \in \Phi} d(\Psi_{\hat{p}}, \Psi_q), \qquad (8)$$

where the distance $(d(\ldots))$ between fragments Ψ_a and Ψ_b is defined as a Sum of Squared Differences (*SSD*) of the pixels that contains information in these fragments (possible already filled pixels). The ideal fragment to fill in one region is that one who minimizes the *SSD*.

Having found the source model, every pixel value $(p'|p' \in \Psi_{\hat{p} \cap \Omega})$ is copied to the correspondent target region inside $\Psi_{\hat{q}}$.

3) Updating the confidence values: After $\Psi_{\hat{p}}$ receive this new values it is redefined the confidence values

$$C(p) = C(\hat{p}) \forall p \in \Psi_{\hat{p}} \cap \Omega. \qquad (9)$$

This simple rule to update the confidence values allows the measure of the relative confidence even without any specific image information. It is expected that the confidence values decay during the filling process, this indicates there are less assurance about the pixels color values that are near the center of the target region.

2.4 Image decomposition

The method proposed by (Vese and Osher 2002) decomposes an image into two sub-images, each representing the components of structure or texture, thus, a better image redefinition can be made. On the structured part, should be applied the technique of inpainting based on DCT, whilst into texture portions and heterogeneous areas is suitable to use texture synthesis.

The generalized model is defined as: $f = u + v$, where f is the input image, u is the structure image and v is the texture image. So, given these sub-images one can reconstruct the original image. However, in practice it is observed that the original image can be only approximately reconstructed. The goal of the method is to have a structure image u that preserves all strong edges with smoothed internal regions, and an image v that contains all the texture and noise information.

The method used to construct the structured image u is based on the assumption that u is a 2D function and in attempt to minimize this function in the space of all Bounded Variation functions (BV). Functions in BV space are functions whose total variation are limited by some constant value less than infinity. Minimizing u in BV space ensures resulting in a stable image and without infinite values. It should be noted, however, that this space allows for functions which have very large derivatives (although non-infinite), thereby ensuring that strong edges are preserved.

Taking in mind the intuition described above, the minimization problem should logically have two terms. One of them will be the fidelity, responsible for maintaining the difference between f and u small. This term ensure that data of the input image are kept on result. The other one imply a smoothing over u, although not necessarily in all u components. The minimization is computed as:

$$F(u) = [(\int |\nabla u|) + (\lambda \int |f - u|^2)] dx dy, \qquad (10)$$

with $u \in BV$ and ∇ representing the gradient operator. The second term is the data term and the first one is a regularization term to ensure a relatively smooth image. λ is a tuning parameter. As can be seen, this seeks find the optimal u and ignores the v image. The reason is that, in Vese and Osher (2002), the authors had considered the v image to be noise, and therefore to be discarded.

There is an unique result to this optimization problem, and methods exist for finding the solution. Noting that $v = f - u$ it is possible to easily modify the above equation to incorporate v:

$$F(u) = \int |\nabla u| + \lambda \int \| v \|^2 dx dy, \qquad (11)$$

(still $u \in BV$). Which yields the Euler-Lagrange equation $u = f + 1/2\lambda div(\nabla u/|\nabla u|)$. Making the right manipulation $v = f - u = -1/2\lambda div(\nabla u/|\nabla u|)$. At this point it is useful to break v into its x and y components respectively. It will be denoted as g_1 and g_2, where:

$$g_1 = -\frac{1}{2\lambda} div(\frac{\nabla u_x}{|\nabla u|}) \; and \; g_2 = -\frac{1}{2\lambda} div(\frac{\nabla u_y}{|\nabla u|}). \; (12)$$

This allows us to write v as: $v = div\vec{g}$ where $\vec{g} = (g_1, g_2)$. It can be seen that $g_1^2 + g_2^2 = 1/2\lambda$, so that $\| \sqrt{g_1^2 + g_2^2} \| = 1/2\lambda$. This allows us to rewrite v as:

$$v(x, y) = div\vec{g} = \partial_x g_1(x, y) + \partial_y g_2(x, y). \qquad (13)$$

And leads to the final minimization problem $(u \in BV)$:

$$G(u, v1, v2) = \int |\nabla u| + \lambda \int |f - u - \partial_x g_1$$

$$-\partial_y g_2|^2 dx dy + \mu \int \sqrt{g_1^2 + g_w^2} dx dy. \qquad (14)$$

Solving the minimization problem (Eq.14) yields the Euler-Lagrange equations:

$$u = f - \partial_x g_1 - \partial_y g_2 + \frac{1}{2\lambda} div(\frac{\nabla u}{|\nabla u|}) \qquad (15)$$

$$\mu \frac{g_1}{\sqrt{g_1^2 + g_2^2}} = 2\lambda[\frac{\partial}{\partial_x}(u - f) + \partial_{xx}^2 g_1 + \partial_{xy}^2 g_2] \quad (16)$$

$$\mu \frac{g_2}{\sqrt{g_1^2 + g_2^2}} = 2\lambda[\frac{\partial}{\partial_y}(u - f) + \partial_{xy}^2 g_1 + \partial_{yy}^2 g_2] \quad (17)$$

156

2.4.1 *Implementation proposed in this work*

In this study was not used the addition of each component obtained above, since this approach results in considerable error generated at the image reconstruction and also due to the difficulty in establishing appropriate parameters for an acceptable image decomposition. Instead, a strategy was proposed for mapping structure and texture areas of an image based on the decomposition process.

This process begins transforming the component that contains the texture information, i.e. image *v*, in a binary image with values 1 for texture heterogeneous areas and 0 for structure areas. Then a nonlinear median filter is applied to make homogeneous (smoothed) areas where small gaps of a given feature are surrounded by predominant regions of another sort. The threshold is defined according the mask (kernel, a square matrix) used in the filter and it will change for each image in an empirical way, depending on the image analyst decision.

This step is critical due to the presence of clouds and shadows in the image, and is performed in order to correctly define the techniques to be employed for each region. This happens because clouds and shadows will always be structure components, so to define which technique use to remove them, one must observe the surrounding regions. Therefore, as a result of application of the filter, those regions to be redefined are mapped in the binary image to the texture or structure regions, to finally apply inpainting or texture synthesis on the input image, respectively.

3 RESULTS AND DISCUSSIONS

A major problem in literature is the lack of quantitative evaluation methods for inpainting algorithms. Along this work was tested the evaluation measures PSNR (local and global), Kappa and the Sum of Absolute Differences (SAD). It was concluded that none of them appropriately evaluate different approaches to redefine regions. For example, certain results from hybrid approach, and even just using texture synthesis, visually performs a region filling which appears more consistently than those resulting from inpainting (sometimes showing large blurs). However, looking to the quantitative evaluations cited, it is common that the inpainting approach achieve better results.

According (Taschler 2006), the only explanation for this discrepancy is that the texture synthesis can reach more faithful results to the goal, but some elements are not located at the corresponding position in the reference image, i.e., if the difference is one or two pixels, it's led to a lower PSNR value for the entire region. Another explanation can be given because these metrics and inpainting are pixel-by-pixel performed, while texture synthesis commonly is block-by-block, taking some disadvantage in evaluation.

With regards to qualitative assessment, Fig. 3 (a) has an image affected by the presence of dense clouds and shadows over texture (urban) and structure (dense

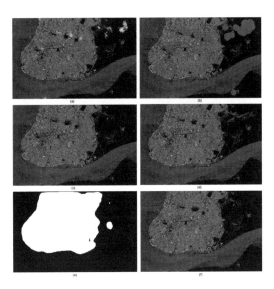

Figure 3. Process of clouds and shadows removal: (a) Original image. (b) Mask from regions detection. (c) Inpainting result. (d) Texture Synthesis result. (e) Binary image representing proposed image decomposition. (f) Proposed hybrid approach results.

vegetation) areas. In Fig. 3 (b) is illustrated a red mask containing these regions to be redefined. This mask was produced by the algorithm cited in Section 2.1. The Figs. 3 (c) and (d) shows the isolated results of inpainting and texture synthesis, respectively. As previously discussed are visible blurs generated by smoothing performed in urban areas (Fig. 3 (c)) and the erroneous replacement of the texture synthesis in areas of dense vegetation (Fig. 3 (d)). From these results, then it is decided to apply the hybrid approach. Fig. 3 (e) shows a binary image after passing *v* image through the median filter, containing texture (white) and structure (black) regions. From this mapping, Fig. 3 (f) shows the hybrid approach result, where becomes clear the union of advantages of the techniques applied in suitable regions thereof, thereby visually overcoming other methods.

4 CONCLUSION

This work aimed to present a new way to perform a hybrid approach toward detection, removal and replacing of clouds and shadows areas in satellite images. The approach proposes a regions decomposition method using a nonlinear median filter in order to map structure and texture regions, where was be applied the methods inpainting by smoothing based on DCT and exemplar-based texture synthesis, respectively.

In qualitative evaluation was evident that the hybrid approach overcomes the use of the techniques in a separated way. In the quantitative tests was not possible to make a fair assessment, due to the non-applicability of the various metrics to evaluate different approaches to redefine regions. Like as in Fig. 3, when applied to the other images, the hybrid method overcome the isolated approaches and due the free-space available these results was omitted.

Actually, we are interested in use quantitative approaches with a more reasonable justification, using not only information of pixel value, but also the context, shape and other attributes that are similar to the subjective evaluation of human eyes.

REFERENCES

Bertalmio, M., G. Sapiro, V. Caselles, and C. Ballester (2000). Image inpainting. In *Proceedings of the 27th annual conference on Computer graphics and interactive techniques*, pp. 417–424. New York:ACM Press/Addison–Wesley Publishing Co.

Buckley, M. (1994). Fast computation of a discretized thin-plate smoothing spline for image data. *81*, 247–258. Oxford: Biometrika.

Bugeau, A. and M. Bertalmio (2009). Combining texture synthesis and diffusion for image inpainting. In VISAPP 2009 – *Proceedings of the Fourth International Conference on Computer Vision Theory and Applications – Volume 1*, pp. 26–33. Lisboa:INSTICC Press.

Criminisi, A., P. Perez, and K. Toyama (2004). Region filling and object removal by exemplarbased image Inpainting. In *IEEE Transactions On Image Processing.*, Volume 13(9), pp. 1200–1212. IEEE Computer Society.

Efros, A. and T. Leung (1999). Texture synthesis by non-parametric sampling. In *International Conference on Computer Vision*, pp. 1033–1038. Washington:IEEE Computer Society.

Garcia, D. (2010). Robust smoothing of gridded data in one and higher dimensions with missing values. *Computational Statistics & Data Analysis 54*(4), 1167–1178. Maryland Heights: Elsevier.

Hau, C. Y., C. H. Liu, T. Y. Chou, and L. S. Yang (2008). The efficacy of semi-automatic classification result by using different cloud detection and diminution method. *The International Archives of the Photogrammetry, Remote Sensing and Spatial Information Sciences.*

Hoan, N. T. and R. Tateishi (2008). Cloud removal of optical image using SAR data for ALOS applications. Experimenting on simulated ALOS data. Beijing: *The International Archives of the Photogrammetry, Remote Sensing and Spatial Information Sciences.*

Liu, H., W. Wang, and X. Bi (2010). Study of image Inpainting based on learning. In *Proceedings of The International MultiConference of Engineers and Computer Scientists*, pp. 1442– 1445. Hong Kong:Newswood Limited.

Liu, Y., A. Wong, and P. Fieguth (2010). Remote sensing image synthesis. In *Geoscience and Remote Sensing Symposium (IGARSS)*, pp. 2467 –2470. Honolulu:IEEE International.

Siravenha, A. (2011). A method for satellite images classification using discrete cosine transform with detection and removal of clouds and shadows. Belem:Federal University of Para.

Taschler, M. (2006). A comparative analysis of image inpainting techniques. Technical report, New York:The University of York.

Vese, L. A. and S. J. Osher (2002). Modeling textures with total variation minimization and oscillating patterns in image processing. *Journal Of Scientific Computing 19*, 553–572. New York:Plenum Press.

Zhang, X., F. Qin, and Y. Qin (2010). Study on the thick cloud removal method based on multitemporal remote sensing images. In *International Conference on Multimedia Technology (ICMT)*, pp. 1–3. Ningbo: IEEE.

Computational Modelling of Objects Represented in Images – Di Giamberardino et al. (eds)
© 2012 Taylor & Francis Group, London, ISBN 978-0-415-62134-2

Wishart classification comparison between compact and quad-polarimetric SAR imagery using RADARSAT2 data

B. Souissi
Infotronics, Physics department, Science Faculty, UMBB University, Boumerdes, Algeria

M. Ouarzeddine & A. Belhadj-Aissa
LTIR Laboratory, Electronics and Computer Science Faculty, UHTHB University, Algiers, Algeria

ABSTRACT: Compact Polarimetry (CP) is a recent technique that affords more target information than a single-pol system, while not suffering as much from the drawbacks of a quad-pol system. In this paper, we present a study of the polarimetric information content of CP- and FP-pol imaging modes. We compare Wishart classifications of the CP data set against the full quad-pol dataset at C-band. Primarily, we use the two dimensional scatter plots to compare between the intrinsic polarimetric parameters which are the entropy (H) and the angle alpha (α) for both modes. Secondly, we generate the polarimetric signatures for different terrain objects from CP- and FP data. Finally, we show overall comparison results and classification accuracy. This CP mode solution has proven to be very attractive but can still be improved.

We illustrate our results by using the polarimetric images of Algiers city, Algeria acquired by the RADARSAT2 satellite in C-band.

1 INTRODUCTION

The standard implementation of full polarimetric SAR (PolSAR) involves the coherent transmission and reception of both vertically (V) and horizontally (H) polarized radar pulses. The PolSAR systems extract the complete polarimetric scattering information from a target scene but suffer from an increase in the pulse repetition frequency by a factor of two and an increase in the data rate by a factor of four over single polarization systems (Raney 2007). Nonetheless several partially polarimetric SAR systems have been proposed and there has been growing interest in dual-pol systems that transmit one polarization (e.g., linear horizontal or circular) and receive two polarizations (Dubois et al. 2008). A dual-pol system has advantage over a full polarimetric system in terms of reductions of pulse repetition frequency, data volume, and system power needs. However, dual-pol SAR systems do not acquire complete information pertaining to the full polarization state of the target. Souyris et al. (2005) proposed a DP mode called compact polarimetry (CP), where the system transmits only one polarization, either H + V ($\pi/4$ mode) or circular (right or left circular, RC or LC). They introduced a radar scattering model that assumes reflection symmetry and a relationship between the linear coherence and the cross-polarization (cross-pol) ratio.

In this paper, we present a study of the polarimetric information content of CP- and FP-pol imaging modes. We compare Wishart classifications (Ferro-Famil et al 2001) based on the polarimetric target decomposition (Cloud & Pottier 1996) of the CP polarimetric dataset against the full quad-pol dataset at C-band low frequency. Primarily, we use the two dimensional (2D) scatter plots to compare between the intrinsic polarimetric parameters which are the entropy (H) and the angle alpha (α) for both modes. Secondly, we generate the 3D-polarimetric signatures based on the Kennaugh matrix (Van zyl et al. 1989) for different terrain objects from CP- and FP data. Finally, we show overall comparison results and classification accuracy of the pseudo-quad-pol data. This CP-mode solution has proven to be very attractive but can still be improved.

We illustrate our results by using the polarimetric SAR images of Algiers city in Algeria acquired by the RADARSAT2 satellite in C-band.

2 DATA USED

The study area is located in the west of the Algiers town which is the capital of Algeria. It consists mainly of urban areas, agriculture fields and sea. The data was acquired on 11th April 2009 by RADARSAT2 in a fully polarimetric C-band at an illumination angle between 38,3 and 39,81°. Figure 1 shows the RGB image of the test site (R:HH, G:HV and B:VV).

3 FULL POLARIMETRY BACKGROUND

The fundamental quantities measured by a polarimetric SAR are the scattering matrix elements S_{TR}, where

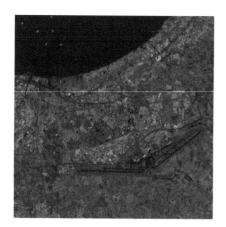

Figure 1. The RGB image of the polarimetric data (Red:HH, Green:HV and Blue:VV).

T and R are the transmit and receive polarizations, respectively. The scattering matrix representation as defined in the linear (H, V) basis is (Cloud & Pottier 1996):

$$S = \begin{bmatrix} S_{hh} & S_{hv} \\ S_{hv} & S_{vv} \end{bmatrix} \quad (1)$$

The scattering vector in the lexicographic basis is given as:

$$\vec{k}_l = \begin{bmatrix} S_{hh} & \sqrt{2}S_{hv} & S_{vv} \end{bmatrix}^T \quad (2)$$

For multilook processed 3×3 positive semi-definite hermitian covariance matrix:

$$\langle [C] \rangle = \langle \vec{k}_l \vec{k}_l^{*T} \rangle \quad (3)$$

Where the superscript $*T$ denotes the conjugate transpose operator. The symbol $<>$ indicates ensemble averaging.

In the case of the target characterized by reflection symmetry, the following relations hold (Nghiem et al 1992):

$$\langle S_{hh}S_{hv}^* \rangle = \langle S_{hv}S_{vv}^* \rangle = 0 \quad (4)$$

Equation (4) can then be written as:

$$\langle [C] \rangle = \begin{bmatrix} \langle |S_{hh}|^2 \rangle & 0 & \langle S_{hh}S_{vv}^* \rangle \\ 0 & 2\langle |S_{hv}|^2 \rangle & 0 \\ \langle S_{vv}S_{hh}^* \rangle & 0 & \langle |S_{vv}|^2 \rangle \end{bmatrix} \quad (5)$$

4 COMPACT POLARIMETRY THEORY

Compact polarimetry is a technique that allows construction of pseudo quad-pol information from dual-polarization SAR systems.

If a single polarization is transmitted, whereas the two canonical orthogonal linear polarizations (H and V) are received, the 2-D measurement vector (or

Table 1. Compact polarimetry modes.

Mode	Trans/Recep	$\vec{k}_{CP}$
π/4	45° */(H,V)	$[S_{hh} + S_{hv}S_{vv} + S_{hv}]^T / \sqrt{2}$
DCP	RC/(RC,LC)	$[S_{RR}S_{RL}]$
CTLR	RC/(H,V)	$[S_{hh} + iS_{hv} - iS_{vv} + S_{hv}]^T / \sqrt{2}$

*45° stands for linear polarization with a 45° inclination.

observable) $\vec{k}_{CP}$ is the projection of the full backscattering matrix S on the transmit polarization state. The relation between $\vec{k}_{CP}$ and S is given by (Dubois et al. 2008):

$$\vec{k}_{CP} = S\vec{J}_t \quad (6)$$

Where, $\vec{J}_t$ represents the transmitted Jones vector.

The scattering vectors $\vec{k}_{CP}$ for the π/4, dual circular polarimetric (DCP), and right circular transmit, linear (horizontal and vertical) receive or hybrid (CTLR) modes are given in Table 1 (Nord et al. 2009).

The measurement compact polarimetric covariance matrix C_{cp} are given by:

$$[C_{cp}] = \langle \vec{k}_{cp} \vec{k}_{cp}^{*T} \rangle = \begin{bmatrix} C_{11} & C_{12} \\ C_{21} & C_{22} \end{bmatrix} \quad (7)$$

Where $C_{11} = \langle |CP_1|^2 \rangle$, $C_{12} = C_{12}^* = \langle CP_1 * CP_2^* \rangle$ and $C_{22} = \langle |CP_2|^2 \rangle$. C_{cp}, is Hermitian and provides four measurements—two real diagonal terms and the real and imaginary parts of one of the off-diagonal terms.

The relevant 2×2 hermitian covariance matrices for the π/4 mode become:

$$\left[C_{\frac{\pi}{4}}\right] = \langle \vec{k}_{\frac{\pi}{4}} \vec{k}_{\frac{\pi}{4}}^{*T} \rangle = \frac{1}{2} \begin{bmatrix} \langle |S_{hh}|^2 \rangle & \langle S_{hh}S_{vv}^* \rangle \\ \langle S_{vv}S_{hh}^* \rangle & \langle |S_{vv}|^2 \rangle \end{bmatrix} + \frac{1}{2}\langle |S_{hv}|^2 \rangle \begin{bmatrix} 1 & 1 \\ 1 & 1 \end{bmatrix}$$
$$+ \frac{1}{2} \begin{bmatrix} 2\Re(\langle S_{hh}S_{hv}^* \rangle) & \langle S_{hh}S_{hv}^* \rangle + \langle S_{hv}S_{vv}^* \rangle \\ \langle S_{hh}^*S_{hv} \rangle + \langle S_{vv}S_{hv}^* \rangle & 2\Re(\langle S_{vv}S_{hv}^* \rangle) \end{bmatrix} \quad (8)$$

The resulting compact polarimetry covariance matrices are expressed as a sum of three terms. The first term contains elements that depend only on S_{hh} and S_{vv} the second term contains $\langle |S_{hv}|^2 \rangle$ elements, and the last term consists only of copolarization (co-pol)/cross-pol correlations.

We come up with an undetermined system of four equations (linked to the two real measurements C_{11}, C_{22} and the complex one C_{12}) and six variables ($H = \langle |S_{hh}|^2 \rangle$, $V = \langle |S_{vv}|^2 \rangle$, $X = \langle |S_{hv}|^2 \rangle$, $P = \langle S_{hh}S_{vv}^* \rangle$, $\langle S_{hh}S_{hv}^* \rangle$ and $\langle S_{vv}S_{hv}^* \rangle$). Additional information is, therefore required to solve it. For this reason, two hypotheses related to the polarimetric behavior of the compact covariance matrix components have been introduced (Souyris et al 2005).

- The first one suppose reflection symmetry as stated in equation ($\langle S_{hh}S_{hv}^* \rangle = \langle S_{hv}S_{vv}^* \rangle = 0$
- The second assumption relates the co-pol correlation coefficient to the relative magnitudes of the cross-pol and co-pol responses.

$$\frac{X}{H+V} \approx \frac{1-|\rho_{h-v}|}{4} \qquad (9)$$

Where $\rho_{(h-v)} = \langle S_{hh}S_{vv}^* \rangle / \sqrt{(|S_{hh}|^2) \cdot (|S_{vv}|^2))}$ is the linear correlation between S_{hh} and S_{vv}.

The assumption of reflection symmetry implies that the last term is null and the covariance matrices become:

$$\left[C_{\frac{\pi}{4}} \right] = \langle \vec{k}_{\frac{\pi}{4}} \vec{k}_{\frac{\pi}{4}}^{*T} \rangle = \frac{1}{2} \begin{bmatrix} \langle |S_{hh}|^2 \rangle & \langle S_{hh}S_{vv}^* \rangle \\ \langle S_{vv}S_{hh}^* \rangle & \langle |S_{vv}|^2 \rangle \end{bmatrix} + \frac{1}{2}\langle |S_{hv}|^2 \rangle \begin{bmatrix} 1 & 1 \\ 1 & 1 \end{bmatrix}$$

$$(10)$$

Here, the mode reduces to a system of four equations from the covariance matrix and five unknowns. The unknowns are $|S_{hh}|^2, |S_{vv}|^2, |S_{hv}|^2, ?(S_{hh}S_{vv}^*)$ and $(S_{vv}S_{hh}^*)$, where the last two unknown are complex.

5 PSEUDO QUAD-POL RECONSTRUCTION

The construction of the pseudo quad-pol covariance matrices from the compact polarimetry modes is based on a pair of equations that are iteratively solved for $\langle |S_{hv}|^2 \rangle$.

The solution of equations starts with the initial values of $\langle |S_{hv}|^2 \rangle$ and the linear co-polarization coherence $\rho = \rho_{h-v}$.

$$\rho_{(0)} = \frac{C_{12}}{\sqrt{C_{11}C_{22}}} \qquad (11)$$

$$\langle |S_{hv}|^2 \rangle_{(0)} = \frac{C_{11}+C_{22}}{2} \left(\frac{1-|\rho_{(0)}|}{3-|\rho_{(0)}|} \right) \qquad (12)$$

and then iterates the following equations:

$$\rho_{(i+1)} = \frac{C_{12} - \langle |S_{hv}|^2 \rangle_{(i)}}{\sqrt{(C_{11} - \langle |S_{hv}|^2 \rangle_{(i)})(C_{22} - \langle |S_{hv}|^2 \rangle_{(i)})}} \qquad (13)$$

$$\langle |S_{hv}|^2 \rangle_{(i+1)} = \frac{C_{11}+C_{22}}{2} \left(\frac{1-|\rho_{(i+1)}|}{3-|\rho_{(i+1)}|} \right) \qquad (14)$$

Given a value for $\langle |S_{hv}|^2 \rangle = \langle |S_{hv}|^2 \rangle_{(n)}$ (where n is the order of iteration), the pseudo quad-pol covariance matrix is then constructed by:

$$[C]_{pseudo\,quad} =$$
$$\begin{bmatrix} C_{11} - \langle |S_{hv}|^2 \rangle & 0 & C_{12} - \langle |S_{hv}|^2 \rangle \\ 0 & 2\langle |S_{hv}|^2 \rangle & 0 \\ (C_{12} - \langle |S_{hv}|^2 \rangle)^* & 0 & C_{22} - \langle |S_{hv}|^2 \rangle \end{bmatrix} \qquad (15)$$

The null components are the characteristic of the reflection symmetry assumption.

6 LINEAR CO-POLARIZATION COHERENCE

Figure 2 displays the mapping of the degree of coherence $|\rho = \rho_{h-v}|$ for consecutive orders of estimation. The test is conducted on the same zone which is the west region of Algiers, with a 5×5 analysis window.

(a)   (b)

(c)   (d)

Figure 2. Degree of coherence $|\rho = \rho_(h - v)|$, (a) reference value of $|\rho_{(h-v)}|$ inferred from FP. (0 = black, 1 = white), (b)-(c) first- and second-order estimates of $|\rho_{(h-v)}|$, (c) reconstructed degree of coherence (3rd order estimates) versus actual degree of coherence (reference value).

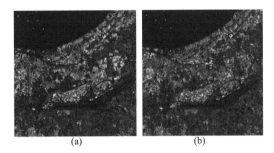

(a)   (b)

Figure 3. Mixed region in Algiers (sea, man-made structures, urban, agriculture fields, etc.), 1600×1600 pixels. RGB color composite (Red:HH), (green:HV), and (blue:VV); (a) Full polarimetric and (b) Compact polarimetric modes.

Figure 2a displays the FP reference value. The first order estimate shown in Figure 2b produces a "milky" impression, which alters the image contrast. However, the third estimated coherence shown in Figure 2c qualitatively very close to the FP coherence. Figure 2d displays the last estimate of the reconstructed degree coherence versus the actual degree of coherence. As expected, this figure shows very good reconstruction performances.

7 QUANTITATIVE ASSESSMENT

For comparison, the original FP data are shown in Figure 3a with H in red, X in green, and V in blue. The $\pi/4$ mode synthesized result is shown in Figure 3b. We notice some differences between these two images, especially in the X intensity in some urban areas, which

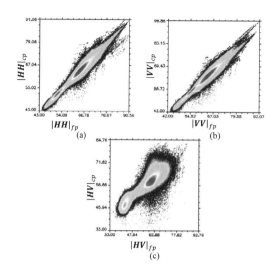

Figure 4. Reconstruction performance for H, V and X channels. Scatter plots of (a) $H_{fp} - H_{cp}$, (b) $V_{fp} - V_{cp}$ and (c) $X_{fp} - X_{cp}$.

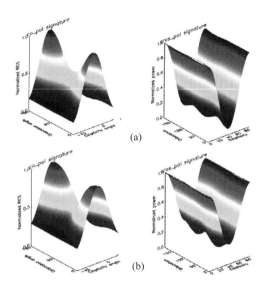

Figure 5. Co- and cross-polarimetric signatures of a sea region using (a) full polarimetric and (b) compact polarimetric modes.

is noticeably lower for the CP mode. However, strong similarity does exist in polarimetric response of most of the rest of test area.

Figure 4 shows scatter plots detailing how well the derived pseudo-quad-pol results fit the original quad-pol. It shows also the performance of the reconstruction algorithm. Most of the points of the scatter plot fall close the one-to-one line with small spread. A small systematic overestimation of the $|S_{hv}|^2$ can be observed over this data. For every channel, an overall agreement is observed between the reconstructed and actual radiometric values.

8 COMPACT POLARIMETRIC SIGNATURE

The concept of the polarization signature of a scatterer was used by Van zyl et al. (1989) to graph the power of a return wave as a function of transmit and receive polarizations. The backscattering radar cross section is given by (Van zyl et al. 1989):

$$\sigma(\psi, \chi) = \frac{4\pi}{k^2} \vec{S_r}[K]\vec{S_t} \qquad (16)$$

The subscripts r and t denote the received and the transmitted polarizations. k is the transmitted wavenumber, ψ and χ are the orientation and the ellipticity angles. The first angle (ψ) ranges between 0° to 180° and the second one (χ) is defined between −45° to 45°. $\vec{S_t}$ and $[K]$ the Stokes vector and the Kennaugh matrix respectively.

The received power, as a function of the ellipticity and orientation angles of the received polarization, completely characterizes the CP-pol response. The two-dimensional surface plot of the received power as a function of polarization ellipticity and orientation provides a simple, graphical way to display this result. Plots of the CP-pol response from a variety of

scattering mechanisms, e.g. rough surface, dihedral, dipole, allow for an easy visual analysis of the CP-pol information.

The selected region of interest is extracted from a sea surface. In Figure 5, we plot the quad- and CP-pol signatures of surface scattering representing the selected sea area.

Of note in these figures is that employing the CP- and quad-pol signatures yield essentially the same results.

9 FULL AND COMPACT POLARIMETRIC CLASSIFICATION

The presented algorithm, is a maximum likelihood classifier based on the complex Wishart distribution for the polarimetric coherency matrix. Each class is characterized by its own coherency matrix $\langle[T]\rangle$ which is estimated using training samples from the ith class ω_i.

According to the Bayes maximum likelihood classification procedure, an averaged coherency matrix $\langle[T]\rangle$ is assigned to the class ω_i, if (Ferro-Famil et al 2001):

$$d(\langle[T]\rangle, \omega_i) \leq d(\langle[T]\rangle, \omega_j) \qquad (17)$$

Where $\sum_i$ and $\sum_i$ are the averaged values of the class centers ω_i and ω_i respectively, with

$$d(\langle[T]\rangle, \omega_i) = Ln(\Sigma_i) + Tr\left(\Sigma_i^{-1}\langle[T]\rangle\right) \qquad (18)$$

The eight classes derived from the H/α decomposition are used as training sets for the initialization of the Wishart classifier. Thereafter, this statistical classifier segments the polarimetric data set into eight

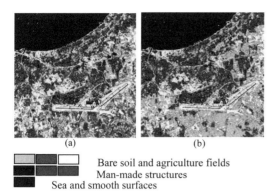

			Bare soil and agriculture fields
			Man-made structures
		Sea and smooth surfaces	

Figure 6. H/Alpha Wishart classification. (a) FP mode, (b) CP mode

classes using iterative technique. The class center coherency matrices are initialized with the results of the cloud/Pottier (or H/α) decomposition.

The combined Wishart classification with H/α decomposition, successfully show a significant improvement of the classification accuracy with respect to the H/α segmented image. The different classes are well established and better discriminated. The bare soil, the agriculture fields, the sea, the urban and man-made targets are all well revealed.

The result of the combined entropy/alpha decomposition with the complex Wishart distribution is given in Figure 6a.

The same processing algorithm as the Wishart classification is applied to the CP-data. The same eight CP-(H/α) training areas were used for all the scene classifications. Once the class centers are defined all image pixels are classified as one of the eleven fixed classes. Classification accuracy is assessed as the percentage of pixels correctly classified for each training area. The classification map for the resulted quad-pol imagery is shown in Figure 6b.

Overall, the CP-pol classification produces goods results in comparison with the FP H/α Wishart classification.

The confusion matrix indicates that the classes C1(57.06%), C3(55.94%) which represent the urban areas, C6(99;79%) and C7(91.99%) which represent the surface scattering regions as the sea and the smooth surfaces are well classified. Overall, the classification accuracies of the CP-mode data are 67%.which is quite high.

10 CONCLUSION

In this study, we have investigated to what extent the CP mode permits to reconstruct the FP information from a single linear transmitted polarization and a reception of two orthogonal polarizations. However, the key property used to estimate the FP information is reflection symmetry, which reveals a complete decorrelation of copolarized and cross-polarized backscattering coefficients.

We've investigated also the CP mode discriminating capability using polarimetric signatures which provide a complete, easy graphic means to analyze the scattering mechanism in any mode. In particular, we showed that the CP-pol signature plots compare well to their corresponding quad-pol.

Finally, the unsupervised combined classification based on the complex Wishart distribution and the most known entropy/Alpha decomposition is derived for the CP- and FP-mode. The results indicate that the generated pseudo quad-pol and the CP-mode classification results compares well to the original quad-pol imagery.

As a conclusion, a compact polarimetry SAR in general cannot be "as good as" a fully polarimetric system. However, in many applications, the results enjoyed from a CP radar are equivalent to those from an FP radar.

ACKNOWLEDGEMENTS

The authors would like to thank the Canadian Space Agency for kindly providing the polarimetric RADARSAT2 data used in this paper.

REFERENCES

Cloude, S. R. & Pottier, E. 1996. A review of target decomposition theorems in radar polarimetry. *IEEE Trans. Geosci. Remote Sens.* 34(2):498–518.

Dubois-Fernandez, P. Souyris, J.-C. S. Angelliaume & Garestier, F. 2008. The compact polarimetry alternative for spaceborne SAR at low frequency.*IEEE Trans. Geosci. Remote Sens.* 46:3208–3222.

Ferro-Famil, L. Pottier, E. & Lee, J. S. 2001. Unsupervised classification of multifrequency and fully polarimetric SAR images based on H/A/Alpha-Wishart classifier. *IEEE Trans. Geosci. Remote Sens.* 39(11): 2332–2342.

Nghiem, S. V. Yueh, S. H. Kwok, R. & Li, F. K. 1992. Symmetry properties in polarimetric remote sensing. *Radio Sci.* 27(5):693–711.

Nord, M. Ainsworth, T. L. Lee, J. –S. & Stacy, N. 2009. Comparison of Compact Polarimetric Synthetic Aperture Radar Modes. *IEEE Trans. Geosci. Remote Sens.*, vol. 47, no. 1, pp. 147–188, Jan. 2009.

Raney, K. 2007. Hybrid polarimetric SAR architecture," *IEEE Trans. Geosci. Remote Sens.* 45(11): 3397–3404.

Souyris, J.-C. Imbo, P. Fjørtoft, R. Mingot, S. & Lee J.-S. Compact polarimetry based on symmetry properties of geophysical media: The $\pi/4$ mode. *IEEE Trans. Geosci. Remote Sens.* 43(3):634–646.

Van Zyl, J. J. Zebker, H. A. & Elachi, C. 1987. Imaging radar polarization signatures: Theory and observation. *Radio Sci.* 22(4):529–543.

Computational Modelling of Objects Represented in Images – Di Giamberardino et al. (eds)
© 2012 Taylor & Francis Group, London, ISBN 978-0-415-62134-2

Land cover differentiation using digital image processing of Landsat-7 ETM+ samples

Y.T. Solano & L. Pencue Fierro
Grupo de Estudios Ambientales (GEA), Grupo de Óptica y Láser (GOL), Universidad del Cauca, Cauca-Colombia

A. Figueroa
Grupo de Estudios Ambientales (GEA), Universidad del Cauca, Cauca-Colombia

ABSTRACT: This study presents the implementation of image processing techniques in satellite bands of the visible and infrared spectrum for the differentiation of land cover in the Colombian Andes, which is dominated by high mountain systems, to do this, we extracted features through texture analysis methods, principal component analysis, discrete cosine transform, combination of color spaces and vegetation indices to be combined together, we have obtained independent zones that guaranteed the separability of 9 classes and correct discrimination in step classification. In the classification module we used two approaches: a neural network and an expert system, with which it was possible to obtain a success rate of 77.38% and 86.36% respectively.

1 INTRODUCTION

The study of land cover that form a region and evaluation of its dynamics, is an important issue to be prerequisite for the sustainable management of natural resources, environmental protection, and basic data monitoring and modeling (Lira 2002), specifically in the case of southwestern Colombia have prominence because it is a high mountain area with a large variability of vegetation cover and land use. Knowing the amount and distribution in spatial and temporal control is important for agriculture, deforestation, environmental degradation and climate change.

Today it is possible to access to all kinds of information from remote sensing satellites providing access to almost anywhere and any data (Gottfried 2003). In this case we have used multispectral satellite images Landsat7 ETM+ from NASA that primarily allow coverage for a considerable spectral range, an appropriate spatial resolution. The images are obtained for the same study area every 16 days.

2 STUDY SITE AND DATA

The study site is located in a region of Cauca – Colombia, between 76°10'17.4818"–76°44'54.7925"W and 2°08'19.8944"–2° 51'32.5787"N (UTM coordinates), Figure 1. Covering around 64 × 80 km to a total area of 5120 km². According to the Landsat standard, this represents an image of 2136 × 2652 pixels (30 m per pixel), in 6 bands (in this study, we didn't use the panchromatic band, or the 2 thermal bands).

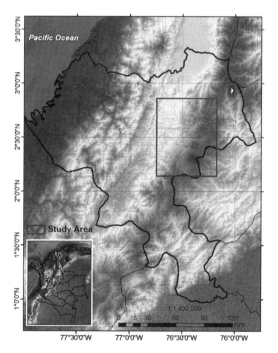

Figure 1. Study site for the land cover classification using *Landsat 7 ETM+*.

To develop this study, we used two Landsat7 ETM+ scenes (30 × 30 m pixel resolution) of September 12 of 2008 and January 21 of 2010, obtained from United States Geological Survey (http://glovis.usgs.gov), in order to perform the correction to the SLC-off problem.

Nine cover types or classes were selected and for each one of them were took the coordinates of 12 in-situ points, these data were supplied by the Grupo de Estudios Ambientales (GEA) and taken on 2011. The classes studied were: natural forest, planted forest, type 1 crop, type 2 crop, grass, city, páramo, water and rock.

3 METHODOLOGY

To perform the classification of land cover, the feature extraction of satellite images was made by creating a specific algorithm developed for this application. Generally the procedure consisted in reducing the number of bands, extracting texture descriptors, color and vegetation index and a final feature space reduction, to proceed to an unsupervised classification using a neural network and an expert system, the procedure is described below.

3.1 Principal component analysis

The objective of principal component analysis (PCA) is summarize a large group of variables in a new (smaller) without losing a significant part of the original information (Hervé & Lynne 2010; Saporta & Niang, 2009).

The study of the relationship between bands can be performed with the matrix of variance-covariance (equation 1) and the correlation matrix (Santa María del Ángel et al. 2011).

$$K_f \cong \frac{1}{MN-1} \sum_{i=1}^{MN} (f_i - \mu_f)(f_i - \mu_f)' \quad (1)$$

where f_i is the value for each pixel in the image, MN is the total number of pixels, μ_f represents the mean of all the pixel in the image and t denote matrix transposed. After applying equation 1 to 6 bands of Landsat7 and by correlation analysis, it was possible to reduce the working space to 3 bands: 3, 4, 5, which were combined in configuration 453 in the RGB color space.

Then we proceeded to perform a linear combination of them to get an image that could represent all the bands, from that image was possible to separate the clouds' shadows, water and rocks of other coverage, by setting a threshold, Figure 2 shows an image on the left side with the presence of a cloud and its respective shade in black and in the right side the same one segmented and displayed in a white color.

From now on, we proceeded with the extraction of other features that allow separability in other classes.

3.2 Texture analysis

For this study, the texture values were calculated using statistical models of first and second order as the mean, standard deviation and descriptors such as contrast, energy and entropy obtained by constructing co-occurrence matrices (Haralick & Shapiro 1992) from a moving window of 5 × 5, of the above, energy

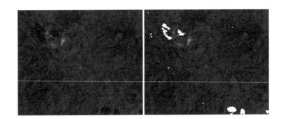

Figure 2. Original image with clouds' shadows (left) and image with segmentate shadows in white color (right).

and the standard deviation gave very good results in the differentiation of type 1 crops, city and clouds, the latter are of great importance because the study area is covered of clouds most of the time, causing difficulties when sorting by ignorance or by identifying areas of erroneous coverage because of their shadows.

3.3 Discrete cosine transform

A discrete cosine transform (DCT), decomposes the signal into spatial frequencies, where each sub-block contains coefficients which tend to give new features in themselves, for the extraction of these, equation 2 was used (Lira 2002) applied in the same way as with the co-occurrence matrices, with a small 5 × 5 window that was shifting along the entire image. This transform was applied to bands 4, 5 and 3 resulting from the correlation with PCA, and the band 3 allowed obtaining large differences in natural and planted forests.

$$C(i, j) = \sum_m \sum_n g(m, n)[\cos(2m+1)i\pi][\cos(2n+1)j\pi] \quad (2)$$

3.4 Color spaces

In this procedure, we used a combination of the bands 453 in RGB space and converted to two new color spaces, YCbCr and HSV in order to highlight some coverage that wouldn't be able to see in the RGB space (Torres 2006). We analyzed each color plane separately and obtained a difference in Cr plane for type 1 and 2 crops (red and blue planes, respectively), which represent potato crops planted on two different dates. For the H plane, it was possible to separate the pántano, characteristic of the Andean region. In addition to this study, we also analyzed each band separately, finding a highly differentiated for grass in band 2.

3.5 Vegetation index

Due to the high presence of vegetation cover in the area of study, vegetation indices were used: NDVI (Normalized Difference Vegetation Index), SAVI (Soil Adjusted Vegetation Index) y PVI (Perpendicular Vegetation Index) (Chuvieco 1996), of which could make a final differentiation to type 2 crops and for the city, the latter due to the non presence of vegetation in the area. Figure 3 shows the results for the 3 indices of a city and vegetation samples image for the NDVI, dark areas

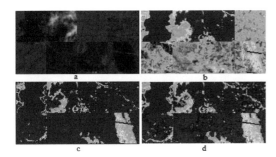

a   b

c   d

Figure 3.   a) Original Image, b) *NDVI*, c) *SAVI*, d) *PVI* .

Table 1.   Neuronal network results for coverage.

Class	Right	Wrong
1. Natural forest	5	7
2. Planted forest	12	0
3. Type 1 crop	4	8
4. Type 2 crop	12	0
5. Grass	12	0
6. City	11	1
7. Páramo	9	3
Total Samples	65	19
	84	

represent samples of town, rock, clouds and water, showing a large separability with other coverage, it does not happen with the other two indexes.

3.6 *Digital image classification*

After applying the different methods, it was possible to reduce the feature space to 8 variables: PCA, standard deviation, energy, DCT for the band 3, Cr and H planes in combination 453, band 2 and NDVI. Based on the above, we evaluated two methods of supervised classification: neural networks and expert system.

For the training of the neural network, were take a set of 4 samples for 7 coverage, the 8 characteristics were extracted for each one of them, and trained 4:16:1 network architecture in configuration multilayer perceptron with tangential and linear activation functions for the output neuron (Bishop 1995).

For the expert system 12 samples were taken for all the covers and additionally samples were taken from areas with cloud and shade, in order to improve the classification, and the 8 features were extracted, which an once plotted, allowed to observe the trend of the coverage to occupy their own space (as already mentioned above) and through this action it was possible to define the ranges in which each branch the decision tree would be allowing the differentiation of all the coverage.

4 RESULTS AND DISCUSSION

Once the neural network trained with the 4 test samples, the classification was performed for 12 samples (the test and other) in order to verify the proper functioning of the network. Table 1 shows the number of correct answers for each class of which it is known that from a total of 84 samples, 19 were classified incorrectly, giving a success rate for the network of 77.38%.

Those cases where the network is wrong classifying correspond to areas in which samples were taken at very small areas, resulting in zones of pixels smaller than 5×5 which were used in the extraction of certain features.

After this procedure, we performed the classification of the entire study area, for which, it required

Figure 4.   Classification of the neuronal network: Salvajina's dam (left) and Popayán (right).

a large time, because for each of the 2136×2652 pixel image, the network had to find the 8 features for classification.

Figure 4 shows the classification made by the neural network for two regions of importance in the area: the Salvajina's dam and the city of Popayán.

Another method implemented, was the use of an expert system, in which the time required for classification decreased to 20%, this because for each class was not necessary to consult on the 8 characteristics.

Table 2 shows the characteristics to differentiate each of the classes with their respective ranges of action, it can be seen that there are 2 additional classes to those set at the beginning of this study, and correspond to areas with presence of clouds and their shadows, for which samples were taken from the image in order to avoid an increase in the classification error.

Additionally it is noted that for classes 1 and 2 (natural and planted forest), the ranges of action are the same, because despite of there is a random and symmetric distribution for each case, it was not possible to differentiate the two types of forests by the methods used principally due to the low resolution images (30 m per pixel).

Table 3 shows the results of the expert system's classification for the 12 samples of each class including clouds and their shadows, where you get that for a total of 132 samples, 18 were classified incorrectly, giving a success rate of 86.36%.

Comparing the results of Tables 1 and 3 clearly shows that for the cases of type 2 crops and grass, the neural network correctly classified all samples, from

Table 2. Features range for expert system classification.

Class	Range	Features
1–2	0.6 < n < 2.0	*DCT* band 3
3	1.3 < n < 3.2	*PCA*
	29 < n < 44	Energy
4	0.42 < n < 0.48	*Cr 453*
	0 < n < 0.2	*NDVI*
5	37 < n < 43	Band 2
6	0.07 < n < 0.115	Standard deviation
	n < 0	*NDVI*
7	0.19 < n < 0.28	*H*453
8	n = 255	*PCA* segmented
11	n < 11	Energy
9	n = 255	*PCA* segmented
10	n > 60	Energy
	n > 0.12	Standard deviation

Table 3. Expert system results for all the coverage.

Class	Rigth	Wrong
1. Natural forest	7	5
2. Planted forest	12	0
3. Type 1 crop	9	3
4. Type 2 crop	10	2
5. Grass	11	1
6. City	12	0
7. Páramo	9	3
8. Water	11	1
9. Rock	10	2
10. Clouds	12	0
11. Cloud shadows	11	1
Total Samples	114	18
	132	

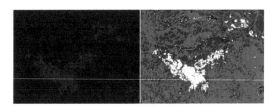

Figure 6. Popayán in 4, 5, 3 combination (left) and classification of expert system (right).

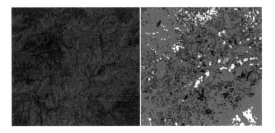

Figure 7. Crops class 1 and 2 in 4, 5, 3 combination (left) and classification of expert system (right).

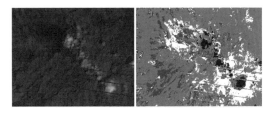

Figure 8. Top of the Puracé's volcano in 4, 5, 3 combination (left) and classification of expert system (right).

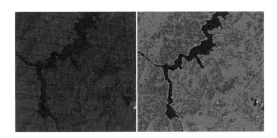

Figure 5. Salvajina's dam in 4, 5, 3 combination (left) and classification of expert system (right).

this result can be deduced that by combining the two methods the classifications could be improved.

The success rate for the expert system is greater than for the neural network, which was reflected in the quality of classification, Figures 5 and 6 show the results for the same areas presented in Figure 4, to Figure 5 the area depicted in the black (right side) represents the water in the Salvajina's dam and light and dark gray areas represents forests and grass respectively.

Figure 5 shows the expert system classification for an additional class to the set, which corresponds to the edge of the dam, which is a mixture of mud and grass, this same characteristic is presented for lakes and ponds in the mountains were classified correctly too.

In Figure 6, the white area corresponds to the city of Popayán and around it can be seen grass and forests again.

The classification of two types of crops was of great interest, since these were in the high mountain area, where programs that are usually used for classification have large errors. In Figure 7 it can be seen in black color the type 2 crops and in light gray, the type 3 crops, to the surrounding there are grazing areas, forests and páramo, this is the topography where crops of potatoes normally grow (corresponding to type 1 and 2 crops)

Finally in Figure 8, there is an image of the top of the Puracé's volcano with some steam coming out of it, which is represented in black in the classification, in a white color it is show the area of páramo and in gray tones the rock, forest and grass present in the area.

The final system of classification allowed to classify 10 classes, where one of them corresponds to natural and planted forest (as already mentioned above),

Table 4. Percentage and area classifified for each coverage.

Class	Area (Ha)	Percentage (%)
Natural and planted forest	381,607.00	74.85
Crop class 1	1780.11	0.35
Crop class 2	8788.68	1.72
Grass	91,315.70	17.91
City	953.28	0.19
Páramo	12,111.60	2.38
Water and Clouds' Shadows	3244.95	0.64
Rock	946.62	0.18
Clouds	2515.68	0.49
Not classified	6556.86	1.29

another correspond to regions with water and cloud shadows, that couldn't be separated because of their response in the frequency spectrum, even when analyzed with textures, a class for unclassified areas and the others mentioned in Table 3.

Table 4 shows the results in area and percentage for each of the 10 classes, where you can see a high presence of forests, grass and pántano, consistent with the topography that can be finds in the Andean areas.

The results were compared with ratings made by the recognized geographic information systems, and verified by experts in the field, allowing corroborate the high degree of improvement in them, in addition to high accuracy for classification of high mountain areas and differentiating areas with presence of clouds. It is thought that by using images with higher resolution should improve the success rate, because although the actual result is very good, there are some areas, especially crop and lakes, which are too small to occupy a good number of pixels that allows them to appear in the image.

5 CONCLUSIONS

We developed a tool that allows the land cover classification in high mountain areas, specifically the Andean region. For which it was use a neural network and expert system combined with answers from the texture analysis, changes in the color space and vegetation indices. While the texture analysis has been used for many years for land cover classification at remote sensing, has shown that when combined with other methods, the result can be even more precise and according to reality.

The development of this new method allows giving a great advantage, since the system is characterized with in-situ data of Colombia, which present much variabil-

ity and can't be compared with the responses obtained by programs that have been characterized with other data regions and also can be easily extrapolated to other countries in the Andean region.

With the use of higher resolution images it could be possible to improve the success rate of classification and even increase the number of classes or coverage selected, this is because having a higher resolution, means a, be greater amount of pixels representing a class than in the case of this study, allowing to obtain a greater number of samples in-situ for the training of the network and for selecting ranges of action in the case of the expert system.

ACKNOWLEDGMENTS

The authors wish to thanks the Universidad del Cauca and the Grupo de Estudios Ambientales for the help and advice given during the course of this investigation.

REFERENCES

Bishop, C. M.. 1995. Neural Networks for Pattern Recognition, Oxford University Press.
Chuvieco Salinero, Emilio. 1996. Fundamentos de Teledetección Espacial. Rialp, Madrid. 3ra Edición.
Haralick, R. M. & L.G. Shapiro. 1992. Computer and robot vision. Addison Wesley. Vol. 1 y 2.
Hervé, Abdi & Lynne J. Williams. 2010. Principal component analysis, Wiley Interdisciplinary Reviews: Computational Statistics,Volume 2, Issue 4, pages 433–459.
Gottfried Konecny. 2003. GEOINFORMATION: Remote Sensing, photogrammetry and geographic information systems. TAYLOR & FRANCIS.
Lira C., Jorge. 2002. Introducción al tratamiento digital de imágenes. Instituto Politécnico Nacional, Universidad Nacional Autónoma de México, Fondo de Cultura Económica, México. ISBN 970-32-0091-5. 1ra Edición.
Santamaría del Ángel, Eduardo., González Silvera a., Millán-Núñez, r., Callejas Jiménez, M. E., Cajal-Medrano, R. 2011. Handbook of Satellite Remote Sensing Image Interpretation Marine Applications: Determining Dynamic Biogeographic Regions using Remote Sensing Data.
Saporta G & Niang N. 2009. Principal component analysis: application to statistical process control. In: Govaert G, ed. *Data Analysis*. London: John Wiley & Sons.
Torres A., Jorge Eliecer. 2006. Determinación de coberturas vegetales a partir del estudio de su repuesta RGB utilizando aerofotografías digitales. Universidad del Cauca. Facultad de Ciencias Naturales Exactas y de la Educación, Ingeniería Física.

Computational Modelling of Objects Represented in Images – Di Giamberardino et al. (eds)
© 2012 Taylor & Francis Group, London, ISBN 978-0-415-62134-2

Comparing land cover maps obtained from remote sensing for deriving urban indicators

T. Santos & S. Freire
e-GEO – Research Centre for Geography and Regional Planning, Faculdade de Ciências Sociais e Humanas, FCSH, Universidade Nova de Lisboa, Lisboa, Portugal

ABSTRACT: From a land planning perspective, it is worthwhile to investigate different levels of information abstraction regarding scale, class detail or minimum mapping unit, that allow characterizing the most common situations land planners have to deal with. The identification of the processes and the characteristics that must be monitored will directly influence the effort and the time spent for land mapping and, consequently, the mapping cost. Accordingly, it is possible to orient the land mapping process towards different applications that require specific products.

This study addresses the extraction of thematic information from remote sensing data for producing urban indicators at local scale in urban environments. Two maps obtained from satellite imagery are compared. The maps differ in methodology, legend and minimum mapping unit. Through the analysis of the different map specifications, it can be concluded that land use classifications can be problematic for estimating urban environmental indicators. For those applications, land cover maps are more suitable.

Keywords: urban indicators, remote sensing, land cover land use, Lisbon

1 INTRODUCTION

Systems based on urban indicators can be used as tools for cities to communicate synthetic information, monitoring and analyzing trends concerning the territory. Many of such indicators require information about the status of land cover. Using the area of each land cover class, indicators on land sealing area, quantification of green area, or the vacant land available in the city, are ecological measures that can be used for monitoring and analyzing trends over the territory. Studies on impacts of urbanization, responses to natural and man-made disasters, vulnerability analysis or housing conditions, all require land-based indicators. The geographical data constitute the base of the spatial representation of the indicators. Urban indicators are designed to measure the quality of life and the nature of development of an urban area. These indicators can be used to make policy and planning decisions, to identify whether policy goals and targets are being met, and sometimes to predict change (Santos et al., 2011).

One major source of information about the urban environment is remote sensing data. Evaluating its efficiency for deriving environmental indicators at local scale in urban environments constitutes the focus of this work.

The extraction of categorical information on the land surface can be accomplished in two ways. The two main methods for information collection from digital images are photo-interpretation and digital

classification. Choosing the best methodology shall take into account economic factors, time and available resources. Moreover, the selection of the appropriate data source for thematic information extraction is crucial to the mapping quality. Indeed, the spatial resolution of images influences the accuracy and the detail of the mapped classes (Benfield et al., 2007).

Traditionally, the cartographic framework is based on the visual analysis of aerial photographs (Herold et al., 2003). More recently, high spatial resolution satellite images have also been used for photo-interpretation, given the recent improvements of the spaceborne digital sensors. The success of the interpretation of images varies with the interpreter's skill and experience, the nature of the objects, the analyzed phenomena and the quality of the images being used (Campbell, 2002). Visual interpretation is a method widely used for production of thematic cartography, not only because it achieves good results, but also because an alternative method is not always available. However, visual interpretation has some associated drawbacks. On one hand, the number of distinguishable grey levels for the human eye (approximately 16) is considerably smaller than the range captured by digital sensors. Similarly, the human eye can only compare three bands simultaneously (in a RGB color composite). But probably the prime disadvantage is that photo-interpretation is based on subjective assumptions. Green and Hartley (2000) found that the subjectivity in placing the boundary between elements which gradually tend to each other is the factor which contributes the most to the positional error on a thematic map. This inconsistency may also create problems in map updating, even if made by the same person (Ahlcrona, 1995), rendering the conclusions based on the analysis of these maps, to be unreliable. In addition to the intrinsic characteristics of human recognition, the entire cartographic framework based on photo-interpretation requires time and allocation of resources.

Alternatively, there are *automatic* methods that are less subjective or time consuming, and require a minimum of resources. Although image classification is mostly performed automatically by the computer in the digital environment, human intervention, either prior to the classification or during post-classification, still plays an indispensable role, even though this intervention is reduced markedly in comparison with manual interpretation (Gao, 2009).

Richards and Jia (2006), comparing two techniques for thematic information extraction from remote sensing data – photo-interpretation and automatic classification – conclude that photo-interpretation, because it involves human interaction and high levels of decision, is a good technique for spatial evaluation but is poor in quantitative accuracy. The higher accuracy of computer analysis originates from its ability on processing every pixel in the image taking into account the full range of spectral, spatial and radiometric characteristics.

2 STUDY AREA AND DATA SET

The study area, where geographic information extraction is tested for building urban indicators, is the city of Lisbon. The municipality occupies an area of 84 Km^2, and is a typical European capital city, with very diverse land use dynamics, ranging from historical neighborhoods (e.g., the downtown area of *Baixa*), where the street network is dense and most of the area is built-up, to modern residential ones (e.g., the area of *Alta*), with ongoing construction of roads and multi-family buildings. Between these two situations, there are more heterogeneous places with land uses that go from built-up, parks, agriculture and vacant land to industrial, utilities, and schools.

Two Land Cover Maps are compared for assessing urban indicators in the city of Lisbon, Portugal. One map – Land Cover Map 2008 (LCM2008) – is obtained from a semi-automatic classification of a Very-High Resolution (VHR) dataset (Santos et al., 2011) (Figure 1).

The spatial database used to produce the LCM2008 included a IKONOS-2 pansharp image of the city acquired in 2008, a derived Normalized Difference Vegetation Index (NDVI) (Rouse et al, 1973), and a normalized Digital Surface Model (nDSM) from 2006. The raster map, resulting from the classification process, was then converted to a vector format.

The LCM2008 nomenclature is organized in two levels of detail. The 1st level includes the classes

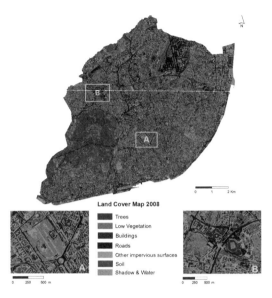

Figure 1. Land Cover Map of 2008 derived from IKONOS imagery for the city of Lisbon (Santos et al., 2011).

"Vegetation", "Impervious Surfaces", "Soil", and "Shadows and Water". On the 2nd level, seven classes were defined: "Trees", "Low Vegetation", "Buildings", "Roads", "Other impervious surfaces", "Soil", and "Shadows and Water".

In the LCM2008, a Minimum Mapping Unit (MMU) of 100 m^2 was adopted for "Impervious surface" and "Soil" classes, and 1 m^2 for "Vegetation".

The other map evaluated in this work is the Urban Atlas, an European product, produced by visual interpretation of satellite imagery, with the support of reference data. The map is available in vector format.

For urban areas, a MMU of 0.25 ha is adopted (Figure 2). This map is available for Lisbon, and was based in ALOS imagery (spatial resolution of 2.5 m), from 2007. The goal of the Urban Atlas is to provide land use information for compiling environmental indicators.

Based on the fact that Urban Atlas and LCM2008 both share the same objective, and are obtained from EO data, a comparison of results is presented.

3 METHODOLOGY

The goal of this study is to compare LCM 2008 and Urban Atlas for the city of Lisbon, outlining major differences and similarities regarding land cover information.

The maps were produced with different goals, methodologies and data sources. Consequently, scale, nomenclature and MMU differ. Therefore, for mapping comparison, a nomenclature compatibilization was performed in order to allow comparing similar classes from the two maps. The area of each class in level 1 of LCM2008 was compared with the corresponding class(es) of the Urban Atlas map, from 2007.

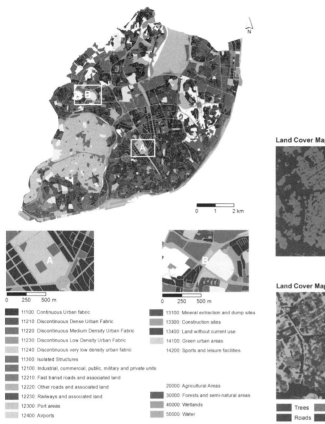

Figure 2. Map from the Urban Atlas for the city of Lisbon.

11100: Continuous Urban fabric
11210: Discontinuous Dense Urban Fabric
11220: Discontinuous Medium Density Urban Fabric
11230: Discontinuous Low Density Urban Fabric
11240: Discontinuous very low density urban fabric
11300: Isolated Structures
12100: Industrial, commercial, public, military and private units
12210: Fast transit roads and associated land
12220: Other roads and associated land
12230: Railways and associated land
12300: Port areas
12400: Airports

13100: Mineral extraction and dump sites
13300: Construction sites
13400: Land without current use
14100: Green urban areas
14200: Sports and leisure facilities

20000: Agricultural Areas
30000: Forests and semi-natural areas
40000: Wetlands
50000: Water

Table 1. Comparing areas from LCM2008 level 1 classes and Urban Atlas, in the city of Lisbon.

LCM2008		Urban Atlas class		
Class	Area (ha)	Class	Area (ha)	Difference Area (ha)
Vegetation	2428	14100; 14200; 2000; 3000	1980	448
Impervious surface	4907	11100; 11200; 11210; 11220; 11230; 11300; 12100; 12200; 12210; 12220; 12230; 12300; 12400	6117	−1210
Soil	839	13100; 13300; 13400	313	526

Regarding the MMU, no harmonization was applied.

The class "Shadows and Water" was excluded from this analysis. This option introduced differences in the total area of each map (Table 1).

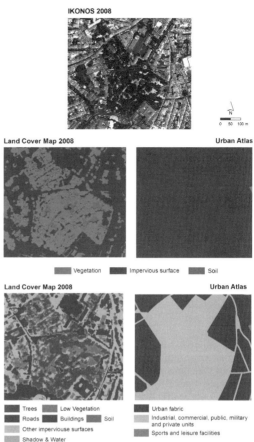

IKONOS 2008

Land Cover Map 2008   Urban Atlas

Vegetation   Impervious surface   Soil

Land Cover Map 2008   Urban Atlas

Trees   Low Vegetation   Urban fabric
Roads   Buildings   Soil   Industrial, commercial, public, military and private units
Other imperviouse surfaces   Sports and leisure facilities
Shadow & Water

Figure 3. Comparison of LCM2008 and Urban Atlas maps in the *Jardim da Estrela*, a public park.

4 RESULTS

The most obvious differences between the two maps are found in "Impervious surface" and "Soil" class areas. The reason for these differences lies in the technical characteristics regarding the mapping process: the MMU and the Urban Atlas's choice of a nomenclature almost exclusively based on land use, rather than a nomenclature that prioritizes land cover.

Urban Atlas uses a nomenclature based on the Corine Land Cover, which prioritizes land use over land cover. LCM2008, on the other hand, is more concerned with land cover. This fact, along with the MMU, also contributes to the different areas occupied by classes, when comparing the two maps. For example, the public park *Jardim da Estrela*, that occupies an area of approximately 6 ha, is classified as "Industrial, commercial, public, military and private units", in the Urban Atlas. Since this class clearly identifies built-up land, it was reclassified as "Impervious surface" for map comparison (Table 1). The same area in LCM2008 it is mainly classified as "Vegetation", "Impervious surface" and "Soil" When using the most disaggregated nomenclature from both maps, it is evident that the estimation of green areas, when using the Urban Atlas map is quite different from the one mapped in LCM2008 (Figure 3).

Differences regarding the "Soil" class are also mainly due to the nomenclature. On one hand, the Urban Atlas applies a land use classification where the class "Land without current use" (class 1.3.4) indicates areas with no constructions, with or without vegetation. On the other hand, in LCM2008, in areas with no built-up elements, the presence or absence of vegetation cover implies classification as "Vegetation" or "Soil", respectively.

The consequences of these differences for urban indicators are evident. For example, when reporting Green area available in the city using LCM2008 as source data, a value of 29% is obtained. But if Urban Atlas is used, the value decreases to 23%.

5 CONCLUSIONS

Remote sensing systems acquire information on land cover and not land use. The comparison between LCM2008 and Urban Atlas, demonstrates that land use classifications can be problematic for estimating urban environmental indicators. In the urban environment, built-up structures are generally impervious surfaces preventing water infiltration and include surfaces such as roof-tops, roads, sidewalks and parking lots and compacted soil and gravel. But these covers coexist with urban vegetation like trees or grass plots. Quantitative assessment of the spatial distribution of the land cover classes constitutes the basic data for building primary surface related indicators.

ACKNOWLEDGEMENTS

This work was conducted in the framework of project GeoSat – Methodologies to extract large scale GEOgraphical information from very high resolution SATellite images, funded by the Portuguese Foundation for Science and Technology (PTDC/GEO/64826/2006).

The authors would like to thank Logica for the opportunity of using the LiDAR data set.

REFERENCES

Ahlcrona, E. 1995. CORINE Land Cover: A pilot project in Sweden. In J. Askne, (Ed), *Sensors and Environmental Applications of Remote Sensing*: 19–22. Rotterdam: Balkema.

Benfield, S.L., Guzman, H.M., Mair, J.M., Young, J.A.T. 2007. Mapping the distribution of coral reefs and associated sublittoral habitats in Pacific Panama: A comparison of optical satellite sensors and classification methodologies. *International Journal of Remote Sensing*, 28: 5047–5070.

Campbell, J. B. 2002. *Introduction to remote sensing*, 3rd edition. New York: The Guilford Press.

Gao, J. 2009. *Digital analysis of Remotely Sensed Imagery*. The McGraw-Hill Companies, Inc.

Green, D.R., Hartley, S. 2000. Integrating photointerpretation and GIS for vegetation mapping: some issues of error. In A. Millington and R. Alexander (Eds), *Vegetation Mapping: From Patch to Planet*, John Wiley and Sons, Chichester, 103–134.

Herold, M., Liu, X.H., Clarke, K.C. 2003. Spatial Metrics and Image Texture for Mapping Urban Land Use. *Photogrammetric Engineering & Remote Sensing*, 69(9): 991–1001.

Richards, J.A., Jia, X. 2006. *Remote Sensing Digital Image Analysis. An Introduction.* 4th Edition, Springer-Verlag New York.

Rouse, J.W., Haas, R.H., Schell, J.A., Deering, D.W. 1973. Monitoring Vegetation Systems in the Great Plains with ERTS. *3rd ERTS Symposium*, 1: 48–62.

Santos, T., Freire, S., Tenedório, J. A, Fonseca, A. 2011. Using satellite imagery to develop a detailed and updated map of imperviousness to improve flood risk management in the city of Lisbon. In J.M. Tavares, R.M. Natal Jorge (Eds.), *Computational Vision and Medical Image – Processing VipIMAGE 2011: 333–336*, CRC Press.

Computational Modelling of Objects Represented in Images – Di Giamberardino et al. (eds)
© 2012 Taylor & Francis Group, London, ISBN 978-0-415-62134-2

Extraction of buildings in heterogeneous urban areas: Testing the value of increased spectral information of WorldView-2 imagery

S. Freire & T. Santos

e-GEO – Research Centre for Geography and Regional Planning, Faculdade de Ciências Sociais e Humanas (FCSH), Universidade Nova de Lisboa, Lisboa, Portugal

ABSTRACT: Large-scale spatial data that include both topographic and thematic information are needed to support decision-making and regular land management activities at municipal level. Buildings are one of the main feature classes of interest for a municipality, whose 'correct' automatic extraction remained a challenging task, even with the advent of Very-High Resolution (VHR) satellite imagery. Using Lisbon as the case study, this work tests the contribution of the increased spectral resolution of the WorldView-2 sensor for building extraction, in an urban planning context. The methodology is based on automated feature extraction and object-based accuracy assessment. Results show that, while the overall accuracy of extraction is relatively high, it has not increased significantly when the newly-available bands were included.

Keywords: WorldView-2, feature extraction, buildings, urban planning, Lisbon

1 INTRODUCTION

Large-scale spatial data that include both topographic and thematic information are needed to support decision-making and regular activities at municipal level. Among these, urban planning and management require faster updating of municipal spatial databases. The combination of widely-available, wide-coverage, cost-effective Very-High spatial Resolution (VHR) satellite imagery and Geographic Object-Based Image Analysis (GEOBIA) hold promise for this purpose (Ehlers 2007, Hay & Castilla 2008). However, until recently, there remained spectral limitations of the former (Herold et al. 2003) and some shortcomings of the latter (Lang 2008). In a detailed evaluation of classification performance, Herold et al. (2003) have identified spectral limitations of IKONOS imagery for pixel-based classification of urban land cover, indicating that four-band VHR satellite imagery are not well suited to capture in detail the unique spectral characteristics of the urban environment (Herold et al. 2003).

The WorldView-2 (WV-2) is the most recent commercial satellite that acquires images with very-high spatial and spectral resolutions, offering an opportunity to test the contribution of the increased spectral resolution for feature extraction. WV-2 acquires a panchromatic band with a pixel size of 0,50 m and multispectral images with eight spectral bands (1-Coastal, 2-Blue, 3-Green, 4-Yellow, 5-Red, 6-Red Edge, 7-Near-Infrared1, and 8-Near-Infrared2), having a 2-m spatial resolution.

Buildings are a major urban element and one of the main feature classes of interest for a municipality, whose 'correct' automatic extraction from imagery remains a challenging task, even with the advent of high spatial resolution. Difficulties include scene complexity, building occlusions (trees, shadows), and heterogeneity of feature class (Freire et al. 2010a), and these increase with refinement of image resolution (Awrangjeb et al. 2010). To obtain a cartographic product from VHR imagery using feature extraction, most of the challenge results from the interplay of several factors, namely: (i) the object and its context, (ii) the nature of the imagery, and (iii) the mapping requirements and its constraints (Freire et al. 2010b). Despite the many methodologies proposed for feature extraction, none has so far proved to be effective in all conditions and for all types of data (Salah et al. 2009). For the image analyst/map producer the challenge may be restricted to handling the necessary stages of image pre-processing, image segmentation, and generalization of features to produce a map. At present, the very quality assessment of extracted buildings for mapping purposes is still a complex endeavor for which there is no optimum, consensual, or standard approach (Rutzinger et al. 2009).

The present work aims at investigating the contribution of the increased spectral resolution of the WV-2 sensor for automated building extraction, within an urban planning context.

2 STUDY AREA AND DATA

For this study, an area located to the northeast of the downtown of the city of Lisbon, Portugal, was selected

Figure 1.   Study area in the city of Lisbon and pansharpened WorldView-2 image (532 RGB).

(Fig. 1). This area occupies 64 ha (800 m × 800 m), and has a diverse land use/land cover (LULC) that varies from urban to open field with and without vegetation. It includes trees, lawns, herbaceous vegetation and agricultural plots, bare soil, a school, industrial properties, roads and rail networks, and residential housing. This latter use includes a mixture of single homes and multi-story apartment buildings, with red tile roofs of varying age and condition. Due to its diversity, this area provided a good testing ground representative of the challenges for building extraction in the city of Lisbon.

Spatial data sets included several band combinations derived from WorldView-2 imagery covering the city of Lisbon, acquired in June 29, 2010. The image has a spatial resolution of 2 m in the multispectral mode (8 bands), a pixel size of 0.5 m in the panchromatic mode, and a radiometric resolution of 11 bits. The imagery was pansharpened to the spatial resolution of the panchromatic band in PCI Geomatica 10, and orthorectified in order to reduce the geometric distortions introduced by the relief and to attribute a national projected coordinate system (ETRS89-PT-TM06). The 0.5 m pixel size is the minimum geometric resolution recommended by Jensen & Cowen (1999) for the detection of building perimeter and area.

To assess the potential benefit brought by the new four WV-2 bands for building extraction, fifteen image datasets were prepared and tested: the first set comprises the four spectral bands (Blue, Green, Red and Near-Infrared1) usually present in optical VHR satellite imagery, and further sets are created by successively adding to this base composite each of the 'new' image bands, in every combination possible.

A vector reference map of building polygons is used for accuracy assessment.

3   METHODOLOGY

The approach includes two stages, namely: (1) the feature (building) extraction, and (2) object-based quality assessment.

3.1   Feature extraction

The extraction of buildings with red tile roofs (polygons) from the 2010 imagery was performed using Feature Analyst 4.2 (Overwatch), as an extension for ArcGIS 9.3 (Esri). Feature Analyst (FA) is an Automated Feature Extraction (AFE) application that utilizes an inductive learning approach for object recognition which allows classifying and extracting only those features belonging to the class of interest. In this process several advanced machine learning techniques can be used, such as artificial neural networks, decision trees, Bayesian learning, and K-nearest neighbor (Opitz & Blundell 2008).

The classification is based on a supervised approach, so the initial step is the identification of training samples for each class, followed by the definition of parameters such as the number of bands to be classified, the type of input representation, and level of aggregation. The classifier uses feature characteristics such as spectral response/color, size, shape, texture, pattern, shadow, and spatial association, for feature classification. After an initial classification, there is the possibility to remove clutter or add missing areas. However, to reduce subjectivity this option was not used and extraction from the image sets was performed in a single pass of the classifier, using the same training set and parameters.

The training set included 35 building samples (about 10% of actual features present), with areas varying between 20.7 and 1232 m². The learning parameters used were Manhattan (representation) 13 pixels wide, with aggregation of areas below 20 m² in order to match the minimum area of the training set. No smoothing or squaring of polygon edges was applied.

3.2   Accuracy assessment

To evaluate the quality of spatial information automatically extracted from images, based on the concept of reference value, it is necessary to measure levels of compliance with information from an independent source. This reference data can be obtained from a field survey (e.g. GPS collection), from an existing map having acceptable accuracy, or from a map created by visual interpretation of the same source data (Congalton & Green 2009). The latter approach is commonly followed in accuracy assessment of building extraction (e.g., Rutzinger et al. 2009, Vu et al. 2009).

Table 1. Accuracy results (in percentage) for each map extracted from the different image data sets (composites).

Image set (bands)	Overall Accuracy	Omission Error	Commission Error
2357	73.33	14.44	16.32
23571	72.42	8.63	22.27
23574	72.33	11.43	20.22
23576	72.52	11.43	19.95
23578	71.72	9.91	22.13
235714	72.86	9.41	21.17
235716	73.44	12.62	17.85
235718	73.44	13.97	16.61
235746	72.59	12.90	18.67
235748	72.83	13.77	17.58
235768	72.48	12.80	18.89
2357146	72.90	10.22	20.50
2357148	72.22	10.93	20.75
2357168	72.72	11.40	19.77
23571468	72.61	12.26	19.19

Figure 2. Detail of extraction result of detached houses for image set 235716 (532 RGB).

Figure 3. Detail of extraction result of adjacent apartment buildings for image set 235716 (532 RGB).

An independent and experienced interpreter created a reference map of building blocks by visual analysis and manual digitizing over the pansharpened image. All the discernible features belonging to the class of interest were digitized, totaling 341 building polygons. Each extracted map was evaluated against the reference data set in terms of its overall thematic accuracy and completeness (lack of errors of omission and commission). The overall thematic quality of building extraction is assessed using the overlap test between classified and reference data (Shan and Lee 2005).

4 RESULTS

The quality assessment metrics for each extracted map are presented in Table 1.

The obtained overall accuracies are very similar, varying between 71.72 and 73.44%. On the other hand, error of omission and commission display significantly more variability. The highest accuracy was obtained with 6-band composites 235716 and 235718, and the lowest using 5 bands (image set 23578), in this case mostly due to high commission error. In fact, in general error of commission was higher that omission, indicating overmapping of features. The lowest omission error was obtained using set 23571.

A detail of the best extraction for an area with detached houses is show in Figure 2.

A common limitation still observed in automated feature extraction (AFE) is that a row of several adjacent buildings can only be extracted as one building block (Fig. 3). This creates problems in using AFE for mapping, since it requires additional editing to obtain a map product where all individual buildings are represented as such.

Taken as a whole, the thematic accuracies are relatively high and significantly above the value obtained by Santos et al. (2009) for the same study area and feature class using a four-band QuickBird pansharp image (60%). This suggests a better performance of the WorldView-2 imagery for this purpose, despite the apparent limited value of the 'new' bands for automated building extraction in the tested conditions. This result is less surprising if one considers that these additional bands are mostly aimed at improving vegetation analysis and applications.

5 CONCLUSIONS

Typical four-band VHR satellite imagery display spectral limitations for urban mapping using automated feature extraction. However, recent space-based sensors possess improved spectral resolution.

This work was an initial attempt at testing the contribution of the additional bands of WorldView-2 imagery for extraction of buildings in a heterogeneous urban area. Results show that, while the overall accuracy of extraction is generally high and above that of previous efforts, it has not increased significantly when the 'new' bands were included. Commission error is more prevalent indicating tendency to overmap.

Future research should consider testing on additional study areas and other building types, as well as using other AFE approaches. A similar study will soon be performed for vegetation extraction.

ACKNOWLEDGEMENTS

We thank Nuno Afonso for kindly performing the pre-processing of the imagery.

REFERENCES

Awrangjeb, M., Ravanbakhsh, M., Fraser, C. 2010. Automatic detection of residential buildings using LIDAR data and multispectral imagery. *ISPRS J. Photogramm. Remote Sens.* 65(5): 457–467.

Congalton, R.G. & Green, K. 2009. *Assessing the Accuracy of Remotely Sensed Data: Principles and Practices.* 2nd ed. Boca Raton, FL, USA: CRC/Lewis Press.

Ehlers M. 2007. New developments and trends for Urban Remote Sensing. In *Urban Remote Sensing*: 357–375. CRC Press.

Freire, S., Santos T., Gomes N., Fonseca A., Tenedório J. A. 2010a. Extraction of buildings from QuickBird imagery – what is the relevance of urban context and heterogeneity? *Proceedings of ASPRS/CaGIS/ISPRS Fall Conference.* Orlando, USA, November 15–18, 2010.

Freire, S., Santos, T., Navarro, A., Soares, F., Dinis, J., Afonso, N., Fonseca, A., Tenedório, J.A. 2010b. Extracting buildings in the city of Lisbon using QuickBird images and LiDAR data. Proceedings of GEOBIA 2010 – GEOgraphic Object-Based Image Analysis, Ghent, Belgium.

Hay, G. J., & Castilla, G. 2008. Geographic Object-Based Image Analysis (GEOBIA): A new name for a new discipline? In T. Blaschke, S. Lang, G. J. Hay (eds), *Object-Based Image Analysis – spatial concepts for knowledge-driven remote sensing applications*: 75–89. Springer-Verlag.

Herold, M., Gardner, M., Roberts, D. 2003. Spectral resolution requirements for mapping urban areas. *IEEE T. Geosci. Remote Sens.* 41(9): 1907–1919.

Jensen, J.R. & Cowen, D.C. 1999. Remote sensing of urban/suburban infrastructure and socio-economic attributes. *Photogramm. Eng. Remote Sens.*, 65(5): 611–622.

Lang, S. 2008. Object-based image analysis for remote sensing applications: modelling reality – dealing with complexity. In T. Blaschke, S. Lang, G. J. Hay (eds), *Object-Based Image Analysis – spatial concepts for knowledge-driven remote sensing applications*: 3–27. Springer-Verlag.

Opitz, D. & Blundell, S. 2008. Object recognition and image segmentation: the Feature Analyst approach. In T. Blaschke, S. Lang, G. J. Hay (eds), *Object-Based Image Analysis – spatial concepts for knowledge-driven remote sensing applications*: 153–167. Springer-Verlag.

Rutzinger, M., Rottensteiner, F., Pfeifer, N. 2009. A comparison of evaluation techniques for building extraction from airborne laser scanning. *IEEE J. of Sel. Top. Appl. Earth Obs. Remote Sens.*, 2(1): 11–20.

Salah, M., Trinder, J., Shaker, A., 2009. Evaluation of the self-organizing map classifier for building detection from LiDAR data and multispectral aerial images. *J. Spat. Sci.*, 54(2): 1–17.

Santos T., Freire S., Tenedório J. A., Fonseca, A. 2009. Classificação de imagens de satélite de alta resolução com introdução de dados LiDAR. Aplicação à cidade de Lisboa. *Proceedings of the VII Congresso da Geografia Portuguesa*, Coimbra, Portugal, 26–28 November 2009.

Shan, J. & Lee, S. D. 2005. Quality of building extraction from IKONOS imagery. *J. Surv. Eng.*, 31(1): 27–32.

Vu, T.T., Matsuoka, M., Yamazaki, F., 2009. Multi-scale solution for building extraction from LiDAR and image data. *International Journal of Applied Earth Observation and Geoinformation,* 11(4): 281–289.

Computational Modelling of Objects Represented in Images – Di Giamberardino et al. (eds)
© 2012 Taylor & Francis Group, London, ISBN 978-0-415-62134-2

An energy minimization reconstruction algorithm for multivalued discrete tomography*

László Varga[†], Péter Balázs & Antal Nagy
Department of Image Processing and Computer Graphics, University of Szeged, Szeged, Hungary

ABSTRACT: We propose a new algorithm for multivalued discrete tomography, that reconstructs images from few projections by approximating the minimum of a suitably constructed energy function with a deterministic optimization method. We also compare the proposed algorithm to other reconstruction techniques on software phantom images, in order to prove its applicability.

Keywords: multivalued discrete tomography; reconstruction; GPGPU; optimization; non-destructive testing

1   INTRODUCTION

Tomography deals with the reconstruction of objects from a given set of their projections. This is usually done by exposing the object to some electromagnetic or particle radiation, and measuring the loss of the energy as the beams pass through it. With this information one can derive the integrals of attenuation coefficients along the path of the beams, and obtain the inner structure of the object.

There are several suitable algorithms for tomography, which can provide satisfactory reconstructions of arbitrary objects, when a sufficiently high amount of information (which usually means hundreds of projections) is available (12).

In *discrete tomography* (DT) (10; 11), one assumes that the object to be reconstructed consists of only a few different materials with known attenuation coefficients. Binary tomography – as a special case of DT – makes the additional restriction that the reconstructed volume contains only two materials. With such prior information, the reconstruction can be performed even from a few projections. DT can be particularly useful, e.g., in non-destructive testing (6), where the goal is to gain some information of the interior of – usually homogeneous – objects without damaging them.

There is a wide range of algorithms for binary and non-binary (called multivalued) discrete tomography. For example, the DART, Discrete Algebraic Reconstruction Technique (4) is capable of producing highly accurate reconstructions by thresholding a continuous reconstruction and then adjusting the object boundaries. Also, there are reconstruction algorithms based on minimizing an energy function by deterministic (13; 15; 16; 18) or randomized (1; 2; 8; 14) optimization strategies.

In this paper we propose a deterministic reconstruction method for multivalued discrete tomography, that solves the problem by minimizing a suitably constructed energy function. The basic idea behind our new method was provided by the algorithm given in (15), that is a highly accurate binary reconstruction algorithm based on D.C. programming – a method for minimizing the difference of convex functions. Unfortunately, the DC algorithm is restricted to the reconstruction of binary images. Our goal is to provide a valuable extension, that is suited for the general case of multivalued DT. Although other simple extensions of the DC algorithm also exist (see, e.g., 13; 16; 18), we propose significant modifications of the original method, to supply an algorithm that is fully adjusted to multivalued DT. We introduce a new energy function for modeling the possible values of the reconstruction, and we also define a new process that can perform a fast approximate optimization of the energy function.

The paper is structured as follows. In Section 2 we give a brief description of the theoretical background of discrete tomography. Then, in Section 3 we describe the proposed method, and in Section 4 we

*This research was in part supported by the támop-4.2.1/ b-09/1/konv-2010-0005 project of the hungarian national development agency co-financed by the european union and the european regional development fund. the work of the second author was also supported by the jános bolyai research scholarship of the hungarian academy of sciences and the pd100950 grant of the hungarian scientific research fund (otka).

[†]The publication is supported by the European Union and co-funded by the European Social Fund. Project title: "Broadening the knowledge base and supporting the long term professional sustainability of the Research University Centre of Excellence at the University of Szeged by ensuring the rising generation of excellent scientists." Project number: TÁMOP-4.2.2/B-10/1-2010-0012

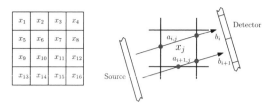

Figure 1. Representation of the parallel beam geometry on a discrete image.

provide experimental results. Finally, in Section 5 we summarize the results.

2 DISCRETE TOMOGRAPHY

For a simple formalism we present our reconstruction algorithm in the case of two-dimensional tomography, but the method can easily be extended to higher dimensions, too. The model we use assumes that a single slice of the reconstructed object is represented by an $n \times n$ size digital image. Moreover, we assume parallel beam projection geometry, i.e., a projection is given by projection values corresponding to parallel projection rays, where each value is given by the integral of the image on a straight line.

With the above considerations the discrete reconstruction problem can be represented by a system of equations

$$\mathbf{Ax} = \mathbf{b}, \quad \mathbf{A} \in \mathbb{R}^{m \times n^2}, \ \mathbf{x} \in L^{n^2}, \ \mathbf{b} \in \mathbb{R}^m \ , \quad (1)$$

where

- $\mathbf{x}$ is the vector of all n^2 unknown image pixels,
- m is the total number of projection lines used,
- $\mathbf{b}$ is the vector of all m measured projection values,
- $\mathbf{A}$ describes the projection geometry with all a_{ij} elements giving the length of the line segment of the i-th projection line through the j-th pixel,
- and $L = \{l_0, l_1, \dots, l_c\}$ is the set of the possible intensities (assuming that $l_0 < l_1 < \dots < l_c$).

An illustration of the applied projection geometry can be seen in Fig. 1.

Note that – as a special case – with $L = \{0, 1\}$ we arrive to the well-known model of binary tomography.

With the above formulation the reconstruction is equivalent to the task of solving the equation system given in (1). Unfortunately, beside the problems arising from the fact that we search a discrete-valued solution, the system of (1) is usually extremely huge, and often underdetermined (owing to the low number of projections) or inconsistent (due to measurement errors). Various techniques have been suggested to overcome these problems, but all of them are heuristic methods. Efficient exact reconstruction algorithms exist only for some special classes of (mostly binary) images (see, e.g., 5; 7)).

3 THE PROPOSED METHOD

Since, even in the binary case, the discrete reconstruction problem is NP-hard if the number of projections is more than two (9), our aim is to provide an approximate solution of the reconstruction task. The algorithm we propose performs the discrete reconstruction by minimizing a suitably constructed energy function.

3.1 The energy function

The energy function consists of two terms. Using the notation of Sect. 2 it can be given as

$$\mathcal{E}_\mu(\mathbf{x}) := f(\mathbf{x}) + \mu \cdot g(\mathbf{x}), \quad \mathbf{x} \in [l_0, l_c]^{n^2} \ . \quad (2)$$

In more detail, the first function

$$f(\mathbf{x}) = \frac{1}{2} \cdot \|\mathbf{Ax} - \mathbf{b}\|_2^2 + \frac{\alpha}{2} \cdot \mathbf{x}^T \mathbf{Sx} \quad (3)$$

is a formulation of the continuous reconstruction problem, where $\mathbf{S}$ is a matrix such that

$$\mathbf{x}^T \mathbf{Sx} = \sum_{i=1}^{n^2} \sum_{j \in N_4(i)} (x_i - x_j)^2 \quad (4)$$

and $N_4(i)$ is the set of pixel indexes 4-connected to the i-th pixel.

Informally, $f(\mathbf{x})$ consists of an $\|\mathbf{Ax} - \mathbf{b}\|_2^2$ projection correctness (or data fidelity) term, and an $\mathbf{x}^T \mathbf{Sx}$ smoothness prior, that is lower if the reconstructed image is smooth, and thus it forces the results to contain larger homogeneous regions.

The second, $\mu \cdot g(\mathbf{x})$, term of (2) is a formulation of the discreteness, which propagates solutions containing values only from the L predefined set of intensities. Here, $\mu \geq 0$ is a constant weight that can be used to balance between the two separate parts of the energy function, and $g(\mathbf{x})$ is constructed to take its minimal values at discrete solutions (i.e., when $\mathbf{x} \in L^{n^2}$) and higher positive values otherwise. The $g(\mathbf{x})$ discretizing function is given in the form

$$g(\mathbf{x}) = \sum_{i=1}^{n^2} g_p(x_i), \quad i \in \{1, 2, \dots, n^2\} \ , \quad (5)$$

where g_p is a one-variable function composed of a set of forth-grade polynomial functions defined over the intervals of L in the way

$$g_p(z) = \begin{cases} \frac{[(z - l_{j-1}) \cdot (z - l_j)]^2}{2 \cdot (l_j - l_{j-1})^2}, & \text{if } z \in [l_{j-1}, l_j] \text{ for each } j \in \{1, \dots, c\}, \\ \text{undefined}, & \text{otherwise.} \end{cases}$$

An illustration of a g_p function can be seen in Fig. 2. Informally, this discretization function assigns a small

180

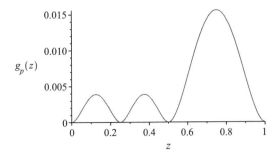

Figure 2. Example of the $g_p(z)$ one-variable discretization function with intensity values $L = \{0, 0.25, 0.5, 1\}$.

value to each pixel if the pixel value in the reconstruction is close to an element of L, and higher values (increasing with the distance) otherwise. There are several other possible functions which could be used for such purposes (see, e.g., 13; 16; 18). We have decided to construct this novel one, since it is easy to handle and can be efficiently computed.

3.2 The optimization process

The process of the optimization in our proposed method is based on breaking the energy function (2) into two parts, and prioritizing between them. The first part is given by the $f(\mathbf{x})$ defined in (3), i.e., two terms responsible for projection correctness and smoothness. The other part is provided by the $\mu \cdot g(\mathbf{x})$ discretization term.

In the beginning, the reconstruction algorithm assumes that the first two terms in the energy function prioritizes the discretization term. Therefore, the process will first focus on finding a continuous reconstruction, and neglect the discretization term. Afterwards, when a good approximation of the continuous reconstruction is found, the weight of the discretization term will be increased, thus the optimization process is steered towards a discrete solution.

The description of the algorithm uses the following notations.

- $\mathbf{A}$, $\mathbf{b}$, $\mathbf{x}$ and n are as defined in Sect. 2,
- $\nabla g_p(x_i)$ denotes the derivate of the discretization term applied for the i-th x_i pixel of the reconstructed image,

$$\nabla g_p(z) = \frac{(z - l_{j-1})(z - l_j)(2 \cdot z - l_{j-1} - l_j)}{(l_j - l_{j-1})^2},$$

$$\text{if } z \in [l_{j-1}, l_j] \,,\tag{6}$$

- $G_{0,\sigma}(z)$ is an unnormalized Gaussian function with 0 mean and σ deviance, that is

$$G_{0,\sigma}(z) = e^{-\left(\frac{z^2}{2 \cdot \sigma^2}\right)} \,,\tag{7}$$

- $\alpha \geq 0$, $\mu \geq 0$, and $\sigma \geq 0$ are predefined constants controlling in the energy function, respectively, the weight of the smoothness term, the weight of the discretization term, and the deviance of the

Gaussian function applying the adaptive weighting of the discretization,

- λ is an upper bound of the largest eigenvalue of the matrix $(\mathbf{A}^T \mathbf{A} + \alpha \cdot \mathbf{S})$, that is used for reasons described in (15).

For obtaining the result, the optimization method uses an adaptive and automatic pixel-based weighting of the discretization term. The detailed description of the algorithm is given in Algorithm 1.

The optimization process makes a connection between the two parts of the energy function (i.e., the formulation of the continuous reconstruction problem, and the discretization term), and assumes that the first part has a higher priority (as our first consideration is to find a reconstruction that satisfies the projections, but we would also like to get a discrete result if possible).

With this, the algorithm is based on optimizing the energy function with a simple projected subgradient method, while applying an automatic weighting between the two terms of the energy function. In each iteration step of the optimization process, one can calculate the gradient of the $\|\mathbf{A}\mathbf{x} - \mathbf{b}\|_2^2$ projection correctness term in the energy function by computing the $\mathbf{A}^T(\mathbf{A}\mathbf{x} - \mathbf{b})$ vector. For each pixel, this vector explicitly contains an estimation of correctness of the pixel in the current solution according to the projections (the higher this value is the more responsible the pixel is for causing incorrect projections). If we apply a Gaussian function on these values we can get a weight, that is smaller when the corresponding pixel needs further adjustments, and higher if the projection rays connected to that specific pixel are more or less satisfied. By weighting the discretization with this value calculated from the gradient of the projection correctness, one can apply an automatic adjustment of the discretizing term for each pixel, omitting it when the projections are not satisfied, and slowly increasing its

---

Algorithm 1. Energy-Minimization Algorithm for Multi-valued DT.

Input: $\mathbf{A}$ projection matrix, $\mathbf{b}$ expected projection values, $\mathbf{x}^0$ initial solution, $\alpha, \mu, \sigma \geq 0$ predefined constants, and L list of expected intensities.

1: $\lambda \leftarrow$ an upper bound for the largest eigenvalue of the $(\mathbf{A}^T \mathbf{A} + \alpha \cdot \mathbf{S})$ matrix.
2: $k \leftarrow 0$
3: **repeat**
4: $\quad \mathbf{v} \leftarrow \mathbf{A}^T(\mathbf{A}\mathbf{x}^k - \mathbf{b})$.
5: $\quad \mathbf{w} \leftarrow S\mathbf{x}^k$.
6: $\quad$ **for** each $i \in \{1, 2, \ldots, n^2\}$ **do**
7: $\quad\quad y_i^{k+1} \leftarrow x_i^k - \frac{v_i + \alpha \cdot w_i + \mu \cdot G_{0,\sigma}(v_i) \cdot \nabla g_p(x_i^k)}{\lambda + \mu}$
8: $\quad\quad x_i^{k+1} \leftarrow \begin{cases} l_0, & \text{if } y_i^{k+1} < l_0, \\ y_i^{k+1}, & \text{if } l_0 \leq y_i^{k+1} \leq l_c, \\ l_c, & \text{if } l_c < y_i^{k+1}. \end{cases}$
9: $\quad$ **endfor**
10: $\quad k \leftarrow k + 1$
11: **until** a stopping criterion is met.
12: Apply a discretization of $\mathbf{x}^k$ to gain fully discrete results.

181

effect as the pixel values get closer to an acceptable reconstruction.

In practice this means that the method starts with an arbitrary initial solution, and first approximates a continuous reconstruction based on the given set of projections. Later, as the projections of the solution get closer to the described vectors, the automatic weighting of the discretizing term begins to increase for each pixel. Thus the pixels will be slowly steered towards discrete values of L.

It is possible that the process will get stuck in a local minimum of the energy function. In this case the process will stop in a semi-continuous solution, where some pixels are properly discretized, and the rest of them are left continuous, since the projection correctness did not allow a full discretization. The μ and σ parameters, are used to control the maximal strength of the discretizing term, and the speed at which the discretizing term gets strengthened during the process, respectively.

Finally, after the optimization process we complete the discretization by simply thresholding the pixel values, to gain a fully discrete reconstruction result. The final thresholding of the result can be performed with values chosen half-way between neighboring intensity levels as

$$x_i = \begin{cases} l_0, & \text{if } x_i^k < (l_0 + l_1)/2, \\ l_j, & \text{if } (l_{j-1} + l_j)/2 \leq x_i^k < (l_j + l_{j+1})/2, \\ l_c, & \text{if } (l_{c-1} + l_c)/2 \leq x_i^k, \end{cases}$$

(8)

where $\mathbf{x}^k$ is the result of the iterative optimization process of Alg. !, $j \in \{1, \ldots, c-1\}$, and i takes each element of the set $\{1, \ldots, n^2\}$ as a value.

4   EXPERIMENTAL RESULTS

We conducted experiments to compare our method to other published algorithms. On one hand, on binary images, we compared our new method to the DC algorithm, to see how the original, and our new approach performs related to each other. Unfortunately, due to the limitations of the DC algorithm (as it is not suited for multivalued tomography), we could only do this comparison for binary images. Also, we ran tests with the recently published DART (4) in order to compare the reconstruction of multivalued images.

We performed the evaluations, by using a set of phantom images (all having a size of 256 by 256 pixels). Three of these phantoms can be seen in Fig. 3. The reconstructions were performed from projection sets containing 2 to 18 projections, distributed equiangularly on the half circle, assuming that the projection with 0° angle corresponds to vertical rays. The angle sets describing the projection directions for a p projection number can be given as

$$S(p) = \left\{ i \cdot \frac{180°}{p} \ \middle| \ i = 0, \ldots, p-1 \right\}.$$

(9)

  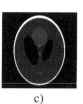

| a) | b) | c) |

Figure 3.   Some of the software phantoms used for testing. a) a binary image; b) a multivalued image from (4); c) the well-known Shepp-Logan head phantom (see, e.g., page 53 of (12)).

As mentioned above, we used a parallel beam projection geometry, where projection values were given by line integrals on the image. The distances between neighboring projection lines were set to be one unit (the width of one pixel on the image), the rotation center of the projections was located in the center of the image half way between two projection lines, and in each projection the rays covered the whole image.

In our tests, the parameters of the DART and DC algorithms were mostly set from the literature, with slight adjustments to get the best performance of all the methods in our tests. The parameters of the DC algorithm were set as given in (17) except that the strength of the smoothness term was $\alpha = 2.5$. In DART, we used 10 iterations of the Simultaneous Iterative Reconstruction Technique (see, e.g., (12)) for performing the continuous reconstructions, applied the same smoothing kernel as described in (3), and terminated the algorithm when the thresholded image did not change in the last 10 DART iterations or the number of iterations reached a limit of 500.

For the parameters of the proposed method, we used the values $\alpha = 2.5$, $\mu = 20$, $\sigma = 1$, and in the $\mathbf{x}^0$ initial solution all the x_i^0 positions were set to the same value in the middle of the range of possible intensities (i.e., $x_i^0 = (l_c - l_0)/2$, for all $i \in \{1, \ldots, n^2\}$). The iteration was stopped when the difference between the solutions of the k-th and $(k+1)$-th iteration steps computed as $\|\mathbf{x}^{k+1} - \mathbf{x}^k\|_2$ became less then 0.001 or the number of iterations reached a limit of 5000. Although, the convergence of the optimization process is not yet proven, we found the algorithm to be convergent in all our tests with these parameter settings.

We implemented the algorithms in C++ with GPU acceleration using the NVIDIA CUDA C sdk. The computation was performed on a PC, with an Intel Q9500 CPU, and an NVIDIA Geforce GTS250 GPU.

After reconstructing the results using all three algorithms, we compared them visually, and by using the error measurement

$$Err = \frac{D(\mathbf{x}, \mathbf{x}^*)}{O(\mathbf{x}^*)} \cdot 100\% \ ,$$

(10)

where $D(\mathbf{x}, \mathbf{x}^*)$ is the number of misclassified pixels on the result, and $O(\mathbf{x}^*)$ is the number of non-zero pixels on the original phantom.

In addition, we also measured the computation times of the algorithms in each case. A summary of the

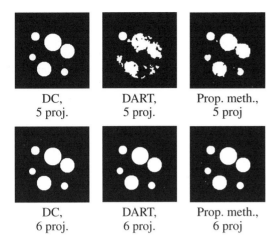

	DART	Prop. meth.
Fig. 3b, 6 proj.	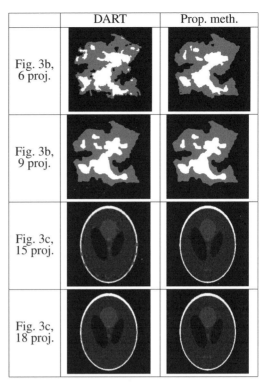	
Fig. 3b, 9 proj.		
Fig. 3c, 15 proj.		
Fig. 3c, 18 proj.		

DC, 5 proj.　　DART, 5 proj.　　Prop. meth., 5 proj

DC, 6 proj.　　DART, 6 proj.　　Prop. meth., 6 proj

Figure 4.　Reconstructions of a binary phantom (Figure 3a), produced by the three compared algorithms, from projection sets containing different numbers of projections.

numerical results can be seen in Table 4, while Fig. 4 and Fig. 5 give some examples of the reconstructed results of binary and multivalued images.

Based on the results we can deduce the following. In case of using very few projections (i.e., 2–3 projections for simple images like the phantoms of figures 3a-b, and up to 5–6 projections for more complex ones like Figure 3c), there was obviously not enough information for the reconstruction algorithms to give accurate solutions. Usually DART produced the best results, but this seems to be irrelevant since the reconstruction error is unacceptably high.

Starting to increase the number of projections, the amount of information in the data was also increasing and the results provided by the algorithms began to improve as well. The optimization based algorithms (DC and the proposed method) showed a faster improvement with the increasing of the projection numbers, therefore after a certain number of projections they started to give better results than the DART. Usually, the advantage of the optimization-based methods caused a sudden drop in the reconstruction error, when the algorithms started to give more accurate results. Thus, we can deduce that these two algorithms can ensure more or less accurate reconstructions from fewer projections than the DART.

Later, when we had even more projections with more than sufficient information for an accurate reconstruction, again the DART provided the best reconstructions, by producing slightly better results than the other two methods.

When comparing the energy minimization based methods, we can observe that on binary images the DC algorithm works better than our proposed method. This might be due to the form of the discretization term in the energy function. The DC algorithm is specialized for binary tomography, and aims a full binarization in the optimization process. The drawback is that the

Figure 5.　Sample of reconstructions of multivalued phantoms of Figure 3b-c, produced by the DART and our proposed algorithm, from different number of projections.

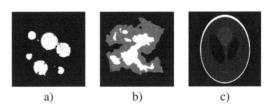

a)　　　　　　b)　　　　　　c)

Figure 6.　Continuous results of the proposed reconstruction algorithm, without the final thresholding. (The images a), b) and c) were reconstructed from 5, 6, and 15 projections, respectively.)

original DC algorithm is not capable of performing multivalued tomography at all.

On the other hand, our algorithm needs a different approach for having the generality to be able to reconstruct multivalued images, and it only makes an approximate discretization. This means that in a later state of the energy minimization process – without the final thresholding – we get a semi-discrete, semi-continuous result. This intermediate result is produced by taking into account that we are looking for a discrete solution, but still contains some uncertainty of the values (some of the examples of such results can be seen in Fig. 6). This kind of soft discretization is necessary for the multivalued reconstruction in our method,

Table 1. Reconstruction errors and computation times of the compared algorithms, reconstructing the phantoms of Figure 3. The error measurement is computed by (10), and the computational time is given in seconds. Reconstructions of the DC algorithm could only be performed on binary test images. In each row, the best result is highlighted in bold.

P. Num.	DC Error	DC Time	DART Error	DART Time	Prop. meth. Error	Prop. meth. Time
Figure 3a						
2	90.7%	12.1 s	**85.6%**	6.6 s	107.4%	10.1 s
3	**22.0%**	12.4 s	52.9%	5.4 s	30.8%	11.2 s
4	**1.2%**	13.6 s	44.9%	8.0 s	22.4%	11.8 s
5	**0.3%**	12.5 s	29.9%	9.5 s	7.9%	12.7 s
6	**0.2%**	8.1 s	**0.2%**	2.7 s	0.8%	7.6 s
9	0.2%	6.5 s	**0.0%**	0.8 s	0.3%	4.6 s
12	0.0%	7.2 s	**0.0%**	0.9 s	0.1%	4.8 s
15	0.0%	8.7 s	**0.0%**	1.2 s	0.1%	5.8 s
18	0.0%	8.7 s	**0.0%**	0.9 s	0.1%	5.8 s
Figure 3b						
2	–	–	62.9%	6.7 s	**52.7%**	10.4 s
3	–	–	45.1%	8.0 s	**41.9%**	11.4 s
4	–	–	43.4%	8.6 s	**35.4%**	12.2 s
5	–	–	36.4%	9.4 s	**26.4%**	13.2 s
6	–	–	27.0%	10.2 s	**11.6%**	13.8 s
9	–	–	**0.7%**	4.5 s	1.9%	15.6 s
12	–	–	**0.4%**	14.9 s	1.0%	11.6 s
15	–	–	**0.3%**	2.3 s	0.8%	11.6 s
18	–	–	**0.1%**	21.3 s	0.6%	10.9 s
Figure 3c						
2	–	–	**84.4%**	6.7 s	85.7%	9.3 s
3	–	–	**77.3%**	8.2 s	82.5%	6.0 s
4	–	–	**75.3%**	8.8 s	81.0%	8.0 s
5	–	–	**73.3%**	9.7 s	74.2%	10.2 s
6	–	–	74.1%	10.2 s	**70.0%**	12.7 s
9	–	–	57.0%	12.6 s	**46.8%**	14.7 s
12	–	–	33.9%	14.5 s	**24.8%**	11.4 s
15	–	–	22.0%	18.0 s	**16.3%**	8.6 s
18	–	–	15.7%	20.8 s	**14.0%**	8.0 s

but it reduces the accuracy of the algorithm on binary images.

Finally, regarding the computational time of the algorithms, we found that depending on the conditions of the reconstruction and the image processed, one or another algorithm gave results faster than the other ones. Still, in general the time requirements showed to be similar.

In summary, the performance of the algorithms were similar on our dataset. All three methods can yield highly accurate reconstructions. Nevertheless, we found that the energy minimization-based methods gave slightly better results when the reconstructions were performed from a low number of projections, but the results of DART were better with more projections. This diversity makes all the algorithms valuable, and in a practical application, we would advise to choose from them based on the conditions of the reconstruction.

5 CONCLUSION AND FURTHER WORK

In this paper we proposed a new algorithm for multi-valued discrete tomography, that is based on the minimization of a suitably constructed energy function. We compared our method to two existing reconstruction algorithms by performing experimental tests on a set of software phantoms. Our results show that the proposed method performs better than the other ones under certain conditions, thus it should be considered a useful alternative for discrete tomography reconstruction.

Also, neglecting the final thresholding step of our algorithm, one can gain reconstruction results which – in some way – might describe the uncertainty in the reconstruction. We think that this property is worth to be investigated in more detail.

In our future work we intend to improve the algorithm by modifying the minimized energy function, and to study the applicability of the technique in different practical fields of discrete tomography. Also, we will make efforts to prove the convergence of the method.

REFERENCES

[1] P. Balázs, M. Gara, *An evolutionary approach for object-based image reconstruction using learnt priors*, Lecture Notes in Computer Science vol. 5575, pp. 520–529 (2009).

[2] K.J. Batenburg, *An evolutionary algorithm for discrete tomography*, Disc. Appl. Math. 151, pp. 36–54 (2005).

[3] K.J. Batenburg, J. Sijbers, *DART: a fast heuristic algebraic reconstruction algorithm for discrete tomography*, IEEE Conference on Image Processing IV, pp. 133–136 (2007).

[4] K.J. Batenburg, J. Sijbers, *DART: a practical reconstruction algorithm for discrete tomography*, IEEE Transactions on Image Processing 20(9), pp. 2542–2553 (2011).

[5] Brunetti, S., Del Lungo, A., Del Ristoro, F., Kuba, A. and Nivat, M. *Reconstruction of 4- and 8-connected convex discrete sets from row and column projections*, Lin. Alg. Appl. 339, pp. 37–57 (2001).

[6] J. Baumann, Z. Kiss, S. Krimmel, A. Kuba, A. Nagy, L. Rodek, B. Schillinger, J. Stephan, *Discrete Tomography Methods for Nondestructive Testing*, Chapter 14 of (11).

[7] M. Chrobak, Ch. Dürr, Reconstructing *hv*-convex polyominoes from orthogonal projections. Information Processing Letters 69(6), pp. 283–289 (1999).

[8] V. Di Gesù, G. Lo Bosco, F. Millonzi, C. Valenti, *A memetic algorithm for binary image reconstruction*, Lecture Notes in Computer Science vol. 4958, pp. 384–-395 (2008).

[9] R. J. Gardner, P. Gritzmann, *Discrete tomography: Determination of finite sets by X-rays*, Trans. Amer. Math. Soc. 349(6), pp. 2271–2295 (1997).

[10] G.T. Herman, A. Kuba (Eds.), *Discrete Tomography: Foundations, Algorithms and Applications*, Birkhäuser, Boston, 1999.

[11] G.T. Herman, A. Kuba (Eds.), *Advances in Discrete Tomography and Its Applications*, Birkhäuser, Boston, 2007.

[12] A. C. Kak, M. Slaney, *Principles of computerized tomographic imaging*, IEEE Press, New York, 1999.

[13] T. Lukić, *Discrete tomography reconstruction based on the multi-well potential*, Lecture Notes in Computer Science vol. 6636, pp. 335–345 (2011).

[14] A. Nagy, A. Kuba, *Reconstruction of binary matrices from fan-beam projections*, Acta Cybernetica 17(2), pp. 359–385 (2005).

[15] T. Schüle, C. Schnörr, S. Weber, J. Hornegger, *Discrete tomography by convex-concave regularization and D.C. programming*, Discrete Applied Mathematics 151, pp. 229–243 (2005).

[16] T. Schüle, S. Weber, C. Schnörr, *Adaptive reconstruction of discrete-valued objects from few projections*, Electronic Notes in Discrete Mathematics vol. 20, pp. 365–384 (2005).

[17] L. Varga, P. Balázs, A. Nagy, *Direction-dependency of binary tomographic reconstruction algorithms*, Graphical Models 73(6), pp. 365–375 (2011).

[18] S. Weber, *Discrete tomography by convex concave regularization using linear and quadratic optimization*, PhD thesis, Heidelberg University, 2009.

Functional near-infrared frontal cortex imaging for virtual reality neuro-rehabilitation assessment

S. Bisconti*, M. Spezialetti*, G. Placidi & V. Quaresima
Department of Health Sciences, University of L'Aquila, Via Vetoio, L'Aquila, Italy

ABSTRACT: Virtual reality-based technologies have great potential for the development of novel paradigms useful for functional neuro-rehabilitation. Functional Near-Infrared Spectroscopy (fNIRS) is an optical brain imaging technique that measures blood oxygenation changes related to brain functions in the cerebral cortex. The present study was aimed at assessing by fNIRS the frontopolar cortex oxygenation response to a balance task in a semi-immersive virtual environment monitored by a depth-sensing camera. An eight-channel fNIRS system was used to measure changes in oxy-hemoglobin (O_2Hb) and deoxy-hemoglobin (HHb) over the frontopolar cortex area. A significant progressive increase of O_2Hb ($p < 0.001$) and a significant progressive decrease of HHb ($p < 0.001$) was found over the frontopolar cortex during the incremental balance task. These changes were modulated by the task difficulty suggesting that the frontopolar cortex is involved in attention-demanding tasks. These results suggest that virtual reality approach could be employed in the neuro-rehabilitation field.

Virtual reality-based technologies have great potential for the development of novel paradigms useful for neuro-rehabilitation (Cameirao et al. 2010; Lucca et al. 2010). Real-time simulations of an environment, scenario, or activity can be created to allow for presentation of complex multimodal sensory information to the user and elicit a substantial feeling of reality and activity, despite its artificial nature (Adamovich et al. 2009). Different advantages can be obtained by the application of virtual reality in motor and sensor-cognitive rehabilitation: 1) motor impairment and recovery can be accurately measured in real time providing the user with the knowledge of the subject performance; 2) interactive virtual environments are flexible and customizable for different therapeutic purposes; 3) online or offline visual feedbacks of the subject performance in the virtual environment can be provided to the patient during the rehabilitation session (Lucca et al. 2010).

Neuroimaging methodologies have been applied in cognitive neuroscience studies for examining the neural networks supporting complex tasks representative of perception, cognition, and action as their occur in natural settings. Functional near-infrared spectroscopy (fNIRS) is an optical brain imaging technique that measures blood oxygenation changes related to brain functions in the cerebral cortex (Ferrari et al. 2004; Wolf et al. 2007; Quaresima et al. 2011, Cutini et al. 2012, Ferrari et al. 2012). fNIRS does not require the subject's head or body to be highly restrained; therefore it represents an optimal cortical imaging monitoring tool to evaluate the subject performance in a virtual reality environment (Holper et al. 2010). fNIRS has been used to study the prefrontal cortex activation during motor tasks as hand grasping (Leff et al. 2011), walking (Atsumori et al. 2010; Holtzer et al. 2011); and balance control (Mihara et al. 2008). However, only few recent studies have investigated the neural response during a real activity realized in a simulated context. The right parietal and occipital areas were found activated during a bisection task in an immersive virtual reality (Seraglia et al. 2011); the superior parietal lobe and superior temporal gyrus were found activated in response to a multimodal dance video game (Tachibana et al. 2011); the prefrontal cortex was found activated during spatial navigation learning in virtual mazes (Ayaz et al. 2011); and the superior temporal gyrus was found activated during a balance task associated to a video game (Nintendo Wii) simulating downhill skiing (Karim et al. 2011).

The present study was aimed at assessing by fNIRS the frontopolar cortex oxygenation response to a balance task in a semi-immersive virtual environment monitored by a depth-sensing camera (that is able to process the depth map of the scenario). This balance task could be applied in neuro-rehabilitative treatment. In particular, the adopted balance task implied an incremental level of difficulty consisting in shifting subject mass center on the right and left to maintain his balance over a board visualized on the computer screen. We predicted a progressive increase of the oxygenation response over the frontopolar cortex in relation to the increase of the balance task difficulty.

---

*Equally contributed to this work

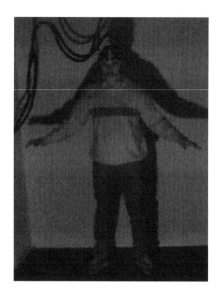

Figure 1. Subject performing the balance task (left panel) and the 150 uniform spheres representing a 3D model standing over the center of a green balance supported by a pivot (right panel).

1 METHODS

1.1 *Experimental protocol*

Six right-handed male volunteers (mean age: 26.3 ± 5.47 y.o.) without self-reported balance or mobility disorders participated to the study. Informed consent was obtained after a full explanation of the protocol to be used and the non-invasiveness of the study. To exclude left-handed subjects, all participants completed the Edinburgh Handedness Inventory assessing hand dominance. The study was conducted in a quiet and dimly lit room. Subjects were standing and watched a 19-inch computer screen at a distance of 3 m. A depth-sensing near-infrared camera (Optrima™, Opticam™130, Brussels, Belgium), that recorded the participant's movements, was positioned near the computer screen. The depth-map of the subject was analyzed and transformed by a home-made software system based on *iisu*™ SDK (Sofkinetic™, Brussels, Belgium), a middleware that enables communication between depth-sensing camera and end-user applications. In particular, the module was utilized to extract a cloud of 150 points from the participant. A sphere was created around each point of the cloud to obtain a 3D subject representation. Then, this information was transferred into a virtual environment and through *Irrlicht* (an open source 3D engine, http://irrlicht.sourceforge.net/) renderized in real time in a virtual environment and represented on the computer screen (Fig. 1). After having placed the fNIRS probe over the skin overlying the frontopolar region, the subject was instructed how to move correctly his body during the protocol and performed a 5-min practice trial to familiarize with the procedure. The duration of the protocol was 9 minutes. During the first 2-min (baseline), the subject was asked to stand still observing his image and the green balance board (supported by a pivot) represented on the screen by the 3D model rendering (Fig. 1).

The 5-min balance task started when an auditory signal was presented; the subject, while watching his image on the screen, had to keep his body balanced enough to avoid that the tilt angle of the green board became greater than $35°$ and the balance board appeared red. The change of the balance board status indicated that the subject was in the wrong position. The participant could control the model by bringing his center of balance on the right or left. As the subject oscillated with respect to his center of gravity, his 3D representation (composed by spheres of equal density) was distributed on his feet proportionally, thus simulating his weight. The participant was informed that the errors were recorded and, for receiving the highest score, one would need to avoid losing the control of the board. The number of times in which the board became red was considered as index of the subject performance. During the task, the difficulty to maintain the own body over the board was increased every 30-s. At each upper level, the board became more sensitive to the right and left movement of the subject. After the end of the task, the subject had to observe his image on the screen without moving his body for a 2-min period. The number of errors performed by the participants was recorded and later analyzed.

After a 10-min rest period, a 9-min control task was performed. During the control task the subject was requested to move his body toward left and right side for 5-min (as done during the proper balance task) without the 3D representation. An auditory stimulus gave the frequency with which the subject had to bring his center of balance on the right or left. During a 2-min recovery period the subject was asked to stand still. In order to assess the perceived exertion during the protocol, subjects completed the Borg Rating of Perception Exertion Scale attributing a value from 0 to 10 to the task (Borg, 1990).

1.2 fNIRS instrumentation

An eight-channel fNIRS system (NIRO-200, Hamamatsu Photonics, Japan) was used to measure changes in oxy-hemoglobin (O_2Hb) and deoxy-hemoglobin (HHb) over the frontopolar cortex area. Two optical fiber bundles carried the light to the left and the right frontal cortex, whereas eight optical fiber bundles collected the light emerging from the frontal area. The optical probe was placed over the head to cover the underlying frontopolar cortex (Brodmann's areas 10 and 11) and the channels #7 and #4 were centered (according to the International 10–20 system for the EEG electrode placement) at the Fp1 and Fp2 for left and right sides, respectively (Fig. 2). The quantification of the concentration changes (expressed in $\Delta\mu M$) of O_2Hb and HHb was obtained by including an age-dependent constant differential pathlength factor ($4.99 + 0.067*Age^{0.814}$) (Duncan et al. 1996). In the 8 fNIRS measurement points (channels), which are defined as the midpoint of the corresponding detector – source pairs (distance set to 3 cm), data were acquired at 1 Hz. The cardiac frequency was monitored by a pulse oximeter (N-600, Nellcor, Puritan Bennett, St. Louis, MO, USA) with the sensor clipped to a ear lobe.

1.3 Data analysis and statistics

The mean value of the concentration changes in O_2Hb and HHb (averaged over the last 30 s of baseline period and over the 300 s of the balance task) was calculated for each measurement point. The O_2Hb and HHb mean values of the balance task were compared with the baseline values by a Student's t-test to evaluate if they were significantly different from the baseline. To examine the effect of the balance task on O_2Hb and HHb changes, the mean value of the baseline period was subtracted from the mean value of the changes in O_2Hb and HHb occurred during the balance task. The resulting data were analyzed by using analysis of variance (ANOVA) to determine the significance of individual changes for channel and hemisphere. To evaluate the relationship between the performance of the subjects and the activation mean values of O_2Hb, the Pearson's correlation coefficient was calculated. Statistical analyses were performed using the SigmaStat 3.5 package (Systat Software, Inc., Point Richmond, CA, USA). The criterion for significance was $p < 0.05$.

2 RESULTS

In all subjects, the increase of the task difficulty was accompanied by a progressive O_2Hb increase and a less consistent HHb decrease over the 8 measurement points of the frontopolar cortex (Fig. 2). O_2Hb and HHb had a tendency to plateau in the last min of the task. After the end of the task, O_2Hb and HHb immediately started to gradually return to their corresponding pre-task values. This fNIRS pattern is generally associated to a cortical activation. In fact, areas of the brain associated with specific cognitive processing may undergo a hemodynamic response (an increase of local blood flow) secondary to cortical neuronal activation (neurovascular coupling) induced by specific stimuli. These blood oxygenation changes lead to an increase in O_2Hb and a decrease in HHb.

The O_2Hb small fluctuations during the baseline, task and recovery period due to the head movements were observed also in the control task, even though no O_2Hb and HHb changes were found during the control task.

The fNIRS data analysis at group level showed a significant increase of O_2Hb ($p < 0.001$) and a significant decrease of HHb ($p < 0.001$) over the frontopolar cortex during the balance task. Figure 3 shows the grand average of the oxygenation response over the subjects in the right (average over 4 measurement points) and in the left (average over 4 measurement points) hemisphere during the incremental balance task. Analysis of variance carried out on O_2Hb and HHb changes did not show any significant main effect for hemisphere and channel. No significant activation was found during the control task. The mean heart rate did not change between the baseline and the task period. The number of the errors committed by the subjects was not correlated with the O_2Hb and HHb change, even though task difficulty was reflected in a greater number of errors and increased O_2Hb levels in the second part of the task. The mean subjective rating of perceived exertion during the balance task was 3 ± 0.9 indicating that the task was evaluated by the participants as moderate.

3 DISCUSSION

To the best of our knowledge this is the first time that fNIRS has been employed to study the frontopolar cortex response to a balance task in a semi-immersive virtual reality environment monitored by a depth-sensing camera. This study highlights the potential advantages offered by the combined use of a virtual reality based neuro-rehabilitation system and fNIRS to assess motor and cognitive tasks induced cortical oxygenation changes. Recently, commercial systems such as Wii by Nintendo have been adopted in neuro-rehabilitation. However, this technology does not allow for a whole body interaction in the virtual space as position and timing information from moving subject are provided by motion sensors. Moreover, these systems need a controller containing motion sensors to simulate the action required by the task: big part of the subject movements are imagined by the system. The use of a depth-sensing camera to capture the whole subject motion preserves geometry of

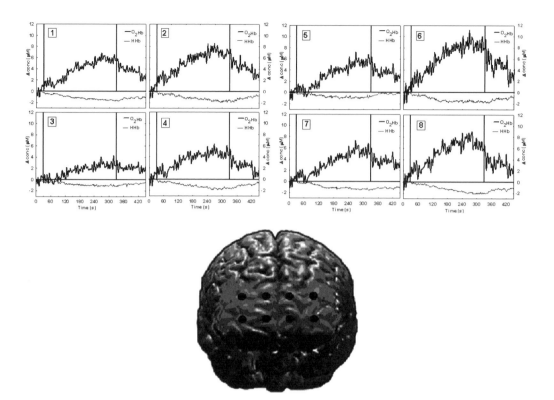

Figure 2. Typical cortical oxygenation response (increase in O_2Hb and decrease in HHb) provoked by the incremental balance task observed in the 8 measurement points over the frontopolar cortex. In each graph, the first vertical line indicates the start of the task after a 30 s rest period in the standing position; the second vertical line indicates the end of the task that was followed by a 120 s rest period. The 8 measurement points over the frontopolar cortex (Brodmann's areas 10 and 11) are shown in the low part of the figure. O_2Hb:oxy-hemoglobin; HHb: deoxy-hemoglobin.

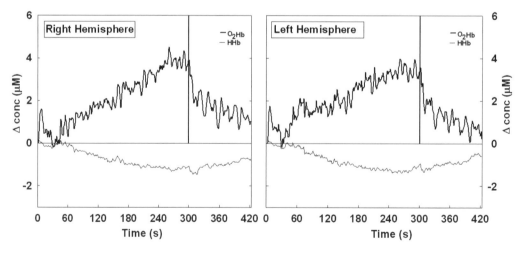

Figure 3. Grand average of the oxygenation response over the subjects in the right (average over 4 measurement points) and in the left (average over 4 measurement points) hemisphere during the incremental balance task. The vertical line indicates the end of the balance task (n = 6). O_2Hb: oxy-hemoglobin; HHb: deoxy-hemoglobin.

the movement with respect to the body and the environment allowing for more accurate mapping of the movement into a virtual environment with respect to commercial systems.

In the present study, a balance task that could be applicable in neuro-rehabilitation was adopted with the aim to evaluate the hemodynamic response over the frontopolar cortex. Although a fNIRS study (Mihara et al. 2008) has already evidenced the role of the prefrontal cortex in the postural balance control after an external perturbation that provoked an horizontal forward and backward translation of the platform on which each subject was asked to stand, our study consisted in controlling the subject balance over the board only by visual feedbacks without a direct contact with it. This task required the ability of the subject to match the movement really performed and the movement represented in real time by the 3D model on the computer screen and to adapt it to the board sensitivity. The results at single subject and group level obtained by the study evidenced that the frontopolar cortex is involved in the balance task. In line with our hypothesis, an increase of the task difficulty was followed by the increase in O_2Hb concentration. The increased oxygenation levels in the frontopolar cortex is consistent with the documented role of the prefrontal region in monitoring and coordinating attention resources to task demands (Stelzel et al. 2009). No frontopolar cortex activation was found in response to control task, suggesting that the observed hemodynamic changes were balance task related. Then, the frontopolar cortex seems to be involved in maintaining subjects balance.

Although the balance task in a semi-immersive virtual environment has not been so far applied in the rehabilitation field, it appears to be a functional task applicable to patients with movement disorders due to neurological dysfunction. Numerous advantages can be obtained by its application in patients to promote a motor (re)learning for body movement: 1) the task requires the ability to maintain one's self over a balance that is not present in the real world, so the patient with a reduced mobility does not risk to fall down from the board; 2) visual feedbacks are shown on the computer screen with the objective of knowing the errors and to sustain the motivation to achieve a higher level of performance than the precedent; 3) the system produces data which can be used online for feedback or offline for analysis which can quantify patient's performance during the motor activity (Kurillo et al. 2011). Moreover, the fNIRS monitoring can be useful to evaluate the effort of the subject during the task.

In conclusion, this study has provided evidence that the oxygenation is increased over the frontopolar cortex in healthy subjects performing an incremental balance task in a semi-immersive virtual environment. Furthermore, this effect was modulated by the task difficulty in all measurement points suggesting that the frontopolar cortex is bilaterally involved in attention-demanding tasks. These results suggest that this task could be employed in the neuro-rehabilitation field.

REFERENCES

Adamovich, S.V., Fluet, G.G., Tunik, E. & Merians, A.S. 2009. Sensorimotor training in virtual reality: a review. *Neuro-rehabilitation* 25(1): 29–44.

Atsumori, H., Kiguchi, M., Katura, T., Funane, T., Obata, A., Sato, H., Manaka, T., Iwamoto, M., Maki, A., Koizumi, H. & Kubota, K. 2010. Noninvasive imaging of prefrontal activation during attention-demanding tasks performed while walking using a wearable optical topography system. *J Biomed Opt* 15(4): 046002.

Ayaz, H., Shewokis, P.A., Curtin, A., Izzetoglu, M., Izzetoglu, K. & Onaral, B. 2011. Using MazeSuite and functional near infrared spectroscopy to study learning in spatial navigation. *J Vis Exp* (56): e3443.

Borg, G. 1990. Psychophysical scaling with applications in physical work and the perception of exertion. *Scand J Work Environ Health* 16(suppl.1): 55–58.

Cameirao, M.S., Badia, S.B., Oller, E.D. & Verschure P.F. 2010. Neuro-rehabilitation using the virtual reality based rehabilitation gaming system: methodology, design, psychometrics, usability and validation. *J Neuroeng Rehabil* 7:48.

Cutini, S., Basso Moro, S. & Bisconti, S. 2012. Functional near infrared optical imaging in cognitive neuroscience: an introdutory review. *J Near Infrared Spectroscopy* 20 (1): 75–92.

Duncan, A., Meek, J.H., Clemence, M., Elwell, C.E., Fallon, P., Tyszczuk, L., Cope, M. & Delpy, D.T. 1996. Measurement of cranial optical path length as a function of age using phase resolved near infrared spectroscopy. *Pediatr Res* 39(5): 889–894.

Ferrari, M., Mottola, L. & Quaresima, V. 2004. Principles, techniques, and limitations of near infrared spectroscopy. *Can J Appl Physiol* 29(4): 463–87.

Ferrari, M. & Quaresima, V. 2012. A brief review on the history of human functional near-infrared spectroscopy (fNIRS) development and fields of application. *NeuroImage* doi:10.1016/j.neuroimage.2012.03.049.

Holper, L., Muehlemann, T., Scholkmann, F., Eng, K., Kiper, D. & Wolf, M. 2010. Testing the potential of a virtual reality neuro-rehabilitation system during performance of observation, imagery and imitation of motor actions recorded by wireless functional near-infrared spectroscopy (fNIRS). *J Neuroeng Rehabil* 7:57.

Holtzer, R., Mahoney, J.R., Izzetoglu, M., Izzetoglu, K., Onaral, B. & Verghese, J. 2011. fNIRS study of walking and walking while talking in young and old individuals. *J Gerontol A Biol Sci Med Sci* 66(8): 879–887.

Karim, H., Schmidt, B., Dart, D., Beluk, N. & Huppert, T. 2012. Functional near-infrared spectroscopy (fNIRS) of brain function during active balancing using a video game system. *Gait Posture* 35 (3): 367–372.

Kurillo, G., Koritnik, T., Bajd, T. & Bajcsy, R. 2011. Real-time 3D avatars for tele-rehabilitation in virtual reality. *Stud Health Technol Inform* 163: 290–296.

Leff, D.R., Orihuela-Espina, F., Elwell, C.E., Athanasiou, T., Delpy, D.T., Darzi, A.W. & Yang, G.Z. 2011. Assessment of the cerebral cortex during motor task behaviours in adults: a systematic review of functional near infrared spectroscopy (fNIRS) studies. *Neuroimage* 54(4): 2922–2936.

Lucca, L.F., Candelieri, A. & Pignolo, L. 2010. Application of virtual reality in neuro-rehabilitation: an overview, Virtual Reality, Jae-Jin Kim (ed.) *In Tech Publisher*.

Mihara, M., Miyai, I., Hatakenaka, M., Kubota, K. & Sakoda, S. 2008. Role of the prefrontal cortex in human balance control, *Neuroimage* 43(2):329–336.

Quaresima, V., Bisconti, S. & Ferrari, M. 2012. A brief review on the use of functional near-infrared spectroscopy (fNIRS) for language imaging studies in human newborns and adults. *Brain Lang* 121(2): 79–89.

Seraglia, B., Gamberini, L., Priftis, K., Scatturin, P., Martinelli, M. & Cutini, S. 2011. An exploratory fNIRS study with immersive virtual reality: a new method for technical implementation. *Front in Hum Neurosci* 5:176. doi: 10.3389/fnhum.2011.00176

Stelzel, C., Brandt, S.A. & Schubert, T. 2009. Neural mechanism of concurrent stimulus processing in dual tasks. *Neuroimage* 48(1): 237–248.

Tachibana, A., Noah, J.A., Bronner, S., Ono, Y. & Onozuka, M. 2011. Parietal and temporal activity during a multimodal dance video game: an fNIRS study. *Neurosci Lett* 503(2): 125–130.

Wolf, M., Ferrari, M. & Quaresima, V. 2007. Progress of near-infrared spectroscopy and topography for brain and muscle clinical applications. *J Biomed Opt* 12(6): 062104.

Computational Modelling of Objects Represented in Images – Di Giamberardino et al. (eds)
© 2012 Taylor & Francis Group, London, ISBN 978-0-415-62134-2

Mandibular nerve canal identification for preoperative planning in oral implantology

T. Chiarelli & E. Lamma
Università di Ferrara, Ferrara, Italy

T. Sansoni
Era Scientific, Cattolica, Italy

ABSTRACT: The identification of relevant anatomies is fundamental for correct preoperative planning in oral implantology. In this setting, few state-of-the-art software products have the option of drawing the mandibular nerve canal, and almost none has a related automatic recognition functionality. We present an algorithm for the automatic recognition and drawing of the mandibular canal for oral implantology 3D-based software systems. The developed algorithm uses two user-identified extremities of the canal to determine a checkpoint at the mandibular foramen ascent base. It identifies the mandibular canal region moving within the radiographic volume, exploiting adaptive thresholds and ROI-intersection-based movement vectors. Finally, the software uses a Catmull-Rom-spline-based drawing feature to produce the canal 3D mesh. Experimental evaluation on scans of 7 human mandibles resulted in 13 successful identifications out of 14, while 1 failed due to extreme scattering.

1 INTRODUCTION

The use of Computed Tomography (CT) images and their three-dimensional (3D) reconstruction has increased over the last decade in the field of dental surgery. When using this kind of software to plan a surgical procedure, apart from achieving fundamental measurements, it is always important to consider relevant related anatomies. In oral implantology the mandibular nerve canal is the most relevant anatomy to take into consideration during the planning stage. The *inferior alveolar nerve*, or mandibular nerve, is the largest of the three branches of the *trigeminal nerve*, passing through the mandibular canal, a longitudinal bone cavity inside the mandible. This canal starts at the *mandibular foramen*, an opening in the inner side of the back of the mandible, and exits at the *mental foramen*, an opening in the outer side of the anterior part of the mandible. The inferior alveolar nerve, including its branches, provides sensory innervation to the lower teeth, the lower lip, and some skin of the lower face. Injury to this area is particularly dangerous: apart from the risk of breaking some of the vessels that pass inside the canal, such as the *inferior alveolar artery*, which would lead to potentially serious blood loss, damaging the mandibular nerve could paralyse the whole mandible. Thus, considering the importance of this anatomy in oral implantology, there is a clear need for an identification system to help the user take this into consideration during planning, which would be even more beneficial if it included an automatic recognition feature. A significant drawback of common planning software for surgery is the lack of automatic identification of this kind of anatomies, and, in the case of oral implantology, this involves mandibular nerve recognition. In fact, there are only a few pre-operative planning software products for oral implantology that are able to highlight the mandibular canal, and, to the best of our knowledge, none is provided with an automatic feature. When the canal identification functionality is present, it is purely manual and requires the user to manually identify points along the slices of a view (generally the cross-sectional view). The canal following the determined path can then be drawn (see for instance SurgiCase® and SimPlant®, Materialise). Although this procedure is certainly valid and necessary, it would be preferable to provide an additional automatic algorithm, in order to better assist implantologists with limited computer skills, thus improving the safety of the software in general.

In (Chiarelli et al. 2009, 2010a, b) we proposed a fully 3D approach to oral implant planning, and a preoperative planning software system in particular, which guarantees better precision in measurements. This paper focuses on the mandibular nerve canal identification functionality, integrated in the above-mentioned planning software, describing in detail both the proposed solution and the developed algorithm, and concluding with an important evaluation by an expert.

2 MANDIBULAR CANAL IDENTIFICATION

2.1 Automatic recognition

Two main types of algorithm on automatic recognition have been identified in literature: template matching algorithms (Rueda et al. 2006) and self-sufficient algorithms (e.g. McInerney & Terzopoulos 2000, Lloréns et al. 2009, Li & Yezzi 2007, Li et al. 2009, Mohan et al. 2010, Benmansour & Cohen 2010). An interesting approach regarding the second category is proposed by Yau et al. (2008). The authors presented an immediate and user-friendly methodology to segment the mandibular nerve canal. Their algorithm is based on the common Dentascan multi-view context (axial, panoramic, and cross-sectional views), and consists of automatically analysing, through the successive applications of a statistical segmentation and a *region growing* filter, each cross-sectional slice along the panoramic curve (to which the cross-sectional planes are orthogonal) between two user-defined extremities. In particular, each new slice *region of interest* (ROI) is achieved intersecting the current slice ROI with an expected nerve region in the following slice. However, conduct irregularities such as holes, bifurcations, and significant canal interruptions, typical of the mandibular canal, strongly affect the proposed algorithm, leading to possible leakages. Additionally, Yau et al. (2008) claim that this problem can be solved by using the ROI itself as the boundary area, however, this does not actually help to a great extent: even when the ROI is used as a boundary constraint, it would allow the new ROI centre to be in the same slice related position as the previous one at most, which, consequently, would then fall outside the canal, losing its track while scrolling forward as a result. Consequently, inspired by the algorithm proposed by Yau et al. (2008), it was decided to work on a new algorithm able to manage the mandibular canal irregularities, while taking advantage of a previously developed functionality that allows run-time custom reconstruction of oblique planes (Chiarelli et al. 2010a).

2.2 Developed algorithm

We developed an automatic recognition algorithm that can be conceptually divided into two parts: the insertion of a virtual magnetic probe or ball to acquire a checkpoint at the bottom of the exit ascent, and an iron-attracted-like rolling probe inserted into the mandibular foramen (Fig. 1). The approach starts by acquiring the starting and end points (i.e. the mandibular and mental foramina) by user identification. Then, the first step consists of inserting the magnetic probe into the canal exit, which moves inside the conduct through the iteration of the following aspects (items 1-8 below):

1 *Soft tissue definition*: as proposed also by Yau et al. (2008), a *region of interest* (ROI) is established around a specific point (e.g. the mandibular foramen), called the *seed point*, and the inner soft tissues greyscale values of the canal are statistically determined using local area based thresholds, thus adaptively:

$$\bar{X} = 1/N \sum_{i=1}^{N} x_i \quad x_i \in ROI \tag{1}$$

$$\bar{x} = 1/n \sum_{i=1}^{n} x_i \quad x_i < \bar{X} \tag{2}$$

$$\sigma = \sqrt{1/n \sum_{i=1}^{n} (x_i - \bar{x})^2} \tag{3}$$

where N is the number of pixels in the ROI, x_i is the greyscale value, $\bar{X}$ is the average of greyscale values in the ROI, and $\bar{x}$ and σ are the average of greyscale values in the ROI (after having removed the influence of the high greyscale values of the bone region), and its standard deviation, respectively. The last two parameters make it possible to achieve the upper and lower thresholds to identify the soft tissues, respectively $\bar{x} + \sigma$ and $\bar{x} - 2\sigma$.

2 *Slice sequence*: the first slice is centred on the second point identified by the user, and its normal aims at the mental foramen entrance, the direction of which has been achieved as interpolation of the related directions of several cases of study. Then, at each slice scrolling step, the direction is updated with the normalisation of the difference vector between the next seed point and the current one. As a consequence, the recognition system is able to move freely and to follow the conduct with enhanced precision as a result.

3 *Nerve canal segmentation*: instead of using a *region growing* filter, a flood labelling filter has been chosen. The first has been replaced since it can cause the main algorithm to fail in certain situations; in particular, it needs to grow a region from a specific point (i.e. the centre of the current ROI) or neighbourhood, analysing its neighbourhood and deciding what to ignore and what to keep, but there is no guarantee that the centre area of the current ROI is characterised by soft tissues. When part of the mandibular canal is characterised by a tiny section, as it usually is, but keeps its surface thickness, there is a good probability of the region growing algorithm to pick a bone area in the ROI centre, preventing the algorithm from continuing.

Instead, a flood labelling algorithm, which consists of labelling closed threshold-defined areas in one full image scan iteration, does not have this drawback. It scans each image pixel and checks if it is not yet labelled and if it satisfies the defined threshold. If the check fails, it moves on, otherwise it uses that pixel as the starting point for a growing filter. Similarly to the region growing algorithm, it then checks the neighbours, labelling those that satisfy the threshold, ignoring the others, and repeating the procedure in the neighbourhood of each labelled one. Then, it continues the scan, in order to find other threshold-fulfilling regions, labelling them with different tags. Finally, this algorithm has been properly adapted. Firstly, the threshold has been

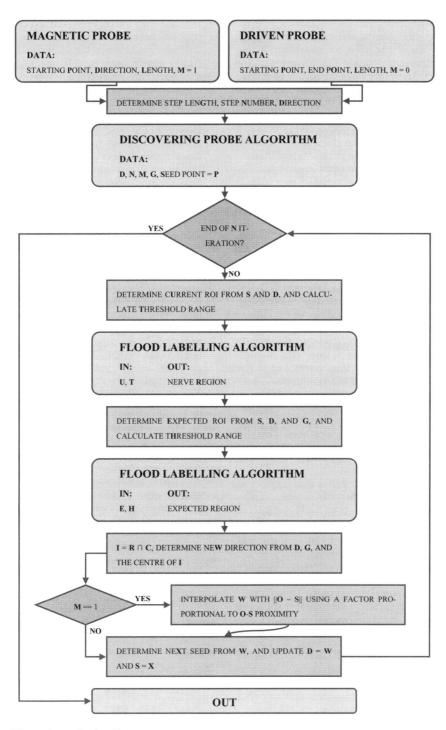

Figure 1.   Discovering probe algorithm.

defined as the range of soft tissue greyscale values. Then, the algorithm has been concluded with the selection of the soft tissues region looked for (labelled area nearer to the ROI centre).

4 *Seed point generation*: as described in Section 2.1: Automatic Recognition, the first seed point is the first point identified by the user (i.e. the mandibular foramen), whereas the subsequent seed points

are achieved immediately before moving on to the following slice. Also in this case, in order to get the next image seed point, an intersection is processed between the current-ROI nerve region and the expected nerve region of the equally positioned ROI in the next slice. However, unlike the approach described by Yau et al. (2008), the expected region is obtained by applying a flood labelling segmentation instead of a threshold filter, using the expected ROI soft tissue greyscale value range as the threshold parameter. This substitution has been performed to prevent the soft tissues out of the canal from influencing the intersection centre (which may occur in case of thin canal edges), and the slice scrolling direction as a result. The centre of the intersection region is the next seed point.

5 *Small noises handling*: a simple system to handle small image noises along the canal has been implemented, in order to improve the robustness of the algorithm. Image artefacts are not unusual in computed tomographies, such as scattering (bright white areas due to radio-reflecting materials) or incongruent densities (e.g. patient moving during the scan), and they may prevent the algorithm from proceeding (e.g. a sudden scattering through the canal would be identified as bone area from the algorithm, preventing it from continuing in that direction); to this regard, two handlers have been implemented. Concerning the nerve canal segmentation, the algorithm may undo an image analysis step, and repeat it using the previously stored direction. While regarding the next-seed-point generation, the algorithm is able to move forward with very short leaps, in order to overcome sudden unexpected noises.

6 *Extent of travel*: the progress of the probe is limited to a short stretch, so it does not go beyond the foramen ascent bottom, identifying the checkpoint before the canal exit. The distance to be travelled has been calculated as the mean of the mental foramen ascents of several cases, which is then converted into the number of progress steps.

7 *Step length*: the step pass has been set so that it is long enough to prevent hairpin turns due to freedom of movement direction (a problem that cannot be encountered in the slice scrolling fixed path).

In this way, the probe rolls down along the descent, only to stop approximately at its bottom, generating the needed checkpoint. After that, the approach aims to insert the second type of probe into the mandibular foramen (pointed out by the first user defined point). This probe works like the magnetic one, but there are a few differences (items 2a and 6a):

2a *Slice sequence*: the first slice is centred on the first point identified by the user (i.e. the mandibular foramen), and its normal is defined by the difference versor between the end and the starting point. Then, at each slice scrolling step, the direction is initially updated with the difference versor between the next seed point and the current

one, and then interpolated with the current-seed-point-to-checkpoint normalised vector. During this interpolation, however, the second vector is modified by a ratio factor $r = 1.0 - |d_{mc}/d_{ms}|$, whereas d_{mc} and d_{ms} are the distances respectively between the magnetic probe and the current seed point, and the magnetic probe and the starting point, consequently achieving different attraction forces in relation to the position within the canal. Finally, the next seed point is updated in relation to the new direction achieved. As a result, the recognition system is able to remain within the lane and to overcome the canal interruption area (common gap near the end of the conduct), reaching the checkpoint pointed out by the magnetic probe.

6a *Extent of travel*: the route length of the probe is calculated by exploiting the distance between the magnetic probe and the mandibular foramen, in order to reach, at least, the interruption of the canal and, at most, the checkpoint. Then, it is converted into the number of progress steps. Even though the obtained length is an approximation, when the probe travel is concluded, the algorithm is able to finish the uncovered path by connecting this second probe to the magnetic one along the straight line between them, which is actually the proper join direction.

As a result, the attraction force of the magnetic probe allows the rolling probe to remain along the canal lane, preventing it from quitting through holes (or soft-tissues-like density edges), and near the end of the conduct that force becomes strong enough to guide the second probe towards the checkpoint through the canal gap. Then, the algorithm calculates the mandibular canal path merging the probes respective routes. Finally, a proper procedure selects the most significant seed points of the path and provides them as the algorithm result. Then, these output points are passed to a Catmull-Rom-spline-based canal mesh builder as cspline control points, obtaining the reconstructed model and its rendering as a consequence.

3 EVALUATION

The algorithm has been implemented in the software previously proposed (Chiarelli et al. 2009, 2010a, b) and tested on a PC with a Pentium D 3.00 GHz CPU, 2 GB of RAM and 512 MB of VideoRAM. The software is written in C++ with the support of libraries: MITK 2.0 to handle the radiographic volume, ProfUIS 2.90 for the frame graphic, and SQLite for the database to store imported cases. Experimental evaluation of the developed functionality has been performed by an implantologist on some relevant cases. Before the evaluation phase, in order to emphasise the canal edges, temporary windowing has been accomplished. The best shared Width/Level (W/L) combination for the seven datasets was 9000/27500. Results are shown in Figures 2–15. For each tested dataset, both the right and the left inferior alveolar nerve identifications have

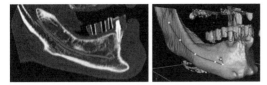

Figure 2. Right mandibular canal automatic recognition (Set 1).

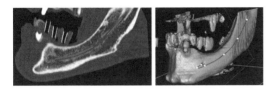

Figure 3. Left mandibular canal automatic recognition (Set 1).

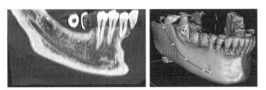

Figure 4. Right mandibular canal automatic recognition (Set 2).

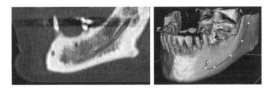

Figure 5. Left mandibular canal automatic recognition (Set 2).

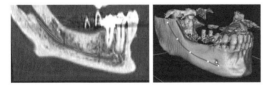

Figure 6. Right mandibular canal automatic recognition (Set 3).

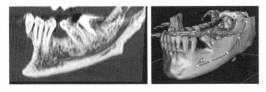

Figure 7. Left mandibular canal automatic recognition (Set 3).

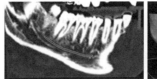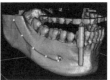

Figure 8. Right mandibular canal automatic recognition (Set 4).

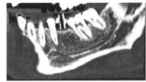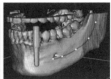

Figure 9. Left mandibular canal automatic recognition (Set 4).

Figure 10. Right mandibular canal automatic recognition (Set 5).

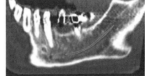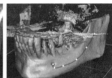

Figure 11. Left mandibular canal automatic recognition (Set 5).

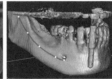

Figure 12. Right mandibular canal automatic recognition (Set 6).

been accomplished, and two views are presented for each one: a 2D oblique view, which approximately cuts the canal longitudinally, and a 3D view, which shows the selected mandibular canal model (solid line plus white control points) within the 3D reconstructed surface of the bone structure.

Notice that Figure 10 is shown with a different W/L pair in order to highlight the shadow effect that implied the recognition algorithm to depart from the canal (rectangle-circumscribed area on the left of Fig. 10). This effect is a consequence of the strong scattering, due to radio-reflective material (see also the explosion-like area on the top of the 3D mandible model on the right of Fig. 10).

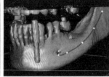

Figure 13. Left mandibular canal automatic recognition (Set 6).

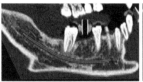

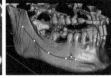

Figure 14. Right mandibular canal automatic recognition (Set 7).

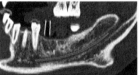

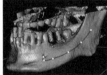

Figure 15. Left mandibular canal automatic recognition (Set 7).

Only one mandibular canal on 14 could not be properly recognised, due to an extreme artefact effect that obscured it. Concerning the 13 successful tests, instead, the assigned implantologist confirmed the correctness of the mandibular canals automatically recognised by the functionality, and reviewed the automatic recognition feature as a very useful tool, allowing the user to highlight a fundamental anatomy, making it possible to perform safe and correct planning, and easing and speeding up the operations needed for its identification.

Given the user-identified canal extremities, time performance tests, concerning the automatic recognition feature on the tested data sets, resulted in an average time of approximately 3 seconds to recognise each conduct.

4 CONCLUSIONS

We presented a software functionality that allows for the automatic identification of the mandibular nerve canal anatomy for oral implantology planning software. The developed algorithm takes advantage of the extremities of the canal, it then proceeds to the identification of the mandibular canal region moving freely within the radiographic volume, exploiting adaptive thresholds and ROI-intersection-based movement vectors modified by the checkpoint position.

The overall performance of the recognition feature has been evaluated by an expert implantologist on 14 mandibular canals of 7 different human mandibles scans (see Section 3). Although 1 identification failed due to the extreme scattering of the related CT images, 13 canals out of 14 were successfully recognised. Finally, the feature has been reviewed as a very fast, useful and user-friendly tool to support the correct preoperative planning in oral implantology.

ACKNOWLEDGMENTS

We thank Dr Federico Franchini for his valuable help in the experiments and evaluation. This work has been partially supported by the Camera di Commercio, Industria, Artigianato e Agricoltura di Ferrara, the project entitled "Image Processing and Artificial Vision for Image Classifications in Industrial Applications".

REFERENCES

Benmansour, F. & Cohen, L.D. 2010. Tubular structure segmentation based on minimal path method and anistropic enhancement. *Int. J. Comput. Vis.* 92(2): 192-210.
Chiarelli, T., Lamma, E., Sansoni, T. 2009. A Tool to Achieve Correct and Precise Measurements for Oral Implant Planning and Simulation. In: IEEE (ed.), *2009 IEEE International Workshop on Medical Measurements and Applications Proceedings*: 246–251.
Chiarelli, T., Lamma, E., Sansoni, T. 2010a. A fully 3D work context for oral implant planning and simulation. *International Journal of Computer Assisted Radiology and Surgery* 5(1): 57–67.
Chiarelli, T., Lamma, E., Sansoni, T. 2010b. Techniques to Improve Preoperative Planning Precision for Oral Implantology. *IEEE Transactions on Instrumentation and Measurement* 59(11): 2887–2897.
Li, H. & Yezzi, A. 2007. Vessels as 4-D curves: global minimal 4-D paths to extract 3-D tubular surfaces and centerlines. *IEEE Trans. Med. Imag.* 26(9): 1213–1223.
Li, H., Yezzi, A., Cohen, L. 2009. 3D multi-branch tubular surface and centerline extraction with 4D iterative key points. *Med. Image Comput. Comput. Assist. Interv. 2009* LNCS 5762: 1042–1050.
Lloréns, R., Naranjo, V., Clemente, M., Alcañiz, M., Albalat, S. 2009. Validation of Fuzzy Connectedness Segmentation for Jaw Tissues. *IWINAC 2009* LNCS 5602: 41–47.
McInerney, T. & Terzopoulos, D. 2000. T-snakes: topology adaptive snakes. *Med. Image Anal.* 4: 73–91.
Mohan, V., Sundaramoorthi, G., Tannenbaum, A. 2010. Tubular surface segmentation for extracting anatomical structures from medical imagery. *IEEE Trans. Med. Imag.* 29(12): 1945–1958.
Rueda, S., Gil, J. A., Pichery, R., Alcañiz, M. 2006. Automatic segmentation of jaw tissues in CT using active appearance models and semi-automatic landmarking. *Med. Image Comput. Comput. Assist. Interv. 2006* LNCS 4190: 167–174.
Yau, H. T., Lin, Y. K., Tsou, L. S., Lee, C. Y. 2008. An adaptive region growing method to segment inferior alveolar nerve canal from 3D medical images for dental implant surgery. *Computer-Aided Design and Applications* 5(5): 743–752.

Computational Modelling of Objects Represented in Images – Di Giamberardino et al. (eds)
© 2012 Taylor & Francis Group, London, ISBN 978-0-415-62134-2

Biomedical image interpolation based on multi-resolution transformations

Guoliang Xu, Juelin Leng & Yanmei Zheng
Institute of Computational Mathematics, Academy of Mathematics and System Sciences,
Chinese Academy of Sciences, Beijing, China

Yongjie Zhang
Department of Mechanical Engineering, Carnegie Mellon University, Pittsburgh, PA, US

ABSTRACT: We present a novel feature-based image interpolation approach. Two continuous maps for the image domain are constructed via an L^2-gradient flow based on their multi-resolution representations so that the features of the given images are matched at multiple scales. The flow equation is efficiently solved using a finite element method in the bicubic B-spline vector-valued function space. The interpolated images are then obtained from the domain maps at any sampling rate. Experimental results show that our interpolation approach is effective, capable of capturing image features from large to small. It yields continuously and uniformly deformed in-between images.

1 INTRODUCTION

Image interpolation is a widely-used operation in image processing, medical imaging and computer graphics. In biomedical imaging, a sequence of image slices of an organ or tissue is obtained with high resolution using CT or MRI devices. However the spacing between slices is much larger than the pixel size. For instance, in CT data, the pixel size is usually in the range of 0.5–2 mm, while the spacing between slices is in the range of 1–15 mm. Direct application of such data for 3D reconstruction often yields step-shaped iso-surfaces and other problems. To obtain volume data with similar spacing in three directions for 3D structure reconstruction (see (Bao and Lin 2003)), images need to be interpolated between slices.

The problem of image interpolation has large randomicity. To make the interpolation problem solvable, several conditions (see (Bao and Lin 2003)) are imposed: 1. The interpolated images should be similar to the original images. 2. The similarity degree of the interpolated image to the original images is inversely proportional to the distance between the interpolated image and the original images. 3. The sequence of interpolated images exhibits a gradual transition from one image to the other.

Many methods have been proposed for solving the image interpolation problem. These methods can be classified into two categories (see (Grevera and Udupa 1998)): scene-based and object-based. For the scene-based method, the interpolated scene intensity values are determined directly from the intensity values of the given scene (see (Chuang, Chen, Yuan, and Yeh 1999)). The earliest method is the linear interpolation, which is simple, fast and widely-used. Afterwards,

cubic spline was used in the medical image interpolation (see (Herman, Rowland, and Yau 1979)). In recent years, the Kringing method in statistic was used in the grey value interpolation (see (Stytz and Parrott 1993)). Basically, all these methods perform an intensity values averaging of the pixels located on the neighboring slices, without considering object shape in the image. Hence, the resultant images have blurring effects at the object boundary (see (Dhawan and Arata 1991; Chuang, Chen, Yuan, and Yeh 1999; Robb 2000)).

As reviewed by Grevera et al. (see (Grevera and Udupa 1998)) and Lehmann et al. (see (Lehmann, Göonner, and Spitzer 1999)), the scene-based method is simple, intuitionistic and easy to use, but the accuracy is usually poor. The object-based method has better accuracy generally, but the computational complexity is higher. In some object-basedmethods, flexible alignment is used to align adjacent slices, and then image interpolation is carried out between the corresponding positions in each slice (see (Goshtasby, Turner, and Ackerman 1992; Williams and Barrett 1993)). These methods, including using the optical flow model, elastic model, fluid model, finite element model and B-spline based free-form deformation, become popular in the 1990s (see (Penney, Schnabel, Rueckert, Hawkes, and Niessen 2004; Penney, Schnabel, Rueckert, Viergever, and Niessen 2004; Rueckert, Sonoda, Hayes, Hill, Leach, and Hawkes 1999; Williams and Barrett 1993)). The success of the alignment-based image interpolation depends on two prerequisites (see (Penney, Schnabel, Rueckert, Hawkes, and Niessen 2004)): (1) the adjacent slices contain similar anatomical features, and (2) the alignment algorithm is capable of finding the transformation whichmaps these similar features correctly. If the first prerequisite is violated, and an anatomical feature disappears from one slice to the next, then the advantages of the alignment-based approach will be lost. The second prerequisite is concerned with the transformation types that the alignment algorithm is capable of. If the transformation between features in adjacent slices is beyond the capabilities of the alignment algorithm, then the results will be sub-optimum.

In this paper, we present a novel and accurate shape-based approach. Two continuous maps for the image domain are constructed based on their multi-resolution representations so that the features of the given images are matched at multiple scales: from large to small. The interpolated images are then obtained from these maps at any sampling rate.

The remainder of this paper is organized as follows. In Section 2, we set up the image interpolation problem and present the algorithm outlines. Sections 3 and 4 are devoted to the computational details of constructing maps between the images to be interpolated. Experimental results are presented in Section 5. Finally, we conclude the paper with a summary in Section 6.

2 PROBLEM DESCRIPTION

Given two similar images $I_0(\mathbf{u})$ and $I_1(\mathbf{u})$ with $\mathbf{u} = [u, v]^T \in \Omega = [0, 1]^2$. We assume $I_0(\mathbf{u})$ and $I_1(\mathbf{u})$ have the same size, $(w + 1) \times (h + 1)$ (w and h are two positive integers). The image interpolation problem is to determine a sequence of in-between images $J_m(\mathbf{u})$, $m = 1, 2, \ldots, M$, such that $J_m(\mathbf{u})$ form a continuous transition from $I_0(\mathbf{u})$ to $I_1(\mathbf{u})$. Here M is a given positive integer.

Since the in-between images can be arbitrary, the image interpolation problem has no unique solution. A natural way to solve the interpolate problem is to determine a continuous (with respect to t in $[0, 1]$) and regular (with respect to $\mathbf{u}$ in Ω) map $\mathbf{y}(t, \mathbf{u})$ from Ω to Ω such that, $\mathbf{y}(0, \mathbf{u}) = \mathbf{u}$ and

$$I_0(\mathbf{y}(1, \mathbf{u})) = I_1(\mathbf{u}), \quad \mathbf{u} \in \Omega. \qquad (1)$$

Then the interpolated images can be defined as $J_m(\mathbf{u}) = I_0(\mathbf{y}(t_m, \mathbf{u}))$ with $t_m = m/(M + 1)$. Since $\mathbf{y}(t, \mathbf{u})$ is continuous with respect to t, so is $J(t, \mathbf{u}) = I_0(\mathbf{y}(t, \mathbf{u}))$. Hence $J(t, \mathbf{u})$ is an interpolation of $I_0(\mathbf{u})$ and $I_1(\mathbf{u})$ that can be sampled for any t in $[0, 1]$. However, to find a map $\mathbf{y}(t, \mathbf{u})$ satisfying the above conditions in general is not easy. The condition (1) is usually satisfied approximately. Then the obtained interpolated images $J_m(\mathbf{u})$ are close to $I_0(\mathbf{u})$ when m is close to 1. But when m gets close to M, $J_m(\mathbf{u})$ may not be close to $I_1(\mathbf{u})$. Hence, the roles played by I_0 and I_1 are unbias. To overcome this unbias, we seek a pair of maps $\mathbf{y}_0(t, \mathbf{u})$ and $\mathbf{y}_1(t, \mathbf{u})$, such that

1. Both $\mathbf{y}_0(t, \mathbf{u})$ and $\mathbf{y}_1(t, \mathbf{u})$ are continuous with respect to t for $t \in [0, 1]$.
2. For each $t \in [0, 1]$, $\mathbf{y}_0(t, \mathbf{u})$ and $\mathbf{y}_1(t, \mathbf{u})$ are regular maps from Ω to Ω. A regular map $\mathbf{y}_k(t, \mathbf{u})$ means

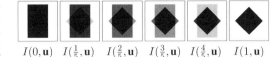

$$I(0, \mathbf{u}) \quad I(\tfrac{1}{5}, \mathbf{u}) \quad I(\tfrac{2}{5}, \mathbf{u}) \quad I(\tfrac{3}{5}, \mathbf{u}) \quad I(\tfrac{4}{5}, \mathbf{u}) \quad I(1, \mathbf{u})$$

Figure 1. Images obtained from the linear interpolation of intensity values of I_0 and I_1. i.e., $I(t, \mathbf{u}) = (1 - t)I_0(\mathbf{u}) + tI_1(\mathbf{u})$, $t \in [0, 1]$.

that the maps are one-to-one and its partial derivatives $\mathbf{y}_{ku}(t, \mathbf{u})$ and $\mathbf{y}_{kv}(t, \mathbf{u})$, with respect to u and v, are linearly independent.
3. $\mathbf{y}_0(0, \mathbf{u}) = \mathbf{y}_1(0, \mathbf{u}) = \mathbf{u}$ and

$$\begin{aligned} & \int_\Omega \|I_0(\mathbf{y}_0(1, \mathbf{u})) - I_1(\mathbf{y}_1(1, \mathbf{u}))\|^2 d\mathbf{u} \\ = & \min_{\mathbf{x}_0, \mathbf{x}_1} \int_\Omega \|I_0(\mathbf{x}_0(\mathbf{u})) - I_1(\mathbf{x}_1(\mathbf{u}))\|^2 d\mathbf{u}. \end{aligned} \qquad (2)$$

It is obvious that the maps $\mathbf{y}_0(t, \mathbf{u})$ and $\mathbf{y}_1(t, \mathbf{u})$ satisfying the above three conditions are not unique. There exist infinite number of many such maps. Each of them yields an interpolation image family. An ideal map-pair should be capable of matching the shapes in the given images from large (low frequencies features) to small (high frequencies features). We achieve this goal by constructing $\mathbf{y}_k(t, \mathbf{u})$ and $\mathbf{y}_1(t, \mathbf{u})$ using a multi-resolution representation. We therefore solve the image interpolation problem in the following steps:

1. Find two maps $\mathbf{y}_k(t, \mathbf{u}) : \Omega \to \Omega, t \in [0, 1], k = 0, 1$, satisfying the above conditions based on their multi-resolution representations.
2. If $M = 2N + 1$ is an odd number, set $J_m(\mathbf{u}) = I_0(\mathbf{y}_0 (t_m, \mathbf{u}))$, $J_{M+1-m}(\mathbf{u}) = I_1(\mathbf{y}_1(t_m, \mathbf{u}))$, for $m = 1, \ldots, N$, with $t_m = m/(N + 1)$, and

$$J_{N+1}(\mathbf{u}) = \frac{1}{2} [I_0(\mathbf{y}_0(1, \mathbf{u})) + I_1(\mathbf{y}_1(1, \mathbf{u}))]. \qquad (3)$$

3. If $M = 2N$ is an even number, set $J_m(\mathbf{u}) = I_0(\mathbf{y}_0 (t_m, \mathbf{u}))$, $J_{M+1-m}(\mathbf{u}) = I_1(\mathbf{y}_1(t_m, \mathbf{u}))$, for $m = 1, \ldots, N$, with $t_m = 2m/(2N + 1)$.

In the above steps, calculating $I_k(\mathbf{y}_k(t, \mathbf{u}))$ is straightforward if $\mathbf{y}_k(t, \mathbf{u})$ are known. The hard problem is how to construct the maps $\mathbf{y}_k(t, \mathbf{u})$. The details for computing $\mathbf{y}_k(t, \mathbf{u})$ are presented in the next section.

Remark 1 As we have discussed in the section 1, directly interpolating intensity values of I_0 and I_1, which results in blurring and ghosting is not an ideal method for image interpolation (see Fig. 1). However, the difference of the images to be averaged in (3) has been minimized by (2). We do not average I_0 and I_1 directly, but their aligned images.

3 COMPUTATION OF $\mathbf{y}_k(t, \mathbf{u})$

To obtain a multi-resolution representation of $\mathbf{y}_k(t, \mathbf{u})$, we represent it as a composition of a sequence of bicubic B-spline vector-valued functions $\mathbf{x}_k^{(\alpha)}(\mathbf{u})$, based on multi-resolution representations of $\mathbf{x}_k^{(\alpha)}(\mathbf{u})$ and $I_k(\mathbf{u})$, $\mathbf{u} \in \Omega$.

Given a non-negative integer L, which represents the number of levels in the multi-resolution representation, let $(w_0, h_0), \ldots, (w_L, h_L)$ be a sequence (we call it N-sequence) of integer pairs satisfying

$$w_0 = w,\ h_0 = h,\ w_\alpha > w_{\alpha+1},\ h_\alpha > h_{\alpha+1},$$

for $\alpha = 0, 1, \ldots, L-1$. Each pair (w_α, h_α) is used to define B-spline representations of $\mathbf{x}_k^{(\alpha)}(\mathbf{u})$ as well as the images I_0 and I_1 with the equally-spaced interval numbers w_α and h_α in the u and v directions, respectively. In our implementation, the N-sequence is given by

$$w_\alpha = E[\lambda_w^\alpha w],\ h_\alpha = E[\lambda_h^\alpha h],\ \alpha = 0,1,\ldots,L, \quad (4)$$

where $\lambda_w < 1$ and $\lambda_h < 1$ are the factors in the geometric proportional sequence, $E[\cdot]$ denotes taking the integer part. In our experiment, both λ_w and λ_h are taken as $5/6$. We refer a bicubic B-spline function in the space spanned by the basis $N_i^{(w_\alpha)}(u)N_j^{(h_\alpha)}(v), i = 0, \ldots, w_\alpha + 2, j = 0, \ldots, h_\alpha + 2$, as (w_α, h_α)-level bicubic B-spline function. The basis function $N_i^{(w_\alpha)}(u)$ and $N_j^{(h_\alpha)}(v)$ defined on the uniform grid can be found in a text book on splines (see (Xu 2008), page 173, for instance). The following algorithm based on multi-resolution representations of $\mathbf{x}_k(\mathbf{u})$ and $I_k(\mathbf{u})$ computes the maps $\mathbf{y}_k(t, \mathbf{u})$.

Algorithm 1 Construct $\mathbf{y}_k(t, \mathbf{u})$

1. Given $\lambda_w < 1$, $\lambda_h < 1$ and $L > 0$. Set $\alpha = L$ and $I_0^{(\alpha+1)} = I_0, I_1^{(\alpha+1)} = I_1$.
2. Iteratively construct two sequences of maps

$$\mathbf{x}_k^{(L)}(\mathbf{u}),\ \mathbf{x}_k^{(L-1)}(\mathbf{u}),\ \cdots,\ \mathbf{x}_k^{(0)}(\mathbf{u}),\ k = 0, 1,$$

as follows. For $\alpha = L, L-1, \ldots, 0$, do the following:

   (a) Compute (w_α, h_α)-level bicubic approximations $\tilde{I}_0^{(\alpha)}$ and $\tilde{I}_1^{(\alpha)}$ of $I_0^{(\alpha+1)}$ and $I_1^{(\alpha+1)}$, respectively, by the following steps:

   i. Smooth $I_0^{(\alpha+1)}$ and $I_1^{(\alpha+1)}$ using the Gaussian filter with deviations (see (Zheng, Jing, and Xu 2012) for explanations why using such deviations) $\sigma_u^{(\alpha)} = \sqrt{\dfrac{h_\alpha^2 - 1}{3}}$, $\sigma_v^{(\alpha)} = \sqrt{\dfrac{w_\alpha^2 - 1}{3}}$.
   ii. Convert the smoothed images to spline representations in the least square sense and obtain the images $\tilde{I}_0^{(\alpha)}$ and $\tilde{I}_1^{(\alpha)}$.

   (b) Compute $\mathbf{x}^{(\alpha)}(\mathbf{u}) = [\mathbf{x}_0^{(\alpha)}(\mathbf{u})^T, \mathbf{x}_1^{(\alpha)}(\mathbf{u})^T]^T \in \mathbb{R}^4$, which are represented as (w_α, h_α)-level bicubic B-spline vector-valued functions, such that

$$
\begin{aligned}
&\mathcal{E}(\mathbf{x}_0^{(\alpha)}, \mathbf{x}_1^{(\alpha)}) \\
&= \frac{1}{2} \int_\Omega \left[\tilde{I}_0^{(\alpha)}(\mathbf{x}_0^{(\alpha)}(\mathbf{u})) - \tilde{I}_1^{(\alpha)}(\mathbf{x}_1^{(\alpha)}(\mathbf{u})) \right]^2 d\mathbf{u} \\
&+ \frac{\lambda}{2} \sum_{k=0}^{1} \int_\Omega [g(\mathbf{x}_k^{(\alpha)}(\mathbf{u})) - 1]^2 d\mathbf{u}
\end{aligned}
\quad (5)
$$

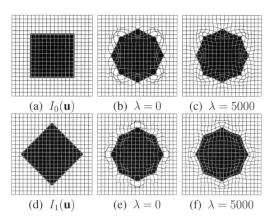

(a) $I_0(\mathbf{u})$	(b) $\lambda = 0$	(c) $\lambda = 5000$
(d) $I_1(\mathbf{u})$	(e) $\lambda = 0$	(f) $\lambda = 5000$

Figure 2. The effect of the regularization term. (b) and (e) are resulting images $I_k(\mathbf{y}_k(1, \mathbf{u}))$ yielded by Algorithm 1 with $\lambda = 0$; (d) and (f) are $I_1(\mathbf{y}_1(1, \mathbf{u}))$ with $\lambda = 5000$. Mesh grids displayed in (b) and (c) show the results of the map $\mathbf{y}_0(1, \mathbf{u})$ deforming an uniform grid; mesh grids displayed in (e) and (f) show the map $\mathbf{y}_1(1, \mathbf{u})$.

is minimized (see the next section for details), where

$$g(\mathbf{x}_k^{(\alpha)}) = g_{k11}^{(\alpha)} g_{k22}^{(\alpha)} - [g_{k12}^{(\alpha)}]^2,$$

$$g_{k11}^{(\alpha)} = (\mathbf{x}_{ku}^{(\alpha)})^T \mathbf{x}_{ku}^{(\alpha)},\quad g_{k12}^{(\alpha)} = (\mathbf{x}_{ku}^{(\alpha)})^T \mathbf{x}_{kv}^{(\alpha)},$$

$$g_{k22}^{(\alpha)} = (\mathbf{x}_{kv}^{(\alpha)})^T \mathbf{x}_{kv}^{(\alpha)},$$

$\mathbf{x}_{ku}^{(\alpha)}$ and $\mathbf{x}_{kv}^{(\alpha)}$ are partial derivatives of $\mathbf{x}_k^{(\alpha)}$ with respect to u and v, respectively.

   (c) Compute $I_0^{(\alpha)}(\mathbf{u}) = I_0^{(\alpha+1)}(\mathbf{x}_0^{(\alpha)}(\mathbf{u}))$, $I_1^{(\alpha)}(\mathbf{u}) = I_1^{(\alpha+1)}(\mathbf{x}_1^{(\alpha)}(\mathbf{u}))$ and then convert $I_0^{(\alpha)}(\mathbf{u})$ and $I_1^{(\alpha)}(\mathbf{u})$ to the (w_0, h_0)-level bicubic B-spline presentations using the least square approximation for the sake of efficient computation of next iteration.

3. Compute

$$\mathbf{x}_k(\mathbf{u}) = \mathbf{x}_k^{(L)}(\mathbf{x}_k^{(L-1)}(\cdots(\mathbf{x}_k^{(1)}(\mathbf{x}_k^{(0)}(\mathbf{u}))))).$$

   Let $\mathbf{y}_k(0, \mathbf{u}) = \mathbf{u}$ and $\mathbf{y}_k(1, \mathbf{u}) = \mathbf{x}_k(\mathbf{u})$, then for any $t \in (0, 1)$, $\mathbf{y}_k(t, \mathbf{u})$ is constructed by linear interpolating $\mathbf{y}_k(0, \mathbf{u})$ and $\mathbf{y}_k(1, \mathbf{u})$. The interpolating formula is written as

$$\mathbf{y}_k(t, \mathbf{u}) = (1-t)\mathbf{y}_k(0, \mathbf{u}) + t\mathbf{y}_k(1, \mathbf{u}). \quad (6)$$

The first term of the right-hand side of (5) aligns the given images. Hence, we call it a fidelity term. If we regard the map $\mathbf{x}_k^{(\alpha)}(\mathbf{u})$ as a parametric surface, then it is easy to see that $g(\mathbf{x}_k^{(\alpha)}(\mathbf{u}))^{1/2}$ is the area element of the surface. Hence, the second term tries to make the mappings $\mathbf{x}_k$ close to an identical map (see Remarks 2 and 3). We therefore call this term a regularizer.

Remark 2 As we discussed previously, $\mathbf{y}_k(t, \mathbf{u})$ is not unique. Our aim is to define $\mathbf{y}_k(t, \mathbf{u})$ so that for each t, $\mathbf{y}_k(t, \mathbf{u})$ is as regular as possible. One role the second term in (5) played is to make $\mathbf{y}_k(t, \mathbf{u})$ have a minimal deformation (close to the identical map). Fig. 2 shows the effect of the regularizer. As shown in Fig. 2, although $I_0(\mathbf{y}_0(1, \mathbf{u}))$ and $I_1(\mathbf{y}_1(1, \mathbf{u}))$ are well matched with either $\lambda = 0$ or $\lambda = 5000$, the constructed maps are much more smooth and regular constrained by the regularization term.

4 COMPUTATION OF $\mathbf{x}_k^{(\alpha)}(\mathbf{u})$

To compute $\mathbf{x}_k^{(\alpha)}(\mathbf{u})$, we minimize $\mathcal{E}(\mathbf{x}_0^{(\alpha)}, \mathbf{x}_1^{(\alpha)})$ by L^2-gradient flows. Performing calculus of variations of $\mathcal{E}(\mathbf{x}_0^{(\alpha)}, \mathbf{x}_1^{(\alpha)})$, we can derive the following (see (Zheng, Jing, and Xu 2012) for details) weak-form L^2-gradient flows for moving $\mathbf{x}_k^{(\alpha)}$

$$
\int_\Omega \frac{\partial \mathbf{x}_k^{(\alpha)}}{\partial t} \phi \, d\mathbf{u}
$$
$$
= (-1)^{k+1} \int_\Omega \left[\tilde{I}_0(\mathbf{x}_0^{(\alpha)}) - \tilde{I}_1(\mathbf{x}_1^{(\alpha)}) \right] \nabla \tilde{I}_k(\mathbf{x}_k^{(\alpha)}) \phi \, d\mathbf{u},
$$
$$
- \lambda \int_\Omega \left(\alpha_k^{(\alpha)} \phi_u + \beta_k^{(\alpha)} \phi_v \right) d\mathbf{u}, \quad k = 0, 1, \tag{7}
$$

where

$$
\alpha_k^{(\alpha)} = (g(\mathbf{x}_k^{(\alpha)}) - 1)(g_{k22}^{(\alpha)} \mathbf{x}_{ku}^{(\alpha)} - g_{k12}^{(\alpha)} \mathbf{x}_{kv}^{(\alpha)}), \tag{8}
$$

$$
\beta_k^{(\alpha)} = (g(\mathbf{x}_k^{(\alpha)}) - 1)(g_{k11}^{(\alpha)} \mathbf{x}_{kv}^{(\alpha)} - g_{k12}^{(\alpha)} \mathbf{x}_{ku}^{(\alpha)}). \tag{9}
$$

Remark 3 From equation (7), we can see that, if $\lambda = 0$ (no regularizer), the motions of $\mathbf{x}_k^{(\alpha)}$ are solely driven by $\nabla \tilde{I}_k(\mathbf{x}_k^{(\alpha)})$. At the region where $\nabla \tilde{I}_k(\mathbf{x}_k^{(\alpha)})$ vanishes, there will be no $\mathbf{x}_k^{(\alpha)}$ motion. This effect makes $\mathbf{x}_k^{(\alpha)}$ very irregular. To reduce the irregularity of $\mathbf{x}_k^{(\alpha)}$, a regularizer is therefore introduced in (5).

4.1 *Spatial discretization*

We solve equation (7) in a bicubic B-spline vector-valued function space. Let

$$
\mathbf{x}_k^{(\alpha)}(\mathbf{u}) = \sum_{i=0}^{w_\alpha + 2} \sum_{j=0}^{h_\alpha + 2} \mathbf{p}_{kij}^{(\alpha)} N_i^{(w_\alpha)}(u) N_j^{(h_\alpha)}(v),
$$

where $\mathbf{p}_{kij}^{(\alpha)} \in \mathbb{R}^2$ are the corresponding two-dimensional control points, $N_i^{(w_\alpha)}(u)$ and $N_j^{(h_\alpha)}(v)$ are one-dimensional cubic B-spline basis functions defined on the uniform grids with spacing $1/w_\alpha$ and $1/h_\alpha$, respectively.

For easy of description, we reorder the two-dimensional control points $\mathbf{p}_{kij}^{(\alpha)}$ of the B-spline mapping into a one-dimensional array and denote its components as $\mathbf{p}_{kl}^{(\alpha)}$, $l = i(w_\alpha + 3) + j = 0, 1, \ldots, n^{(\alpha)}$, where $n^{(\alpha)} = (w_\alpha + 3)(h_\alpha + 3) - 1$. The B-spline basis functions $N_i^{(w_\alpha)}(u) N_j^{(h_\alpha)}(v)$ are correspondingly

reordered and denoted as $\phi_0^{(\alpha)}, \ldots, \phi_{n^{(\alpha)}}^{(\alpha)}$. Using this ordering of the basis functions and control points, the mapping $\mathbf{x}_k^{(\alpha)}(\mathbf{u})$ can be represented as

$$
\mathbf{x}_k^{(\alpha)}(\mathbf{u}) = \sum_{j=0}^{n^{(\alpha)}} \mathbf{p}_{kj}^{(\alpha)} \phi_j^{(\alpha)}(\mathbf{u}). \tag{10}
$$

Substituting $\mathbf{x}_k^{(\alpha)}(\mathbf{u})$ into (7), and then taking the test function ϕ as $\phi_i^{(\alpha)}$ for $i = 0, \ldots, n^{(\alpha)}$, we can discretize (7) as systems of ordinary differential equations (ODEs) with the control points $\mathbf{p}_{kj}^{(\alpha)}, j = 0, \ldots, n^{(\alpha)}$, as unknown vectors.

$$
\sum_{j=0}^{n^{(\alpha)}} m_{ij}^{(\alpha)} \frac{d\mathbf{p}_{kj}^{(\alpha)}(t)}{dt} = \mathbf{q}_{ki}^{(\alpha)}, \quad i = 0, \cdots, n^{(\alpha)}, \quad k = 0, 1,
$$
$$
\tag{11}
$$

where

$$
m_{ij}^{(\alpha)} = \int_\Omega \phi_i^{(\alpha)} \phi_j^{(\alpha)} d\mathbf{u}, \tag{12}
$$

$$
\mathbf{q}_{ki}^{(\alpha)} = (-1)^{k+1} \int_\Omega \left[\tilde{I}_0(\mathbf{x}_0^{(\alpha)}) - \tilde{I}_1(\mathbf{x}_1^{(\alpha)}) \right] (\nabla \tilde{I}_k) \phi_i^{(\alpha)} d\mathbf{u}
$$
$$
- \lambda \int_\Omega \left(\alpha_k^{(\alpha)} \phi_{iu}^{(\alpha)} + \beta_k^{(\alpha)} \phi_{iv}^{(\alpha)} \right) d\mathbf{u} \in \mathbb{R}^2. \tag{13}
$$

Equation (11) can be written in the matrix form,

$$
\mathbf{M}^{(\alpha)} \frac{d\mathbf{P}_k^{(\alpha)}(t)}{dt} = \mathbf{Q}_k^{(\alpha)}, \quad k = 0, 1, \tag{14}
$$

with

$$
\mathbf{M}^{(\alpha)} = [m_{ij}^{(\alpha)}]_{ij=0}^{n^\alpha} \in \mathbb{R}^{(n^\alpha + 1) \times (n^\alpha + 1)},
$$
$$
\mathbf{P}_k^{(\alpha)}(t) = [\mathbf{p}_{k0}^{(\alpha)}(t), \cdots, \mathbf{p}_{kn^{(\alpha)}}^{(\alpha)}(t)]^T \in \mathbb{R}^{(n^\alpha + 1) \times 2},
$$
$$
\mathbf{Q}_k^{(\alpha)} = [\mathbf{q}_{k0}^{(\alpha)}, \cdots, \mathbf{q}_{kn^{(\alpha)}}^{(\alpha)}]^T \in \mathbb{R}^{(n^\alpha + 1) \times 2}.
$$

4.2 *Temporal discretization*

For the temporal direction discretization of the ODE systems (14), we use the following forward Euler scheme

$$
\frac{d\mathbf{P}_k^{(\alpha)}(t)}{dt} \approx \frac{\mathbf{P}_k^{(\alpha,s)} - \mathbf{P}_k^{(\alpha,s-1)}}{\tau_s}, \tag{15}
$$

where τ_s is a given temporal step-size, s is the iteration number. The initial value $\mathbf{P}_k^{(\alpha,0)}$ is given by the coefficients of $\mathbf{x}_k^{(\alpha,0)}(\mathbf{u}) = \mathbf{u}$. Substitute (15) into (14), we obtain $\mathbf{P}_k^{(\alpha,s)}$ iteratively from the previous $\mathbf{P}_k^{(\alpha,s-1)}$.

Using the inverse of the matrix $\mathbf{M}^{(\alpha)}$ (see (Zheng, Jing, and Xu 2012) for the fast computation of the inverse matrix) to solve the linear systems (14) for $k = 0, 1$, we obtain $d\mathbf{P}_k^{(\alpha)}(t)/dt = (\mathbf{M}^{(\alpha)})^{-1} \mathbf{Q}_k^{(\alpha)}$ and the new control points of $\mathbf{P}_k^{(\alpha)}$ from (15). We treat the right-hand terms in (14) as the known quantities. After $\mathbf{Y}_k^{(\alpha)} := (\mathbf{M}^{(\alpha)})^{-1}\mathbf{Q}_k^{(\alpha)}$ is computed, we project the control points at the boundaries to the boundaries of Ω so that the domain Ω of $\mathbf{x}_k$ maps to Ω.

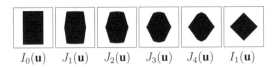

$I_0(\mathbf{u}) \qquad J_1(\mathbf{u}) \qquad J_2(\mathbf{u}) \qquad J_3(\mathbf{u}) \qquad J_4(\mathbf{u}) \qquad I_1(\mathbf{u})$

Figure 3. Interpolation of two images with consistent topology.

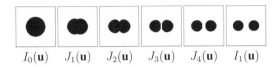

$I_0(\mathbf{u}) \qquad J_1(\mathbf{u}) \qquad J_2(\mathbf{u}) \qquad J_3(\mathbf{u}) \qquad J_4(\mathbf{u}) \qquad I_1(\mathbf{u})$

Figure 4. Interpolation of two images with distinct topology.

5 ILLUSTRATIVE EXAMPLES

We verify the performance of our image interpolation method through several numerical experiments. In each of the experiments, we construct the mapping $\mathbf{y}_k(t, \mathbf{u})$ so that $I_0(\mathbf{y}_0(1, \mathbf{u})) = I_1(\mathbf{y}_1(1, \mathbf{u}))$ approximately.

At first, our proposed interpolation method is validated on simple synthetic images. Fig. 3 and Fig. 4 show the experimental results, where I_0 and I_1 are the given images to be interpolated, four in-between images are produced in each experiment. It can be seen that, the deformation between two given images are continuous and uniform. Moreover, the interpolation process is topology changeable, as shown in Fig. 4.

Next, we present two examples to show the performances of our method for interpolating real images as well. As shown in Fig. 5 and Fig. 6, $I_0(\mathbf{u})$ and $I_1(\mathbf{u})$ are two similar images. From the figures in the second rows, we can see that the in-between images generated by our interpolation method describe the evolution process as t varies from 0 to 1. Compared with the images produced by the conventional linear

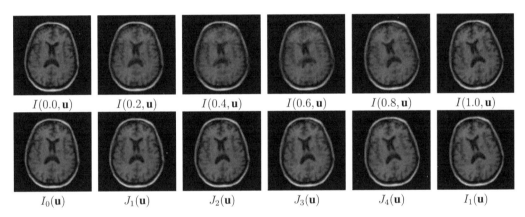

$I(0.0, \mathbf{u}) \qquad I(0.2, \mathbf{u}) \qquad I(0.4, \mathbf{u}) \qquad I(0.6, \mathbf{u}) \qquad I(0.8, \mathbf{u}) \qquad I(1.0, \mathbf{u})$

$I_0(\mathbf{u}) \qquad J_1(\mathbf{u}) \qquad J_2(\mathbf{u}) \qquad J_3(\mathbf{u}) \qquad J_4(\mathbf{u}) \qquad I_1(\mathbf{u})$

Figure 5. First row: Images obtained from the linear interpolation of intensity values of I_0 and I_1. Second row: Interpolation images for different t for I_0 and I_1 by our method.

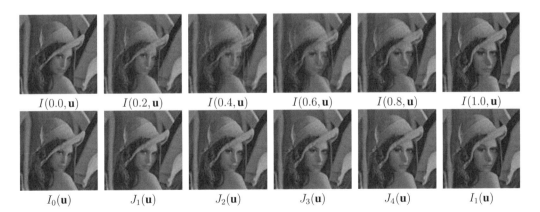

$I(0.0, \mathbf{u}) \qquad I(0.2, \mathbf{u}) \qquad I(0.4, \mathbf{u}) \qquad I(0.6, \mathbf{u}) \qquad I(0.8, \mathbf{u}) \qquad I(1.0, \mathbf{u})$

$I_0(\mathbf{u}) \qquad J_1(\mathbf{u}) \qquad J_2(\mathbf{u}) \qquad J_3(\mathbf{u}) \qquad J_4(\mathbf{u}) \qquad I_1(\mathbf{u})$

Figure 6. First row: Images obtained from the linear interpolation of intensity values of I_0 and I_1. Second row: Interpolation images for different t for I_0 and I_1 by our method.

interpolation method which are shown in the first rows of Fig. 5 and Fig. 6, the in-between images produced by our approach have no ghosting, and shape features are preserved very well. Hence, we can conclude that our interpolation method is effective and satisfies the requirements mentioned in section 1.

6   CONCLUSIONS

A novel feature based image interpolation approach has been presented. Two continuous and regular maps for the image domain are constructed based on their multi-resolution representation so that the features of the given images are matched from large to small. The interpolate images then are obtained from these maps at any sampling rate. The maps are represented by B-splines. L^2-gradient flows are used to construct the maps. The L^2-gradient flow is efficiently solved in the B-spline finite element space. The proposed interpolation method is validated by several experiments. The experimental results demonstrate that the proposed method is robust, effective, and capable of generating clear-cut in-between images.

ACKNOWLEDGEMENTS

G. Xu, J. Leng and Y. Zheng were supported by NSFC under the grant (11101401), NSFC key project under the grant (10990013) and Funds for Creative Research Groups of China (grant No. 11021101). Y. Zhang was supported in part by a research grant from the Winters Foundation.

REFERENCES

Bao, S. and B. Lin (2003). Weighted interpolation algorithm based on contours of the research. *Journal of Chongqin Normal University 20*(3), 29–32.

Chuang, K., C. Chen, L. Yuan, and C. Yeh (1999). Shape-based grey-level image interpolation. *Physics in medicine and biologyg 44*(6), 1565–1577.

Dhawan, A. and L. Arata (1991). Knowledge-based 3D analysis from 2D medical images. *IEEE Engineering in Medicine and Biology Magazine 10*(4), 30–37.

Goshtasby, A., D. Turner, and L. Ackerman (1992). Matching of tomographic slices for interpolation. *IEEE Transactions on Medical Imaging 11*(4), 507–516.

Grevera, G. and J. Udupa (1998). An objective comparison of 3-D image interpolation methods. *IEEE Transactions on Medical Imaging 17*(4), 642–652.

Herman, G., S. Rowland, and M. Yau (1979). A comparative study of the use of linear and modified cubic spline interpolation for image reconstruction. *IEEE Transactions on Nuclear Science 26*(2), 2879 – 2894.

Lehmann, T., C. Göonner, and K. Spitzer (1999). Survey: Interpolation methods in medical image processing. *IEEE Transactions on Medical Imaging 18*(11), 1049–1075.

Penney, G., J. Schnabel, D. Rueckert, D. Hawkes, and W. Niessen (2004). Registration-based interpolation using a high-resolution image for guidance. In *Proceedings of Medical Image Computing and Computer-Assisted Intervention*, Volume 3216-I, pp. 558–565.

Penney, G., J. Schnabel, D. Rueckert, M. Viergever, and W. Niessen (2004). Registration-based interpolation. *IEEE Transactions on Medical Imaging 23*(7), 922–926.

Robb, R. (2000). *Biomedical imaging, visualization, and analysis*. John Wiley & Sons.

Rueckert, D., L. I. Sonoda, C. Hayes, D. L. G. Hill, M. O. Leach, and D. J. Hawkes (1999). Nonrigid registration using free-form deformations: Application to breast MR images. *IEEE Transactions on Medical Imaging 18*(8), 712–721.

Stytz, M. and R. Parrott (1993). Using kriging for 3-D medical imaging. *Computerized Medical Imaging and Graphics 17*(6), 421–442.

Williams, W. and W. Barrett (1993). Optical flow interpolation of serial slice images. In *Proceedings of SPIE Medical Imaging 1993: Image Processing*, Volume 1898, pp. 93–104.

Xu, G. (2008). *Geometric Partial Differential Equation Methods in Computational Geometry*. Science Press, Beijing, China.

Zheng, Y., Z. Jing, and G. Xu (2012). Flexible Multi-scale Image Alignment Using B-Spline Reparametrization. In Y. Zhang (Ed.), *Image-based Geometric Modeling and Mesh Generation*, pp. 21–51. New York: Springer (to appear).

Computational Modelling of Objects Represented in Images – Di Giamberardino et al. (eds)
© *2012 Taylor & Francis Group, London, ISBN 978-0-415-62134-2*

A spectral and radial basis function hybrid method for visualizing vascular flows

Boris Brimkov, Jae-Hun Jung, Jim Kotary, Xinwei Liu & Jing Zheng
Department of Mathematics, State University of New York at Buffalo, Buffalo, NY, US

ABSTRACT: In this paper we develop a hybrid between the spectral method and Radial Basis Function (RBF) method in order to quickly and efficiently analyze vascular flows. Our hybrid method allows us to adopt the complex geometry of a blood vessel while maintaining high-order accuracy. The method produces streamline graphs and pressure gradients which can easily be visualized by open-source programs, and can be used in the diagnosis of defective blood vessels.

We also consider the concept of a vascular library of a large number of pre-computed full 3D solutions with a variety of parameters, and propose an interpolating algorithm to obtain the full flow fields of an unknown geometry from a linear combination of existing solutions with similar parameters. Finally, we provide numerical experiments to show that the obtained solutions are satisfactory.

1 INTRODUCTION

Vascular disease is one of the main causes of human mortality world-wide and affects more than a million people in the United States every year (Roger et al. 2011). To properly treat this disease, a reliable hemodynamic analysis is crucial; however, the medical assessment instruments used today are intrusive and yield data with low accuracy. An alternative assessment tool is a patient-specific computational fluid dynamics (CFD) analysis (Berthier et al. 2002), (Antiga et al. 2008), (Mut et al. 2010). Unfortunately, full 3D CFD typically takes too long to be done in a clinically relevant time frame due to its high computational complexity, and is therefore not usable by clinicians. In this work, we propose a faster hemodynamic analysis through a new CFD method, and expand the concept of a vascular library, which is based on our previous work (Jung et al. 2012), (Lee 2011).

The first goal of our research is to develop an extremely fast CFD method that can be used in a clinical time frame, by introducing an innovative hybrid method between the high-order spectral method – which needs a strict grid restriction and is therefore used to model the smooth areas of a blood vessel – and the radial basis function (RBF) method, which does not require a strict grid and is used to model the irregular parts of a blood vessel. Hence, with this new way to use CFD, we can adopt a complex geometry while maintaining a high-order accuracy. The data points produced by our method can be visualized by open-source programs such as *Paraview*, and presented in the form of pressure graphs and streamline fields.

We also create a vascular library of a large number of pre-computed full 3D CFD solutions with a variety of parameters based on the previous work (Jung et al. 2012; Lee 2011), and we use an interpolating algorithm to obtain the full flow fields of an unknown geometry from a linear combination of existing solutions with similar parameters without requiring an extra CFD run. This makes the patient-specific full 3D CFD assessment available to clinicians within a clinical time frame.

The paper is organized as follows. In the next section, we recall some basic definitions of human vasculature and fluid dynamics. In Section 3, we investigate a hybrid method between the spectral and RBF methods and support our discussion with illustrations and numerical results. In Section 4, we consider the advantages of creating a vascular library. We conclude with some final remarks in Section 5.

2 PRELIMINARIES

Most of the human cardiovascular network has a streamline flow with no lateral mixing, but occasional recirculation can be witnessed (see Figure 1) near irregular geometric areas such as stenoses (narrowings in the blood vessels), aneurysms (dilations in the blood vessels), and bifurcations (divisions of a main vessel into two smaller vessels). These can cause serious health problems which often require immediate intervention (Mark 2003).

We want to be able to provide an accurate model of a segment of human vasculature and the corresponding blood flow pattern in optimal time. Our first goal is to produce a library of vascular flows for different blood vessel geometries using a Spectral-RBF hybrid

Figure 1. Left: laminar blood flow, which is witnessed in healthy blood vessels. Right: turbulent blood flow, often occurring in unhealthy blood vessels, can cause health problems.

method. The second is to produce a flow field for any vascular geometry by high-order interpolation from these pre-existing solutions.

To obtain a blood flow field, we solve the 3D incompressible Navier-Stokes equation,

$$\frac{\partial \vec{V}}{\partial t} + (\vec{V} \cdot \nabla)\vec{V} = -\nabla P + \frac{1}{Re}\nabla^2 \vec{V}$$

with the incompressibility condition

$$\nabla \cdot \vec{V} = 0,$$

where $\vec{V} = \vec{V}(u, v, w)$ is the blood velocity, and P is the pressure. We assume that blood flow is incompressible, which means that density is homogeneous in space and time. We also adopt a no-slip boundary condition $u = v = w = 0$ along the wall to make calculations easier, even though in reality the boundary has Navier conditions. We also assume that there is no change in pressure, i.e. $\partial P/\partial t = 0$ at the walls of the vessel, and w.l.o.g. that the blood flows from left to right, making the pressure at the left boundary $P = P_0$ and the pressure at the right boundary $P = P_1 < P_0$. Finally, we assume that blood is a Newtonian fluid, i.e. that it continues to flow regardless of the forces acting on it.

There are two different methods to solve the Navier Stokes equations using RBFs: the Artificial Compressibility Method and the Projection Method. The former introduces an artificial compressibility δ into the equations of motion in such a way that the steady state does not depend on δ. The latter forms an intermediate velocity field and updates it to obtain a flow field for the irregular portion of the blood vessel. There are several ways to generate the nodes for the RBF method; they can either be extracted from a 3D mesh generator like *ISO2Mesh*, or selected using a random number generator from a program like *Matlab* (Fang & Boas 2009). Both approaches yield similar results.

3 HYBRID METHOD

Our proposed hybrid method consists of a spectral method for the cylindrical regions of the vessel and a RBF method for the irregular regions (see Figure 2). RBFs are relatively little-known, but they are introduced thoroughly in (Flyer & Fornberg 2011) and (Driscoll & Heryudono 2007). There are various types of commonly used RBFs; we use the Multiquadric

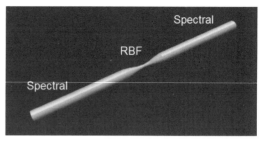

Figure 2. The spectral method is used in the smooth areas of the vessel, and the RBF method is used in the irregular portion which may feature a stenosis, aneurysm, or bifurcation.

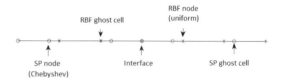

Figure 3. An example of Chebyshev nodes patched with uniform RBF nodes.

radial basis function (MQ-RBF), $\phi_i(x) = \sqrt{r^2 + \epsilon_i}$, which is one of the most widely studied RBFs. There are several reasons why we choose to use RBFs in the first place. Firstly, the RBF methods possess spectral accuracy for smooth functions. Second, RBFs are meshless methods, having high flexibility for irregular geometries such as stenoses, aneurysms, and bifurcations. Finally, they are very easy to implement, and can save computation time by being used locally (Sanyasiraju & Chandhini 2008).

We use a 2nd order finite difference patching algorithm at the interface between the portion of the vessel solved by the spectral method and the portion solved with the RBF method. The first step of the algorithm is to locate the ghost cells for the spectral and RBF regions around the interface (see Figure 3).

Next, the algorithm interpolates function values at the ghost cells using Lagrange interpolation,

$$I_N f(x) = \sum_{i=1}^{n} f(x_i) \prod_{i=1, i \neq j}^{n} \frac{x - x_j}{x_i - x_j}$$

with given data $(x_1, f(x_1)) \cdots (x_n, f(x_n))$ respectively on the spectral and RBF regions. Finally, the algorithm updates $u_{interface}$ and $v_{interface}$ using a 5-point central scheme separately, and then takes the average of the two. For example, for the first derivatives u_x and v_x, $x \in Interface$, the general formula is as follows:

$$f'(x) =$$
$$\frac{1}{2\lambda(\lambda+1)(\lambda+2)h_1}f(x - h_1 - h_2) - \frac{(\lambda+1)^2}{2\lambda(\lambda+2)h_1}f(x - h_1) +$$
$$\frac{(\lambda+1)^2}{2\lambda(\lambda+2)h_1}f(x + h_1) - \frac{1}{2\lambda(\lambda+1)(\lambda+2)h_1}f(x + h_1 + h_2),$$

where $\lambda = h_2/h_1$. The second derivatives u_{xx} and v_{xx}, $x \in Interface$, are found in a similar way. As an illustration, in Figure 4, we present the spectral-RBF patching

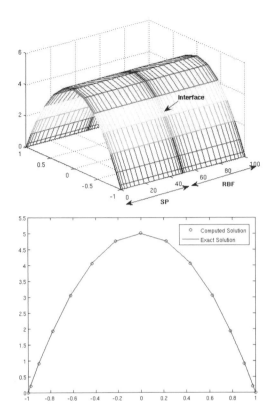

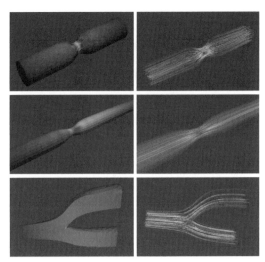

Figure 5. Top Row: Surface distribution of pressure field for a 50% stenosis at steady state obtained using MQ-RBF, and corresponding streamline field. Middle Row: Surface distribution of pressure field for a 50% stenosis at steady state obtained using Spectral Element method, and corresponding streamline field. Bottom Row: Cross sectional distribution of pressure field for a bifurcation at steady state obtained using MQ-RBF, and corresponding streamline field. Red-colored areas represent regions of high pressure or velocity, while blue-colored areas represent regions of low pressure or velocity.

Figure 4. Top: Velocity Field. Bottom: Comparison between the computed value and the exact solution at the interface.

algorithm for the 2D Incompressible Navier-Stokes Equations, where the numerical results are at a steady state.

Using our method, we have generated pressure distributions and streamlines for simple stenotic and bifurcating blood vessels. We also generated similar solutions using only a spectral method, and we compare the quality and accuracy of our results with the control spectral results. The figures are visualized using the open-source program *Paraview*. In the example below, the MQ-RBF method is using only 1500 centers, while the spectral method is using 3600 elements and 230400 nodes. Even though the MQ-RBF method is using roughly a hundred times fewer data points than the spectral method, it can be seen from Figure 5 that it displays comparable output.

4 VASCULAR LIBRARY

While effective, the methods described thus far are time-consuming and cannot be expected to give accurate solutions in a clinical timeframe (Jung et al. 2012), (Lee 2011). One of our goals is to create a method which can produce useful results in real-time. To this end, we use the Spectral-RBF method to create a library of vascular flows for all different blood vessel geometries.

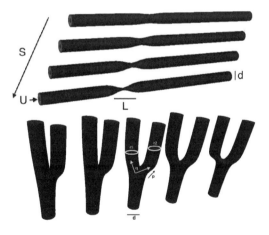

Figure 6. A schematic illustration of different parametrized blood vessels.

Each vessel with a different geometry is parameterized by several physical and geometric parameters such as S (percentage of stent), U (maximum blood velocity), d (vessel diameter), and q (bifurcating angle). The total number of parameters determines the dimension of the library (see Figure 6). Given an image of a patient's blood vessel, we use interpolation of known solutions from a library to approximate a solution to the problem at hand.

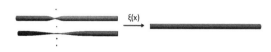

Figure 7. A schematic illustration of a sample linear interpolation.

The basic idea of the interpolation scheme is as follows; first, solutions in the CFD library are grouped according to topological equivalence so that all elements of similar type may be mapped to the same reference domain. Next, a rapid interpolation method finds the unknown solution based on the nearby solution elements in the library. The unknown solution U is then given by the linear sum of nearby solutions as a vector interpolation such that $\sum_i A_i(P_{i1}, \ldots, P_{iN}) \cdot U_i$ where U_i are the nearby basis solutions, A_i are the expansion coefficients, and $P_{i1}, P_{i2}, \ldots, P_{iN}$ are the hemodynamic parameters of U_i. Figure 7 shows a homeomorphic transformation of the computational mesh into the uniformized reference domain for the cases with stenosed vessels. The unknown flow fields are found in the reference domain using similar solutions in the existing library.

5 CONCLUDING REMARKS

We have developed a stable and accurate hybrid between the RBF and spectral methods. We have solved the incompressible Navier Stokes equations for an arbitrary vascular arrangement using this method, and we have developed a stable patching algorithm to hybridize the RBF and spectral methods. The RBF method is a highly efficient way to solve laminar and turbulent dynamic flows due to its high order accuracy and capability to adapt to a flexible geometry, whereas the spectral method performs accurately in the smooth geometric areas.

Future work will be based on optimizing the hybrid method to improve its speed even further, and developing a mobile device application which can exploit the flexibility of the proposed method.

ACKNOWLEDGEMENTS

This work was supported by the National Science Foundation CSUMS grant, No. 0802994. We thank the referees for their comments.

REFERENCES

Antiga L., Piccinelli M., Botti L., Ene-Iordache B., Remuzzi A. & Steinman D.A. 2008. An image-based modeling framework for patient-specific computational hemodynamics. *Med Biol Eng Comput.* 46: 1097–112.

Berthier B., Bouzerar R. & Legallais C. 2002. Blood flow patterns in an anatomically realistic coronary vessel: influence of three different reconstruction methods. *J Biomech.* 35: 1347–56.

Driscoll, T.A. & Heryudono, A.R.H. 2007. Adaptive residual subsampling methods for radial basis function interpolation and collocation problems, *Comput. Math. Appl.* 53: 927–939.

Fang, Q. & Boas, D. 2009. Tetrahedral mesh generation from volumetric binary and gray-scale images, Proceedings of *IEEE International Symposium on Biomedical Imaging*: 1142–1145.

Flyer, N. & Fornberg, B. 2011. Radial basis functions: Developments and applications to planetary scale flows, *Computers and Fluids* 46: 23–32.

J.-H. Jung, K. Hoffmann, J. Lee, & T. Dorazio, *A rapid vascular CFD through interpolation with spectral collocation methods*, submitted, 2012.

J. Lee, Rapid determination of flow parameters for arbitrary vessel geometries using a database of CFD solutions, MS Thesis, University at Buffalo, 2011.

Mark, R. G. 2003. Physiological fluid mechanics, HST.542J: *Quantitative Physiology: Organ Transport Systems*. MIT.

Mut, F., Aubry, R., Lhner, R. & Cebral, JR. 2010. Fast Numerical Solutions of Patient-Specific Blood Flows in 3D Arterial Systems. *Int. J. Numer. Method Biomed. Eng.* 26: 73–85.

Roger et al. 2011. Heart Disease and Stroke Statistics – 2012 Update: A Report From the American Heart Association, *Circulation* (10.1161/CIR.0b013e31823ac046).

Sanyasiraju, Y.V.S.S. & Chandhini, G. 2008. Local radial basis function based gridfree scheme for unsteady incompressible viscous flows, *Journal of Computational Physics* 227: 8922–8948.

Identification of models of brain glucose metabolism from ^{18}F – Deoxyglucose PET images

Paola Lecca

The Microsoft Research – University of Trento Centre for Computational and Systems Biology, Italy

ABSTRACT: The signals detected during physiological activation of the brain with ^{18}F – deoxyglucose (FDG) Positron Emission Tomography (PET) reflect predominantly uptake of this tracer into astrocytes. This notion provides a cellular and molecular basis for the FDG PET technique. In this paper, we present two new computational models of the molecular interactions governing the brain energy metabolism. The first model describes the glutamate-stimulated glucose uptake and use into astrocytes. The second includes also the effects of inter-cellular waves of Na^+ and Ca^{2+} generated by astrocytes on the glucose metabolism. The kinetic rates constants of the models have been identified by fitting the sets of ordinary differential equations to dynamic PET scans of 31 patients.

Keywords: astrocyte, glucose metabolism, ^{18}FDG-PET images, functional imaging, kinetic analysis.

1 INTRODUCTION

We present two ordinary differential equation models of the glutamate-triggered glucose uptake and metabolism by focusing on the emerging central role of the reactions occurring within astrocytes. Astrocytes are sub-types of the glial cells in the brain. These cells play a central role in brain function by affecting the activity of neurons, by taking an active part in the distribution of energy substrates from the circulation to neurons (Takano, Tian, Peng, Lou, Li-bionka, Han, & Nedergaard 2006).

Glutamate uptake into astrocytes is driven by the electro-chemical gradient of Na^+; it is an Na^+-dependent mechanism with a stoichiometry of three Na^+ ions cotransported with one glutamate molecule. A consequence of the glutamate uptake into astrocytes is the stimulation of glucose uptake and aerobic glycolysis in these cells (Pellerin & Magistretti 1997). Glutamate-stimulated increase in glucose uptake into astrocytes is abolished in the absence of Na^+ in the extracellular medium, consistently with the necessity for an electro-chemical gradient for the ion to drive glutamate uptake. A central role in the coupling between glutamate transporter activity and glucose uptake into the astrocytes is the activation of the Na^+/K^+-ATPase. The astrocytic Na^+/K^+-ATPase responds to increases in intracellular Na^+ concentration. Well established experimental observations (Pellerin & Magistretti 1997) show that glutamate activates Na^+/K^+-ATPase. There is also an ample evidence from studies in a variety of cellular systems including brain, kidney, vascular smooth muscle and erythrocytes, that increases in the activity of the Na^+/K^+-ATPase stimulates glucose compartments uptake and glycolysis (Pellerin & Magistretti 1997). Finally, the specific glutamate transporter inhibitor β-threohydroxyaspartate ((β_t)) inhibits the glutamate-stimulate glucose use (Magistretti & Pel-lerin 1999, Pellerin & Magistretti 1997).

Recently it has been shown that, in parallel with its uptake into the astrocyte, glutamate is released in association with Ca^{2+} waves (Pasti, Zonta, Pozzan, Vicini, & Carmignoto 2001, Schipke, Boucsein, Ohlemeyer, Kirkhoff, & Kettenmann 2002),, which represent a form of multicellular bidirectional communication with neurons. Astrocytes can communicate with each other by the propagation of Ca^{2+} elevation. Astrocytes can also generate Na^+ waves whose behavior is driven by the Ca^{2+} waves. It has been observed that inhibiting $Na^+/$glutamate transporters blocks the Na^+ waves without affecting the Ca^{2+} waves (Bernardinelli, Magistretti, & J. 2004). On the contrary each process that inhibits Ca^{2+} also inhibits the Na^+ waves. Since the glutamate released in the synaptic cleft during neuronal activity is rapidly taken up by astrocyte with three Na^+ ions, the uptake of glutamate results in an increase of the astrocytic Na^+ concentration. Glutamate is thus involved in the generation of the Na^+ wave, as the inhibition of glutamate transporter TOBA causes a strong inhibition of the wave without affecting the Ca^{2+} wave. Recent experiments proved that glutamate release and its subsequent Na^+ dependent uptake mediate the regenerative Na^+ wave (Bernardinelli, Magistretti, & J. 2004).

2 MODEL 1: BASIC GLUCOSE CONSUMPTION MECHANISM

In section 2 we present a model (*Model 1*) not including the waves of $[Na^+]_i$ and $[Ca^{2+}]_i$. This model is an extension of a previous work of the author (see (Lecca & Lecca 2007)). The rate equation of the concentration of glucose in the astrocyte ($[Gluc]_i$) is composed by three terms (Eq. (1)). The first term models the glutamate-stimulated glucose increase as a direct proportionality between the time derivative of glucose astrocytic concentration and the glutamate astrocytic concentration ($[Glut]_i$). The second term is the product of the rate of glucose uptake and its astrocytic concentration. This term expresses the proportionality between the time change of astrocytic glucose and both the flux of incoming glucose ($Gluc_{flux}$) and the glucose astrocytic concentration. Finally, the third term in Eq. (1) represents the decrease of glucose in astrocyte due to the Na^+/K^+-ATPase – stimulated glycolysis. Since the astrocytic Na^+/K^+-ATPase is activated by glutamate in response to increases in intracellular Na^+ concentration, the rate equation for the concentration of this enzyme, $[Na^+/K^+$-ATPase] in Eq. (3), is given by a term proportional to the concentration of Na^+ in the astrocyte ($[Na^+]_i$ and by a negative term proportional to the amounts of β-threohydroxyaspartate and Na^+/K^+-ATPase. This term models the inhibition of glutamate-stimulated glucose use performed by β-threohydroxyaspartate. In Eq. (3) the inhibition of glycolysis is modeled by a decrement term in the rate equation of $[Na^+/K^+$-ATPase]. In fact a decrement of the amount of this enzyme causes a decrement of the glycolytic events.

The rate equation for the astrocytic glutamate concentration (Eq. (2)) is the product of the glutamate amount in the cell and the flux of incoming glutamate. The fluxes of glutamate and glucose entering the astrocyte ($Gluc_{flux}$) and $Glut_{flux}$, respectively) have been modeled as functions of time. Experimentally the rate at which glucose is transported into the cell is determined by the rate at which the concentration of glucose accumulates inside the cell in the absence of metabolism (Marland & Keizer 2004). Thence, the temporal derivatives of the glucose and glutamate fluxes ($[Glut]_{flux}$, and $[Glut]_{flux}$) are given by Eq. (5) and Eq. (6) respectively. Eq. (5) contains a term accounting for the number of glutamate transporters in an open state ($[Glut\_t]_{open}$), i.e. transporters that are facing the exterior of the cell and ready to receive a glucose molecule. Similarly, Eq. (6) contains a term proportional to the fraction of two types of glutamate transporters Glt_1 and Glast and a term proportional to the difference between the internal and external concentration of Na^+ ($[Na^+]_i$ and $[Na^+]_e$, respectively).

Glucose entering into the astrocyte

$$\frac{d[Gluc]_i}{dt} = k_1\ [Glut]_i + k_2\ Gluc_{flux} \times [Gluc]_i - k_3\ [NA/K\_ATPase]\cdot[Gluc]_i \tag{1}$$

Glutamate entering into the astrocyte

$$\frac{d[Glut]_i}{dt} = k_4\ [Glut]_{flux}\cdot[Glut]_i \tag{2}$$

Na^+_K^+_ATPase

$$\frac{d[NA/K\_ATPase]}{dt} = k_5\ [Na^+]_i - k_6[\beta\_t]\cdot[NA/K\_ATPase] \tag{3}$$

β_threohydroxyaspartate

$$\frac{d[\beta\_t]}{dt} = -k_7\ [Na^+]_i\cdot[\beta\_t] \tag{4}$$

Rate of glucose uptake into the astrocyte

$$\frac{dGluc_{flux}}{dt} = k_8 Gluc\_t_{open}\cdot Gluc_{flux} \tag{5}$$

Rate of glutamate uptake into astrocyte

$$\frac{dGlut_{flux}}{dt} = (k_9\ Glt\_1 + k_{10}\ Glast)\cdot Glut_{flux} + k_{11}([Na^+]_i - [Na^+]_e) \tag{6}$$

Na^+ uptake into astrocyte

$$\frac{d[Na^+]_i}{dt} = k_{12}\ [Glut]_i \tag{7}$$

Eq. (4) is the rate equation for the β-threohydroxyaspartate concentration. The time derivative of this inhibitor is given by the product of its concentration and the concentration of Na^+. Namely, the inhibitory activity of the β-threohydroxyaspartate is consequent to the increase of the concentration of Na^+, that in turn is also responsible for the activation of the glycolytic activity of Na^+/K^+-ATPase. Finally, Eq. (7) describes the time behavior of the astrocytic concentration of Na^+. Its time derivative is proportional to the astrocytic concentration of glutamate.

3 MODEL 2: Ca^{2+} AND Na^+ WAIVES

In section 3 we modify Model 1 by including the generation of $[Na^+]_i$, the $[Ca^{2+}]_i$ waves (*Model 2*). IN particular, this second model describes how the Na^+ waves depend on the Ca^{2+} waves and quantifies the influence of these oscillations on the rate of glucose consumption. Unlike the model in section 2, the model presented in this section contains two equations more, defining the time behavior of the astrocytic concentration of calcium ions $[Ca^{2+}]_i$, intracellualr and extracellualar sodium concentrations, respectively $[Na^+]_i$ and $[Na^+]_e$. The $[Ca^{2+}]$ oscillations are modeled by a cosine waveform (Eq. (16)).

The dependency of the $[Na^+]_i$ wave from the $[Ca^{2+}]_i$ wave has been modeled in the following way: the release of glutamate in Eq. (9) is represented by the negative term proportional to $[Ca^{2+}]_i$. Moreover, in Eq. (14), the rate of change of $[Na^+]_i$ contains a term proportional to $[Glut]_i$, so that the oscillatory behavior of $[Ca^{2+}]_i$ is passed on to the rate of $[Glut]_i$, that, in turn transmits it to the rate of $[Na^+]_i$. Thus, glutamate is the mediator that transfers the $[Ca^{2+}]$ oscillations to the intracellular sodium. In this model, the rate equation for the glutamate flux that fits the FDG PET data (Eq. (12)) differs from the correspondent Eq. (6) in *Model 1*, as it is not proportional neither to $[Glut]_{flux}$ or to $[Na]_e^+ - [Na]_i^+$. The introduction of the calcium and sodium waves in the model results in a simplification of Eq. (6), that assumes the form of Eq. (12).

Glucose entering into the astrocyte

$$\frac{d[Gluc]_i}{dt} = k_1'[Glut]_1 + k_2' \, Gluc_{flux}[Gluc]_i - k_3'[Na/K\_ATPase][Gluc]_i \quad (8)$$

Glutamate entering into the astrocyte

$$\frac{d[Glut]_i}{dt} = k_4' \, Glut_{flux} \, [Glut]_i - [Ca^{2+}]_i [Glut]_i \quad (9)$$

$Na^+\_K^+\_ATPase$

$$\frac{[Na/K\_ATPase]}{dt} = k_5'[Na^+]_i - k_6'[\beta\_t][Na/K\_ATPase] \quad (10)$$

Rate of glucose uptake into the astrocyte

$$\frac{Gluc_{flux}}{dt} = k_{11}' \, Glut\_t_{open} - k_{12}' Glut\_t_{total} \quad (11)$$

Rate of glutamate uptake into the astrocyte

$$\frac{dGlut_{flux}}{dt} = (k_7' Glt + k_8' Glast) \cdot [Na]_i^+ \quad (12)$$

$\beta\_threohydroyaspartate$

$$\frac{[\beta\_t]}{dt} = -k_{13}'[Na^+]_i[\beta\_t] \quad (13)$$

Na^+ uptake into astrocyte

$$\frac{d[Na^+]_i}{dt} = k_{10}'[Glut]_i - k_{14}'[Na^+]_i \quad (14)$$

Na^+ release from astrocyte

$$\frac{d[Na^+]_e}{dt} = -[Na^+]_i' \quad (15)$$

Ca^{2+} wave

$$\frac{d[Ca^{2+}]_i'}{dt} = A \cos(0.5t) \quad (16)$$

4 PET IMAGE PROCESSING AND PARAMETER ESTIMATION

The dynamic FDG PET data used in this work have been provided by the Neurobiology Research Group, Rigshospitalet of Copenhagen. These data consist of 31 three-dimensional gray level images of the brain of a normal subject. The scans have been taken with

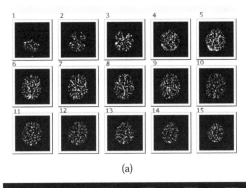

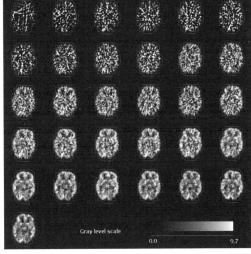

Figure 1. (a) A view of a set of brain slices. (b) The red boundary encloses the regions R_j^k for $j = 1, \ldots 31$ and $k = 7$. The values on the gray level scale are measured in Bq/cc.

a Scanditronix 4096 scanner on a time range of 3429 seconds. Figure 1(a) shows a set of brain slices of the database used in this work. To identify the kinetic rates k-s of the model we used a standard fit procedure of the time-dependency of glucose concentration obtained from the PET images. For the fit we used a simple least squares cost function. Before obtaining the measured time-dependence of glucose concentration, the images have been processed in order to eliminate noise and border effects and identify exclusively the region corresponding to the brain. The identification of the brain region and the elimination of the noisy parts on the borders of the skull have been performed with the following procedure. Let I_j denote the 3D-scan taken at time t_j and $\{I_j^1, \ldots, I_j^{15}\}$ with $j = 1, \ldots, 31$, the set of 15 slices of the j-th scan. For each scan I_j and for each slice I_j^k ($k = 1, \ldots, 15$), we calculated the smallest polygon P_j^k, enclosing the pixels, whose grey-level is greater than zero (i. e. the pixels which do not belong to the background). The boundary of this polygon has been smoothed by a simple procedure of

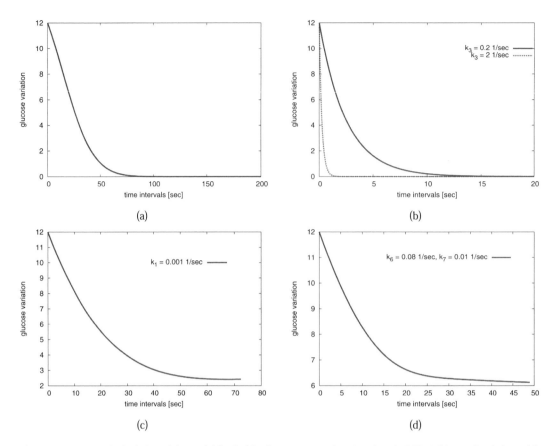

Figure 2. (a) Numerical solution of the model for the kinetic parameters of set 1 as show in Table 1.(b) Two simulations with different values of k_2 (0.2 and 2 sec^{-1}). The other parameters maintain the values shown in Table 1). (c) Simulations with $k_1 = 0.001$ sec^{-1} (the other parameters are fixed as in Table 1). (d) Two simulations with increased values of k_6 and k_7 (0.08 and 0.01 sec^{-1}). (see in the text for more details)

elimination of its parts having thickness larger than one pixel. Hence, for each slice I_j^k, the region R_j^k, we estimated as region effectively corresponding to the brain, is given by the topological internal part of P_j^k and the P_j^k boundary itself (see Figure 1(b)). Moreover, we defined $R_k \equiv \cup_{h=1}^{31} R_h^k$, $k = 1, \ldots, 15$, and we calculated the glucose concentration variation slice by slice using the following formula

$$\frac{dG_k}{dt}(t_i) = \frac{1}{\text{Area}(R_k)} \sum_{p \in R_k} \frac{|\sigma_p(t_i) - \sigma_p(t_{i+1})|}{t_{i+1} - t_i} \quad (17)$$

where $\sigma_p(t_i)$ is the intensity of the pixel $p \in R_k$ at time t_i ($i = 1, \ldots, 31$).

Two kinds of analysis has been carried out pixel by pixel to reveal a possible partitioning of the brain slice in activation areas.

In Table 1 the values of the initial concentrations of the reactants and the kinetic rates constants are shown. Figures 2 shows the simulations of glucose concentration for different values of the rate constants. We observed that the glucose concentration is mainly affected by the change of the values of k_1, k_2, k_3, k_6, and

k_7. Increasing k_1 and k_2 (as well as increasing simultaneously k_6 and k_7 or k_7 only) means to decrease the speed of glucose metabolism (Figure 2b-c), whereas increasing k_3 speeds up the glucose use (Figure 2b).

1. For each slice I_j^k of an image I_j we defined a *frequency map* $M : R_k \longrightarrow \mathbb{N}$ of pixel activation in the following way:

$$M[p] = \begin{cases} 0 & \text{if } p \notin R_k \\ m & \text{otherwise} \end{cases}$$

where m is the number of intensity's changes occurred in pixel p. This analysis showed that almost all the pixels exhibit the same frequency of intensity's change.

2. We also computed pixel by pixel for each slice of each scan the average glucose variation to detect possible clusters of pixels characterized by different levels of changes in glucose concentration variation. Also this kind of analysis showed that the time changes in glucose concentration are homogeneously distributed.

Table 1. Parameter space of the *Model 1.*	
Species	Initial concentration ($\times 0.0379016$ Bq/cc)
$[Gluc]_i$	12.00
$[Glut]_i$	11.00
NA/K_ATPase	2.0
$[\beta\_t]$	0.01
$Gluc_{flux}$	0.10
$Glut_{flux}$	0.10
$[Na^+]_i$	0.70
Constants	**Values**
$[Na^+]_e$	0.1
$Glut\_t_{open}$	0.1
Glt_1	0.1
$Glast$	0.1
Rate constant	**Value (sec^{-1})**
k_1	0.00003
k_2	0.00003
k_3	0.02000
k_4	0.00100
k_5	0.00100
k_6	0.00800
k_7	0.00100
k_8	0.00100
k_9	0.01000
k_{10}	0.01000
k_{11}	0.01000
k_{12}	0.10000

Table 2. Parameter space of the *Model 2.*	
Species	Initial concentration ($\times 0.0379016$) Bq/cc
$[Gluc]_i$	12.00
$[Glut]_i$	11.00
Na/K_ATPase	2.0
$[\beta\_t]$	0.01
$Gluc_{flux}$	0.10
$Glut_{flux}$	0.10
$[Na^+]_i$	0.10
$[Na^+]_e$	1.00
$[Ca^+]_i$	1.00
Constants	**Values**
$Glut\_t_{open}$	0.1
$Glut\_t_{total}$	0.1
Glt_1	0.1
$Glast$	0.1
Rate constants	**Value (sec^{-1})**
k'_1	0.0003
k'_2	0.0003
k'_3	0.0100
k'_4	0.0100
k'_5	0.1000
k'_6	0.0080
k'_7	0.1000
k'_8	0.0100
k'_9	0.01000
k'_{10}	0.1000
k'_{11}	0.1000
k'_{12}	0.1000
k'_{13}	0.0100

5 SIMULATIONS OF MODEL 1

The simulated glucose curves are in agreement with the solutions of the set of ordinary differential equations of the Sokoloff's model and they also reproduce the typical behavior of the tracer density in arterial blood as in (Svarer, I., Holm, Mørch, Paulson, Hansen, & Fog 1995). However, simulations – not reported here – demonstrate that the time behavior of the other species involved in the glucose metabolisms reveals that this first model needs to be refined. We obtained a linear increase of $[Na^+]_i$, $Gluc_{flux}$ and $Glut_{flux}$. This behavior points out that the model does not include any mechanism responsible for the oscillatory behavior of the Na^+, for an equilibrium of the glutamate and glucose fluxes and for the glutamate release from astrocyte. Therefore The Model 2 takes into account these aspects.

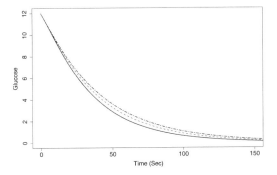

Figure 3. Time course of the glucose concentration in Model 2. Black solid line refers to A = 0.0345, dashed red line refers to A = 1, dotted green line refers to A = 2.94, and dot-dash blue line to A = 3.91.

6 SIMULATIONS OF MODEL 2

The amplitude A of the $[Ca^{2+}]$ wave has been increased from the value of 0.0345 (black solid line) inferred from the PET images up to 3.91 (dot-dash blue line) to evaluate changes in glucose consumption (see Figure 3). The kinetic rates of the *Model 2* are shown in Table 2: *Model 2* fit almost the same kinetic parameters both for the kinetics in *Set 1* and *Set 2* of brain slices.

As from experimental observations (Bernardinelli, Magistretti, & J. 2004), the simulations show that increasing the amplitude of the calcium wave only slightly affects the glucose flux and the rate of glycolysis, while the rate of glutamate release significantly increases and the amplitude of the $[Na^+]$ wave strongly decreases. Table 3 shows the average deviation between the reference simulation curves obtained with $A = 0.0034$, i. e. the amplitude of $[Ca^{2+}]$ waves inferred from the PET images.

Table 3. Average deviation of the simulated time behavior of the different species foor different values of A from the reference simulation curves obtained with $A = 0.00345$.

	$A = 1$	$A = 2.94$	$A = 3.91$
$\langle \Delta [\text{Gluc}]_i \rangle$	~0.00	~0.00	~0.00
$\langle \Delta [\text{Glut}]_i \rangle$	0.10	0.18	0.20
$\langle \Delta [\text{Na/K\_ATPase}] \rangle$	0.01	0.01	0.02
$\langle \Delta [\beta\_t] \rangle$	1.23	3.69	4.93
$\langle \Delta \text{Gluc}_{flux} \rangle$	0.30	0.47	0.51
$\langle \Delta \text{Glut}_{flux} \rangle$	1.67	1.50	1.46
$\langle \Delta [\text{Na}^+]_i \rangle$	1.04	1.78	1.99
$\langle \Delta [\text{Na}^+]_e \rangle$	0.01	0.01	0.02

7 CONCLUSIONS

Functional neuro-imaging techniques such as PET have provided valuable insights into the working brain. However, fundamental questions related to the cellular and molecular aspects of neurometabolic coupling are unresolved. Our computational models describe the molecular origin of the neuro-metabolic coupling and attempts also to build a theoretical framework to understand and experiment the glucose metabolism by the initial conditions and rate parameters. The model simulations, performed with the kinetic parameters derived from the PET images, are consistent with the blood activity curves observed in the PET studies on normal subjects (Svarer, I., Holm, Mørch, Paulson, Hansen, & Fog 1995).

REFERENCES

Bernardinelli, Y., P. Magistretti, & C. J. (2004). Astrocytes generate na+-mediated metabolic waves. 101(41), 14937–14942.

Lecca, P. & M. Lecca (2007). Molecular mechanism of glutamate-triggered brain glucose metabolism: a parametric model from fdg pet-scans. LNCS BVAI 2007 4729, 350–359.

Magistretti, P. & L. Pellerin (1999). Cellular mechanisms of brain energy metabolism and their relevance to functional brain imaging. Phil. Trans. R. Soc. Lond. B 354, 1155–1163.

Marland, E. & J. Keizer (2004). Computational Cell Biology, Chapter Transporters and Pumps. C. P. Fall, E. S. Marland, J. M. Wagner and J. J. Tyson Editors, Springer-Verlag.

Pasti, L., M. Zonta, T. Pozzan, S. Vicini, & G. Carmignoto (2001). Cytosolic calcium oscillations in astrocytes may regulate exocytotic releaseof glutamate. The J. of Neuroscience 21(2), 477–484.

Pellerin, L. & P. Magistretti (1997). Glutamate uptake stimulates na+/k+-atpase activity in astrocytes via an activation of the na+/k+-atpase. J. Neurochem. 69, 2132–2137.

Schipke, C., C. Boucsein, C. Ohlemeyer, F. Kirkhoff, & H. Kettenmann (2002). Astrocyte ca22+ waves trigger responses in microglial cells in brain slices. FASEB J. 16, 255–257.

Svarer, C., L. I., S. Holm, N. Mørch, O. Paulson, L. K. Hansen, & T. Fog (1995). Estimation of the glucose metabolism from dynamic pet-scan using neural networks. In Neural Networks for Signal Processing V. Proceedings of the 1995 IEEE Workshop.

Takano, T., G. Tian, W. Peng, N. Lou, W. Libionka, X. Han, & M. Nedergaard (2006). Astrocytes mediated control of cerebral blood flow. Nat. Neurosci. 9, 260–267.

Computational Modelling of Objects Represented in Images – Di Giamberardino et al. (eds)
© 2012 Taylor & Francis Group, London, ISBN 978-0-415-62134-2

Dynamic lung modeling and tumor tracking using deformable image registration and geometric smoothing

Yongjie Zhang, Yiming Jing & Xinghua Liang
Department of Mechanical Engineering, Carnegie Mellon University, Pittsburgh, PA, US

Guoliang Xu
LSEC, Institute of Computational Mathematics, Academy of Mathematics and System Sciences, Chinese Academy of Sciences, Beijing, China

Lei Dong
Scripps Proton Therapy Center, San Diego, CA, US

ABSTRACT: A greyscale-based fully automatic deformable image registration algorithm, based on an optical flow method together with geometric smoothing, is developed for dynamic lung modeling and tumor tracking. In our computational processing pipeline, the input data is a set of 4D CT images with 10 phases. The triangle mesh of the lung model is directly extracted from the more stable exhale phase (Phase 5). In addition, we represent the lung surface model in 3D volumetric format by applying a signed distance function and then generate tetrahedral meshes. Our registration algorithm works for both triangle and tetrahedral meshes. In CT images, the intensity value reflects the local tissue density. For each grid point, we calculate the displacement from the static image (Phase 5) to match with the moving image (other phases) by using merely intensity values of the CT images. The optical flow computation is followed by a regularization of the deformation field using geometric smoothing. Lung volume change and the maximum lung tissue movement are used to evaluate the accuracy of the application. Our testing results suggest that the application of deformable registration algorithm is an effective way for delineating and tracking tumor motion in image-guided radiotherapy.

1 INTRODUCTION

Today, cancer is the second most common cause of death in the United States. In 2011, about 571,950 Americans are expected to die of cancer. It is estimated that 221,130 people will be diagnosed with lung and bronchus cancer, and approximately 156,940 of them will die from this disease (Siegel, Ward, Brawley, and Jemal 2011). Approximately 45,000-50,000 of these patients will be diagnosed with locally advanced non-small-cell lung cancer (NSCLC) with an expected 5-year survival of only 10–20%. The poor results of radiotherapy for medically inoperable NSCLC may be due to various deficiencies in conventional radiation treatment techniques. One of such deficiencies is the respiratory-induced organ motion, which limits further reduction in treatment margins, and consequently also limits further dose escalation without significantly increasing treatment-related toxicities. There have been numerous studies demonstrating significant respiration motions and their dosimetric effects. However, it was not until recently that 4D CT scans became available. 4D CT images allow for quantitative modeling of internal organ motion for both treatment targets (primary tumors and involved lymph nodes) and normal tissues and organs that may be at risk due to radiation related toxicity. The internal organ motion determined from 4D CT provides evidence-based strategies to improve treatment plans. In addition, CT imaging modality provides the necessary tissue density information needed for radiation transport calculations, which are critical in designing accurate radiation treatments taking into account of density variations due to breathing. However, quantitative dosimetric studies using 4D CT are scarce at the present time. One reason was the need for deformable image registration to track dose deposited in the same target volume in multiple CT images at different time. This technique is still under intense research and development.

As reviewed in (Zitová and Flusser 2003), four main deformable registration techniques were developed for medical image data: elastic registration, level-set method, diffusion-based registration, and optical flow method. In elastic registration (Bajcsy and Kovacic 1989; Davatzikos, Prince, and Bryan 1996), external forces are introduced to stretch the image while internal forces defined by stiffness or smoothness constraints are applied to minimize the amount of bending and stretching. One of its advantages is that the

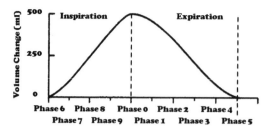

Figure 1.   Ten phases during respiration.

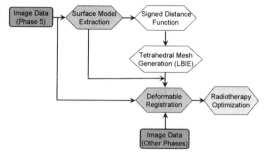

Figure 2.   Pipeline of dynamic lung modeling and tumor tracking for radiotherapy optimization.

feature matching and mapping function design can be done simultaneously. The level-set method (Osher and Fedkiw 2003) is a numerical technique for tracking interfaces and shapes, which can easily track topology change and combine segmentation together with registration (Moelich and Chan 2003; Droske and Ring 2006). The diffusion-based registration (Thirion 1998; Andresen and Nielsen 2001) considers the contours and other features in one image as membranes, and the other image as a deformable grid model, with geometrical constraints. This approach relies mainly on the notion of polarity, as well as the notion of distance. The optical flow method (Horn and Schunck 1981; Barron, Fleet, and Beauchemin 1994) assumes that the corresponding intensity value in the static image and the moving image stays the same, and then estimates the motion as an image velocity or displacement. This method is suitable for deformations in temporal sequences of images. Optical flow and diffusion registrations can be combined to have better matching results.

In this paper, we develop a systematic computational framework for dynamic lung modeling and tumor tracking using an optical flow registration together with geometric modeling techniques. In our computational processing pipeline, the input data is a set of 4D CT images with 10 phases. The triangle mesh of the lung model is directly extracted from one stable phase (Phase 5). In addition, we represent the lung surface model in 3D volumetric format by applying a signed distance function and then generate tetrahedral meshes. Our registration algorithm works for both triangle and tetrahedral meshes. In CT images, the intensity value reflects the local material density. For each grid point, we calculate the displacement from the Phase 5 image to match with images at other phases by using merely intensity values. The optical flow computation is followed by a regularization of the deformation field using geometric smoothing. Lung volume change and the maximum lung tissue movement are used to evaluate the accuracy of the application. Our testing results suggest that the application of the deformable image registration is an effective way for delineating and tracking tumor motion for image-guided radiotherapy.

The remainder of this paper is organized as follows. Section 2 overviews the systematic computational framework and then the following sections explain details. Section 3 describes an optical flow approach together with geometric smoothing. Section 4 shows testing results, and finally Section 5 draws conclusions.

2   COMPUTATIONAL FRAMEWORK

The respiration process can be divided into ten phases (Fig. 1), and the CT image data at each phase were obtained automatically from a 4D CT machine. During respiration, the motion of the lung results in the movement of the tumor inside the lung. It is important to study the movement of the lung and find out the exact position of the tumor at each phase of respiration. The ultimate goal is to use the lung and tumor tracking results for dose calculation and non-active lung tissue identification during the lung cancer treatment planning. Fig. 2 shows a computational framework of dynamic lung modeling and tumor motion tracking for the optimization of radiation therapy. During the respiration, Phase 5 in Fig. 1 is relatively stable due to its very small volume change. We construct a surface model of the lung as well as the tumor directly from the CT data at Phase 5, and define it as the reference.

The triangular surface mesh of the lung model is generated using our in-house software named LBIE-Mesher (Level-set Boundary Interior and Exterior Mesher) (Zhang, Bajaj, and Sohn 2005; Zhang and Bajaj 2006). Noise may exist in the constructed 3D surface models, therefore geometric flows (or geometric partial differential equations) (Zhang, Xu, and Bajaj 2006; Zhang, Bajaj, and Xu 2009) are adopted to smooth the surface and improve the aspect ratio of the surface mesh, while preserving surface features. Fig. 3 shows one constructed lung model with tumor. The constructed surface is then converted to volumetric grid data using the signed distance function method, which puts the surface into grids and calculates the shortest distance from each grid point to the surface, and finally assigns different signs to grids inside and outside the boundary. Then the volumetric data will be used as input to generate tetrahedral meshes for the lung-tumor model. Both triangular and tetrahedral meshes will be used in the following dynamic lung modeling and tumor tracking.

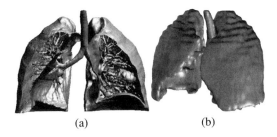

(a)   (b)

Figure 3. The extracted surface model of the lung (a) and the triangular mesh after smoothing (b).

3 DEFORMABLE IMAGE REGISTRATION

CT images are used in radiation dose calculation because Hounsfeld units (CT pixel values) are calibrated to the attenuation coefficient of water and therefore the pixel values are well defined. CT images directly reflect tissue density, therefore we choose an intensity-based algorithm for radiotherapy application (Wang, Dong, O'Daniel, Mohan, Garden, Ang, Kuban, Bonnen, Chang, and Cheung 2005). Our algorithm is primarily based on an optical flow method, also known as the "demons" algorithm (Thirion 1998), together with a geometric smoothing technique.

3.1 OPTICAL FLOW

Given one static image S and one moving image M, the "demons" algorithm evaluates the demons force using the gradient of the intensity field from S to match these two images. Usually, the optical flow formula is applied to calculate one passive force $\vec{f}_s$ at grid point on a greyscale image,

$$\vec{f}_s = \frac{(m-s)\vec{\nabla}s}{|\vec{\nabla}s|^2 + (s-m)^2}, \tag{1}$$

where $\vec{\nabla}s$ is the gradient on the static image. This algorithm may not be efficient especially when image varies little among neighboring grid points in one local region. Based on Newton's third law of motion, an active force f_m was introduced to speed up the rate of convergence by making use of information from both static and moving images,

$$\vec{f}_m = -\frac{(s-m)\vec{\nabla}m}{|\vec{\nabla}m|^2 + (s-m)^2}. \tag{2}$$

The term "passive" force denotes the contribution to the force from the static image. Similarly, the term "active" force denotes the influence from the moving image, in which the deformation is iteratively calculated to match with the moving image and it is active to track the corresponding point on the moving image.

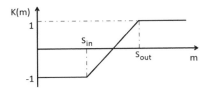

Figure 4. The $K(m)$ function.

Combining f_s and f_m, the total force at a specific grid point can be calculated as

$$\vec{f} = \vec{f}_s + \vec{f}_m$$

$$= (m-s)(\frac{\vec{\nabla}s}{|\vec{\nabla}s|^2 + (s-m)^2} + \frac{\vec{\nabla}m}{|\vec{\nabla}m|^2 + (s-m)^2}). \tag{3}$$

Eqn (3) is suitable for 3D image analysis with a complete grid of demons, and can deal with large deformation between two images. However, it may not be able to capture the boundary precisely. Therefore, we introduce "demons 2" to improve the performance along the boundary. For each contour point P in S, the "passive" and "active" forces are obtained using

$$\vec{f}_s' = K(m)\vec{n}_s \quad and \quad \vec{f}_m' = K(m)\vec{n}_m, \tag{4}$$

where $\vec{n}_s$ is the oriented normal of the contour point in S, and $\vec{n}_m$ is the oriented normal of the contour point in M (both from inside to outside). $K(m)$ is the demon function, see Fig. ??. Here $s_{in} = s(P - 2\vec{n}_s)$, and $s_{out} = s(P + 2\vec{n}_s)$. Combining both "passive" and "active" forces, we can obtain the total demon force,

$$\vec{f}' = K(m)(\vec{n}_s + \vec{n}_m). \tag{5}$$

3.2 GEOMETRIC SMOOTHING

To obtain smoothed geometry after registration, we also include geometric smoothing into our algorithm besides the Gaussian filter on the image domain. Here, we minimize an energy functional

$$\epsilon(x) = \int_\Omega (g(x(u,v,w)) - 1)^2 dudvdw, \tag{6}$$

where $\Omega = [0,1]^3$, $x(u,v,w)$ is the position vector of one grid point, and

$$g(x) = g_{11}g_{22}g_{33} + 2g_{12}g_{23}g_{13} \\ - (g_{13}^2 g_{22} + g_{23}^2 g_{11} + g_{12}^2 g_{33}) \tag{7}$$

with $g_{11} = x_u^T x_u$, $g_{12} = x_u^T x_v$, $g_{13} = x_u^T x_w$, $g_{22} = x_v^T x_v$, $g_{23} = x_v^T x_w$, and $g_{33} = x_w^T x_w$. For the existence of the solution of Eqn (6), see (Xu 2008). Here, we construct an L^2-gradient flow to minimize the energy functional $\epsilon(x)$,

$$\bar{x}(u,v,w,\epsilon) = x + \epsilon\Phi(u,v,w) : \Phi \in C_0^1(\Omega)^2. \tag{8}$$

Then, we have

$$\delta(\epsilon(x), \Phi) = \frac{d}{d\epsilon}\epsilon(\bar{x}(\cdot, \epsilon))|_{\epsilon=0}$$

$$= 2\int_\Omega (g(x(u,v,w)) - 1)\delta(g)d\Omega. \tag{9}$$

Hence, the equation becomes

$$\delta(\epsilon(x), \Phi) = 2\int_\Omega (\Phi_u^T \alpha + \Phi_v^T \beta + \Phi_w^T \gamma)dudvdw. \tag{10}$$

After applying Green's theorem, we have

$$\delta(\epsilon(x), \Phi) = -2\int_\Omega \Phi^T(\alpha_u + \beta_v + \gamma_w)dudvdw$$

$$+ \ const, \tag{11}$$

where $\alpha_u = 2 * (\frac{\partial g(x)}{\partial u}(x_u g_{22}g_{33} + x_v g_{23}g_{13} + x_w g_{12}g_{23}$
$-x_w g_{13}g_{22} - x_v g_{12}g_{33} - x_u g_{23}^2) + (g(x)-1)(x_{uu}g_{22}g_{33}$
$+x_u \frac{\partial g_{22}}{\partial u}g_{33} + x_u \frac{\partial g_{33}}{\partial u}g_{22} + x_{vu}g_{23}g_{13} + x_v \frac{\partial g_{23}}{\partial u}g_{13}$
$+x_v \frac{\partial g_{13}}{\partial u}g_{23} + x_{wu}g_{12}g_{23} - x_w \frac{\partial g_{12}}{\partial u}g_{23} + x_w \frac{\partial g_{23}}{\partial u}g_{12}$
$-x_{wu}g_{13}g_{22} - x_w \frac{\partial g_{13}}{\partial u}g_{22} - x_w \frac{\partial g_{22}}{\partial u}g_{13} - x_{vu}g_{12}g_{33}$
$-x_v \frac{\partial g_{12}}{\partial u}g_{33} - x_v \frac{\partial g_{33}}{\partial u}g_{12} - x_{uu}g_{23}^2 - 2x_u \frac{\partial g_{23}}{\partial u}g_{23}))$,
$\beta_v = 2 * (\frac{\partial g(x)}{\partial v}(x_v g_{11}g_{33} + x_u g_{23}g_{13} + x_w g_{12}g_{13}$
$-x_u g_{12}g_{33} - x_w g_{23}g_{11} - x_v g_{13}^2) + (g(x)-1)(x_{vv}g_{11}g_{33}$
$+x_v \frac{\partial g_{11}}{\partial v}g_{33} + x_v \frac{\partial g_{33}}{\partial v}g_{11} + x_{uv}g_{23}g_{13} + x_u \frac{\partial g_{23}}{\partial v}g_{13}$
$+x_u \frac{\partial g_{13}}{\partial v}g_{23} + x_{wv}g_{12}g_{13} + x_w \frac{\partial g_{12}}{\partial v}g_{13} + x_w \frac{\partial g_{13}}{\partial v}g_{12}$
$-x_{uv}g_{12}g_{33} - x_u \frac{\partial g_{12}}{\partial v}g_{33} - x_u \frac{\partial g_{33}}{\partial v}g_{12} - x_{wv}g_{23}g_{11}$
$-x_w \frac{\partial g_{23}}{\partial v}g_{11} - x_v \frac{\partial g_{11}}{\partial v}g_{23} - x_{vv}g_{13}^2 - 2x_v \frac{\partial g_{13}}{\partial v}g_{13}))$, and
$\gamma_w = 2 * (\frac{\partial g(x)}{\partial w}(x_w g_{11}g_{22} + x_v g_{12}g_{13} + x_u g_{12}g_{23}$
$-x_u g_{13}g_{22} - x_v g_{23}g_{11} - x_w g_{12}^2) + (g(x)-1)(x_{ww}g_{11}g_{22}$
$+x_w \frac{\partial g_{11}}{\partial w}g_{22} + x_w \frac{\partial g_{22}}{\partial w}g_{11} + x_{vw}g_{12}g_{23} + x_v \frac{\partial g_{12}}{\partial w}g_{23}$
$+x_v \frac{\partial g_{23}}{\partial w}g_{12} + x_{uw}g_{12}g_{23} + x_u \frac{\partial g_{12}}{\partial w}g_{23} + x_u \frac{\partial g_{23}}{\partial w}g_{12}$
$-x_{uw}g_{13}g_{22} - x_u \frac{\partial g_{13}}{\partial w}g_{22} - x_u \frac{\partial g_{22}}{\partial w}g_{13} - x_{vw}g_{23}g_{11}$
$-x_v \frac{\partial g_{23}}{\partial w}g_{11} - x_v \frac{\partial g_{11}}{\partial w}g_{23} - x_{ww}g_{12}^2 - 2x_v \frac{\partial g_{12}}{\partial w}g_{12}))$.

We define a term $G = -2\lambda(\alpha_u + \beta_v + \gamma_w)$ (λ is an input parameter), and merge it with Eqns (3) and (5). Then, we obtain

$$\vec{f} = \vec{f_s} + \vec{f_m} + \vec{f_s'} + \vec{f_m'} - G. \tag{12}$$

To apply the displacement field from image grid points to the mesh model, a trilinear interpolation on 3D regular grids is used to calculate the corresponding displacement of the mesh model. All the information required is the intensity value of each grid point on the static and moving images. As shown in Fig. 5, Eqn (12) is calculated iteratively. In each iteration, the regularization of the deformation field and geometric smoothing follow this optical flow calculation, using a Gaussian filter with a variance of σ^2 (here we choose $\sigma = 1.0$) and the G term in Eqn (12). The regularization plays an essential role as a smoothing

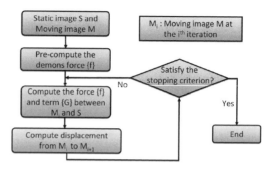

Figure 5. The iterative scheme for the optical flow and geometric smoothing.

operation to remove noise and preserve the geometric continuity, when this algorithm calculates displacement merely using the local information. After each iteration a stopping criterion is required. For each mesh vertex, if the maximum displacement difference is smaller than a given threshold as which we set 0.01, which is roughly 10% of the minimum span among X, Y and Z coordinates, the program stops.

4 TESTING RESULTS

We have applied our algorithm to 2D lung images. For example, we took the same cross section (slice 48) from the lung images at Phase 5 and Phase 9. In Fig. 6, (a) shows the contour curve overlaid with the static image at Phase 5. In (b), the green curve denotes the contour curve in (a) overlaid with the moving image at Phase 9, the red curve denotes the deformed contour using demons 1&2, and the blue curve denotes the deformed contour using demons 1&2 and with G term influence. The deformed grids are shown in (c) and (d). It is obvious that the deformed isocontour matches with the moving image very well, and with G terms the deformed grids is much smoother.

We also applied our algorithm to tetrahedral lung mesh and calculated the volume at each time phase. We generated the tetrahedral mesh directly from Phase 5 image data utilizing our meshing tool LBIE-Mesher. In Fig. 7, the blue line denotes the registration result using our algorithm, which always takes Phase 5 as the reference phase to generate the targeted tetrahedral meshes. The green line denotes the registration results, which takes the adjacent phase mesh (obtained from registration) as the reference to generate the targeted tetrahedral mesh in two different directions, one is Phase 5-4-3-⋯-7-6-5 and the other one is Phase 5-6-7-⋯-3-4-5. After finishing both computations, we took the average of these two sets of results. From the results, we can observe that the gradual registration yields a better match with Fig. 1, with the volume change reaching the maximum at Phase 0 through the ten phases of one cycle.

Furthermore, we applied our algorithm to the triangle surface mesh and calculated the maximum

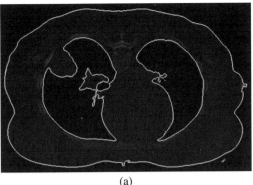

(a)

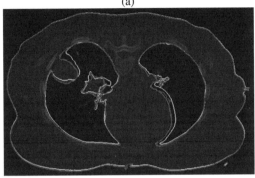

(b)

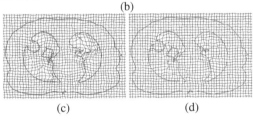

(c)   (d)

Figure 6.   Registration results of 2D lung images at Phases 5 and 9. (a) The contour curve overlaid with the static image at Phase 5; (b) the green curve denotes the contour in (a) overlaid with the moving image at Phase 9, the red curve denotes the deformed contour without G term influence, and the blue one denotes the deformed contour with G term influence; (c) the deformed grids without G term influence; and (d) the deformed grids with G term influence.

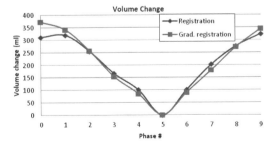

Figure 7.   The volume change using various methods.

displacement of the left lung, the right lung and the tumor during the breath, see Fig. 8. The green line denotes the maximum displacement of the left lung, where the tumor was. The red and blue lines denote

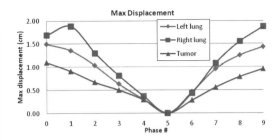

Figure 8.   The maximum displacement of the left lung, the right lung and the tumor.

the maximum displacement of the right lung and the tumor, respectively. Ideally, both the left and right lungs should deform roughly in the same displacement range. From this figure, it is obvious that the left lung deforms less and the tumor can move as large as 1 *cm*. This is because the lung tumor is one kind of abnormal mass consisting of non-active lung tissues, it has lost the functionality of the regular lung cell during inspiration and expiration. The moving trajectory of the tumor can be used to control the movement of the radiation probes, and thus to optimize the radiation therapy for lung cancer treatment planning.

Discussion. Compared to the standard tidal lung volume in Fig. 1, the maximum volume change in Fig. 7 is much smaller. This is reasonable because each patient is different. From Fig. 8, we can observe that during the respiration, the tumor movement can be large, even reach 1–2 cm. In addition, the physical condition of this patient is not good due to the large size tumor, which contributes to the shortness of breath. Since we do not know much about the health situation of that patient, we merely deduce that the lung tumor leads to the reduced lung displacement, especially for the left lung with the tumor.

5   CONCLUSIONS

We have developed an effective deformable image registration technique using the optical flow method together with geometric smoothing, which was validated using a set of 4D CT images of the lung. However, we only tested our technique on limited samples. As part of our future work, we will test more datasets and compare with other state-of-art techniques.

There are several potential developments which could improve this technique, such as multi-resolution registration. To effectively register two images from large structure features to fine details, in the future we will investigate new techniques supporting multi-resolution alignment. In addition, we will study how to identify the detailed level for each grid point based on the intensity gradient information. In this way, we can skip quite a lot grid points whose neighboring points vary little, and thus to improve the effectiveness of this algorithm.

ACKNOWLEDGEMENT

This research was supported in part by a research grant from the Winters Foundation.

REFERENCES

Andresen, P. R. and M. Nielsen (2001). Non-rigid registration by geometry-constrained diffusion. *Medical Image Analysis. 5*(2), 81–88.

Bajcsy, R. and S. Kovacic (1989). Multiresolution elastic matching. *Computer Vision, Graphics, and Image Processing. 46*(1), 1–21.

Barron, J. L., D. J. Fleet, and S. S. Beauchemin (1994). Systems and experiment: performance of optical flow techniques. *International Journal of Computer Vision. 12*, 43–77.

Davatzikos, C., J. Prince, and R. Bryan (1996). Image registration based on boundary mapping. *IEEE Transactions on Medical Imaging. 15*(1), 112–115.

Droske, M. and W. Ring (2006). A mumford–shah level-set approach for geometric image registration. *SIAM Journal on Applied Mathematics. 66*(6), 1–19.

Horn, B. K. P. and B. G. Schunck (1981). Determining optical flow. *Artifical Intelligence. 17*, 185–203.

Moelich, M. and T. Chan (February 2003). Joint segmentation and registration using logic models. In *UCLA CAM Report 03-06*.

Osher, S. and R. Fedkiw (2003). *Level set methods and dynamic implicit surfaces*. Springer-Verlag.

Siegel, R., E. Ward, O. Brawley, and A. Jemal (2011). Cancer statistics, 2011. *CA: A Cancer Journal for Clinicians 61*(4), 212–236.

Thirion, J. P. (1998). Image matching as a diffusion process: an analogy with Maxwell's demons. *Medical Image Analysis. 2*(3), 243–260.

Wang, H., L. Dong, J. O'Daniel, R. Mohan, A. Garden, K. Ang, D. Kuban, M. Bonnen, J. Chang, and R. Cheung (2005). Validation of an accelerated demons algorithm for deformable image registration in radiation therapy. *Physics in Medicine and Biology 50*(12), 2887–2905.

Xu, G. (2008). Geometric partial differential equation methods in computational geometry. *Scientific Publishing Press*.

Zhang, Y. and C. Bajaj (2006). Adaptive and quality quadrilateral/hexahedral meshing from volumetric data. *Computer Methods in Applied Mechanics and Engineering 195*(9-12), 942–960.

Zhang, Y., C. Bajaj, and B.-S. Sohn (2005). 3D finite element meshing from imaging data. *Computer Methods in Applied Mechanics and Engineering 194*, 5083–5106.

Zhang, Y., C. Bajaj, and G. Xu (2009). Surface smoothing and quality improvement of quadrilateral/hexahedral meshes with geometric flow. *Communications in Numerical Methods in Engineering 25*(1), 1–18.

Zhang, Y., G. Xu, and C. Bajaj (2006). Quality meshing of implicit solvation models of biomolecular structures. *Computer Aided Geometric Design 23*(6), 510–530.

Zitová, B. and J. Flusser (2003). Image registration methods: a survey. *Image and Vision Computing. 21*, 977–1000.

Computational Modelling of Objects Represented in Images – Di Giamberardino et al. (eds)
© 2012 Taylor & Francis Group, London, ISBN 978-0-415-62134-2

Motion and deformation estimation from medical imagery by modeling sub-structure interaction and constraints

Ganesh Sundaramoorthi
Department of Electrical Engineering & Department of Applied Mathematics and Computational Science
King Abdullah University of Science and Technology (KAUST), Saudi Arabia

Byung-Woo Hong
Computer Science Department, Chung-Ang University, Seoul, Korea

Anthony Yezzi
School of Electrical and Computer Engineering, Georgia Institute of Technology, Atlanta, US

ABSTRACT: This paper presents a novel medical image registration algorithm that explicitly models the physical constraints imposed by objects or sub-structures of objects that have differing material composition and border each other, which is the case in most medical registration applications. Typical medical image registration algorithms ignore these constraints and therefore are not physically viable, and to incorporate these constraints would require prior segmentation of the image into regions of differing material composition, which is a difficult problem in itself. We present a mathematical model and algorithm for incorporating these physical constraints into registration / motion and deformation estimation that does not require a segmentation of different material regions. Our algorithm is a joint estimation of different material regions and the motion/deformation within these regions. Therefore, the segmentation of different material regions is automatically provided in addition to the image registration satisfying the physical constraints. The algorithm identifies differing material regions (sub-structures or objects) as regions where the deformation has different characteristics. We demonstrate the effectiveness of our method on the analysis of cardiac MRI which includes the detection of the left ventricle boundary and its deformation. The experimental results indicate the potential of the algorithm as an assistant tool for the quantitative analysis of cardiac functions in the diagnosis of heart disease.

1 INTRODUCTION

Registration of medical images is of utmost importance for a number of applications in medicine. Typically, medical images are composed of many interacting parts and sub-structures, and each of these parts have different physical laws governing their motion and deformation (or at least different parameters from the same governing equations). For example, in the case of cardiac MRI registration, the deformation of ventricles and the myocardium (heart muscle) satisfy different governing equations because of differing material composition. Therefore, in order to correctly register medical images, one must first segment or detect each of the sub-structures where different governing laws hold, and then register the sub-structures based on the equation that holds for the particular sub-structure. Of course, now the problem becomes performing the segmentation. Current methods for medical image registration assume that the segmentation is given (usually done by hand or segmentation techniques in hardware, see for example, [5, 2, 12]) or completely ignore this problem, register the images

directly, and hope that good choices of regularization can overcome this predicament ([17, 6, 9]). Even when manual segmentation is done perfectly, and when the registration is performed correctly (which is highly unlikely due to noise and modeling errors) in each sub-structure, there is no guarantee that physical constraints for the deformation across sub-structures are satisfied.

In this work, we propose a novel registration method that models materials of different composition and therefore differing laws of deformation in each of the sub-structures composing the image, and the physical constraints of the deformation that must occur across sub-structures. We accomplish this by automatically detecting (segmenting) sub-structures using their *discriminating dynamics or deformation*. The deformation in each sub-structure is simultaneously computed while the sub-structures are estimated. Further, physical constraints across sub-structures are naturally incorporated into a unified optimization framework for the deformation and the sub-structure segmentation. Although our framework is valid in a number of registration scenarios (e.g., brain registration), we demonstrate the idea on the analysis of cardiac MRI, of critical importance [14, 2, 12], which includes the detection of the left ventricle boundary, its deformation, and the deformation in the surrounding area, i.e., the myocardium. For simplicity, we illustrate the formulation on an image composed of two sub-structures, although the formulation may be similarly extended to any number of sub-structures.

2 MATHEMATICAL METHODOLOGY

In this section, we formulate a mathematical model for registering medical images (and determining motion and deformation of objects within the image) under the assumption that several objects and/or sub-parts that are composed of different materials are imaged. This is the case, for example, in brain MRI (white/gray matter) and cardiac MRI (ventricles/myocardium). We model differing dynamics among sub-parts and the constraints that are imposed across the common boundaries of sub-parts.

2.1 Interaction between inhomogeneous materials

We start by considering the most basic assumption governing deforming objects typical in medical imagery, that is, the conservation of mass. This assumption is reasonable for cardiac MRI, and is a good approximation in other cases of interest. The assumption is not central to the thesis of the work, which is to model the constraints and interactions among neighboring sub-parts, and our method is general and can be used with other governing equations. We will see however, that the mass conservation property leads to the standard equations that are typically used in image registration.

We assume that the domain of the object(s) of interest is $\Omega \subset \mathbb{R}^n$ (where $n = 2$ or $n = 3$ depending on whether we wish to model in two or three dimensions). Within a homogeneous material (e.g., the myocardium or ventricle in cardiac MRI) of an object, the differential form of the conservation of mass implies the continuity equation:

$$\rho_t + \mathrm{div}\,(\rho v) = 0 \tag{1}$$

where $\rho : [0, 1] \times \Omega \to \mathbb{R}^+$ denotes the density of the material and $v : [0, 1] \times \Omega \to \mathbb{R}^n$ denotes the infinitesimal velocity of the material, ρ_t denotes differentiation with respect to the first variable, and div () indicates the divergence operator. The first parameter of ρ and v is time denoted by t, and the second parameter x indicates spatial location. Note that the above equation holds within regions of the same material characteristics, but *not across* material boundaries. Noting that tissue is composed mostly of water and therefore, can be considered as an incompressible fluid [15], we have that

$$\mathrm{div}\,(v) = 0, \tag{2}$$

and therefore, (1) reduces to

$$\rho_t + \nabla \rho \cdot v = 0 \tag{3}$$

within materials of the same material composition. Here $\nabla \rho$ denotes the spatial gradient (derivative with respect to x). We may replace ρ in (3) with the image intensity $I : [0, 1] \times \Omega \to \mathbb{R}^+$ since in many imaging modalities such as MRI and CT, the intensity represents a conserved quantity [15] via roughly proportionality to the density, and thus

$$I_t + \nabla I \cdot v = 0. \tag{4}$$

The above equation is equivalent to the differential form of the *brightness constancy equation* [11] for two-dimensional images which is commonly used in the computer vision literature.

Assuming that the input image is composed of two differing material regions (for example, the myocardium and ventricles in cardiac MRI), we have that

$$\begin{cases} I_t(t, x) + \nabla I(t, x) \cdot v_{in}(t, x) = 0 & x \in \mathrm{int}\,(R_t) \\ I_t(t, x) + \nabla I(t, x) \cdot v_{out}(t, x) = 0 & x \in \mathrm{int}\,(\Omega \backslash R_t) \end{cases} \tag{5}$$

where Ω is the domain of the image, $R_t \subset \Omega$ denotes the first material region (as a function of t), and v_{in} and v_{out} (defined on R_t and $\Omega \backslash R_t$, respectively) denote the velocities inside R_t, the first region, and $\Omega \backslash R_t$, the second material region. The notation int R_t indicates the interior of the set R_t (not including the boundary). The formulation may be similarly extended to any number of material regions, but we choose to illustrate our technique on two regions for simplicity.

Note that the differing statistics of v_{in} and v_{out} due to material differences make determining the unknown R_t possible. The solution of (5) is not unique and therefore, regularization is required, as typical in determining optical flow in computer vision. Even with typical regularization (for example, assumption of spatial smoothness of the velocity fields), the equations (5) still do not yield a unique solution without specification of a boundary condition on ∂R_t, the boundary of R_t, which is an extra condition not needed in optical flow problems. However, this can be resolved by noting the physical constraint that the velocity (not the density ρ) must be continuous across the boundary ∂R_t of material regions:

$$v_{in}(t, x) = v_{out}(t, x),\ x \in \partial R_t. \tag{6}$$

The constraint above makes it possible to solve (5) directly on ∂R_t when the image is two-dimensional. Indeed, one can show, by solving (5) and (6) directly on ∂R_t, that

$$v_b(t, x) = I_t(t, x) \frac{(\nabla I_{in}(t, x) - \nabla I_{out}(t, x))^\perp}{\nabla I_{in}(t, x) \cdot \nabla I_{out}(t, x)^\perp},\ x \in \partial R_t \tag{7}$$

where $v_b(t, x) = v_{in}(t, x) = v_{out}(t, x),\ x \in \partial R_t,\ w^\perp$ denotes a vector perpendicular to w, and ∇I_{in} and

∇I_{out} denote the limits of the gradients approaching the boundary ∂R_t from within int R_t and int $\Omega \backslash R_t$, respectively. The expression holds when ∇I_{in} and ∇I_{out} are non-parallel, and are non-zero. In other words, the expression breaks down when there is no discontinuity of the gradient angle of the image across the material boundary. The expression also holds in three-dimensions except the expression no longer yields a unique solution.

2.2 Energy-based formulation for registration

In order to determine the registration between images in the sequence, one would need to determine the velocity field v for each time t at every spatial location. The registration (cumulative motion/deformation between two images) $\phi : [0, 1] \times \Omega \to \Omega$ between image at time zero and time t can be obtained by integration:

$$\phi(t, x) = x + \int_0^t v(\tau, \phi_\tau(x)) \, d\tau, \qquad (8)$$

where $v|R_t = v_{in}$, $v|\Omega \backslash R_t = v_{out}$, and $v|\partial R_t = v_b$. Therefore, we need to determine v_{in}, v_{out}. Note that since v_{in} is supported on R_t and v_{out} is supported on $\Omega \backslash R_t$, R_t either must be known or it must be estimated from the data. Our approach is to jointly estimate v_{in}, v_{out} and R_t as part of an optimization problem.

We setup an optimization problem incorporating the conditions (5) and the constraint (6). The energy that we propose is

$$E(R, v_{in}, v_{out}, v_b) = \int_0^1 \int_{R_t} f_{in}(v_{in}(t, x)) \, dx \, dt$$
$$+ \int_0^1 \int_{\Omega \backslash R_t} f_{out}(v_{out}(t, x)) \, dx \, dt$$
$$+ \int_0^1 \int_{\partial R_t} f_{bndry}(v_b(t, x)) \, dS(x) \, dt \qquad (9)$$

where v_b is the velocity defined on ∂R_t for all t, and dS denotes the surface area element (or arclength element if $n = 2$). Further, the optimization is subject to the constraint $v_b(t, x) = v_{in}(t, x) = v_{out}(t, x)$, $x \in \partial R_t$. An auxiliary variable v_b has been added to the energy in order to simplify imposing the constraint (6). The functions f are defined as follows:

$$f_{in}(v_{in}) = (I_t + v_{in} \cdot \nabla I)^2 + \text{Reg}(v_{in}) \qquad (10)$$

$$f_{out}(v_{in}) = (I_t + v_{out} \cdot \nabla I)^2 + \text{Reg}(v_{out}) \qquad (11)$$

$$f_{bndry}(v_b) = (I_t + v_b \cdot \nabla I_{in})^2 + (I_t + v_b \cdot \nabla I_{out})^2$$
$$+ \text{Reg}(v_b) \qquad (12)$$

where Reg indicates a regularization term that includes spatial and time regularity, e.g., $\text{Reg}(v_{in}) = |\nabla v_{in}|^2 + |\partial_t v_{in}|^2$. Further, another constraint that is imposed is

$$R_t = \phi(t, R), \qquad (13)$$

that is, the region at time t is obtained by warping the initial region R along the velocity field. Note that we exclude weighting in (9) each of the terms for ease of presentation, but they are used in practice.

The use of regularization of v_{in} and v_{out} as seen above is to regularize the brightness constancy equation (such a motion prior is realistic since the motion within homogeneous regions in medical imagery is smooth spatially and in time). The sophisticated deformation in many organs (e.g., the heart) makes the use of TV regularization [4], which favors piecewise constant motion and therefore popular in computer vision, unsuitable. We choose not to incorporate the closed form solution (7) directly, but instead incorporate the constraint (6) and (5) into the term f_{bndry}. Therefore, the portion of the energy due to the term f_{bndry} is small when the constraints are satisfied. Due to noise, pointwise estimates of v_b using (7) may not provide an accurate estimate of v_b and when ∇I_{in} and ∇I_{out} are close to parallel, numerical problems may arise, and therefore, we work with the constraints directly and further add regularity of the velocity along the boundary as in f_{bndry}.

In order to simplify matters for optimization, we assume $n = 2$ (the image is two-dimensional), omit regularization of the vector fields v_{in}, v_{out}, v_b and R_t in time and the constraint $R_t = \phi(R)$, and in this case, the region and velocities can be computed by optimizing the following energy for each time t:

$$E(R, v_{in}, v_{out}, v_b) = \int_R \left[(I_t + v_{in} \cdot \nabla I)^2 + |\nabla v_{in}|^2 \right] dx$$
$$+ \int_{\Omega \backslash R} \left[(I_t + v_{out} \cdot \nabla I)^2 + |\nabla v_{out}|^2 \right] dx + \int_{\partial R} ds$$
$$+ \int_{\partial R} \left[(I_t + v_b \cdot \nabla I_{in})^2 + (I_t + v_b \cdot \nabla I_{out})^2 + |(v_b)_s|^2 \right] ds$$

subject to $v_b = v_{in} = v_{out}$ on ∂R. $\qquad (14)$

The variable s denotes the arclength parameter of ∂R, and $(v_b)_s$ denotes the derivative with respect to the arclength. We suppress the time variable in R_t and the velocities in the above energy for convenience. The integral over time has been eliminated since R_t' and the velocities at time t' are treated as independent of R_t and the velocities at time t. The formulation assumes that the images are sampled fine enough in time so that the motion/deformation between frames is small. This is realistic in cardiac MRI, which we test in the experiments. Larger deformations between consecutive frames can be handled using multiscale methods and/or ideas in [5], but we forgo this to illustrate the main concept of modeling interactions between different materials and constraints across material boundaries.

2.3 Relation of the model to existing work

The first two terms of the energy (14) are reminiscent of the energy used in Motion Competition [8] in computer vision where the objective is to segment objects that move with differing velocities. Each of the objects are described by parametric motions. We do not restrict the model of motion to simple parametric models such as translations or affine motions, as the motion of the heart is much more sophisticated, and further the velocity in our model is continuous across ∂R (although the derivatives need not be), something that has not been considered before and requires added reasoning and sophistication. Each of the first two terms of the energy (14) are identical to Horn & Schunck optical flow [11], however, we extend the model to model discontinuities in the gradient of the velocity field across (unknown) material boundaries. One can think of our model as an extension of the Mumford and Shah model [13] to deformations with an added physical constraint of continuous (and not differentiable) motion across boundaries.

The energy (9) can be thought of as a generalization of the popular Large Deformation Diffeomorphic Metric Mapping (LDDMM) [5] framework used in medical image registration. In LDDMM, only one region $R = \Omega$ is assumed, and therefore, constraints across boundaries of differing materials are not considered. The first term of (9) with the constraint (8) is similar to the energy considered in LDDMM, and regularization across material boundaries is performed – an undesirable property. Our framework can include multiple differing material regions, and these regions are estimated as part of the optimization process. Our energy (9) is also related to the Riemannian formulation of optimal mass transport methods for image registration [10]. In those methods, $R = \Omega$ and the first term of (9) that contains the term $I_t + \nabla I \cdot v_{in}$ is the mass preservation constraint used in [10] with the additional constraint of incompressible motion. The regularization of v_{in} differs in mass transport methods, however, and the penalty is on the magnitude of v_{in} rather than ∇v_{in}, as in our case.

2.4 Optimization method

We now present a method to optimize the energy in (14). The optimization of E consists of an alternating minimization in v_b, v_{in}, v_{out}, and R. First, fixing R, v_{in}, v_{out} and optimizing in v_b through variational calculus, we find the following condition, which is an ODE defined on ∂R, and ensures a global optimum for a given R:

$$-(v_b)_{ss} + (\nabla I_{in} \nabla I_{in}^T + \nabla I_{out} \nabla I_{out}^T) v_b =$$
$$-I_t (\nabla I_{in} + \nabla I_{out}) \quad (15)$$

with circular boundary conditions. The subscripts in the expression $(v_b)_{ss}$ denotes the second derivative with respect to the arclength parameter of ∂R. The above equation can be discretized as a sparse linear system and can be solved efficiently.

The vector field v_b that is defined on ∂R is the boundary condition for v_{in} and v_{out}, and with this condition, we may optimize E for v_{in} and v_{out} holding v_b and R fixed. The global optimum solution for the given R and v_b is given by the solution of the following decoupled PDE:

$$\begin{cases} -\Delta v_{in} + \nabla I \nabla I^T v_{in} = -I_t \nabla I & \text{on int}(R) \\ -\Delta v_{out} + \nabla I \nabla I^T v_{out} = -I_t \nabla I & \text{on int}(\Omega \backslash R) \\ v_{in} = v_{out} = v_b & \text{on } \partial R, \end{cases}$$
$$(16)$$

We impose the standard Neumann boundary conditions for v_{out} on $\partial \Omega$, as typical in optical flow problems. These PDE can be solved efficiently using a conjugate gradient solver as (16) forms a positive definite linear system.

Finally, holding v_b, v_{in}, v_{out} fixed, one can solve for R using an iterative technique, and for simplicity, we use a gradient descent to optimize E in R (see [6, 7]), although other methods may be used (e.g., [3]). Setting $c = \partial R$, the gradient descent equation is

$$\partial_\tau c = \left[(I_t + v_b \cdot \nabla I_{in})^2 - (I_t + v_b \cdot \nabla I_{out})^2 \right.$$
$$\left. + |\nabla v_{in}|^2 - |\nabla v_{out}|^2 + \kappa \right] \mathcal{N} \quad (17)$$

where $\mathcal{N}$ denotes the inward normal vector of c, κ denotes the curvature of $c = \partial R$, and τ denotes an artificial variable that parameterizes the evolution of c. Note that the variation of the last term in E (14) is neglected by weighting that term arbitrarily small in this last step of optimizing with respect to R.

The optimization algorithm, in summary, is given by the following steps:

1. Initialize R
2. Solve the ODE (15) on ∂R for v_b
3. Solve the PDE (16) with boundary condition v_b
4. Update the region R by an iteration of (17)
5. Repeat Steps 2–4 until convergence of R

3 EXPERIMENTAL RESULTS

In this section, we present experimental results on real cardiac MRI data to illustrate the performance of our method in determining motion and deformation of the left ventricle (LV) and surrounding areas, and also the left ventricle boundary. Region R will represent the LV and $\Omega \backslash R$ represents everything other than the LV. Therefore, we model only the material differences across the LV boundary. In these experiments, images I_1 and I_2 at two consecutive time instances are taken from a sequence of cardiac MRI images, and $I_t = I_2 - I_1$, $\nabla I = \nabla I_1$ in the energy E. The results obtained by our method are shown in Figure 1. We initialize the region R as a small seed point within the

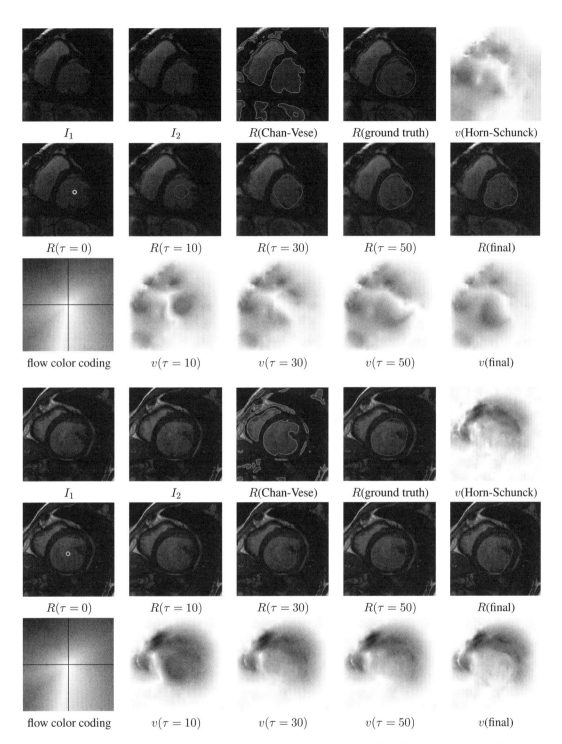

Figure 1. This figure shows the segmentation of the left ventricle from a sequence of cardiac MRI data. (Top tow) image I_1, image I_2, segmentation result by the Chan-Vese method, ground truth for the left ventricle region, optical flow by Horn-Schunck model. (Middle row) segmenting curve initialization, snapshots of the segmenting curve evolution in time and the final segmentation of the left ventricle by our method. (Bottom row) standard color coding scheme for the optical flow, snapshot of the deformation field evolution in time, and the final deformation field by our method. See text for assessment of the results.

LV, and this is hand initialized. Snapshots of the evolution of $c = \partial R$ representing the LV boundary and the deformation field in the interior, v_{in}, and the exterior, v_{out}, of the left ventricle are shown in Figure 1 for two different example pairs of images.

We provide comparison to a standard segmentation method [7], the Chan-Vese method, which assumes a piecewise constant distribution of the image intensity and two regions. The segmentation is performed on the image I_1. Also, for comparison, we provide standard optical flow results using the Horn-Schunck method [11] to estimate the deformation between the images I_1 and I_2 (not using the region R at all in the computation). In Figure 1, the top row displays images I_1 and I_2, the segmentation result obtained by the Chan-Vese method, ground truth for the left ventricle boundary, and the optical flow obtained by the Horn-Schunck model in the standard color coding scheme [1]. The curve evolution for the left ventricle boundary from the initial (left) to the final (right) is presented on the middle row. The registration result obtained by the deformation field at different various time steps, τ, is shown on the bottom row.

The experimental results demonstrate that our model provides an accurate boundary for the left ventricle and the obtained deformation field agrees with the physical motion characteristics of the left ventricle and its surrounding region. The fact that the LV is detected quite accurately indicates also that the deformation is accurate (since inaccurate deformation would lead to errors in estimation of the LV). As can be seen from the deformation recovered by our method, the discontinuity of the gradient of the deformation across the detected LV is clearly sharp and visible, unlike the standard Horn-Schunck method. Our method is shown to be robust to the inhomogeneity of the intensity distribution within the left ventricle. Indeed, the dark regions within the LV are valves, and our method captures them more accurately than segmentation based on intensity such as the Chan-Vese method. This desirable property is inherited from our model that considers the physical laws of motion and deformation and the interactions and constraints across materials of different chemical composition.

which the deformation and boundary of the substructures are simultaneously estimated. This not only achieves a more physically plausible registration (estimation of deformation), but also yields the detection of sub-structures, which are of clinical interest in many applications. In our method, the characteristic feature employed in determining the object/ sub-part boundaries is the inhomogeneity of deformation across material boundaries, and thus our method is applicable to objects with sophisticated appearance in which distributions of intensity are not discriminative to distinguish object(s)/sub-parts(s). Objects with such sophisticated appearance are typical in medical images.

To illustrate proof-of-concept, we have implemented a simplified version of our general model (9) which is given by the energy described in (14). In this simplification, we have made the assumption that the deformation across the sequence is uncorrelated, and the region describing an object/sub-part is uncorrelated in time. The simplification leads to a rather simple alternating optimization scheme. The algorithm has been demonstrated on cardiac MRI in the application of detecting the left ventricle and estimating its deformation, although the technique is general for many medical registration scenarios. We have shown that our algorithm accurately detected the left ventricle, provided the deformation that agreed with physical constraints, and estimated the deformation outside the left ventricle in the myocardium region. In comparison to standard intensity-based image segmentation, our method achieves better results, and a more physically plausible deformation. The results demonstrate its effectiveness and potential for quantitative analysis of cardiac functions. Future work includes implementing the full model (9), and testing it on 3D data.

ACKNOWLEDGEMENT

This work was supported by the Korea Research Foundation Grant (NRF-2010-220-D00078).

4 CONCLUSION

We have proposed a novel general framework for medical image registration in which several objects or sub-parts described by differing material properties are explicitly modeled and the constraints of their motion/deformation across parts are incorporated. This is ignored in prior work on medical image registration to the best of our knowledge, where a common deformation field obeying smoothness properties is assumed on the entire domain of the image, and thus ignoring physical constraints of the deformation field across substructures. As a by-product, our algorithm also automatically determines the substructures in a unified optimization framework in

REFERENCES

[1] S. Baker, D. Scharstein, J. Lewis, S. Roth, M. Black, and R. Szeliski. A database and evaluation methodology for optical flow. *International Journal of Computer Vision*, 92(1):1–31, 2011.

[2] A. Bistoquet, J. Oshinski, and O. Skrinjar. Left ventricular deformation recovery from cine mri using an incompressible model. *Medical Imaging, IEEE Transactions on*, 26(9):1136–1153, 2007.

[3] X. Bresson, S. Esedoglu, P. Vandergheynst, J. Thiran, and S. Osher. Fast global minimization of the active contour/snake model. *Journal of Mathematical Imaging and Vision*, 28(2):151–167, 2007.

[4] T. Brox, C. Bregler, and J. Malik. Large displacement optical flow. In *Computer Vision and Pattern Recognition, IEEE Conference on*, 2009.

[5] Y. Cao, M. Miller, R. Winslow, and L. Younes. Large deformation diffeomorphic metric mapping of vector fields. *Medical Imaging, IEEE Transactions on*, 24(9):1216–1230, 2005.

[6] V. Caselles, R. Kimmel, and G. Sapiro. Geodesic active contours. *International journal of computer vision*, 22(1):61–79, 1997.

[7] T. Chan and L. Vese. Active contours without edges. *Image Processing, IEEE Transactions on*, 10(2):266–277, 2001.

[8] D. Cremers and S. Soatto. Motion competition: A variational approach to piecewise parametric motion segmentation. *International Journal of Computer Vision*, 62(3):249–265, 2005.

[9] P. Ghosh, M. Sargin, and B. Manjunath. Generalized simultaneous registration and segmentation. In *IEEE Conference on Computer Vision and Pattern Recognition*, pages 1363–1370, Jun 2010.

[10] S. Haker, L. Zhu, A. Tannenbaum, and S. Angenent. Optimal mass transport for registration and warping. *International Journal of Computer Vision*, 60(3):225–240, 2004.

[11] B. Horn and B. Schunck. Determining optical flow. *Artificial intelligence*, 17, 1981.

[12] X. Liu, K. Abd-Elmoniem, and J. Prince. Incompressible cardiac motion estimation of the left ventricle using tagged mr images. *Medical Image Computing and Computer-Assisted Intervention*, 2009.

[13] D. Mumford and J. Shah. Optimal approximations by piecewise smooth functions and associated variational problems. *Communications on pure and applied mathematics*, 42(5):577–685, 1989.

[14] N. Paragios, M. Rousson, and V. Ramesh. Knowledge-based registration & segmentation of the left ventricle: a level set approach. In *Workshop Applications of Computer Vision*, 2002.

[15] S. Song and R. Leahy. Computation of 3-d velocity fields from 3-d cine ct images of a human heart. *IEEE Trans. Medical Imaging*, 10(3):295–306, 1991.

[16] G. Unal and G. Slabaugh. Coupled pdes for nonrigid registration and segmentation. In *Computer Vision and Pattern Recognition. Conference on*, 2005.

[17] A. Yezzi, L. Zollei, and T. Kapur. A variational framework for joint segmentation and registration. In *Mathematical Methods in Biomedical Image Analysis. IEEE Workshop on*, 2001.

Computational Modelling of Objects Represented in Images – Di Giamberardino et al. (eds)
© 2012 Taylor & Francis Group, London, ISBN 978-0-415-62134-2

Automated and semi-automated procedures in the assessment of Coronary Arterial Disease by Computed Tomography

M.M. Ribeiro
Scientific Area of Radiology from Higher School of Health Technology, Polytechnic Institute of Lisbon PT
New University of Lisbon- Medicine Faculty, Human Anatomy Department, Lisbon PT

I. Bate & N. Gonçalves
Scientific Area of Radiology from Higher School of Health Technology, Polytechnic Institute of Lisbon PT

J. O'Neill
New University of Lisbon, Medicine Faculty, Head of Human Anatomy Department, Lisbon PT
CEFITEC, New University of Lisbon, Science and Technology Faculty, Lisbon PT

J.C. Maurício
Euromedic, Medical Imaging Centre, Tomar PT

ABSTRACT: The aim of this paper is to explain the flow of procedures to quantify the Coronary Arterial Disease (CAD), and the mean differences to the clinical results when automated or semi-automated method is used. Will be described the disease concerning its epidemiology, imaging, the risk factors and prognosis.

The software steps as well as the values that allow predicting the risk to develop CAD will be presented.

Due the ionizing radiation damage, one experience with pig hearts was developed and used to demonstrate advantages and disadvantages of the Agatston method applying two different protocols in Multidetectors CT scan and its post processing concerning the operator-dependent variable.

1 INTRODUCTION

The Coronary Arterial Disease (CAD) is currently the leading cause of death in the developed countries [1,2,3]. Although the verified effectiveness of new available therapies to treat and to reduce the progression of CAD, still are of high variability in the risk prediction and hence its complex identification due the confluence of the risk factors such as: family history, age, sex, cholesterol levels, smoking, diabetes and hypertension. However, these risk factors cannot by itself explain more than 50% of morbidity and mortality caused by coronary artery disease [1,4].

One consequence from the risk factors is the presence of calcium into the coronary arteries and therefore the development of CAD [1,5]. The amount of calcium in the coronary arteries allow predicting the risk likelihood as well as the disease development [1,5,6]. So, there is a need of a reliable tool to identify an increase of the predictive risk for developing heart disease in the future [2]. In 1990, Agatston *et al* described, for the first time, one technique that allows assessing the amount of calcium into arteries through an electron beam CT scan [7].

With the arising of Multi-Detector Computed Tomography (MDCT) and Dual Source Computed Tomography (DSCT) techniques, the diagnostic of CAD has been facilitated in the quantification of cardiovascular risk. Calcium Score (CS) is considered the most reliable quantification technique known, regarding its high temporal resolution in the newer generation of equipments. This property allows acquiring best cardiac images free of motion artifacts. Nevertheless, there is wide evidence that the presence of high levels of calcium in coronary arteries by itself is enough to predict the risk of CAD. [4,8–10].

Thus, the clearest potential of cardiac CT is to provide CAD screening in asymptomatic patients, as well as to assure CAD risk prevention for now and in ten years (with correlation with other risk factors), to determine the prognosis in patients with established disease and for therapeutic guidance [4].

With the development of new MDCT equipments, the previous technique to detect CS has been replaced. Even though the temporal resolution is lower due to configuration, type and number of detectors, their accurate spatial resolution allows detection of smaller lesions [8–9,14].

The increase of Temporal Resolution reduces motion artifacts of the heart and High Spatial Resolution allows to evaluate a large number of segments, through a detailed explanation of the coronary anatomy [14,16–17].

Table 1. CS values and risk of CAD correspondence, according to Agatston method.

Calcium Score units	Risk level of DAC
0	Without risk evidence
1–10	Minimum risk
11–100	Middle risk
101–400	Moderate risk
>401	High risk

The new technology MCDT with 64 detectors and dual source (two xRay tubes) without the use of beta-blockers helped to keep the values of spatial and temporal resolution in patients with high heart frequency. [4,14–16].

CS provides a quantitative assessment of the calcium plaques in coronary arteries [1]. Also provides information and helps the decision to choose the eligible's candidate for Angio CT with contrast media injection. [4,17].

Risk prediction of CAD by Agatston method is in accordance with the follow reference values shown in table 1.

1.1 Calcium score application: semi automated and automated methods

The CS software identifies areas with calcium, with a density of 130 HU or more, and gives the total of CS amount based on the values of each branch coronary artery (automated option); however this value can be changed by operator in the called semi automated method.

In this last option the areas in the range of this density are pink-colored, (Fig. 1 – top left) and automatically the bone density is excluded enabling a fast quantification of areas likely to injury. In the following step, through free hand mark tool appliance, the operator selects a ROI (Region Of Interest) around the vessel in study and close the circle. Immediately the ROI becomes yellow to differentiate from the other anatomic structures with the similar HU. It is intended that the yellow zone (zone of assessment) is filled and embedded in the calcic densities as most as possible. Subsequently, there are available tools to select the main areas of injury at the coronary branches (Right Coronary Artery-RCA, Left Main-LM, Left Anterior Descending-LAD and circumflex-CX). To perform an accurate assessment and include the total amount of calcium, the operator should observe the images one by one and adjust (enlarging, reducing or replacing) the covered area. Each vessel is labeled and identified with a different color.

The figure 1 represents the screen where we can see the general workflow. Through images, we are able to select the vessels to include (top left), two images to plan the procedure and locate the coronary arteries branches way, one in coronal view (lower left) and other in transversal view (top right). The table

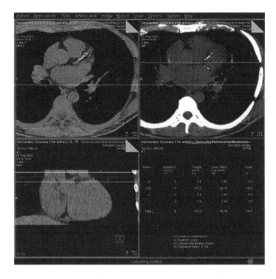

Figure 1. Image of the screen during the procedure.

Figure 2. Example of summary table values.

summary with values of CS obtained in this case is enlarged in figure 2.

The images to plan the procedure may be replaced by 3D MIP or VRT (Volume Rendering Technique) images to give extra information.

In case of smaller lesions, it is allowed by the operator to change the limit of 130 HU to a lower value, in order to include lesions with less density.

The CS application, after to measure the areas with calcium, automatically selects and shows the amount of calcium associated with these preset areas (Fig. 2) and determines the risk of CAD for a period of 5 to 10 years. Furthermore, it is possible to add other risk factors such as cholesterol levels, blood pressure and body mass index, among others, to accurately estimate the risk of scale GRF (Glomerular Filtration Rate) [19]. CS is a simple technique, relatively inexpensive, as well safe. The radiation dose is low and does not involve the use of iodinated contrast agents.

The automated CS application has the advantage that automatically selecting the anatomic areas, generally with a density of 130 HU and enables a fast quantification of areas likely to injury. Despite the

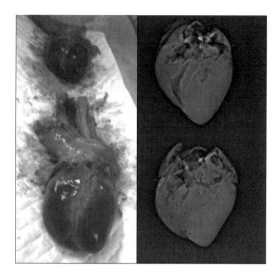

Figure 3. Pig hearts models are shown on the left. (Courtesy of Simões & Galapito, Lda). The test image by Xray is shown on the right side.

semi automated option be more time-consuming is considered be more reliable.

It is intended that the assessment zone should be filled as evenly as possible and this effect only can be efficient when clinically assessed by the operator.

In figure 2, we can observe the summary values of the lesion's number, in which the total volume under study was based in: a) the calcium score based on the Agatston equivalent; b) the isotropic interpolated volume and c) the calibration factor in equivalent of mass from mg of Calcium Hydroxyl Apatite to each coronary branches.

2 METHOD

There was a quantitative, quasi-experimental and cross-sectional study. Was evaluated the Acquisition Protocol measured by the mAs – independent variable.

As dependent variables, we considered CS and CTDI (Computed Tomography Dose Index) values measured in mGy. In this case, factors such as thickness and/or gap between slices that we know that could influence CTDI and CS values were fixed.

The research questions were: 1- "The variation in the protocol acquisition, with the aim of reducing the dose of radiation received by the patient, through manipulation of mAs, can modify the values of the CS while keeping the reliability of the technique in assessing the risk of Coronary Heart Disease?; 2- Is there a similarity in semi-automated and the automated method concerning the values of CS obtained?".

Concerning the ethical procedures safeguard and biological effects from the ionizing radiation of the CT scanners, this study was developed with pig hearts as phantoms (Fig. 3) – Heart 1 and Heart 2. Recently research studies in cardiovascular disease have shown success with an experimental pig heart [13].

Table 2. Protocols' parameters applied by each heart.

Protocols	Electric Current	Voltage in the Rx tube	Acquisitions	Applied to the hearts
A	128 mAs	120 kVp	1	1 and 2
B	64 mAs		2	
C	256 mAs		3	

Each phantom weighted 1.234 kg (Heart 1) and 1.118 kg (Heart 2). The two hearts were carefully observed. Criteria such as the cardiac anatomy unchanged and arterial branches easily identified, were considered. An initial x-ray was obtained to assure that both hearts haven't any calcium at the beginning of experiment.

Swinddle (1983), cited by Mariano (2003), [21] states that the use of pigs in experimental research is a practice already quite old. From the literature reviewed it seems well accepted that the cardiovascular anatomy of pig is identical to human. [13.21].

It was found that the pig heart has an apex and base of the heart with characteristics comparable with those of the human heart.

In this study, it was felt appropriate to evaluate the effectiveness and reliability of the CS in the prediction of CAD, trying to find a compromise between these values and the radiation dose to the patient, measured by CTDI. The CTDI is the most commonly used dose indicator required by EU regulations [11].

For each protocol acquisition resulted of 48 images. Thus, and for the three protocols were performed three acquisitions, resulting a total of 864 analyzed images. For data collection a MDCT scanner Dual Source of 64 row detectors from Siemens Medical Systems, model SOMATOM® Definition, was used, and equipped with the software Syngo Calcium Score®.

To simulate the calcium plaques inside vessels, two Calcioral® pills crushed in powder and mixed with 2 ml of physiological saline were injected into each of the major coronary branches to simulate calcic lesions. The pills dose was 1250 mg of calcium carbonate, equivalent to 500 mg of calcium. The density of the compound was tested for beam X-rays attenuation through the acquisition of a radiogram as well as a CT scan. Only the variable mAs of the acquisition protocol were modified, from 128 mAs values – protocol A, 64 mAs – protocol B and 256 mAs – protocol C. (Table 2). The thickness has 3 mm, and 3 mm gap. The two hearts in same time were placed in the longitudinal direction (Fig.3). The other fixed parameters of the protocols are: the rotation time (0.33 s); Acquisition (6 × 3.0 mm); Slice collimation (3.0 mm); Slice width (3.0 mm); Feed/scan (18.0 mm); Kernel (B35f) and Temporal Resolution (165 ms).

Heart number 1 was injected in the arteries LM, LAD and RCA. Images were acquired thought the protocols A, B and C.

The CS value was determined by the Agatston method using a limit HU 130. The process was replicated to give more reliability to the study and the

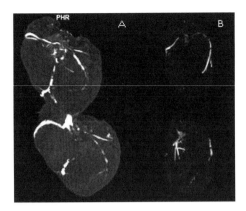

Figure 4. A – MPR in two different views, and B – 3D with VRT.

Table 3. CTDI values by each protocol (A, B, C) and heart 1 and 2.

CTDI (mGy)					
Protocol A (128 mAs)		Protocol B (64 mAs)		Protocol C (256 mAs)	
Heart		Heart		Heart	
1	2	1	2	1	2
0.43	0.43	0.44	0.44	0.89	0.87

values shown are the ones found by average concerning the three acquisitions. The same procedure was performed for the heart 2, being, however, the vessels groups studied: LAD, CX and RCA.

To calculate the variability intra-observer, we conducted a second reading for each image. Finally, there was acquired a Cardiac CT in two hearts simultaneously for a possible 3D images reconstruction, Multiplanar Reconstructions (MPR) and Volume Rendering Technique (VRT) (Figures 4A and 4B).

In CS calculation with semi-automated method is the operator that selects and analyzes the images of the coronary arteries, by visual observation of calcium density. In this case the application of the ROI(s), to determinate the HU values can be necessary to help in the decision about the lesion's composition. When the automatic method is used, the computer automatically selects the images and applies the value (usually 130 HU) to calculate the amount of calcium present. In 50 of 864 images, randomly selected, the automatic method was applied to determine whether there were significant statistics differences between the two methods.

The data were analyzed using Microsoft Excel 2007® and SPSS® for windows. Bar graphs and tables were presented; allowing to establish a better evaluation between CS, CTDI and mAs. Finally, were analyzed the presence and type of relationship between the CS and CTDI variables. The aim was to determine what is the variation, into the reliable limits, of CS obtained by protocols B and C related to Protocol A, and at the same time, looking for a decreasing in the of CTDI value.

3 RESULTS

Only the images when the density values were ≥130 HU were analyzed for CS evaluation. Regarding the assessment of intra-observer variability, the obtained values showed differences in average, lower than 1 unit and therefore these differences were not considered.

The biggest difference between observers was 7 units and was verified only in one image.

Trough graph 1, we can visualize that CS total value corresponding to heart 1 has a gradual increase, with the increasing of mAs. Using Protocol B, CS values were lower, assuming an inverse behavior concerning the Protocol A and still exceeding Protocol C values. Thus, the total amount of CS applying Protocol A obtained a mean of 1069.5; 997.3 using Protocol B and 1103.73au, using Protocol C.

In the heart 2 (graph 2), CS highest value was obtained through Protocol A. Protocol B and C values were quite similar and lower than A (averaging 1311.8, 1058.8, and 1064.8au respectively). In the heart 1, CS values in LM and LAD showed a maximum difference of 8.1 and 27.96au, respectively.

Concerning CTDI values (Table 3), there are some differences, although not significant between protocol A and B when applied in the hearts 1 and 2. Using Protocol A, 0.43 mGy in both hearts were readable. Using Protocol B, 0.44 mGy values were recorded in both hearts, a little bit more than using Protocol A but however negligible. Using Protocol C, higher differences were recorded than using protocols A and B (0.89 mGy and 0.87 mGy) in the heart 1 and heart 2, respectively) For CS/CTDI analysis, this study found that this relation hasn't a linear variation. In the heart 1, for 0.43 mGy (Protocol A) corresponded higher values of CS, compared to a radiation dose of 0.44 mGy (using protocol B). In turn, CS values increased again when using Protocol C, which corresponds to a higher value of CTDI – 0.89 mGy. There is not a linear tendency in the relationship between these variables in both hearts. When CTDI values increase, CS values oscillate corresponding in the start, decreasing after and then increasing back again.

Concerning semi-automatic versus automatic analysis, of the 50 in 864 images analyzed, with this last method it was revealed that the differences obtained in CS values were statistically significant with $p = 0.0038$.

4 DISCUSSION

Concerning the relationship between mAs and CTDI it would be expected that by halving the mAs, the dose

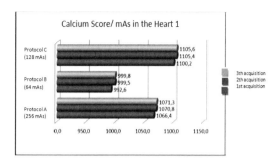

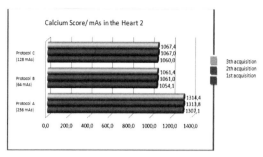

Graph 1. In the Heart 1: CS/mAs by acquisitions in the tree protocols (A,B, C).

Graph 2. In the Heart 2: CS/mAs by acquisitions in the tree protocols (A,B, C).

would be significantly lower. However, using Protocol B, 0.44 mGy values were obtained in both hearts, higher values compared to the results obtained using Protocol A.

This may be due to CareDose4D software, an automatic exposure control which calculates the optimal radiation dose from the mAs, defined by the operator.

The CareDose4D automatically adjusts these values to the effective's mAs taking into account the volume of the anatomic area under study and the different dose attenuation. According to these data, the CareDose4D, based on a profile or face scout view previously acquired, automatically handles mAs with each tube rotation. Its function is mainly to protect the patient from over-exposure, while maintaining a commitment to image quality [23].

After analyzing CS values, a considerable variation between the two hearts is observed. It is also possible to assign many reasons for these results. In the case of the heart 1, using Protocol B, with less mAs, it was found less calcium measured such as expected. Furthermore, when using Protocol C, calcium values were higher. This may be explained by the fact that to an increase of mAs corresponds an increase in the spatial resolution of the image and a better definition of the lesion, thus allowing more effective detection of the calcium [22,23].

However, summarizing data obtained from heart 2, the three protocols application contradict this hypothesis, since the protocol A shows higher value of CS. The results of this study are against to the results of Shemesh et al. (2005) [10]. The authors conducted a research to compare the CS values obtained with two protocols of 55 mAs and 165 mAs in a sample of 51 individuals which was underwent to cardiac CT. They found that there were no significant differences in the amount of CS, as well as for each artery, when comparing the two protocols applied. Although the results did not meet the expected, based on the CS values obtained, if the study were applied to patients, it would be possible to allocate all high risk of CAD in the next 5–10 years, according to the classification of Agatston, since in both hearts and in all acquisition protocols, the values of CS obtained were higher than 400.

Also the presence of an automatic exposure control CareDose4D- can constitute a limitation by the influence in the dependent variables.

Similarly, the absence of factors such as the heart rate, respiratory motion, bone and surrounding soft tissue may influence the results obtained when a comparison with established tests is performed on humans. In contradiction the fact that there is no motion by the hearts may contributes to the reliability of experience.

5 CONCLUSIONS

The presence of calcium in coronary arteries is a reliable indicator to the coronary artery disease [1,5] and is very important to have a consistent tool to detect and quantify the calcium, predicting the risk of this disease. The current equipment for DSCT have an application, the CS, which allows to evaluate the calcium present in the coronary branches, fast and noninvasively [4,8–10]. We concluded that the CS values increased linearly with mAs values. It can be used as a screening test and this is very important due the radiation dose is reduced to the minimum as possible (ALARA principle) by keeping a balance between the reliable quantification of calcium, allowing a correct diagnosis. By changing parameters such as kVp and mAs, it is possible to reduce the dose levels. [10,11,12].

We found that is not beneficial for the patient to be exposed to the Protocol C (64 mAs), since the radiation dose tends to increase and it is not favorable to CS detection. Thus, as we said above, the CS's values decreases with the decrease of the mAs. Having regard to the value obtained in the RCA of the heart 2, the protocol could have been applied in a 3rd heart to complement the study, to stabilize the sampling error and corroborate the results of heart 1.

Regarding the change of variables, it is now clear that the manipulation of the technical parameters, namely the mAs, aims to protect the patient, however, we can contribute to a higher dose and even to a decrease in diagnostic image quality as we found in Protocol B, which resulted in a higher radiation dose.

We found that the protocol A (128 mAs), used as a standard in this hospital, is the most efficient one.

With trying to test the influence of the variables that determine the radiation dose, we can unwittingly harming the patient by the radiation dose parameter or due a misjudgment in the CS values. Regarding automatic and semiautomatic methods from this evaluation, we concluded that the semi-automated method produces more reliable results but nevertheless requires great and more experience by the operator while spends more time.

This study has a great limitation of the CTDI variation may be higher when we scan human cardiac CT scan due to the synchronization by tracking retrospective ECG which leads to an increase in the total examination time.

We suggest to carry out more studies without the interference of Care Dose4D and to compare the automatic and semiautomatic methods in larger samples with 10 observers at a minimal.

DISCLOSURE

The authors declare no conflicts of interest that may influence the results obtained in this study.

REFERENCES

[1] Boyar, A. EBT Coronary Calcium Scoring Guide. In: Advance Body Scan Newport's. 2004.

[2] Jakobs T, Wintersperguer B, Herzog P, Flohr T, Suess C, Knez A, Reiser M, Becker C. Ultra-low-dose coronary artery calcium screening using multislice CT with retrospective ECG gating. Eur Radiol. 2003; 13:1923–1930.

[3] Fishman E. Multidetector-Row Computed Tomography to Detect Coronary Artery Disease: The Importance of Heart Rate. Center for Bio-Medical Communication. 2005.

[4] Wann S. Cardiac CT for risk stratification. Applied Radiology. 2006; 40–47.

[5] Jasinowodolinski D, Szarf G. Escore de cálcio na Avaliação Cardiovascular Paciente com diabetes. Arq Bras Endocrinol Metab.2007;51/2: 294–98;

[6] Schmermund A, Mohlenkamp S, Erbel R. Coronary artery calcium and its relationship to coronary artery disease. Cardiol Clin. 2003; 21(4): 521–534.

[7] Agatston A, Janowitz W, Hildner F, Zusmer F, Vilamonte M, Detrano R. Quantification of Coronary Artery Calcium Using Ultrafast CT. J Am Coll Cardiol. 1990;15:827–832.

[8] Groen J, Greuter M, Vliegenthart R, Suess B, Schmidt B, Zijlstra F, Oudkerk M. Calcium scoring using 64-slice MDCT, dual source CT and EBT: a comparative phantom study. Int J Cardiovasc Imaging. 2008; 24: 547–556.

[9] Sousa N. Estudo das correlações entre tecido adipose visceral abdominal e epicárdico, aterosclerose coronária e níveis circulantes de células progenitoras endoteliais. Mestrado em Medicina e Oncologia Molecular. FMPorto. 2009.

[10] Shemesh J, Evron R, Koren-Morang N, Apter S, Rozenman J, Shaham D, Itzchak Y, Motro M. Coronary Artery Calcium Measurement with Multi-Detector Row CT and Low Radiation Doses: Comparison between 55 and 165 mAs. Radiology. 2005; 236: 810–814.

[11] Lee C, Goo F, Lee H, Ye S, Park C, Chun E, Im F. Radiation Dose Modulation Techniques in the Multidetector CT Era: From Basics to Pratice. RadioGraphics. 2008; 28(5): 1451–1459.

[12] Mahnken A, Wildberger J, Simon J, Koss R, Flohr T, Schaller S, Gunther R. Detection of Coronary Calcifications: Feasibility of Dose Reduction With a Body Weight-Adapted Examination Protocol. AJR. 2003; 181:533–538.

[13] Crick S, Sheppard M, HO S, Gebstein L, Anderson R. Anatomy of the pig heart: comparisons with normal human cardiac structures. J. Anat. 1998; 193:105–119.

[14] Fishman E. Cardiac CT: Where are we today and where are we going?. Applied Radiology. 2006:5–9.

[15] Serra D, Martins J, Barata S, Magalhães J, Ouro M, Janeiro L. Valor prognóstico da angiografia por tomografia computorizada na avaliação da doença coronária. Saluitis Scientia-Revista de Ciências da Saúde de ESSCVP. 2009; 1: 41–52.

[16] Mesquita A. A acurácia da TC de múltiplos detectores dual-source no dignóstico da doença arterial coronariana: revisão sistemática. Universidade do Estado do Rio de Janeiro, Centro Biomédico- Instituto de Medicina Social. 2010.

[17] Jacobs J. How to perform coronary CTA: A to Z. Applied Radiology. 2006; 10–21.

[18] Radiological Society of North America. Cardiac CT for CS. In: RadioloyInfo.org. 2011.

[19] Siemens AG. Syngo Calcium Scoring. In: Siemens AG Medical Solutions. 2002–2011: Available from: http://www.siemens.com/entry/cc/en/.

[20] Schanaider A, Silva P. Uso de animais em cirurgia experimental. Acta Cir Bras. 2004; 19(4):441–447.

[21] Mariano M. Minisuíno (minipig) na pesquisa biomédica experimental. O Minipig br1. Acta Cirúrgica Brasiliera. 2003; 18(5): 387–391.

[22] Siemens AG. Cardiac CT Apllication Guide. In: Siemens AG Medical Soluitions. 2007.

[23] Siemens AG. SOMATOM Definition. Application Guide. In: Siemens AG Medical Solutions. 2007.

Computational Modelling of Objects Represented in Images – Di Giamberardino et al. (eds)
© 2012 Taylor & Francis Group, London, ISBN 978-0-415-62134-2

Image-analysis tools in the study of morphofunctional features of neurons

D. Simone & A. Colosimo
Department SAIMLAL – Sapienza, University of Rome, Borelli, Roma, Italy

F. Malchiodi-Albedi
I.S.S. – Deptartment of Mol. Biol. and Neurosciences, Regina Elena, Rome, Italy

A. De Ninno, L. Businaro & A.M. Gerardino
CNR-Institute for Photonics and Nanotechnology, Cineto Romano, Rome, Italy

ABSTRACT: We focused on the quantitative assessment of shape changes observed in neuronal populations upon chemical stimulation. In particular, we studied pure cultures of rat hippocampal neurons, after fixation, by fluorescence microscopy. The number of dendrites in CNF1 treated hippocampal cells was substantially reduced as compared to controls, and significant changes were observed in the shape of the cells body. Our results, obtained by systematic use of image analysis software tools, underline the importance of quantitative cell morphometry to clarify detailed structure-function relationships, a critical step in the design of reliable mechanistic models. The quite similar and promising approach concerning the analysis of high-resolution images of alive and mobile cells grown in micro-devices, is briefly discussed.

1 INTRODUCTION

The morphometric techniques mainly used in Cellular Biology to characterize cells and subcellular organelles have been often limited to the estimate of surface area and perimeter (Petty 2007). However, the huge amount of knowledge accumulated in the last twenty years on both morphoanatomical features and physiological/pathological behaviour of cell populations, has been largely due to some substantial improvements in the image processing techniques. Such improvements, besides the resolution power of the optical and microscopical instrumentation, concern the reliability, power and friendliness of the data analysis algorithms/procedures (Shamir et al. 2011). In this context, a number of specialized software tools have been made recently available, with particular emphasis on the image analysis of different cell types from the central nervous systems, namely neurons and microglia. If, on top of that, one adds the continuously growing popularity of measurements carried out in the finely controlled and realistic conditions provided by nanodevices (De Ninno et al. 2010), it becomes worth of attention any experimental strategy taking advantage of the synergic combination of both sets of tools.

In the present contribution we exemplify the use of algorithms and morphometric parameters appropriate to analyze some typical shape modifications of excitable cells during chemical or physical stimuli. We evaluate the circularity of the cell body (the

increase of circularity is a key morphological index of tumoral cells) and the body elongation (a common shape change often associated to increased motility). The study has been carried out on the basis of cellular images obtained by immunofluorescence techniques, since a measure of the fluorescence intensity also provides a direct indication of the average cellular density and of its spatial distribution in mobile cellular populations. As a matter of fact, a quite similar techniques can be used to characterize cell movements in terms of velocity and directional migration (Solecki et al. 2009).

It is worth considering our approach also in the light of the increasing number of problems tackled by seeding and stabilizing cells in specific microdevices. The aim is to follow the morphological changes ensuing to carefully planned chemical and physical stimulation even at the level of single (or small number of) cells (Lindstrom & Andersson-Svahn 2010).

2 MATERIALS AND METHODS

2.1 Cell cultures

Pure neuronal cultures were obtained from the hippocampus of Wistar rat (Charles River) embryos at gestational day 18, as previously described (Charles River). The cultures were treated with CNF1[1] for 14

---

[1] Courtesy of Dr. Carla Fiorentini (ISS, Rome)

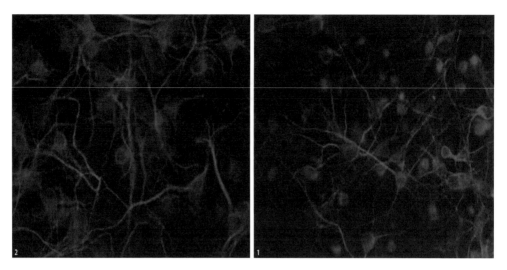

Figure 1. *Primary culture of hippocampal neurons.* Left: Hippocampal cells subjected to the action of CNF1 for 14 days. Right: Control cells labeled with MAP2.

days and fixed on day 14 of growth. After fixation, pure hippocampal cultures were permeabilized in Triton X-100 (0.2%) and incubated with monoclonal antibodies anti Microtubule Associated Protein 2 (MAP-2). After washing in PBS, Alexafluor 546-conjugated goat anti-mouse IgG were used as secondary antibodies (Molecular Probes).

2.2 *Image analysis*

The cultures were observed by a fluorescence microscope (Eclipse 80i Nikon; Nikon), equipped with a Video Confocal system1. The images of the MAP-2 marked cells (Figure 1) were imported into the Image J software, and analyzed as 8-bit images (Figure 2-top). Background artifacts (if present), were digitally removed and images appropriately thresholded in order to put in evidence the binary images of neurons. The images were subjected to further analysis focusing on the size of the soma and the branching complexity of the dendritic arbor. The size of the soma area (SA) was estimated after disconnecting the primary dendrites from the cell bodies (Figure 2-bottom). The area occupied by the dendrites (DA) was measured by subtracting the total area of cells soma from the total area marked with MAP-2. The histograms shown in Figure 3 refer to the means of SA (a), DA (b) and the whole cell surface, namely SA + DA (c).

3 RESULTS

In order to check the efficacy of our quantitative estimate of cellular morphometric features under different conditions, we considered some typical images of neuron populations (see figure 1) produced in the study of the effect of a toxic agent named CNF1 (Malchiodi-Albedi et al. 2012). Malchiodi and collaborators recently demonstrated that CNF1 profoundly

remodeled the cytoskeleton of hippocampal and cortical neurons, which showed philopodia-like, actin-positive projections, thickened and poorly branched dendrites, larger neuronal cell bodies, and a decrease in synapse number, suggesting a block of neuronal differentiation.

In a study of the CNF1 effects on neuronal differentiation, a dendritic tree growth and synapse formation – which, in general, are strictly modulated by Rho GTPases – were also observed. Thus, in the analysis of images of neurons treated with CNF1 we focused on the size of the soma (SA) and on the branching complexity of the dendritic arbor (DA). At least 30 cells in each image and at least 20 images for both control and treated cells were considered. Under these conditions we first determined in each image the values of the following parameters: average Soma Area (SA), average Dendritic Area (DA) and average total area (positive to the MAP-2). The mean values calculated over the analyzed images are shown in Figure 3. Treatment with the toxin produced significant changes in the immunoreactivity for the MAP-2 through an increase of the total SA and of the total area positive to the MAP-2. The Mann Whitney test indicates a significant difference for the SA ($p < 0.03$) but not for the whole cell area ($p = 0.11$).

4 DISCUSSION AND CONCLUSIONS

4.1 *Biological considerations*

Our analysis based on images of neuronal cells culture treated for 14 day with CNF1 and then fixed, was based on the sophisticated plug-ins included in the ImageJ software (Abramoff et al. 2004). The results confirmed the significant difference in the soma area ($p < 0.03$). On the contrary, no significant differences were found in the whole cell area ($p = 0.11$) and dendritic area

Figure 2. *Analysis of the fluorescence image of Figure 1- Right.* Left: Binarization of the image. Right: ROI (Region Of Interest) extraction.

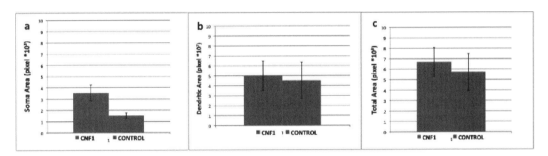

Figure 3. *Shape changes induced by CNF1 in primary cultures of hippocampal neurons.* Color code: blue = control cells; red = treated cells. The histograms rappresent the mean values of: a) soma area (SA); b) dendritic area (DA); c) total area (positive to the MAP2); The significance levels of differences assessed by the Mann-Withney test were: $p < 0.03$ (SA); $p = 0.12$ (SA + DA); $p = 0.88$ (DA).

($p = 0.88$) by the increased diameter of the dendrites induced by the toxin. These findings, apparently in contrast with the effect of CNF1 described in the literature (Malchiodi-Albedi et al. 2012, De Filippis et al. 2011), can be explained by the increased diameter of the dendrites induced by the toxin.

4.2 *Shape and motility changes of single cells on microdevices*

The opportunities offered by micro- and nano-technology to biological investigations are significantly amplified by reliable and robust techniques able to analyze high resolution images. Moreover, it is difficult to overestimate the importance of time-resolved observations in order to understand the physico-chemical mechanisms modulating the behavior of excitable cells (e.g. neurons and miocytes) down to the single-cell level. Figure 4 shows the basic architecture of a microdevice designed to explore the

role of controlled diffusion of chemical effectors (e.g. drugs) from their own reservoir to the cell compartment through a grid of microchannels. Microchannels play a central role in such an architecture, since: 1) the diffusion-driven flux of chemical effectors from the reservoir to the cell-hosting compartment, can be finely tuned by concentration, temperature and density gradients; 2) mechanically confined microenvironment seem to influence both shape and motility of migratory cells such as cancer cells (Irimia & Toner 2009) even in the absence of any chemical effector.

The latter point, in particular, is clarified by Figure 5, in which the shapes of individual cells is apparently depending on their location over the microchannels grid. To provide a quantitative estimate of the shape changes the following two indexes can be used: Circularity Ratio (CR) and Aspet Ratio (AR). CR is defined as the ratio of the area of the considered shape over a circle having the same perimeter. From a previous estimate of the Soma Area (see the Results

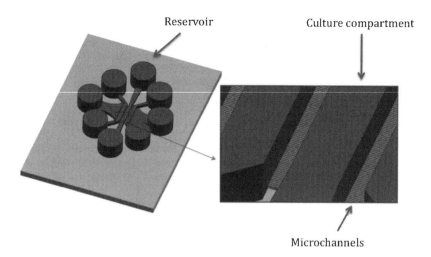

Reservoir   Culture compartment

Microchannels

Figure 4. *Multichamber microfluidic culture platform design for real-time cell analysis.* The platform is fabricated in Poly-dimethylsiloxane (PDMS), a biocompatible elastomer, following well standard lithography and replica molding procedures (De Ninno 2010). Cell culture compartments are 1 mm wide, 8 mm in length and 100 µm high. They are connected via sets of channels each with dimensions: width = 12 µm, length = 500 µm, height = 10 µm. The circular wells (8 mm in diameter) serve as loading inlets and cell medium reservoirs for nutrient and gas exchange.

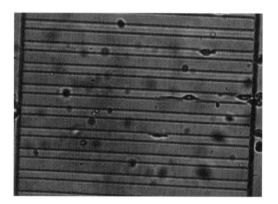

Figure 5. *High resolution image of microchannels' grid shown in Figure 4 (in green).* For the channels size, see the legend to Figure 4. The dark spots of different shape above the channels, are single glial cells from the CNS.

and Figure 2) this parameter is given by $CR = 4AP^{-2}$ where A is the soma area and P is the perimeter. AR is defined as the ratio between the major axis and the minor axis of the particleÕs fitted ellipse (Jelinek et al. 2011) and represents a measure of how the analyzed shape is elongated.

Finally, it should be underlined that an experimental set-up of the type shown in Figure 4 opens the door to quantifying the motility of alive cells from time-lapse recordings at the level of individual somata. Time-lapse images acquired at fixed time intervals are now being analyzed by both manual and automatic tracking tools available in ImageJ, for a reliable assessment of the physical and chemical parameters influencing different cell types with distinct migratory phenotypes.

REFERENCES

Abramoff, M., P. Magalhaes, & S. Ram (2004). Image processing with imagej. *Biophotonics International 11(7)*, 36–42.

De Filippis, B., A. Fabbri, D. Simone, R. Canese, L. Ricceri, F. Malchiodi-Albedi, G. Laviola, & F. C. (2011). Modulation of rhogtpases improves the behavioral phenotype and reverses astrocytic deficits in a mouse model of rett syndrome. *Neuropsychopharmacology 37*, 1152–63.

De Ninno, A., A. Gerardino, G. Birarda, G. Grenci, & L. Businaro (2010). Top-down approach to nanotechnology for cell on chip applications. *Biophysics and Bioengineering Letters 3*, 1–22.

Irimia, D. & M. Toner (2009). Spontaneous migration of cancer cells under conditions of mechanical confinement. *Integr Biol. 1*, 506–512.

Jelinek, H. F., D. Ristanovic, & N. Milosevic (2011). The morphology and classification of alpha ganglion cells in the rat retinae: A fractal analysis study. *Journal of Neuroscience Methods 201*, 281–287.

Lindstrom, S. & H. Andersson-Svahn (2010). Overview of single-cell analyses: microdevices and applications. *Lab Chip 10*, 3363–3372.

Malchiodi-Albedi, F., S. Paradisi, M. D. Nottia, L. F. M. G. C. A. C. F. A. D. Simone, S. Travaglione, & F. C. (2012). Cnf1 improves astrocytic ability to support neuronal growth and differentiation in vitro. *PlosOne 7*, 1–12.

Petty, H. (2007). Fluorescence microscopy: Established and emerging methods, experimental strategies, and applications in immunology. *Microscopy Research And Technique 70*, 687–709.

Shamir, L., J. Delaney, N. Orlov, M. Eckley, & I. Goldberg (2011). Pattern recognition software and techniques for biological image analysis. *PLoS Comp. Biology 6*, 1–10.

Solecki, D., N. Trivedi, E. Govek, R. Kerekes, S. Gleason, & H. M.E. (2009). Myosin-2 motors and f-actin dynamics drive the coordinated movement of the centrosome and soma during cns glial-guided neuronal migration. *Neuron 63(1)*, 63–80.

GPU computing for patient-specific model of pulmonary airflow

T. Yamaguchi, Y. Imai, T. Miki & T. Ishikawa
Tohoku University, Sendai, Japan

ABSTRACT: We propose an implementation of Lattice Boltzmann (LB) method on GPUs for simulating airflow in pulmonary airways with complex branches. An adaptive meshing method is developed for optimizing memory accessing, where the global domain comprises unstructured subdomains, while the local subdomain consists of a structured grids. We also develop a multi-GPU computing method based on a domain decomposition. For strong scaling tests with a subject-specific geometry (12 million LB nodes), the performance on 8 GPUs is approximately 200 GFLOPS, which is 100 times faster computation than 8 CPU cores.

1 INTRODUCTION

Advancement in medical imaging technology has enabled us to easily obtain precise geometry data of pulmonary airways. Computational fluid dynamics (CFD) with such geometry data provides subject-specific simulation of pulmonary airflow. It would be very helpful for the diagnosis and treatment of respiratory diseases. For example, it can be used to identify airflow features in each patient, and also be applied to designing inhaled drug delivery.

Numerical simulations of pulmonary airflow and inhaled particle transport have been performed by several researchers, using either an idealized model (Kleinstreuer & Zhang 2009, Zhang et al. 2009) or a realistic model based on multi-slice computer tomography (CT) images (van Ertbruggen et al. 2005, Lin et al. 2007, Gemci et al. 2008, Ma & Lutchen 2008, Inthavong et al. 2009, De Backer et al. 2010, Comeford et al. 2010, Yin et al. 2010, Miki et al. 2011, Lambert et al. 2011). However, pulmonary airways are very complex geometry with multi-level bifurcations, then a huge number of computational mesh is needed to resolve the flow field in all bronchi. Computational time is a barrier to using such simulations more practically. A major problem is the convergence of a Poisson equation. Finite volume and finite element methods are often used for simulating pulmonary airflow. These methods employ semi-implicit time integration, where Poisson equation for pressure field is solved by an iterative solver. However, in the case of flow in vessels, the convergence of the iterative solver is very slow because of Neumann boundary condition at the wall.

To avoid the convergence problem, we use an explicit method, lattice Boltzmann method (LBM) (Chen & Doolen 1998, Aidun and Clausen 2010) in this study. We proposed a novel patient-specific method of modeling pulmonary airflow using the LBM on graphics processing unit (GPU) (Miki et al.,

in press). A GPU has hundreds of streaming processors and is attached to very wide bandwidth device memory. Recent studies have reported the capacity of GPU computing with the LBM (Zhao 2008, Kuznik et al. 2010, Tölke 2010, Wang & Aoki 2011). For further efficient implementation, we developed an adaptive meshing method to optimize memory accessing. This method was implemented on multi-GPUs with message passing interface library (MPI). In this paper, we present the method for multi-GPU computing of pulmonary airflow.

2 METHODS

2.1 Lattice Boltzmann method

In the LBM, a lattice BGK equation is solved:

$$f_i(t + \Delta t, \mathbf{x} + \mathbf{c}_i \Delta t) =$$
$$f_i(t, \mathbf{x}) + \frac{1}{\tau}\left\{ f_i^{eq}(t, \mathbf{x}) - f_i(t, \mathbf{x}) \right\}, \quad (1)$$

where f_i refers to the particle distribution, t is the time, Δt is the time interval, $\mathbf{x}$ is the position vector, c_i is the the particle velocity, τ is the relaxation time, and the superscript eq denotes equilibrium.

2.2 Implementation on GPU

Computational performance of the LBM is mainly restricted by memory accessing. In particular, for implementation on GPU, it is important to consider coalesced memory access and less access to GPU device memory. This may be achieved easily on a simple rectangular domain with Cartesian grids. However, it is not economical for pulmonary airways, because the pulmonary airways are a set of long branches,

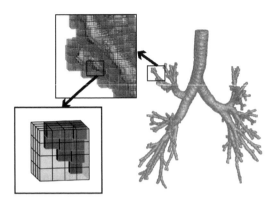

Figure 1. Adaptive meshing method. An airway model is decomposed by isotopic subdomains. Each subdomain consists of a structured Cartesian mesh.

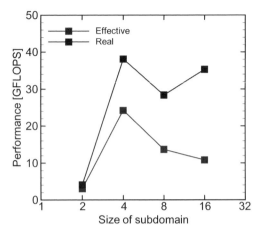

Figure 2. Effect of subdomain size on performance.

which occupies a very small volume in the rectangular domain. Therefore, we propose an adaptive mesh method as shown in Fig. 1. The airway geometry is decomposed into small isotropic subdomains consisting of Cartesian grids. Each subdomain includes information on its connectivity to the neighboring subdomains. Consequently, the global domain comprises unstructured subdomains, while the local subdomain consists of a structured grids. A GPU code is developed with the compute unified device architecture (CUDA). We compute each subdomain in a CUDA block, in which the block is a group of processes consisting of CUDA threads. Based on this, rather than one thread, several threads in a block co-operatively access each direction of the subdomain connectivity information at once via coalesced memory access, which accelerates the computation.

2.3 Multi-GPU computation

For multi-GPU computations, we employ a domain decomposition method with MPI library. In multi-GPU computing, first, the data transfer is needed from the GPU device memory to the host computer memory through a PCI Express bus. Then, the data is sent and received among cluster nodes through a computer network using the MPI library. Finally, the received data is transferred from the host memory to the device memory through the PCI Express bus. This can be a bottleneck in terms of overall computational time. To reduce the bottleneck, the subdomains near the processor boundary are computed first, then the data transfer and computation of non-processor boundary subdomains are processed simultaneously.

3 RESULTS AND DISCUSSION

A subject-specific geometry is extracted from multi-slice CT images of a 41-year-old-male. The geometry model is constructed as shown in Fig. 1, where the total number of grid points is 7746378 with the grid size of $0.215\,\mu\text{m} \times 0.215\,\mu\text{m} \times 0.215\,\mu\text{m}$.

First, we investigate the effect of the size of subdomains on the computational performance. Larger subdomains has an advantage for memory accessing, while smaller ones reduce unnecessary grid points. We examine the performance on a single GPU with various subdomain sizes, ranging from $2 \times 2 \times 2$ to $16 \times 16 \times 16$. The performance is evaluated as

$$\text{Performance [FLOPS]} = \frac{ON}{TS}, \qquad (2)$$

where O is the number of floating-point operations per one LBM node ($O = 229$ in the case of D3Q19 LBGK model), N is the total number of LBM nodes, T is the computational time, S is the time steps. We define "real performance" for $N = N_{total}$, and "effective performance" for $N = N_{fluid}$. Figure 2 shows the performance on NVIDIA Tesla C1060. Both the real performance and effective performance peak at $4 \times 4 \times 4$ subdomains. Note that the memory usage is also the smallest for $4 \times 4 \times 4$ subdomains among the cases. These results suggest that $4 \times 4 \times 4$ subdomains is the most effective size for the current airway model. The resultant domain consists of 176024 subdomains and the total number of LBM nodes is 11265536.

The performance on multi-GPUs is evaluated with a large GPU cluster, TSUBAME1.2 (Tokyo Institute of Technology). Figure 3 shows the performance of CPUs or GPUs as a function of the number of CPU cores or GPUs. The performance on 8 GPUs is approximately 200GFLOPS, while 8 CPU cores is approximately 2GFLOPS, indicating 100 times faster computation on GPUs. The performance almost linearly increases with increasing the number of GPUs up to 8 GPUs. However, further increase in the number of GPUs shows smaller performance than ideal performance. This is because smaller number of LBM nodes per one GPU causes lower speed for the memory accessing and data transfer. Moreover, the time for the data transfer becomes longer than that for the computation at non-processor boundary subdomains; the data transfer is the bottleneck of the computation. In the other

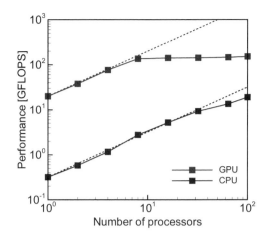

Figure 3. Comparison of parallel computing performance between GPU and CPU on TSUBAME 1.2.

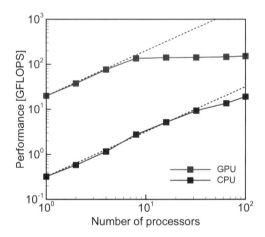

Figure 4. Comparison of parallel computing performance between GPU and CPU on a small cluster.

words, the computation with 1.2 million LBM nodes can be performed even with 8 GPUs, whereas at least 800 CPU cores are needed for parallel CPUs.

We also examine the performance on a small GPU cluster. The GPU cluster has 8 nodes connected with QDR Infiniband, in which each node consists of a CPU (Intel Core i7-930) and a GPU (NVIDIA Geforce GTX480). It costs 3 million Japanese yen (Spring, 2010), and the dimension is $0.7\,m \times 1.0\,m \times 2.0\,m$. The performance on GPUs shows again, almost ideal strong scaling as shown in Fig. 4. Although the performance on CPUs increases almost linearly with the number of CPU cores up to 16 cores, the speedup is decreased for further increase in the CPU cores. The peak performance may be estimated as 2–3 GFLOPS, which is comparable to the performance on single GPU.

4 CONCLUSIONS

We proposed an implementation of LBM on GPUs for simulating airflow in pulmonary airways with complex branches. We developed an adaptive meshing method for optimizing memory accessing, where the global domain comprises unstructured subdomains, while the local subdomain consists of a structured grids. We also developed a multi-GPU computing method based on a domain decomposition. To overlap the data transfer and the computation, the subdomains near the processor boundary are computed first, then the data transfer and computation of non-processor boundary subdomains are processed simultaneously. For strong scaling tests with a subject-specific geometry (12 million LBM nodes), the performance on 8 GPUs was approximately 200 GFLOPS, which was 100 times faster computation than 8 CPU cores.

Our method is a fully-automated method, from the airway extraction from CT images, the topological analysis of airways, and to the fluid dynamics simulations. It can also be applied to simulate the deposition of inhaled particle drugs (Imai et al. in press). The GPU computing method used in this study is an innovative technique for accelerating simulations, providing 40–100 times faster computations than CPU. Advantages for the use of our method in medical applications include a short computational time, as well as a low cost and small footprint to install. This method has a potential to break a barrier against achieving a computer-aided, subject-specific drug delivery system.

REFERENCES

Aidun, C.K. & Clausen, J.R. 2010. Lattice-Boltzmann method for complex flows. *Annual Review of Fluid Mechanics* 42: 439–472.

Chen, S. & Doolen, G.D. 1998. Lattice Boltzmann method for fluid flows. *Annual Review of Fluid Mechanics* 30: 329–364.

Comeford A., Förster C. & Wall, W.A. 2010. Structured tree impedance outflow boundary conditions for 3D lung simulations. *Journal of Biomechanical Engineering* 132: 081002.

De Backer, J.W., Vos, W.G., Vinchurkar, S.C., Claes, R., Drollmann, A., Wulfrank, D., Parizel, P.M., & De Backer W. 2010. Validation of computational fluid dynamics in CT-based airway models with SPECT/CT. *Radiology* 257: 854–862.

Gemci, T., Ponyavin, V., Chen, Y., Chen, H. & Collins, R. 2008. Computational model of airflow in upper 17 generations of human respiratory tract. *Journal of Biomechanics* 41: 2047–2054.

Imai, Y., Miki, T., Ishikawa, T., Aoki, T., & Yamaguchi, T. in press. Deposition of micrometer particles in pulmonary airways during inhalation and breath holding. *Journal of Biomechanics*.

Inthavong, K., Choi, L.-T., Tu, J., Ding, S., & Thien, F. 2010. Micron particle deposition in a tracheobronchial airway model under different breathing conditions. *Medical Engineering and Physics* 32: 1198–1212.

Kleinstreuer, C. & Zhang, Z. 2009. An adjustable triple-bifurcation unit model for air-particle flow simulations in human tracheobronchial airways. *Journal of Biomechanical Engineering* 131: 021007.

Kuznik, F., Obrecht, C., Rusaouen, G. & Roux, J. 2010. LBM based flow simulation using GPU computing processor. *Computers and Mathematics with Applications* 59: 2380–2392.

Lambert, A.R., O'Shaughenessy, P.T., Tawhai, M.H., Hoffman, E.A. & Lin, C.-L. 2011. Regional deposition of particles in an image-based airway model: large-eddy simulation and left-right lung ventilation asymmetry. *Aerosol Science and Technology* 45: 11–25.

Lin, C.L., Tawhai, M.H., Mclennan, G. & Hoffman, H.A. 2007. Characteristics of the turbulent laryngeal jet and its effect on airflow in the human intra-thoracic airways. *Respiratory Physiology and Neurobiology* 157: 295–309.

Ma, B., & Lutchen, R. 2008. CFD simulation of aerosol deposition in an anatomically based human large-medium airway model. *Annals of Biomedical Engineering* 37. 271–285.

Miki, T., Imai, Y., Ishikawa, T., Wada, S., Aoki, T. & Yamaguchi, T. 2011. A fourth-order Cartesian local mesh refinement method for the computational fluid dynamics of physiological flow in multi-generation branched vessels. *International Journal for Numerical Methods in Biomedical Engineering* 27: 424–435.

Miki, T., Wang, X., Aoki, T., Imai, Y., Ishikawa, T., Takase, K. & Yamaguchi, T. in press. Patient-specific modelling of pulmonary airflow using GPU cluster for the application in medical practice. *Computer Methods in Biomechanics and Biomedical Engineering.*

Tölke, J. 2010. Implementation of a lattice Boltzmann kernel using the compute unified device architecture developed by nVIDIA. *Computing and Visualization in Science* 13: 22–39.

van Ertbruggen, C., Hirsch, C. & Paiva, M. 2005. Anatomically based three-dimensional model of airways to simulate flow and particle transport using computational fluid dynamics. *Journal of Applied Physiology* 98: 970–980.

Wang, X., & Aoki, T. 2011. Multi-GPU performance of incompressible flow computation by lattice Boltzmann method on GPU cluster. *Parallel Computing* 37: 521–535.

Yin, Y., Choi, J., Hoffman E.A., Tawhai, M.H. & Lin, C.L. 2010. Simulation of pulmonary air flow with a subject-specific boundary condition. *Journal of Biomechanics* 43: 2159–2163.

Zhang, Z., Kleinstreuer, C., & Kim, C.S. 2009. Comparison of analytical and CFD models with regard to micron particle deposition in a human 16-generation tracheobronchial airway model. *Journal of Aerosol Science* 40: 16–28.

Zhao, Y. 2008. Lattice Boltzmann based PDE solver on the GPU. *The Visual Computer* 24: 323–333.

Computational Modelling of Objects Represented in Images – Di Giamberardino et al. (eds)
© 2012 Taylor & Francis Group, London, ISBN 978-0-415-62134-2

Bone volume measurement of human femur head specimens by modeling the histograms of micro-CT images

F. Marinozzi, F. Zuppante & F. Bini
*"Sapienza" University of Rome, Faculty of Engineering, Department of Mechanical and Aerospace Engineering.
Mechanical & Thermal Measurement Lab. Eudossiana, Rome, Italy*

A. Marinozzi
Orthopaedics and Traumatology Area, University "Campus Bio-Medico", Rome, Italy

R. Bedini & R. Pecci
Department of Technology and Health of Istituto Superiore di Sanità, Rome, Italy

ABSTRACT: Micro-CT analysis is a powerful technique for bone characterization. It allows to obtain histo-morphometric parameters of bone sample. To do this, some different levels of data processing are required, a binarization of the images is usually performed as first step. Binarization consists in establishing a gray-level threshold in order to assign each pixel to bone or void spaces. A wrong choice of this threshold induces an overestimation or underestimation of the parameters. The main problem is represented by the quantization error due to spatial sampling, causing partial volume artifacts. Each voxel which samples the external surface of the tissue represents an average of both air and bone and the simple thresholding process operates an incorrect discrimination of the two materials. To overcome this limitation, the aim of this study is the extraction of bone volumetric information by fitting and reconstruction of the grey-levels histogram with a suitable set of functions.

1 INTRODUCTION

Morphometric characterization of bone specimens is an important issue since it can be useful for the study and early diagnosis of pathologies such as osteoarthritis or osteoporosis. In the last years many studies worked on the bone characterization by micro-Computed Tomography (Khun 1980, Odgaard 1997, Parfitt 1983). This technique allows for a nonde-structive histomorphometric characterization of the samples.

One of the main problem of bone analysis by micro-CT is the processing of the scans for sample reconstruction. This usually requires a process named binarization which consists on the definition of a threshold value of grey-level, necessary to distinguish the bone from the background.

The choice of this value is a crucial task since a standard method doesn't exist, so it has been the subject of several studies and different solutions (Tan 2011). M. Ding, et al. (1999) calculate the threshold using an adaptive method. The graylevel data set is segmented at different thresholds. The level, where the volume fraction changes the least is chosen as the threshold for the data set. C. H. Kim, et al. (2007) define threshold value considering two different parameters, density connectivity and structure model index. Noticing that in a certain range of grey levels, density connectivity have a constant value and structure model index is linear,

as threshold value is chosen as the mean value of the range. In a previous study (Bedini 2010), a different method was implemented in order to calculate threshold value of bone samples. This method are based on the histogram analysis and require the splitting of the histogram in three different populations: pixels marked as bone, pixels marked as air and a third population of pixels that can not be statistically referred to bone or air. A statistical analysis was then used in order to define a threshold value.

Another problem arise from the resolution spatial sampling of bone elaboration of microCT system is that bone is a inhomogeneous material. Binarization process associate each voxel to bone or air, not considering that each voxel of the acquired image can be composed by both of them. Laidlaw (Laidlaw 1998) presented a new algorithm for identifying the distribution of different material types in volumetric datasets computing the relative proportion of each material in the voxels and introducing also information from neighboring voxels.

This method allows to reconstruct a continuous function that represents the distribution of values that the function takes on, also within the region of a single voxel.

In ideal condition of homogeneous material, each voxel contains only one material, so histogram will be composed of two peaks, one for each material. In this case bone volume could be calculated as the product between voxel volume and number of bone voxels. In actual conditions, histogram is the result of two ideal pulses spread over the entire range of greylevels. Histogram can be split into two base functions corresponding to the materials and a third function representing voxel that contain a mixture of bone and air. This third contribution can be considered such as an estimation of the so called "partial volume effect". Under these hypotheses, the number of bone voxels is the sum of voxels in the bone distribution plus the 50% of voxels of the third distribution. In order to verify this method, volumetric data were compared with data obtained by a standard method for the bone volume measurement.

2 MATERIALS AND METHOD

2.1 Data acquisition

Nineteen trabecular bone specimens were extracted from femoral heads of eight patients subject to a hip arthroplasty surgery. After being catalogued, specimens were preserved in a freezer at $10°C$ for about a month. From the middle of the femur capita of each specimen was obtained a slice of about 10 mm of thickness corresponding to the frontal plane and then these slices were kept 10 hours in a freezer.

Subsequently, slices were subjected to three complete cycles of dehydration with aqueous solutions having crescent percentage of ethanol, severally 70%, 90% and 99.9%, in order to defatted them. Between the dehydratation cycles, specimens were kept in a refrigerator for about 10 hours. Slices were then cut with a diamond belt saw (EXTEC Labcut 1010, Enfield CT) in order to obtain cubic specimens of about 6 mm. These specimens were further dehydrated and defatted by other three cycles with ethanol as previously described.

Trabecular bone specimens were acquired using a Skyscan 1072® micro-CT Scanner. The acquisition parameters were set at 100 kV and 98 μA, with 1 mm of aluminum filter with an isotropic voxel size in the range of 11.24 μm–14.66 μm. An angular step of 0,45° was used and 400 projections were acquired over an angular range of 180°. Using Cone Beam Reconstruction, software based on the Feldkamp algorithm, from these projection images in TIFF format, transversal sections of the specimen in BMP format can be obtained (Fig. 1). The distance between two consecutive sections is the double of the resolution chosen for the acquisition.

After the reconstruction, data were processed by a software of Skyscan®, named Ct-Analyzer®. This software allows to obtain histomorphometric parameters of the samples. The main interesting parameters for the description of bone architecture are (Parfitt et al 1987):

– Trabecular Separation (Tb.Sp) which is a measurement of the average distance between trabecula and is calculated as the diameter of the largest

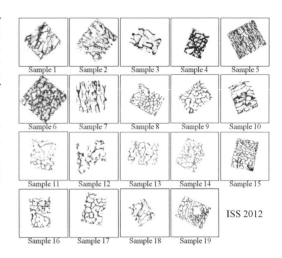

Figure 1. A single slice of each bone sample.

sphere which is entirely bounded within the solid surfaces.
– Trabecular Number, expressed as the number of trabeculae per unit length.

In ideal condition of homogeneous material, each voxel represents only one material, so histogram will be composed of two peaks, one for each material. In actual conditions, noise, inhomogeneity of the sample and finite resolution of the CT apparatus spread each peak into a Gaussian distribution. The actual histogram is then roughly composed by two gaussian-like functions, one centered on the mean gray level produced by the bone x-ray absorption, while the other represents the portions of the image occupied by air.

A third function represents the voxels that contain a mixture of bone and air, since they are located across the bone/air interface. This contribution can be considered such as an estimation of the partial volume effect.

The volume of the bone specimens were also measured by an helium pycnometer AccuPyc II 1340 pycnometers (Micromeritics-GA, USA).

2.2 Data processing

For each sample the histogram was computed.

After Laidlaw, the histogram can be fit with two gaussian basis functions:

$$f_{\sin gle}(v;c,s) = \left(\prod_{i=1}^{n_c} \frac{1}{s_i\sqrt{2\pi}}\right)\exp\left(-\frac{1}{2}\sum_{i=1}^{n_c}\left(\frac{v_i - c_i}{s_i}\right)^2\right) \quad (1)$$

where v_i, c_i, s_i are scalar component of v, c, s. The symbols c and s represent mean and standard deviation, respectively. The histogram component relative to the mixture along boundaries between two materials (bone and air, in our case) is calculated with this third basis function:

$$f_{mixture}(v;c_1,c_2,s) = \int_0^1 f_{\sin gle}(v,(1-t)c_1 + tc_2,s)dt \quad (2)$$

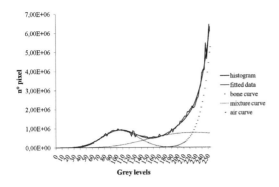

Figure 2. Histogram and the functions approximating bone, air and the mixture bone-air for one sample.

Considering the actual histogram obtained by the acquired images and the equations of the three basis functions $f_{bone}, f_{air}, f_{mixture}$, the fitting process was based on the Levenberg-Marquardt algorithm (Manolis & Lourakis 2005), which is an iterative method to solve non-linear least squares problems.

Finally, the bone volume was evaluated as the sum of all the voxels pertaining the estimated f_{bone}, plus the 50% of voxels that belongs to $f_{mixture}$. This latter assumption was justified considering the distribution of the mixture meanly composed of 50% of bone and 50% of air.

In order to verify this method, volumetric data were compared with the volume measured by the helium pycnometer (Aguilar-Chávez 2004). This method, in our measuring conditions, allowed to determine the volume of samples with a reproducibility of $2\,mm^3$.

3 RESULTS

The data processing allowed to fit each histogram with two Gaussian distributions and a distribution relative to the mixture according to eq. (1) and (2). In Figure 2 are reported a typical histogram and the distributions for one of the analyzed samples.

In table 1 are reported the volumes measured by the pycnometer (mean and standard deviation), together with those computed by the proposed method.

Comparing these values, an excellent correlation can be established, as reported in the graph of Figure 3.

Angular coefficient of the linear equal to 0.956 shows a slight underestimation of computed vs. measured volume.

In Figure 4a and 4b are reported the relation between values of trabecular separation and trabecular density with the bone volume values.

Observing these graphs, none significant correspondance between the percentage error and the parameters that describe the bone architecture was found.

In order to verify the accuracy of this new method bone volume calculated is also compared with bone volume obtained with the method implemented in

Table 1. Measured and calculated bone volume of the samples.

Samples	Measured volume mm³	Dev.st mm³	Computed volume mm³	Difference %
Sample1	192.3	10,4	238,6	−24,1
Sample2	241,4	3,1	254,2	−5,3
Sample3	68,5	2,9	59,1	13,7
Sample4	280,2	2,6	291,8	−4,1
Sample5	480,7	4	547,8	−14,0
Sample6	436,9	2,8	311,5	28,7
Sample7	55,9	3,3	65,8	−17,8
Sample8	111,6	4,7	130,7	−17,1
Sample9	112,8	3,1	120,7	−7,0
Sample10	146,9	2,4	165,8	−12,9
Sample11	93,7	3	110,8	−18,3
Sample12	65,1	1,5	52,6	19,1
Sample13	70,3	1,2	85,4	−21,4
Sample14	105,9	1,5	100,4	5,2
Sample15	161,5	4	165,5	−2,5
Sample16	84,1	2,9	99,4	−18,2
Sample17	93,7	0,9	82,6	11,9
Sample18	67	0,4	66,2	1,2
Sample19	96,7	1,7	93,7	3,2

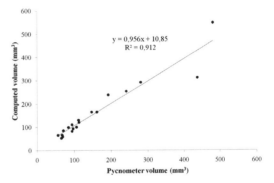

Figure 3. Comparison of volume values calculated with both the methods.

the previous study. An example of this comparison is reported in the graph below.

Considering the value obtained by pycnometer as reference, it can be observed that the volume calculated with this new method is more accurate than the previous method. Both of the methods focus on the histogram analysis, however the first method leads to define a threshold for the image processing, this new method avoid the problem of the threshold definition because it allows to extract bone volumetric information directly from the histogram.

4 CONCLUSION

Bone characterization is a crucial research issue because of its importance in the prediction of diseases such as osteoarthritis or osteoporosis.

Microtomographic technique is a good investigation tool since it allows a noninvasive analysis. However, for the calculation of several parameters,

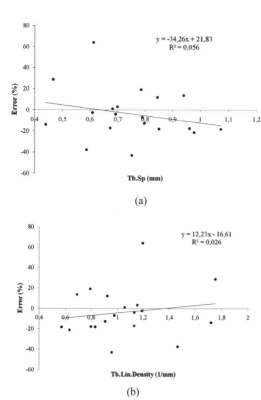

(a)

(b)

Figure 4. Relation between the percentage error and parameters characteristic of the bone architecture, Trabecular separation (a) and Trabecular linear density (b).

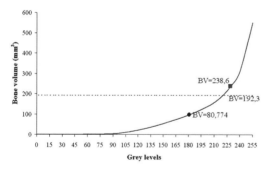

Figure 5. Comparison of volume values calculated with pycnometer (dotted line), method of the previous study (black circle) and the method object of this study (grey square).

though the choice of a proper grey level threshold is a concern, no accurate procedure is yet available, despite this topic has been the subject of many studies in the recent years. In this work a method for the bone volume calculation was investigated in order to overcome the threshold definition problem in CT images. Nineteen samples of bone cube, extracted from femoral heads of eight patients subject to a hip arthroplasty surgery, were acquired by a microCT apparatus. The hystograms of reconstructed images were analyzed in order to calculate bone volumetric information, i.e. the number of voxel which are internal, external

and on the bone surface. The volume of the samples was also measured by a Helium pycnometer, taken as a reference. The comparison of data obtained by both of the methods demonstrates a good correlation between them and so the histogram processing can be defined a viable method for an accurate and non destructive bone volume calculation.

ACKNOWLEDGEMENTS

The Authors wish to thank Dr. P. Scarlato end Dr. S. Mollo from Italian National Institute of Geophysics and Volcanology (INGV) for providing the Helium pycnometer and for their assistance during the measurements.

REFERENCES

Bedini R, Marinozzi F, Pecci R, Angeloni L, Zuppante F, Bini F, Marinozzi A. 2010. Analisi microtomografica del tessuto osseo trabecolare: influenza della soglia di binarizzazione sul calcolo dei parametri istomorfometrici.*Roma: Istituto Superiore di Sanità.* (Rapporto Istisan 10/15).

Ding M., Odgaard A., Hvid I. 1999. Accuracy of cancellous bone volume fraction measured by micro-CT scanning. *J. Biomechanics* 32(3):323–6.

Kim C. H., Zhang H., Mikhail G., von Stechow D., Muller R., H. Kim H., Guo X. E. 2007. Effects of thresholding techniques on micro-CT-based finite elemet model of trabecular bone. *Journal of Biomechanical Engineering* 129(4):481–6.

Kuhn J. L., Goldstein S. A., Jesion, 1990. Evaluation of a Microcomputed Tomography System to Study Trabecular Bone Structure. *Journal of Orthopaedic Research* 8(6):833–842.

Laidlaw D. H., Fleisher K. W., Barr A. H., 1998. Partial Volume Bayesian Classification of Material Mixtures in MR Volume Data Using Voxel Histograms. *IEEE transaction on medical imaging* 17(1): 74–86.

Manolis I. A. Lourakis, 2005. A Brief Description of the Levenberg-Marquardt Algorithm Implemened by levmar. *Institute of Computer Science Foundation for Research and Technology,* Crete, GREECE.

Odgaard A. 1997. Three-Dimensional Methods for Quantification of Cancellous Bone Architecture.*Bone,* 20(4): 315–28.

Parfitt A. M., Mathews C. H. E., Villanueva A. R, Kleerekoper M., Frame B., Rao D. S. 1983. Relationships between Surface, Volume, and Thickness of Iliac Trabecular Bone in Aging and in Osteoporosis Implications for the microanatomic and cellular mechanisms of bone loss. *J Clin Invest.*; 72(4): 1396–1409.

Parfitt A., Drezner K., Glorieux H., Kanis A., Recker, 1987. Bone histomorphometry: Standardization of nomenclature, symbols and units. *Journal of bone and mineral research* 2(6):595–610.

Tamari S., Aguilar-Chávez A., 2004. Optimum design of the variable-volume gas pycnometer for determining the volume of solid particles. *Meas. Sci. Technol.*15:1146–1152.

Tan Y., Kiekens K., Kruth, Voet A., Dewulf W. 2011. Material Dependent Thresholding for Dimensional X-ray Computed Tomography. *International Symposium on Digital Industrial Radiology and Computed Tomography* – Mo.4.3.

Computational Modelling of Objects Represented in Images – Di Giamberardino et al. (eds)
© 2012 Taylor & Francis Group, London, ISBN 978-0-415-62134-2

Stress and strain patterns in normal and osteoarthritic femur using finite element analysis

F. Marinozzi, A. De Paolis, R. De Luca & F. Bini
*"Sapienza" University of Rome, Faculty of Engineering, Department of Mechanical and Aerospace Engineering,
Mechanical & Thermal Measurement Lab. Eudossiana, Rome, Italy*

R. Bedini
Department of Technology and Health of Istituto Superiore di Sanità, Rome, Italy

A. Marinozzi
Orthopaedics and Traumatology Area, University "Campus Bio-Medico", Rome, Italy

ABSTRACT: Bone tissue stress and strain levels play a crucial role in bone remodeling. In fact, abnormal remodeling is documented in pathological conditions, such as in subject affected by osteoarthritis, even if this aspect seems to be poorly explored. We used FE analysis to investigate how the stress and strain distributions can be correlated with the loading conditions of the osteoarthritic femur and how a weird stimulation of trabecular tissue could lead to anomalous rearrangements of its pattern. For our analysis we developed, from X-ray images, 2D isotropic, homogeneous and linearly elastic models. The results show that stress and strain for the healthy femur are consistent with literature, whereas in osteoarthritic hips the physiological distribution of strain results altered and, consequently, anomalous remodeling processes take place: zones with very low (geodes) and/or very high (eburnation) bone density could appear, due to an unbalance in the activity of osteoblasts and osteoclasts.

1 INTRODUCTION

Osteoarthritis of the hip is the most common chronic-degenerative disease that can strike adults and it is characterized by severe pain (Nancy & Lane 2007, Murray, 1965). It causes the progressive thinning of the articular cartilage covering the femoral head and the acetabulum as well as the formation of new born at the joint surface.

This involves the formation of osteophytes in the attempt to compensate the stress increase by augmenting the contact area. (Elkholy 2005, McCarty 1989) A typical manifestation of osteoarthritis is the growth of geodes, i.e. spherical voids, located just below the articular cartilage whose etiology is currently not well understood even if authors connected them with bone resorption. (Collins 1953, Resnick 1977, Trueta 1953, Zirattu 2005) Bone, in fact, has the ability to remodel itself and change its morphology in accordance with the mechanical demands as stated by Wolff's law. (Bullough 1981, Pustoc'h & Cheze 2009, Wolff 1986) The purpose of this study is to show, by means of the Finite Element (FE) analysis, the consequences that the load changes, due to osteoarthritis, can have on the stress and strain distributions within the femur head. These alterations in turn are likely to be the cause of the (anomalous) re-arrangement of the trabecular tissue. In particular, massive bone resorption (geodes) may appear where stress and strain reach very low values with respect to the physiological ones; in this state the osteoclasts activity is favored. On the contrary, where the trabecular tissue experiments high value of stress and strain, osteoblasts activity is likely to be promoted with consequent increase of bone density (eburnation).

2 MATERIALS AND METHODS

Improved computed and modeling techniques render the FE method a very reliable and accurate approach in biomechanical applications. In this work, FE static analyses were performed on a 2-D model of the proximal half of a human femur in order to address the problems mentioned in the Introduction. The geometry of the model was reconstructed from an anterior-posterior X-rays image of a healthy femur. First, the contour of the femur was manually drawn through a CAD software (Fig. 1) and suitably filtered in MATLAB environment and then it was exported in COMSOL Multiphysics for the FE analysis.

The radius and center of the femoral head were found by calculating the radius and the centre of the circle that best fits the contour of the femoral head (diameter femur, 45 mm) (Brinckmann 2002, Van Rietbergen 2003).

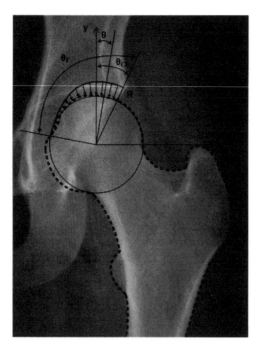

Figure 1. Radiographic image utilized for modeling the healthy femur together with the load distribution and angles according to (Brinckmann 2002, Nordin & Frankel 2001, Vengust 2001, Van Rietbergen 2003). The dashed line indicates the femur's contour drawn for the FE analysis.

Even if the structure of bone is non-homogeneous anisotropic, in this study, a simplified model of bone was used: the femur was considered isotropic, homogeneous and linearly elastic. Therefore, the trabecular tissue was approximated with a compact one with the following mechanical properties: apparent Young's modulus, 1 GPa, Poisson's ratio, 0.3 (Cowin 2001, Elkholy 2005, Vengust 2001).

The FE model was implemented for both the healthy and the osteoarthritic femur in three different stages of the disease.

In each numerical analysis, the distal ends of the models were fully constrained to prevent any movement in the plane. (Beauprè 1990, Van Rietbergen 2003)

The healthy hip joint was assumed to be a friction-free joint and a physiological distributed load was applied normal to the femoral head surface (Brinckmann 2002, Shive 1995, Van Rietbergen 2003).

These external forces were chosen to represent the monopodal stance phase and were obtained from biomechanical calculations: the load changes over the weight-bearing area according to a cosine function of the polar angle(Fig. 1) (Brinckmann 2002, Nordin & Frankel 2001, Vengust 2001, Van Rietbergen 2003).

To distinguish the healthy case from the pathological one, the center-edge angle of Wiberg (θ_{CE} in Figure 1) was used. Previous study, in fact, proved the importance of θ_{CE} as the main radiographic parameter

for the assessment of the hip dysplasia. The range of θ_{CE} between 20° and 25° is considered as the lower limit for normal hips, while the value of θ_{CE} below 20° is pathognomic for the hip dysplasia (Legal 1987, Pawels 1976, Wiberg 1933, Wiberg 1953). The size of the angle θ_{CE} correlates with the size of the weight bearing surface (θ_F in Figure 1) (Brinckmann 2002) and may therefore serves as an indirect measure of the hip joint contact stress which determines the state of health or disease of the joint (Bombelli 1984, Ipavec 1999, Maquet 1999, Pompe 2003, Vengust 2001). It was demonstrated that, in presence of osteoarthritis, the contact angle between the acetabulum and the femur head decreases. As a consequence, elevated forces are transmitted through a smaller surface and therefore the area experiments higher stress gradient. In addition, in the osteoarthritic femur the development of osteophytes can be observed with consequent augmenting in the number of contact points between the femur and the acetabulum. For these reasons we repeated the biomechanical calculations for the osteoarthritic femur and parabolic symmetric load distributions were implemented to reproduce the points loads due to the presence of osteophytes (Elkholy 2005, Vengust 2001).

In agreement with the X-ray data which suggest the progressive fusion between acetabulum and femur due to osteoarthritis, in the present study various grades of the disease were reproduced varying the number and the position of the contact points between femur and acetabulum and the amplitude of θ_{CE}. In particular, a θ_{CE} of 30° and 10° was considered respectively for the healthy femur and the osteoarthritic one.

Four simulations were performed: one for healthy femur and three for osteoarthritic femur at different stages of the disease (corresponding to two, three and four contact points between femur head and acetabular surfaces, respectively).

A triangular mesh was created resulting in 6026 nodes with 24104 degrees of freedom (Lagrange quadratic elements).

3 RESULTS AND DISCUSSION

The distribution of strain in the healthy femur (Fig. 2) agrees with the literature: the load is conveyed to the femur along the so-called Wolff's lines, in analogy with the actual case. In this condition the area of the femoral head is stimulated in an even manner consistent with the absence of poorly stimulated zones subjected to resorption (Elkholy 2005, Van Rietbergen 2003).

The portions of scarcely loaded trabecular tissue, on the contrary, are subjected to bone resorption. It seems consistent with what stated by Cox (2011) about the growth of fluid filled cavity (geodes).

The simulations for the pathological stages considered in this study show the presence of extremely low stress portion of trabecular tissue (Figs 3–5), right below the articular surface (Bombelli 1984, Zirattu 2005).

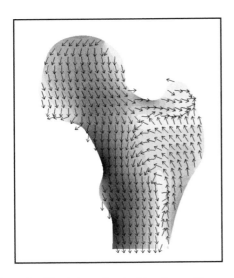

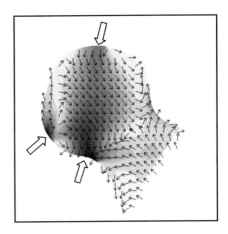

Figure 2. Simulation of a healthy femur. The arrows represent the direction of strain while the shaded gray map depicts its modulus. Dark grey relates to high strain values and vice-versa.

Figure 4. Simulation of a pathologic condition (osteoarthritis) in which three contact points (white arrows) are present between femur and acetabulum. The arrows represent the direction of strain while the shaded gray map depicts its modulus. Dark grey relates to high strain values and vice-versa. Only the femur head is shown.

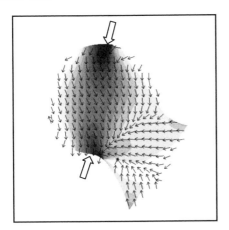

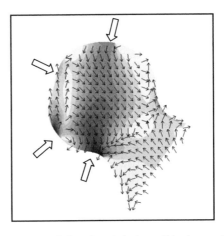

Figure 3. Simulation of a pathologic condition (osteoarthritis) in which only two contact points (white arrows) are present between femur and acetabulum. The arrows represent the direction of strain while the shaded gray map depicts its modulus. Dark grey relates to high strain values and vice-versa. Only the femur head is shown.

Figure 5. Simulation of a pathologic condition (osteoarthritis) in which four contact points (white arrows) are present between femur and acetabulum. The arrows represent the direction of strain while the shaded gray map depicts its modulus. Dark grey relates to high strain values and vice-versa. Only the femur head is shown.

The low stress areas, according to the theory of bone remodeling, result in bone atrophy and produce voids which can act as areas of excessive stresses on the surrounding bone resulting in microfractures and pain (Elkholy 2005).

The vector graph in Figure 2 shows the strain in the x-y plane of the healthy femur, while in Figures 3–5 the results of the three stages of the osteoarthritis are reported. Each vector represents the direction and orientation of strain but has the same module for clarity. The strain magnitude of the femur is visualized in maps with different gray levels.

From the analysis of the graphs, compressed and stretched fibers result evident. For the healthy femur (Fig. 2) in particular, compression is found for trabeculae running from the femoral head to the medial cortex and tension for trabeculae in the perpendicular direction. In osteoarthritic hips (Figs 3–5), on the contrary, the vectors seem to form some kind of "vortices", indicating that the physiological distribution of strain has been lost, together with the normal balance between osteoblast and osteoclast activity.

4 CONCLUSIONS

This work is aimed to perform 2D FE modeling of the median surface of the human femur head taken from

X-ray projections. Two different scenarios were simulated, corresponding to loading conditions of normal and osteoarthritic patients. FE analysis allowed for an immediate visualization of stress and strain distributions and provided information to localize the areas in which anomalous bone formation and/or resorption are likely to take place. We believe that this approach could be useful to explore the various loading conditions that can arise in pathologic state and with the progression of the disease.

REFERENCES

Beauprè, G.S., Orr, T.E., Carter, D.R., 1990. An approach for time-dependent bone modeling and remodeling-theoretical development. *Journal of Orthopaedic Research* 8:651–661.

Bombelli, R., Santore, R. F., Poss, R. 1984. Mechanics of the normal and osteoarthritic hip. A new per-spective. *Clin. Orthop.* 182:69–78.

Brinckmann, P., Frobin, W., Leivseth, G. 2002. *Musculoskele-tal Biomechanics*. Thieme.

Bullough, P. G. 1981. The geometry of diarthrodial joints, its physiological maintenance, and the possible significance of age-related changes in geometry-to-load distribution and the development of osteoarthritis. *Clin. Orthop. Relat. Res.*: 156:61-6.

Collins, D. 1953. Osteoarthritis. *The Journal of Bone and Joint Surgery* 35-B: 518–520.

Cowin, S. 2001. *Bone Mechanics Handbook*. CRC Press.

Cox, L.G., Lagemaat, M.W., van Donkelaar, C.C., van Riet-bergen. B., Reilingh, M.L., Blankevoort, L., van Dijk, C.N., Ito, K. 2011. The role of pressurized fluid in subchondral bone cyst growth. *Bone* 49(4):762–8.

Elkholy, A.H., Ghista, D.N., D'Souza, F.S., Kutty, M.S. 2005. Stress analysis of normal and osteoarthritic femur using finite element analysis. *J. Computer Applications in Technology* 22(4): 205–211.

Ipavec, M., Brand, R.A., Pedersen, D.R., Mavcic, B., Kralj-Iglic, V., Iglic, A. 1999. Mathematical modeling of stress in the hip during gate. *Journal of Biomechanics* 32(11): 1229–35.

Legal, H. 1987. Introduction to the biomechanics of the hip. *Congenital dysplasia and dislocation of the hip joint*. Springer-Verlag.

Maquet, P. 1999. Biomechanics of the hip. *Acta Orthopaedica Belgica* 65: 302–314.

McCarty, D. J. 1989. *Arthritis and allied conditions*. Lea and Febiger.

Murray, R. O. 1965. The aetiology of primary osteoarthritis of the hip. *Br. J. Radiol.* (38): 810–824.

Nancy, E. & Lane, M. D. 2007. Osteoarthritis of the hip. *The New England Journal of Medicine* 357: 1413–1421.

Nordin, M. & Frankel V.H. 2001. *Basis Biomechanics of the Musculoskeletal System*. Lippincott Williams&Wilki; 3th edition.

Pawels, F. 1976. *Biomechanics of the normal and diseased hip*. Springer-Verlag.

Pompe, B. 2003. Gradient of contact stress in normal and dysplastic human hips. *Medical Engineering & Physics* 25: 379–385.

Pustoc'h, A. & Cheze, L. 2009. Normal and osteoarthritic hip joint mechanical behaviour: a comparison study. *Med. Biol. Eng. Comput.* 47(4): 375–383.

Resnick, D., Niwayama, G., Coutts, R.D. 1977. Subchon-dral cysts (geodes) in arthritic disorders: pathologic and radiographic appearance of the hip joint. *Annual meeting of the American Roentgen Ray Society* 128(5): 799–806.

Trueta, J. 1953. Osteoarthritis of the hip: a study of the nature and evolution of the disease. *The Journal of Bone and Joint Surgery* 15(3): 174–192.

Van Rietbergen, B., Huiskes, R., Eckstein, F., Rüegsegger, P. 2003. Trabecular bone tissue strains in the healthy and osteoporotic human femur. *Journal of Bone and Mineral Research* 18(10): 1781–1788.

Vengust, R., Daniel, M., Antolic, V., Zupanc, O., Iglic , A., Kralj-Iglic, V. 2001. Biomechanical evaluation of hip joint after Salter innominate osteotomy: a long-term follow-up study. *Arch. Orthop. Trauma Surg.* 40: 511–516.

Shive, N. G. 1995. Articular cartilage. *Biomechanics of the Musculo –Skeletal System*, Wiley.

Wiberg, G. 1933. Studies on the dysplastic acetabulum and congenital subluxation of the hip joint with special reference to the complication of osteoarthritis. *The Journal of Bone and Joint Surgery* 15:6.

Wiberg, G. 1953. Shelf operator in congenital dysplasia of the acetabulum and in subluxaton and dislocation of the hip. *The Journal of Bone and Joint Surgery* 35-A:65–80.

Wolff, J. 1986. *The law of bone remodeling*. Springer-Verlag.

Zirattu, G., Fadda, M., Manunta, A., Marras, F., Zirattu, F., Bandiera, P. 2005. I geodi nell'artrosi dell'anca. *Minerva Ortopedica Traumatolgica* 56(2): 115–9.

Computational Modelling of Objects Represented in Images – Di Giamberardino et al. (eds)
© 2012 Taylor & Francis Group, London, ISBN 978-0-415-62134-2

A new index of cluster validity for medical images application

Amel.Boulemnadjel & Fella.Hachouf
Laboratoire d'automatique et de robotique, Département d'électronique, Université Mentouri de Constantine, Constantine, Algérie

ABSTRACT: In this paper a new index of cluster validity is proposed. It is based on the density of clusters. The proposed index operates at two levels; the first level is dedicated to the clustering estimation and the second one concerns a correction of the classification step. Clustering is performed using Kohonen Self-Organizing Maps algorithm (SOM). Maximal number of obtained clusters in the training step allows selecting a threshold which represents the minimal density in a cluster. This threshold divides the clusters in two categories; accepted and refused clusters. Weights representing the refused clusters are eliminated. The proposed index can evaluate the clustering in one iteration. Obtained results on medical images demonstrate that the proposed index is able to help in separating correctly objects than the existing indexes.

1 INTRODUCTION

Clustering is one of the most useful tasks in data mining process. It allows discovering groups and identifying interesting distributions and patterns in the underlying data. There are two major types of clustering approaches: supervised (Daumé III. & Marcu, 2005) and unsupervised Clustering (Bouveyron 2009, a,b, Erman, et al, 2006, c, Lung Wu, & Shen Yang 2007, d, Ozdemir, R et al, 2007). It is a mechanism where a set of objects usually multidimensional in nature, are classified into groups (classes or clusters) such that members of one group are similar according to a predefined criterion. Clustering methods have been successfully used to segment an image into a number of clusters (segments). However, clustering-based segmentation techniques have used several control parameters, e.g., the predefined number of clusters to be found or some tunable thresholds. These parameters should be adjusted to obtain the best image segmentation. The choice of values for the various parameters is a nontrivial task. Almost every clustering algorithm depends on the characteristics of the dataset and on the input parameters. Incorrect input parameters may lead to clusters that deviate from those in the dataset. In order to determine the input parameters that lead to clusters that fit best a given dataset, we need reliable guidelines to evaluate the clusters. Quantitative evaluation function (known under the general term of cluster validity indexes) has been used. Many criteria have been developed to determine clusters validity (Rendón, et al. 2011, a,b, Stein et al. 2003, c, Ammor et al. 2007, d, Yu Yen, & Cios, 2008, e, Pakhiraa et al. 2005, f, Xu et al. 2005, j, Wu & Yang. 2005). All of which have a common goal to find the clustering which results in compact clusters that are well separated. In this work, a new index validity is proposed and compared to Davies-Bouldin (Davies & Bouldin, 1979) and Dunn (Dunn, 1974) indexes. Davies-Bouldin and Dunn indexes are based on two accepted concepts: a cluster's compactness and a cluster's separation. The number of clusters that minimizes DB is taken as the optimal number of clusters. Dunn index is limited to the interval [0, 1] and should be maximized. These two indexes cannot improve the clustering result. They just evaluate the clustering quality. In addition, they do not take in to consideration the notion of cluster density. This property has been used in the proposed index allowing correction in classification results. After a training step by SOM algorithm, the first level of the proposed index consists in clustering estimation where the clustering correction will be performed in the second step. This index has been applied to medical images, especially on mammography images. Mammography is one of the most reliable methods for early detection of breast carcinomas. However, it is difficult for radiologists to provide both accurate and uniform evaluation for the enormous number of mammograms generated in widespread screening. There are some limitations of human observers: 10–30% of breast lesions are missed during routine screening (Chabriais, 2001). The quality of image interpretation depends on the clustering result.

This paper is organized as follows. Section 2 presents the method of clustering based on the SOM algorithm. Section 3 describes the cluster validity measures and the proposed index. The proposed index validity measure is compared qualitatively and quantitatively to the other indexes. The results are shown and discussed in Section 4, while conclusions are given in Section 5.

2 CLUSTERING USING THE SOM

In this paper, clustering is performed using the Kohonen Self-organising Maps algorithm (SOM) (Wu. & Chow. 2004,a,b, Azzag1 & Lebbah. 2008) which has been widely used in industrial applications such as pattern recognition, biological modeling, data compression, signal processing and data mining (AROUI et al. 2009). It is an unsupervised and nonparametric neural network approach. SOM is trained iteratively. At each training step, a sample vector is randomly chosen from the input data set. The SOM consists of a regular, usually two-dimensional (2-D), grid of map units. Each unit i is represented by a prototype vector $w_i = \{w_{i1}, \ldots, w_{id}\}$ where d is the dimension of an input vector. The units are connected to adjacent ones by a neighborhood relation. The number of map units which typically varies from a few dozen up to several thousand, determines the accuracy and generalization capability of the SOM. During training, the SOM forms elastic net that folds onto the "cloud" formed by the input data. Data points lying near each other in the input space are mapped onto nearby map units. Thus, the SOM can be interpreted as a topology preserving mapping from input space onto the 2-D grid of map units

3 INDEX VALIDITY MEASURES

Cluster validity is a very important issue in clustering analysis because the result of clustering needs to be validated in most applications. In most clustering algorithms, the number of clusters is set as user parameter. There are a lot of approaches to find the best number of clusters. In general terms, there are three approaches to investigate cluster validity. The first one is based on external criteria. This implies that we evaluate the results of a clustering algorithm based on a pre-specified structure. It is imposed on a data set and reflects our intuition about the clustering structure on it. The second approach is based on internal criteria. We may evaluate the results of a clustering algorithm in terms of quantities that involve the vectors of the data set themselves (e.g. proximity matrix). The third approach of clustering validity is based on relative criteria.

There are two criteria proposed for clustering evaluation and for the selection of an optimal clustering scheme:

a. Compactness: members of each cluster should be as close to each other as possible. A common measure of compactness is the variance which should be minimized.
b. Separation: the clusters themselves should be widely spaced. There are three common approaches measuring the distance between two different clusters:

- Single linkage: It measures the distance between the closest members of the clusters.

- Complete linkage: It measures the distance between the most distant members.
- Comparison of centroids: It measures the distance between the centers of the clusters.

This section introduces two validation methods known as the Dunn's and the Davies-Bouldin indexes. They have shown to be robust strategies for the prediction of optimal clustering partitions.

3.1 Davies-Bouldin index

This index (Davies & Bouldin, 1979) is defined as follows:

$$DB = \frac{1}{n}\sum_{i=1}^{n} \max_{i \neq j}\left\{\frac{s_x(Q_i) + s_x(Q_j)}{s(Q_i, Q_j)}\right\} \tag{1}$$

$$s_x = \frac{\sum_{i=1}^{k}\|x_i - Q_k\|}{n_k} \tag{2}$$

where n represents the number of clusters, S_x: is the average distances of all patterns in cluster $\mathbf{i}$ to their cluster center, $s(Q_i, Q_j)$ distance between clusters centers. Hence the ratio is small if the clusters are compact and far from each other. Consequently, Davies-Bouldin index will have a small value for a good clustering.

3.2 Dunn index

Dunn index (Dunn, 1974) aims to identify dense and well-separated clusters. It is defined as the ratio between the minimal inter-cluster distances to maximal intra-cluster distance. For each cluster partition, the Dunn index can be computed by the following formula:

$$I(c) = \frac{\min_{i \neq j}\{\delta(c_i, c_j)\}}{\max_{1 \leq l \leq k}\{\Delta(c_l)\}} \tag{3}$$

$$\delta(c_i, c_j) = \min_{x \in c_i, y \in c_j} d(x, y) \tag{4}$$

$$\Delta(c_i) = \max(\max_{x, y \in c_i} d(x, y)) \tag{5}$$

where $\delta(c_i, c_j)$ represents the distance between clusters i and j, and $\Delta(c_i)$ measures the intra-cluster distance of cluster c_i. Inter-cluster distance $\delta(c_i, c_j)$ between two clusters may be any number of distance measures, such as the distance between the centroids of the clusters. Similarly, the intra-cluster distance $\Delta(c_i)$ may be measured in a various ways, such as the maximal distance between any pair of elements in cluster c_i. Since internal criterion seek clusters with high intra-cluster similarity and low inter-cluster similarity, algorithms that produce clusters with high Dunn index are more desirable.

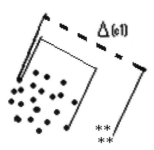

Figure 1. Clustering example.

3.3 *The proposed index*

Previously described indexes use a distance between objects as a measure to evaluate the clustering quality but in high dimensional spaces, distances between points become relatively uniform.

Dunn index becomes prohibitive in the case of high dimensions. It does not include any information about the cluster density which influences the quality of clustering result. As shown in figure 1, it is clear that the objects presented in stars are more distant from the others. Although they belong to the same cluster. The distance Δ is computed by equation (5) which represents the diameter of the cluster. It is represented by the discontinuous line, while the real distance is represented by the continuous line. The difference between the two distances is clearly seen. It gives a false imagination of the shape of the cluster. This is the result of the negligence of the density. Some other indexes need several partitions to select the best clustering which leads to a considerable augmentation of the runtime. Generally the main goal of a cluster validity technique is to identify the clusters partition for which a measure of quality is optimal. B our index is also able to offer more than the other indexes classifying more again badly classified elements.

This is the great advantage of this index. The proposed index IP solves these problems operating at two levels, the first level consists in estimating the clusters validity by computing is the index defined as follows:

$$IP = \frac{\sum_{i=1}^{nmx} N_i}{\sum_{j=1}^{n} N_j} \qquad (6)$$

where:
nmx: maximal number of the obtained clusters in the training step.
n: total number of clusters obtained by final clustering step.
N: cluster density.

Threshold SL represents the minimal density of cluster which is defined by equation (7) :

$$SL = \min_{1 \le l \le nmx} (N_l) \qquad (7)$$

This threshold divides the clusters in two categories: one category establish the correct clusters and the other

one the refused clusters. The weights representing the refused cluster are eliminated (w = 0).

The second level deals with the classification correction clustering data with the remaining weights.

These steps can be summarized by the following algorithm:

Algorithm
- Input: Number of neurons, input data for training step, number of iterations.
   Step 2. Training:
- Train input data by SOM
- Output: **nmx** number.
   Step 3. Clustering estimation:
- Compute the clusters density N
- Compute the index IP value by using (6)
- Compute the Threshold SL by using (7)
   Step 4. Classification correction
       For i=1 to n **do**
           If N (i) < SL **then**
              W(i)=0
       End for
   Step 5. Clustering:
- Cluster image with the remaining weight values w_i
End.

4 RESULTS AND DISCUTIONS

The proposed index has been evaluated on synthetic and real datasets. The technique consists of running the clustering algorithm (SOM) several times and obtaining different partitions, and the clustering partition that optimizes a validity index is selected as the best one. The various images obtained by the implementation of this algorithm are shown in Figures 2 , 4. For selecting the best partition, the three indexes are used. The number of clusters that maximizes Dunn index (D) is taken as the optimal, and the cluster configuration that minimizes Davies_Bouldin (DB) is taken as the optimal number of clusters. The value of the proposed index, Davies_Bouldin and the Dunn indexes are listed in table 1, the minimum value of DB index and the maximum value of Dunn show that the best partition is in Figure 2.C. Also our index selects the figure 2.c as the best partition. For a synthetic image, it is easy to select the best partition. we notice that the Figure 2.C is classified by comparison other images. The image background is almost uniform and the clusters are separated. These observations are confirmed by the three indexes. The comparison between the images obtained by SOM and the correction by the proposed index (Fig. 3,A) confirm the improvement. The background has become more homogeneous than before and edges are limited. Davies_Bouldin and Dunn indexes have also confirmed that the best partition is the partition obtained after correction by our index, according to the histogram (Fig. 3. B, C) which shows the number of cluster before and after correction.

This algorithm has been applied on a medical images such the image of the breast (mammography). The different images obtained by the

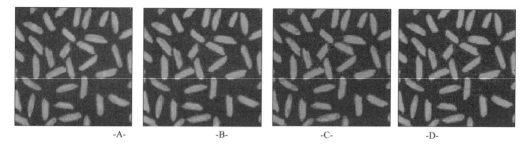

-A- -B- -C- -D-

Figure 2. A. Original image, B,C,D: Different partitions obtained by SOM.

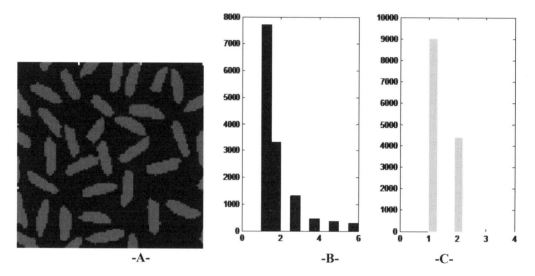

-A- -B- -C-

Figure 3. A: Result obtained after correction by the proposed index. The histogram B: before correction, C: after correction.

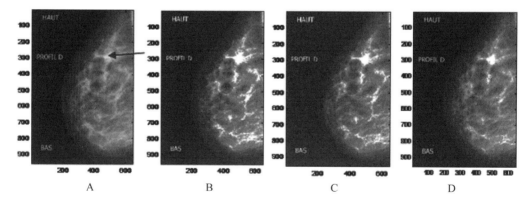

A B C D

Figure 4. A: Original image, B,C,D: different partitions obtained by applying SOM.

implementation of the clustering algorithm and the proposed index are given in Figures 4, 5.

According to calculations in table 1, the Figure 4.C is the best partition. It is noticed that the Figure 4.C does not represent all parts of the breast, and opacity that is not detected completely and the clusters are not homogeneous. After application of the proposed index we find that the opacity edges are limited, they are well separated compared to other clusters (Fig. 5.A),

according to the histogram (Fig. 5.B.C). The clusters are more homogeneous

5 CONCLUSION

A new index for cluster validity has been proposed. It allows easily the clustering estimation and gives an evaluation value in one iteration. Unsupervised classification guided by this index turns out effectively for

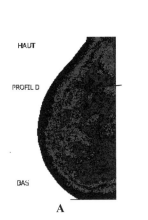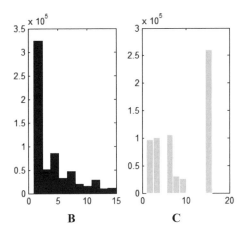

Figure 5. A: Result obtained after correction by the proposed index. The histogram of the breast image B: Before correction. C: After correction.

Table 1. The index values.

Images		DB	D	IP
Synthetic image	B	0.54	0.10	0.88
	After correction	0.33	0.11	/
	C	0.36	0.37	0.97
	After correction	0.16	0.79	/
	D	0.46	0.02	0.92
	After correction	0.31	0.10	/
	B	0.45	0.10	0.79
	After correction	0.36	0.12	
	C	0.44	0.37	0.86
Breast image	After correction	0.40	0.39	/
	D	0.46	0.14	0.73
	After correction	0.42	0.13	/

different types of images and especially the medical images. We are continuing efforts for improving this index taking into account smallclusters density representing objects in an image. Further information is needed to select the threshold.

REFERENCES

Ammor, O et al. 2007. Détermination du nombre optimal de classes présentant un fort degré de chevauchement. *REVUE MODULAD*, no. 37: 32–45.
Aroui, T et al. 2009. Clustering of the Self-Organizing Map based Approach in Induction Machine Rotor Faults Diagnostics. *Leonardo Journal of Sciences* Issue 15: 1–14.
Azzag1, H & Lebbah, M. 2008. Clustering of Self-Organizing Map. *ESANNbruges (Belgium)*: 209–214.
Bouveyron1, C, & Brunet, C.2009. Classification automatique dans les sous-espaces discriminants de Fisher. Manuscrit auteur: *publié dans 41èmes Journées de Statistique, SFdS*, Bordeaux.

Chabriais, J. 2001. *Les* standards en Imagerie Médicale. *Série documents d'information.*
Daumé III, H. & Marcu, D. 2005. A Bayesian Model for Supervised Clustering with the Dirichlet Process Prior. *Journal of Machine Learning Research, vol. 6,* pp.1551–1577.
Davies, D.L & Bouldin, D. W. 1979. A cluster separation measure. *IEEE Trans. Patt. Anal Machine Intell*: 224–227.
Dunn, J.C.1974. Well separated clusters and optimal fuzzy partitions. *J. Cybern.* Vol 4: 95–104.
Erman, J et al. 2006. Traffic Classification Using Clustering Algorithms. *SIGCOMM'06 Workshops*: Pisa, Italy.
Lung Wu, K & Shen Yang, M. 2007. Mean shift-based clustering. *Pattern Recognition, Elsevier*: 3035–3052.
Ozdemir, R et al. 2007. The modifier fuzzy art and a two-stage clustering approach to cell design. *Information science,* vol. 177: 5219–5236.
Pakhiraa, M. K. et al. 2005. Astudy of some fuzzy cluster validity indices, genetic clustering and application to pixel classification. *Fuzzy Sets and Systems,* 155: 191–214.
Rendón, E, et al. 2011. Internal versus External cluster validation Indexes. *international journal of computers and communications,* vol. 5, issue 1: 27–34.
Stein, B et al. 2003. On Cluster Validity and the Information Need of Users. ACTA Press: 216–221.
Wu, K.L. & Yang, M.S. 2005. A cluster validity index for fuzzy clustering. *Pattern Recognition Letters*, vol. 26: 1275–1291.
Wu, S. & Chow, T.W.S. 2004. Clustering of self-organising map using a clustering validity index on inter-cluster and intra-cluster density. *Pattern recognition,* vol. 37: 175–188.
Xu, Y. et al. 2005. A comparative study of cluster validation indices applied to genotyping data. *Chemometrics and Intelligent Laboratory Systems*, vol. 78: 30–40.
Yu Yen, C.H. & Cios, K. J. 2008. Image recognition system based on novel measures of image similarity and cluster validity. *Neurocomputing, Elsevier.*

Computational Modelling of Objects Represented in Images – Di Giamberardino et al. (eds)
© 2012 Taylor & Francis Group, London, ISBN 978-0-415-62134-2

Knowledge sharing in biomedical imaging using a grid computing approach

M. Castellano & R. Stifini
Politecnico di Bari, Bari, Italy

ABSTRACT: This paper will present the Knowledge Grid based system model, the architecture and the design principles focusing the discussion on the biomedical imaging process.

1 INTRODUCTION

The imaging is a process to investigate a reality that escapes directly to the human eye producing a visual representation. Several applications have been developed up to now, that span from biological investigations to the validation of scientific hypotheses (Hudson et al. 2007). A field of investigation that makes an increasing use of the imaging process is the biomedical one. Biomedical imaging makes possible to observe an area of a body no visible from the outside with the aim to use advanced diagnostic and therapeutic techniques in order to improve the effectiveness of biomedical process (Marion et al. 2011). According to different investigative objectives some examples of important technology based on biomedical imaging are: endoscopy, ultrasound, eco-doppler, computed tomography, magnetic resonance imaging, angiography, mammography, positron emission tomography (PET), single photon emission tomography (SPECT).

The general objectives of the studies, on innovative management, of the imaging process results, resides in the improvement of efficiency and effectiveness of a diagnostic and therapeutic process. The development of the accuracy of the result and the quality of the biomedical process, operate on the improvement of user satisfaction (Elmroth et al. 2005). For example, based on telemedicine solutions approach the performance of diagnostic therapeutic process the patient and make transparent the site where the process takes place against the expert evaluator. In terms of automated systems are known the Picture Analisys and Communication Systems (PACS) and the Computer Aided Detection and Diagnostics (CAD) (Bellotti et al. 2004).

The results of a process of biomedical imaging are generally used in decision making according to a hierarchical model in which the final outcome is expressed on the basis of observations (or opinions), or knowledge sharing multiple (produced by most experts) (Cabrera et al. 2002) on the same representation of information (image).

This work, with reference to the image or, more generally, to the process of biomedical imaging will discuss the benefits that can result from a system design based on knowledge and distribution of resources. The technology of the knowledge grid is proposed and discussed in the last section where is possible to develop a layer of services oriented to the biomedical imaging.

2 KNOWLEDGE IN BIOMEDICAL IMAGING

Biological imaging investigations are normally aimed at validating scientific hypotheses. The first step, therefore, is hypothesis definition, followed by experiment design. The experiment is then conducted and results are analyzed to understand the biological processes. Owing to the complexity of biological organisms, it is evident that the research in biological sciences, the diagnosis formulation, the development of therapies and the drug development will require more and more collaboration among research groups (Smal et al. 2011).

A decision process based on several observations requires intensive searches for related knowledge that often is available in different locations. As the data in this case happen to be in the form of large collections of images, a key requirement is to have automated systems and formalisms for representation of extracted knowledge so that software agents can search for relevant knowledge (Earl et al. 2001).

The Knowledge Engineering deals solutions to complex problems which usually ask for a high levels of human skill. The goal of K-Engineer is to build up a knowledge model useful for an algorithmic description by a structured approach. Through the discipline of knowledge engineering (KE) is possible the definition of a specific model of knowledge with a decomposition of the atomic elements of knowledge itself, in terms of tasks and activities. The result of this approach is a formal description of knowledge, but nothing infers about the technologies that could translate in terms of computer system (Anya et al. 2009).

The using of a method of Knowledge Engineering is useful for modeling the medical knowledge relating to illustrations drawn out by imaging processes. To produce so some schematics, understandable both by human and both by machines, through which the collaboration of imaging experts can create an automatic process focalized on the patient. In other words, the Knowledge Model contains all the schematic models and the mechanisms for the construction of the interoperable and collaborative processes (Vaezi et al. 2009).

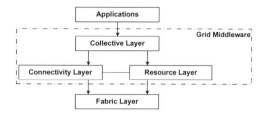

Figure 1. A reference architecture for grid computing systems.

3 COLLABORATIVE ENVIRONMENTS USING A GRID COMPUTING APPROACH

Grid computing is a promising information technology which meets collaborative requirements, and has great potential to become a standard cyber-infrastructure for life sciences. Medical informatics is one of the areas in which these technologically revolutionary advances could bring significant benefit both for scientists' research study and clinicians' everyday work. Recently there has been much excitement in the distributed and parallel systems community as well as that of distributed database applications in the emergence of Grids as the platform for scientific and medical collaborative computing (Stockinger et al. 2006). Grid computing promises to resolve many of the difficulties in facilitating medical informatics and medical image analysis by allowing radiologists and clinicians to collaborate without having to co-locate. Grid technology can potentially provide medical applications with an architecture for easy and transparent access to distributed heterogeneous resources (like data storage, networks, computational resources) across different organizations and administrative domains. The Grid offers a configurable environment whereby structures can be reorganized dynamically without affecting any overall active Grid processing (Zhao et al. 2008). In particular the Grid can address the following issues relevant to biomedical imaging:

- Data distribution: The Grid provides a connectivity environment for medical data distributed over different sites. It solves the location transparency issue by providing mechanisms which permit seamless access to and the management of distributed data. These mechanisms include services which deal with virtualization of distributed data regardless of their location.
- Heterogeneity: The Grid addresses the issue of heterogeneity by developing common interfaces for access and integration of diverse data sources. Such generic interfaces for consistent access to existing, autonomously managed databases that are independent of underlying data models are defined by the Global Grid Forum Database Access and Integration Services (GGF-DAIS) working group. These interfaces can be used to represent an abstract view of

data sources which can permit homogeneous access to heterogeneous medical data sets.
- Data processing and analysis: The Grid offers a platform for transparent resource management in medical analyses. This allows the virtualization and sharing of resources connected to the grid. For handling computationally intensive procedures (e.g. CADe), the platform provides automatic resource allocation and scheduling and algorithm execution, depending on the availability, capacity and location of resources.
- Security and confidentiality: Enabling secure data exchange between hospitals distributed across networks is one of the major concerns of medical applications. The Grid addresses security issues by providing a common infrastructure for secure access and communication between grid connected sites (Tourino et al. 2005). This infrastructure includes authentication and authorization mechanisms, amongst other things, supporting security across organizational boundaries.
- Standardization and compliance: Grid technologies are increasingly being based on a common set of open standards (such as OGSI, OGSA, XML, SOAP, WSDL, HTTP etc.).

In the Grid Computing Systems no assumptions are made concerning hardware, operating systems, networks, administrative domains, security policies, etc.

A key issue in a grid computing system is that resources from different organizations are brought together to allow the collaboration of a group of people or institutions. Such a collaboration is realized in the form of a virtual organization. The people belonging to the same virtual organization have access rights to the resources that are provided to that organization (Taswell et al. 2008). Typically, resources consist of compute servers (including supercomputers, possibly implemented as cluster computers), storage facilities, and databases. Given its nature, much of the software for realizing grid computing evolves around providing access to resources from different administrative domains, and to only those users and applications that belong to a specific virtual organization. For this reason, focus is often on architectural issues. An architecture proposed by Foster et al. (2001) is shown in Fig. 1.

The architecture consists of four layers. The fabric layer that provides interfaces to local resources at a specific site. The connectivity layer that consists of

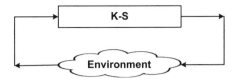

Figure 2. Conceptual architecture for a Knowledge System.

communication protocols for supporting grid transactions that span the usage of multiple resources and contain security protocols to authenticate users and resources. The resource layer which is responsible for managing a single resource. The collective layer that deals with handling access to multiple resources and typically consists of services for resource discovery, allocation and scheduling of tasks onto multiple resources, data replication, and so on.

4 KNOWLEDGE SYSTEM IN BIOMEDICAL IMAGING

In this section is discussed the design of a Knowledge System (K-S). It consists into transform the knowledge model requirement specifications into technical solutions (Castellano et al. 2011). Here, the design model describes the software specifications in terms of computing model, architecture and computational mechanisms. Any knowledge system can function satisfactorily only if it is properly integrated in the organization where the system is operational (Zhuge et al. 2004) (Abdullah et al. 2008). Fig. 2 shows the knowledge system acting as an agent that can cooperating with many others, human or software, and it carries out just a fraction of the many tasks that are performed in the organization. The knowledge systems must be viewed as supporting components within the business processes of the organization.

One of the requirements that emerged from the phase of design of the knowledge model is the collaboration (Kalabokidis et al. 2009). This is a very important requirement from which you can pull out the technical specification of collaborative computing. The collaborative computational mechanisms require to break down the existing barriers between different organizations and deal with issues concerning the sharing of resources.

Fig. 3 present an architecture for building a Knowledge Grid. It is based on a tiered approach in which each level focuses on a specific functionality and is organized in such a way as to provide services at a higher level. To fulfill its role the same level may require the services at lower level.

The core of the architecture consists of three layers discussed here:

- The Data Layer – The data to which it refers can normally be contained in files or databases, will then be the task of the service grid associated with each specific resource to process the data and to extract

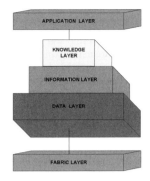

Figure 3. Knowledge Grid.

a representation used by the grid system. The application of the health service could also be seen as a data-intensive: for example, a statistical analysis of patient data. Moreover, even the data to be processed may be a great deal, or data files may be large. This layer is designed to provide higher level functionality through exposure of a range of services is of general type (Data Grid) that a particular type which is related to the application.

- The Information Layer that deals with the way that information, metadata, tables (informative structures) and databases are represented, stored, accessed, shared, and maintained (Zettsu et al. 2009).
- The Knowledge Layer that deals with the way that knowledge (managed by complex tools such as Ontologies – Data-Mining – Text Mining) is acquired, used, retrieved, published and maintained to achieve goals and objectives. In other words the grid layer is a virtual knowledge or experience sharing environment for innovation (Song et al. 2009). Experts, software, physician are all viewed as the resource nodes of the grid framework.

Each levels are characterized by a strong focus and functional specialization, has to perform specific tasks and homogeneous, is able to communicate with other layers and leaves it to them possible actions not of own pertinence (Zhan et al. 2009). Among the different layers there is a certain hierarchy in the sense that the relationship between the layers is not equal footing in general, but is governed by a set of dependencies that permit to identify an order of relation. This means that each level "is leaning" to one or more layers in order to perform its tasks, depends on them and only communicates with them.

From being each level of grid-based type, you will have to provide solutions on the major features of grid computing, for each, such as those of resource sharing, virtualization and communication.

5 CONCLUSIONS

This paper, discusses a grid based system to develop collaborative solutions to approach the results of biomedical imaging process. A pool of experts can implement a decision making with knowledge sharing only if both the systems role and its potential impact on the organization are made explicit. A collaborative model for biomedical imaging based on knowledge engineering and knowledge grid must be designed. A knowledge system model design from the computational point of view in term of computing models, architecture and computational mechanism is finally produced. Moreover, in images analysis, important aspects have been reviewed as collaboration, resource sharing, data, information and knowledge.

REFERENCES

Abdullah A, Othman M, Sulaiman MD N, Ibrahim H, Abu T. Othman, Malaysia (2008). "Towards A Scalable Scientific Data Grid Model and Services", Proceedings of the International Conference on Computer and Communication Engineering 2008, IEEE, Vol. 1, No. 1, pp. 20–25.

Anya O, Tawfik H, Nagar A, (2009). "Task-Centred Knowledge Sharing in e-Collaboration", Second International Conference on Developments in eSystems Engineering, Vol. 2 No. 1, pp. 403–408.

Bellotti R, Cerello P, De Carlo F, Bagnasco S, Bottigli U, Castellano M et al (2004) The MAGIC-5 Project: Medical Applications on a Grid Infrastructure Connection In: Proceedings of IEEE-NSS/MIC 2004, Vol.3, Roma, Italy, October 16–22, 2004 ISBN 0-7803-8700-7

Cabrera A, Cabrera E F, (2002). "Knowledge-sharing Dilemmas". Organization Studies 23, pp. 687–710.

Castellano M, Stifini R, (2011). "Biomedical Knowledge Engineering Using a Computational Grid", Biomedical Engineering, Trends in Electronics, Communications and Software, InTech, Vol. 1, No. 1, pp. 601–625.

Earl M, (2001). "Knowledge management strategies: Toward taxonomy", Journal of Information Systems, Vol. 18, No. 2; pp. 215– 233.

Elmroth E, Nylen M, Oscarsson R, (2005). "User-centric cluster and grid computing portal". Proceedings of the *International Conference on Parallel Processing Workshops*, pp. 103–112.

Hudson D L, Cohen M E, (2007). "Technologies for patient-centered healthcare", *Life Science Systems and Applications Workshop*, 2007. LISA 2007. IEEE/NIH Digital Object Identifier, pp. 148–151.

Kalabokidis, Kostas, Vaitis, Michail, Soulakellis, Nikolaos, (2009). "Towards a semantics-based approach in the development of geographic portals". Computers and Geosciences, Vol. 35, No. 2, pp. 301–308.

Marion A. et al, (2011). Multi-modality medical image simulation of biological models with the Virtual Imaging Platform (VIP), Computer-Based Medical Systems (CBMS), 2011 24th International Symposium on, pp. 1–6

Smal I., Carranza-Herrezuelo N. , Klein S., Niessen W., Meijering, E., (2011), Quantitative comparison of tracking methods for motion analysis in tagged MRI, Biomedical Imaging: From Nano to Macro, 2011 IEEE International Symposium on, pp. 345–348

Song W, Geldart J, (2009). "An Analysis Model for Knowledge Grid and Flows", An Analysis Model for Knowledge Grid and Flows, Vol. 1, pp. 240–249.

Stockinger H, (2006). "Grid Computing in Physics and Life Sciences", Physics Vol 18, pp. 1–6.

Taswell C, (2008). "DOORS to the semantic web and grid with a PORTAL for biomedical computing". IEEE Transactions on Information Technology in Biomedicine, v 12, n 2, pp. 191–204.

Tourino J, Martin M, Tarrio J, Arenaz M, (2005). "A grid portal for an undergraduate parallel programming course". IEEE Transactions on Education, v 48, n 3, pp. 391–399.

Vaezi S K, (2009). "Critical Success Factors for Implementing Knowledge Based Models for Electronic Public Services (EPS)", ACM Proceedings of International Conference on Theory and Practice of Electronic Governance (ICEGOV 2008), (Dec 2008), pp 471–473.

Zettsu K, Nakanishi T, Iwazume M, Kidawara Y, Kiyoki Y, (2008). "Knowledge cluster systems for knowledge sharing, analysis and delivery among remote sites," Information Modelling and Knowledge Bases, vol. 19, pp. 282–289.

Zhan H, Gu X, (2009). "Process Model Based Scale-free Knowledge Grid Modeling Technology" Factory of Engineering, Ningbo University, Zhejiang, China.

Zhao J, Xu J, Dong X, Zhu Z, Wang Z, (2008). "A scalable and low-cost grid portal", Proceedings of 7th *International Conference on Grid and Cooperative Computing*, GCC 2008, pp. 570–576.

Zhuge H, (2004). "The Knowledge Grid", World Scientific Publishing Co. Pte. Ltd., Chinese Academy of Sciences, China.

Computational Modelling of Objects Represented in Images – Di Giamberardino et al. (eds)
© 2012 Taylor & Francis Group, London, ISBN 978-0-415-62134-2

Modeling anisotropic material property of cerebral aneurysms for fluid-structure interaction computational simulation

Hong Zhang
Department of Mechanical Engineering, Carnegie Mellon University, Pittsburgh, PA, US

Yuanfeng Jiao
Department of Biomedical Engineering, Carnegie Mellon University, Pittsburgh, PA, US

Erick Johnson
Water Power Technologies Department, Sandia National Laboratories, Albuquerque, NM, US

Yongjie Zhang & Kenji Shimada
Department of Mechanical Engineering, Carnegie Mellon University, Pittsburgh, PA, US
Department of Biomedical Engineering, Carnegie Mellon University, Pittsburgh, PA, US

ABSTRACT: Fluid-Structure Interaction (FSI) simulation is a useful tool to estimate the wall stress distribution and predict the rupture risk of a cerebral aneurysm. Accurate simulation is important to assist the treatment of cerebral aneurysms. Although the accuracy has been increased in recent years, the process still remains physically non-reasonable with most simulation studies relying on a uniform wall thickness and isotropic material property. In this paper, we present an anisotropic material property modeling scheme with a non-uniform wall thickness distribution for patient-specific cerebral aneurysms. The non-uniform wall thickness distribution and material anisotropy are obtained through deforming the mesh of a healthy blood-vessel onto an aneurysm model, where the mesh deformation simulates the formation of the aneurysm by stretching the mesh elements. With the addition of material anisotropy, the simulation results can be improved and become more physically meaningful.

1 INTRODUCTION

Cerebral aneurysms (CAs) (Brisman, Song, and Newell 2006) are defined as brain blood-vessel dilations at spots with weakened walls and the formation of bulges. Compared with other kinds of aneurysms such as abdominal aortic aneurysms, CA ruptures are harder to predict due to their smaller size. The mean diameter of CAs is 5.47 ± 2.536 mm in anterior cerebral artery, 6.84 ± 3.941 mm in inferior cerebral artery, 7.09 ± 3.652 mm in middle cerebral artery and 6.21 ± 3.697 mm in vertebrobasilar artery (Jeong, Jung, Kim, Eun, and Jang 2009). Though not all ruptures are fatal and some patients suffer no complications, the mortality rate is still relatively high. Based on the statistical data, there is a 40% chance of mortality within 24 hours and an additional 25% chance of death due to complications within the next 6 months (NINDS 2008). The mortality rate could be reduced if the growth and rupture of CAs or even surgical outcomes, in the future, become predictable.

Fortunately, computational simulation is a useful tool to help physicians in diagnosing and treating CAs potentially. Physically, the vessel wall changes in both its thickness and material properties when an

aneurysm develops. During these changes, the cerebral aneurysm wall is stretched and the primary, load-bearing component becomes the collagen layer as the elastic and smooth muscle cell layers degrade. Thus the vessel tissue starts to lose the linear-elastic properties and enters the non-linear and hyperelastic region after high-strain loading based on its stress-strain curves (Scott, Ferguson, and Roach 1972). In the region of high strain, collagen growth is increased to reinforce the vessel wall and some modifications are needed in the simulation to model the increase (Kroon and Holzapfel 2008). The additional mass of collagen may change the aneurysm wall thickness and the thickness distribution may vary over the cerebral aneurysm dome. The average thickness of the dome decreases to 50–100 μm (Abruzzo, Shengelaia, Dawson, Owens, Cawley, and Gravanis 1998).

Thus in the computational simulation of cerebral aneurysms these changes are important, and the simulation results may be inaccurate and even wrong if the changes are not reflected in the simulation (Bazilevs, Hsu, Zhang, Wang, Liang, Kvamsdal, Brekken, and Isaksen 2010). However, the changes are unmeasurable in vivo. Even if the vivisection is possible, the changes are impossible to observe due to the small size of CAs. For this reason, most of the previous simulation studies use uniform wall thickness and isotropic material model, lacking meaningful physical properties (Baek, Rajagopal, and Humphrey 2006). These two restrictions severely decrease the simulation accuracy.

In this paper, we provide an algorithm to overcome the two restrictions by estimating the thickness distribution and the material directions from a healthy model, which are generated from an aneurysm geometry by removing the bulge and replacing it with a pipe-like object. Then the vertex correspondence between the healthy model and the aneurysm model is built via parameterization. The wall mesh is generated with an anisotropic spring relaxation based on the material directions of each element. Finally, the anisotropic mesh with anisotropic material property are used in the FSI simulation.

2 ANISOTROPIC MESH GENERATION

Compared with other existing methods, our algorithm improves two main aspects: estimating the wall thickness and building an anisotropic material model as shown in Figure 1. The basic idea is to deform the surface mesh from a healthy blood vessel onto the aneurysm model. In this way the wall thickness (related to the area change) and the material direction (related to the mesh deformation) are calculated on each mesh element. The algorithm contains three main steps:

1. Deform the mesh from the healthy model, S_0, onto the patient-specific aneurysm model, S_1, through surface parameterizations and build up an element-to-element correspondence;
2. Calculate surface curvature, and relax the nodes of the mesh, S_2, using an anisotropic non-linear spring model based on the material directions;
3. Calculate the thickness at each node as a result of the mesh deformation.

Given the healthy and aneurysm meshes, both of them are parameterized onto planar discs of the same size, and nodes on these discs are moved until the outlet-boundaries from one map line up with the corresponding ends in the other. Then each node on the healthy model finds a location on the aneurysm surface through a barycentric coordinate system, see (Johnson, Zhang, and Shimada 2011; Johnson 2010) for details.

2.1 Isotropy versus anisotropy

Though an isotropic spring model is easy to implement, an anisotropic model is closer to real strain-energy curves shown in Figure 2 (Fung, Fronek, and Patitucci 1979). From the curves, the non-linear stress-strain relationships in the axial and circumferential directions are significantly different. The circumferential material property is stiffer than the axial material

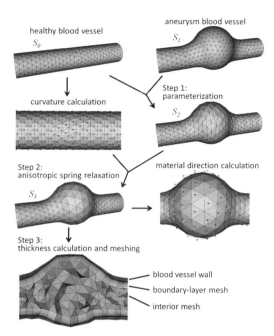

Figure 1. Overview of anisotropic mesh generation.

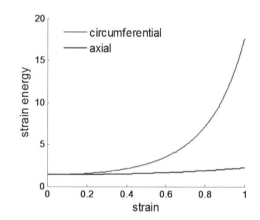

Figure 2. Strain-energy curve for anisotropic tissue (Fung, Fronek, and Patitucci 1979).

property and the mechanical response should be different in each direction. It means the vessel behaves differently depending on how it is stretched.

To induce the anisotropic characteristics of the vessel wall into the meshing step, we relax the mesh with an anisotropic spring model based on the material directions. Thus the primary difficulty in adding material anisotropy to an aneurysm model becomes how to determine the material directions.

The most straightforward way to calculate the material direction is to align the axial vector following the centerline and define the circumferential direction as the vector perpendicular to the plane of axial and normal vectors on each element. The solution is valid when the vessel grows from the centerline, similar to a pipe with variable diameters (Rissland, Alemu, Einav,

Ricotta, and Bluestein 2009). However, it depends largely on the quality of the centerline, and in the case of aneurysms the surface is complex and the centerline is hard to detect. In this way, the axial direction on an aneurysm is nearly identical to its healthy model and the solution is not applicable for aneurysms.

Another solution is to use the principal curvatures on surface to define the material directions. However, this approach only utilizes geometry information, while not reflecting the physical process of aneurysm formation. In this paper, we propose to calculate the anisotropic material directions on the healthy model first, based on the principal curvatures, and then map that directions to the aneurysm model by using anisotropic non-linear springs.

2.2 Surface-curvature calculation

On the healthy model the principal curvature is defined locally at every vertex. The principal curvature directions are found through the extended quadric method (Sahni, Muller, Jansen, Shephard, and Taylor 2006). The extended patch modification from Garimella and Swartz is also used to improve the algorithm near boundaries (Garimella and Swartz 2003). For each vertex P on S_0, a quadric surface is fit over the local patch of elements,

$$\bar{Z} = a\bar{X}^2 + b\bar{X}\bar{Y} + c\bar{Y}^2 + d\bar{X} + e\bar{Y}, \qquad (1)$$

where a, b, c, d and e are the coefficients being solved for, and $\bar{X}$, $\bar{Y}$, and $\bar{Z}$ are the component directions of the local coordinate frame at the node translated from the global coordinate frame X, Y, and Z to make the current vertex as the origin. The local coordinate frame is defined using P as the origin:

- $\bar{Z}$ is the regularized surface normal, $\hat{n}$, at P, and can be approximated as the average of the neighboring vertex normals.
- $\bar{X}$ takes the direction of the global X-axis projected against the plane defined by $\hat{n}$,

$$\bar{X} = \frac{\hat{i} - (\hat{i} \cdot \hat{n})\hat{n}}{||\hat{i} - (\hat{i} \cdot \hat{n})\hat{n}||}, \qquad (2)$$

where $\hat{i}$ is the global X-axis $[1, 0, 0]^T$.
- $\bar{Y}$ is orthogonal to both $\bar{X}$ and $\bar{Z}$, $\bar{Y} = \bar{Z} \times \bar{X}$.

In addition to the origin, the 2-ring neighbors are required to solve the least-square problem in Equation 1. At the boundary of a surface, only part of the patch is present and some vertices are missing. One possible solution is to mirror the incomplete patch along P. The neighbors are transformed into the local frame using

$$\bar{x} = R(x - x_P), \qquad (3)$$

where x_P is the position vector of P in the global frames; x and $\bar{x}$ are position vectors in the global and local frames, respectively; and the rotation matrix is $R = [\bar{X}, \bar{Y}, \bar{Z}]^T$. The linear terms of the extended

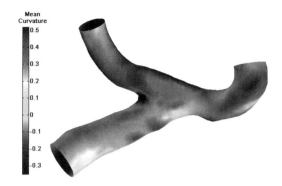

Figure 3. The mean curvature on a healthy model.

quadric in Equation 1 allow the surface normal to be updated using

$$\hat{n} = \frac{[-d, -e, 1]^T}{\sqrt{(d^2 + e^2 + 1)}} \qquad (4)$$

and will in turn update $\bar{X}$ and $\bar{Y}$. After updating $\bar{X}$ and $\bar{Y}$, $\hat{n}$ is updated accordingly, and the iteration will stop until $\hat{n}$ converges and yields a more accurate estimation of the principal curvature directions. The quadratic terms of Equation 1 define a shape operator for the patch (Petitjean 2002),

$$S = \begin{bmatrix} 2a & b \\ b & 2c \end{bmatrix}, \qquad (5)$$

and the eigenvectors of this matrix are the principal curvature directions. The eigenvalues are the principal curvature magnitude. These vectors need to be transformed back into the global frame by multiplying the inverse of the updated rotation matrix. The mean curvature on a healthy model is shown in Figure 3.

2.3 Using anisotropy for spring relaxation

Based on the principal curvature directions on the healthy model, the material directions are estimated and embedded in the springs. After building up the correspondence between the healthy model and the aneurysm as in Step 1 in Figure 1, the anisotropic springs are set up on the aneurysm surface. In practice, the springs are split into two components along the axial and circumferential directions through the dot-product with the material directions shown in Figure 4. Now the spring stiffness can be viewed as a combination of the two components,

$$K = \frac{C}{l} e^{a_1 \epsilon_{\theta\theta}^2 + a_2 \epsilon_{zz}^2 + 2a_3 \epsilon_{\theta\theta}\epsilon_{zz}}, \qquad (6)$$

where ϵ_{zz} and $\epsilon_{\theta\theta}$ are the axial and circumferential strains for each spring, respectively, a_1, a_2, a_3, and C are material constants, and l is the current length of the spring. Using Equation 6 as the stiffness equation, the relaxation is both non-linear and anisotropic along the material directions.

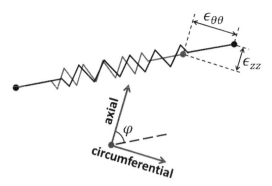

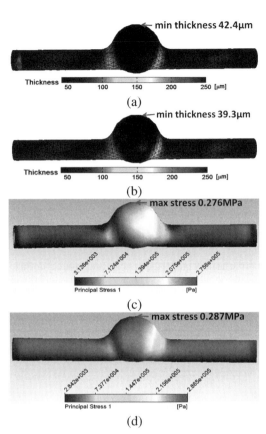

Figure 4. The split of a spring in the axial and the circumferential directions.

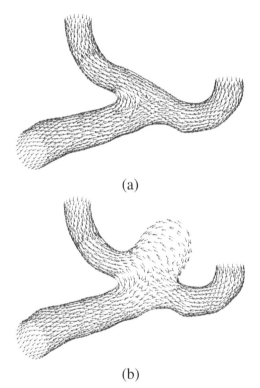

(a)

(b)

Figure 5. The axial material direction distribution on the healthy model (a) and the aneurysm (b).

Figure 6. Thickness and principal stress distribution. (a) Isotropic mesh; (b) anisotropic mesh; (c) isotropic mesh with isotropic material property; and (d) anisotropic mesh with anisotropic material property (t = 0.32 s).

2.4 Transforming material directions

After relaxation, a group of correct vertex-to-vertex correspondences is established. Given the material directions at each vertex on the healthy model, the material directions at each vertex on the aneurysm are calculated from the springs connected to this vertex.

As shown in section 2.3, the springs before relaxation are split into two components along the axial and the circumferential directions and we assume the angles between the springs and the axial and the circumferential directions on the springs do not

change after relaxation. Thus the axial and circumferential directions can be estimated by rotating the new spring direction back according to the stored φ in Figure 4. The mean of estimated axial and circumferential directions on springs are used as the transformed material directions on this vertex.

The axial material directions on the healthy model and the aneurysm after transformation are shown in Figure 5. We can observe that the material directions follow the axial directions consistently.

3 ANISOTROPIC MATERIAL PROPERTY AND FSI SIMULATIONS

With the assumption that the wall volume does not change during the aneurysm formation, the wall thickness of an element should be inversely proportional to the change of surface area and calculated from the correspondence between the healthy model S_0 and the deformed model S_3, see Step 3 in Figure 1. This assumption estimates the weakening of the wall, even if the wall does physically increase in thickness. The advancing front method is utilized to grow the wall

264

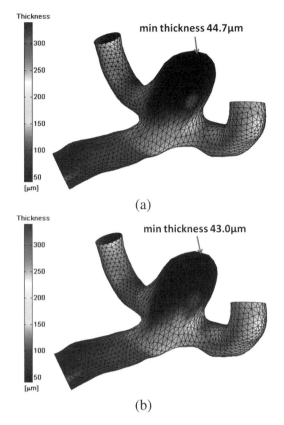

(a)

(b)

Figure 7. Thickness distribution over isotropic (a) and anisotropic (b) meshes.

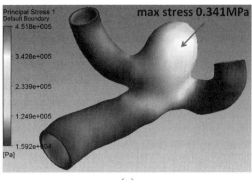

(a)

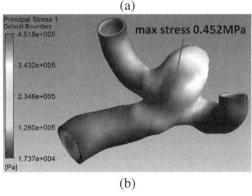

(b)

Figure 8. Principal stress distribution. (a) Isotropic mesh with isotropic material property; and (b) anisotropic mesh with anisotropic material property (t = 0.32s).

into the solid and fluid meshes (Johnson, Zhang, and Shimada 2011; Johnson 2010).

The generated mesh with material anisotropy and non-uniform wall thickness presents a potential improvement and has more physical meaning, while the comparison of simulation results between an isotropic mesh and an anisotropic mesh of the same aneurysm can reveal this improvement. First of all, from the comparison in Figures 6 and 7 we can observe the element size on the anisotropic mesh is adaptive according to the material direction and the elements on the bulge are slightly and reasonably stretched in the circumferential direction.

The FSI simulations were run with ANSYS® v13.0 and ANSYS®, CFX^{TM} v13.0. The blood flow was approximated as a Newtonian fluid with a dynamic viscosity of $\mu = 3.85$cP. A simplified representation of the cardiac cycle (Torii, Oshima, Kobayashi, Takagi, and Tezduyar 2010) was used to provide boundary conditions giving an inlet velocity that varies between 4 and 41 cm/s and a pressure with a range of 75121 mmHg, where an approximate phase shift of $0.08\,s$ is used (Ogoh, Fadel, Zhang, Selmer, Jans, Secher, and Raven 2005). In the isotropic model, the vessel wall was modeled with a MooneyRivlin material model with a density of $\rho = 1055\,$kg/m^3, and material coefficients

of $C_1 = 0.174\,MPa$, $C_2 = 1.88\,MPa$ (Johnson, Zhang, and Shimada 2011). In the anisotropic model, the vessel wall was modeled as an anisotropic hyperelastic material. Borrowing the idea from (Scherer, Treichel, Ritter, Triebel, Drossel, and Burgert 2011), we transfer the Holzapfel model in (Eriksson, Kroon, and Holzapfel 2009) to the AHYPER model in ANSYS, with a density of $\rho = 1055$ kg/m^3, an initial shear modulus of material $c = 0.3$ MPa, polynomial parameters $c_2 = e_2 = 120$ kPa, $c_4 = e_4 = 50.358$ kPa, $c_6 = e_6 = 14.088$ kPa with Taylor expansion.

The simulation results of the two meshes are shown in Figures 6 and 8. Compared to the isotropic mesh with isotropic material property, (1) the anisotropic mesh and the anisotropic material properties do not change the location and the value of the maximum stress much and the location is still on one side of the dome; (2) from Figures 6(c) and 6(d), we can observe that over the dome the stress changes faster along the axial direction than along the circumferential direction on the anisotropic mesh, while it is prone to be the same in each direction on the isotropic mesh. We can also obtain the same insight from Figure 8 by comparing it with the axial material direction distribution in Figure 5(b). Therefore, our simulation results suggest that the anisotropic mesh with anisotropic material properties generates a more reasonable stress distribution than the isotropic mesh with isotropic material properties.

4 CONCLUSIONS

The algorithm we propose includes the anisotropic mesh, non-uniform wall thickness and anisotropic material properties. The process embeds the anisotropic material characteristics through transforming the material directions of the healthy model onto the aneurysm and applying the spring relaxation and anisotropic material properties. From the simulation results, the anisotropic model demonstrates directional stress distribution along the material directions while the isotropic model remains the same in each direction. Future work will need to test more patient-specific models.

ACKNOWLEDGEMENTS

We would like to thank thank University of Pittsburgh Medical Center for providing us the patient-specific cerebral aneurysm models. This research was supported in part by the University of Pittsburgh Medical Center's Healthcare Technology Innovation Grant.

REFERENCES

Abruzzo, T., G. Shengelaia, R. Dawson, D. Owens, C. Cawley, and M. Gravanis (1998). Histologic and morphologic comparison of experimental aneurysms with human intracranial aneurysms. *American Journal of Neuroradiology 19*(7), 1309–1314.

Baek, S., K. Rajagopal, and J. Humphrey (2006). A theoretical model of enlarging intracranial fusiform aneurysms. *Journal of Biomechanical Engineering 128*, 142–149.

Bazilevs, Y., M. Hsu, Y. Zhang, W. Wang, X. Liang, T. Kvamsdal, R. Brekken, and J. Isaksen (2010). A fully-coupled fluid-structure interaction simulation of cerebral aneurysms. *Computational Mechanics 46*(1), 3–16.

Brisman, J., J. Song, and D. Newell (2006). Cerebral aneurysms. *New England Journal of Medicine 355*(9), 928–939.

Eriksson, T., M. Kroon, and G. Holzapfel (2009). Influence of medial collagen organization and axial in situ stretch on saccular cerebral aneurysm growth. *Journal of biomechanical engineering 131*, 101010 (7 pages).

Fung, Y., K. Fronek, and P. Patitucci (1979). Pseudoelasticity of arteries and the choice of its mathematical expression. *The American Journal of Physiology 237*(5), H620–H631.

Garimella, R. and B. Swartz (2003). Curvature estimation for unstructured triangulations of surfaces. *Tech. Rep. LA-UR-03-8240, Los Alamos National Laboratory*.

Humphrey, J. and C. Taylor (2008). Intracranial and abdominal aortic aneurysms: similarities, differences, and need for a new class of computational models. *Annual Review of Biomedical Engineering 10*, 221–246.

Jeong, Y., Y. Jung, M. Kim, C. Eun, and S. Jang (2009). Size and location of ruptured intracranial aneurysms. *Journal of Korean Neurosurgical Society 45*(1), 11–15.

Johnson, E. (2010). *Improving the accuracy of Fluid-Structure Interaction analyses of patient-specific cerebral Aneurysms*. Ph. D. thesis, Department of Mechanical Engineering, Carnegie Mellon University.

Johnson, E., Y. Zhang, and K. Shimada (2011). Estimating an equivalent wall-thickness of a cerebral aneurysm through surface parameterization and a nonlinear spring system. *International Journal for Numerical Methods in Biomedical Engineering 27*(7), 1054– 1072.

Kroon, M. and G. Holzapfel (2008). Estimation of the distributions of anisotropic, elastic properties and wall stresses of saccular cerebral aneurysms by inverse analysis. *Proceedings of the Royal Society A: Mathematical, Physical and Engineering Science 464*(2092), 807–825.

NINDS (2008). Cerebral aneurysm fact sheet. *NIH Publication* (08-5505).

Ogoh, S., P. Fadel, R. Zhang, C. Selmer, . Jans, N. Secher, and P. Raven (2005). Middle cerebral artery flow velocity and pulse pressure during dynamic exercise in humans. *American Journal of Physiology-Heart and Circulatory Physiology 288*(4), H1526–H1531.

Petitjean, S. (2002). A survey of methods for recovering quadrics in triangle meshes. *ACM Computing Surveys (CSUR) 34*(2), 211–262.

Rissland, P., Y. Alemu, S. Einav, J. Ricotta, and D. Bluestein (2009). Abdominal aortic aneurysm risk of rupture: patient-specific fsi simulations using anisotropic model. *Journal of Biomechanical Engineering 131*, 031001–1– 031001–10.

Sahni, O., J. Muller, K. Jansen, M. Shephard, and C. Taylor (2006). Efficient anisotropic adaptive discretization of the cardiovascular system. *Computer Methods in Applied Mechanics and Engineering 195*(41-43), 5634– 5655.

Scherer, S., T. Treichel, N. Ritter, G. Triebel, W. Drossel, and O. Burgert (2011). Surgical stent planning: simulation parameter study for models based on dicom standards. *International Journal of Computer Assisted Radiology and Surgery 6*(3), 319–327.

Scott, S., G. Ferguson, and M. Roach (1972). Comparison of the elastic properties of human intracranial arteries and aneurysms. *Canadian Journal of Physiology and Pharmacology 50*(4), 328–332.

Torii, R., M. Oshima, T. Kobayashi, K. Takagi, and T. Tezduyar (2010). Influence of wall thickness on fluid–structure interaction computations of cerebral aneurysms. *International Journal for Numerical Methods in Biomedical Engineering 26*(3-4), 336–347.

Analysis of left ventricle echocardiographic movies by variational methods: Dedicated software for cardiac ECG and ECHO synchronizations

Massimiliano Pedone

InfoSapienza ICT Center, Sapienza Università di Roma, Roma, Italy

ABSTRACT: The image sequences elaboration, which represent our main topic, concerns the identification of contours of an object for segmentation and study of its movement over time. A newer dynamic approach to the well-known static variational method for the time-series medical echocardiographic images is presented, and a graphic software applications for synchronization between cardiac movement and electric signals (ECG) are developed. Many approaches have been proposed to process time-series of digital images, and it is difficult establish the most effective one. Here we focus on PDE-based and variational methods.

1 INTRODUCTION

The image segmentation is a core component for medical application. We have studied the feasibility of applying some segmentation techniques for the determination of cardiac efficiency parameters, on echocardiographic image sequence of the left ventricle. To face this first issue in order to structure not invasive medical technique of analysis and to build up an automatic protocol for image sequence analysis. Standard snake models of closed curve evolution, pertaining to the Level-Set Method, have been implemented to characterize the internal ventricular area over time. We have applied standard finite difference techniques for the approximation solution of the involved eikonal equations for front evolution. Well-known techniques of Gaussian regularization, such as heat equation, have highlighted the loss of definition of ventricular edge Fig. 1a.

Thus we now present a numerical frame preprocessing technique, based on a variational model, that consists of functional minimization with a Mumford-Shah (M-S) time-dependent energy term, which is suitable to enhance ventricular edge and regularize initial data for curve evolution. Our goal is to retrieve the ventricular area frame by frame as in the Fig. 1b.

2 MODEL PROBLEMS

Standard snake models of closed curve evolution, pertaining to the Level-Set Method due to S. Osher and J. A. Sethian et al., have been implemented to characterize the internal ventricular area over time. Let us now recall some basic definition for a curve as isovalues of surface and its speed of propagation to which we refer the edge detection model problem. For the recognition of contour, the front of a parametrized curve

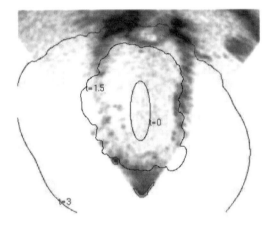

Figure 1a. Curve evolution fails on smoothed image.

Figure 1b. Recognized ventricular area.

evolves from an allocated initial configuration, of the domain $\Omega \subset \mathbb{R}^2$, to stops its evolution on the internal edge of the object.

2.1 Edge-detection

Given an image, in particular an echocardiographic image, we need to automatically detect the internal contour of object for clinical purposes. We focus ours attention to a ventricular cavity. We consider a closed subset Ω of $\mathbb{R}^2$, having $\partial\Omega$ as the external edge of the rectangular image and being $I: \Omega \to [0,1]$ the brightness intensity. The edge-detector (detector of contours) for segmentation of the image is a positive real coefficient, which is dependent of $I(x)$'s gradient at every point of the curve. In particular, the model is represented by a filter function such that

$$g : \mathbb{R}^+ \longrightarrow \mathbb{R}^+$$

where

$$g(z) = \frac{1}{1+z}$$

decreasing with z and

$$\lim_{z \to \infty} g(z) = 0$$

such that $0 \leqslant g$ and $g(0) = 1$.

A curve in $\mathbb{R}^2$ can be represented as the zero-level line of a function in higher dimension. More precisely, let us suppose that there exists a function $u: \mathbb{R}^2 \times \mathbb{R}^+ \to \mathbb{R}$ solution of the initial value problem:

$$\begin{cases} u_t(x,t) - g(|\nabla I(x)|)|\nabla u(x,t)| = 0; \\ \quad x \in \Omega \subset \mathbb{R}^2 \times [0,T] \\ u(x,0) = u_0(x); \\ \quad x \in \Omega \subset \mathbb{R}^2 \end{cases}$$

which is the model of **eikonal equation for front evolution**. We have applied standard finite difference techniques for the approximation solution of the involved eikonal equations for front evolution. We have to note that the evolving function $u(x,t)$ always remains a function as long as $I(x)$ is smooth. So we choose a different expression for the speed terms in order to limit the loss of image definition of Gaussian smoothing and by new approach with functional minimization to preserve this property.

2.2 Speed choice for image processing

The speed term is dependent of the brightness intensity at every pixel and is directed along the outward normal, starting from an initial elliptic profile E centered in Ω. Insofar we define the composite function

$$g(|\nabla(G_\sigma * I(x))|) = \frac{1}{1 + |\nabla(G_\sigma * I(x))|},$$

where $G_\sigma * I(x)$ is the convolution of the image $I(x)$ with a Gaussian regularizing operator G_σ. The operator G_σ is a filter, implemented by discrete heat equation, that allows to calculate the brightness-intensity gradient of the image in the presence of discontinuous data (Camilli and Siconolfi 2003). The discretization of the problem is performed by building a rectangular lattice, which is made fine-grained according to image's pixel definition. In the numerical tests the algorithm predicts a stop of the evolving curve at a threshold value th for the speed term, such that, if $g_I(x) \leqslant th \Rightarrow g_I(x) = 0$, then the threshold parameter allows curve evolution to stop in the presence of different gradient values for the regularized image.

2.3 Left ventricle echographic image

The ventricular contour of the cardiac muscle behaves like an elastic gum, so inducing a good performance of the model for a large part of image frames, but, at its maximum expansion, the ventricle reveals low echo response from thin tissue at various locations, so causing the image-frame to be affected by the presence of a hole in the edge delimitation and the curve to go out. The standard choice of threshold-parameter value at variable number of smoothing iterations, as we can see, is insufficient to give a correct stop of the curve within a small range of the pixels neighboring the ventricle edge. Indeed, this kind of image, with its amount of noise so overlapping with the studied signal, suggests us to apply a different technique in the frame analysis. Many methods and different approaches are currently available and lots of papers were written about this specific topic. An exhaustive work, due to Stanley Osher et al., is (OSher and Paragios 2003). As a large part of other scientists, expert in image analysis, we choose here to investigate a newer approach including a pre-processing step, based on functional minimization, to enhance the border, following the idea that image noise is quite static over time, while studied signal (ventricle border, in our case) is rhythmically moving. Thus, let us try to apply the Mumford-Shah functional to this dynamic problem.

2.4 Regularization and evolution choice

The standard convolution with a regularizing operator, allows to compute the bright-intensity gradient of the image for velocity term of the curve evolution.

In the continuous model problem the image has treated as a surface, the speed term $V(x,t,\nabla(I(x))) = g_I$ in the equation depends of its gradient.

If exists a fracture the gradient jumps to infinity value. Then a regularization is needed! Regularization choice

- **Convolution with Mollifier:** *level-set* standard model

$$g_I(x) = g(|\nabla(G_\sigma * I(x))|) = \frac{1}{1 + |\nabla(G_\sigma * I(x))|}$$

268

- **Regularization and Edge Enhancing** by functional minimization

$$g_I(x) = \frac{1}{1 + |\nabla u_{k_\epsilon}(x)|},$$

u_{k_ϵ} minimum of an energy functional

3 APPROXIMATION AND ANALYSIS

Analysis of images through variational methods finds application in a number of fields such as robotics, elaboration of satellite data, biomedical analysis and many other real-life cases. By segmentation is meant a search for constituent parts of an image, rather than an improving of its quality or characteristics. Our aim is to develop a criterium for enhancing movie frame in two possible ways by functional minimization. First, we adopt the M-S functional (Mumford and Shah 1989) and its approximated form proposed by Ambrosio Tortorelli (Ambrosio and Tortorelli 1990) to regularize data frame for curve evolution method, instead of Gaussian regularization. We call it "classic functional" since the function space the variational integral converges over contains functions that have the regularity needed for image gradient calculus. Second, we present a numerical scheme, where a time-dependent parameter is inserted, precisely in the second integral part of functional distinguished by a "time gradient", for enhancing the internal moving parts of the object. A detailed analytical treatment and a numerical scheme for minimization of the functional, which involves some delicate conjectures and refined mathematical steps, can be found in (Aubert and Kornprobst 2002). In the following section we recall in brief the essential formulation of the model problem used to regularize and enhance image. The reader can look up, for a complete review, the book by Morel and Solimini (Morel and Solimini 1995). Thus we now present a numerical frame preprocessing technique, based on a variational model, that consists of functional minimization

$$F_\epsilon(u, S) = \mu \int_\Omega (u - g)^2 \, dx + \int_\Omega S^2 \mid \nabla u \mid^2 dx +$$

$$+\alpha \int_\Omega \left(\epsilon \mid \nabla S \mid^2 + \frac{1}{4\epsilon} (1 - S)^2 \right) dx$$

with a Mumford-Shah (M-S) time-dependent energy term $|\nabla u|^2$, which is suitable to enhance ventricular edge and regularize initial data for curve evolution.

3.1 Approximation of the functional

The numerical scheme is made by dividing in two coupled parts with $u_0 = g$ and $S_0 = 1$. At every step we calculate u_1 for $S_0 = 1$ solving a linear elliptic equation and, this way, we find S_1 from the second equation; this process is repeated for a fixed number of iterations.

3.2 System discretization

We use a numerical scheme based on explicit finite differences, over the rectangle Ω with step h; this way we obtain $(x, y) = (ih, jh)$ for $0 \leqslant i, j \leqslant N$. Reducing Ω to a square of side 1 and taking $h = 1/N$, discrete coordinates will become: $u(ih, jh) \cong u_{i,j}$ and $S(ih, jh) \cong S_{i,j}$. We use an approximated scheme that is enough to enhancing little areas, characteristic of the echographic image. Some scheme for the Ambrosio-Tortorelli segmentation problem can be found on Spitaleri et al. article (Spitaleri et al. 1999). In order to determine the minimum of the functional we adopt the schema given, for a finite element, in Birindelli and Finzi Vita (Birindelli and Finzi-Vita 1998) to a finite difference mesh grid through the following iterative scheme:

given a maximum number of iteration Nit and a tolerance ϵ, then we construct:

– $S_0 = 1, u_0 = g$
– for $n = 1, 2, \ldots, Nit$ find u_n, by solving:

$$\begin{cases} div(2(S_{n-1}^2 + K_\epsilon)\nabla u_n) = \mu(u_n - g) & \text{in } \Omega \\ \frac{\partial u_n}{\partial \mathbf{n}} = 0 & \text{in } \partial\Omega \end{cases}$$

and S_n by solving:

$$\begin{cases} \alpha\epsilon\Delta S_n = S_n \mid \nabla u_n \mid^2 - \frac{\alpha}{4\epsilon}(1 - S_n) & \text{in } \Omega \\ \frac{\partial S_n}{\partial \mathbf{n}} = 0 & \text{in } \partial\Omega \end{cases}$$

– stop for $n = Nit$.

From minimization theorem 3.1 Proposition 2.1 in (Birindelli and Finzi-Vita 1998), it respectively follows that:

S_u is a piecewise C^2 submanifolds of $\mathbb{R}^2$

for any $n > 1$ there exists u_n an S_n solution of the respective system which satisfy the bounds:

$$\|u_n\|_{L^\infty} \leq \|g\|_{L^\infty}.$$

Then we discretize the equations by finite differences.

3.3 Applicability conjecture

The applicability of the evolutionary method based on curve evolution is conditioned, in the presence of discontinuous data, by restrictive hypotheses the punctual dependence of the speed term. Then the existence of the spatial gradient of the brightness-intensity function requests the convolution, in Ω-domain, with a standard mollifier. In the approximation of the functional, the sequence of functional minima are iterated by the system by Ambrosio Tortorelli method. The technique to fix every equation arise to a quasi-linear class of elliptic systems explained in the paper (Birindelli and Finzi-Vita 1998).

Conjecture Bounded gradient at given iteration If u_{k_ϵ} represents a solution at k^{th} iteration of the Ambrosio-Tortorelli sequence (Ambrosio and Tortorelli 1990) given from alternate solution of the elliptic system for fixed number of iteration, then u_{k_ϵ} is enough smooth to calculate $|\nabla u_{k_\epsilon}(x)|$ i.e. $\|\nabla u_{k_\epsilon}(x)\|_\infty < \infty$; $\forall x \in \Omega^\circ$.

Sketch of Proof. In the internal points of the domain Ω the amplitude of the fracture is ϵ's proportional, then we can found, at every step of the iterative solution of AT_ϵ algorithm, a constant C_{Nit} such that

$$|\nabla u_{k_\epsilon}(x)| \le \frac{C_{Nit}}{\epsilon}$$

then $v(x) = \dfrac{1}{1 + |\nabla u_{k_\epsilon}(x)|} \ge \dfrac{1}{1 + \dfrac{C_{Nit}}{\epsilon}} > 0.$

It is then possible to calculate the speed term in the eikonal equation ☐.

As an extension of one dimensional results by (Francfort et al. 2009).

3.4 *Approximation algorithm*

At every iteration we obtain a regularized pair of approximated functions dependent from ϵ. This suggests to leave Gaussian regularization aside, with smaller loss of details, using the function of the K_ϵ item of the iteration sequence as the brightness-intensity function for curve evolution. u-function can directly be used in the eikonal eq. for front evolution. This technique is used in (Chan and Vese 2001), but optical flow theory is left aside (see (Aubert and Kornprobst 2002)); This choice provides the needed regularity in the fracture (see (Francfort et al. 2009)) that represents the critical points of brightness-intensity function g_I of the given image.

A further step of technique refinement, in order to make ventricular walls get emphasized, consists, as mentioned above, of making the second term of the functional dependent of the gradient over the time (Chan and Shen 2005).

3.5 *Time dependence in ∇*

Different approach in Gradient calculus on a Lattice of h spacing: Standard spatial way respectively for two or three points stencil:

$$|\nabla u|^2_{i,j} = \frac{1}{4h^2}\left((u_{i+1,j} - u_{i-1,j})^2 + (u_{i,j+1} - u_{i,j-1})^2\right).$$

$$|\nabla u|^2_{i,j} = \frac{1}{4h^2}\left(\frac{(u_{i+1,j} - u_{i,j})^2 + (u_{i,j} - u_{i-1,j})^2}{2}\right) +$$

$$+ \frac{1}{4h^2}\left(\frac{(u_{i,j+1} - u_{i,j})^2 + (u_{i,j} - u_{i,j-1})^2}{2}\right).$$

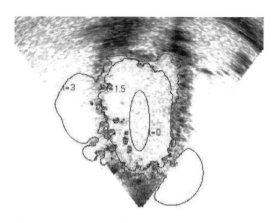

Figure 2c. Curve evolution on image regularized by M-S functional.

Following the **optical flow** model (see page 184 (Aubert and Kornprobst 2002)), we introduce dependence on time in the gradient term $(|\nabla u|^{(f)})^2$ respectively for two or three points stencil.

$$(|\nabla u|^{(f)}_{i,j})^2 = |\nabla u|^2_{i,j} + \frac{1}{4h^2}\left((u^{(f-1)}_{i,j} - u^{(f+1)}_{i,j})^2\right)$$

$$(|\nabla u|^{(f)}_{i,j})^2 = |\nabla u|^2_{i,j} + \frac{1}{4h^2}\left(\frac{(u^{(f-1)}_{i,j} - u^{(f)}_{i,j})^2 + (u^{(f)}_{i,j} - u^{(f+1)}_{i,j})^2}{2}\right).$$

Represents a new formulation of the model with a dynamic "mean" between frame. The square gradient is point by point calculated over function u as the difference in brightness intensity between the preceding and the following frame. Globally, the variability of this term is mostly due to movements of ventricular walls. In the presence of a continuous movement in the same "scene", we are in condition to assume regularity for function g_I over time.

4 APPLICATION AND RESULTS

We show the formulation of the standard model problem for image segmentation by curve evolution and a newer approach of pre-treating model by functional minimization. The regularization and edge enhancing obtained by pre-processing give back an image enhanced. This image frame of the movie presents a better resolution. In particular the pixels nearest the ventricular edge are emphasized than the other Fig. 2c. That choice produce a good results so we can compute the ventricular area over the frames Fig. 2d.

4.1 *Protocol for movie processing*

We describes the digital format of movies, the adaptation of single frames, the step which characterize the elaboration protocol Fig. 2e. We report the quality of the results obtained with the preprocessing on a real movie.

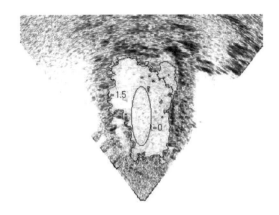

Figure 2d. Curve evolution on image frame pre-treated by time dependence.

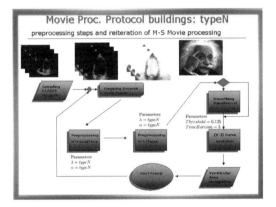

Figure 2e. Frame to frame elaboration protocol.

4.2 *Medical Parameter*

We compare results obtained by various preprocessing methods on identified areas,the curve evolution beyond the edge, ventricular area his trends over time and the evaluated Ejection Fraction. The Fig 3.2 shows results from the elaboration representing the main goal of this work: the trend of ventricular area over time. The max and min peak of expulsion for the diastolic cycle and the calculated Ejection Fraction are shown as well. We want to remark that a non-conventional information is provided: time needed for expulsion, which offers a possible field of investigation about internal ventricle pressure. In cardiovascular physiology, ejection fraction (E_f) is the fraction of blood pumped out of a ventricle with each heart beat. By definition, the volume of blood within a ventricle immediately before a contraction is known as the end-diastolic volume. Similarly, the volume of blood left in a ventricle at the end of contraction is end-systolic volume. The difference between end-diastolic and end-systolic volumes is the stroke volume, the volume of blood ejected with each beat. Ejection fraction (E_f) is the fraction of the end-diastolic volume that is ejected with each beat; that

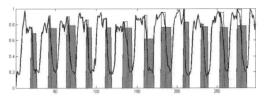

Figure 2f. Ventricular AREA trends and ejection fraction.

is, it is stroke volume (SV) divided by end-diastolic volume (EDV):

$$E_f = \frac{SV}{EDV} = \frac{EDV - ESV}{EDV}$$

In the last figure the values of E_f is represented by the dark green line, the red one is the maximal value of EDV and the cyan one is ESV, These values are only an appraisal, because referable to an area and not to a volume information. There we show a static reproduction of the implemented graphic Matlab[©] application. The input consists of three file. The echographic movie in avi format, the area trend with the Ejection Fraction and the twelve tracks of Electrocardiography simultaneous data trends. The principal use of the signal-synchronizing software consists of the reproduction of the data files and the possibility to establish a delay time between Echo and ECG, see Fig 2g. Once the delay time is recognized the medical staff can appreciate, especially for the arrhythmic beat, a detailed innovative and non invasive analysis of the patient. In facts they see the dynamic evaluation in time of the single ejection fraction beat to beat related to the electrics signals. This work is a first step of an extended project whose final target is the analysis of Echocardiographic image sequences for non-invasive and a-posteriori medical diagnostics of heart left-ventricle diseases. The medical protocol requires the determination of local pressure and internal volume of the left ventricle during its cyclic work.

In this first step, which has many up-gradable features, a specific method has been developed, based on mathematical models of image processing, for the recognition of ventricular area in different frames of Echocardiographic images.

5 CONCLUSIONS AND DEVELOPMENTS

The exposed methods and their applications are an approach to answer to requests from medical diagnostics.

From the point of view of applied mathematics, numerical techniques have been used in order to obtain results which fit the problem. In echocardiographic images very ragged contours are usually available and pixel definition is low, but the described sophistication of the M-S method does not require the use of onerous numerical algorithms for convergence to the

271

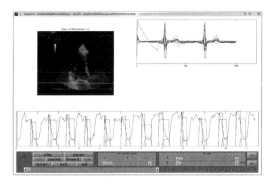

Figure 2g.   Echo ECG graphic signal synchronizer.

functional minimum. In agreement with cardiologists involved in this project, future developments could be:

- The elaboration of ventricular areas by a ventricular "short and long side" approach, in order to obtain a better volumetric reconstruction and a 3D analysis.
- The validation of theories which allow the interpretation of correspondences between the electric signal (ECG) and ventricular volume filling (ejection fraction).
- The implementation of this protocol for an Internet service for reconstruction of the ventricular area profile, which would prove useful for modern e-health-care.
- The study of mobility profiles of ventricular walls to draw speed vector-field curves during cardiac periodical movement; this would prove useful for the individuation of muscular portions affected by diseases.
- The possibility to model, by finite-element analysis, the ventricular pump to make internal pressure (Urheim et al. 2002) and volume fit experimental results in the comparison between real data and simulation.

From a computational point of view, Fast Marching Methods are a possible improvement to reduce CPU time for curve evolution by eikonal eq.

REFERENCES

Ambrosio and Tortorelli (1990). Approximation of functionals depending on jumps by elliptic functionals via γ-convergence. *Comm. Pure Appl. Math. 43*, 999–1036.

Aubert and Kornprobst (2002). *Mathematical Problems in Image Processing*. Springer Verlag.

Birindelli and Finzi-Vita (1998). A class of quasi-linear elliptic systems arising in image segmentation. *Nonlinear Differential Equations and Applications NoDEA 5*, 445–449.

Camilli and Siconolfi (2003). Hamilton-jacobi equation with a measurable dependence on the state variable. *Advances in Differential Equation. 8(6)*, 733–708.

Chan and Shen (2005). Variational image inpainting. *Comm. on Pure and Appl. Math. LVIII*, 579–619.

Chan and Vese (2001). Active contour without edges. *IEEE Transaction on Image Processing 10*, 266–276.

Francfort, Le, and Serfaty (2009). Critical points of ambrosio-tortorelli convergence to critical points of m-s in the one dimensional dirichlet case. *ESAIM: Control, Optimisation and Calculus of Variation 15*, 576–598.

Morel and Solimini (1995). Variational methods in image segmentation. *Progress in Nonlinear Differential Equations and Their Applications..*

Mumford and Shah (1989). Optimal approximation by piecewise smooth functions and associated variational problems. *Comm. Pure Appl. Math. 42*, 577–685.

Osher and Paragios (2003). *Geometric Level Set Methods in Imaging, Vision, and Graphics*. Springer.

Spitaleri, March, and Arena (1999). Finite difference solution of euler equation arising in variational image segmentation. *Numerical Algorithms*, 353–365.

Urheim, Bjornerheim, and Endersen (2002). Quantification of left ventricular diastolic pressure-volume relations during routine cardiac catheterization by two-dimensional digital echo quantification and left ventricular micromanometer. *Journal of the American Society of Echocardiography 15*, 225–232.

A survey of echographic simulators

A. Ourahmoune
USTHB Faculty of Electronic and Computer Science, Algiers, Algeria
Telecom Bretagne, Brest, Bretagne, France

S. Larabi
USTHB Faculty of Electronic and computer science, Algiers, Algeria

C. Hamitouche-Djabou
Telecom Bretagne, Brest, Bretagne, France

ABSTRACT: Echographic training simulators are becoming prolific. They are deployed in a multitude of different forms, from full scale responsive mannequins with dummy probe to computer-based interactive with virtual representation of the body, the probe, a detailed 3D anatomic structures and haptic feedback. This paper presents a survey of different echographic simulators in the scope of proposing a virtual environment for echographic training.

1 INTRODUCTION

Echography is a very useful diagnostic tool in almost every medical field. This technique is real-time, non-invasive and less expensive but it is operator-dependant, echographic images don't give direct correspondence to anatomy representation and their quality is poor due to artifacts presence. All of these drawbacks make image interpretation difficult.

Nowadays, echographic training still follows the classical two stage model: theoretical knowledge comes with using the major textbooks in the field and practical knowledge (experience) is only generated by performing ultrasound examination as many as possible which implies a long training technique. This training method depends on the access to the patients, chance of pathologies' met and the availability of the teachers. In (Köhn et al 2004), the authors show that after completing one-year training course, students will have seen only 80 % of the possible findings. Consequently, every day, medical personnel are asked to diagnose ultrasound images without appropriate ultrasound training.

To overcome these obstacles and improve training efficiency, the scientists introduced simulation as new way of training; where neither patients nor ultrasound equipment would be necessary. The first medical simulators were simple models of human patients in clay used to demonstrate clinical features of disease states since antiquity (Meller 1997). In 1968, one of the earliest medical simulators for trainees was produced by university of Miami; the mannequin performs more than 27 different cardiac functions of the human body like varying blood pressure and pulse (Gordon 1974). More recently, interactive models have been developed that respond to actions taken by a physician for fiber-endoscopic training (Gillies & Williams 1987). The last generation of medical simulators is computer-based which are able to reproduce touch and force feedback feelings using specific joysticks such as the phacoemulcification simulator (Choi et al. 2009).

In this paper an overview of echographic simulators will be presented and discussed. We will present in section 2 the state of art of existing echographic simulators. In section 3 we will discuss the lacks in the existing echographic simulators and finally we will conclude in section 4 and present our perspectives and future work in section 5.

2 EXISTING ECHOGRAPHIC SIMULATORS

Nowadays, echographic simulators have blossomed we present the most interesting ones.

2.1 UltraSim training echographic simulator: (www.medsim.com)

It uses a mannequin and a dummy probe equipped with a 3D motion sensor to register the relative position and orientation of the probe. The ultrasound image is created using a 3D ultrasound volume acquired on real patient as described in (Aiger & Cohen-Or 1998). This simulator is unrealistic since it doesn't represent echographic image deformation under probe compression. Moreover, the haptic feedback feeling using the dummy probe on the mannequin is far from to be equivalent to the one on human.

2.2 Ultratrainer: (Stallkamp & Wapler 1998)

This simulator uses a 3D ultrasound database taken from real ultrasound examinations where the shape and orientation of the mannequin which replace the real body are registered. The matching between the cross section of the dummy probe on the scan field area and the selected slice from the ultrasound volume is realized using a 3D tracking system on the dummy probe. The student can move the dummy probe freely on the scan field area but the feeling of haptic feedback is not realistic and no image deformation is provided under probe compression.

2.3 Sonosim3D: (Ehricke 1998)

It uses a virtual probe and the human body is represented by a virtual patient model through 3D image volumes MRI and CT scans. 2D ultrasound images are real time generated from a 3D ultrasound data which are calculated from a set of 2D echographic images using a tri-linear interpolation method. This simulator doesn't represent echographic image deformation under probe compression and doesn't permit to trainees feeling haptic feedback.

2.4 A virtual reality training system for pediatric sonography: (Arkhurst et al. 2001)

It simulates an ultrasound examination of baby's head. It allows the exploration of detailed 3D anatomical model based on MRI selected by the cross section of the virtual probe and matched with the real ultrasound and MRI images registered. This simulator has only two degrees of freedom which limits movements of the probe. It doesn't allow users feeling haptic feedback and doesn't represent echographic image deformation under probe compression.

2.5 A virtual simulator for thrombosis detection: (Laugier et al. 2001)

It uses a virtual 3D representation of the probe and a virtual model of the thigh. The ultrasound image is generated in real time using 3D data acquired from patient using linear interpolation method over a set of 2D echographies. The deformation of the 2D ultrasound under probe pressure is done by using an octree spline image deformation method following the arterial deformation but ignores the non homogenous anatomical reaction under probe compression. The haptic feeling is real time calculated using a simplified masse spring model; the biomechanical data are registered using an indentation method. This simulator uses a 3D haptic joystick to manipulate the virtual probe.

2.6 Echocardiography simulator: (Maul et al. 2004)

This simulator uses a mannequin and a dummy probe equipped with a magnetic 3D tracking system which registers the relative position and orientation of the probe to the mannequin. These information are used to real time calculate the 2D ultrasound image from a pre enquired real 3D ultrasound data. This simulator doesn't offer realistic haptic feedback due to the use of a plastic mannequin and doesn't represent echographic image deformation under probe compression.

2.7 SonoTrainer: (Weidenbach et al. 2004)

It simulates normal and abnormal fetal scans under real-time conditions of a late-first or early second trimester pregnancy and fetal echocardiography, providing full motion of the fetal heart. The system provides also al measurement functions such as distance. The real body is replaced with a mannequin with a soft, pliable, rubber surface which provides a non realistic haptic feedback. A 3D tracking system is used to register the position of the dummy probe, corresponding to the ultrasound cross section image calculated from 3D ultrasound data. This simulator doesn't represent echographic image deformation under probe compression.

2.8 TRUS: (Persoon et al. 2010)

It is a simulator for teaching transrectal ultrasound. It uses a mannequin equipped with an electromagnetic localization system. Real-time 3D ultrasound data are obtained from ultrasound prostate imaging of real patients using 3D ultrasound machine and transferred to the simulator using a connection between the computer of the simulator and the ultrasound machine. The simulation of tissue resistance is done only by simulating the opening in the mannequin with foam plastic, but the external and internal anal sphincter are not simulated and don't represent echographic image deformation under probe compression.

2.9 Echocardiogram simulator: (Sun & McKenzie 2011)

It combines a virtual simulator including a virtual probe and a 3D model of the human heart with mannequin simulator using a tracker based system. The ultrasound image is synthesized in real time based on 3D model of the heart according to the position and the orientation of the dummy probe which is combined with a precompiled artificial ultrasound texture based on the linear convolution *Bamber* and *Dickinson* model (Bamber & Dickinson 1980) and mapped to the 2D image using a basic scan-fill algorithm, finally a background noise is added to the resulting image without representing deformation under probe compression.

2.10 Schallware: (www.schallware.com)

It is an abdominal, cardiac and obstetrical simulator which uses a mannequin and a dummy probe with unknown tracking and calibration system. The 2D ultrasound image is synthesized using a 3D ultrasound data acquired from real patients using unknown interpolation and re-slicing methods. This simulator also doesn't allow realistic haptic feeling.

Table 1. The set of simulators.

Simulator	2D US simulation	Body representation	Probe representation
UltraSim	3D US D	mannequin	Dummy probe 3D MS
UltraTrainer	3D US D	mannequin	Dummy probe 3D MS
SonoSim3D	3D US D	Virtual body	Virtual probe
Pediatric Simulator	3D US D	Virtual body	Virtual probe 2DOF
Echocardio-graphy simulator	3D US D	mannequin	Dummy probe 3D MS
Thrombosis simulator	3D US D	Virtual body	Virtual probe 3DOF
SonoTrainer	3D US D	mannequin	Dummy probe 3D MS
TRUS	3D US D	mannequin	Dummy probe 3D MS
Echocar-diogram simulator	Linear convolution	Mannequin Virtual body	Dummy probe 3D MS & virtual
Schallware	3D US D	mannequin	Dummy probe
TEE	Videos US	Virtual body	Virtual probe
Scan trainer	NoI	Virtual body	Virtual probe

US: indicates (Ultra Sound), NoI: indicates (No information available), DOF: indicates (Degrees Of Freedom), MS: indicates (Motion Sensor), D: indicates (Data).

2.11 TEE echocardiogram training: (pie.med.utoronto.ca/TEE)

It is an online virtual transesophageal echocardiography based on video clips TEE views and those of the transitional movements between them. It includes a 3D rotational model of the heart and the internal structures that the echo plane passes through, a virtual probe and the ultrasound plane position. There is no haptic feedback represented.

2.12 Scan trainer: (www.scantrainer.com)

It is a completely virtual endovaginal echographic simulator which includes a virtual 3D model of the anatomy, a realistic 2D ultrasound image and allows haptic feedback with 6 degrees of freedom haptic joystick.

A summary of the presented simulators is presented in table 1.

All the presented simulators are discussed in the next section.

3 DISCUSSION

The use of the simulator in complement of clinical experiment can help the student to improve his clinical competence. Over our study, we noted that the haptic feedback on the existing simulators is far from to be cross to reality; in the simulators with a mannequin and a dummy probe (Sun & McKenzie 2011) (Stallkamp & Wapler 1998) (Maul et al. 2004) (Weidenbach et al. 2004) (www.schallware.com)

(Persoon et al. 2010), the haptic feeling doesn't match the one in real case because the elasticity of the plastic mannequin and human structures are completely different. In (Arkhurst et al. 2001) haptic feeling isn't incorporate at all but in (Laugier et al. 2001) the calculated haptic feedback is based on a simplified biomechanical model to make the calculi frequency equivalent to these of the joystick which is about 1 KHZ, so the haptic feeling becomes no realistic. Also deformation of the ultrasound image under probe pressure is not realistic because the author deforms the entire image following the deformation of only the arteries which is not true because each anatomical structure has a specific deformation. As far as the simulated 2D ultrasound images are concerned, two approaches exist in literature. The first one is the generative approach, it synthesized echographic texture using methods based on ray tracing model (Jensen 1996) and methods based on a linear convolution model (Bamber & Dickinson 1980), this approach can be used in the case of dynamic structures but it is time consuming, it was adopted in (Persoon et al. 2010). The second one (Rohling et al. 1999) is interpolation based using the 3D ultrasound volume acquired off–line from real patients, this method is quiet fast to be used in real time but it needs to choose the adequate interpolation method and it is not adapted for completely dynamic structures, most of the presented simulators have used it. In terms of training, the existing simulators resolve the problem of training on patients, and preserve their safety but don't present solution for the non availability of the supervisors. No simulator discusses cost and adaptability for large use in trainee community; the presented simulators use a 3D tracking motion system or expensive haptic joystick which makes them no adapted for trainees. We note that there is a lack of research showing effectiveness, transfer of training to the clinical environment and cost-effectiveness study, these are barriers to diffusion of the technology. We can just note the INIST-CNRS study's (Monsky et al. 2002), it evaluates the effectiveness of an ultrasound simulator over 2 years of study and concludes that self-assessment of resident knowledge and their abilities for image scanned interpreting are also significantly improved.

4 CONCLUSION

In this paper, we have presented the state of art of echographic simulators. We have detailed each existing echographic simulator and discussed their functionalities. We conclude that although the existing simulators make possible the students training safely, the physical presence of the expert is still necessary.

5 TOWARDS AN OPTIMAL ECHOGRAPHIC SIMULATOR

In order to offer to trainees an optimal training method, we propose to realize not a simple simulator but a whole virtual environment to avoid the physical presence of the instructor. Both the instructor and the trainees have access to the virtual simulator on a local network allowing the supervision of trainees by instructor. The virtual simulator includes a virtual scene which represents a virtual body and a virtual probe manipulated with accessible haptic joystick. The simulator displays the ultrasound 2D image simulated in real time based on real ultrasound data and the corresponding detailed anatomy structures in 3D with each probe manipulation. The ultrasound data are registered dynamically off simulation with a corresponding position and orientation of the probe using a 3D tracking system and the adequate pressure calculated using biomechanical features of human structures. Furthermore, we propose to incorporate to the system an intelligent agent to supervise the student avoiding a virtual presence of the instructor.

REFERENCES

Aiger, D. & Cohen-Or, D. 1998. Real Time Ultrasound Imaging Simulation. *Real time Imaging* 4: 263–274.

Arkhurst, W. Pommert, A. Richter, E. Frederking, H. Kim, S. I. Schubert, R & Hohne K. H. 2001. A virtual reality training system for pediatric sonography. *Proc. Computer Assisted Radiology and Surgery (CARS)*: 453–457.

Bamber, J. C. Dickinson, R. J. 1980. Ultrasonic B scanning: a computer simulation. *Physics in Medicine and Biology* 25: 463–479.

Choi, Kup-Sze. Soo, Sophia & Chung, Fu-Lai. 2009. A virtual training simulator for learning cataract surgery with phacoemulsification. Elsevier 39 (11): 1020–1031.

Ehricke, H. H. 1998. Sonosim3d: A multimedia system for sonography simulation and education with an extensible case database. *European Journal of Ultrasound* 7: 225–300.

Gordon, MS. 1974. Cardiology patient simulator: development of an automated manikin to teach cardiovascular disease. *The American Journal of Cardiology* 34 (3): 350–355.

Gillies, DF & Williams, CB. 1987. An interactive graphic simulator for the teaching of fibrendoscopic techniques. In: G. Marechal (ed), *Eurographics'87*: 127–138.

Jensen, J. A. 1996. Field: a program for simulating ultrasound systems. *Medical & Biological Engineering & Computing,* 34: 351–353.

Köhn, S. van Lengen, R.H. Reis, G. Bertram, M. & Hagen, H. 2004. VES: Virtual Echocardiography System. *4th IASTED International Conference on Visualization Imaging and Image Processing (VIIP-04)*: 465–471.

Laugier, C. Mendoza, C. & Sundaraj, K. 2001. Towards a Realistic Medical Simulator Using Virtual Environments and Force Feedback. *International Symposium in Research Robotics ISRR-2001. Australia.*

Maul, H. Scharf, A. Baier, P. Wustemann, M. Gunter, H. H. Gebauer, G. & Sohn, C. 2004 . Ultrasound simulators: Experience with the sonotrainer and comparative review of other training systems. *Ultrasound ObstetGynecol* 24: 581–585.

Meller, G. 1997. A Typology of Simulators for Medical Education. *Journal of Digital Imaging* 10 (3, S1): 194–196.

Monsky, W L. Levine, D. Mehta, T. S. Kane, R. A. Ziv, A. Kennedy, B & Nisenbaum, H. 2002. Using a Sonographic Simulator to Assess Residents Before Overnight Call. *American Roentgen Ray Society* 178: 35–39.

Persoon, Marjolein C. Schout, Barbara M.A. Martens, Elisabeth J. Tjiam, Irene M. Tielbeek, Alexander V. Scherpbier, Albert J.J.A. Witjes, J. Alfred & Hendrikx, Ad J.M. 2010. A Simulator for Teaching Transrectal Ultrasound Procedures TRUS. *Society of simulation in Healthcare* 5: 311–314.

Rohling, R. N. Gee, A. H & Berman, L. 1999. A comparison of freehand three-dimensional ultrasound reconstruction techniques. *Med Image Anal* 3(4):339–359.

Stallkamp, J & Wapler, M. 1998. Ultratrainer: a training system for medical ultrasound examination. *Studies in Health Technology and Informatic* 50: 298–301.

Sun Bo & McKenzie Frederic D. 2011. Real-Time Sonography Simulation for Medical Training. *International journal of education and information technologies* 5(3):328–335.

Weidenbach, M. Trochim, S. Kreutter, S. Richter, C. Berlage, T. & Grunst, G. 2004. Intelligent training system integrated in an echocardiography simulator. *Computers in Biology and Medicine* 34:407–425.

http:// www.medsim.com

http://www.schallware.com

http://pie.med.utoronto.ca/TEE

http://www.scantrainer.com

Computational Modelling of Objects Represented in Images – Di Giamberardino et al. (eds)
© 2012 Taylor & Francis Group, London, ISBN 978-0-415-62134-2

Hip prostheses computational modeling: Mechanical behavior of a femoral stem associated with different constraint materials and configurations

I. Campioni
Istituto Superiore di Sanità, Rome, Italy

U. Andreaus
Sapienza University of Rome, Rome, Italy

A. Ventura
Orthopedic Trauma Center "A. Alesini", Rome, Italy

C. Giacomozzi
Istituto Superiore di Sanità, Rome, Italy

ABSTRACT: The paper deals with finite element (FE) models of femoral stems aimed at supporting fatigue mechanical tests according to Medical Devices Standards. A basic model was created in agreement with the ISO7206-4:2010 requirements and used to investigate a certain number of varied configurations: the abduction and the flexion angle were varied in the range $2° \div 15°$ and $2° \div 13°$ respectively, and the constraint level in the range $(30 \div 80)$mm. Once the most critical configuration had been identified, femur-like FE models were created to investigate it in a context closer and closer to the in-vivo scenario: the models, in fact, were based on a femur-like fixture, and on various materials to simulate cement and cancellous bone. Both Titanium and Co-Cr-Mo alloys were used for the stem. For both alloys the highest stresses were found in correspondence with a 80 mm constraint and a more soft cancellous bone; higher risky deflection was estimated for the Titanium stem. The great potential of FE methodology and analysis is here commented, as a valid support to experimental tests.

1 INTRODUCTION

Fatigue behavior of hip joint prostheses represents a crucial aspect to be investigated before introducing the devices on the market. Classified as high risk Medical Devices (MD) for permanent replacement of the hip joint, they need to prove to be adequate for replacing the anatomical joint function for quite a long period, under different loading conditions depending on patient's activity, body mass, age, clinical status. Manufacturers and Notified Bodies all over the world mainly design and implement their assessments on the basis of International Standards, harmonized or not, which help to demonstrate compliance with MD Regulations. Being mechanical fatigue tests destructive, simulation of loading conditions by means of proper computational modeling seems to be extremely attractive: the implementation of static analyses under varied conditions may thus help to identify the most critical configurations to focus the mechanical tests on.

In the present paper, simple Finite Element (FE) models of femoral stems are implemented, based on ISO7206-4:2010 testing requirements and on a previously validated FE model (Campioni et al, in press);

main results and their clinical relevance are here reported and commented.

2 METHODS

2.1 Fatigue experimental tests

In EU, prosthetic femoral stems are class III MD. Usually, Manufacturers and Notified Bodies voluntarily apply the harmonized Technical International Standard EN ISO 21535 to be compliant with MD Directives 93/42/EC and 2007/47/EC. The above Standard makes reference to Operating Procedures, Testing Apparatus and Testing Conditions described in the Technical Standard ISO7206, last revised in 2010. In particular, Part 4 of the ISO7206 is dedicated to mechanical fatigue tests on femoral stems. The rationale for the Standard testing methodology relies on the hypothesis that, once correctly inserted in the femoral channel and following a proper function recovery process, the stem reaches a full, solid and durable integration. During time wear, misuse, pathological conditions or other factors may induce a certain mobilization in the upper part of the implant. Solid contact usually remains in the distal part of the stem, and the main aim of the fatigue mechanical tests is just to verify that this "residual" constraint is sufficient to prevent the stem fracture and the implant failure for at least 10 years of regular use.

To this purpose, and trying to replicate the expected *in-vivo* condition, the Standard suggests to test the stem as follows: i) it must be embedded in a cement mantle – the cement elasticity being comparable with that of surgical bone cement – up to a pre-defined distance D from the centre of the head which corresponds to 80 mm for stems in the range (120 to 250) mm; ii) stem adduction angle α must be set to 10°, stem flexion angle β to 9°, and a rotation equal to the eventual anti-version angle must be performed before cementing the stem; iii) a sinusoidal load in the range (300 to 2300) N must be applied with a frequency in the range (1 to 30) Hz. The fatigue test has to be continued until the deflection of the head exceeds 5 mm, or fracture of the specimen occurs, or the maximum number of requested cycles (5×10^6 cycles) is reached.

2.2 FE simulation: the Basic Model

At ISS (Italian National Institute of Health) there is a laboratory dedicated to the above mechanical tests being ISS a Notified Body. To perform a deeper investigation of the femoral stem mechanical behavior, FE models have also been created and validated to carry out the FE analysis (COMSOL Multiphysics, COMSOL AB, Sweden). To investigate a specific stem, a digital acquisition of the 3D model of the stem is done in the Lab with a 3D-laser scanner (SCANNY3D-base, SCANNY3D s.r.l., Italy) and a process based on reverse-engineering techniques (Rhinoceros 4.0, McNeel, USA) is then applied to obtain the proper COMSOL Basic Model.

The Model consists of the following elements: head and stem of the investigated hip prosthesis system; cement mantle; an external metallic fixture dimensionally equal to the specimen holder used in the experimental test (Fig. 1A). For the numerical simulations and the analysis performed in the present study, a straight small femoral stem for uncemented implants has been considered (115 mm); the stem is made of a Ti-Al-V Titanium alloy (E = 114 GPa, Poisson's ratio 0.24, $\rho = 4400$ kg/m^3), has a neck-shaft angle of 125° and a 12/14 cone. The stem is embedded with the cement mantle (E = 3 GPa, Poisson's ratio 0.30, $\rho = 1190$ kg/m^3) inside the metallic stainless steel fixture (E = 200 GPa, Poisson's ratio 0.33, $\rho = 7850$ kg/m^3), oriented 10° and 9° respectively in the anatomical frontal and sagittal plane, and rigidly fixed at distance D = 80 mm in accordance with the ISO Standard. A slighlty augmented load of 3 kN is applied at the centre of the metallic head (E = 220 GPa, Poisson's ratio 0.33, $\rho = 8830$ kg/m^3) on a surface of 311 mm^2 (Campioni et al, in press). Finally, the base of the metallic fixture is fully bound, as well as the prosthesis-cement and the cement-metallic fixture interfaces. A static elastic-linear analysis is then

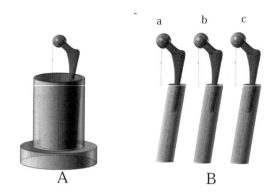

Figure 1. A. Basic Model with the stem positioned according to the ISO 7206-4:2010. B. Femur-like Models: (a) metallic fixture-cement; (b) bone-cement; (c) bone-cancellous bone.

performed and attention is focused on the maximum value of the von Mises stress (σ) on the stem at the constraint level.

Varied configurations of the same basic model have been then investigated as follows:

– variation of the angle α in the range 2° ÷ 15°;
– variation of the angle β in the range 2° ÷ 13°;
– variation of constraint level (30 to 80) mm;
– investigation of two different stem materials, Co-Cr-Mo alloy and Titanium alloy.

2.3 FE simulation: the femur-like Models

To investigate the major effects of varied loading configurations and of varied implant and interface materials in a context closer to the real *in-vivo* scenario, further FE Models have been created, based on the same sample stem, and variations have been identified on the basis of the ISS own experimental background, most critical results of the Basic Model analysis, reported clinical risky conditions, information about occurred implant failures *in-vivo*. In these femur-like Models, the stem is fully bonded in a 3 mm cement mantle (Mann et al. 1995), and the external fixture of the basic Model is replaced by a femur-like fixture (Fig. 1B). The following FE Femur-like Models have been created:

– Model with an external metallic fixture and a cement mantle, to perform static elastic-linear analysis in the ISO Standard configuration in the most critical configuration;
– Model with an external bone fixture and a cancellous bone (CB) to simulate a non-cemented condition in the same analysis as above but with three different E values in the range (0.03 to 3) GPa (Turner et al, 1990).

Finally, each of the above configurations has also been investigated by replacing the Titanium alloy with the Co-Cr-Mo alloy.

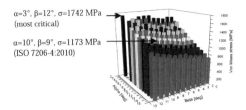

α=3°, β=12°, σ=1742 MPa
(most critical)

α=10°, β=9°, σ=1173 MPa
(ISO 7206-4:2010)

Figure 2. Basic Model: max σ for each angular combination in the range $\alpha = 2° \div 15°$ and $\beta = 2° \div 13°$.

3 RESULTS

3.1 Basic Model: identification of critical configurations

Figure 2 summarizes the max σ at the constraint level D = 80 mm for each angular combination in the range $\alpha = 2° \div 15°$ and $\beta = 2° \div 13°$. Due to the relative position between the stem and the metallic fixture and to their specific geometry, some extreme combinations have not been allowed for the specific stem. The most critical configuration MCC ($\alpha = 3°$; $\beta = 12°$) showed a σ increase of 48.5% with respect to the standard configuration SC indicated in the ISO Standard. In general, the maximum σ was well above the yield strength of the Titanium alloy (860 MPa) but for combinations in the range $\alpha = 13° \div 15°$ and $\beta = 2° \div 5°$.

SC and MCC have been then analyzed by varying the constraint level D in the range (30 to 80) mm. As expected, the more proximal the constraint level, the lower the stress. It is here only interesting to compare the results obtained with respect to D1 = 80 mm, which is the current indication of the ISO Standard, and to D2 = 52 mm which, for the specific stem, corresponds to the parametric indication of the 2002 version of the same Standard: i) as for SC, max σ was 480 MPa at D2 and 1173 MPa at D1; ii) as for MCC, it was 650 MPa at D2 and 1742 MPa at D1.

The 3D representation of stress distribution on the stem greatly help to investigate the stress concentration area: as shown in Figure 3, this area is located in the medial-anterior portion of the stem, at the upper surface of the cement. More specifically, Figure 3 represents Von Mises stress spatial distribution on the stem surface in correspondence with the standard configuration at the constrain level D1 and D2.

3D stress distribution did not significantly change when replacing the Titanium alloy with Co-Cr-Mo alloy; σ only showed a negligible reduction, from 1173 MPa to 1165 MPa for SC, and from 1742 MPa to 1692 MPa for MCC.

3.2 Numerical simulations with femur-like Models

Table 1 shows the max σ and the head vertical deflection δ obtained when performing the static elastic-linear analysis of the femur-like Models with the

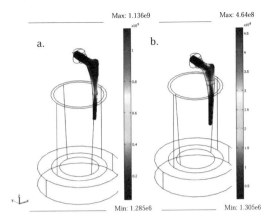

Figure 3. Basic Model: Von Mises stress distribution in correspondence with the standard configuration at D1 (a) and D2 (b).

Table 1. max σ and head deflection δ for two identical Titanium and Co-Cr-Mo stems, in several femur-like FE Models analysis (D = 80 mm, load 3 kN).

Configuration (SC/MCC) and materials	Titanium stem		CoCrMo stem	
	σ MPa	δ mm	σ MPa	δ mm
SC: steel + cement	**1225**	**−2.8**	**1194 (−3%)**	**−1.6**
MCC: steel + cement	1679 (+37%)	−4.6	1632 (+33%)	−2.6
SC: bone + cement (E = 3 GPa)	1227 (0%)	−3.4	1193 (−3%)	−2.1
MCC: bone+ cement (E = 3 GPa)	1677 (+37%)	−5.6	1633 (+33%)	−3.6
SC: bone + CB(E = 0.3 GPa)	1184 (−3%)	−4.3	1202 (−2%)	−2.8
MCC: bone + CB(E = 0.3 GPa)	1658 (+35%)	−6.1	1676 (+37%)	−3.4
SC: bone + CB(E = 0.03 GPa)	1352 (+10%)	−7.4	1430 (+17%)	−5.5
MCC: bone + CB(E = 0.03 GPa)	1791 (+46%)	−11.1	1809 (+48%)	−7.9

combinations of materials and material properties indicated in Paragraph 2.3, under SC and MCC. For each stress, the percentage of increase or decrease is also reported with respect to the reference configuration of the femur-like Model including the original Titanium stem, the metallic fixture and the cement interface as indicated in the ISO Standard.

It is interesting to observe that for both alloys the highest stresses are found in correspondence with a softer cancellous bone and that the Titanium stem showed a significantly higher deflection – about 40%

Max: 3.584e⁻⁴   Max: 1.92e⁻⁴

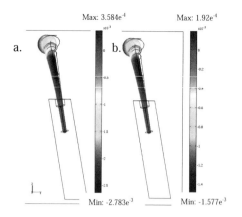

Min: -2.783e⁻³   Min: -1.577e⁻³

Figure 4. Femur-like Model: deflection δ in reference configuration for the stem in Ti alloy (a) and Co-Cr-Mo (b).

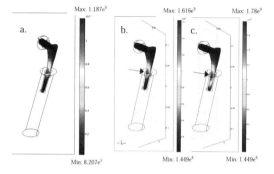

Max: 1.187e⁹   Max: 1.616e⁹   Max: 1.78e⁹

Min: 8.207e⁵   Min: 1.449e⁶   Min: 1.449e⁶

Figure 5. Femur-like Model, Von Mises stress distribution for the stem in Ti alloy in the following configurations: (a) reference configuration (SC, steel + cement); (b) MCC with cement (E = 3 GPa); (c) MCC with cancellous bone (E = 0.03 GPa).

greater – with respect to the corresponding Co-Cr-Mo stem behavior despite comparable stresses.

Figure 4 shows deflection values, and the deformed shape, for both alloys in the same reference configuration, greater for Titanium alloy.

In Figure 5 Von Mises stress distribution is showed on the Titanium stem surface in the reference configuration; it is also possible to note how the high-stress area is more expanded in MCC with a softer cancellous bone than for the stem embedded with cement in the same configuration.

4 DISCUSSION AND CONCLUSIONS

Being mechanical fatigue tests on femoral stems destructive, simulations of loading conditions by means of digital acquisition of 3D geometrical models and FE modeling seems to be extremely attractive to perform structural analyses under varied geometrical configurations and loading conditions.

The Basic Model analysis conducted in this study showed that, besides the ISO Standard requested testing configuration, other, more critical configurations should be tested which might occur in a clinical scenario. Also, as expected, the more proximal the constraint level, the lower the stress. It is here worth to be noted that the constraint level corresponding to the ISO 2010 standard causes stress peaks which are about twice those caused by the constraint level corresponding to the indication of the 2002 version of the same Standard. Furthermore, it is interesting to observe that, in the femur-like Models, for both alloys the highest stresses occur in correspondence with a softer cancellous bone. Finally, despite comparable stresses, the Titanium stem undergoes to significantly higher deflection with respect to the corresponding Co-Cr-Mo stem.

As a final remark, it is worth to highlight the great potential of FE methodology and computational images in better understanding 3D stress distribution in both prosthetic stems and bony components, hopefully resulting in a more effective design and implementation of experimental fatigue tests.

REFERENCES

Campioni, I. Notarangelo, G. Andreaus, U. Ventura, A. Giacomozzi, C. Chapter VI: Hip Prostheses Computational Modeling: FEM simulations integrated with fatigue mechanical tests in Springer Book U. Andreaus & D. Iacoviello (eds), *Biomedical Imaging and Computational Modelling in Biomechanics*. In press.

Mann, K.A. Bartel, D.L. Wright, T.M. Burstein, A.H. (1995). Coulomb frictional interfaces in modelling cemented total hip replacements: a more realistic model. *J Biomech* 28: 1067–1078.

Turner, C.H. Cowin, S.C. Young Rho, J. Ashman, R.B. and Rice, J.C. (1990). The fabric dependence of the orthotropic elastic constants of cancellous bone. J Biomech 23(6): 549–561.

Histogram based threshold segmentation of the human femur body for patient specific acquired CT scan images

D. Almeida, J. Folgado & P.R. Fernandes
IDMEC – Instituto Superior Técnico, Technical University of Lisbon, Lisbon, Portugal

R.B. Ruben
ESTG, CDRSP, Polytechnic Institute of Leiria, Portugal

ABSTRACT: In this paper, we propose a fully automated computational procedure, based on histogram based threshold segmentation, which generates a volume rendered CT scan of a specific patient femur. Foremost, the developed procedure will read the stack of images and locates the dominant peaks in the image histogram. Subsequently, a correspondence between the CT scan specter and the image histogram is established, which is used to bias the segmentation of the bone tissue. A line following algorithm is then used to obtain the coordinates of the points that define the bone region contour of each acquired image so that, finally, a 3D Delaunay triangulation may be used to render the obtained volume.

1 INTRODUCTION

Image segmentation techniques are based on the division of an image or a set of images in multiple regions or pixel sets according to some determined characteristic of the image or any of its computational properties, such as the intensities of the pixels in the color map or image gradient. Therewith, it is possible to obtain a more or less precise definition of the border between these regions and therefore proceed to its isolation for a more simple and precise analysis of the region of interest. A glimpse on the state of the art of this matter makes clear that, among the several different segmentation techniques developed, it is not obvious to state which one provides the best results for a generic use (Lakare, 2000, Unter et al. 2008). Depending on the image type, the segmentation approach must be previously studied in order to take full advantage of the image's specific characteristics. However, even taking in account the most recent segmentation techniques developed, the user's experience is still a preponderate factor that clearly influences the obtained results. Therefore, the development of an algorithm that can perform a fully automatic segmentation without taking in account some previous information about the images to segment and therefore constraining its use to a very specific application is still a work in progress. In Medical Imaging, segmentation is of major interest because it allows the location and tracking of tumors or other pathologies as well as real time guidance for computer assisted orthopedic surgery or the acquisition and computational representation of an organ or tissue, providing information about its properties and morphology. The fully automation of this process will save significant amount of time both in the identification of the pathology and in surgery preparation. In addition, the integration of such an algorithm in a commercial software package will allow less qualified users to perform such kind of analysis without having to fully comprehend the relying problems behind the graphic user interface.

Magnetic resonance (MRI), x-ray computed tomography (CT) and ultrasounds are the most widely used methods to acquire in vivo medical images from the human body. These methods acquire a sequenced pile of images that are ideally parallel among them. This does not occur in most cases, due to the patient's movements during the exam or, in the case of ultrasonography, the force used to press the device against the patient's body. Therefore, the process of image segmentation and registration of a stack of medical images is slow and complex depending on the complexity of the anatomic structure and the quality of the images acquired. Moreover, the better the quality desired, the more resolution per slice is required, as well as smaller distance between slices resulting in an increase of the number of slices for the same segmentation volume. There is also the probability of image artifacts to appear on the images, which increases the complexity of the process. These artifacts are generally related to the patient's natural movements during the acquisition process or even the functionality of the machine itself.

Image Segmentation techniques of medical volumes can be divided into three major groups: structural, stochastic and hybrids (Lakare, 2000). The structural segmentation techniques are essentially based in the structure and morphology of the region to segment

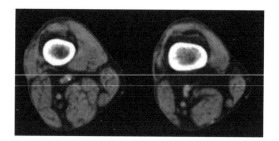

Figure 1. Two examples of CT scan acquired images.

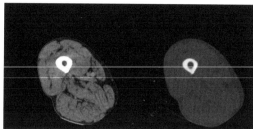

Figure 2. Example of an unedited image on the left side and the correspondent rescaled slice on the right side.

while the stochastic approach takes in account a discrete analysis over every pixel characteristics or the characteristics of the pixel and the nearby region. The combination of both of these techniques gives rise to the hybrid approach which, in most cases, provides better results in an inferior computational time.

Therefore, a computational procedure was implemented in order to select and map the region of each image which corresponds to the bone tissue which will define the shape of the three dimensional representation of the body of the femur. This representation will allow the definition of a finite element mesh which can be used in several kinds of analysis to generate information about the femur morphology prior to the surgical intervention.

2 METHODS

The stack of images used in the present work were acquired by X-ray computed tomography and gently given to us in the standard DICOM format. This acquisition method was chosen since it allows the acquisition of the anatomical structures in all three dimensions and, as shown in Figure 1, because the fact that it uses X-rays, hard tissues such as bone tissue are easily identified given such difference in physical density. This medical image format contains a set of images equally spaced from each other. Every image has the same amount of pixels and the same color map. Each pixel has an intensity value of the gray scale color map according to the tissue's physical density.

Such segmentations can be very challenging to obtain since osseous tissue toes not always yield readily distinguishable from soft tissue regions in CT images. Indeed, osseous tissue on the surface of the bones is very dense and strong, and therefore has a larger electromagnetic absorption coefficient than the weaker osseous tissue in the interior of the bones. Hence, cortical bone appears as a thin bright rim surrounding the darker region corresponding to the cancellous bone on CT images. Complicating matters further, cortical bone contrast with soft tissue and with cancellous bone can be highly uneven for the same bone, leading to faint intensity boundaries between the regions, as is notorious in Figure 1, and consequently difficult to identify its specter peak in the image histogram.

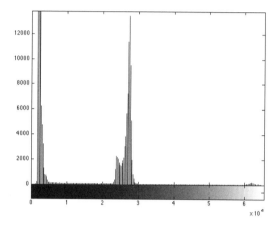

Figure 3. Example of a DICOM histogram. Notice the local maximum in the highest intensities region that corresponds to the bone tissue pixels.

The values of the image pixels are limited to a narrow band in the total possible dynamic range of the image, so the data had to be rescaled. Due to the fact that all the images in the series have the same dimensions and bit-depth, and that information is easily accessible on the DICOM info data, it is possible to perform a linear combination in order to rescale the image data to fill the entire 16-bit dynamic range. The linear combination to rescale the grayscale map values is

$$y = (x - b)m \qquad (1)$$

where b is the minimum x value and m is a constant ratio derived from the input and output ranges. The described procedure is visible in Figure 2. The image on the right side provides a more precise segmentation due to the increase of the number of bits.

Looking at the image, it is notorious that the pixels corresponding to the bone tissue region have a characteristic intensity in the gray scale specter. Therefore, it was chosen to approach the problem through a pixel by pixel analysis in which a band-pass filter was implemented that sets the non-matching pixels' intensity to zero (black). The limits of this filter are estimated by an automatic analysis on the histogram of the image. An example of a histogram of such kind is presented in Figure 3. On the horizontal axis are the different

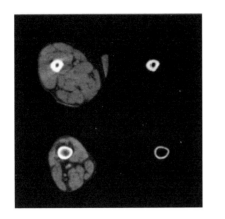
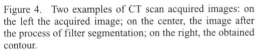

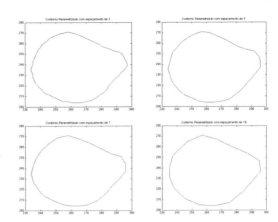

Figure 4. Two examples of CT scan acquired images: on the left the acquired image; on the center, the image after the process of filter segmentation; on the right, the obtained contour.

Figure 5. Various examples of the contour parameterizations obtained. From left to right and top to bottom, the numbers of points were reduced by factors of 3, 5, 7 and 10 respectively.

values for the intensities of the gray scale image pixels. Since the image has been rescaled to 16 bits, the values are therefore comprised between 0 and 65536. On the vertical axis are shown the frequencies on which these values occur, i.e., the number of pixels that share the same intensity value. The developed code will easily calculate the frequency peaks on the histogram. Due to the characteristic pixel intensity of bone, significantly higher than the surrounding tissues, the correspondence to frequency peak and the levels of tolerance are easily predicted. This approach is called histogram based threshold segmentation (Weszka, 1997). This technique is very effective in getting segmentation done in volumes with a very good contrast between regions and has a low computational cost, as is the femur body. On the other hand, it is a technique very sensitive to noise and intensity inhomogeneities. In order to avoid this last drawback, the image rescaling proved to be of good use since it smoothens the boundaries between different tissue regions.

At this point, a line following algorithm was implemented which returns the position of each point that defines the contour and allows the contour parameterization and consequently the level of detail of the structure 3D representation. It is important to keep in mind that the decrease of the number of points that define the contour due to the parameterization will result in a loss of the detail level of the obtained 3D surface. It is up to the user to balance this equilibrium based on the final purpose of the segmented volume. The 3D representation was achieved using 3D Delaunay triangulation. The first step towards a three dimensional representation of the segmented volume was the plot in a Cartesian coordinate system of the points that define the contours of every slice. The distance between each slice of the stack is available in the DICOM file, in a field called *SliceThickness*. This process results in a 3D representation of the body of the femur for any CT scan of this region.

Figure 6. Example of the 3D representation of the body of the femur.

3 RESULTS

Figure 4 shows the detail of the most important steps of the method and stands as a proof to the correct estimation of the band-pass filter limits based on the histogram data. We can observe the correct segmentation of our region of interest on both examples. On the left, the raw image imported to the workframe, on which cortical bone tissue is clearly distinguishable;

283

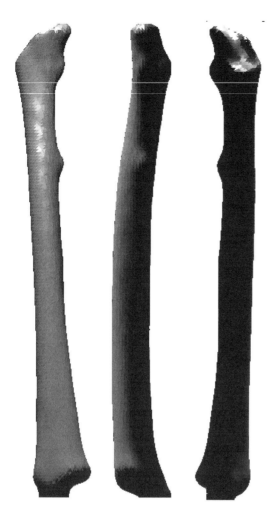

Figure 7. Example of the 3D representation of the body of the femur.

4 CONCLUSIONS AND FUTURE WORK

An automated, patient specific, segmentation approach was developed based in threshold segmentation. The presented procedure proved to be robust and computationally effective, generating the 3D volume in less than 30 seconds on an Intel Core i7 2600K processor with 8GB of RAM memory without taking in account the possibility of the contour parameterization which in decrease the computational cost of the procedure. Additionally, it has not a single step that requires the intervention of the user, making the procedure suitable for everyone, even those who are not deeply familiarized with such medical segmentation methods. Therefore, it is possible to infer some information about the bone's morphology prior to the surgical intervention. In a near future, it is our will to make the procedure capable of a fully segmentation of the femur. Due to the proximity of bone tissue in the region of the head of the femur, this method proved not to be the effective at all. A new approach is being developed based in energy minimizing snakes, more specifically an adaptation of the active contours without edges (Chan, 2001), taking in account some *a priori* information (Leventon, 2003). This new approach does not overrule the presented work at all, since it is significantly slower without presenting any benefits on the obtained results. A combination of both techniques seems to be the most effective to achieve our purpose. This will open doors to numerical experimentations and finite element analysis which will hopefully provide a better preparation of the surgery from the physicians.

ACKNOWLEDGEMENTS

This work was supported by Portuguese Foundation for Science and Technology (FCT) through the project: PTDC/SAU-BEB/103408/2008. D. Almeida would also like to thank FCT for the PhD scholarship SFRH/BD/71822/2010.

REFERENCES

Chan, T., Vese, L. 2001. Active Contours Without Edges. *IEEE Transactions on Image Processing,* Vol. 10, No. 2, 266–277.
Lakare, S. 2000. *3D Segmentation Techniques for Medical Volumes.* Center of Visual Computing: New York.
Leventon, M. et al. 2003. Knowledge-Based Segmentation of Medical Images. *Geometric Level Set Methods in Imaging, Vision, and Graphics,* Springer, New York, 401–420.
Unter, M., Pock, T. & Bischof, H. 2008. *Interactive Globally Optimal Image Segmentation.* Institute for Graphics and Vision, Graz University of Technology: Graz.
Weszka, S. 1997. A Survey of Thresholding Techniques. *Computer Graphics and Image Processing,* 259–265.

in the middle the segmented bone tissue; on the right, finally, the obtained contour.

The parameterization of the contour was also tested and it is possible to obtain a less detailed structure resulting on a decrease of the computational time of the process, as shown in Figure 5. The purpose of this parameterization process is to reduce the computational cost of the segmentation process, keeping in mind that a parameterization by a factor too large would be translated in significant losses on the geometry of the segmented volume.

The 3D plot of the anatomical structure using the Delaunay triangulation is also shown in Figure 6. The lines that define each tetrahedron are shown in the representation and some of the morphological characteristics of the body of the femur are identifiable, such as the non-circular section as it approaches the femoral head. Finally, in Figure 7 we have a surface plot of the obtained results, where it is visible the correct segmentation of the whole body of the femur.

Computational Modelling of Objects Represented in Images – Di Giamberardino et al. (eds)
© 2012 Taylor & Francis Group, London, ISBN 978-0-415-62134-2

A hybrid sampling strategy for Sparse Magnetic Resonance Imaging

L. Ciancarella, D. Avola, E. Marcucci & G. Placidi
Department of Health Sciences, University of L'Aquila, Via Vetoio Coppito, L'Aquila, Italy

ABSTRACT: A hybrid acquisition sequence for Sparse 2D Magnetic Resonance Imaging (MRI) is presented. The method combines random sampling of Cartesian trajectories with an adaptive 2D acquisition of radial projections. It is based on the evaluation of the information content of a small percentage of the k-space data collected randomly, to identify radial blades of k-space coefficients having maximum information content. An entropy function is defined on the power spectrum of the projections for evaluating the information content of each direction. The method has been tested on MRI images and it was also compared to the weighted Compressed Sensing. Some results are reported and discussed.

1 INTRODUCTION

Magnetic Resonance Imaging (MRI) has become a major non invasive imaging modality over the past 25 years, due to its ability to provide structural details of human body, like Computed Tomography, and additional information on physiological status and pathologies, like nuclear medicine. The reconstruction of a single MR image usually involves collecting a series of trajectories. The measurement of a trajectory is a sampling process of a function evolving with time in a 2D or 3D space domain, referred to as "k-space". The raw data from this sequence of acquisitions are then used to reconstruct an image, through Fourier Transform after gridding (O'Sullivan 1985, Jackson et al. 1991).

The most popular k-space trajectories are straight lines from a Cartesian grid, in which each line of k-space corresponds to the frequency encoding readout at each value of the phase encoding gradient (Spin Warp Imaging, (Edelstein et al. 1980)). The lines in the grid are parallel and are equally separated. Although the acquisition of Cartesian trajectories allows easier image reconstruction, recent advances in MR hardware allow other acquisition patterns, such as spirals (Meyer 1998), or radial trajectories (Projection Reconstruction, PR (Lauterbur 1973)). PR, for example, has many advantages over the conventional Cartesian k-space trajectory, because of its robustness to the motion artifacts, such those due to blood flow or respiration.

A fundamental limitation of MRI is the linear relation between the number of acquired trajectories and net scan time: to collect a complete data set, minutes are often required. This acquisition time is considered to be too high especially when dynamic processes have to be observed at high temporal resolution, such as in fMRI studies (Bernstein et al. 2004). The acquisition time for each trajectory is limited by the slow natural relaxation processes that are beyond the control of the acquisition sequence. Therefore, the only way to speed up acquisition is to reduce the acquisition time by using fewer trajectories, that is by using undersampling of the k-space.

Undersampling is the violation of the Nyquist's criterion where images are reconstructed by using a number of data lower than that theoretically required to obtain a fully-sampled image. One of these methods (Placidi et al. 2000) presented a k-space adaptive acquisition technique for MRI from projections. The method defined the entropy function on the power spectrum of the collected projections, to evaluate their information content, thus driving the acquisition where data variability is maximum. The choice of the projections was made during the acquisition process; this allowed the reduction of acquisition time, by reducing the scanned directions. Candès et al. (Candès et al. 2006) and Lustig et al. (Lustig et al. 2007) presented the theory of Compressed Sensing (CS) and the details of its implementation for rapid MRI. They demonstrated that if the underlying image exhibits sparsity in some transform domain, then the image can be recovered from randomly undersampled frequency domain data, providing that an appropriate nonlinear recovery scheme is used. Besides, the optimization of the nonlinear sparse MRI reconstruction can be achieved by increasing samples in the central part of the k-space (low frequency data have greater information content, as demonstrated in the CS, weighed by a Gaussian function (Lustig 2008, Wang et al. 2010)).

In this paper a hybrid MRI acquisition sequence to reduce scan time while preserving image quality is presented. It is based on some significant characteristics of CS (random sampling of spectral coefficients and L_1−minimization), combined with a radial adaptive acquisition criteria. Numerical simulations are

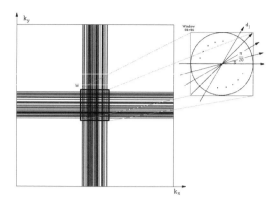

Figure 1. The entropy function is measured for a set of 20 equispaced radial projections inside the window W (size M/6 × M/6).

reported and compared with weighted CS to show its performances.

2 THE PROPOSED METHOD

Consider the k-space image support as an $M \times M$ matrix. In a first phase, the acquisition process consists of the random collection of Cartesian trajectories, through Spin Warp Imaging, in a central region of the k-space whose width is M/6, both along the rows and along the columns. Each of these trajectories is completely sampled but the number of collected rows and columns is lower than M/6. Lines are collected by randomizing the phase-encoding direction, the columns are collected by reversing the phase-encoding gradient with the frequency-encoding gradient (also in this case, the randomization process involves the phase-encoding direction). On the central region W, where rows and columns crossed (whose size is M/6 × M/6), the acquisition algorithm provides the foundation for radial adaptive sampling.

In fact, we consider an equispaced set of 20 radial projections in the window W, whose angular separation was $\pi/20$ (Fig. 1).

The entropy function of the *j-th* direction is defined as:

$$E_j = \frac{1}{q_j} \sum_{i=1}^{q_j} v_i \log v_i \quad \forall \, j = 1,\ldots,20 \qquad (1)$$

where q_j is the number of measured coefficients in W falling along the *j-th* radial projection and v_i represents the power spectrum of *i-th* coefficient allowing to the *j-th* projection. Once calculated the entropy of each projection, the average value is chosen as a threshold value T. If there are projections whose entropy is above the threshold value, then a "blade", composed by 10 parallel lines of k-space coefficients, is collected around each of these projections (the acquisition resembles a sort of PROPELLER (Pipe 1999)).

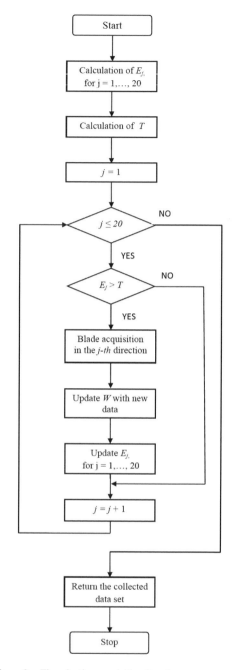

Figure 2. The adaptive acquisition iterative process.

The process is repeated until entropy values above T are present. The adaptive choice of the most significant projections is summarized in Figure 2. The method is suitable for nonlinear L_1-norm reconstruction, in line with the theory of CS (Placidi 2012). It is important to note that the radial directions involved into the entropy calculation are confined inside the window W, while the blades of coefficients are collected into the whole image support.

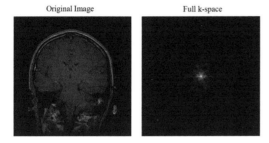

Figure 3.   A complete MRI image and its k-space spectrum.

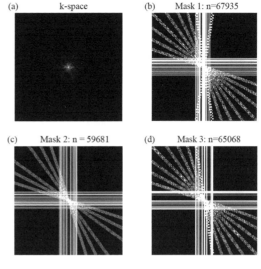

Figure 4.   Complete k-space spectrum of the image reported in Figure 3 (a) and three different sampling masks ((b), (c) and (d)) obtained with three different applications of the proposed method. The number of collected data, n, is slowly different for the three cases.

3   NUMERICAL SIMULATIONS

To evaluate the performance of proposed method, simulations on different MRI images have been performed. The considered images were complete 512×512 images with different orientations and weighting parameters. To simulate MRI acquisition, the Fast Fourier Transform (FFT) of each image was performed and the obtained coefficients were treated as the k-space collected data, assuming the image and the whole k-space data set are not known a priori (we found this procedure useful to compare the under-sampled reconstructed images with the complete, theoretical, image).

The obtained results have been compared both with the completely sampled original image and with the image reconstructed by using compressed sensing undersampling weighted with a Gaussian distribution function having zero mean and $\sigma = 0.1$.

The σ value has been optimized experimentally: $\sigma < 0.1$ produced severe losses of resolution (small details were lost), because the mask resulted too con-centrated on the central part of the k-space; $\sigma > 0.1$ produced severe artifacts due to undersampling.

In what follows two of these simulations are reported and discussed: the results of each simula-tion have been compared both visually and numeri-cally, by using the *peak signal-to-noise ratio* (PSNR), with the complete image. Images of enlargements of small details and difference between the original and reconstructed images are also shown.

In the first reported simulation (Fig. 3), the proposed acquisition method has been preliminary applied three times. This was done to demonstrate the reproducibility of the proposed method. In fact, the starting point of the adaptive method is a set of randomly collected data: if this data set is too sparse, the resulting adaptively collected data set is strongly dependent upon the k-space initial positions. In this case, different results could occur for different starting sets.

As can be noted from Figure 4, though a certain variability due to the initial random choice, the corre-sponding images, reported in Figure 5, are very similar each other, both visually and in PSNR values.

Based on PSNR values, the worst-case result was considered with the mask #3 (Fig. 5, PSNR =

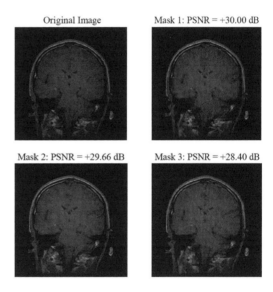

Figure 5.   Images reconstructed by using the three masks reported in Figure 4. In particular, the original image (a) is compared with the images obtained with mask 1 (Fig. 4b), mask 2 (Fig. 4c), and mask 3 (Fig. 4d), respectively.

+28.40 dB, n = 65,068). This result was compared with the weighed CS result. The comparison is reported in Figures 6, 7, and 8. In particular, CS allowed better PSNR value, but the proposed method conserved better resolution, as can be observed from Figure 7, and produces incoherent artifacts, as can be seen from Figure 8.

In the second reported simulation (Fig. 9a), the sampling mask shown in Figure 9b, is obtained for the proposed method (73,566 samples are acquired).

(a)

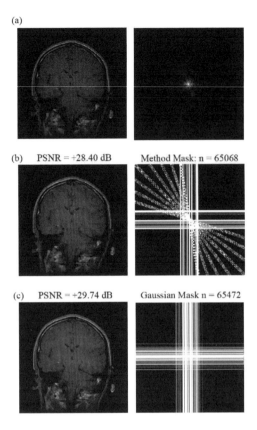

(b)   PSNR = +28.40 dB   Method Mask: n = 65068

(c)   PSNR = +29.74 dB   Gaussian Mask n = 65472

Figure 6.   Visual comparison between the original image (a), the reconstructed image with proposed method (b), and the reconstructed image with weighted CS (c). On the right part the used corresponding masks are reported (for the complete image the whole k-space spectrum is reported).

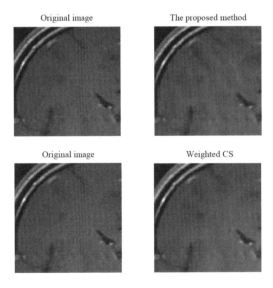

Original image   The proposed method

Original image   Weighted CS

Figure 7.   Enlargement of a particular for the original image (left column), for proposed method (right top), and for the weighted CS (right bottom).

(a)   (b)

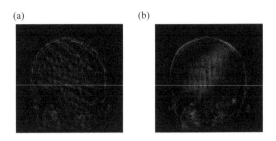

Figure 8.   Difference between the original image and the reconstructed image with the proposed method (a) and with the weighed CS (b). Differences are shown in an expanded gray-scale.

(a)

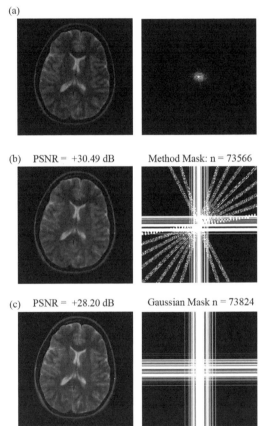

(b)   PSNR = +30.49 dB   Method Mask: n = 73566

(c)   PSNR = +28.20 dB   Gaussian Mask n = 73824

Figure 9.   Test image and its power spectrum coefficients (a). Reconstruction obtained by using the proposed method (b), and the weighted CS (c). The corresponding masks are reported on the right.

The corresponding weighted CS result is reported in Figure 9c. In this case, the hybrid adaptive method overcame weighted CS both visually and in PSNR. This is also evident by looking at the enlargement (Fig. 10). The better conservation of the image edges with our method is evident also from the difference images. In fact, the difference image referred to the weighted CS maintained a huge part of useful information.

Original image   The proposed method

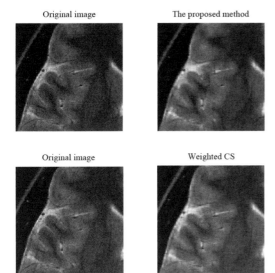

Original image   Weighted CS

Figure 10. Enalrgements of a particular for the original image (left column), for proposed method (right top), and for the weighted CS (right bottom).

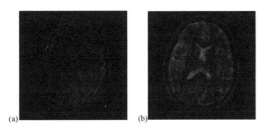

(a)   (b)

Figure 11. Difference between the original and the reconstructed image with the proposed method (a) and with the weighed CS (b). Differences are shown in an expanded gray-scale.

4 CONCLUSIONS AND FUTURE WORK

In the present work, a hybrid acquisition method for MRI has been described, that makes it possible to reduce the total acquisition time, with minimum loss of resolution. It combined two acquisition techniques: Spin Warp and Adaptive Projection Reconstruction.

The method was tested on different MRI images and two of them were reported. The use of the hybrid adaptive technique allowed good quality reconstruction, though a lower set of "most informative" data than the theoretical data set required to meet the Nyquist's

criterion was used. Our method has given comparable results compared with weighted CS, but it furnished a criterion to estimate the near optimal number of data to reconstruct a good image: this is impossible for CS, being it a blind method (for CS, the number of k-space coefficients has to be chosen in advance, independently of the image shape).

The values of the peak signal-to-noise ratio and the visual comparison confirmed the proposed method was slightly better than CS.

In the future we aim to demonstrate that our method, associated with a weighted random sampling in the first phase of acquisition, can improve its results while reducing the used samples.

REFERENCES

Bernstein, M. A., King, K. F. & Zhou X. J. 2004. *Handbook of MRI Pulse Sequences*: 468–469. USA: Elsevier.

Candès, E. J., Romberg, J. & Tao, T. 2006. Robust Uncertainty Principles: Exact Signal Reconstruction From Highly Incomplete Frequency Information. *IEEE Transactions on Information Theory* (2) 52: 489–509.

Edelstein, W. A., Hutchison, J. M., Johnson, G. & Redpath, T. 1980. Spin warp NMR imaging and applications to human whole-body imaging. *Phys. Med. Biol*. 25: 751–756.

Jackson, J. I., Meyer, C. H., Nishimura, D. G. & Macovski, A. 1991. Selection of a convolution function for Fourier inversion using gridding. *IEEE Trans. Med. Imaging* 10 (3): 473–478.

http://overcode.yak.net/15?width=1600&size=XS.

Lauterbur, E. C. 1973. Image formation by induced local interactions: Examples employing nuclear magnetic resonance. *Nature* 242: 190–191.

Lustig, M., Donoho, D. & Pauly, J. M. 2007. Sparse MRI: The Application of Compressed Sensing for Rapid MR Imaging. *Magnetic Resonance in Medicine* 58: 1182–1195.

Lustig, M. 2008. SPARSE MRI Thesis.

Meyer, C. H. 1998. Spiral echo-planar imaging. *Echo-planar imaging:* 633–655. Berlin: Springer.

O'Sullivan, J. 1985. A fast sinc function gridding algorithm for Fourier inversion in computer tomography, *IEEE Trans. Med. Imaging* 4 (MI-4): 200–207.

Pipe, J. 1999. Motion Correction With PROPELLER MRI: Application to Head Motion and Free-Breathing Cardiac Imaging. *Magnetic Resonance in Medicine* 42: 963–969.

Placidi, G., Alecci, M. & Sotgiu, A. 2000. ω-Space Adaptive Acquisition Technique for Magnetic Resonance Imaging from Projections. *Journal of Magnetic Resonance* 143: 197–207.

Placidi, G. 2012 *MRI: Essentials for Innovative Technologies*: 111–160. CRC Press Inc.

Wang, Z. & Arce, G. R. 2010. IEEE, Variable Density Compressed Image Sampling. *IEEE Transactions on Image Processing* (1) 19: 264–270.

Computational Modelling of Objects Represented in Images – Di Giamberardino et al. (eds)
© 2012 Taylor & Francis Group, London, ISBN 978-0-415-62134-2

Two efficient primal-dual algorithms for MRI Rician denoising

A. Martin
Fundación CIEN – Fundación Reina Sofía, Madrid, Spain
Neuroimaging lab. (UPM – URJC) – Center for Biomedical Technology, Madrid, Spain
Departamento de Matemática Aplicada, Universidad Rey Juan Carlos, Madrid, Spain

J.F. Garamendi
INRIA/INSERM U746/CNRS, UMR6074/University of Rennes I, VisAGeS Research Team, Rennes, France

E. Schiavi
Neuroimaging lab. (UPM – URJC) – Center for Biomedical Technology, Madrid, Spain
Departamento de Matemática Aplicada, Universidad Rey Juan Carlos, Madrid, Spain

ABSTRACT: In this work we consider the numerical resolution of a variational Rician denoising model recently proposed for Magnetic Resonance Images (MRI) restoration. Two new different primal-dual algorithms are presented. In a first setting we adapt to the resolution of the Rician denoising minimization problema a primal-dual algorithm originally applied for Total Variation based Gaussian restoration and denoising. As an alternative, a gradient flow approach combined with a semi-implicit numerical scheme is used to convert the problem into a sequence of Rudin, Osher and Fatemi (ROF) model problems which we solve iteratively in a primal-dual framework. For comparison purposes we also solve the above sequence of ROF problems using a well established dual descent method. Validation of the above methods is performed considering synthetic noisy MR images which are obtained from a ground truth solution (a phantom). Our results show the effectiveness of the two proposed numerical primal-dual formulations. Some modelling assumptions are also highlighted.

1 INTRODUCTION

Medical image processing is an established research field where advanced mathematical methods can be applied to improve the quality of images used in the statistical analysis of large amounts of noisy data sets. A proper and accurate modelling of the acquisition process is particularly important for low Signal to Noise Ratio (SNR) magnetic resonance images, e.g. diffusion weighted images (DWI), functional magnetic resonance images (fMRI) or perfusion images, restoration. These data are severely affected by Rician noise and their denoising improves the quality results of the images post-processing such the tensor estimation from DWI (Basu, Fletcher, and Whitaker 2006) or the activity maps from fMRI.

For this purpose, two new numerical schemes to solve the Rician denoising model equations proposed (independently) in (Martín, Garamendi, and Schiavi 2011) and (Getreuer, Tong, and Vese 2011) are presented. The effectiveness of the model for real DWI images and FA reconstruction has been recently assessed in (Martín, Garamendi, and Schiavi 2012).. The numerical schemes presented are based on a primal-dual algorithm originally proposed for gaussian denoising and restoration in (Zhu and Chan 2008) which was further analysed in a more general

framework in (Chambolle and Pock 2011). This novel approach allows to consider, in both algorithms, the exact, not approximated, total variation operator (in a dual maximization step) together with the original non-linear and (possibly) non-convex likelihood term associated to Rician noise (in a primal minimization step).

In Section 2 the model equations for MRI Rician denoising are reviewed. Then we describe the first new primal-dual scheme in Section 3 and present the second algorithm in Section 4: a semi-implicit scheme applied to the Rician denoising gradient flow equations leading to a sequence of well posed quasilinear problems solved with a primal-dual algorithm. Finally, in section 5, we compare the numerical results in artificially contaminated MR images, using as ground-truth solution a fully dual approach on the sequence of problems seen in (Martín, Garamendi, and Schiavi 2012) and we discuss the advantages of the above settings and the suggested future work.

2 MODEL EQUATIONS

Let Ω be a bounded open subset of $\mathbb{R}^d$, $d = 2, 3$ defining the image domain and let $f : \Omega \to \mathbb{R}$ be a given noisy image representing the data, with $f \in L^\infty(\Omega)$.

Let $BV(\Omega)$ be the space of functions with bounded variation in Ω equipped with the seminorm $|u|_{BV}$ defined as

$$|u|_{BV} = \sup \left\{ \int_{\Omega} u(x) \operatorname{div} \bar{\xi}(x)\, dx : \right.$$
$$\left. \bar{\xi} \in C_c^1(\Omega, \mathbb{R}^d), |\bar{\xi}(x)|_{\infty} \leq 1,\, x \in \Omega \right\} \quad (1)$$

where $|\cdot|_{\infty}$ denotes the l_{∞} norm in $\mathbb{R}^d$, $|\bar{\xi}|_{\infty} = \max_{1 \leq i \leq d} |\xi_i|$ (details on this space and the related geometric measure theory can be found in (Ambrosio, Fusco, and Pallara 2000)). Following a Bayesian modelling approach we consider the minimization problem

$$\min_{u \in BV(\Omega)} J(u) + \lambda H(u, f) \quad (2)$$

where $J(u)$ is the convex nonnegative Total Variation regularization functional

$$J(u) = |u|_{BV} = \int_{\Omega} |Du| \quad (3)$$

being $\int_{\Omega} |Du|$ the Total Variation of u with Du its generalized gradient (a vector bounded Radon measure). When $u \in W^{1,1}(\Omega)$ we have $\int_{\Omega} |Du| = \int_{\Omega} |\nabla u| dx$. The λ parameter in (2) is a scale parameter tuning the model.

The data term $H(u, f)$ is a fitting functional which is nonnegative with respect to u for fixed f. To model Rician noise the form of $H(u, f)$ has been deduced in (Basu, Fletcher, and Whitaker 2006) in the context of diffusion tensor MR images. The Rician likelihood term is of the form:

$$H(u, f) = \frac{1}{2\sigma^2} \int_{\Omega} u^2 dx - \int_{\Omega} \log I_0 \left(\frac{uf}{\sigma^2} \right) dx \quad (4)$$

where σ is the standard deviation of the Rician noise of the data and I_0 is the modified zeroth-order Bessel function of the first kind. It is shown in (Getreuer, Tong, and Vese 2011) that functional 4 is possibly nonconvex depending on the data f. Using (2), (3) and (4) the minimization problem is formulated as: Fixed λ and σ and given a noisy image $f \in L^{\infty}(\Omega)$ recover $u \in BV(\Omega) \cap L^{\infty}(\Omega)$ minimizing the energy:

$$J(u) + \lambda H(u, f) =$$
$$\int_{\Omega} |Du| + \frac{\lambda}{2\sigma^2} \int_{\Omega} u^2 dx - \lambda \int_{\Omega} \log I_0 \left(\frac{uf}{\sigma^2} \right) dx \quad (5)$$

Due to the fact that the functional in (3) (hence in (5)) is not differentiable at the origin we introduce the subdifferential of $J(u)$ at a point u by

$$\partial J(u) = \{ p \in BV(\Omega)^* |\, J(v) \geq J(u) + <p, v - u> \}$$

for all $v \in BV(\Omega)$, to give a (weak and multivalued) meaning to the Euler-Lagrange equation associated to the minimization problem (5). In fact the first order optimality condition reads

$$\partial J(u) + \lambda \partial_u H(u, f) \ni 0 \quad (6)$$

with (Gâteaux) differential

$$\partial_u H(u, f) = \frac{u}{\sigma^2} - \frac{I_1 \left(uf/\sigma^2 \right)}{I_0 \left(uf/\sigma^2 \right)} \frac{f}{\sigma^2} \quad (7)$$

where I_1 is the modified first-order Bessel function of the first kind and verifies $0 \leq I_1(\xi)/I_0(\xi) < 1, \forall \xi > 0$. This model, first proposed in (Martín, Garamendi, and Schiavi 2011), differs from (Basu, Fletcher, and Whitaker 2006) because of the geometric prior (the TV-based regularization term) which substitutes their Gibb's prior model based on the Perona and Malik energy functional (Perona and Malik 1990). Also it differs from the classical gaussian noise model because of the nonlinear dependence on the solution of the ratio I_1/I_0. In the following we introduce the non-linear function $r(u, f) = I_1(uf/\sigma^2)/I_0(uf/\sigma^2)$ for notacional simplicity.

3 PRIMAL-DUAL ALGORITHM FOR RICIAN DENOISING

In this section we adapt to the Rician denoising a primal-dual algorithm which was first presented in (Zhu and Chan 2008) for Total Variation Image Restoration and which can be traced back to the classical Arrow-Hurwicz method for saddle-point problems. This algorithm has been studied in a general framework in (Esser, Zhang, and Chan 2010) and it was also included in a yet more theoretical work about first order primal-dual algorithms in (Chambolle and Pock 2011). In all these studies this algorithm was proved to be one of the most efficient methods when the adaptive time steps proposed in (Zhu and Chan 2008) are used, which promptly we do.

3.1 Discrete framework

In order to present the primal-dual algorithm we introduce the discrete setting of the problem. For this, we consider a regular Cartesian grid of size $N \times M$: $\{(ih, jh) : 1 \leq \leq N, 1 \leq j \leq M\}$ where h denotes the size of the spacing. The matrix $(u_{i,j})$ represents a discrete image where $u_{i,j}$ denotes the intensity pixel values. In what follows, we shall choose $h = 1$ because it causes only a rescaling of the energy and it does not change the solutions obtained. Also we assume $N = M$. The discrete gradient is defined as

$$(\nabla u)_{i,j} = \begin{pmatrix} u_{i+1,j} - u_{i,j} \\ u_{i,j+1} - u_{i,j} \end{pmatrix} \text{ for } i, j = 1, ..., N-1 \quad (8)$$

with $(\nabla u)_{i=N, j=1,...N} = 0$ and $(\nabla u)_{i=1,...N, j=N} = 0$. Then ∇ is a linear map from X to $Y = X \times X$, and if we endow both spaces with the standard Euclidean scalar product its adjoint ∇^*, denoted by $-\text{div}$, is defined by

$$< \nabla u, p >_Y = - < u, \text{div } p >_X \qquad (9)$$

for any $u \in X$ and $p = (p_{i,j}^x, p_{i,j}^y) \in Y$, and it is given by the following formulas

$$(\text{div } p)_{i,j} = (p_{i,j}^x - p_{i-1,j}^x) + (p_{i,j}^y - p_{i,j-1}^y)$$

for $2 \leq i, j \leq N - 1$. The term $(p_{i,j}^x - p_{i-1,j}^x)$ is replaced with $p_{i,j}^x$ if $i = 1$ and with $-p_{i-1,j}^x$ if $i = N$, while the term $(p_{i,j}^y - p_{i,j-1}^y)$ is replaced with $p_{i,j}^y$ if $j = 1$ and with $-p_{i,j-1}^y$ if $j = N$. Defining the discrete version of the Total Variation semi-norm:

$$||\nabla u||_{\ell^1} = \sum_{i,j} |(\nabla u)_{i,j}|,$$

$$|(\nabla u)_{i,j}| = \sqrt{((\nabla u)_{i,j}^1)^2 + ((\nabla u)_{i,j}^2)^2}$$

we can write the discrete version of the energy of the Rician denoising problem (5):

$$\sum_{i,j} |(\nabla u)_{i,j}| + \lambda \sum_{i,j} \left[\frac{u_{i,j}^2}{2\sigma^2} - \log I_0 \left(\frac{u_{i,j} f_{i,j}}{\sigma^2} \right) \right] \quad (10)$$

where the matrix $(f_{i,j})$ represents the discrete noisy image with pixel values $f_{i,j}$ at nodes (i,j).

3.2 Primal-dual algorithm

Following (Chambolle and Pock 2011), we need to write the discrete version of the minimization problem (10) in the form

$$\min_{u \in X} F(Ku) + G(u) \qquad (11)$$

In our problem $K = \nabla : X \to Y$ is the discrete gradient defined in (8), so $F(Ku) = \sum_{i,j} |(\nabla u)_{i,j}|$ and $G(u) = \lambda \sum_{i,j} \left[u_{i,j}^2 / (2\sigma^2) - \log I_0 \left(u_{i,j} f_{i,j} / \sigma^2 \right) \right]$ is obtained discretizing (4). Now we can consider the saddle-point problem equivalent to (11) which results from applying the Fenchel-Legendre Transform to the term $F(Ku)$:

$$\min_{u \in X} \max_{p \in Y} < Ku, p >_Y - F^*(p) + G(u)$$

and problem (11) is:

$$\min_{u \in X} \max_{p \in Y} < Ku, p >_X - I_P(p) + G(u) \qquad (12)$$

where $p \in Y$ is the dual variable. The convex set P is given by

$$P = \{ p \in Y : ||p||_\infty \leq 1 \}.$$

The function $I_P(p)$ denotes the indicator function of the set P and is defined as

$$I_P(p) = \begin{cases} 0 & \text{if } p \in P, \\ \infty & \text{if } p \notin P. \end{cases}$$

The algorithm consists of an alternative maximization and minimization, where the dual variable p is updated in the maximization step and then the solution u is set in the minimization step. These two steps are repeated iteratively until the convergence of the solution. Given the previous solution (u^k, p^k):

• Maximization step:
 Fix u^k and apply a semi-implicit gradient ascent method to maximize (12) using a fixed time step t_d (we denote it with the subscript d because the dual variable is updated in this step):

$$(p^{k+1} - p^k)/\tau_d = Ku^k - \partial F^*(p^{k+1})$$

So we obtain p^{k+1} as:

$$p^{k+1} = (I + \tau_d \partial F^*)^{-1}(p^k + \tau_d K u^k)$$

Since here F^* is the indicator function of the convex set P, the operator $(I + \tau_d \partial F^*)^{-1}$ is the Euclidean projector into the set P, and we can write this updating step explicitly:

$$p^{k+1} = \frac{p^k + \tau_d \nabla u^k}{\max(1, |p^k + \tau_d \nabla u^k|)}$$

• Minimization step:
 Fixed p^{k+1}, apply a gradient descent method to minimize (12) using a fixed time step τ_p:

$$(u^{k+1} - u^k)/\tau_p = -K^* p^{k+1} - \partial G(u^k)$$

with K^* the adjoint operator of K. So using the identity (9) and the discretized form of (4) and (7) the updating of the variable u can be explicitly written as:

$$u^{k+1} = (1 - \tau_p) u^k + \tau_p \left[\frac{\sigma^2}{\lambda} \text{div } p^{k+1} + r(u^k, f) f \right]$$

4 THE RICIAN PROBLEM AS A SEQUENCE OF PRIMAL-DUAL ROF PROBLEMS

In (Martín, Garamendi, and Schiavi 2012) a framework which allows to solve the Rician denoising problem using the exact Total Variation operator was presented. It consists of converting the Rician problem into the resolution of a sequence of ROF problems (Rudin, Osher, and Fatemi 1992) and each one of them were solved with a dual gradient descent following (Chambolle 2004). However we can apply the same primal-dual algorithm used in the previous section at each ROF problem resolution as a way to solve the original Rician problem using a primal-dual approach. For this purpose we need to write first the original Euler-Lagrange equation associated to the Rician

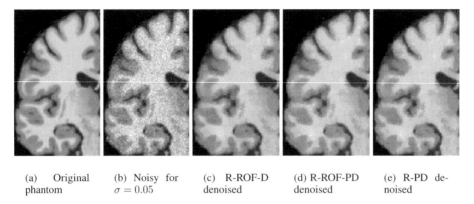

(a) Original phantom   (b) Noisy for $\sigma = 0.05$ (c) R-ROF-D denoised   (d) R-ROF-PD denoised   (e) R-PD denoised

Figure 1. The original image and the contaminated phantom are shown in a) and b). The denoised images obtained with the R-ROF-D, R-ROF-PD, R-PD algorithms and for the parametric values $\sigma = 0.05$, $\lambda = 0.075$ are presented in the sub-plots c), d) and e) respectively.

energy (5) (with abuse of notation for the diffusive term):

$$-\mathrm{div}\left(\frac{\nabla u}{|\nabla u|}\right) + \frac{\lambda}{\sigma^2}\left[u - r(u,f)f\right] = 0 \qquad (13)$$

Then a gradient descent scheme combined with forward finite differences for the temporal derivative of the gradient descent and a semi-implicit scheme lead to the following equation:

$$\left(1 + \tau\frac{\lambda}{\sigma^2}\right)u^{k+1} =$$

$$= u^k + \tau\left(\mathrm{div}\left(\frac{\nabla u^{k+1}}{|\nabla u^{k+1}|}\right) + \frac{\lambda}{\sigma^2}r(u^k,f)f\right) \qquad (14)$$

Defining

$$\beta = (\tau\lambda)/\sigma^2 \,,\, \gamma = (1+\beta)/\tau \qquad (15)$$

$$g^k = (1+\beta)^{-1}\left(u^k + \beta\,r(u^k,f)f\right) \qquad (16)$$

the equation (14) can be written as the Euler-Lagrange equation of the ROF energy functional ((Rudin, Osher, and Fatemi 1992)):

$$E(u) = \int_\Omega |Du| + \left(\frac{1}{2\gamma}\right)\int_\Omega (u - g^k)^2 dx \qquad (17)$$

Hence, at each gradient descent step τ, we can solve a ROF problem associated to the minimization of the energy (17) in the space $BV(\Omega)$ with the primal-dual algorithm presented in (Zhu and Chan 2008), which was denoted by PDHG. Fixed λ, σ, τ and given a noisy image f and an initial condition $u_0 \neq 0$, with $f, u_0 \in L^\infty(\Omega)$ the proposed Rician denoising algorithm is:

1. Calculate β, λ with (15) and function g_k using (16).
2. Compute u^{k+1} as the minimum of the ROF problem (17) with the PDHG algorithm.
3. Repeat until the convergence.

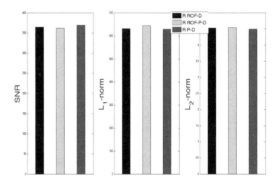

Figure 2. In the colored bars the quality measures used for comparison among the different numerical methods. In terms of accuracy and denoising power the results are very similar with a better global performance of R-PD (in red).

5 RESULTS AND DISCUSSION

In this section we shall validate the two new methods based in primal-dual formulations proposed in sections 3 and 4. We also consider the method proposed in (Martín, Garamendi, and Schiavi 2012) which was shown to be more accurate than the ε-approximating problems solved in (Martín, Garamendi, and Schiavi 2011) and (Getreuer, Tong, and Vese 2011). Notice that in the dual step the exact Total Variation operator is considered and solved.

In our study we used synthetic brain images from the BrainWeb Simulated Brain Database[1] at the Montreal Neurological Institute. The original phantoms were artificially contaminated with Rician noise considering the data as a complex image with zero imaginary part and adding random Gaussian perturbations to both the real and imaginary part, before computing the magnitude image.

This process allows to control the amount and distribution of the Rician noise. In the following we shall

_____

[1] available at http://www.bic.mni.mcgill.ca/brainweb

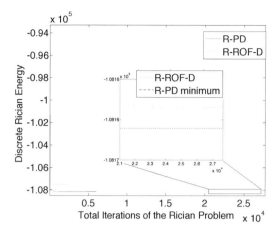

Figure 3. R-PD needs much less iterations (22%) to fulfill the same convergence criterium that R-ROF-D. We zoom in the plot to show that R-PD reaches also a lower energy than R-ROF-D.

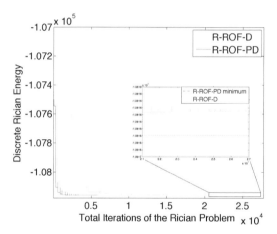

Figure 4. R-ROF-PD also needs much less iterations (21%) to fulfill the same convergence criterium that R-ROF-D. Notice the vertical bars in the light blue curve indicating the number of inner iterations in the dual gradient descent of R-ROF-D.

denote as R-ROF-D the method presented in (Martín, Garamendi, and Schiavi 2012), R-PD the new algorithm from section 3 and R-ROF-PD the algorithm from the section 4. In order to compare the solutions we have used the same convergence criterium for all the algorithms. The results are shown in Figure 1 where a ROI of the denoised images is considered. It can be seen, by visual inspection that the computed solutions are very similar, the details (edges) have been preserved and the noise has been partially removed. To quantify the numerical performances we compute the Signal-to-Noise-Ratio (SNR) defined as $\mathrm{SNR}(u, \hat{u}) = \sum_{i,j} u_{ij}^2 / \sum_{i,j} (u_{ij} - \hat{u}_{ij})^2$ and the ℓ^1, ℓ^2 norms given by $||u - u_k||_{\ell^1} = \sum_{i,j} |u_{ij} - \hat{u}_{ij}|$ and $||u - u_k||_{\ell^2} = \left[\sum_{i,j} (u_{ij} - \hat{u}_{ij})^2 \right]^{1/2}$ where u denotes the original image and $\hat{u}$ a numerical solution.

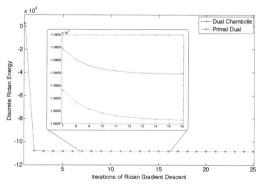

Figure 5. The energy curves computed for the R-ROF-PD vs R-ROF-D algorithm showing the effectiveness of the primal-dual approach compared with the pure dual approach.

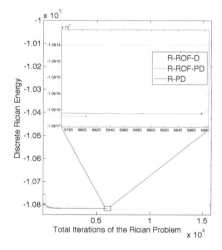

Figure 6. A comparison of all the methods together. Both primal-dual algorithms R-PD and R-ROF-PD outperform clearly R-ROF-D.

DISCUSSION

Our discussion is based on the quality measures in Figure 2. They present very similar values indicating convergence to the same numerical solution. The solution of R-PD (in red) provides the most accurate (denoised) image having the highest SNR and the lowest ℓ^1, ℓ^2 residual norms. This indicate that the ROF iteration, which is the modeling assumption leading to R-ROF-D (in blue) and R-ROF-PD (in green) can be avoided when high quality denoised images are a fundamental issue. The computational aspects are illustrated in Figures 3, 4, 5 and 6. If we compare R-PD vs R-ROF-D, the R-PD algorithm needs much less iterations (6000 vs 28000) to fulfil the same convergence criterion, with R-PD also reaching smaller energy (as it is shown in Figure 3). In R-PD the bottleneck in the numerics relies in the evaluation of the Bessel's ratio (see (Getreuer, Tong, and Vese 2011) for an approximation of this term which sensibly reduces the computational load but also the accuracy of the solution). Comparing R-ROF-PD vs R-ROF-D, which gives insight into the primal-dual vs dual approach the R-ROF-PD algorithm needs much less iterations (5800 vs 28000) to converge (Figure 4). Once again the R-ROF-PD reaches a smaller energy. Notice that after the same number of steps of the Rician gradient descent the energy is always smaller in the R-ROF-PD as it can be seen in Figure 5. This is consistent with the results obtained in (Chambolle and Pock 2011) where the supremacy of the primaldual approaches was established for TV-based Gaussian denoising. In Figure 6 the computational speed is analysed, the R-PD, R-ROF-PD need a similar number of iterations but much less evaluations of the Bessel's ratio (40 vs 6000) are done in R-ROF-PD so each iteration is less costly. The R-ROF-PD also reaches the smallest energy but R-PD is very close. Summarizing, our results indicate that the Rician denoising minimization problem can be efficiently and better solved with primal-dual approaches instead of pure dual iterations and that the ROF iterations are outperformed in terms of quality measures by the pure Rician iteration of R-PD. This has interesting consequences whenMRI medical images are considered for statistical analysis and future work shall be devoted to the application of these schemes to real MR-DW images for DTI denoising.

REFERENCES

Ambrosio, L., N. Fusco, and D. Pallara (2000). *Functions of Bounded Variation and free discontinity Problems*. The Clarendon Press, Oxford University.

Basu, S., T. Fletcher, and R. Whitaker (2006). Rician noise removal in diffusion tensor mri. *Medical Image Computing and Computer- Assisted Intervention* 9, 117–125.

Chambolle, A. (2004). An algorithm for total variation minimization and applications. *Journal Mathematical Imaging and Vision* 20, 89–97.

Chambolle, A. and T. Pock (2011, May). A firstorder primal-dual algorithm for convex problems with applications to imaging. *J. Math. Imaging Vis. 40*, 120–145.

Esser, E., X. Zhang, and T. F. Chan (2010). A general framework for a class of first order primaldual algorithms for convex optimization in imaging science. *SIAM Journal on Imaging Sciences 3*(4), 1015–1046.

Getreuer, P., M. Tong, and L. Vese (2011). A variational model for the restoration of mr images corrupted by blur and rician noise. In *Advances in Visual Computing*, Number 6938 in LNCS, pp. 686–698. Springer Berlin/Heidelberg.

Martín, A., J. Garamendi, and E. Schiavi (2011). Iterated rician denoising. In *Proceedings of the International Conference on Image Processing*, Computer Vision, and Pattern Recognition (IPCV'11), Las Vegas, USA, pp. 959–963. CSREA Press.

Martín, A., J. Garamendi, and E. Schiavi (2012, May). An efficient numerical resolution for mri rician denoising. In *Proceedings of the International Conference on Bio-inspired Systems and Signal Processing*, Villamoura, Portugal, pp. 15–24.

Perona, P. and J. Malik (1990). Scale-space and edge detection using anisotropic diffusion. *Pattern Analysis and Machine Intelligence, IEEE Transactions on 12*(7), 629–639.

Rudin, L. I., S. Osher, and E. Fatemi (1992, November). Nonlinear total variation based noise removal algorithms. *Physica D Nonlinear Phenomena 60*, 259–268.

Zhu, M. and T. Chan (2008). An efficient primaldual hybrid gradient algorithm for total variation image restoration. *UCLA CAM Report*, 08–34.

Computational Modelling of Objects Represented in Images – Di Giamberardino et al. (eds)
© 2012 Taylor & Francis Group, London, ISBN 978-0-415-62134-2

Graph based MRI analysis for evaluating the prognostic relevance of longitudinal brain atrophy estimation in post-traumatic diffuse axonal injury

Emanuele Monti
Dipartimento di Biotecnologie e Scienze della Vita Università dell'Insubria Varese, Italy

Valentina Pedoia & Elisabetta Binaghi
Dipartimento di Scienze Teoriche e Applicate – Sezione Informatica, Università dell'Insubria Varese, Italy

Alessandro De Benedictis & Sergio Balbi
Dipartimento di Biotecnologie e Scienze della Vita Università dell'Insubria Varese, Italy

ABSTRACT: The brain volume is a biomarker for neurodegenerative diseases therefore quantitative measurement of change in brain size and shape (e.g., to estimate atrophy) is an important current area of research. New methods of change detection are proposed in an attempt to improve accuracy, robustness, and level of automation. In this work a new strategy for estimating the prognostic relevance of longitudinal brain atrophy in post-traumatic diffuse axonal injury is presented. A key point in the strategy is the correlation of the changes in the brain volume and the patient response to neuropsychological assessment; an accurate and robust brain segmentation method is required. In this work a graph-based segmentation framework is adopted within the overall procedure to extract the volume changes of White Matter (WM), Gray Matter (GM) and CSF Cerebral Spinal Fluid (CSF).

1 INTRODUCTION

Diffuse axonal injury (DAI) (Meythaler et al. 2001) is one of the most common and important pathological features of traumatic brain lesion, which is characterized by widespread disruption of white matter fibers induced by shearing forces originated by the mechanism of rapid acceleration-deceleration of the head. The clinico-pathological result is a sudden coma, which lasts conventionally more than six hours. In addition to this clinical state there is the diagnostic criterium of radiological evidence of micro-hemorrhages on CT or MRI scan. This particular kind of brain lesion does not require surgical treatment because the brain damage underlying the described syndrome occurs only at a microscopic level and throughout the brain with prevalent involvement of some areas, such as cortico-subcortical junction of frontal and temporal lobes and the upper dorsal brainstem. The clinical course of these patients is characterized by severe cognitive impairments conditioning a persistent disability. However, despite white matter disruption represents a critical factor in pathophysiology of post-traumatic cognitive consequences, conventional neuroimaging underestimates its extent. In fact, the greatest part of subcortical damage, called delayed atrophy, occurs in a non-immediate stage and is driven by neuronal deafferentation and axonal damage under apoptotic cellular inputs. On a macroscopic level, the resulting effect is progressive white matter atrophy, which can

be quantified as brain volume reduction. Interestingly memory, executive functions and speed processing are the most common affected cognitive domains (Scheid and Walther 2006). This is likely due to the fact that these brain functions depend on the coherent activity of widely distributed brain networks constituted by nodes connected by short and long white matter pathways which may be damaged in traumatic brain injury. Clinical scales like Glasgow Coma Scale (GCS) are useful in general in brain traumas because they are related to the outcome, but in the particular subgroup of DAI patients, as we experiment, it does not seem to be true, because for the same level of depth of coma at the arrival in Intensive Care Unit, sometimes different trajectories of recovery are followed by different patients. A common, but not accurate method for evaluating the gravity of DAI consists in considering the distributions of the lesions on Magnetic Resonance Imaging (MRI). In fact, deeper damages are correlated with worse prognosis, but even this kind of consideration is not so accurate in clinical settings. As most survivors are young and have nearly normal life expectancy, the possibility of predicting the final outcome of a patient with diffuse axonal injury assumes great importance because this information could give a feedback on the rehabilitative process, with increasing scientific evidence of outcome modulation, if conducted in a proper temporal window. Therefore there is the need of clinical biomarker related to the amount of diffuse brain damage which could be measured in the subacute period when the patient is still recovering. The importance of such an indicator is not only to predict the possibility of consciousness restoration, but also to infer about the entity of recovery of higher cerebral functions not contemplated by conventional outcome scales, but dramatically related to social functioning of an individual. We know that recovery trajectory is under the influence of different moderators, such as age, pre-morbid I.Q., years of school education; parameters which are related to cerebral plasticity and connectivity and more in general to the concept of functional reserve. The concept of brain volume as a biomarker for neurodegenerative diseases emerged in relatively recent times and represents a challenging instrument for clinical research. Quantitative measurement of change in brain size and shape (e.g., to estimate atrophy) is an important current area of research (Ng et al. 2008). Several methods of change analysis are proposed in attempt to ensure accuracy, robustness, and a good level of automation. The basic idea is that to characterize quantitatively post-traumatic alterations of the brain in such a way may allow us to understand which kind of variation will affect the neurobehavioural functioning of the patient in the early period. Based on these assumptions, we set up a longitudinal multicentric study oriented to track the trajectory of brain volume reduction in MRI with respect to the functioning of the brain as assessed by neuropsychological tests. Our aim is to collect and analyze a great amount of clinical and morphometric data of patients who suffered from brain injury (DAI type) in order to find out statistical correlations that could allow us to infer about clinical prognosis; it would lead us to the creation of a prognostic algorithm, helping us decide the correct timing and intensity of rehabilitation process. In a previous retrospective study conducted at our institution, we analyzed perspectives of using brain volume as a predictor of clinical recovery after traumatic brain injury with pure DAI pattern, and we found statistically significant correlations between cognitive performance scores on tests and reduction of brain white matter between two time points.

2 BRAIN TISSUES VOLUMETRY

The overall procedure to extract the volume changes of white matter (WM), gray matter (GM) and cerebral spinal fluid (CSF) is composed of two essential steps. The aim of the first phase is to segment the whole brain from the head MRI and the aim of the second phase is to cluster WM, FM and CSF for examining separately the changes in volume over time of the three classes.

2.1 *Whole brain segmentation*

In the present contest the whole brain volume estimation requires an accurate and robust MRI volume segmentation procedure. The problem of segmenting all or part of the brain in MRI imagery continues to be investigated giving rise to a variety of approaches attempting to satisfy the high accuracy demand in diversified clinical and neuroimaging application (Clarke et al. 1995; Bomans et al. 1990; Dellepiane 1991). Graph searching has become one of the best investigated segmentation tool for 2D medical image data (Martelli 1972; Pope et al. 1985; Sonka et al. 1995) and 3D volume (Thedens 1995; Sonka et al. 1995; Li et al. 2006). The segmentation strategy based on graph searching principles (Martelli 1972) has the typical advantage of the optimization methods of embedding global information about edges and the structure of object in a figure of merit allowing accurate border detection from noisy pictures (Montanari 1971). This approach has been widely experimented on medical image data with applications to coronary angiography, ultrasound and cardiac MRI imaging (Thedens 1995).

In a previous work we investigate the use of graph searching technique for MRI brain segmentation (Pedoia et al. 2011). To make the present work self contained, we briefly describe the key points of this strategy.

A contour of radial shape can be conveniently treated by using polar coordinates. Working in polar space, the radial boundary of a given object can be represented by a transformation $\rho = f(\theta)$ characterized by the following feasibility constraints:

Boundary as a Function – $f(\theta)$ is single valued and the value ρ exists for each θ with $1 \leq \theta \leq N$.
Connectivity Constraint – $|f(\theta + 1) - f(\theta)| \leq 1$ for $1 \leq \theta \leq N - 1$.
Closing Constraint – $|f(1) - f(N)| \leq 1$ imposing that the first and last pixels satisfy the connectivity constraint.

Each feasible function $f(\theta)$ is a candidate object boundary. The goal is then to find the minimum cost boundary subject to the feasibility constraints. The boundary detection task within the graph searching framework is modeled by embedding the properties of the boundary in a cost function and formulating the boundary extraction as the problem of minimizing this function subject to the feasibility constraints. The boundary cost is defined as follows:

$$B_c = \sum_{\theta=1}^{N} C(\theta, f(\theta)) \qquad (1)$$

where $C(\theta, f(\theta))$ is a cost image. The value of each pixel in the cost image must be inversely related to the likelihood that an edge is present at that point. The likelihood is usually determined by the application of a low-level local edge operator (Thedens 1995). In general, the definition of this cost function depends on a priori knowledge on the object to be segmented. The boundary cost allows to express both local and global information that can be incorporated within the constrained minimization strategy for optimal boundary detection. The 2D brain segmentation is accomplished in the three following phases.

Phase 1 Polar Conversion – The image is converted in polar coordinates, both the correct number of sampling angles both the brain's centroid are automatically computed

Phase 2 Skull Boundary Detection – The cost image is computed using vertical Sobel filter and the graph searching technique is applied, finding the skull boundary. A binary mask is computed distinguishing between pixels with ρ less and greater than the edge. This mask is applied to cost image and it is inverted for finding a new cost image for the second step: the actual brain segmentation.

Phase 3 Brain Boundary Detection – The strategy is again applied on this new cost image. The minimal path in the graph is the brain boundary in the polar space. The last step is the conversion in cartesian coordinates of the detected boundary.

The value of graph search method lies both in robustness and improvements in computing time. Robustness derives from the fact that it inherit the property of optimization methods of embedding global information about edges and the structure of object in a figure of merit allowing accurate border detection.

2.2 White matter, gray matter, cerebral spinal fluid segmentation

After the whole brain segmentation phase the WM, GM and CSF are segmented in the entire volume. We address this task choosing within the graph searching framework the min-cut technique (Boykov 2001).

We assume that each voxel v of the brain volume V has three cost values concerning the association of the specific voxel to the White Matter $R_v(O_{WM})$ to the Gray Matter $R_v(O_{GM})$ and to the Cerebral Spinal Fluid $R_v(O_{CSF})$. Moreover, we assume each pair of adjacent voxels $v, w \in N$ has a cost $B_{v,w}$ associated with the fact that the pair of voxel has different labeling, where N is a set of pairs of 3D neighboring voxel. Our goal is to find the optimal labeling $L = (L_{WM}, L_{GM}, L_{CSF})$ in assigning each voxel v to the brain tissues by the minimization of the following cost function:

$$E(L) = \lambda R(L) + B(L) \qquad (2)$$

where $R(L) = \sum_{v \in D} R_v(O_v)$ is called Regional Term and $B(L) = \sum_{v,w \in N} B_{v,w} \delta_{L_i, L_j}$; (where δ_{L_i, L_j} is 0 if $L_i = L_j$ and is 1 if $L_i \neq L_j$) is called Boundary Term. In our application the Regional Term is the euclidian distance between each voxel and the centroid of the three clusters (WM, GM and CSF) computed with the K-means algorithm and the Boundary Term is computed with gradient function. The segmentation problem can be formulated as finding the optimal set of voxel belonging to a desired 3D objects; considering the translation of the segmentation problem in an optimization problem solvable with 3D graph $G\{V, E\}$, we have to look for the partition of the graph that minimizes the cost of the cut.

Each $v \in D$ becomes a graph node V that is connected to the 26, 3D neighbors with a link called

n-link. There are also three specially designated terminal nodes, that represent WM, GM and CSF prototypes. Each node is also connected to both the terminal node with a link called t-link. To every link both t-links and n-links is given a weight. Weight assigned to n-link represents the distance between two neighbor nodes, thus the Boundary term $B_{v,w}$ and the weight assigned to t-link represents the distance between nodes and both the terminal nodes, thus the Regional terms $R_v(O_{WM})$, $R_v(O_{WM})$ and $R_v(O_{CSF})$. An s-t cut is a subset of edges such that the terminal nodes become completely separated on the graph. The cost of a cut is sum of all the weight crossed by the cut. The segmentation goal is to identify the cut that minimize the cost using the max-flow/min-cut algorithm (Boykov 2001; Boykov 2004). The min-cut Graph Cuts framework offers a globally optimal object extraction method for N-dimensional images. A fairly general cost function is described that can include both region and boundary properties of segments and certain types of topological constraints that naturally fit into our global optimization framework.

In the present work a specific problem dependent solution has been introduced within the general min-cut technique which is suggested by regularity of the spatial distribution of WM, GM, CSF. By exploiting the flexibility in the description of the energy function that allows to take into account a-priori information, work we consider a modified version of equation 2 by including a new term:

$$E(L) = \lambda R(L) + \mu A(L) + B(L) \qquad (3)$$

where $A(L) = \sum_{v \in D} A_v(O_v)$ is an Atlas Term. We assume that each voxel v of the brain volume V has three a-priori cost values concerning the association of the specific voxel to the White Matter $A_v(O_{WM})$, to the Gray Matter $A_v(O_{GM})$ and to the Cerebral Spinal Fluid $A_v(O_{CSF})$. These costs are derived using a brain tissue probabilistic atlas, made available from the LONI (Laboratory of Neuroimaging) of the UCLA University (Shattuck et al. 2008). The atlas consists of three volumes, which the value of each voxel indicates the probability that this belongs to the WM GM and CSF respectively. In the 3D lattice structure, besides the classic n-links and n-links, the a-links for each tissue are added. The weight of these connections is given by the a priori probability that the voxel belongs to a class, which is found in the probabilistic atlas. The min-cuts of this new graph are the 3D boundaries of the WM, GM anf CSF.

3 EXPERIMENTAL RESULTS

Our multidisciplinary study is recruiting patients with pure DAI lesion pattern on MRI, being in a state of coma of sudden onset, with no history of psychiatric disease, no alcohol or drug abuse, no previous head trauma or brain intervention, and with no incompatibility with MRI examination. The first MRI is

Table 1. Quantitative evaluation and comparison of graph based segmentation strategy.

WB Volume	K-MEANS		GRAPH CUT		ATLAS GRAPH CUT	
	WM Voume	GM Voume	WM Voume	GM Voume	WM Voume	GM Voume
0.59%	15.73%	12.89%	8.08%	8.43%	1.08%	2.33%
1.24%	16.45%	16.99%	8.07%	10.19%	1.36%	3.91%
0.15%	17.36%	11.89%	8.61%	6.21%	1.86%	1.22%
2.14%	14.16%	15.26%	8.76%	11.33%	1.16%	5.12%
1.42%	23.92%	21.00%	8.84%	10.43%	1.54%	4.20%
2.66%	10.79%	13.28%	7.35%	1.70%	0.23%	5.17%
3.16%	25.28%	11.50%	5.99%	0.97%	0.38%	6.05%
2.44%	13.61%	16.90%	3.25%	8.09%	1.16%	3.77%
3.15%	23.80%	12.67%	8.21%	1.27%	0.69%	5.45%
1.96%	20.57%	11.97%	9.93%	4.55%	2.15%	1.80%

performed for the clinical need of diagnosing the DAI, which could explain the consciousness impairment in the absence of focal brain lesion on CT examination. Once the relatives submit the written informed consent, the patient is MRI-scanned at 4 time points: 1 month post-trauma, 6 months, 1 year and 2 years later. The acquisition sequence is 3D T1 MP-RAGE isotropic with voxel of 1 mm^3. The cognitive recovery is monitored by neuropsychologic evaluation performed 7 months, 13 months and 25 months post-trauma (according to patient clinical condition). Our expectation is that the neuropshycological characterization of brain pathology by means of a clinical biomarker will lead us to better interpret the patient's clinical condition and its potential evolution.

The evaluation phase is composed of qualitative and quantitative assessment of our graph based segmentation strategy, the analysis of the white matter atrophy trend and the study of the coupling between the atrophy and the neuropsychological evaluation.

Figure 1 shows three example of the WM/GM/CSF segmentation; the slices have been randomly chosen in our dataset. Our expert team judged the results satisfactory in general. The quantitative evaluation is performed using, as ground truth, a set of 10 MRI volumes segmentation already evadable positively evaluated by experts. Table 1 shows the percentage deviation between the whole brain volume (WB Volume) computed with our strategy and the reference data; the deviations are always low with an average of 1.8%. As regards the brain tissue clustering, the atlas Graph-Cut results are compared with those obtained by the K-means algorithm and by the classical Graph-Cut. The results are expressed in terms of percentage deviation in WM and GM volume estimation. The proposed strategy prevails in all cases with maximum deviation in the white matter volume estimation equal to 2.15%. this results are acceptable in the application we are dealing with.

Figure 2 shows the trend of the atrophy across the time for 4 patients of our data set. Analyzing the atrophy rate, computed for the 4 time point scans, we note that this phenomenon is grossly influencing the early seven months period; once past this point the curve

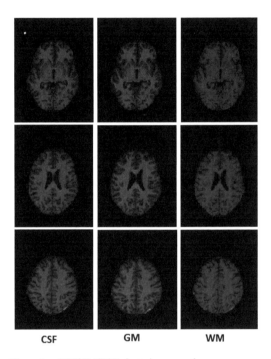

CSF   GM   WM

Figure 1. WM/GM/CSF clustering example.

slopes down as atrophy rate progressively reduces, tending to normal value and atrophy rate progressively reduces according to its average age-dependent trend. Therefore, we realized the importance of focusing on this particular period of post-traumatic recovery, in order to characterize white matter decrease trajectory in the tract where the curve may greatly vary from one patient to another according to the damage sustained. This is the most informative tract because the patient is usually still recovering in. In our opinion if we introduced one ore more further time point between 1st and 2nd ones (i.g. 2 months post-trauma) we would be able to better infer about the recovery progress during the phase in which we generally have no proper instruments to predict the clinical evolution

Table 2. Evaluation of the coupling between the white matter atrophy and the patients tests performance.

PTS	Time	WM Atropry	TMTA	TMTB	RAVLTi	RAVLTd	fReyc	fReyi	RavenCPM	phon. flu.	sem.flu.
1	6 M.	0.21%	14.0	278.0	NP	NP	24.4	12.6	23.8	21.0	30.0
2	6 M.	1.83%	48.0	126.0	25.2	0.0	31.2	1.8	27.6	NP	NP
3	6 M.	1.00%	59.0	137.0	54.1	9.5	33.4	25.9	30.2	32.0	16.0
4	6 M.	7.30%	NP	NP	15.1	19.0	29.7	NP	31.0	7.8	31.5
4	12 M.	8.46%	89.0	190.0	14.3	3.2	33.0	26.0	31.0	7.8	31.5
Pearson Correlation			**0.813**	−0.084	**−0.820**	0.290	0.380	0.465	0.720	**−0.885**	0.587

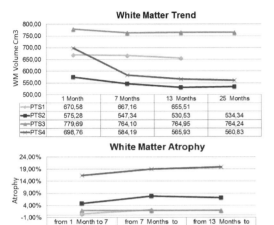

Figure 2. WM/GM/CSF clustering example.

of the patient. The basis of such a study consist in systematically studying the atrophy phenomenon with respect to its timing, and correlation to patient prognosis. We decided to perform first MRI examination 1 month post-trauma, not to take into account early brain volume variations related to inflammation and affine phenomena. We set up a battery of neuropsychological tests tailored on these patients to evaluate the cognitive domains mainly affected by traumatic DAI. In a preliminary retrospective work conducted at our institution we applied these tests to patients finding agreement with what medical literature reports. So now we submitted just the most significant ones from them. The key concept is to get some objective measurement of patient cognitive status allowing us to define numerically the recovery process and its relation with the loss of white matter fibers caused by head trauma.

Here is reported a synthetic description of the tests performed by the patients. Trail making test (TMTA, TMTB): is a test of visual attention and task switching. It can provide information about visual search speed, scanning, speed of processing, mental flexibility, as well as executive functioning. Rey-Osterrieth Complex Figure: this test permits the evaluation of different functions, such as visual-spatial abilities, memory, attention, planning, and working memory (executive functions). The Rey Auditory Verbal Learning Test (RAVLT) evaluates a wide diversity of functions: short-term auditory-verbal memory, rate of learning, learning strategies, retroactive, and proactive interference, retention of information, and differences between learning and retrieval. Raven's CPM measures clear-thinking ability; is used to assess the degree to which children and adults can think clearly, or the level to which their intellectual abilities have deteriorated. Verbal fluency tests are a kind of test in which participants have to say as many words as possible from a category (semantic or phonemic) in a given time. Regarding the brain areas used in this task, neuropsychological investigations implicate both frontal and temporal lobe areas. Analyzing the initial data coming from this study we observed a trend of correlation between white matter volume reduction and neuropshycological performance assessed by tests. Considering the small size of the sample, we are not allowed to perform a robust parametric statistics. However, we report in Table 2 the Pearson correlation values for each test. We are aware of the fact that only a larger number of cases may confirm or refute the thesis made on these data. These encouraging results agree with statistically significant correlation between TMT A, RAVLT, phon. flu. and WM volume as we found in our previous study.

4   CONCLUSION

The present work addresses the problem of providing a quantitative strategy for brain volume tracking and use it as biomarker for neurodegenerative diseases. In this contest the assessment of an accurate prognosis has both ethic and clinico-economical values because it could regulate the intensity of therapeutic rehabilitative intervention in function of what it is assumed to be achievable on the basis of a sophisticated characterization of brain damage, substrate of patient disability. An early clinico-radiological follow up represents therefore an important prognostic moment. We record the satisfaction of patients, relatives and caregivers, and this highlights the social value of this kind of intervention. Preliminary results are presented in this work. Future plan contemplates further investigation with more robust medical cases.

REFERENCES

Bomans, M., K.-H. Hohne, U. Tiede, and M. Riemer (1990, jun). 3-D segmentation of MR images of the head for 3-D display. *Medical Imaging, IEEE Transactions on 9*(2), 177–183.

Boykov, Yuri; Veksler, O. . Z. R. (2001, November). Fast approximate energy minimization via graph cuts. *IEEE Trans. Pattern Anal. Mach. Intell. 23*, 1222–1239.

Boykov, Yuri; Kolmogorov, V. (2004, September). An experimental comparison of min-cut/max-flow algorithms for energy minimization in vision. *PAMI 26*, 1124–1137.

Clarke, L., R. Velthuizen, M. Camacho, J. Heine, M. Vaidyanathan, L. Hall, R. Thatcher, and M. Silbiger (1995). MRI segmentation: Methods and applications. *Magnetic Resonance Imaging 13*(3), 343–368.

Dellepiane, S. (1991, oct-3 nov). Image segmentation: Errors, sensitivity, and uncertainty. In *Engineering in Medicine and Biology Society, 1991. Vol. 13: 1991, Proceedings of the Annual International Conference of the IEEE*, pp. 253–254.

Li, K., X. Wu, D. Z. Chen, and M. Sonka (2006, January). Optimal surface segmentation in volumetric images-a graph-theoretic approach. *IEEE Trans. Pattern Anal. Mach. Intell. 28*, 119–134.

Martelli, A. (1972). Edge detection using heuristic search methods. *Computer Graphics and Image Processing 1*(2), 169 – 182.

Meythaler, J. M., J. D. Peduzzi, E. Eleftheriou, and T. A. Novack (2001). Current concepts: Diffuse axonal injury–associated traumatic brain injury. *Archives of Physical Medicine and Rehabilitation 82*(10), 1461–1471.

Montanari, U. (1971, May). On the optimal detection of curves in noisy pictures. *Commun. ACM 14*, 335–345.

Ng, K., D. J. Mikulis, J. Glazer, N. Kabani, C. Till, G. Greenberg, A. Thompson, D. Lazinski, R. Agid, B. Colella, and R. E. Green (2008). Magnetic resonance imaging evidence of progression of subacute brain atrophy in moderate to severe traumatic brain injury. *Archives of Physical Medicine and Rehabilitation 89* (12, Supplement), S35–S44.

Pedoia, V., E. Binaghi, S. Balbi, A. De Benedictis, E. Monti, and R. Minotto (2011). 2d MRI brain segmentation by using feasibility constraints. In *Proceedings on Computational Vision And Medical Image Processing, VipIMAGE 2011*, pp. 251–256.

Pope, D. L., D. L. Parker, P. D. Clayton, and G. D. E (1985). Left ventricular border recognition using a dynamic search algorithm. *Radiology 155*, 513–518.

Scheid, R. and K. Walther (2006). Cognitive sequelae of diffuse axonal injury. *Archives of Neurology 63*, 418–424.

Shattuck, D. W., M. Mirza, V. Adisetiyo, C. Hojatkashani, G. Salamon, K. L. Narr, R. A. Poldrack, R. M. Bilder, and A. W. Toga (2008). Construction of a 3d probabilistic atlas of human cortical structures. *NeuroImage 39*(3), 1064–1080.

Sonka, M., M. Winniford, and S. Collins (1995, mar). Robust simultaneous detection of coronary borders in complex images. *Medical Imaging, IEEE Transactions on 14*(1), 151–161.

Sonka, M., X. Zhang, M. Siebes, M. Bissing, S. Dejong, S. Collins, and C. McKay (1995, dec). Segmentation of intravascular ultrasound images: a knowledge-based approach. *Medical Imaging, IEEE Transactions on 14*(4), 719–732.

Thedens, D.R.; Skorton, D. F. S. (1995). Methods of graph searching for border detection in image sequences with applications to cardiac magnetic resonance imaging. *IEEE Transaction On Medical Imaging 14*, 42–55.

Computational Modelling of Objects Represented in Images – Di Giamberardino et al. (eds)
© 2012 Taylor & Francis Group, London, ISBN 978-0-415-62134-2

Hemispheric dominance evaluation by using fMRI activation weighted vector

Valentina Pedoia
Dipartimento di Scienze Teoriche e Applicate – Sezione Informatica Università dell'Insubria Varese, Italy

Sabina Strocchi
U.O. Fisica Sanitaria, Ospedale di Circolo e Fondazione Macchi, Varese, Italy

Renzo Minotto
U.O. Neuroradiologia, Ospedale di Circolo e Fondazione Macchi, Varese, Italy

Elisabetta Binaghi
Dipartimento di Scienze Teoriche e Applicate – Sezione Informatica Università dell'Insubria Varese, Italy

ABSTRACT: In this work, a new quantitative method for the evaluation of the hemispheric dominance using fMRI data has been studied and experimentally evaluated. The quantitative evaluation of the Statistical Parametric Map (SPM) could be a valuable tool supporting the radiologist in the interpretation of the map and in the localization of the active area related to a specific task, avoiding a presumptive attitude in the analysis. This quantitative analysis is useful in the evaluation of the functional modifications, due to neuronal plasticity phenomena. The aim of this analysis is to extract a synthetic but comprehensive representation, the Activation Weighted Vector (AWV) from which to derive from the SPM indexes which describe in an objective way the distribution of the cerebral activations. Relevant information may be extracted from the clinical point of view, such as the hemispheric dominance.

1 INTRODUCTION

Functional MRI is a noninvasive technique for studying brain activity. During the course of an fMRI examination, series of brain images are acquired while the subject performs a set of tasks. Changes in the measured signal between individual images are used to make inferences regarding task activations in the brain, and the localization of relevant functional area. Before the use of fMRI the hemispheric dominance evaluation for the presurgical planning involves the use of invasive techniques as the Wada test to identify the hemisphere of the areas of language (Bruce 2005). fMRI allows to obtain similar results in a non-invasive manner and an index of hemispheric dominance has been introduced (Lateralization Index) (Seghier 2008):

$$LI = 100 * \frac{Q_l - Q_r}{Q_l + Q_r} \qquad (1)$$

where Q_l and Q_r are representative quantities measured by fMRI for the left and right contributions, respectively. Recently the use of this index has highlighted some limitations essentially due to its low reliability causing artifacts and to a new definition of hemispheric dominance that makes the test inadequate. The concept of hemispheric dominance stems

from the hypothesis of functional asymmetry identified by Broca in 1861. Systematic studies on the subject begun in the sixties. Initially the left hemisphere was considered dominant, indeed it is the sites of language and reasoning in right-handed. fMRI studies allow to exploit new complex aspects. Lateralization is less clear (the areas of language are mainly localized in the left hemisphere for about 95% of the right-handed and for about 75% of the left-handed), and both hemispheres contribute to brain functions with interpersonal variability. Moreover, for different functions there may be a different degree of lateralization. Currently, hemispheric specialization, (functional difference detected in the hemispheres with certain tasks and the subsequent specialization in certain skills) is preferred.

The brain is considered '*asymmetric but integrated*'. Hemispheric specialization is mostly recognized and accepted because supported by several studies, but at the same time the importance of cooperation between the hemispheres in a healthy brain it is emphasized. In the presurgical studies is essential to identify topologically and quantitatively, the location of eloquent areas in order to preserve them as far as possible. A complete analysis requires the use of DTI in an attempt to identify the white matter bundles and the corresponding networks.

In this work, a new quantitative method for the evaluation of the hemispheric specialization using fMRI data has been studied. The aim of our method is to assess hemispheric dominance for different functional areas, emphasizing the importance of the activation clustering with respect to the singular activation value. The method is based on synthetic indexes for hemispheric dominance useful for statistical purpose and a new kind of synthetic description derived from Statistical Parametric Map (SPM). The aim of this method, called Activation Weighted Vector, is to support the qualitative inspection of the activation map by the radiologist.

2 DATA ACQUISITION MODE

Magnetic resonance imaging device used is a Philips Achieva® 1.5T. The evaluation framework is developed in MATLAB® release 2010a. Stimulation paradigms are those more often used in neuroradiology: motor (finger tapping) and language (verb generation, word generation, object naming) tasks. All tasks are block designs.

We assessed the studies with functional MRI BOLD activation in 10 patients, 6 males and 4 females, mean age 50.5 years and age range between 22 to 70 years 9 right-handed and 1 left-handed. In our series, functional tests were performed preoperatively in patients awaiting surgery for removal of brain tumors. From the acquired volumes, activation maps were calculated as statistical parametric maps (SPMs) through using IViewBOLD software supplied by Philips®. Statistical analysis is carried out applying t-test to each voxel time-course, after retrospective image registration for motion correction. SPMs are function of t, minimum t value 3, maximum t value as results from analysis. Contiguous voxels = 4 and default mask were set.

3 ACTIVATION WEIGHTED INDEX AND HEMISPHERIC DOMINANCE EVALUATION

In this section we describe the sequence of steps involved in the evaluation of the hemispheric dominance accomplished by our strategy.

3.1 Brain segmentation

Some software tools for fMRI time series analysis do not include brain segmentation in the processing pipeline, therefore all MRI voxels are analyzed and may be active, with the result that active zones can be found outside the brain. Brain Segmentation allows to restrict the quantitative analysis in the brain. Moreover, segmentation allows to extract brain surface that will be used in the next registration phase. The brain segmentation method used in this work is a fully automatic and unsupervised method (Pedoia et al. 2011); this a 2D boundary-based strategy that use the graph searching principles (Martelli 1972; M. Sonka and

Figure 1. Statistical parametric map of a word generation task.

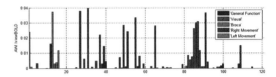

Figure 2. Activation weighted vector of a word generation task.

Boyle 1993; Thedens et al. 1995). The brain boundary is described using a set of feasibility constraints and the segmentation problem is described in terms of constrained optimization obtaining solvable as the search of a minimal path in a weighted graph.

3.2 Cerebral surface registration

During fMRI exam, a set of brain volumes is acquired and SPM is the result of the time series analysis. The maps are shown in relation with one of the acquired volumes, usually the first or the last. The aim of this step is to align this volume to a standard one to which an anatomical atlas is linked. Our strategy makes use of a surface registration method based on spherical harmonic decomposition has been used (Pedoia et al. 2011). This method performs affine 3D surface registration using shape modeling based on **SPH**erical **HARM**onic (SPHARM).

3.3 Activation weighted vector extraction

The objective of this step is to briefly describe the volumetric map in an attempt to give to the radiologist a synoptic view of the SPM in a simply "features vector". For this purpose an index that describe the information contained in the map was studied in our previous work (Pedoia et al. 2011). The SPM is normalized with respect to its own maximum and each active voxel is assigned to the relevant atlas anatomical area and the *Activation Weighted Index (AWI)*, that summarizes the activation of the specific brain portion, is computed as follows:

$$AWI_J = \frac{1}{N_J} \sum_{i=1}^{N_J} w_i \qquad (2)$$

where N_J is voxel total number in the Jth area and w_i is the normalized voxel value in the Jth area. *AWI* value can vary between zero and one and has greater value in areas where activations are clustered. Collecting these indexes for every zones, a feature vector can be defined. *AWV (Activation Weighted Vector)* is brief but comprehensive. Figure 1 shows an example of SPM extracted from the execution of a Word

Generation task, the map includes widespread activations in both hemispheres with high levels of noise that creates difficulties in determining which activities are significant and which are merely acquisition artifacts. Moreover, the volumetric nature of the data makes difficult to assess the general activation distribution forcing the radiologist to proceed through each SPM axial plan. In Figure 2 the *Activation Weighted Vector* computed for the above example is shown. Each bar in the histogram represents an anatomical zone in the atlas chosen as reference. This representation gives a synoptic view of all the Map offering to the expert a complete overview of the SPM. The *Activation Weighted Vector* can be used to recognize particular patterns, extract indexes, as the hemispheric dominance, or as a guide to the full exploration of the map. Analyzing the activation map by means of *AWV*, makes the analysis more objective, avoiding a presumptive approach in which the expert only verifies a priori hypothesis of activation. Qualitative analysis can not be exhaustive on the map; thus the expert will intend only to verify that the activation occurs in areas classically related to the task performed. But it is well known that the brain is a plastic organ particularly in presence of diseases. Moreover, functional areas locations are very individual-dependent. In these terms comprehensive and quantitative analysis of the map is essential.

3.4 *Hemispheric dominance evaluation*

The *AWV* can be separated in two vectors AWV_l and AWV_r considering the activation of only the left and right hemisphere respectively. The hemispheric specialization can be simply evaluated as follows:

$$LI_{AWV} = 100 * \frac{Q_l^{AWV} - Q_r^{AWV}}{Q_l^{AVW} + Q_r^{AWV}} \qquad (3)$$

$$Q_l^{AWV} = \sum_{i \in AWV_l} AWI_i; \quad Q_r^{AWV} = \sum_{i \in AWV_r} AWI_i$$

For the example above the LI_{AWV} value is 60.4 indicating a clear overall dominance of the left hemisphere; this quantitative assessment agrees with that made in qualitative way by looking at the map. It is worth to note that the calculation of hemispheric dominance as here proposed is very different from the classical one in which the left and right hemisphere activations are directly compared. LI_{AVW} allows to attribute less importance to isolated active voxels which are probably artifacts and gives significance to voxels belonging to clusters. Indeed, in the example above, the classical LI doesn't detects a predominant hemisphere ($LI = 0.86$) due to the accumulation of noise voxels in the right hemisphere. Moreover, LI_{AVW} can be computed also to evaluate a quantitative hemispheric specialization considering separately the brain areas classically linked to some of the more relevant functions.

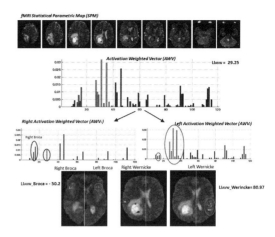

Figure 3. Example of the Hemispheric Dominance evaluation using *AVW*: Verb Generation task.

4 EXPERIMENTAL RESULTS

A set of experiments have been conducted to assess the capability of the AWV analysis to describe the hemispheric specialization. The hemispheric dominance is evaluated on all the 10 patient in our dataset. SPMs were qualitatively assessed by an expert, the classic LI is computed and the AWV hemispheric dominance is performed considering the global LI_{AWV} and the Broca and Wernicke language zones LI_{AWV}^{Borca} and $LI_{AWV}^{Wernicke}$ respectively. The fMRI exam pipeline includes three language tasks: Object Naming (L_1), Verb Generation (L_2) and Word Generation (L_3). Object Naming is performed by showing simple color images (ON block) alternating with a white cross on black background (OFF block). During the image show (an image every 3 seconds) the patient has to think of the word representing the shown object. During the OFF block the patient has to think nothing. Each block lasts 30 seconds, during which 10 volumes are acquired. The block sequence is: 4 times OFF-ON plus a last OFF block. Verb Generation is performed showing of simple black and white drawings representing an action (ON block) alternating with a white cross on black background (OFF block). During the drawings show (an image every 3 seconds) the patient has to think of the verb representing the shown action. During the OFF block the patient has to think nothing. Each block lasts 30 seconds, where 10 volumes are acquired. The block sequence is: 6 times OFF-ON plus a last OFF block. Word Generation is performed by the show of single white letters on black background (ON block) alternating with a white cross on black background (OFF block). During the letters show (a letter every 6 seconds) the patient has to think of all the words beginning with that letter that he/she can. During the OFF block the patient has to think nothing. Figure 3 shows an example of the hemispheric dominance evaluation performed using the *AWV* on one case of the dataset. The patient is left handed and has an expansive mass in the right parietal lobe. The example in Figure 3 is

Table 1. Hemispheric dominance evaluation.

idx	Task	LI	LI_{AWV}	LI^B_{AWV}	LI^W_{AWV}	Report
	L_1	−84.6	−2.1	−54.4	−7.5	
1	L_2	−2.5	20.1	−27.0	72.5	Bilateral
	L_3	−33.6	64.2	74.8	88.4	
	L_1	−47.0	−39.4	18.8	−29,0	
2	L_2	40.2	−14,2	−8.5	−22.7	Bilateral
	L_3	−9.1	35.9	36.2	59.9	
	L_1	−28.8	−18.0	−27.0	−60.1	
3	L_2	13.2	29.6	−50.2	81.0	Bilateral
	L_3	1.8	55.8	71.9	37.6	
	L_1	−44.4	0.3	−35.8	57.9	Bilateral with
4	L_2	−63.7	25.5	19.6	71.9	tendency to
	L_3	−36.4	25.9	39.9	49.7	the Left
	L_1	−85.9	19.2	21.4	35.3	
5	L_2	−78.1	45.9	71.0	85.4	Left Dominance
	L_3	−26.4	25.2	26.6	15.9	
	L_1	−16.7	25.7	50.3	20.7	
6	L_2	−26.9	15.4	38.2	16.0	Bilateral
	L_3	−14.4	2.3	21.0	3.6	
	L_1	−57.1	15.8	12.0	69.8	
7	L_2	−50.4	46.2	41.6	100.0	Left Dominance
	L_3	36.3	64.9	59.8	77.7	
	L_1	−63.4	1.5	−3.3	52.9	Left Dominance
8	L_2	−87.1	21.1	−25.0	68.1	with Controlateral
	L_3	−24.5	18.5	−11.9	55.1	Attivation
	L_1	−76.3	−9.4	−36.7	46.7	
9	L_2	−62.8	25.9	77.3	76.4	Left Dominance
	L_3	0.8	60.1	46.7	88.8	
	L_1	−70.6	40.5	19.2	75.1	
10	L_2	−52.9	33.1	17.3	45.3	Left Dominance
	L_3	20.9	43.2	−7.4	100.0	

that the analysis for specific areas is essential for a correct evaluation. The activation distribution can be very complex, therefore appears to be inadequate describe the hemispheric specialization with a single index.

5 CONCLUSION

In this paper a strategy for quantitative evaluation of the hemispheric dominance, based on the analysis of fMRI activation Map, is presented. The key point of our method is to describe the SPM in a brief but comprehensive manner avoiding the presumptive attitude of the expert in the qualitative analysis. A new index for evaluating the hemispheric dominance that uses this SPM description is proposed. The aim is to assess hemispheric dominance for different functional areas, emphasizing the importance of the activation clustering with respect to the singular activation value. The experimental results and the experts evaluation are encouraging with respect to the use of this method in clinical practice.

REFERENCES

Bruce, H. (2005). Wada test failure and cognitive outcome. *Epilepsy Currents 5*(2), 61–62.

Sonka, M. V. H. and R. Boyle (1993). *Image Processing, Analysis and Machine Vision* (3 ed.). London: Chapman and Hall.

Martelli, A. (1972). Edge detection using heuristic search methods. *Computer Graphics and Image Processing 1*(2), 169 – 182.

Pedoia, V., E. Binaghi, S. Balbi, M. E. De Benedictis, Alessandro and, and R. Minotto (2011). 2d MRI brain segmentation by using feasibility constraints. In *Proceedings Vision And Medical Image Processing, VipIMAGE 2011*.

Pedoia, V., E. Binaghi, and I. Gallo (2011). Affine SPHARM registration: Neural estimation of affine transformation in spherical domain. In *Proceedings of International Conference on Computer Vision Theory and Applications (VISAPP)*. INSTICC Press.

Pedoia, V., V. Colli, S. Strocchi, C. Vite, E. Binaghi, and L. Conte (2011). fMRI analysis software tools: an evaluation framework. In *Proc. SPIE 7965*.

Seghier, M. L. (2008). Laterality index in functional mri: methodological issues. *Magnetic Resonance Imaging 26*(5), 594–601.

Thedens, D., D. Skorton, and S. Fleagle (1995). Methods of graph searching for border detection in image sequences with applications to cardiac magnetic resonance imaging. *IEEE Transaction On Medical Imaging 14*(1), 42–55.

related to the task L_2. The SPMs obtained are very noisy and difficult to be interpreted. The qualitative analysis of the SPM performed by a radiologist didn't highlight a clear hemispheric dominance in this language task; thus, the patient is judged with bilateral dominance. The *AWV* analysis allows to quantify the observation and to study the SPM comprehensively, faithfully and quickly. As regards the Broca's area the right hemisphere is dominant ($LI^{Broca}_{AWV} = -50.2$). Instead, as regards the Wernicke's area, located in the inferior parietal lobe, the left hemisphere is dominant ($LI^{Wernicke}_{AWV} = 80.97$). The *AVW* automatic analysis agrees with the expert comments, with the additional merit of giving a quantitative evaluation thus, comparable with other cases or usable for the clinical follow-up. Table 1 shows the results obtained for all the cases analyzed. The *AVW* analysis is always in agreement with the radiologist opinion. Otherwise the classical *LI* is often in disagreement. It is worth to note

Computational Modelling of Objects Represented in Images – Di Giamberardino et al. (eds)
© 2012 Taylor & Francis Group, London, ISBN 978-0-415-62134-2

An automatic unsupervised fuzzy method for image segmentation

Silvana G. Dellepiane, Valeria Carbone & Sonia Nardotto
Università degli Studi di Genova, DITEN, Genova, Italy

ABSTRACT: Starting from the fuzzy intensity-connectedness definition (χ-connectedness) and the related growing mechanism, a new method is here proposed for the unsupervised, automatic, global region segmentation of digital images, hereinafter referred to as the "Automatic Fuzzy Segmentation" (AFS). One of the major advantages lies in the strict and very simple integration between the analysis of topological connectedness and grey level similarities of the pixels belonging to the same region.

By overcoming the previous drawback due to the need of some seed points selection, an iterative processing is here developed, able to adapt to the image content. The automatic selection of seed points is driven by intermediate connectedness results which alternates the analysis of inter-region similarities with inter-region separation measurements. The robustness of the method with respect to the three required parameters is discussed. Example cases related to the biomedical application domain are here presented and discussed.

1 INTRODUCTION

Image segmentation is widely recognized as a difficult and challenging problem in computer vision with a wide variety of applications such as medical imaging. The aim of the segmentation is to partition an image into regions with homogeneous properties that faithfully correspond to the objects or parts of the objects of interest [1]. Among the methods and solutions suggested in the literature, the ones exploiting fuzzy logic have proved to be very promising. As a matter of fact, fuzziness is an inherent feature of real images, which are created through imaging devices, naturally affected by defects such as resolution limitations, blurring, noise, etc. Another hindrance to image analysis is represented by the heterogeneity of the materials which make up the objects under examination. Even when acquired from a single homogeneous class, the imaging signal may be characterized by various levels of intensity.

Fuzzy logic has been widely used in image segmentation, mainly as an appropriate approach to clustering based on the signal intensity. Indeed, even though the spatial and topological properties of images are of major importance, only a few methods have been developed able to exploit such an additional dimension.

In this context, fuzzy connectedness [2][3] may help to solve these problems by a precise identification of the elements composing an image and belonging to the same region of interest, despite the existing uncertainties and variations in intensity.

The aim of the proposed method is to extend the fuzzy intensity-connectedness approach to more seed points which are automatically selected with the goal of a global segmentation of the image.

Such an iterative procedure starts from a random selection of a few seed points, and ends with the complete segmentation of the image. Each iterative step alternates addition of new significant seeds and deletion of useless ones.

Since it is mainly a random process driven by fuzzy operators, it is unsupervised and makes use of only three parameters, robustness was very deeply investigated. An extensive experimental session has proved a very strong robustness in addition to good performances, when compared with other methods.

The outline of the paper is as follows. In the next section, we shall briefly review the segmentation approaches that are related to our method. Section 3 presents the formulation of the proposed segmentation method. In Section 4 results will be reported and compared with already published methods. The conclusions section ends the paper.

2 RELATED WORK

In the snake and active contour models [4–6], an edge detection function is used to evolve the curve on the sharp boundary of the desired object. Since these methods are sensitive to noise the image needs to be smoothed, thus blurring the edges. A novel approach to obtain smooth segmentation using edge information was proposed in [7]: the idea is to extract edge points and use them as interpolation points of the threshold surface. In addition to the edge-based methods, region-based methods incorporate region and edge information and are generally more robust to noise. The best known and most influential approach is the Mumford-Shah model [8], which approximates the image by piecewise smooth functions with regular boundaries. The piecewise constant Mumford-Shah model has been studied by Chan and Vese [9]. The merit of this model is that it can detect object contours with or without gradient.

Another variational method for image segmentation is the so-called region competition. Zhu and Yuille [10] proposed a statistical and variational region competition model that unifies snake, region growing and Bayesian statistics. The model penalizes the length of the boundaries and the Bayes error in each region that is characterized by a parametric probability distribution. Curve evolution is used in the implementation of region competition.

Inspired by [10], Tang and Ma [11] proposed a general scheme of region competition based on scale space clustering for image segmentation. A nonparametric statistical approach has also been studied and reported in [12] and [13].

Apart from the aforementioned approaches, which give hard segmentation results, fuzzy segmentation methods have been widely used in data mining and medical images [14]–[15].

In such circumstances, the processing of images that possess ambiguities is better performed using fuzzy segmentation techniques, which are more adept at dealing with imprecise data. In Tizhoosh's state-of-the-art work [16] fuzzy techniques are broadly classified into five main categories: fuzzy thresholding, fuzzy clustering, fuzzy rule-based, fuzzy geometry, and fuzzy integral based segmentation techniques. Of these, the most widely used are fuzzy clustering and fuzzy rule-based segmentation.

The two most popular fuzzy clustering techniques are the fuzzy c-means (FCM) [17] and possibilistic c-means (PCM) algorithms [18].

While both these methods have been applied extensively, neither integrates human expert knowledge nor includes information about pixel spatial relations. Image segmentation which relies upon only feature based information without considering inter-pixel relationships, does not generally produce good results, because there are usually a large number of overlapping pixel values between different regions.

In contrast, fuzzy rule-based image segmentation techniques are able to integrate expert knowledge and are less computationally expensive compared with fuzzy clustering. They are also able to interpret linguistic as well as numeric variables [19]. However, in many applications, the performance of fuzzy rule-based segmentation is sensitive to both the structure of the membership functions and of the associated parameters used in each membership function.

Paper [20] presents a new generic fuzzy rule-based image segmentation (GFRIS) algorithm, which addresses a number of the aforementioned issues, most crucially by incorporating spatial pixel information and automatically data-mining.

In general, even though the spatial and topological properties of images are of major importance, they are only partially exploited by literature methods. In addition to the five cited categories, a few approaches

have been developed able to exploit the topological information in a fuzzy-logic framework [2][3].

These approaches are based on a close examination of the connectedness between pairs of image pixels and the relevant paths, in addition to the actual signal intensity.

Starting from [20], in [21] Udupa et al. have proposed the application of what they call "the relative fuzzy connectedness" where various objects in the image "are let to compete among themselves in having pixels as their members". In the experiments there proposed, some seeds have been utilized to specify a class of objects and some others correspond to different co-objects in the background. The user makes a distinction between the seeds corresponding to the class of interest and the seeds corresponding to the background.

In the present paper, starting from the fuzzy intensity-connectedness [2], new image processing steps are defined and introduced to realize an automatic unsupervised fuzzy image segmentation.

3 PROPOSED METHOD

As it is well known, the purpose of image segmentation is the partition of an image M into a set of regions, $\{R_k\}$, where each R_k is a connected component and it holds that:

$$\bigcup_k R_k = M$$
$$R_k \cap R_h = \Phi \quad for\ k \neq h \tag{1}$$

This method is divided into five steps described in sections from 3.1 to 3.5:

- *Fuzzy χ – connectedness map:*
  Unlike the MSMC method [25] where seeds are manually selected by the user, in this method the seed points are randomly placed.
- *Redundant seed:*
  Obviates the possibility that more than one seed is associated to the same region.
- *Total connectedness Map:*
  The seeds-related membership maps are merged into a single "total-connectedness map" through a fuzzy union operator.
- *Residual and Seed Generation:*
  Automatically finds new seeds when the previously selected ones are not sufficient to achieve a correct segmentation result.
- *Stopping Criterion and Labeling*
  The stopping criterion and the labeling step are defined in subsection 3.5.

3.1 *Fuzzy χ – connectedness map*

Let V be a generic 2D square lattice, where $v_i \in V$ with $i = 1, \ldots, |V|$, is the i-th pixel. The intensity map M is a function from V to the scalar domain, representing the random field of the grey levels of an original digital image. Given an appropriate rationale number, r, a fuzzy mapH is derived from M where $\eta(v_i) = r \cdot M(v_i)$ such that $\eta(v_i)$ is a fuzzy value within the unit real interval $[0,1]$. Following paper [2], as a preliminary step, when a seed point is chosen, named v_a, the function χ^{v_a} from V to the scalar fuzzy domain is defined as

$$\forall v_i \subset V \quad \chi^{va}(v_i) = 1 - |\eta(v_i) - \eta(v_a)| \quad (2)$$

From [25], the fuzzy-intensity connectedness (or χ-connectedness) map, $C_{\chi a}$, is defined as

$$\forall v_i \in V \quad c_{\chi^{v_a}}(v_i) = \max_{P(v_a, v_i)} \left[\min_{z \in P(v_a, v_i)} \chi^{v_a}(z) \right] \quad (3)$$

where $P(v_a, v_i)$ is a path, i.e., a connected sequence of pixels from v_a to v_i (four-connectivity holds).

Following the method described in [2] and [22], the above connectedness map is generated for each seed point.

3.2   Redundant seeds

At the beginning, a few seed points are randomly placed, then it is required to determine the eventual cases where more than one seed is associated to the same region. It is thus necessary to define and search for these situations, in order to collapse more seeds into one. To this aim, we define the so-called "redundant seeds".

For this purpose, a distance between the connectivity map generated from seed v_a (by applying equations 2 and 3) and the one generated from seed v_b is defined as:

$$d(v_i) = |c_{\chi^{v_a}}(v_i) - c_{\chi^{v_b}}(v_i)| \quad (4)$$

and associated to the distance energy value:

$$\alpha = \sum_i d(v_i) \quad (5)$$

A decreasing sigmoidal fuzzy membership function "low_s", with flex point in s, is also defined:

$$low_s : \alpha \to [0,1] \quad \begin{cases} \lim_{\alpha \to 0} low_s(\alpha) = 1 \\ \lim_{\alpha \to +\infty} low_s(\alpha) = 0 \end{cases} \quad (6)$$

Two seeds v_a and v_b are considered to be redundant if the energy of the distance field is low, given the original grey level of the two seeds are similar enough. In such a case only one of the two seeds is preserved, while the other is removed from the seeds list.

3.3   Total connectedness map

As more seed points are considered at each iteration step, connectivity information independently derived from each generic seed t, and contained in the related membership maps $C_{\chi t}$, are merged into a single "Total-Connectedness Map" through a process of fuzzy union:

$$C = \bigcup_{t=1}^{T} C_{\chi t} \quad (7)$$

where

$$c(v_i) = \bigcup_{t=1}^{T} c_{\chi t}(v_i) = \max_t \{c_{\chi t}(v_i)\} \quad (8)$$

3.4   Residuals and seeds generation

On the contrary of redundant seeds, when a significant region is not located by any of the already selected seeds, a new seed point must be appropriately looked for. Once it is found it is added to the already existing seeds list and the related χ-connectedness map is generated. To this end, the above defined intermediate total connectedness map is used. In fact, when a region has not been properly identified, some "residuals" appear in the map and can be automatically located.

Residuals are defined as those pixels that have low values in the total connectedness map. On the basis of equation 8, the residual map, P, is then derived by assigning to each pixel site the value:

$$\rho(v_i) = low_\rho \{c(v_i)\} \quad (9)$$

From this last map, a residual seed is added to the seeds list and its connectedness map is generated, giving start to a new iteration step.

3.5   Stopping criterion and labeling

The process is iteratively repeated by checking for the redundancy of the new seed and by looking for new residuals.

The stopping criterion is defined on the basis of residuals disappearance. When such a condition is verified, no more seeds are added or deleted and an intermediate label map is generated from the final total-connectedness map. It associates each image pixel to the most-connected seed:

$$\lambda_0(v_i) = \arg\{\max_t (c_{\chi t}(v_i))\} \quad (10)$$

Such a labeling usually gives rise to a larger number of connected components than the seed number. To avoid any loss of information, each connected component is finally labeled independently of the seed that has originated it, and the final label map, Λ, is created.

In all the processing steps only three parameters have to be set: the value s of the function low_s in Equation 6, the value ρ of the function low_ρ in Equation 9, and the number, ϕ, of final residuals accepted in the stop condition. In the following paragraph, the robustness of the AFS method with respect to these parameters is discussed, together with a preliminary comparison with other methods.

4 RESULTS

In order to evaluate the AFS performances, MRI brain images have been considered, addressing the segmentation of intracranial brain tissues, i.e., Cerebrospinal Fluid (CSF), White Matter (WM) and Gray Matter (GM). To achieve a quantitative numerical evaluation and to make the test results comparable with other works in the literature, the Brainweb dataset [23], which provides simulated MRI volumes for normal brain has been used. The datasets have complete volume phantoms available to represent the correct segmentation result.

An under-segmentation error occurs when a segmented region is actually corresponding to more than one region of the phantom. On the contrary, over-segmentation errors occur when more than one connected component is found corresponding to a single region of the phantom. With the aim of a classification step to be applied to the region segmentation result, under-segmentation errors are penalized and evaluated, while over-segmentation errors are not considered since the classifier can easily merge small regions to find the correct final phantom's regions.

The segmentation performances reported in the present section refer to Brainweb T1-slices with 1% noise level and Intensity Uniformity – RF = 0%. The used image size is 181×217 pixels, spatial resolution being 1 mm^2.

Figure 1(a) shows, as an example, the 100th brain slice of a considered volume.

The final total-connectedness-map C, generated by the proposed algorithm is presented in Fig. 1(b), and the segmentation result (i.e., the label map) Λ is shown in Fig. 1(c). For a visual comparison, the related phantom slice is shown in Figure 1(d).

By describing the parameters as a percentage value of the related dynamics, the setting of the reported processing session was:

- $s = 4\%$
- $\rho = 5\%$
- $\varphi = 0.3\%$.

To numerically evaluate the performances of the proposed segmentation, different metrics have been computed, referring to the true segmentation provided by the Brainweb phantom. In particular, the following parameters have been extracted for each class: Specificity, Sensitivity, Accuracy [24], as presented in Table 1.

By leaving unchanged the parameters' configuration different tests have been executed, to evaluate the repeatability of the method. The values reported in Table 1 are an average over the various executed tests, showing a very low standard error.

It was considered important to also report the ROC scatter plot for the six tests in the same parameters' setting ($s = 8\%$, $\rho = 10\%$, $\varphi = 0.3\%$) for a T1-weighted image with 0% RF, 3% noise (Figure 2). Each class is represented by a color: CFS is red, GM is blue, and WM is green.

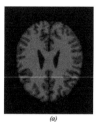

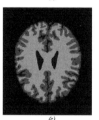
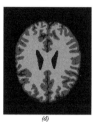

(a) (b)

(c) (d)

Figure 1. Simulated T1-weighted MR image, 100th slice: 1% noise (a) Original image; (b) Final total-connectedness map; (c) Label map; (d) Phantom.

Table 1. Accuracy results for the T1-weighted MRI slice of Fig. 1.

	Specificity	Sensitivity	Accuracy
CSF	98.68%	96.32%	98.45%
GM	98.94%	99.10%	99.07%
WM	98.73%	95.24%	97.52%

All three classes present a scatter plot very concentrated in the range 0-0.045 in terms of complement of specificity and in the range 0.85-1 in terms of sensitivity . This is certainly a positive factor because it validates the hypothesis of robustness and excellent performance of the method despite the randomness of seed selection phase.

For comparison, in Figure 3, the ROC scatter plot related to three methods is shown: AFS (red), MSMC (green) and Fuzzy Vectorial (blue). The graph refers to images with 0% RF, 1% noise and it presents all three classes CSF (circle), GM (triangle), WM (diamond). AFS method is run on T$_1$-weighted images with $s = 3\%$, $\rho = 5\%$, $\varphi = 0.3\%$. The MSMC method refers to the fuzzy segmentation described in paper [25], where seeds have been manually placed. The Fuzzy Vectorial method, reported in [26] makes use of the multiparametric volumes, by exploiting T1-weigthed, T2-weigthed and PD at the same time.

For the AFS method the average of the values obtained by the six tests has been considered

Despite using only one parametric image, AFS performs similarly or even better than Fuzzy Vectorial method, both in terms of specificity and sensitivity. In order to prove the application generality of the proposed AFS method some tests have been performed on

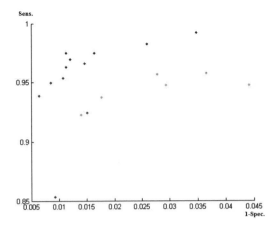

Figure 2. ROC scatter plot ($s = 8\%$, $\rho = 10\%$, $\varphi = 0.3\%$) for a T1-weighted image with 0% RF, 3% noise for CFS (red), GM (blue) and WM (green).

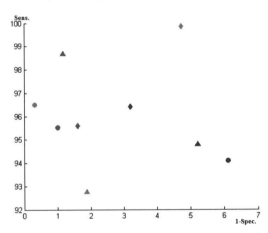

Figure 3. ROC scatter plot of AFS (red), MSMC (green) and Fuzzy Vectorial (blue) for the three different classes: CSF (circle), GM (triangle), WM (diamond) (0% RF, 1% noise with $s = 3\%$, $\rho = 5\%$, $\varphi = 0.3\%$)

real MRI datasets and on satellite images. A real T1-weighted MR brain image in coronal view provided by the Internet Brain Segmentation Repository (IBSR) [3] is reported in Figure 4, together with the regions obtained by the segmentation procedure.

5 CONCLUSIONS

In the current work, an automatic unsupervised fuzzy method for image segmentation has been proposed whose aim is the total segmentation of a digital images without user intervention.

In Section 4 performance evaluation of the method has been presented. The major advantage is that the present method does not require any a-priori information neither a training phase.

As shown above, good results have been achieved, often comparable or superior to those presented in the

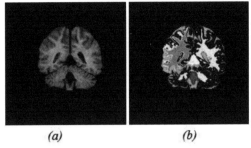

Figure 4. Real T1-weighted MR brain image (a) Original image; (b) Randomly colored segmented image.

literature but with the advantage that the present AFS method does not make use of any parameter estimation through a supervised phase.

Since the AFS method is also independent on the image content and on the image resolution, it works well for any kind of images.

The robustness of the method with respect to the three required parameters has been deeply investigated and has been here briefly discussed, showing very promising behaviour.

As a final advantage, the proposed method is very simple and transparent to the user. Its extension to the multiparametric exploitation of more source channels is straightforward and should be related to better performance quality.

REFERENCES

[1] S. K. Choy, M. L. Tang, and C. S. Tong, *Image Segmentation Using Fuzzy Region Competition and Spatial/Frequency Information Member*, IEEE Trans. on Image Processing, vol. 20, no. 6, June 2011, pp. 1473-1484.

[2] Dellepiane, S., Fontana, F., Vernazza, G.: *Nonlinear Image Labelling for Multivalued Segmentation.* In: IEEE Trans. on Image Processing, vol. 3, pp. 429–446, 1996.

[3] Udupa, J.K, Samarasekera, S.: *Fuzzy Connectedness and Object Definition: Theory, Algorithms, and Applications in Image Segmentation.* In Graphical Models and Image Processing, vol. 58, pp. 246–261, 1996.

[4] V. Caselles, F. Catte, T. Coll, and F. Dibos, "*A geometric model for active contours in image processing,*" Numer. Math., vol. 66, pp. 1–31, 1993.

[5] V. Caselles, R. Kimmel, and G. Sapiro, "*Geodesic active contours,*" Int. J. Comput. Vis., vol. 22, no. 1, pp. 61–79, 1997.

[6] M. Kass, A. Witkin, and D. Terzopoulos, "*Snakes: Active contour models,*" Int. J. Comput. Vis., vol. 1, pp. 321–331, 1988.

[7] C. S. Tong, Y. Zhang, and N. Zheng, "*Variational image binarization and its multi-scale realizations,*" J. Math. Imaging Vis., vol. 23, no. 2, pp. 185–198, Feb. 2005.

[8] D. Mumford and J. Shah, *"Optimal approximations by piecewise smooth functions and associated variational problems,"* Commun. Pure Appl. Math., vol. XLII, no. 4, pp. 577–685, 1989.

[9] T. Chan and L. Vese, *"Active contours without edges,"* IEEE Trans. Image Process., vol. 10, no. 2, pp. 266–277, Feb. 2001.

[10] S. C. Zhu and A. Yuille, *"Region competition: Unifying snake/balloon, region growing and Bayes/MDL/energy for multi-band image segmentation,"* IEEE Trans. Pattern Anal. Mach. Intell., vol. 18, no. 9, pp. 884–900, Sep. 1996.

[11] M. Tang and S. Ma, *"General scheme of region competition based on scale space,"* IEEE Trans. Pattern Anal. Mach. Intell., vol. 23, no. 12, pp. 1366–1378, Dec. 2001

[12] B. Mory, R. Ardon, and J.-P. Thiran, *"Variational segmentation using fuzzy region competition and local non-parametric probability density functions,"* in Proc. ICCV, 2007, pp. 1–8.

[13] J. Kim, J. Fisher, A. Yezzi, M. Cetin, and A. Willsky, *"A nonparametric statistical method for image segmentation using information theory and curve evolution,"* IEEE Trans. Image Process., vol. 14, no. 10, pp. 1486–1502, Oct. 2005.

[14] M. Ahmed, S. Yamany, N. Mohamed, A. Farag, and T. Moriarty, *"A modified fuzzy c-means algorithm for bias field estimation and segmentation of MRI data,"* IEEE Trans. Med. Imag., vol. 21, no. 3, pp. 193–199, 2002

[15] K. Ni, X. Bresson, T. Chan, and S. Esedoglu, *"Local Histogram Based Segmentation Using the Wasserstein Distance,"* UCLA CAM report 08-47, 2008.

[16] Tizhoosh, H.R., 1998. Fuzzy image processing. http://pmt05.et.uni-magdeburg.de/hamid/segment.html.

[17] Bezdek, J.C., 1981. Pattern Recognition With Fuzzy Objective Function Algorithms. Plenum, New York.

[18] Krishnapuram, R., Keller, J., *A possibilistic approach to clustering.* IEEE Trans. Fuzzy Syst. 1, 98–110, 1993.

[19] Chang, C.W., Ying, H., Hillman, G.R., Kent, T.A., Yen, J., 1998. *A rule-based fuzzy segmentation system with automatic generation of membership functions for pathological brain MR images.* Computers and Biomedical Research. http://gopher.cs.tamu.edu/yen/publications/index.html.

[20] Shih, F.Y., Cheng, S.: *Automatic Seeded Region Growing for Color Image Segmentation.* In: Image Vision and Computing, vol. 23, pp. 877–886, 2005.

[21] Udupa, J.K., Saha, P.K., Lotufo, R., *Relative Fuzzy Connectedness and Object Definition: Theory, Algorithms, and Applications in Image Segmentation.* In: IEEE Trans. on Pattern Analysis and Machine Intelligence, Vol. 24, No. 11, pp. 1485–1500, Nov. 2002

[22] S. Dellepiane, F. Fontana: *Extraction of Intensity Connectedness for Image Processing.* In: Pattern Recognition Letters, vol. 16, pp. 313–324, 1995.

[23] BrainWeb: www.bic.mni.mcgill.ca/brainweb/

[24] Richards, J., Jia, X. *Remote Sensing Digital Image Analysis*, Springer, 2005.

[25] E. Angiati, I. Minetti, S. Dellepiane, *Multi-Seed Segmentation of Tomographic Volumes Based on Fuzzy Connectedness*, ICIC 2010, Sixth Int. Conf. on Intelligent Computing, Changsha, China, August 18–21, pp. 360–367, 2010.

[26] Zhuge, J. K. Udupa, P. K. Saha: *Vectorial scale-based fuzzy-connected image segmentation.* In: Computer Vision and Image Understanding, vol. 101, no. 3, pp. 177–193, 2006.

Machining process simulation

Claudia H. Nascimento, Alessandro R. Rodrigues & Reginaldo T. Coelho
University of Sao Paulo, Brazil

ABSTRACT: This research estimated the temperature field in the cutting zone and compared it to the experimental results when machining AISI 5135 steel applied in the automotive industry. The simulation was performed by employing ABAQUS/Explicit® version 6.10-1 which uses a finite element method for spatial discretization of momentum equation and an explicit integration scheme for discretization relative to time. Through this numerical technique various machining parameters can be predicted such as force and temperature. A 2D orthogonal cutting has been modeled using Arbitrary Lagrangian-Eulerian (ALE formulation). The thermo-viscoplastic behavior of the workpiece material is modeled by Johnson-Cook constitutive work flow stress model. The numerical results presented a good agreement to the experimental ones validating them by an error of only ~3%.

1 INTRODUCTION

Since the first approaches in the early 1970s (Okushima; Kakino, 1971; Tay; Stevenson; Davis, 1974), the Finite Element Modeling (FEM) of the cutting process has evolved toward the implementation of more realistic and complex phenomena in the cutting models like elasto-viscoplastic material behavior (Llanos et al., 2009). However, in that time, precision simulations were not possible due to the scarse resources and high errors caused by simplifications of analitical models. Therefore, with the advent of computers the number of researches on this type of simulation has increased significantly.

Cutting processes simulation is effective for improving cutting tool design and selecting optimum working conditions, especially in advanced applications such as moulds and dies machining, high-performance machining and nano-micromachining (Rodrigues et al., 2012). This technique reduces time consumption and expensive experimental testing (Ceretti; Lucchi; Altan, 1997). In addition, many parameters involved in the machining process are very difficult or impractical to measure experimentally.

The machining process simulation involves the development of the material removal model, workpiece mesh and remesh model and force prediction model. Elevated temperature is a significant problem in metal cutting because it has visible influences on the physical property changes of the workpiece in cutting and the machined surface quality (Zong et al., 2007; Rodrigues et al., 2010). The researches about it can optmize the manufacturing process and consequently bring low costs for industries.

In this paper, an orthogonal cutting model was developed to simulate the temperature field in

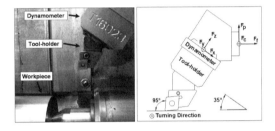

Figure 1. Dynamometer setup and schematic drawing for force calculations (front view).

cutting zone when machining AISI 5135 steel (37CrS4) applied in the automotive industry.

2 MATERIAIS AND METHODS

2.1 *Experimental procedure and geometrical model*

Experimental tests were carried out in a Boehringer VDF 180C CNC lathe. An acquisition system connected to a Kistler 9067 3-component piezoelectric dynamometer (Figure 1) was used to measure the machining forces by which the specific friction energy and cutting temperature were calculated and simulated.

Valenite CNMA120408 ceramic tools with rake angle $\gamma = -6°$, clearance angle $\alpha = 6°$, tool cutting edge angle $k_r = 95°$, corner radius $r_\epsilon = 0.8$ mm, Titanium Nitride coating and no chip breaker were used in the experiments. The cutting parameters were cutting speed $V = 1.3$ m/s, feed $f = 0.15$ mm/rev, depth of cut $d = 0.54$ mm, cutting length $l_c = 20$ mm and no lubricant/cooling. The experiment had two replications with identical conditions and a new cutting edge was

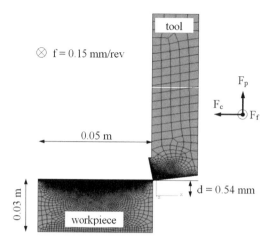

Figure 2. Geometrical model for simulation (side view).

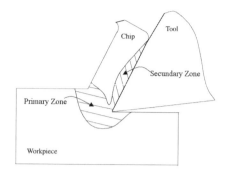

Figure 3. Temperature zones in machining process (Dirikolu; Childs, 2000).

Table 1. Material Properties.

AISI 5135 steel (workpiece)	
Thermal conductivity (W/m°C)	38
Coefficient of thermal expansion (μm/m°C)	0.000032
Density (kg/m^3)	7838
Young's modulus (GPa)	200
Poisson's ratio	0.29
Specific heat (J/kg°C)	477
Ceramic (tool)	
Thermal conductivity (W/m°C)	20
Density (kg/m^3)	14950
Young's modulus (GPa)	400
Poisson's ratio	0.21
Hardness (HV)	240

used for each test to assure the equal initial conditions. The cutting (F_c), feed (F_f) and radial (F_p) forces were determined by considering the dynamometer components (F_x), (F_y) and (F_z) as well as the dynamometer setup angle of 35°. Equation 1 and 2 were deduced using tool angles and dynamometer setup and applied to obtain the tangential (F_t) and normal (F_n) forces acting over tool rake face.

$$F_t = F_c \sin\gamma + (F_f \sin k_r + F_p \cos k_r) \cos\gamma \quad (1)$$

$$F_n = F_c \cos\gamma + (F_f \sin k_r + F_p \cos k_r) \sin\gamma \quad (2)$$

The specific friction energy (u_f) was calculated by numerical integration of Equation 1, considering the removed chip volume (Vol_{chip}) and cutting length (l_c).

$$u_f = \frac{1}{Vol_{chip}} \int_0^{l_c} \left\{ F_c \sin\gamma + \left[F_f \sin k_r + F_p \cos k_r \right] \cos\gamma \right\} dl_c \,(3)$$

The cutting temperature was determined by applying the Cooks model (Groover, 2007), based on the specific friction energy and physical properties of the workpiece material (Incropera; Witt; Bergman, 2008). The workpiece was modeled in rectangular geometry with 0.05 m length and 0.03 m height using ABAQUS® (Figure 2). The cutting parameters for simulation input are the same used in machining tests. ABAQUS® adopts Coulomb friction model, which was determined experimentally as $\mu = F_t/F_n = 0.66$.

The friction model establishes a shear stress τ_{max} who generates sliding friction if the magnitude of shear stress reaches this value. A reasonable estimation value of upper bound is $\tau_{max} = \delta_y/\sqrt{3}$, where δ_y is the Von Mises Yield stress (Abaqus, Inc., 2006).

2.2 Heat generation in machining process

Practically all mechanical energy envolved in machining process is converted in thermal energy.

Theoretically this energy is divided into two main deformation zones: primary zone caused by the work of inner shearing of material and secundary zone due to high stress in chip-tool contact conditions. The primary shear zone is where the major shearing of work material takes place (Özel, 2006). Figure 3 shows these temperature zones.

The experimental measurement of the cutting temperature in these regions is very complicated and FEM allows obtaining estimations very close to the experimental ones.

2.3 Material model

Table 1 shows the input data for the ABAQUS® simulation, i.e., the properties of material used in the processes of machining.

During cutting process large plastic deformation occurs and consequently the workpiece material reaches high temperature, high strain rate and material strengthening phenomenon. Flow stress is taken as a function of strain, strain rate and temperature. The Johnson-Cook model (Equation 4) has been chosen to implement its mechanical behavior. This model integrates strain hardening (I), sensitivity to the strain

Table 2. Johnson-Cook constants for stress and strain.

AISI 4340	
A	792
B	510
C	510
n	0.26
m	1.03
d_1	0.05
d_2	3.44
d_3	−2.12
d_4	0.0022
d_5	0.61
T_m	1793
T_0	305

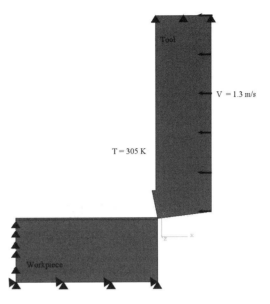

Figure 4. Boundary conditions.

speed (II) and thermal softening (III) (Valiorgue et al., 2008).

$$\sigma = \underbrace{[A + B\,(\bar{\epsilon})^n]}_{I} \underbrace{\left[1 + C\ln\left(\frac{\dot{\bar{\epsilon}}}{\dot{\bar{\epsilon}}_0}\right)\right]}_{II} \underbrace{\left[1 - \left(\frac{T - T_0}{T_m - T_0}\right)^m\right]}_{III} (4)$$

Where σ is the equivalent plastic flow stress, A is the yield stress, B is the strength coefficient, $\bar{\epsilon}$ the equivalent plastic strain, C is the strain rate constant, $\dot{\bar{\epsilon}}$ is the equivalent plastic strain rate, $\dot{\bar{\epsilon}}_0$ is the reference equivalent plastic strain rate (set to 1) T is the temperature, T_m is the melting temperature, T_0 is the reference temperature, n is the strain-hardening exponent and m is the thermal softening exponent.

Johnson-Cook Failure model (Equation 5) was adopted to describe the chip formation mechanism mathematically. It states that material fails according to a certain amount of strain as a function of strain-rate, temperature and the stress triaxiality. The criterion calculates the strain for failure according to:

$$\epsilon^f = [d_1 + d_2 \exp d_3 \sigma^*] \left[1 + d_4 \ln \frac{\dot{\bar{\epsilon}}}{\dot{\epsilon}_0}\right] \left[1 - d_5 \left(\frac{T - T_0}{T_f - T_0}\right)^m\right] (5)$$

The specific constants are listed in Table 2 (Johnson; Cook, 1985). Constants of the AISI 4340 steel were considered for simulation due to the proximity of chemical composition between the materials.

Not only the failure criterion is necessary, but also some kind of damage progress and separation criterion used by the FEM code to separate the chip. That has to be incorporated into the software code to allow a real simulation. The one adopted at the present work is given by:

$$D = \sum \frac{\Delta \epsilon}{\epsilon^f} \qquad (6)$$

Separation is then allowed when $D = 1.0$ and the elements matching such criterion are removed from the mesh (Johnson; Cook, 1985).

The FEM technique requires the use of either implicit or explicit solution method. Explicit methods can deal with highly non-linear systems but need small steps. On the other hand, implicit methods can deal with mildly non-linear problems but with large steps. Machining process simulation involves large deformation, multiple nonlinearities, complex contact problem and high strain rates. Thus, the most appropriate solution method is the explicit integration.

Therefore, the cutting process is treated as a coupled thermo mechanical process (HAIRUDIN; AWANG, 2011).

2.4 Adaptive meshing and boundary conditions

The boundary conditions imposed for this simulation is described in Figure 4. The botton surface was constrained along the X-direction and left surface in Y-direction. The velocity vector was applied on the right side of the tool and the top surface was constrained along the X-direction. The system was considered adiabatic (no heat loss to the environment) and set at 305 K initially for the thermal boundary conditions.

An approach for developing finite element modelling of machining process is to use the Lagrangian formulation, generally adopted for transient and non-steady processes with a dependency on the loading path and the materials history (Rakotomalala; Joyot, 1993).

Lagrangian approach is default in ABAQUS® and the meshes have to change with the flowing material (Lu et al., 2011). This formulation is advantageous for satisfying less complex governing equations compared with the pure Eulerian approach.

3 ANALYSIS

The computational time for solving the temperature parameters was approximately within 6 hours, using

315

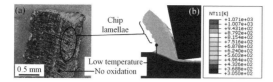

Figure 5. (a) Real chip and (b) comparison with simulated chip formation and temperature field.

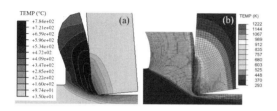

Figure 6. Temperature simulation of AISI 4340 steel with (a) $V = 3.0$ m/s, $f = 0.07$ mm/rev, 595 HV (material hardness), negative tool geometry (TiNAl PCBN) and (b) $V = 5.0$ m/s, $f = 0.2$ mm/rev, 230 HV (hardness material), positive tool geometry (uncoated carbide).

INTEL XEON 3.2 GHz, 32 GB RAM, 8 microprocessors workstation. Figure 5 shows the simulation result.

The temperature calculated in the cutting zone was 247.4°C and the simulated temperature in the related region was 254.4°C. So the error was found in the simulation result of 2.83%. This result (percentile error) agrees with (Grzesik, 2006) when mentioning that FEM accuracy is around 3 to 5%.

Figure 6 shows that the results of temperature simulated by this work present close agreement with (Coelho; Eu-Gene; Elbestawi, 2007; Arrazola; Özel, 2010). As referências esto abaixo. Despite some cutting parameters, tool geometrical features and mechanical properties of workpiece material being little different, the temperature fields in the chip and machined workpiece surface are very similar. The differences more significant are the tool temperature and the region where it is maximum. Most likely the substrate and coating of tools are the cause of temperature difference.

Figures 6(b) indicates that temperature in the chip inner reaches approximately 460°C while the machined workpiece surface and tool rake face attain about 280 and 160°C, respectively. In other words, even with low cutting speed and ceramic tool (which conduct less heat than carbide insert) the increase of temperature can affect the workpiece integrity and wear the tool especially for a long machining time. The chip-tool interface (secondary deformation zone) is the region more critical given the temperature of ~800°C obtained by simulation. This considerable increase of temperature takes place distant from tool tip and it is in this location that crater wear usually occurs

The maximum strain energy density curve is found to be located within the primary deformation zone and lies almost parallel to the so-called shear plane (Lin; Pan, 1993).

4 CONCLUSIONS

In summary, this research showed that the simulation is efficient to describe the machining process because the obtained results are very closed to the experimental ones. These errors can be caused by the simplification of the physical model and numerical approximations, but the simulations allow to reduce costs and processing time. With the FEM turning simulation, it is possible to estimate the values of process variables that are very difficult to measure by experiment or are not measurable, such as cutting temperatures at the toolchip and toolworkpiece interfaces.

REFERENCES

Abaqus, Inc. *ABAQUS Analysis*: User's manual abaqus. United States of America, 2006.

Arrazola, P.; Özel, T. Investigations on the effects os friction modeling in finite element simulation of machining. *International Journal of Mechanical Sciences*, v. 52, p. 31–42, 2010.

Ceretti, E.; Lucchi, M.; Altan, M. Fem simulation of orthogonal cuttin: serrated chip formation. *Journal of Material Processing Technology*, v. 95, p. 17–26, 1997.

Coelho, R.; Eu-Gene, N.; Elbestawi, M. Tool wear when turning hardened aisi 4340 with coated pcbn tools using finishing cutting conditions. *International Journal of Machine Tools & Manufacture*, v. 32, p. 263–272, 2007.

Dirikolu, M. H.; Childs, T. H. C. Modelling requirements for computer simulation of metal machining. *Turk J Engin Environ Sci*, v. 46, n. 24, p. 81–93, 2000.

Groover, M. P. *Fundamentals of Modern Manufacturing*. [S.l.]: Wiley, 2007.

Grzesik, W. Determination of temperature distribution in the cutting zone using hybrid analytical-fem technique. *International Journal of Machine Tools and Manufacture*, v. 46, n. 6, p. 651 – 658, 2006.

Hairudin, W. M. B.; Awang, M. B. Preliminary study on thermomechanical modelling of cutting process. *Advanced Materials Research*, v. 383, p. 6741–6746, 2011.

Incropera, F.; Witt, D.; Bergman, T. *Fundamentals of Heat and Mass Transfer*. [S.l.]: LTC, 2008.

Johnson, G.; Cook, W. Fracture characteristics of three metals subjected to various. *Engineering Fracture Mechanics*, v. 21, n. 1, p. 31–48, 1985.

Lin, Z. C.; Pan, W. C. A thermo-elastic-plastic model with special elements in a cutting process with tool flank wear. Int. *J. Manufact.*, v. 34, n. 24, p. 757–770, 1993.

Llanos, I. et al. Finite element modeling of oblique machining using an arbitrary lagrangianeulerian formulation. *Machining Science and Technology*, v. 13, n. 3, p. 385–406, 2009.

Lu, D. et al. Finite element modeling for high speed machining of ti-6al-4v using ale boundary technology. *Applied Mechanics and Materials*, v. 66, p. 1509–1514, 2011.

Okushima, K.; Kakino, Y. Residual stress produced by metal cutting. *Annals of the CIRP*, v. 20, n. 1, p. 13–14, 1971.

Özel, T. The influence of friction models on finite element simulations of machining. *International Journal of Machine Tools & Manufacture*, v. 46, p. 518–530, 2006. ISSN 0890-6955.

Rakotomalala, R.; Joyot, P. Arbitrary lagrangian-eulerian thermomechanical finite-element model of material cutting. *Communications in numerical methods in engineering*, v. 9, p. 975–987, 1993.

Rodrigues, A. et al. Surface integrity analysis whem milling ultrafine-grained steels. *Materials Research*, v. 15, n. 1, p. 1–6, 2012.

Rodrigues, A. et al. Effects of milling condition on the surface integrity of hot forged steel. *Journal of the Brazilian Society of Mechanical Sciences and Engineering*, v. 47, n. 1, p. 37–43, 2010.

Tay, A.; Stevenson, M.; Davis, G. Using the finite element method to determine temperature distributions in orthogonal machining. *Proceedings of the Institution of Engineers*, v. 188, p. 627–638, 1974.

Valiorgue, F. et al. Modelling of friction phenomena in material removal processes. *Journal of Materials Processing Technology*, v. 20, p. 450–452, 2008.

Zong, W. et al. Finite element optimization of diamond tool geometry and cutting-process parameters based on surface residual stresses. *Int J Adv Manuf Technol*, v. 32, p. 666–674, 2007.

Computational Modelling of Objects Represented in Images – Di Giamberardino et al. (eds)
© *2012 Taylor & Francis Group, London, ISBN 978-0-415-62134-2*

A least square estimator for spin-spin relaxation time in Magnetic Resonance imaging

Fabio Baselice, Giampaolo Ferraioli & Vito Pascazio
Università degli Studi di Napoli Parthenope, Dipartimento per le Tecnologie, Italy

ABSTRACT: Magnetic Resonance (MR) imaging techniques are widely used for medical examination of biophysical properties of tissues. Clinical diagnoses are mainly based on the evaluation of contrast in MR images, evaluating the water component of involved tissues. Moreover, weighted images are considered in order to evaluate the difference response time in order to detect pathologies such as cancer. In this paper we propose some some first results of a statistical technique able to estimate the Spin-spin Relaxation Time of observed tissues exploiting both real and imaginary parts of MR images. Working in the complex domain instead of the amplitude one allows us to write the joint probability distribution of real and imaginary signals. Considering the optimal estimator for the considered case, we are able to evaluate the tissues relaxation times with the lowest possible variance. This estimation technique can lead to a different scenaries in medical diagnostic. The method has been tested on real data, showing interesting results.

1 INTRODUCTION

Magnetic Resonance Imaging (MRI) is a technique used in medical environment to produce high quality images of tissues inside human body. MRI is based on the principles of nuclear magnetic resonance (NMR), a spectroscopic technique used to obtain microscopic chemical and physical information about molecules.

To obtain an MR image, tissues have to be positioned inside a magnetic field and excited by proper Radio Frequency (RF) signals. Depending on such signal, many different imaging sequences are defined. The final acquired image will depend both on tissues under investigation parameters and on imaging sequence kind and configuration.

In particular, from an MRI point of view, tissues are defined by three intrinsic values: the proton density ρ, the Spin-lattice Relaxation Time $T1$ and the Spin-spin Relaxation Time $T2$. The first parameters is related to the intensity of the signal generated by each voxel (i.e. imaged tissue represented by one pixel in the final image), while $T1$ and $T2$ define the decay time and the speed of phase lost of the signal (Slichter 1996)

Considering a Spin Echo imaging sequence, there are two system parameters that can be tuned id order to produce images with different characteristics: the Echo Time TE and the Repetition Time TR. Depending on the considered TE/TR couple, we can obtain images that empathize one of the intrinsic tissues values (i.e. obtaining the so called weighted images). In particular, a quantitative analysis of Spin-Spin Relaxation Time $T2$ can give useful information for cancer discrimination (Roebuck et al. 2009).

Conventional relaxation parameter estimation techniques work on the amplitude of MR images. In this paper we propose an approach for Spin-Spin Relaxation Time estimation that works directly on the real and imaginary decomposition of the MR images. Exploiting this decomposition, since the noise is Gaussian both on real and imaginary parts, we implement a Least Square Estimator which is the optimal one under Gaussian noise conditions.

In Section 2 the theoretical background is presented, while an analysis of the performances of the proposed method based on a real data set is presented in Section 3. Finally, we draw some conclusions.

2 CONCEPTUAL MODEL

The MR complex signal can be modeled as (Baselice et al. 2010):

$$m = g(\boldsymbol{\theta}) + n \tag{1}$$

where $g(\boldsymbol{\theta})$ is the noise free MR complex signal, $\boldsymbol{\theta} = [\rho\ T_1\ T_2]$ is the vector of tissues parameters we are interested in and $n = n_R + in_I$ is the noise. In the MRI framework, the noise is a zero mean jointly circular Gaussian random variable (Wang et al. 1996), i.e. both real and imaginary parts are independent Gaussian distributed. The probability density function (pdf) of the real noise component n_R is given by

$$f_{N_R}(n_R) = \frac{1}{\sqrt{2\pi\sigma^2}} e^{-\frac{n_R^2}{2\sigma^2}} \tag{2}$$

The same distribution applies for the imaginary part (i.e. n_I instead of n_R).

Considering a Spin Echo imaging sequence, the complex noise free MR signal becomes (Slichter 1996):

$$g(\boldsymbol{\theta}) = ke^{-\frac{TE}{T2}}e^{i\phi} = ke^{-\frac{TE}{T2}}\cos(\phi) + ike^{-\frac{TE}{T2}}\sin(\phi)\,(3)$$

where k takes into account the ρ and $T1$ parameters while ϕ is the phase of the signal $g(\boldsymbol{\theta})$ (Baselice et al. 2010). This notation lead to the following statistical distribution of real (m_R) and imaginary (m_I) components of the acquired signal m:

$$f_{M_R}(m_R) = \frac{1}{\sqrt{2\pi\sigma^2}}e^{-\frac{\left(m_R - ke^{-\frac{TE}{T2}}\cos(\phi)\right)^2}{2\sigma^2}}$$

$$f_{M_I}(m_I) = \frac{1}{\sqrt{2\pi\sigma^2}}e^{-\frac{\left(m_I - ke^{-\frac{TE}{T2}}\sin(\phi)\right)^2}{2\sigma^2}} \quad (4)$$

Considering these distributions, a Non Linear Least Square Estimator (NL-LSE) is implemented in order to reconstruct the $T2$ values for each pixel of the tissues under investigation. The reconstruction of k, as it is not meaningful from a physical point of view, is ignored. We recall that, in case of Gaussian distributed noise, the NL-LSE represents the Minimum Variance Unbiased (MVU) estimator, which is the best possible, with respect to variance minimization, estimator in the classical estimation theory. In particular, we propose a fast non linear estimator able to dramatically reduce the computational time with respect to a non optimized implementation (Kay 1993).

3 ANALYSIS

The performances of the estimator have been, first, tested on simulated data. A signal generated by a single voxel with $k = 78.5$ and $T2 = 65$ ms and with a phase of $\phi = 60$ degrees has been considered. The simulated data has been corrupted by circular Gaussian complex noise.

First, an evaluation of $T2$ estimator performances, using Monte Carlo simulation, has been conducted for different Signal to Noise Ratio (SNR) values, considering 7 acquisitions with $TE = [30\ 45\ 70\ 85\ 110\ 120\ 145]$ ms, respectively. The number of Monte Carlo simulation was set to 10^3 for each SNR configuration.

The relative error of the estimator is reported in figure 1(a). It can be noted that its value is always lower than 2%, and it approaches zero for SNRs higher than 10 dB. With respect to the standard deviation of the estimator, reported in figure 1(b), we can asses that it is deeply affected by the SNR value, as expected. Again, high SNRs (i.e. SNR higher than 10) can assure effective results.

An evaluation of $T2$ estimator performances has also been conducted for different number of acquisitions (from 3 to 10), considering an SNR of 15 dB. In

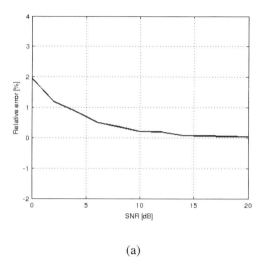

(a)

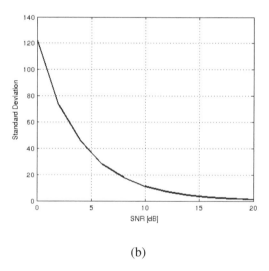

(b)

Figure 1. Relative error (a) and standard deviation (b) of the proposed estimator with respect to SNR.

case of 3 acquisitions, TEs have been set to $[30\ 45\ 70]$ ms, while in case of 10 acquisitions the considered TEs were $[30\ 45\ 70\ 85\ 110\ 120\ 145\ 155\ 173\ 200]$ ms. The results are reported in figure 2. Both estimator relative error and standard deviation, as expected, show a positive trend increasing the number of acquisitions.

For the evaluation of performances of the proposed approach a case study of four complex images m_1, m_2, m_3, m_4 has been considered. The images have been acquired in a 7 tesla MR machine with a resolution of 256×256 pixels and are representative of a lemon slice. The four images have been acquired with different Echo Times, $TE = [10\ 20\ 30\ 40]$ ms and a Repetition Time $TR = 2$ s. The amplitude of the acquisition with $TE = 10$ ms is reported in figure 3.

The algorithm, implemented in Matlab© environment, has been applied to the available complex

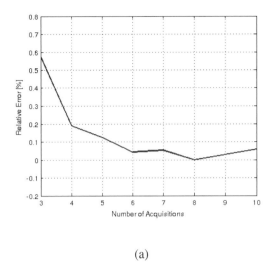

(a)

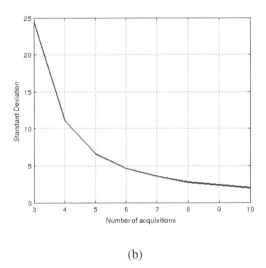

(b)

Figure 2. Relative error (a) and standard deviation (b) of the proposed estimator with respect to the number of acquisitions.

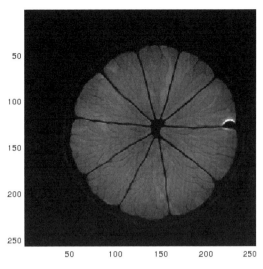

Figure 3. Amplitude of image acquired with $TR = 2$ s, $TE = 10$ ms.

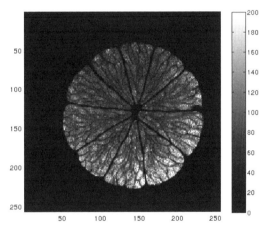

Figure 4. Estimated Spin-spin Relaxation Time ($T2$).

images, generating the estimate $T2$ map reported in figure 4.

The algorithm is able to retrieve different $T2$ values across the image. In particular, some regions show a Spin-spin Relaxation Time $T2$ higher than 150 ms, indicating a different chemical composition compared to other areas. This distinction cannot directly be done from the amplitude image of figure 3, where all areas show similar contrast. Of course the possibility of classify areas with different chemical composition can be a very interesting application in medical diagnosis. For example, it is known that cancer tissues exhibit different $T2$ values with respect to surrounding healthy tissues.

Clearly, since we do not have the true $T2$ map, the algorithm cannot be quantitatively evaluated. A simulated case, with quantitative analysis, can be found in (Baselice et al. 2010).

Moreover, an interesting application of this estimation method is the possibility to form, after $T2$ estimation, an a posteriori $T2$ weighted image with any TE, i.e. it is able to simulate an MR image acquired with any specific Echo Time. For example, we report the a posteriori synthesized image for $TE = 100$ ms. It has to be stressed that, as this image is computed from the estimated $T2$ map, does not present a very low Signal to Noise Ratio (SNR), as a real acquired image would do. From a computational point of view, the algorithm, working pixel wise can be easily parallelized, providing the estimated $T2$ map in almost real time.

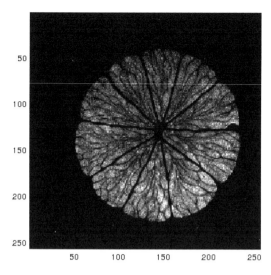

Figure 5.   Synthesized image with $TE = 100$ ms.

4   CONCLUSIONS

In this paper we have shown the potentialities of a novel Spin-Spin Relaxation Time estimation method. The technique, based on Non Linear Least Square approach, is able to provide an estimated $T2$ map of the tissues under investigation. This makes the method an interesting tool for medical imaging application, such as tissue classification, ill-healthy voxel detection and

a posteriori weighted MR images formation. We tested the effectiveness of the method using a real data set at 7 T. The approach has proven to be not excessively time demanding and computationally heavy.

ACKNOWLEDGEMENT

Authors would like to acknowledge Johan Van Audekerke of the Departement Biomedische Wetenschappen, Antwerpen University, for supplying the MR images.

REFERENCES

Baselice, F., G. Ferraioli, and V. Pascazio (2010). Relaxation time estimation from complex magnetic resonance images. *Sensors 10*(4), 3611–3625.

Baselice, F., G. Ferraioli, and A. Shabou (2010). Field map reconstruction in magnetic resonance imaging using bayesian estimation. *Sensors 10*(1), 266–279.

Kay, S. M. (1993). *Fundamentals of Statistical Signal Processing, Volume 1: Estimation Theory*. Prentice Hall.

Roebuck, J. R., S. J. Haker, D. Mitsouras, F. J. Rybicki, C. M. Tempany, and R. V. Mulkern (2009). Carr-purcell-meiboom-gill imaging of prostate cancer: quantitative t2 values for cancer discrimination. *Magnetic Resononance Imaging 27*, 497–502.

Slichter, C. P. (1996). *Principles of Magnetic Resonance, 3rd ed.* Springer.

Wang, Y. J., T. Lei, W. Sewchand, and S. K. Mun (1996). Mr imaging statistics and its application in image modeling. In *Proc. SPIE 2708, 706*.

Computational Modelling of Objects Represented in Images – Di Giamberardino et al. (eds)
© 2012 Taylor & Francis Group, London, ISBN 978-0-415-62134-2

Numerical modelling of the abdominal wall using MRI. Application to hernia surgery

B. Hernández-Gascón, A. Mena, J. Grasa, M. Malve, E. Peña & B. Calvo
Aragón Institute of Engineering Research. University of Zaragoza, Spain
CIBER-BBN. Centro de Investigación en Red en Bioingeniería, Biomateriales y Nanomedicina, Spain

G. Pascual & J.M. Bellón
Departament of Surgery, Faculty of Medicine, University of Alcalá, Spain
CIBER-BBN. Centro de Investigación en Red en Bioingeniería, Biomateriales y Nanomedicina, Spain

ABSTRACT: The aim of this work is to present a methodology to model the human abdomen to study its mechanical response and simulate hernia surgery. For this purpose, a realistic geometry of the human abdomen using magnetic resonance imaging is obtained. The model defines the different anatomical structures of the abdomen and the anisotropic mechanical properties are assigned depending on the considered structure. The finite element model obtained from the human model is used to simulate the passive mechanical behaviour of the healthy and totally herniated human abdomen under physiological loads using different surgical meshes. Concretely, *Surgipro®*, *Optilene®* and *Infinit®* were modelled. Results suggest the linea alba is the most susceptible part to fail. Besides, just after surgery, surgical repair procedure does not fully restore normal physiological conditions.

1 INTRODUCTION

Routine hernia repair surgery involves the implant of synthetic mesh. However, this type of procedure may give rise to several problems, causing considerable patient disability. Nowadays, there is a great variety of meshes available on the market. However, the "ideal prosthesis" which best adapts to the mechanical conditions of the host tissue has not yet been achieved.

The abdominal wall contains four expiratory muscles: the rectus abdominis (RA), the external oblique (EO), the internal oblique (IO), and the transverse abdominis (TA). Anatomically, the IO lies internal to the EO muscle in the lateral abdominal wall, whereas the TA is the innermost abdominal muscle. Furthermore, the linea alba (LA) is a fibrous structure that runs down the midline of the abdomen and is composed mostly of collagen connective tissue. Particularly, it is formed by the fusion of the aponeuroses of the abdominal muscles (anterior rectus sheath (ARS) and posterior rectus sheath (PRS) which cover the anterior and posterior faces of the rectus abdominis, respectively), and it separates the left and right RA muscles. Another anatomical structure is the fascia transversalis (FT) which covers the internal surface of the abdominal cavity.

In the case of the abdominal muscle it is necessary to distinguish between the passive mechanical response, given by the collagen fibres (CF), and the active mechanical response, associated with the muscular fibres (MF). Regarding the constitutive behaviour, the purely passive response is often modelled within the framework of hyperelasticity by means of a strain energy function (SEF).

To the authors' knowledge, there is no computational study which defines a realistic geometry of the human abdomen to simulate the behaviour under physiological loads. Thus, with the aim of improving surgical procedures, this study presents a finite element (FE) model of the human abdomen obtained using magnetic resonance imaging (MRI). The model defines the different anatomical units of the abdomen and the anisotropic mechanical properties are assigned depending on the considered structure. This numerical human model is used to simulate the passive behaviour of the healthy and herniated human abdomen. Different surgical meshes, characterized by different pore size and materials are simulated.

2 MODEL OF THE HUMAN ABDOMEN

For the reconstruction of the real geometry of the human abdomen, a reference model from a healthy man aged 38 was created. To obtain the abdomen geometry, DICOM files from MRI were used. Using the commercial software MIMICS® a manual segmentation of the 3D DICOM images was made. This procedure permitted obtaining the geometrical model of the internal and the external surfaces of the abdomen as well as differentiating the different anatomical structures using different masks: linea alba, rectus abdominis and oblique muscles. Despite oblique muscles comprise three different muscles (EO, IO and TA), our work considers these muscles as a composite, a monolayer. This assumption is valid to simulate the passive behaviour since the layers of the abdomen do not express separate movements but only change their thickness. Although the different muscle layers in the flat muscles are considered as a monolayer, the number of elements through the thickness was chosen according to the work published by Norasteh et al. (Norasteh et al. 2007). Thus, 5 elements were defined in the thickness so that approximately 2 elements represent the EO, 2 elements represent the IO and the last one represents the TA. Thus, using the surfaces previously obtained with MIMICS® and the software ABAQUS®, a structured volumetric mesh was made employing a total of 13200 hexahedral elements.

The abdomen is not only comprised of muscles but also of aponeuroses (namely, fascias or tendons as well). Apart from the linea alba, aponeuroses usually appear wrapping muscles. Besides, some tendinous structures appear inside the rectus abdominis (rectus tendon, RT) and oblique muscles (oblique muscle tendon, OMT) which are aponeuroses as well (See Fig. 1.a). Thus, these anatomical structures forming the aponeuroses are defined over the mesh previously created, using the software ABAQUS®, taking into account the human anatomy (W. Moore 2008). Those aponeuroses wrapping muscles are modelled using membrane elements so 2092 membrane elements are defined. On the other hand, RT and OMT are part of the structure volumetric mesh previously made with hexahedral elements (See Fig. 1.a).

Finally, despite our work is only focussed on analyzing the results in the front of the abdomen (where abdominal hernias appear), the diaphragm and pelvis are modelled in order to close the abdominal cavity. For that purpose, 432 shell elements are added to the model using the software FEMAP® (Siemens Software) (See Fig. 1.a).

Summing up, hexahedral elements are subdivided into different groups: lina alba, rectus abdominis muscle and rectus tendon, oblique muscles (including EO, IO and TA) and oblique muscle tendon, chest, shoulder and pelvis support (See Fig. 1). On the other hand, membrane elements are subdivided into fascia transversalis, anterior and posterior rectus sheath and, finally, shell elements are subdivided into diaphragm and pelvis (see Fig. 1).

3 CONSTITUTIVE MODELLING

Some common features are usually found in the mechanical response of soft tissues. Those features are non linear mechanical response, large deformations, incompressibility, anisotropy and the presence of residual stresses or initial strains that can be relieved by selective cutting of the living tissue and removal of its internal constraints. Thus, regarding the constitutive modelling, the purely passive response of the

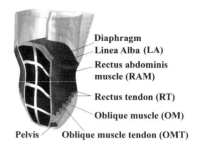

(a)

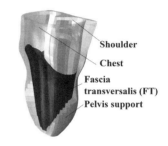

(b)

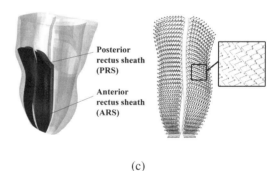

(c)

Figure 1. Different groups defined in the model. (a) RAM, LA, RT, OM, OMT, diaphragm and pelvis. (b) FT, chest, shoulder and pelvis support. (e) ARS and PRS. Collagen fibre disposition in the ARS.

abdominal muscle is modelled within the framework of the large deformation anisotropic hyperelasticity by means of a strain energy function (SEF) (Hernández et al. 2011).

In order to more easily handle the quasiincompressibility constraint, a multiplicative decomposition into volume-changing and volumepreserving parts is usually established (Flory 1961). Then, we postulate the existence of a unique decoupled representation of the SEF which is expressed in terms of kinematic invariants (Spencer 1971) which depends on the orientation of the collagen fibres (Weiss et al. 1996):

$$\Psi(\mathbf{C}, \mathbf{M}) = \Psi_{\text{vol}}(J) + \bar{\Psi}(\bar{I}_1, \bar{I}_2, \bar{I}_4) \quad (1)$$

Table 1. Material model parameters obtained from the fitting procedure of the experimental *in vitro* results. The angle α is relative to the craneo-caudal direction. *These properties are assigned to the following sets of elements: LA, RAT, OMT, FT, ARS, PRS (See Fig. 1).

	c_1 [MPa]	c_2 [−]	c_3 [MPa]	c_4 [−]
*Sets**	0.2434	0.8	0.0064	9.63

	c_5 [MPa]	c_6 [MPa]	c_7 [MPa]	I_{40} [−]	ε [−]
*Sets**	31.8214	−36.9188	−31.4118	1.0	0.1483

	c_1 [MPa]	c_2 [−]	c_3 [MPa]
Oblique muscles	0.16832	0.6319	0.01219
Rectus abdominis muscle	0.10445	6.86123	0.001
Chest − Shoulder − Pelvis	0.16832	0.6319	0.01219

	c_4 [−]	α [°]	ε [−]
Oblique muscles	5.68158	87.8	0.17873
Rectus abdominis muscle	0.00491	0.0	0.10923
Chest − Shoulder − Pelvis	5.68158	−	0.17873

In order to define the SEF, material properties were taken from the literature (Hernández et al. 2011; Martins et al. 2012). Due to the fact that reported experimental data on the mechanical properties of human abdominal wall are limited, experimental animal data are used when there is no data from humans. After an exhaustive analysis to fit original data (Hernández et al. 2011; Martins et al. 2012), the best agreement between the experimental data and the modelled one was achieved using different SEF for each tissue. Thus, regarding the human aponeurosis (Martins et al. 2012), the isotropic response was modelled by means of the Demiray's SEF (Demiray et al. 1988) while the anisotropic response was represented by Calvo's SEF (Calvo et al. 2009). On the other hand, muscles (Hernández et al. 2011) were modelled using the SEF proposed by Demiray et al. (Demiray et al. 1988) and Holzapfel et al. (Holzapfel et al. 2000) to fit the isotropic and the anisotropic response, respectively. Those SEF were implemented in a user-material subroutine for ABAQUS®.

This study only considers the passive mechanical behaviour. However, the abdominal tissue is anisotropic so this response is modelled by introducing a preferential direction of anisotropy (PDA) (Arruda et al. 2006). For that purpose, the PDA is included at each integration point of the finite element model through a unit vector.

The material parameters that fit the experimental behaviour in each tissue are shown in Table 1. Lastly, the behaviour of the diaphragm and pelvis was modelled using a Neo-Hookean model (Ogden 1996) whose constant c_1 was equal to 0.18 (Note that since we are interested in the frontal part of the abdomen,

the results in the chest, shoulder, diaphragm and pelvis are not analized so an isotropic model is considered).

The preferential direction of anisotropy is coincident with the direction of the collagen fibres when considering the aponeuroses so it can be obtained from the literature. On the contrary, muscles are characterized by a preferential direction of anisotropy that is obtained from the fitting procedure (See Table 1). Thus, the fitting procedure indicates that the PDA in the rectus abdominis runs longitudinally to the abdomen whereas it runs transversally in the oblique muscles. On the other hand, with the aim of introducing the preferential direction of anisotropy in the aponeuroses or collagenous structures, some guidelines previously published are followed (W. Moore 2008; Park et al. 2006). Thus, the PDA in the fascia transversalis is disposed transversally to the abdomen. On the other hand, we assume in our work that all aponeurotic fibres from the external and internal obliques are included in the anterior lamina of the rectus sheath and the aponeurotic fibres from the TA are included in the posterior lamina of the rectus sheath. Therefore, the PDA in the posterior rectus sheath run transversally to the abdomen whereas the fibre patron in the anterior one is that shown in Figure 1.c. Since the LA is formed by the junction of the aponeuroses coming from the anterior and posterior lamina of the rectus sheath (W. Moore 2008), the PDA is defined following the same disposition. Therefore, three preferential directions of anisotropy are defined in the linea alba.

4 BOUNDARY CONDITIONS

MRI data came from an alive subject in a supine position. Thus, this geometry includes the initial strains and the additional deformation due to the intraabdominal pressure (IAP) in the supine position (3.6 mmHg) (Cobb et al. 2005). Since the unloaded configuration is unknown, we followed an iterative process (Lanchares et al. 2008) to approximately determine the initial strain distribution, $\mathbf{F}_0^n$, that produce an autoequilibrated state by means of balancing the IAP in the reference geometry.

IAP has been considered responsible some abdominal diseases such as formation or recurrence of hernias. With the aim of studying its effects, firstly, the change from the supine position to the standing position is modelled. For that purpose, both the body mass (gravity equal to $4.64 \cdot 10^{-6}$ Kg mm^{-3}), which includes the weight of the viscera and the muscles, and the IAP in this position (20 mmHg) are considered (Cobb et al. 2005). After that, the IAP corresponding to the jumping movement (P3 = 168.76 mmHg) was applied (Cobb et al. 2005).

In a physiological state under IAP, the ribs in the shoulder prevent from any movement of the back of the abdomen. Thus, boundary conditions were defined in the model as follows. The constraint imposed by the ribs in the shoulder and lower part of the abdomen is included by fixing the nodes at the back

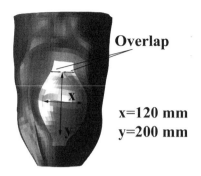

Overlap

x=120 mm
y=200 mm

Figure 2. FE model of the repaired abdomen including a large incisional hernia along the LA. The FE model of the surgical mesh is placed over the defect. The overlap between prostheses and tissue is indicated.

of the abdomen and those corresponding to the pelvis support (See Fig. 1.b).

5 APPLICATION TO MODELLING HERNIA SURGERY

To simulate hernia surgery just after surgery (at time-zero without tissue in-growth), a FE model based on the healthy one previously obtained was defined. For that purpose, some nodes from the linea alba are separated and internal pressure is applied until the abdomen is opened so that the hernia defect measures 120×200mm. After this procedure, the deformed orientation of the PDA is considered in the hernia FE model.

After generating the hernia in the front of the abdomen, a prosthesis that covers the defect is modelled. Instead of considering the fabrics of the prostheses, a membrane model is proposed, where the geometry and weave of the surgical mesh is homogenized, using a 3D anisotropic hyperelastic formulation. The modelling of the prosthesis was developed using the software FEMAP® (Siemens Software) and 279 membrane elements were added to the model. Besides, an overlap is considered between the prosthesis and the abdominal wall so that the suture can be modelled as well (See Fig. 2). The thickness of the membrane elements was 1 mm.

Concretely, three non-absorbable, biocompatible surgical meshes are modelled in the defect: Surgipro® (SUR), Optilene® (OPT) and Infinit® (INF). The material properties of these prostheses, obtained within the framework of large deformation anisotropic hyperelasticity, were taken from a previous work (Hernández-Gascón et al. 2011). For the anisotropic meshes, OPT and INF, the preferential direction of anisotropy was defined at each integration point of the membrane elements. In all cases, the stiffest direction of the mesh was transversally aligned in the abdomen (Hernández-Gascón et al. 2011). To model the running sutures used to fix the surgical mesh, the nodes in the overlap were matched with the abdominal

nodes. Finally, an internal pressure corresponding to the jumping movement was applied (168.76 mmHg).

6 RESULTS

In the healthy abdomen, maximal displacements (MD) and maximal principal stresses (MPS) appear in the front of the belly. As previously described, in the frontal part of the abdomen, there are different anatomical structures through the whole thickness of the model (see Fig. 1). The MPS and MD are obtained in the linea alba due to its high collagen content, that makes this structure to be the stiffest one. Besides, the rectus tendon and anterior and posterior rectus sheaths return high values in stresses as well. The values obtained in the rectus abdominis muscle and oblique muscles are lower since the stiffness of the muscle is not as high as in the aponeuroses (Hernández et al. 2011; Martins et al. 2012). The fascia transversalis, in the area corresponding to the frontal part of the abdomen, barely supports stresses whereas the areas under the oblique muscles address higher values. In the frontal part of the abdomen, stiffest structures (LA) or structures with similar stiffness (ARS and PRS) exist over the fascia transversalis. Thus, the FT does not support high stresses whereas the other structures do. On the contrary, when the oblique muscles and the FT appear together, the stiffest structure is the last one. As a consequence, the highest stresses are supported by the fascia transversalis. Close to the boundaries (ribs), the displacements and stresses values are practically zero.

Figures 3.a and 3.b show MD and MPS obtained in the healthy abdomen when jumping, respectively, along the line AB, defined in Figure 3.a. MD and MPS take place in the central part of the belly. However, due to the asymmetry of the abdomen, total symmetry is not appreciated. There is a perturbation on the MPS around the hernia defect since these areas correspond to the transition area between the rectus abdominis and linea alba where the material properties change.

After simulating hernia surgery, results in the abdomen area are similar to that obtained in the healthy abdomen whereas those obtained in the area of the defect show differences (See Fig. 4). Along line AB, MD with SUR and OPT in the defect decrease lightly because the stiffness of the prostheses is higher than that of the abdomen (See Fig. 4.a). Thus, the surgical meshes restrict the movement. However, MD with INF in the defect increase notably compared to the other meshes (See Fig. 4.a). This is due to the low stiffness in this mesh in the longitudinal direction (Hernández-Gascón et al. 2011). Besides, SUR mesh shows the stiffest mechanical response compared to the others.

Regarding stresses, the high stiffness of the prostheses compared to that of the tissue provokes a notably increment of the MPS in the area of the defect (See Fig. 4.b). Furthermore, a stress concentration appears in the overlap as depicted in Fig. 4.b around the 30% and 65% of the normalized distance. Since the

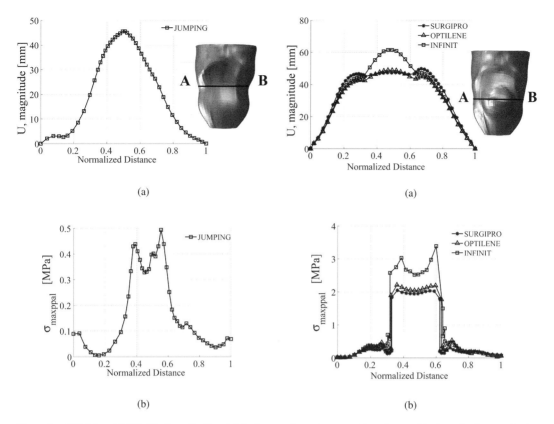

Figure 3. MD (a) and MPS (b) along the line AB in the model of the healthy abdomen. The abscissa shows the normalized distance of the lines AB. $x = 0$ and $x = 1$ correspond to points A and B, respectively.

Figure 4. MD (a) and MPS (b) along the line AB in the model of the herniated abdomen just after surgery. The abscissa shows the normalized distance of the line AB. $x = 0$ and $x = 1$ correspond to points A and B, respectively.

stiffness of the INF mesh in the transversal direction is higher than that of the OPT and SUR meshes (Hernández-Gascón et al. 2011), the MPS appear with INF mesh.

7 DISCUSSION

A methodology to model the human abdomen and study its mechanical response after physiological movements as well as to simulate hernia surgery is presented. Besides, the purely passive response is modelled within the framework of large deformation hyperelasticity and mechanical properties and the preferential direction of anisotropy are considered.

In the healthy abdomen, MD and MPS appear in the front of the belly. Thus, this distribution suggests that the highest probability of hernia appearance is in the LA. This area becomes stiffer since it is composed mostly of collagen connective tissue. Our results agree with other studies which report that the most important structure for the stability of the abdominal wall from a mechanical point of view is the linea alba (Axer et al. 2001).

Just after surgery, no mesh was able to match the displacements of the frontal part of the abdomen. Thus, surgical repair procedure does not fully restore normal physiological conditions and the risk of hernia recurrence by the suture is high due to the stress concentration. This is due to the low stiffness of the wall next to the the defect because, just after mesh implantation, the regeneration process has not yet occurred (Hernández-Gascón et al. 2012).

Our study includes some simplifications in the geometry and in the definition of the preferential direction of anisotropy due to their complexity (W. Moore 2008; Park et al. 2006). Furthermore, experimental animal data are used when there is no data from human tissue. Regarding hernia surgery approach, the study does not consider tissue in-growth over the prosthesis so the results are applicable only just after surgery.

From now on, future studies can be developed to automatically simulate the mechanical behaviour of the abdominal wall in personalized studies. The analysis of the anatomy of a specific patient would provide some guidelines to predict the risk of failure of the abdominal wall. Furthermore, simulating the surgical procedure before the real one would predict the outcomes depending on the prostheses and the best results could be achieved.

8 CONCLUSIONS

A methodology to model the human abdomen and study its mechanical response after physiological movements is presented as well as to simulate hernia surgery is presented.

In the healthy abdomen, the distribution of MD and MPS suggest that the highest probability of hernia appearance is in the LA.

Just after surgery, no mesh was able to match the displacements of the frontal part of the abdomen and the risk of hernia recurrence by the suture is high.

REFERENCES

Arruda, E. M., K. Mundy, S. Calve, and K. Baar (2006). Denervation does not change the ratio of collagen I and collagen II mRNA in extracellular matrix of muscle. *American Journal of Physiology – Regulatory, Integrative and Comparative Physiology 292*, 983–987.

Axer, H., D. G. Keyserlingk, and A. Prescher (2001). Collagen fibers in linea alba and rectus sheaths: I. General scheme and morphological aspects. *Journal of Surgical Research 96*, 127–134.

Calvo, B., E. Peña, P. Martins, T. Mascarenhas, M. Doblare, R. Natal, and A. Ferreira (2009). On modelling damage process in vaginal tissue. *Journal of Biomechanics 42*, 642–651.

Cobb, W. S., J. M. Burns, K. W. Kercher, B. D. Matthews, H. J. Norton, and B. T. Heniford (2005). Normal intra-abdominal pressure in healthy adults. *J Surg Res 129*, 231–235.

Demiray, H., H. W. Weizsacker, K. Pascale, and H. Erbay (1988). A stress-strain relation for a rat abdominal aorta. *Journal of Biomechanics 21*, 369–374.

Flory, P. J. (1961). Thermodynamic relations for high elastic materials. *Trans Faraday Soc 57*, 829–838.

Hernández, B., E. Peña, G. Pascual, M. Rodriguez, B. Calvo, M. Doblaré, and J. M. Bellón (2011). Mechanical and histological characterization of the abdominal muscle. A previous step to model hernia surgery. *J Mech Behav Biomed Mater 4*, 392–404.

Hernández-Gascón, B., E. Peña, H. Melero, G. Pascual, M. Doblaré, M. P. Ginebra, J. M. Bellón, and B. Calvo (2011). Mechanical behaviour of synthetic surgical meshes. Finite element simulation of the herniated abdominal wall. *Acta Biomater 7*, 3905–3913.

Hernández-Gascón, B., E. Peña, G. Pascual, M. Rodríguez, J. M. Bellón, and B. Calvo (2012). Long-term anisotropic mechanical response of surgical meshes used to repair abdominal wall defects. *J Mech Behav Biomed Mater 5*, 257–271.

Holzapfel, G. A., T. C. Gasser, and R. W. Ogden (2000). A new constitutive framework for arterial wall mechanics and a comparative study of material models. *Journal of Elasticity 61*, 1–48.

Lanchares, E., B. Calvo, J. A. Cristóbal, and M. Doblaré (2008). Finite element simulation of arcuates for astigmatism correction. *J Biomech 41*, 797–805.

Martins, P., E. Peña, R. M. N. Jorge, A. Santos, L. Santos, T. Mascarenhas, and B. Calvo (2012). Mechanical characterization and constitutive modelling of the damage process in rectus sheath. *Journal of the Mechanical Behavior of Biomedical Materials 8*, 111–122.

Norasteh, A., E. Ebrahimi, M. Salavati, J. Rafiei, and E. Abbasnejad (2007). Reliability of B-mode ultrasonography for abdominal muscles in asymptomatic and patients with acute low back pain. *Journal of Bodywork and Movement Therapies 11*, 17–20.

Ogden, R. W. (1996). *Non-linear Elastic Deformations.* Dover, New York.

Park, A. E., J. S. Roth, and S. M. Kavic (2006). Abdominal wall hernia. *Current Problems in Surgery 43*, 326–375.

Spencer, A. J. M. (1971). Theory of invariants. In *Continuum Physics*, pp. 239–253. Academic Press, New York.

W. Moore (2008). *Gray's Anatomy celebrates 150th anniversary.* The Telegraph (Telegraph Media Group).

Weiss, J. A., B. N. Maker, and S. Govindjee (1996). Finite element implementation of incompressible, transversely isotropic hyperelasticity. *Computer Methods in Applied Mechanics of Engineering 135*, 107–128.

Computational Modelling of Objects Represented in Images – Di Giamberardino et al. (eds)
© 2012 Taylor & Francis Group, London, ISBN 978-0-415-62134-2

A splitting algorithm for MR image reconstruction from sparse sampling

Zhi-Ying Cao & You-Wei Wen
Department of Mathematics, Kunming University of Science and Technology, Yunnan, China
Research supported in part by NSFC and KMUST Grant

ABSTRACT: In this paper, we proposed a new splitting algorithm for MR image reconstruction from random variable density sample. Due to the sparsity in the transform domain and piecewise smoothness in the spatial domain of MR images, the reconstruction can be obtained by performing total variation and wavelet L_1 regularization optimization. By introduce an auxiliary variable, we derive a new quadratic majorizing function for data fitting term in the objective function. Alternative minimization approach is applied to find the minimizer of the objective function. For the auxiliary variable, the minimum has a closed form solution, and for the original variable, the minimum is a proximity operator of the hybrid regularizers. We develop an efficient algorithm to compute the proximity operator. We compare the proposed algorithm with gradient methods in term of signal-to-noise ratio. Numerical results demonstrate that the proposed algorithm is very efficient and outperforms that of gradient methods.

1 INTRODUCTION

Magnetic Resonance (MR) imaging is primarily a medical imaging technique. Because of its non-invasive manner and excellent depiction of soft tissue changes, it is most commonly used in radiology to visualize the internal structure and functions of the body. In the acquisition process, a radio frequency excitation field produces a magnetization component transverse to the static field, and points in the Fourier domain (so-called the k-space) are sampled. In order to avoid the aliasing artifacts in the reconstruction image and obtain perfect reconstruction, the amount of acquired data must at least match the amount of information needed for recovery. During scanning, the patients are required to lie motionless in the machine for serval ten minutes, which is uncomfortable.

To shorten the acquisition time, one can undersample the points in the k-space. This results in artifacts such as coherent aliasing–a superposition of shifted replicas. But recent developments in compressive sensing theory show that it is possible to accurately reconstruct the MR images from random undersampled data in k-space and therefore surely alleviate the long and expensive MR imaging diagnosis process.

In [7], Lustig *et al* developed the formwork of sparse MRI. They gave an example to illustrate the sparsity in the transform domain of MR images. They also showed the piecewise smoothness in the spatial domain of angiogram images. Since the MR images have these two characteristics, Lustig *et al* proposed to reconstruct the MR image $\mathbf{u}_0$ by performing total variation (TV) and wavelet L_1 regularization optimization:

$$\mathbf{u}_0 = \arg\min \mathcal{J}(\mathbf{u}) \qquad (1)$$

where

$$\mathcal{J}(\mathbf{u}) = \frac{1}{2}\left\| \mathcal{F}_p\mathbf{u} - \mathbf{b} \right\|_2^2 + \beta_1 \left\| \nabla\mathbf{u} \right\|_1 + \beta_2 \left\| \boldsymbol{W}\mathbf{u} \right\|_1. \quad (2)$$

Here $\mathbf{b}$ is the observed data in k-space, $\mathcal{F}_p$ is a Fourier transform evaluated only at a subset of frequency domain samples (corresponding to one of the k-space undersampling schemes), $\beta_i(i = 1, 2)$ is the regularization parameter, ∇ is the discrete gradient operator defined by

$$(\nabla\mathbf{u})_{r,s} = \left((\nabla_x\mathbf{u})_{r,s}, (\nabla_y\mathbf{u})_{r,s} \right)$$

with $(\nabla_x\mathbf{u})_{r,s} = u_{r+1,s} - u_{r,s}$ and $(\nabla_y\mathbf{u})_{r,s} = u_{r,s+1} - u_{r,s}$. and W is an orthogonal wavelet transform matrix.

The main challenge to solve the problem [1] is that both TV regularizer and L_1 regularizer are non smooth. Lustig *et al* [7] applied the conjugate gradient method to compute the minimizer of [1], while He *et al* considered PDE approach to handle it [6]. The main drawback of these approaches is that their convergence are very slow for practical.

Since the data fitting term has a Lipschitz-continuous gradient as required in [4], it is possible to use the forward-backward splitting proximal algorithm to solve the optimization problem. One important task in forward-backward splitting proximal algorithm is to compute the proximal operator of the regularizers. Chambolle [1] proposed a project algorithm to compute the proximal operator of TV regularizer. It is well known that the proximal operator of wavelet L_1 regularizer is a shrinkage operator [5]. Combettes and Pesquet developed an iterative method to compute the proximity operator of the composite regularizers by performing the proximity operator of each regularizer independently [3]. However, when the regularizers are the linear combination of TV norm and wavelet L_1 norm, to the best of our knowledge, there is no papers studying how to compute the proximal operator of the hybrid regularizers. This will prevent us from using the forward-backward splitting proximal algorithm to solve [1].

In this paper, we will develop an efficient algorithm for MR image reconstruction using optimization transform. We derive a new quadratic majorizing function for data fitting term in the objective function, then the new objective function can be minimized by alternating minimization. We develop an efficient algorithm to compute the proximal operator of hybrid regularizers.

The outline of the paper is as follows. In Section 2, we reformulate MR image reconstruction problem as a Fourier domain inpainting problem and propose a bivariate functional together with an alternating minimization algorithm to find its minimizer. A dual approach is developed to compute the proximal operator of hybrid regularizers. Experiment results are reported in Section 3 to demonstrate the effectiveness of our algorithm. Finally, a short conclusion is given in Section 4.

2 ITERATIVE METHOD FOR MR IMAGE RECONSTRUCTION

The main objective in this paper is to propose an efficient iterative algorithm to compute the minimizer of objective function $\mathcal{J}(\mathbf{u})$, see [1]. Notice that the under sampled Fourier transform matrix $\mathcal{F}_p$ can be rewritten as $\mathcal{F}_p = PF$ with P be a sampling matrix and F be the full Fourier transform. Therefore, MR image reconstruction problem can be regarded as a Fourier domain inpainting problem: filling in the missing sampled points in the k-space.

Since the rows of the downsampling matrix are orthogonal, we have $PP^T = I$. Using certain matrix operation and optimization transform, we derive a quadratic majorizing function for data fitting term in the objective function (2)

$$\|PF\mathbf{u} - \mathbf{b}\|_2^2 = \frac{1+\alpha}{\alpha} \min_{\mathbf{x}} \|P\mathbf{x} - \mathbf{b}\|_2^2 + \alpha\|\mathbf{x} - F\mathbf{u}\|_2^2.$$

Define the bivariate function $\mathcal{J}_1(\mathbf{x}, \mathbf{u})$ as

$$\mathcal{J}_1(\mathbf{x}, \mathbf{u}) = \frac{1}{2}\|P\mathbf{x} - \mathbf{b}\|_2^2 +$$

$$\frac{\alpha}{2}\|\mathbf{x} - F\mathbf{u}\|_2^2 + \alpha\lambda_1\|\nabla\mathbf{u}\|_1 + \alpha\lambda_2\|W\mathbf{u}\|_1.$$

Here $\lambda_i = \beta_i / 1 + \alpha$. By the convexity of $\mathcal{J}(\mathbf{u})$ and $\mathcal{J}_1(\mathbf{x}, \mathbf{u})$, the minimization problem for $\mathcal{J}(\mathbf{u})$ is equivalent to $\mathcal{J}_1(\mathbf{x}, \mathbf{u})$, i.e.,

$$\min_{\mathbf{u}} \mathcal{J}(\mathbf{u}) = \frac{1+\alpha}{\alpha} \min_{\mathbf{x}, \mathbf{u}} \mathcal{J}_1(\mathbf{x}, \mathbf{u}).$$

Hence, instead of computing the minimizer of the objective function $\mathcal{J}(\mathbf{u})$, we would like to calculate the minimizer of objective function $\mathcal{J}_1(\mathbf{x}, \mathbf{u})$.

We propose an alternating minimization algorithm to find a minimizer of $\mathcal{J}_1(\mathbf{x}, \mathbf{u})$. Starting from an initial guess $\mathbf{u}^{(0)}$, we use an alternating minimization algorithm to generate the sequence:

$$\begin{cases} \mathbf{x}^{(k)} &= \operatorname{argmin}_{\mathbf{x}} \mathcal{J}_1(\mathbf{x}, \mathbf{u}^{(k-1)}), \\ \mathbf{u}^{(k)} &= \operatorname{argmin}_{\mathbf{u}} \mathcal{J}_1(\mathbf{x}^{(k)}, \mathbf{u}). \end{cases}$$

2.1 Subproblem for the variable $\mathbf{x}$

We see that the objective function $\mathcal{J}_1(\mathbf{x}, \mathbf{u})$ is quadratic with respect to $\mathbf{x}$. Hence $\mathbf{x}^{(k)}$ can easily be computed by the formula

$$\mathbf{x}^{(k)} = \left(P^T P + \alpha I\right)^{-1}\left(P^T \mathbf{b} + \alpha F\mathbf{u}^{(k-1)}\right). \quad (3)$$

Because P is a random undersampled matrix, $P^T P$ is a diagonal matrix and its diagonal entries are 0 and 1. Therefore, we can compute $\mathbf{x}^{(k)}$ easily.

We remark that $\mathbf{x}^{(k)}$ can be regarded as an average of $P^T\mathbf{b}$ and $F\mathbf{u}^{(k-1)}$. If the point in k-space is unsampled, its value is filled by the Fourier coefficient of the restored image.

2.2 Subproblem for the variable $\mathbf{u}$

Due to the convexity of both TV norm and wavelet L_1 norm, we know that there exists a unique solution for the variable $\mathbf{u}$. Notice that F is a Fourier transform matrix, which implies that $\|\mathbf{x} - F\mathbf{u}\|_2^2 = \|F^{-1}\mathbf{x} - \mathbf{u}\|_2^2$, we have

$$\mathbf{u}^{(k)} = \operatorname{prox}_{\lambda_1, \lambda_2}(F^{-1}\mathbf{x}^{(k)}),$$

where $\operatorname{prox}_{\lambda_1, \lambda_2}(\cdot)$ is the proximity operator of the hybrid regularizers:

$$\operatorname{prox}_{\lambda_1, \lambda_2}(\mathbf{g})$$

$$= \operatorname*{argmin}_{\mathbf{u}} \frac{1}{2}\|\mathbf{g} - \mathbf{u}\|_2^2 + \lambda_1\|\nabla\mathbf{u}\|_2 + \lambda_2\|W\mathbf{u}\|_1.$$

We will discuss how to compute $\operatorname{prox}_{\lambda_1, \lambda_2}(\mathbf{g})$ in Section 2.3.

Now, we summarize the resulting algorithm in Algorithm 1.

2.3 Proximity operator of hybrid regularizers

Now we consider how to compute the proximity operator $\operatorname{prox}_{\lambda_1, \lambda_2}(\mathbf{g})$. We propose a dual approach to compute the proximity operator, where the TV norm is represented using Legendre-Fenchel's duality to represent the TV norm.

Algorithm 1. Iterative Method for MR Image Reconstruction

Ensure: $\mathbf{u} = \text{IterMethod}\,(P, \mathbf{b}, \beta_1, \beta_2)$.
Require: $P, \mathbf{b}, \beta_1, \beta_2$.
  1: Initialize $\mathbf{u}^{(0)}$. Set the parameter α.
  2: $\lambda_i = \frac{\beta_i}{1+\alpha}\,(i = 1, 2)$.
  3: **while** stopping criterion is not satisfied **do**
  4:    $\mathbf{x}^{(k)} = (P^T P + \alpha I)^{-1}(P^T \mathbf{b} + \alpha \mathbf{u}^{(k-1)})$;
  5:    $\mathbf{u}^{(k)} = \text{prox}_{\lambda_1, \lambda_2}\,(F^{-1}\mathbf{x}^{(k)})$;
  6: **endwhile**
  7: **return** $\mathbf{u} = \mathbf{u}^{(k+1)}$.

Let

$$p_{r,s} = \begin{bmatrix} p_{r,s}^x \\ p_{r,s}^y \end{bmatrix}$$

the dual variable of the (r, s)-th pixel, $\mathbf{p}$ be the concatenation of all $p_{r,s}$, and define the discrete divergence of $\mathbf{p}$ as:

$$(\text{div}\mathbf{p})_{r,s} \equiv p_{r,s}^x - p_{r-1,s}^x + p_{r,s}^y - p_{r,s-1}^y$$

with $p_{0,s}^x = p_{r,0}^y = 0$ for $r, s = 1, \ldots, n$. The vector $\text{div}\mathbf{p}$ is the concatenation of all $(\text{div}\mathbf{p})_{r,s}$. The TV norm can be represent by using the dual formulation:

$$\|\nabla \mathbf{u}\|_2 = \max_{\mathbf{p} \in A} \mathbf{u}^T \text{div}\mathbf{p}.$$

Here $A = \{\mathbf{p} : |p_{r,s}| \leq 1, \forall r, s\}$.

We represent the TV norm using the dual formulation and define the objective function $\mathcal{Q}(\mathbf{u}, \mathbf{p})$ as

$$\mathcal{Q}(\mathbf{u}, \mathbf{p}) = \frac{1}{2}\|\mathbf{u} - \mathbf{g}\|^2 + \lambda_1 \mathbf{u}^T \text{div}\mathbf{p} + \lambda_2 \|W\mathbf{u}\|_1. \quad (4)$$

Now the optimization problem is turn to

$$\min_{\mathbf{u}} \max_{\mathbf{p} \in A} \mathcal{Q}(\mathbf{u}, \mathbf{p})$$

By using arguments of duality theory for convex programming, a pair $(\mathbf{u}^*, \mathbf{p}^*)$ is a saddle point for $\mathcal{Q}(\mathbf{u}, \mathbf{p})$ if and only if

$$\mathcal{Q}(\mathbf{u}^*, \mathbf{p}) \leq \mathcal{Q}(\mathbf{u}^*, \mathbf{p}^*) \leq \mathcal{Q}(\mathbf{u}, \mathbf{p}^*)$$

for any $\mathbf{u}$ and $\mathbf{p} \in A$, that also means

$$\min_{\mathbf{u}} \max_{\mathbf{p} \in A} \mathcal{Q}(\mathbf{u}, \mathbf{p}) = \mathcal{Q}(\mathbf{u}^*, \mathbf{p}^*) = \max_{\mathbf{p} \in A} \min_{\mathbf{u}} \mathcal{Q}(\mathbf{u}, \mathbf{p}). \quad (5)$$

Before describing our iterative method to calculate the sadddle point, we first introduce two definiation: soft-thresholding opertor and orthogonal projection.

For a scalar b and the λ, define the operator $S_\lambda(b)$ as

$$S(b) = \begin{cases} b - \lambda_2, & b \geq \lambda_2 \\ 0, & |b| < \lambda_2 \\ b + \lambda_2, & b \leq -\lambda_2 \end{cases}.$$

For the vector $\mathbf{b}$, the operator $\mathcal{S}_\lambda(\mathbf{b})$ is defined by $(\mathcal{S}_\lambda(\mathbf{b}))_i = S_\lambda(\mathbf{b}_i)$. It is known that the optimal solution of the minimization problem $\min_{\mathbf{x}} \frac{1}{2}\|\mathbf{x} - \mathbf{b}\|^2 + \lambda\|\mathbf{x}\|_1$ is given by $\mathbf{x} = \mathcal{S}_\lambda(\mathbf{b})$.

We use the notation $\mathcal{P}_A$ to denote any function for which $\mathcal{P}_A(\mathbf{x})$ is a projection of $\mathbf{x}$ on the set A, i.e., for all $\mathbf{x}$, we have

$$\mathcal{P}_A(\mathbf{x}) = \underset{\mathbf{z}}{\text{argmin}}\{\|\mathbf{z} - \mathbf{x}\|^2 | \mathbf{z} \in A\}.$$

We have $(\mathcal{P}_A(\mathbf{x}))_i = \mathbf{x}_i$ if $|\mathbf{x}_i| \leq 1$, and $(\mathcal{P}_A(\mathbf{x}))_i = \mathbf{x}_i / \|\mathbf{x}_i\|$ if $|\mathbf{x}_i| > 1$.

We establish the following theorem.

Theorem 1. $(\mathbf{u}^*, \mathbf{p}^*)$ *is the saddle point of* $\mathcal{Q}(\mathbf{u}, \mathbf{p})$ *if and only if* $(\mathbf{u}^*, \mathbf{p}^*)$ *satisfies*

$$\mathbf{u}^* = W^T \mathcal{S}_{\lambda_2}(W(\mathbf{g} - \lambda_1 \text{div}\mathbf{p}^*)). \quad (6)$$

and

$$\mathbf{p}^* = \mathcal{P}_A(\mathbf{p}^* - \tau \lambda_1 \nabla \mathbf{u}^*). \quad (7)$$

Here $\tau > 0$ is the parameter. Moreover, the equation (6) and (7) can be reformulated to a more compact form

$$\mathbf{p}^* = \mathcal{P}_A(\mathbf{p}^* - \tau \lambda_1 \nabla W^T \mathcal{S}_{\lambda_2}(W(\mathbf{g} - \lambda_1 \text{div}\mathbf{p}^*))). (8)$$

Theorm 1 states that the saddle point of $\mathcal{Q}(\mathbf{u}, \mathbf{p})$ can be calcuated by seeking dual variable $\mathbf{p}$ statisfies (8), then replacing the dual variable into (6) to compute the primal variable. However, the equation (8) involves the expression of soft thresholding operator and the projection operation, the equation is non-differentiable, which poses serious difficulties in numerical solution. To remove the difficulty, Chen et.al. [2] introduced a concept of slant differentiablity in view of its Lipschitz continuity and proposed semismooth methods to solve this type equations.

We will describe a smiple iterative method. Theorm 1 also suggests that the computation of dual variable $\mathbf{p}$ in (8) might be done by performing an iterative scheme. Starting from some initial pair $\mathbf{p}^{(0)}$, the iteration scheme has the form

$$\mathbf{p}^{(k+1)} = \mathcal{P}_A(\mathbf{p}^{(k)} - \tau \lambda_1 \nabla W^T \mathcal{S}_{\lambda_2}(W(\mathbf{g} - \lambda_1 \text{div}\mathbf{p}^{(k)}))). \quad (9)$$

The following theorem shows that the sequence of $\mathbf{p}^{(k)}$ converges to some point $\mathbf{p}^*$ satisfying (7) under the assumption on the stepsize τ.

Theorem 2. *Suppose that the sequence of step sizes obeys $0 < \tau < 1/4\lambda_1^2$. Then the sequence $\mathbf{p}^{(k)}$ generated by (9) converges, i.e., there exists $\mathbf{p}^*$ such that*

$$\lim_{k \to \infty} \mathbf{p}^{(k)} = \mathbf{p}^*.$$

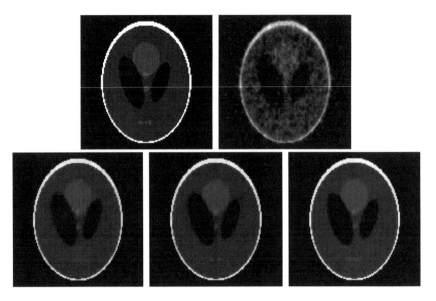

Figure 1. The original 128×128 image, the observed image with 25% k-space data and the restored image by conjugate gradient method, nonlinear conjugate gradient method, and the proposed method respectively.

Let $\mathbf{u}^*$ is computed using (6), then the pair $(\mathbf{u}^*, \mathbf{p}^*)$ is the saddle point of $\mathcal{Q}(\mathbf{u}, \mathbf{p})$.

The resulting algorithm for proximity operator of the compound regularizers is summarized in Algorithm 2.

Algorithm 2. Dual Approach for Proximity Operator of Hybrid Regularizers

---

Ensure: $\mathbf{u} = \text{prox}_{\lambda_1, \lambda_2}(\mathbf{g})$.
Require: g.
  1: Initialize $\mathbf{p}^{(0)}$. Set the step sizes τ.
  2: **while** stopping criterion is not satisfied **do**
  3:     Update $\mathbf{p}^{(k+1)}$ by (9);
  4: **end while**
  5: **return** $\mathbf{u} = \mathbf{u}^{(k+1)}$.

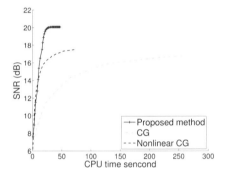

Figure 2. SNR versu CPU time for 128×128 Shepp-Logan phantom image.

3 NUMERICAL RESULTS

We illustrate the performance of the proposed algorithm for sparse MRI reconstruction problems and compare it with the conjugate gradient method and the non-linear conjugate gradient method [7]. Our codes are written in Matlab R2009a.

The experiments were performed under Mac OS X10.7.3 and MATLAB R2011a on a MackBook Air Laptop with a 1.7GHz Intel Core i5 processor and 4GB of RAM. The Signal-to-Noise Ratio (SNR) is used to measure the quality of the restoration results. It are defined as follows: $\text{SNR} = 10 \log_{10} \left(\|\mathbf{u}\|_2^2 / \|\mathbf{u} - \widehat{\mathbf{u}}\|_2^2 \right)$, where $\mathbf{u}$ and $\widehat{\mathbf{u}}$ are the original image and the restored image respectively.

The observed image is chosen as the initial image. The sample rate in the tests is 25%, which is selected with polynomial variable density sampling. Gaussian white noise with standard deviation $\sigma = 0.01$ is added in the sampling data. We choose the parameter as follows: $\beta_1 = \beta_2 = 0.01$ and $\tau = 0.248$. Figure 1 shows the original 128×128 Shepp-Logan phantom image, the observed image and the restored image by conjugate gradient method, nonlinear conjugate gradient method, and the proposed method respectively. We plot the SNRs of the restored images versus CPU times, see Figure 2. We observe that the proposed method produces the best SNRs.

Besides the Shepp-Logan phantom image, we also tested Cameraman, Lena images. The size of both images is 256×256. All parameters are kept unchanged. The original image, the observed image and the restored image by conjugate gradient method, nonlinear conjugate gradient method, and the proposed method are presented in Figure 3 and Figure 5, respectively. While the SNRs of the restored images

Figure 3. The original 256×256 image, the observed image with 25% k-space data and the restored image by conjugate gradient method, nonlinear conjugate gradient method, and the proposed method respectively.

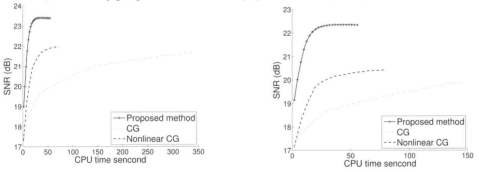

Figure 4. SNR versu CPU time for 256×256 cameraman image.

Figure 6. SNR versu CPU time for 256×256 lenna image.

Figure 5. The original 256×256 image, the observed image with 25% k-space data and the restored image by conjugate gradient method, nonlinear conjugate gradient method, and the proposed method respectively.

333

versus CPU times are shown in Figure 4 and Figure 6, respectively. These results demonstrate that the proposed method is competitive with gradient methods.

4   CONCLUSION

In this paper, we proposed an efficient algorithm to reconstruct MR images from highly random under sampled k-space data. Since the reconstruction problem is very ill-conditioned problem, a regularization method should be used in the image reconstruction process. A hybrid TV and wavelet L_1 norm is applied to represent the sparsity and piecewise smoothness of MR images. Numerical difficulties arise due to the non smoothness of the regularization term. To handle this problem, a quadratic majorizing function for data fitting term in the objective function is derived. Alternative minimization approach was proposed to find the minimizer. We develop an efficient algorithm to compute the proximity operator of the hybrid regularizers. Our experimental results show that the proposed algorithm is very efficient and outperforms the gradient descent method.

REFERENCES

[1]   A. Chambolle. An algorithm for total variation minimization and applications. *J. Math. Imag. Vision*, 20(1–2):89–97, 2004.

[2]   X. Chen, Z. Nashed, and L. Qi. Smoothing methods and semismooth methods for nondifferentiable operator equations. *SIAM J. Numer. Anal.*, 38(4):1200–1216, 2000.

[3]   P. Combettes and J. Pesquet. A proximal decomposition method for solving convex variational inverse problems. *Inverse Problems*, 24(1):1–27, 2008.

[4]   P. Combettes and V. Wajs. Signal recovery by proximal forward-backward splitting. *Multiscale Model. Simul.*, 4(4):1168–1200, 2005.

[5]   D. Donoho. Nonlinear solution of linear inverse problems by wavelet-vaguelette decompositions. *Appl. Comput. Harmon. Anal.*, 1:100–115, 1995.

[6]   L. He, T.C. Chang, S. Osher, T. Fang, and P. Speier. Mr image reconstruction by using the iterative refinement method and nonlinear inverse scale space methods. *UCLA CAM Report*, 6(35), 2006.

[7]   M. Lustig, D. Donoho, and J.M. Pauly. Sparse MRI: The application of compressed sensing for rapid MR imaging. *Magnetic Resonance in Medicine*, 58(6):1182–1195, 2007.

Computational Modelling of Objects Represented in Images – Di Giamberardino et al. (eds)
© *2012 Taylor & Francis Group, London, ISBN 978-0-415-62134-2*

A look inside the future perspectives of the digital cytology

D. Giansanti, S. Morelli, S. Silvestri, G. D'Avenio & M. Grigioni
Istituto Superiore di Sanità, Dipartimento di Tecnologie e Salute, Roma, Italy

MR. Giovagnoli, M. Pochini, S. Balzano, E. Giarnieri & E. Carico
Università Sapienza, Facoltà di Medicina e Psicologia, Roma, Italy

ABSTRACT: The work approaches the new technological scenario relevant to the introduction of the Digital-Cytology (D-CYT) in the National Health Service. A detailed analysis of the state of the art on the introduction of D-CYT in the hospital and more in general in the dispersed territory has been conducted. The analysis focused on the ICT technologies available for both the remote diagnosis and the cooperative diagnosis in the Hospital. In particular the work critically revised the client/server technologies, the tablet solutions, the 3D Virtual reality solutions. From a global point of view the paper showed that the today's technologies have a great potentiality in the digital-cytology.

1 INTRODUCTION

The virtual microscopy is an alternative solution between the static and dynamic tele-pathology. Telemedicine applied to pathology (T-P) is considered a valid aid to pathologists (Weisz-Carrington et al. 1999, Dunn et al. 2000, Leong et al. 2000); in fact it is supposed to allow the remote exchanging of information about a tissue on glass and in particular:

1) The tele-diagnosis;
2) The audit of complex cases, by means of shared virtual desktop;
3) The minimization of resources (more hospitals could share the professionals).

There are two basic methods of T-P. *Static* T-P consists in the capture and digitalization of images selected by a pathologist or pathologist assistant, which are then transmitted through electronic means to a tele-pathologist. *Dynamic* T-P consists in the direct communication between two different centres by using microscopes equipped with a telerobotic system oriented to explore the glass, remotely operated by the tele-pathologist who makes the diagnosis.

As an alternative solution between the two methods, widely increasing today is the diffusion of virtual microscopy (VM) (Çatalyürek et. al 2003a, Aferwork et al. 1998) The latter does not refer to the tele-control of microscopes, whilst the glass is scanned as a whole, producing a "virtual glass", and a pathologist can navigate remotely (via internet) inside this virtual glass or virtual slide in a manner similar to a real microscope.

1.1 *The evolution of digital-cytology*

Until recently the management of the information on the glasses (virtual slides) in tele-pathology

applications was principally based on the design and construction of a few identical and expensive platforms with microscope units and software tools for both the display and the tele-control (zooming, moving cutting of pieces of images) (Giansanti et al. 2007a,b). In the first applications of these methodologies the latency of information during the transmission caused displacement errors in the positioning of the microscope's mechanics on the glass. For these reasons the need of investigation based on the so called virtual microscopy with virtual navigation on scanned images without using tele-control provided new solutions. Severe problems were also noticed for example in the remote information exchange. The lacking of both a wide-band channel and an ad-hoc visualization strategy (VS) strongly delayed the image transmission. It has to be considered that a single file representing a virtual glass for cytology applications could reach several tens of gigabytes, more than in the case of applications of tele-echography (Giansanti et al. 2007a,b). Thus the design of an appropriate VS is a basic core aspect. Clearly it is not feasible and reasonable to fully manage a single file of several GB in the World Area Network!

Today the Scenario is completely changed thanks to the introduction of the VM. The principal changes in the world of the Information Technologies affecting Tele Pathology were the following:

– Availability of wide band channels.
– Diffusion of new VSs.
– Availability of new power image scanners.
– Availability of free visualization software.

The first point was driven by the diffusion of the Information Communication technologies (ICT) for Internet/Intranet/Extranet connections. The second point was driven by the pressing request of very large image exchange by internet. New methodologies today allow the archiving of an image arranged in layers assigned to different magnification factors (Giansanti et al. 2007a) and answers to the so called internet need "I give you something to see before you become angry!!", as for example in Google Earth (Giansanti et al. 2007a, b, c) and allow a remote information exchange using a reasonable wide-band-channel. The third point was driven by the exceptional changes in photonics applied to medicine. Many producers, leaders in photonics, are using their skill to design scanners for virtual microscopy. The fourth point was driven by the diffusion of free web-viewers and by new commercial strategies of the producers of tele-pathology systems.

1.2 Digital-pathology and digital-cytology

In order to face deeply the introduction of the use of digital cytology in the VS we have to make the following basic considerations.

The Tele-pathology mainly relates to the world of histology; the histologist navigates the slide without the need to use the focus function. The virtual slide in histology does not require the emulation of the focus function; in histology it is sufficient a flattened vision, in other words the histologist does not need to "smash" in the sample.

When coming to the D-CYT which relates to the world of cytologist, basic problems arise. At first, the cytologist uses a way to navigate completely different from the histologist: she or he in fact widely uses the focus function. Then, we should consider that the cytologist on the contrary to the histologist, uses a lateral area of the eye while navigating (the same used by the primitives to avoid animals attacks); as a consequence of this second aspect a third aspect arises, i.e. the stereo-vision is much more important for the cytologist. The focus can be emulated by software, by means of specific functions using also interpolation algorithms not opened to the user. This implies that, at a defined zoom, several focal planes are captured (till 100) for a given sample (depending on the thickness of the sample and the chosen level of magnification) (Giansanti et al. 2010,ba) (Figure 1 and 2). This leads to the generation of very large virtual slides during the digitization, and the focus function at a defined zoom should be emulated by generating different images ready for interpolation; in the VM this functionality is called by manufacturers "Z-stack". This implies the generation of virtual slides tens time larger than in the case of pathology. Furthermore the cytologist's needs are different from the ones of the pathologist when he or she interacts with the VM.

2 METHODS

An analysis has been conducted on the possibility to introduce the D-CYT technology in the Hospital and more in general in the dispersed territory. The

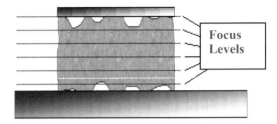

Figure 1.   Example from scanning process in the case of the Z-stack for emulating the focus function.

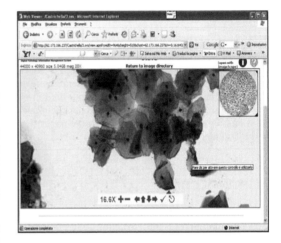

Figure 2.   Example of a view of a layer of a Virtual Slide.

analysis conducted in a form of a review was arranged with consideration of the recent developments of the technologies. In particular the analysis considered that today a system for the D-CYT:

- embeds a Web-based server as repository of the Virtual Slides and clients with a light software to display, often furnished with no costs
- is thus a client-server web-based architecture.

3 RESULTS

The study focused on the available ICT technologies allowing client-server web-based architecture and in particular the tablet technology, the integration with the 3D technologies and the standardization with particular care to the integration with DICOM standard.

3.1 Integration to tablet technology

Among the available promising technologies to share image information, the following have been investigated, focusing on D-CYT:

- Smart-phones (wearable-tablet)
- A4-tablet (portable-tablet) such as the Apple Ipad.
- Large touch table (not-portable table) such as the XDesk or Microsoft Surface.

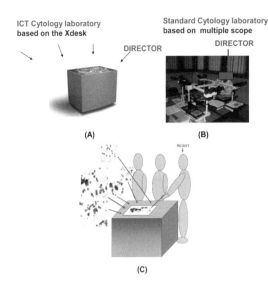

ICT Cytology laboratory based on the Xdesk

Standard Cytology laboratory based on multiple scope

DIRECTOR

DIRECTOR

(A)

(B)

REGIST

(C)

Figure 3. The Virtual Table. (A) The xDesk. (B) The optical multiple scope in standard cytology (C) A scenario of cooperative decision with xDesk.

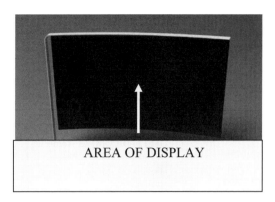

AREA OF DISPLAY

Figure 4. Example of Holography monitor.

3D glasses

Figure 5. Example of 3 D glasses.

The first two systems, that are widely used for many different purposes, allow to reach everyone in the world, therefore representing a chance for the remote consulting in D-CYT.

Regarding the last technology, we focused on the Epson XDesk.

The Epson XDesk (www.epson.it) represents a powerful ICT solution for cooperative analysis and discussion of cases of virtual cytology. In detail, the Epson XDesk is an interactive table; some call it a coffee table because you can put anything on the surface of the table, it works by projections, with the very latest technology on that. This desk is also compatible with Bluetooth communication protocol and as soon as you put your phone or camera on the surface of the table, the XDesk will be able to see all your files and pictures on the desk. By natural interface pictures on the table can be managed freely and resized, zoomed in and out by finger movements as the iPod touch does, only on a fair larger scale. The Epson XDesk has a 52-inch screen and a 1024x768 touch screen display. It represents the appropriate high technology solution for cooperative discussions, clinical audit (D'Avenio et al. and Grigioni 2007), and ultimately the future direction of cooperative virtual microscopy environment. Furthermore it could represent a tool suitable to recover the inheritance of cytology solutions for large screening abandoned because the technology was not so ready for the application, such as Pap-Net (Neuromedical Systems, Suffern, NY) for computer assisted cervical/vaginal cytology diagnosis (Giansanti et al. 2010b). Figure 3 A shows the xDdesk, emulating in D-CYT the multiple-scope equipment (B). Figure 3C shows a scenario of cooperative decision with XDesk.

3.2 Integration to 3 D Technologies

Different studies have been proposed to approach the introduction of the potentialities of 3D/holographic techniques (Fig. 4 and 5) in cytology vision (Giansanti et al. 2010b). However these studies have only dealt with the local diagnosis conducted in stand-alone equipments and have not been conducted over the WAN/LAN in telemedicine applications, apart for cardiovascular purpose (the specific targeted research project Collaborative Holographic Environments for Networked Tasks, COHERENT). The use of these techniques in consideration to the importance of the stereovision of the cytologist, could be of help both to emulate the stereovision thanks to the use of 3D/holographic techniques and to minimize the data transfer among remote nodes as a partial solution for the area occupancy caused by the Z-stack.

3.3 Integration to DICOM

Two Work Groups (WG)s of the NEMA are facing the integration of the D-CYT into DICOM.

The WG26 operates in synergy with the NEMA WG6 (Giansanti et al. 2010 a, b), the group responsible for the adequacy to the standard, reviewer of the activity of the WG26. The initial objectives of the WG26 were to extend minimal capabilities to describe specimens in DICOM and to create a mechanism to allow exchange and use of whole slide microscopic images within DICOM. The long term goals were to face the integration of also other imaging modalities, such as multi-spectral images, electron microscopy, flow cytometry (another field of D-CYT). The recent approval of the DICOM supplement 145 (Supplement 145: Whole Slide Microscopic Image IOD and SOP

Table 1. Issues investigated by the WG 26.

ISSUE	Comment
CHARACTERISTICS OF WHOLE-SLIDE IMAGES (WSI)	Faces the characteristics of the WSI in terms of image dimensions, data size, access patterns, data organization, image data compression, sparse image data
WSI IN DICOM	Faces the core aspect to import the WSI in DICOM: i.e. storing an image pyramid as a series.
DESCRIPTION OF THE WSI STORAGE AND ACCESS	Details the storage and access procedures.
THE WSI INFORMATION OBJECT DEFINITION (IOD)	Details aspects relevant to image orientation, assumptions and data Interpretation.
WSI FRAME OF REFERENCE	Gives explanations on the frame of reference.
DIMENSIONS, Z-PLANES, AND MULTISPECTRAL IMAGING	Goes in deep on aspects particularly correlated to the digital-cytology.
WSI SPARSE ENCODING	The sparse encoding is also an important issue.
WSI ANNOTATION AND ANALYSIS RESULTS	Goes in deep on aspects relevant to the annotation allowed, the presentation states, the segmentation, the structured reporting.
WSI WORKFLOW	Goes in deep on aspects relevant to the workflows associated to the WSI.
IMPACT ON THE NEMA PUBBLICATIONS	Faces the impact and the modifications introduced by WG26 in comparison to the previous NEMA publications and considers the needed revisions.

Classes) in the 2010/08/24, by the DICOM WG 26 allows today to store the digital pathology images also from D-CYT in a form that is compatible with the same DICOM archive systems used by hospitals and opens new chances for D-CYT. The D-CYT in fact uses very large files in D-CYT (> 2GB), the Whole Slide Images (WSI)s produced by digitizing microscope slides at diagnostic resolution. In addition these WSIs have a different type of access regarding the other digital images stored in the PACS systems, due to the fact that the pathologist needs to rapidly pan and zoom images. The WSI representation allows to also consider the Z-stack problem of the D-CYT. The 145 supplement addresses also the considerations available in a previous supplement, the 122 supplement (Supplement 122: Specimen Module and Revised Pathology SOP Classes), about the digital microscopic images. The 145 supplement furnishes now the definitions of the WSI as DICOM Information Object Definitions and describes the WSI image characteristics. Vendors of D-CYT products will make in the next years

efforts to adequate their products to the new standards and probably will suggest revisions on the basis of their analysis. Table 1 resumes the issues currently investigated by WG 26 relevant to D-CTY

4 DISCUSSION AND CONCLUSION

Today thanks to the development of the information technology, the diffusion of new visualization strategies and the availability of low cost or free visualization proprietary tools, the scenario of the tele-pathology has radically changed: the Virtual Microscopy offers new promising opportunities oriented to the application of D-CYT. This study has reviewed the technologies to promote the diffusion of the D-CYT via the web-based-client-server technologies. Three lines of actions relevant to the consolidation of the D-CYT are now strongly needed:

- *Direction of research*, proposing to investigators new fields of interest requiring strong scientific efforts.
- *Direction of Health Technology Assessment.* The new technologies dedicated to the D-CYT, now available based on the *virtual slides*, have in fact radically changed the scenario of work-place, increasing the need to develop a specific HTA centred on the cytologist.
- *Direction of institutional bodies*, asking from the so called "stake-holders" proposals relevant to the destination of funds to be directed towards University and Health Care Systems for the necessary changes to investigate and include the D-CYT in the Public Administration procedures.

REFERENCES

Afework, A. Beynon, MD. Bustamante, F. Cho, S. Demarzo, A. Ferreira, R. Miller, R. Silberman, M. Saltz, J. Sussman A, Tsang H. 1998. Digital dynamic telepathology the virtual microscope. *Proc AMIA Symp* 23:912–6.

Ash, WM. Krzewina, L. Kim, MK. 2009. Quantitative imaging of cellular adhesion by total internal reflection holographic microscopy. *Appl Opt* 48(34)144–52.

Bashshur, RL. 2000.Editorial: telemedicine nomenclature: What does it mean? *Telemed J* 6(1):1–2.

Bondi, A. 2007. CQ con immagini digitali nello screening dei tumori del collo dell'utero: il progetto della Regione Emilia Romagna. *Proc Congresso nazionale SICI* 2007;459–62.

Bondi, A. Pierotti, P. Crucitti, P. Lega, S. 2010. The virtual slide in the promotion of cytologic and hystologic quality in oncologic screenings. *Ann Ist Super Sanita* 46(2): 144–50.

Catalyurek, U. Beynon, MD. Chang, C. Kurc, T. Sussman, A. Saltz, J. 2003b.The virtual microscope. *IEEE Trans Inf Technol Biomed* 2003;7(4):230–48.

Çatalyürek, Ü. Beynon, MD. Chang, C. Kurc, T. Sussman, A. Saltz, J. 2003a. The virtual microscope. *IEEE Trans Inf Technol Biomed* 2003;7(4):14–22.

Cheong, FC. Sun, B. Dreyfus, R. Amato-Grill, J. Xiao, K. 2009. Dixon L, Grier DG. Flow visualization and flow cytometry with holographic video microscopy. *Opt Express* 17(15):1371–9.

Cosentino, A. Ghidoni, D. Salemi, M. Folicaldi, S. Amadori, A. Zani, J. Grasso, G. Bondi, A. 1999. An interlaboratory study of the use of PapNet in the quality control of cervico-vaginal cytology *Pathologica* 91(2):101–6.

D'Avenio, G. Balogh, T. Grigioni. M. 2007. Holographic display as innovative training tool. ESAO 2007, XXXIV Annual Congress European Society for Artificial Organs, Sept. 5–8, 2007 Krems, Austria.

Della Mea, V. Demichelis, F. Viel, F. Dalla Palma, P. Beltrami, C. 2006. User attitudes in analyzing digital slides in a quality control test bed: A preliminary study. *Comp Meth Progr Biomed* 82(2):177–86.

Demichelis, F. Barbareschi, M. Dalla Palma, P. Forti, S. 2002. The virtual case: a new method to completely digitize cytological and histological slides. *Virchows Arch* 441(2):159–64.

Demichelis, F. Barbareschi, M. Dalla Palma, P. Forti, S. 2002. The virtual case: a new method to completely digitize cytological and hislo cal slides. *Virchows Arch* 441(2):159–64.

Dunn, BE. Almagro, UA. Recla, DL. Davis CW. 2000. Telepathology Networking in VISN-12 of the veterans health administration. *Telemed J E Health* 6(3):349–54.

Dunn, BE. Choi, H. Almagro, UA. Recla, DL. Krupinski, EA. Weinstein, RS. 1999. Routine Surgical Telepathology in the department of Veterans Affairs: Experience-Related improvements in Pathologist Performance in 2200 Cases. *Telemed J* 5(4):323–37.

Ferreira, R. Moon, B. Humphries, J. Sussman, A. Saltz, J. Miller, R. Demarzo, A. 1997. The virtual microscope. *Proc AMIA AnnuFall Symp* 22:449–53.

Frankewitsch T, Söhnlein S, Müller M, Prokosch HU. Computed quality assessment of MPEG4-compressed DICOM video data. *Stud Health Technol Inform* 2005;116: 447–52.

Giansanti, D. Castrichella, L. Giovagnoli, MR. 2007c New models of e-learning for healthcare professionals: a training course for biomedical laboratori technicians. *J Telemed Telecare* 13(7):374–6.

Giansanti, D. Castrichella, L. Giovagnoli, MR. 2008a. Telepathology training in a master of cytology degree course. J *Telemed Telecare* 14(7): 338–41.

Giansanti, D. Castrichella, L. Giovagnoli, MR. 2008b. Telepathology requires specific training for the technician in the biomedical laboratory. *Telemed J E Health* 14(8) : 801–7.

Giansanti, D. Castrichella, L. Giovagnoli, MR. 2008c.The design of a health technology assessment system in telepathology. *Telemed J E Health* 14(6):570–5.

Giansanti, D. Cerroni, F. Amodeo, R. Filoni, M. Giovagnoli, MR. 2010c. A pilot study for the integration of cytometry reports in digital cytology telemedicine applications. *Ann Ist Super Sanita.* 46(2):138–43.

Giansanti, D. Grigioni, M. D'Avenio, G. Morelli, S. Maccioni, G. Bondi, A. Giovagnoli, MR. 2010b Virtual microscopy and digital cytology: state of the art. *Ann Ist Super Sanita.* 46(2):115–22.

Giansanti, D. Grigioni, M. Giovagnoli, MR. 2010a. Virtual microscopy and digital cytology: fact or fantasy? Preface. *Ann Ist Super Sanita.*46(2):113–4.

Giansanti, D. Morelli, S. and Macellari, V. 2007. A protocol for the assesmentof diagnostic accuracy in tele-echocardiography imaging. *Telemed J E Health* 13(4):669–7.

Giansanti, D. Morelli, S. and Macellari, V. 2007. Telemedicine technology assessment part 2: tools for a quality control system. *Telemed J E Health* 13:456–63.

Giansanti, D. Morelli, S. Macellari, V. 2007a. Telemedicine technology assessment part II: tools for a quality control system. *Telemed J E Health* 13(2):130–40.

Giansanti, D. Morelli, S. Macellari, V. 2007b. Telemedicine technology assessment part I: setup and validation of a quality control system. *Telemed J E Health* 13(2):118–29.

Gilbertson, JR. Ho, J. Anthony, L. Jukic, DM. Yagi, Y. Parwan, AV. 2006. Primary histologic diagnosis using automated whole slide imaging: a validation study. *BMC Clin Pathol* 27:16–24.

Giovagnoli, MR. Giarnieri, E. Carico, E. Giansanti, D. 2010. How do young and senior cytopathologists interact with digital cytology? *Ann Ist Super Sanita* 46(2):123–9.

Hunter, DC. Brustrom, JE. Goldsmith, BJ. Davis, LJ. Carlos, M. Ashley, E. Gardner, G. 1999. Gaal teleoncology in the department of defense: a tale of two systems. *Telemed J* 5(3):273–82.

Hunter, DC. Brustrom, JE. Goldsmith, BJ. Davis, LJ. Carlos, M. Ashley, E. Gardner, G. Gaal, I. 1999. Teleoncology in the department of defense: a tale of two systems. *Telemed J* 1999;5(3):273–82.

ITU Work Group. 2007. ITU-R Recommandation F.390-4 (07/82). Definitions of terms and references concerning hypothetical reference circuits and hypthetical reference digital paths for radio-relay systems. Geneva: Recommendations of the International Telecommunication Union/Radiocommunication Sector; July 1982. Suppressed on 10/10/07 (CACE/435).

Krupinski, EA. Webster, P. Dolliver, M. Weinstein, RS. Maria Lopez, A. 1999. Efficiency analysis of a multi-specialty telemedicine service. *Telemed J* 5(3):265–71.

Lee, ES. Kim, IS. Choi, JS. Yeom, BW. Kim, HK. Ahn, GH. Leong, AS. 2002. Practical telepathology using a digital camera and the internet. *Telemed J E Health* 8(2):59–165.

Leong, F. Graham, AK. Schwarzmann, P. McGee, J. 2000.Clinical trial of telepathology as an alternative modality in breast histopathology quality assurance. *Telemed J E Health* 6(4): 373–7.

Martelli, F. Giordano, A. Giansanti, D. Morelli, S. Macellari, V. 2005. Evaluation of a new metric for telemedicine video quality asment.In: *Proceeding of 2nd Annual Meeting on Health Technology Assessment International.* Rome: 2005.

Miyahara, S. Tsuji, M. Iizuka, C. Hasegawa, T. Taoka, F. 2006 On the evaluation of economic benefits of japanese telemedicine and factors for its promotion. *Telemed J E Health* 12(6):691–7.

Molnar, B. Berczi, L. Diczhazy, C. Tagscherer, A. Varga, SV. Szende, B. Tulassay, Z. 2003. Digital slide and virtual microscopy Based routine and telepathology evaluation of routine gastrointes Tinal biopsy specimens. *J Clin Pathol* 56(6):433–8.

Moore, PT. O'Hare, N. Walsh, KP. Ward, N. Conlon, N. 2008. Objectivvideo quality measure for application to tele-echocardiography *Med Biol Eng Comput* 2008;46(8):807–13.

Morelli S, Giordano A, Giansanti D. 2009. Routine tests for both planning and evaluating image quality in tele-echocardiography. *Ann Ist Super Sanità* 45(4):378–91.

Mori, I. Nunobiki, O. Ozaki, T. Taniguchi, E. Kennichi, K. 2008. Issues for application of virtual microscopy to cyto-screening, perspetivesbased on questionnaire to Japanese cytotechnologists. *Diagn Path* 3(1):S11–15.

Palombo, A. La Scaleia, M. Silvestri, S. D'Avenio, G. Daniele, C. Grigioni. M. 2010. Collaborative environment for Clinical Audit by integration of Surface Computing, 3D Dynamical Visualization and Imaging Fusion technique. Proceedings of the II National Congress of

Bioengineering (GNB 2010), July 8–10, 2010, Torino, Italy

Pinson, M. Wolf, S. Austin, PG. Allhands, A. 2002. *Video Quality Measurement PC User's Manual*. Santa Clara, CA: Intel Corporation; Nov 2002.

Seet, KY. Nieminen, TA. Zvyagin, AV. 2009. Refractometry of melanocyte cell nuclei using optical scatter images recorded by digital Fourier microscopy. *J Biomed Op* 14(4):135–42.

Settakorn, J. Kuakpaetoon, T. Leong, F. Thamprasert, K. Ichijima, K. 2002. Store-and-forward diagnostic telepathology of small biopsies by E-mail attachment: A feasibility pilot study with a view for future application in thailand diagnostic pathology services. *Telemed J E Health* 8(3):333–41.

Tanriverdi, H. Iacono, CS. 1999. Diffusion of telemedicine: a knowledge barrier perspective. *Telemed J* 5(3):223–44.

Tuominen, VJ. Isola, J. 2009. Linking Whole-Slide Microscope Images with DICOM by Using JPEG2000 Interactive Protocol. *J Digit Imaging* 13:451–3.

Weisz-Carrington, P. Blount, M. Kipreos, B. Mohanty, L. Lippman, R. Todd, WM. Trent, B. 1999 Telepathology between richmond and beckley veterans affairs hospitals: report on the first 1000 cases. *Telem J* 5(4):367–73.

Wolf, S. Pinson, M. 2002. *NTIA Report 02-392* – Video Quality Measurement Techniques. Boulder, Colorado: US Department of Commerce – National Telecommunications and Information Administration; June 2002; available from: www.its.bldrdoc.gov/n3/video/documents.htm; last visited 11/12/08.

Wolf, S. Pinson, MH. 2002. *The relationship between performance and spatial-temporal region size for reduced-reference, in-service video quality monitoring systems*. Institute for Telecommunication Sciences, National Telecommunications and Information Administration.325 Broadway, Boulder, CO 80305, USA.

Yearwood, J, Pham, B. 2000. Case-Based Support in a cooperative medical diagnosis environment. *Telemedicine Journal* 6(2):243–50.

Yellowlees, PM. 2000. Intelligent health systems and third millennium medicine in Australia. *Telemed J* 6(2): 197–200.

Yu, L. Mohanty, S. Zhang, J. Genc, S. Kim, MK. Berns, MW. Chen, Z. 2009. Digital holographic microscopy for quantitative cell dynamic evaluation during laser microsurgery. *Opt Express* 17(14):12031–8.

Computational Modelling of Objects Represented in Images – Di Giamberardino et al. (eds)
© 2012 Taylor & Francis Group, London, ISBN 978-0-415-62134-2

Radial Basis Functions for the interpolation of hemodynamics flow pattern: A quantitative analysis

R. Ponzini
CILEA, Milan, Italy

M.E. Biancolini
Università di Tor Vergata, Rome, Italy

G. Rizzo
IBFM-CNR, Milan, Italy

U. Morbiducci
Politecnico di Torino, Turin, Italy

ABSTRACT: Non-invasive quantitative map of blood flowing in the cardiovascular system is now feasible thanks to imaging techniques (PC MRI). The potentiality of PC MRI data are only poorly exploited due to the limited spatial and temporal accuracy (pixel size, slice-thickness, number of time frames per cardiac cycle). As a consequence, advanced in vivo quantitative hemodynamics cannot be fully exploited as well. A possible approach to overcome some of the limitations consists in the application of interpolation strategies. In this work we test for the first time the reliability of Radial Basis Function theory (RBF) to interpolate blood flow fields in the human aortic arch. Thanks to a previously validated synthetic PC MRI data generator, a quantitative analysis of the percentage error distribution with respect to a gold standard flow field has been performed. The obtained results clearly show that the new method is well suited for this kind of application limiting the error values below 5% in almost every zone of the bulk flow.

1 INTRODUCTION

1.1 *Background*

Non-invasive quantitative map of blood flowing in the cardiovascular system is now feasible thanks to imaging techniques such as Phase Contrast MRI (PC MRI). Actually the potentiality of PC MRI data are only very little exploited due to the limited spatial and temporal accuracy (pixel size, slice-thickness and number of time frame per cardiac cycle) of the data and thus limiting the reliability of advanced processing based on computational fluid dynamics (CFD) algorithms. A possible approach consist on build more suitable images by mean of an interpolation algorithm. Nevertheless generation of new flow field data of complex fluid dynamics patterns with standard interpolation algorithms usually involve also noise and error generation. For this reason a possible new choice could be the application of well-established and robust data interpolator based on Radial Basis Function theory (RBF). In this work we tested for the first time the reliability of RBF to interpolate the blood flow of the human aortic arch. This was done using a recently validated synthetic MRI-like data generator based on CFD modeling (Morbiducci et al., 2012).

The work is divided in three sections: (i) the material and method briefly describe the synthetic data generator, the complex flow pattern present in the human aortic arch and the radial basis function method and setup; (ii) the results section describe the main results of the work including spatial and temporal interpolation analysis; (iii) in the last section the results are discussed along with the perspectives and the future works.

2 MATERIAL AND METHOD

2.1 *In silico MRI data generator*

To mimic the PC MRI flow data complexity by taking advantage of synthetic dataset generation, we mapped selected sets of velocity values taken from a patient specific image-based CFD model into equally spaced grids of imposed pixel size, slice thickness and number of time frames. Considering the temporal axis, the CFD data were wrapped together in a subset of time instant datasets according to an arithmetic mean filter while for the space sampling, in analogy to MRI signal averaging, the data on the k-th voxel was synthesized from CFD data using a weighting Gaussian

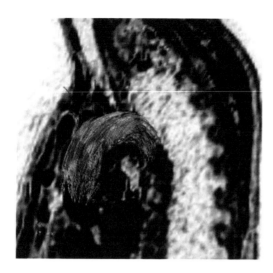

Figure 1.   Complex flow pattern measured *in vivo*.

function based on the distance of the i-th CFD velocity point with the respect to the center of the voxel (see Morbiducci et al. 2012). According to this processing the output of the synthetic PC MRI data generator is a set of images differing for slice thickness, in plane pixel size and number of time frame per cardiac cycle. Differently to in-vivo measurements, the proposed synthetic ones, are noise free data and they can conveniently be used for quantitative algorithms evaluation.

2.2   Aortic flow patterns

The human aortic arch present a complex spatial/temporal evolving fluid dynamics. More in details during the acceleration phase and up to the peak systole the flow pattern are mainly aligned with the aortic axis, the main component of the velocity being driven by the feet-to-head component (namely v component). During the deceleration phase helical flow patterns predominate in the flow field as given by the composition oftranslational and rotational fluid motion, this last mainly generated by the increases of the right-to-left velocity component and of the anterior-posterior velocity component (namely u and w respectively). It must be also noticed that the magnitude of the three velocity components spans a wide range of values within the cardiac cycle. This complexity in the aortic hemodynamics, recently observed *in vivo* and quantitatively described (Morbiducci et al., 2011), is well reproducedalso by the synthetic dataset.

2.3   Synthetic data: spatial and temporal sampling

In what follows, high spatial/temporal accuracy synthetic images (Morbiducci et al., 2012) are considered for the evaluation RBF interpolation.They will be refernced as: gold standard (GS), i.e., the reference dataset, coarser images (N_i, i = 1, . . . , 3), representative of a range of lower resolution images used in

the clinical practice and against which the accuracy of image interpolation is tested. The spatial accuracy of the GS is equal to 0.5×0.5 [mm^2], with a 0.01 [s] temporal accuracy. The coarser Niimages have a spatial and temporal definition that ranges from 1.0×1.0 [mm^2] to 4.0×4.0 [mm^2] and from 0.02 [s] to 0.07 [s] respectively.

2.4   Radial Basis Function interpolation

Radial Basis Functions (RBF) are a very powerful tool created for the interpolation of scattered data; RBF are successfully used for a wide range of applications including mesh morphing (see Biancolini 2011) and interpolation of velocity data set (see Vennell & Beatson 2009). They are able to interpolate everywhere within the space a function defined at discrete points giving the exact value at original points. The behaviour of the function between the points depends on the kind of radial function adopted that can be fully or compactly supported;some functions requirea polynomial correction to guarantee compatibility for rigid modes.

Alinear system of equations (of order equal to the number of source point introduced) needs to be solved for the coefficients calculation. Once the unknown coefficients are calculated, the dataat an arbitrary point inside or outside the domain (interpolation/extrapolation) is expressed as the summation of the radial contribution of each source point (if the point falls inside the influence domain). The interpolation function composed of a radial basis φ and a polynomial h is defined as follows:

$$s(\mathbf{x}) = \sum_{i=1}^{N} \gamma_i \phi(\|\mathbf{x} - \mathbf{x}_i\|) + h(\mathbf{x}) \qquad (1)$$

The degree of the polynomial has to be chosen depending on the kind of radial function adopted. A radial basis fit exists if the coefficients γ and the weights of the polynomial can be found such that the desired function values are obtained at source points and the polynomial terms gives 0 contributions at source points, that is:

$$s(\mathbf{x}_{k_i}) = g(\mathbf{x}_{k_i}) \quad 1 \le i \le N \qquad (2)$$

$$0 = \sum_{i=1}^{N} \gamma_i q(\mathbf{x}_{k_i}) \qquad (3)$$

for all polynomials q with a degree less or equal to that of polynomial h. The minimal degree of polynomial h depends on the choice of the basis function. A unique interpolant exists if the basis function is a conditionally positive definite function. If the basis functions are conditionally positive definite of order m <= 2, a linear polynomial can be used:

$$h(\mathbf{x}) = \beta + \beta_1 x + \beta_3 y + \beta_4 z \qquad (4)$$

The values for the coefficients γ of RBF and the coefficients β of the linear polynomial can be obtained by solving the system:

$$\begin{pmatrix} \mathbf{M} & \mathbf{P} \\ \mathbf{P}^{\mathrm{T}} & \mathbf{0} \end{pmatrix} \begin{pmatrix} \gamma \\ \beta \end{pmatrix} = \begin{pmatrix} \mathbf{g} \\ \mathbf{0} \end{pmatrix} \qquad (5)$$

where $\mathbf{g}$ are the known values at the source points.

$$M_{ij} = \phi\left(\left\| \mathbf{x}_{k_i} - \mathbf{x}_{k_j} \right\|\right) \quad 1 \le i \quad j \le N \qquad (6)$$

$\mathbf{M}$ is the interpolation matrix defined calculating all the radial interactions between source points and $\mathbf{P}$ is a constraint matrix that arises to balance the polynomial contribution and contains a column of "1" and the x y z positions of source points in the others three columns:

$$\mathbf{P} = \begin{pmatrix} 1 & x_{k_1}^0 & y_{k_1}^0 & z_{k_1}^0 \\ 1 & x_{k_2}^0 & y_{k_2}^0 & z_{k_2}^0 \\ \vdots & \vdots & \vdots & \vdots \\ 1 & x_{k_N}^0 & y_{k_N}^0 & z_{k_N}^0 \end{pmatrix} \qquad (7)$$

Radial basis interpolation works for scalar fields. For the flow interpolation each component of the flow field prescribed at the source points is interpolated as follows:

$$\begin{cases} v_x = s_x(\mathbf{x}) = \sum_{i=1}^{N} \gamma_i^x \phi\left(\left\| \mathbf{x} - \mathbf{x}_{k_i} \right\|\right) + \beta_1^x + \beta_2^x x + \beta_3^x y + \beta_4^x z \\ v_y = s_y(\mathbf{x}) = \sum_{i=1}^{N} \gamma_i^y \phi\left(\left\| \mathbf{x} - \mathbf{x}_{k_i} \right\|\right) + \beta_1^y + \beta_2^y x + \beta_3^y y + \beta_4^y z \\ v_z = s_z(\mathbf{x}) = \sum_{i=1}^{N} \gamma_i^z \phi\left(\left\| \mathbf{x} - \mathbf{x}_{k_i} \right\|\right) + \beta_1^z + \beta_2^z x + \beta_3^z y + \beta_4^z z \end{cases}$$

The radial basis method has several advantages that make it very attractive. The key point is that being a meshless method it's able to interpolate everywhere regardless the position of points used to generate the field. So it can be easily used for particle tracking because the field values can be recovered continuously in the space. Of course a regular grid can be used to place source points, but points with null contribution (a lot of them can arise when a regular domain is used as original dataset) can be excluded from the fit (in this case a new scattered distribution will result). Another interesting option is that extra points can be added at desired positions to better track wall behaviour of the flow. Scattered dataset can be also generated from the original fine dataset using a distance based decimation algorithm.

It is worthwhile to notice that the closed form interpolator can be differentiated analytically (the order depend on the class of continuity of the radial function adopted).

2.5 RBF setup

Presented study has been conducted using direct fit of RBF by means of a MathCAD worksheet (it comes with Lapack solver for linear systems). This approach

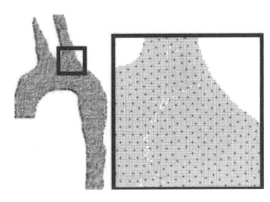

Figure 2. RBF reference points (in red) and generated points (in black). In this case about 2400 velocity points are used to evaluate the flow fields at about 58000 new velocity points. Complete slice on the left, enlarged detail on the right.

allows to evaluate the feasibility of the method very quickly (see Biancolini 2011). The biharmonic kernel ($\varphi(r) = r$) has been here considered for two reasons: it is the smoothest interpolant and it can be accelerated using fast RBF algorithms. The latter observation is crucial to extend the method to real life applications; the direct method is limited to about 10.000 RBF points (its complexity grows as N^3) while a fast solver can manage very large data set (complexity grows as $N^{1.4-1.6}$; the solver used in the commercial software RBF Morph can fit 2.600.000 points in about 2 hours). Although the original dataset is structured, RBF centers have been processed as a generic distribution; RBF points set is defined excluding first all zero velocity points outside the domain boundary and then a subset is selected applying a sub-sampling method based on prescribed radius.

In figure 2 the RBF points selection is shown. The red dots are the point selected as input for the RBF interpolation (about 2400) while the black dot represent the point excluded from the original dataset in which the flow field is evaluated by the interpolation process (about 58000). The fine original dataset is used to check the interpolation quality because at that points the flow interpolated on the small set can be compared with the original value (GS).

The sub-sampling radius has been defined to obtain a spacing of the RBF dataset typical of *in vivo* measurements as explained above; the synthetic dataset is available at an higher resolution and can be used as a reference to understand whether an RBF interpolation at typical *in vivo* spacing can be used for the reconstruction of the actual flow field. The reliability of the interpolation has been evaluated computing the percentage error between interpolated data and GS data on the high resolution distribution.

3 RESULTS

3.1 Spatial interpolation

As a first test we studied the systolic peak instant. Figure 3 shows the results of the mapping procedure

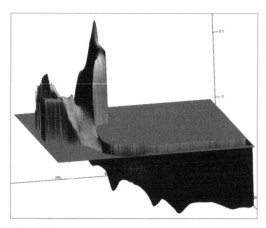

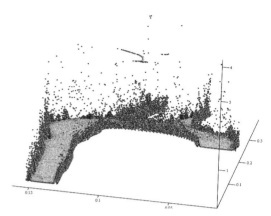

Figure 4. Error distribution on the 2D slice represented as a scatter plot; height represent the error. Notice how the bulk flow is properly recovered by the interpolator.

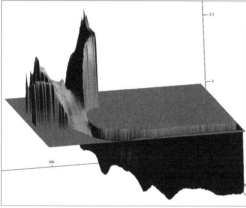

Figure 3. Qualitative comparison of reference data (top) with interpolated data (bottom).

at peak systole for a single 2D slice centered on the geometry model. The overall features of the velocity component are in good agreement as expected. Notably using the synthetic data generator we are able to perform quantitative and very detailed point wise analysis and quantify percentage error in each point of the slice excluding the effect of any eventual external source of noise.

The larger part of the point population have a very low percentage error value; in details:

- 82% below 5%
- 5% in the range 5%–10%
- 8% in the range 10%–20%
- 4% in the range 20%–50%
- 1% over 50%

Figure 4 localize within the 2D slice the values of the percentage error; the most interesting fact is that in the bulk points the flow field is very well described by the interpolated data while at wall vessel location the errors are relevant. This kind of error distribution should be kept in mind when *in vivo* computation are performed at the wall vessel location such as for wall shear stress evaluation (WSS).

3.2 Temporal interpolation

RBF spatial interpolation is available at each sampled time frame. The flow information at an arbitrary position in the domain for a given intermediate time is recovered computing first the value at a certain number of time frames using RBF interpolation. Extracted data are time interpolated; the quality and order of interpolation depends on the number of interpolated frames, the best is achieved using the full dataset. At present this kind of analysis is at the very early stage but could represent the most relevant application due the intrinsic timedependent nature of hemodynamics problems.

4 DISCUSSION AND CONCLUSION

This is the first attempt to use well established RBF technique to perform blood flow field data interpolation being this kind of procedure involved in *in vivo* image processing analysis (see Busch et al. 2012). The robustness of the results is very promising showing a very low percentage of generated point with error over 10%. Further investigations will span the sensitivity of the new approach to other image parameters such as pixel size, slice thickness and number of time frame per heart cycle. All these parameters are involved in MRI *in vivo* data sampling and must be kept under consideration in the evaluation of the performance of any image processing algorithm.

It is very important to remark that: (i) the meshless nature of the methods allows to easily improve the method; (ii) the predicted flow field exhibits a very low percentage error in the bulk; (iii) at the wall location pixels velocity values are poorly resolved. A substantial improvement of the interpolation can be achieved by using a non-uniform RBF points seeding strategy and therefore placing extra RBF points at wall location that impose to the flow the desired wall values. Furthermore the analytic nature of the fit allows to evaluate spatial derivatives in closed form enriching substantially the information extracted from the experimental data flow and allowing in theory to directly compute wall shear stress values (and therefore the oscillating shear index (OSI) and the time averaged wall shear stress (TAWSS)) 'analytically' directly from *in vivo* measurements.

REFERENCES

Biancolini, M. E., 2011 MeshMorphing and Smoothing by Means of Radial Basis Functions (RBF): A Practical Example Using Fluent and RBF Morph, Handbook of Research on Computational Science and Engineering: Theory and Practice (in press).

Busch J et al 2012, Construction of divergence-free velocity fields from cine 3D phase-contrast flow measurements, Magnetic Resonance in Medicine.

Morbiducci et al. 2011. Mechanistic insight into the physiological relevance of helical blood flow in the human aorta: An *in vivo* study. (2011) Biomechanics and Modeling in Mechanobiology

Morbiducci et al. 2012. Synthetic dataset generation for the analysis and the evaluation of image-based hemodynamics of the human aorta. (2011) Medical and Biological Engineering and Computing

Vennell, R., and R. Beatson 2009, A divergence-free spatial interpolator for large sparse velocity data sets, J. Geophys. Res., 114, C10024, doi:10.1029/2008JC004973.

Computational Modelling of Objects Represented in Images – Di Giamberardino et al. (eds)
© 2012 Taylor & Francis Group, London, ISBN 978-0-415-62134-2

A software for analysing erythrocytes images under flow

G. D'Avenio
Department of Technology and Health, Istituto Superiore di Sanità, Rome, Italy

P. Caprari
Department of Hematology, Oncology and Molecular Medicine, Istituto Superiore di Sanità, Rome, Italy

C. Daniele
Department of Technology and Health, Istituto Superiore di Sanità, Rome, Italy

A. Tarzia
Department of Hematology, Oncology and Molecular Medicine, Istituto Superiore di Sanità, Rome, Italy

M. Grigioni
Department of Technology and Health, Istituto Superiore di Sanità, Rome, Italy

ABSTRACT: Red Blood Cells (RBCs) have a critical role in the behaviour of blood under flow. Normally the RBCs have the capacity to aggregate in the presence of plasma proteins but the aggregates are dispersed by the shear stress generated by the blood flow; propensity to aggregation is critical in determining blood rheology. This study presents a software tool capable to determine, by means of analysis of erythrocytes images under flow, the aggregation – disaggregation patterns of RBCs, either obtained from controls or pathological patients. The results of the analysis of actual images of physiological RBC under different shear rates confirm the effectiveness of the proposed method.

Red blood cells (RBCs) have a critical role in the behaviour of blood under flow: cell deformability and aggregation properties may influence blood rheology. Haematological diseases characterized by morphological and structural alterations in RBCs may be associated with hemorheological abnormalities. Thus, the availability of image processing tools capable to evaluate quantitatively how RBCs reacts to mechanical loading induced by increasing shear rates could be useful in highlighting the pathological profile of the patient.

Normally the RBCs have the capacity to aggregate in the presence of plasma proteins but the aggregates are dispersed by the shear stress generated by the blood flow (Neu, 2007). Structural alterations in RBCs might induce the formation of larger and stronger aggregates that are resistant to dispersion by the blood flow or smaller and weaker aggregates that are easily disaggregated.

This study presents a software tool capable to determine the aggregation – disaggregation patterns of RBCs, either obtained from controls or pathological patients.

1 MATERIALS AND METHODS

A custom software was written in the Matlab environment to analyze video sequences of RBCs under flow.

The software calculates, for each frame, the number and size of the clusters of RBCs, by nonlinear filtering of the images. Thus, the time course of the mean cluster area could be identified, for the operating conditions of the recording (i.e., shear rate).

Each video frame is partitioned in smaller domains, and the histogram of intensity values in each subsection of the image is calculated. Thus, a local optimum intensity threshold is determined as the average value of the intensities associated to the peak values of the background's and of the erythrocytes' distributions (see example in fig. 1). This local thresholding allows dealing with non-uniform illumination of the image, such as in fig. 2.

After binarizing each subimage with the relative threshold and reconstituting the binarized version of the original frame, spurious groups of pixels are removed with morphological opening (Dougherty, 1992), using structuring elements of lower size than that of a single RBC.

Erythrocytes' images under microscope have a much brighter center than their border, due to their biconcave shape. Thus, upon binarization the image can present ring-like structures, instead of simply connected domains of points (fig. 2). This effect is dealt with by removing from the image all connected components (objects) that have fewer than N pixels, where N is a suitable fraction of the RBC size. This is

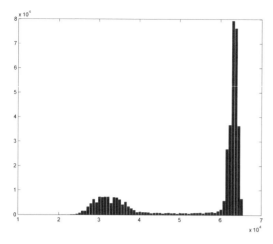

Figure 1. Histogram of the intensities in the image in fig. 4 (top). A roughly Gaussian bimodal distribution can be noticed, given by the erythrocytes (left) and background (right) contribution.

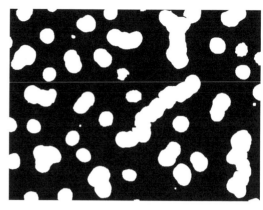

Figure 3. Result of the operation of morphological opening on Fig. 2. The ring-like structures are replaced by correct shapes.

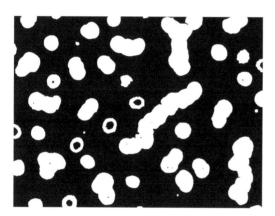

Figure 2. Image of aggregating erythrocytes, after local thresholding: there are a number of segmented RBCs with only their periphery marked as belonging to a cellular structure. Ring-like structures are yielded by the low value of thickness at the center of the RBCs.

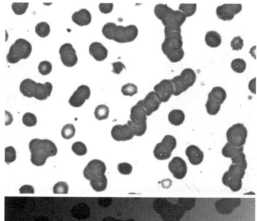

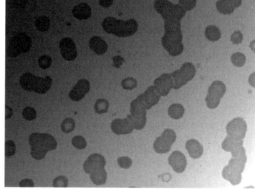

Figure 4. Typical images of aggregating erythrocytes, in ideal illumination conditions (top) and with illumination gradients (bottom).

done with a suitable morphological opening of the image.

Finally, all remaining connected components in the image (i.e., single or aggregated erythrocytes) are labeled, making it possible to measure the centroid and area of each labeled cluster. The total area of aggregates, as well as the average area of the aggregates and the number of clusters in each image are then computed.

The result of the procedure is shown in fig. 5, referring to the example in fig. 4: notwithstanding the not ideal illumination conditions, the difference in the number of pixels identified as belonging to a cluster of RBCs, between the two images is very small (0.6%).

The experimental part of this study has been carried out using the Rheo-Microscope (Anton Paar), constituted by a glass parallel-plate rheometer, and an optical microscope. Whole blood was subjected at 37°C to increasing shear rates (between 1 and 250 s-1) and simultaneously imaged with a CCD camera. Video sequences of the flowing RBCs were subsequently analyzed.

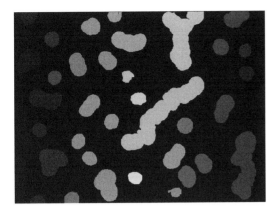

Figure 5. Result of the segmentation of the image in Fig. 4 (top). Each labeled cluster is shown with a different colour.

Table 1. Rate of increase of cluster number (RI), after analysis of a sample of physiological blood under rheo-microscope.

SR [s^{-1}]	50	100	200
RI [s^{-1}]	0.66	1.18	−0.20

2 RESULTS

The algorithm was tested with synthetic images of aggregating RBCs, randomly disposed in the image. N = 1000 synthetic images were generated, and added with gaussian white noise of increasingly higher level. Having defined the image SNR as

$$SNR = 10 \log_{10} \frac{\sigma_s}{\sigma_n} \qquad (1)$$

$\sigma_s^2 \sigma_n^2$ and being the variance of the original and the noise image, respectively, the algorithm was found to be fairly insensitive to additive white noise, at least up to SNR = 10 dB.

The software was then tested with RBC video sequences from the Rheo-Microscope. As shown in Fig. 6, the number of clusters was found to be steadily increasing, after the start of the rheometer's plate movement (vertical line). The rate of increase of clusters (1/s) is calculated by least–square-fitting the data comprised between two limit points (red circles).

The results of the analysis are reported in table I. At lower shear rate, 50 s^{-1}, the mechanical load caused the aggregates to break down (table I and Fig. 7). This applied for 50 as well as for 100 s^{-1} SR values, the latter causing a rate of increase of cluster number which was almost double of that associated to the former.

Finally, at 200 s^{-1}, the applied shear was high enough to cause immediate fragmentation, and a low, negative value of the rate of increase of cluster number was found (table I and Fig. 8).

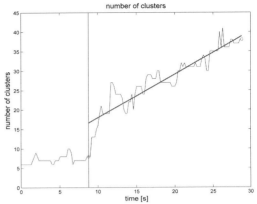

Figure 6. Analysis of the video recording of flowing erythrocytes, at shear rate = 100 s^{-1}. The number of clusters is shown against time. The number of clusters is steadily increasing, after the start of the rheometer (vertical line).

3 DISCUSSION

Other studies about automatic processing of hematological images have been proposed in the past. For instance, Gering et al (2004) report a method for counting nucleated erythrocytes from digital micrographs of thin blood smears, which enables to estimate intensity of hematozoan infections in nonmammalian vertebrate hosts. This method, which subjects blood images to automatic thresholding and particle analysis, uses ImageJ, a Java-based open source software tool, developed at the U.S. National Institutes of Health (http://rsb.info.nih.gov/ ij/index.html).

The image enhancements operations provided by ImageJ are:

smoothing, sharpening, edge detection, median filtering and thresholding on both 8-bit grayscale and RGB color images, brightness and contrast interactive adjustment.

The presented software supports local automatic thresholding, as opposed to ImageJ. Moreover, it uses morphological operators to post-process blood images after thresholding, whereas ImageJ does not allow morphological operations, in its basic version.

It must be underlined that the blood smears considered by Gering et al. (2004) had been previously subjected to staining, in order to enhance the contrast between nuclei and RBC cytoplasm, whereas our study addresses the analysis of unstained blood images.

The problem of uneven illumination can hinder automatic processing of microscopic images (Leong et al, 2003), since it is something that is not always recognized by the operator, who generally compensates inhomogeneities of this sort in the perceived image. The software hereby proposed can deal with such a problem (figure 4), allowing for local thresholding of images.

Previous works in related areas have also used morphological operators: in (Di Ruberto et al., 2002),

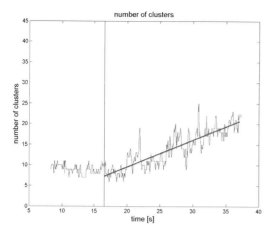

Figure 7. Analysis of the video recording of flowing erythrocytes, at shear rate $= 50\,s^{-1}$. The number of clusters is shown against time.

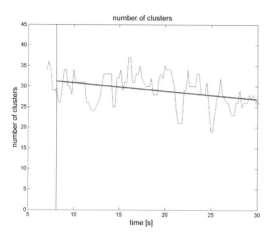

Figure 8. Analysis of the video recording of flowing erythrocytes, at shear rate $= 200\,s^{-1}$. The number of clusters is shown against time.

grey scale granulometry based on opening with disk-shaped elements (flat and hemispherical) was applied. These structuring element were used to enhance the roundness and the compactness of the red cells, in conjunction with the classical watershed algorithm, while a disk-shaped flat structuring element was used to separate overlapping cells. These methods make use of previous knowledge of the red blood cell structure, as opposed to traditional watershed-based algorithms.

In our study, morphological operators were not used to alter the morphological features (e.g., roundness/compactness) of RBCs (this would have not been useful, since the RBCs were generally found in aggregates), but to perform filtering of cell shapes after thresholding (i.e., pruning of stray pixels related to image noise, and correction of ring-like structures due to the biconcavity of RBCs).

An often cited method for monitoring RBC aggregability under shear was proposed by Chen et al. (1994). In this work, image processing was applied to differentiate aggregates by segmentation of the background. Such operation was not clearly described, though, hence it is not possible to rigorously compare the approach with that hereby presented. In particular, it is not possible to assess if an uneven illumination can impact on the accuracy of the analysis. Anyhow, all of the parameters yielded by the method of (Chen et al., 1994), i.e., projected aggregate area, number of cells per aggregate, aggregate distribution, etc., are also provided by our method. Moreover, the center of mass of single aggregates is also computed, so that the relative trajectory as a function of time can be calculated.

4   CONCLUSIONS

Image analysis was found to be capable to yield useful information on the erythrocyte aggregation and could be used to correlate hemorheological profile of healthy and pathological subjects with RBCs flow behaviour. Enhanced RBC aggregation has been observed and implicated in the pathophysiology of many pathological states (cardiovascular diseases, sickle cell disease, thalassemia) (Baskurt, 2007), so the determination of the aggregation – disaggregation patterns of RBCs, by means of the proposed method, might be a useful tool to evaluate the presence of rheological alterations affecting the microcirculation.

REFERENCES

Baskurt OK. Mechanisms of blood rheology alterations. O.K. Baskurt et al. (Eds) 2007. Handbook of Hemorheology and Hemodynamics. Amsterdam. IOS press pp.170–190.
Chen S, Barshtein G, Gavish B, Mahler Y, Yedgar S. 1994. Monitoring of red blood cell aggregability in a flow-chamber by computerized image analysis. Clinical Hemorheology, 14, 497–508.
Di Ruberto C, Dempster A, Khan S, Jarra B. 2002. Analysis of infected blood cell images using morphological operators. Image and Vision Computing 20:133–146
Dougherty ER 1992. An Introduction to Morphological Image Processing. ISBN 0-8194-0845-X
Gering E and Atkinson CT. 2004. A Rapid Method for Counting Nucleated Erythrocytes on Stained Blood Smears by Digital Image Analysis. Journal of Parasitology, 90 (4): 879–881.
Leong FJWM, Brady M, McGee J. 2003. Correction of uneven illumination (vignetting) in digital microscopy images. J Clin Pathol, 56:619–621.
Neu, B. Red blood cell aggregation. O.K. Baskurt et al. (Eds) 2007. Handbook of Hemorheology and Hemodynamics. Amsterdam. IOS press pp. 114–136.

Computational Modelling of Objects Represented in Images – Di Giamberardino et al. (eds)
© 2012 Taylor & Francis Group, London, ISBN 978-0-415-62134-2

Flow patterns in aortic circulation following Mustard procedure

G. D'Avenio
Department of Technology and Health, Istituto Superiore di Sanità, Rome, Italy

S. Donatiello & A. Secinaro
Pediatric Hospital "Bambino Gesù", Rome, Italy

A. Palombo
Department of Clinical and Molecular Medicine, 'Sapienza' University of Rome, Rome, Italy

B. Marino
Department of Pediatrics, 'Sapienza' University of Rome, Rome, Italy

A. Amodeo
Pediatric Hospital "Bambino Gesù" , Rome, Italy

M. Grigioni
Department of Technology and Health, Istituto Superiore di Sanità, Rome, Italy

ABSTRACT: The transposition of the great arteries (aorta and main pulmonary artery) is a pathological condition which is faced with negative outcome in the very first years of life, if left untreated. A surgical procedure for restoring a more physiological situation is the Mustard procedure, which allows total correction of transposition of the great vessels, by means of surgical redirection of caval blood flow to appropriate atria. In the Mustard procedure, pulmonary and systemic circulation are driven by the left and right ventricle, respectively. The present study is meant to characterize, by means of computational simulation, the aortic hemodynamics in patients operated on with a Mustard procedure, in order to highlight whether the native connection of the aorta to the right ventricle can cause particular concerns, with respect to the physiological case.

1 INTRODUCTION

A not negligible series of cardiac pathologies of the newborn implies the transposition of the great arteries (aorta and main pulmonary artery). Since the functional difference between right and left ventricle, the right ventricle, which in such a case is destined to sustain the systemic circulation, will be overloaded. This type of circulation is thus faced with negative outcome in the very first years of life, if left untreated.

A surgical procedure for restoring a more physiological situation is the Mustard procedure, which allows total correction of transposition of the great vessels, by means of surgical redirection of caval blood flow to the left atrium and then to the left ventricle, which pumps the deoxygenated blood to the lungs (it must be recalled that in a normal heart the deoxygenated blood is pumped instead by the right ventricle into the lungs). In the Mustard procedure, pulmonary and systemic circulation are driven by the left and right ventricle, respectively.

The present study is meant to characterize the aortic hemodynamics in patients operated on with a Mustard procedure, in order to highlight whether the native connection of the aorta to the right ventricle can cause particular concerns, with respect to the physiological case.

2 MATERIALS AND METHODS

MRI images of a patient operated on with Mustard procedure at the Bambino Gesù Pediatric Hospital, Rome, were processed with Mimics software (Materialise, Belgium), allowing the construction of a 3D model of the aorta, from the aortic root downstream to the abdominal aorta, including the supra-aortic arteries. The model was built by means of suitable (upper and lower) thresholds for intensity of the image.

The 3D model was subsequently imported into 3-Matic software (Materialise, Belgium), in order to prune the parts not belonging to the aortic circulation.

A volume mesh of about 2.75 millions tetrahedral was then created and imported in Ansys Fluent 12.1, a flexible general-purpose computational fluid dynamics software package used to model flow, turbulence, heat transfer, and other physical phenomena (Carroll et al., 2011; Redaelli et al., 2004).

Suitable boundary conditions were set at the input and outputs of the model. The flow (at the inlet section) and pressure (at the outflow sections, i.e., the cerebral arteries and the descending aorta) values were imposed according to the typical values in the subjects at the age of the candidate for the Mustard procedure.

The flow motion was described by the principle of momentum conservation, expressed by the Navier-Stokes equations:

$$\frac{\partial U_i}{\partial t} + U_j \frac{\partial U_i}{\partial x_j} = \frac{1}{\rho} \frac{\partial \sigma_{ij}}{\partial x_j} \qquad (i, j = 1,2,3) \quad (1)$$

where

$$\sigma_{ij} = -P\delta_{ij} + \mu \left(\frac{\partial U_i}{\partial x_j} + \frac{\partial U_j}{\partial x_i} \right) \qquad (2)$$

is the stress tensor, and by the continuity equation, representing the principle of mass conservation:

$$\frac{\partial U_i}{\partial x_i} = 0 \qquad (3)$$

Steady, laminar flow was assumed in the calculation. The blood was modeled as a Newtonian fluid, with 1.06 g/cc density and 3.5 cP dynamic viscosity.

We used the PISO scheme as solution method. The Pressure-Implicit with Splitting of Operators (PISO) pressure-velocity coupling scheme is a non-iterative method for solving the implicity discretised, time-dependent, fluid flow equations (Issa, 1986). This technique allows to decouple operations on pressure from those on velocity. At each time-step, the procedure yields solutions which approximate the exact solution of the difference equations.

The PISO scheme is especially recommended for transient calculations, but it was also demonstrated that it is stable for fairly large time steps, which renders it useful for steady-state calculations as well (Issa et al., 1986). Its usefulness is even more appreciable in case of steady-state simulations with distorted meshes. Actually, the unstructured tetrahedral mesh provided by the 3D reconstruction of the aortic model from MRI images was somewhat distorted, hence the PISO scheme allowed an improvement over faster but less accurate algorithms (e.g., SIMPLE).

As for pressure interpolation, the PRESTO! Scheme was used, as recommended for flows in strongly curved domains. The PRESTO! (PREssure STaggering Option) scheme uses the discrete continuity balance for a "staggered" control volume about the face to compute the "staggered" pressure (face pressure).

A custom software was written in the Matlab environment, in order to process the velocity and pressure data yielded by the CFD analysis for each section, and saved by the user in an ASCII file. The software enables

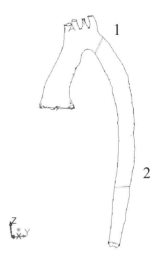

Figure 1. Outline of the mesh used in the CFD simulations. The borders of sections 1 and 2 are depicted.

to do a suitable transformation of the common coordinate system, in order to decompose the velocity vectors in two components, parallel and perpendicular to the given section. This feature enabled, in particular, an easy calculation of secondary flows.

Similarly to previous studies (Wood et al., 2001), we chose to evaluate the results in two sections, immediately downstream of the left subclavian artery (LSA) and in a section positioned in the descending aorta, at 14 cm below the center of the latter. These two sections will be denoted henceforth as section 1 and 2, respectively.

Figure 1 shows the outline of the mesh for CFD simulation, with the 2 sections referred to above.

Moreover, we evaluated the results also on a sagittal plane. The latter was defined here as the plane passing through the centers of section 1 and 2, and parallel to the vertical direction.

3 RESULTS

Figure 2 shows the static pressure distribution in section 1, downstream of the LSA. The inner part of the bend is towards the bottom part of the figure. The section is viewed form the upstream side, and the patient's right-hand side is towards observer's left.

The pressure values are higher towards the convex part of the aortic arch, the pressure difference in the section is 0.66 mmHg.

Figure 3 shows the axial velocity with respect to the plane of section 1, downstream of LSA.

Referring again to section 1, fig. 4 shows the in-plane velocity vectors, as seen from the upstream side.

Fig. 5 reports the axial velocity at section 2, close to the exit of the fluid domain, as seen from above. The antero-posterior direction is along y, whereas patient's right hand side is towards observer's left.

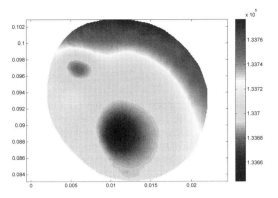

Figure 2. Static pressure [dyn/cm^2] in section 1, downstream of LSA.

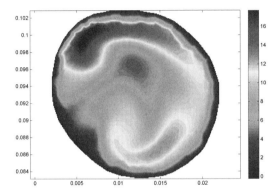

Figure 3. Axial velocity [cm/s] (positive towards the descending aorta) in section 1, downstream of LSA.

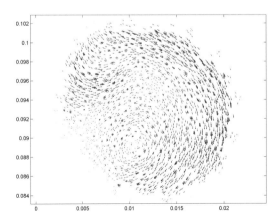

Figure 4. In-plane velocity in section 1, downstream of LSA.

The static pressure distribution in the sagittal plane is shown in fig. 6. It was not possible to define a plane intersecting all cerebral arteries, due to the anatomical constraints. It can be seen that there are pressure minima just ahead of and after the aortic bend, which is a condition promoting flow separation. This notwithstanding, the axial velocity did not show flow reversal, even though a remarkable low flow velocity zone was

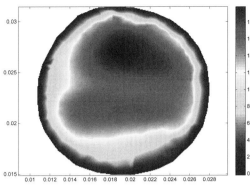

Figure 5. Axial velocity [cm/s] (positive towards the descending aorta) in section 2.

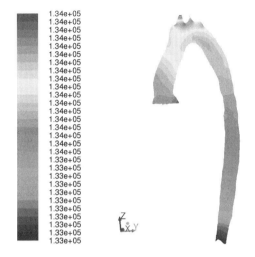

Figure 6. Static pressure [dyn/cm^2] in the sagittal plane.

observed close to the concave side of the aortic arch (Fig. 7).

4 DISCUSSION

The pressure distribution downstream of the LSA (fig. 2) is very similar to the result in Wood et al (2001). Generally, the results hereby presented share the characteristic patterns of physiological aortic flow, even though a zone of separated flow was not observed, contrary to, e.g., Kilner et al. (1993) and Wood et al. (2001). These studies, however, reported results relative to flow in the unstable regime (late systole), when flow separation and, possibly, disturbances can be expected. In the present study a stationary condition was instead considered, which should be referred more realistically to peak systole, at minimal inertial effects (in fact, at peak systole the derivative of flow rate is zero). As already recalled in Wood et al (2001), retrograde flow appears in zones of initially slower antegrade velocities, under the retarding pressure gradient in late systole, therefore the low velocity zone

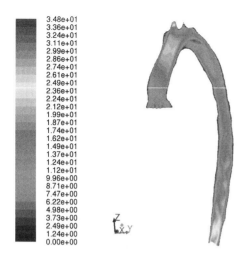

Figure 7. Velocity magnitude [cm/s] in the sagittal plane.

at the bend of the arch in fig. 7 can be associated to the already investigated flow reversal zone in the physiological aorta (Kilner et al. 1993).

It must be underlined that many of the characteristic signatures of aortic flow can be appreciated. For instance, secondary flow is clearly seen in section 1 (fig. 4). The rotation is mainly right-handed, as already observed by Kilner et al. (1993) for the ascending and upper aortic arch in late systole.

This clockwise rotation can also explain the rotation of the location of peak axial velocity, from section 1 to section 2, which can be seen in figs. 3 and 5, respectively.

Recent investigations with time-resolved three-dimensional (3D) phase-contrast magnetic resonance imaging (Hope et al., 2010) enabled to compare different anatomic structures in the ascending aorta with regard to the associated secondary flows. In particular, the incidence of (mainly right-handed) secondary flows was found to be more pronounced in subjects with dilated ascending aorta, compared with subjects presenting a normal aorta. A similar difference was also found in patients with a dilated ascending aorta, comparing patients with a bicuspid valve to those with a tricuspid valve. These findings suggest that the features of aortic flows may be indicative of pathological states, and possibly useful for predicting the evolution of the patient's conditions in the long term. More generally, the identification of aortic flow structures can

be a tool for the classification of the patient's state, allowing the evaluation of the effectiveness of, e.g., a surgical reconstruction such as the Mustard operation.

5 CONCLUSIONS

The results of the study indicate that the aortic flow after the Mustard procedure, even though driven by the right ventricle, retains the properties of the physiological aortic flow. In particular, the lower curvature of the ascending part of the aorta, immediately downstream of the ventricle, in comparison to the physiological case (i.e., ascending aorta connected to the left ventricle), was not found to be associated to significant differences with respect to the previously characterized aortic flow patterns in the physiological case, confirming the effectiveness of the operation on the hemodynamical point of view.

REFERENCES

Carroll G.T., McGloughlin T.M., Burke P.E., Egan M., Wallis F., Walsh M.T. 2011. Wall shear stresses remain elevated in mature arteriovenous fistulas: a case study. J Biomech Eng. 133(2):021003.

Hope M.D., Hope T.A., Meadows A.K., Ordovas K.G., Urbania T.H., Alley M.T., Higgins C.B. 2010. Bicuspid aortic valve: four-dimensional MR evaluation of ascending aortic systolic flow patterns. Radiology. 255(1):53–61.

Issa R.I. 1986. Solution of the implicitly discretised fluid flow equations by operator-splitting. Journal of Computational Physics. 62(1):40–65

Issa R.I., Gosman A.D., Watkins A.P. 1986. The computation of compressible and incompressible recirculating flows by a non-iterative implicit scheme. Journal of Computational Physics. 62(1):66–82

Kilner P.J., Yang G.Z., Mohiaddin R.H., Firmin D.N., Longmore D.B. 1993. Helical and retrograde secondary flow patterns in the aortic arch studied by three-directional magnetic resonance velocity mapping. Circulation 88(5 Pt 1):2235–47.

Redaelli A., Bothorel H., Votta E., Soncini M., Morbiducci U., Del Gaudio C., Balducci A., Grigioni M. 2004. 3-D simulation of the St. Jude Medical bileaflet valve opening process: fluid-structure interaction study and experimental validation. J Heart Valve Dis. 13(5):804–13.

Wood N.B., Weston S.J., Kilner P.J., Gosman A.D., Firmin D.N. 2001. Combined MR imaging and CFD simulation of flow in the human descending aorta. J Magn Reson Imaging 13(5):699–713.

Computational Modelling of Objects Represented in Images – Di Giamberardino et al. (eds)
© 2012 Taylor & Francis Group, London, ISBN 978-0-415-62134-2

A new workflow for patient specific image-based hemodynamics: Parametric study of the carotid bifurcation

M.E. Biancolini
Università di Tor Vergata, Rome, Italy

R. Ponzini
CILEA, Milan, Italy

L. Antiga
Orobix, Bergamo, Italy

U. Morbiducci
Politecnico di Torino, Turin, Italy

ABSTRACT: Engineering applications involving biological fluids have highly transversal requirements in terms of domain definition from clinical images, complex flow conditions, rheological properties of fluids, structure motion and deformation, visualization and post-processing of the results. For these reasons, a properly tailored computer-aided–engineering workflow represents an elective environment where to perform realistic hemodynamics studies. Nowadays a large part of the technological requirements needed to tackle these problems in a computational environment are already available in open source and/or commercial codes. Nevertheless, success still strongly depends on technical knowledge and best practice. In other words, the design of the workflow must be translated into a stable and usable framework. Here we present a new workflow based on three fully validated software used to effectively fulfill the requirements related to hemodynamics: the **Vascular Modeling Toolkit (VMTK)** for the pre-processing step (i.e., from clinical images-to-anatomic models); the mesh morphing tool **RBF Morph** to impose changes to the vascular anatomy; **Ansys Fluent** as solver of the governing equations of fluid motion. As a first test case we focused our attention on the study of a realistic model of carotid bifurcation, where geometrical factors such as bifurcation angle and the bulb flare are deformed starting from image-based models. In perspective the herein proposed workflow could be a powerful tool supporting image-based surgical planning optimization in several arterial districts.

1 INTRODUCTION

1.1 Background

Engineering applications involving biological fluids have highly transversal requirements in terms of domain definition from clinical images, complex flow conditions, fluid rheological properties, structure motion and deformation, visualization and post-processing of the results.

In particular, in the last decade, it has been demonstrated that the coupling of medical imaging and computational fluid dynamics (CFD) allows to calculate highly resolved four-dimensional blood flow patterns in anatomically realistic models of the cardiovascular district, thus obtaining information, for example on quantities such as the distributions of the friction and pressure forces at the luminal surface, which are difficult to be measured in vivo [Boussel et al., 2009]. In detail, image-based CFD is becoming a powerful tool to quantify and classify local hemodynamic conditions, thus enabling the study of disease mechanisms [Friedman et al., 2010; Taylor and Steinman, 2010]. However, the increasing reliance on CFD for hemodynamic studies necessitates a close look at the various assumptions required by in silico modeling. For example, much effort has been spent to assess the sensitivity to assumptions regarding boundary conditions [Moyle et al., 2006; Wake et al., 2009; Morbiducci et al., 2010; Hoi et al., 2010; Spilker and Taylor, 2010], blood rheology [Morbiducci et al., 2011], geometric uncertainties [Thomas etal., 2003], vascular compliance [Perktold and Rappitsch, 1995]. Moreover, great effort has been put in studying the links between disturbed flow, responsible for the onset and development of atherosclerosis and other vascular wall pathologies [Ku et al., 1983] and geometric features such as tortuosity of the vessel, bifurcation angle, flaring etc [Lee et al., 2008; Bijai et al., 2012].

In this scenario, linking fluid structures to wall deformations is a challenging task, due to the need to (1) have detailed knowledge of the highly complex mechanical behavior of the arteries or (2) impose realistic vessel wall deformation/motion, as extracted from clinical imaging, when only the hemodynamics is of interest.

In this work we present a new workflow based on three fully validated software used to effectively fulfill the requirements related to hemodynamics:

- the **Vascular Modeling Toolkit (VMTK)** for the pre-processing step. VMTK can take clinical images as an input and give anatomic models of cardiovascular districts as output;
- the mesh morphing tool **RBF Morph** to impose changes to the vascular anatomy;
- **Ansys Fluent** as finite volume-based solver of the governing equations of fluid motion.

As a first test case we focused our attention on the study of a subject-specific carotid bifurcation being this anatomical site of major interest in hemodynamics and its relation to atherosclerosis. The present study is focused on the impact that the shape of the carotid bulb and the bifurcation angle have on the resulting flow patterns, which are thought to be tightly related to the focal development of atheromatous plaques at that site, or could be related to the output of remodeling surgical procedures. The effect of the shape is unrealistically magnified in order to emphasize the potency of the approach.

2 MATERIAL AND METHOD

2.1 The VMTK toolkit

The Vascular Modeling Toolkit is a collection of open-source libraries and tools for 3D reconstruction, geometric analysis, mesh generation and surface data analysis for image-based modeling of blood vessels (see Antiga et al. 2008). It has been designed to provide seamless integration with downstream CFD codes. For the present application, it has been used to segment the carotid bifurcations from clinical Magnetic Resonance (MR) images, to obtain and characterize the anatomy (centerlines, radius, bifurcation geometry, branch geometry, section areas, curvature, tortuosity), to generate a suitable triangulated, superficial mesh for CFD and export it directly to Fluent (.msh) format.

2.2 The RBF Morph add-on

RBF Morph (see Biancolini 2011) is a morpher code that combines a very accurate control of the geometrical parameters with an extremely fast mesh deformation, fully integrated in the CFD solving process. RBF Morph is the meeting point between state-of-the-art scientific research and top-level industrial needs.

For the present application it has been used to impose clinically relevant modifications on the carotid bifurcation geometry such as changes on relative internal/external angle between internal carotid artery (ICA) and external carotid artery (ECA), named ICA/ECA angle, deformation of the carotid bulb and presence of a stenosis at the ICA.

2.3 Ansys Fluent solver

The finite volume method was applied to solve the governing equations of the fluid motion. The general purpose CFD code Fluent ANSYS12 (ANSYS Inc., USA) was used on mesh-grids generated using the Gambit mesh generator software: the fluid domain was divided into about $1.40 \cdot 000$ tetrahedral cells. In order to solve the nonlinear system of matrix equations derived from the discretization of the flow equations on the computational grid, a second order upwinding method was used to obtain the solution at each time step of the time-dependent problem, respectively. The backward Euler implicit time integration scheme was implemented with a fixed time increment (time step equal to 2 ms). Fluid was modelled as homogeneous, isotropic and Newtonian (density equal to $1060 \, \text{kg} \, \text{m}^{-3}$, dynamic viscositiy equal to 3.5 cP). As for the boundary conditions, a measured flow rate waveform was applied at the inlet section of the model in terms of flat velocity profile and a fixed 60/40 flow split repartition was imposed between ICA and ECA outlet sections (exhaustive details can be found in Morbiducci et al 2010).

2.4 Changes to shape of the carotid bifurcation

In order to perform a parametric study of the carotid bifurcation, in the present application, starting from the acquired carotid bifurcation, relevant modifications have been considered. Mesh morphing allows to continuously change the shape of the vessels, two specific shape modifications are herein explained in detail: the first one consists in the change on external angle (ECA angle), the second one consists in the deformation of the carotid bulb shape.

2.5 Parameter setup for vascular morphing

RBF Morph allows to set-up and store several shape modifiers. It's important to notice that only the information required for mesh updating (i.e. RBF coefficients) are stored and there is no need to save the morphed mesh. This approach makes the CFD model parametric; the desired set of amplifications for all the shape modifications can be applied at calculation stage just before the iterations.

The set-up of one of the two shape modifiers (ECA angle) is described in detail. A cylinder is used to limit the morphing action to the part of the model that has to be bent. Points at the branch root are extracted as surface points defining all the points on the wall and an auxiliary selection cylinder (with the outside option). The bending of the vessel is imposed applying a rigid rotation to all the points on the ECA outlet section. Extracted RBF points are represented in Figure 2 where the prescribed movement is highlighted.

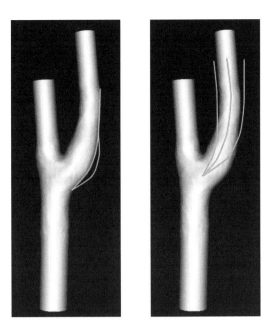

Figure 1. Addressed shape modifications: change of carotid bulb shape (left), changing of the ECA angle (right).

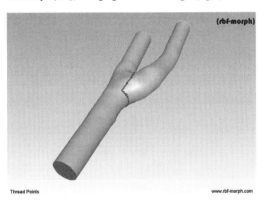

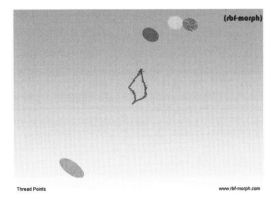

Figure 2. RBF Morph set-up for ECA angle variation. A zero movement is imposed to the red points at the branch root whilst a rotation around the branch is imposed to the points at the ECA surface (top). Rotation axis in blu and a preview of the ECA points in the rotated configuration (bottom).

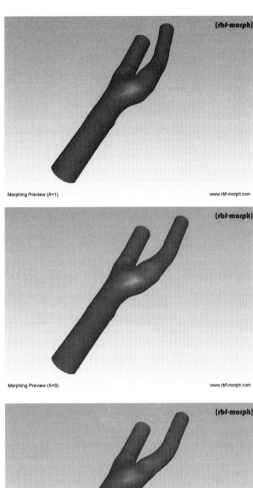

Figure 3. The effect of shape modification (ECA angle) is previewed at 3 values of the amplification (−1,0,1).

It is important to notice that even if a rigid rotation produces a non-linear deformation field, the software has the ability to deal with it and so even rotation are properly amplified with respect to the baseline value used for set-up.

The effect of shape modifier is clearly represented in Figure 3 where various amplifications are explored (all previewed configuration were also checked for the quality of resulting volume mesh). Whilst a preview is advisable at set-up stage the best practice is to morph the model just before the CFD calculation; the morphing module works in Fluent parallel.

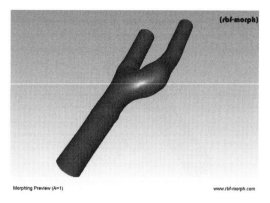

Morphing Preview (A=1)   www.rbf-morph.com

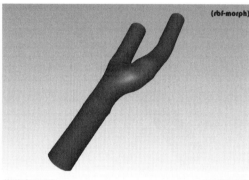

Morphing Preview (A=0)   www.rbf-morph.com

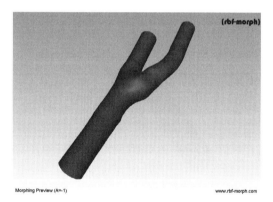

Morphing Preview (A=-1)   www.rbf-morph.com

Figure 4.   The effect of shape modification (bulb shape) is previewed at 3 values of the amplification ($-1,0,1$).

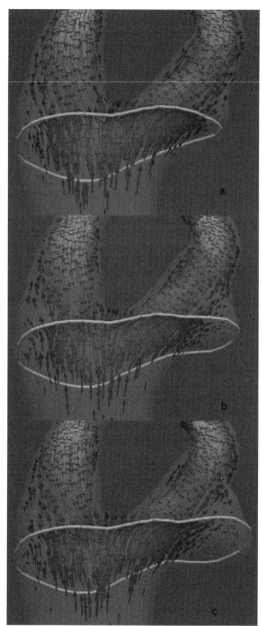

Figure 5.   Effect of the increasing bulb dimension. The bulb dimension is increased from panel (a) to (c) and accordingly the velocity flow patterns show relevant changes highlighted by the vectors distribution.

The second shape modifier is defined limiting the action of the morpher with a sphere surrounding the carotid bulb. A second sphere is used to select vessel points (with outside option); this allows to define points at the boundary of the bulb area that are fixed.

The bulb shape is then controlled adding individual points on the bulb surface, each one with a movement normal to the surface defined so that the combined action gently deform the bulb surface in a meaningful fashion. The action of the shape modifier is represented in Figure 4.

3   RESULTS

The new parametric model of the carotid bifurcation allows to investigate several new shapes. The two shape modifiers can be combined; other shape modifications (for instance vessel bending in other planes) can be easily added and combined as well.

The case of bulb shape is here presented to demonstrate how the parametric model can be exploited.

Three different amplifications have been considered (see Figure 5). The bulb shape has a relevant effect on the shape of the flow field. Qualitative and quantitative hemodynamics parameters could be extracted via user-defined functions (UDF) to understand how they are related to the shape parameters, as in [Morbiducci et al., 2010; Morbiducci et al., 2011].

4 DISCUSSION AND FUTURE WORKS

Here we show that thanks to the proposed new workflow, a comprehensive fluid dynamics study of a patient-specific carotid bifurcation with several relevant anatomical modifications can be performed successfully. The workflow takes clinical images as input and it is capable to generate and visualize flow fields in reshaped vascular districts. The power, the usability and the full level of integration demonstrated by the three computational bricks, together with the possibility of exploiting their features in High Performance Computing environments make this new workflow very attractive for future applications on a wide range of clinically relevant hemodynamics problems. The parametric nature of the model allows an easy integration of the workflow with additional computational tools (modeFRONTIER, Ansys DesignXplorer) for shape optimization problems, with particular reference to virtual testing of surgical connections, virtual surgical training/planning and medical devices design.

REFERENCES

Antiga L, Piccinelli M, Botti L, Ene-Iordache B, Remuzzi A, Steinman DA. *An image-based modeling framework for patient-specific computational hemodynamics*. Med Biol Eng Comput. 2008 Nov;46(11):1097-112. Epub 2008 Nov 11. Review. PubMed PMID: 19002516.

Biancolini, M. E., 2011 *Mesh Morphing and Smoothing by Means of Radial Basis Functions (RBF): A Practical Example Using Fluent and RBF Morph*, Handbook of Research on Computational Science and Engineering: Theory and Practice, IGI Global, ISBN13: 9781613501160.

Bijari, P.B., Antiga, L., Gallo, D., Wasserman, B.A., Steinman, D.A. *Improved prediction of disturbed flow via hemodynamically-inspired geometric variables*. Journal of Biomechanics 2012 (in press).

Boussel, L., V. Rayz, G. Acevedo-Bolton, M. T. Lawton, R. Higashida, W. S. Smith, W. L. Young, and D. Saloner. *Phase-contrast magnetic resonance imaging measurements in intracranial aneurysms in vivo of flow patterns, velocity fields, and wall shear stress: comparison with computational fluid dynamics*. Magn. Reson. Med. 61(2):409–417, 2009.

Ku, D.N., Giddens, D.P., Zarins, C.K., Glagov,S., 1985. *Pulsatile flow and atherosclerosis in the human carotid bifurcation. Positive correlation between plaque location and low and oscillating shear stress*. Arteriosclerosis 5, 293–302.

Lee, S.W., Antiga, L., Spence, J.D., Steinman, D.A., 2008. *Geometry of the carotid bifurcation predicts its exposure to disturbed flow*. Stroke 39, 2341–2347.

Morbiducci et al (2010). *Quantitative analysis of bulk flow in image-based hemodynamic models of the carotid bifurcation: The influence of outflow conditions as test case* (2010). Annals of Biomedical Engineering.

Morbiducci, U., D. Gallo, D. Massai, R. Ponzini, M. A. Deriu, L. Antiga, A. Redaelli, and F. M. Montevecchi, 2011. *On the importance of blood rheology for bulk flow in hemodynamic models of the carotid bifurcation*. Journal of. Biomechanics 44:2427–2438.

Moyle, K.R., Antiga, L., Steinman, D.A., 2006. *Inlet conditions for image-based CFD Models of the carotid bifurcation:is it reasonable to assume fully developed flow?* Journal of Biomechanical Engineering 128, 371–379.

Perktold, K., Rappitsch, G., 1995. *Computer simulation of local blood flow and vessel mechanics in a compliant carotid artery bifurcation model*. Journal of Biomechanics 28, 845–856.

Spilker, R.L., Taylor, C.A., 2010. *Tuning multidomain hemodynamic simulations to match physiological measurements*. Annals of Biomedical Engineering 8, 2635–2648.

Taylor, C.A., Steinman, D.A., 2010. *Image-based modeling of blood flow and vessel wall dynamics: applications, methods and future directions*. In:Proceedings of the Sixth International Bio-Fluid Mechanics Symposium and Workshop, Pasadena, California, March 28–30, 2008. Annals of Biomedical Engineering 38, 1188–1203.

Thomas, J.B., Milner, J.S., Rutt, B.K., Steinman, D.A., 2003. *Reproducibility of image-based computational fluid dynamics models of the human carotid bifurcation*. Annals of Biomedical Engineering 31, 132–141.

Wake, A.K., Oshinski, J.N., Tannenbaum, A.R., Giddens, D.P., 2009. *Choice of in vivo versus idealized velocity boundary conditions influences physiologically relevant flow patterns in a subject-specific simulation of flow in the human carotid bifurcation*. Journal of Biomechanical Engineering 131, 021013.

Computational Modelling of Objects Represented in Images – Di Giamberardino et al. (eds)
© 2012 Taylor & Francis Group, London, ISBN 978-0-415-62134-2

Radial Basis Functions for the image analysis of deformations

M.E. Biancolini & P. Salvini

Università di Tor Vergata, Rome, Italy

ABSTRACT: In this work we tested the reliability of Radial Basis Function theory (RBF) for the image analysis of deformations. RBF conformal transformation allows to map the original image onto the deformed one using only a few RBF centres to control the field. The backward transformation (i.e. with centres located at target positions and a displacement field that move them in the original positions) allows to obtain a deformed image from the original one using a few control points. A proper points' placement allows to generate a parametric morphed image that can carefully overlap the actual deformed one. Whilst the overall procedure consists in the minimisation of the matching error acting on the control parameters (i.e. unknown displacements at control points), in this study we have successfully verified the feasibility of the method feeding the image morpher with the actual displacements at RBF points.

1 INTRODUCTION

1.1 *Background*

In present time electronically controlled sensors are experiencing an astonishing progress. Beside their availability in almost all applications – according to several physical properties, their prices are rapidly decreasing. Among all recently available sensors, digital cameras based on CCD or CMOS technologies are currently very precise (if adequate optics are used). They offer high resolution together with a very attractive economical cost. Even if inexpensive optics are used, it is easy to introduce procedures for the calibration of the system so that an optimal precision is reached (Torres & Menéndez 2004).

In the application of digital images to the structural analysis many approaches have been recently proposed (Sutton & et al. 2009). All of them are based on the comparison of sequential frames, taken during the progressive application of loads. Generally, the surfaces are prepared so that speckles images appear on them, by previous coloring, or by laser speckle. Most of the methods proposed are based on the recognition of individual points on the pictures through correlation functions (Sutton & et al. 1991). These methods are very fast but the resulting displacement field can be affected by some discontinuities.

Recently, interesting and smoothed analyses have been conducted with the application of Finite Element mesh on the acquired regions. The identification is thus performed as a whole, ensuring the continuity of the computed displacement field (Sutton & et al. 2009, Amodio & et al. 2005, Besnard & et al. 2006).

However, these methods increase their suitability if they are preceded by algorithms able to identify some

points of the region of interest (considered as reference points). As a consequence of this previous recognition, they can reach convergence even if large displacements are encountered (Flusser & et al. 2009). As a matter of fact, in many applications the applied loads cause a strain field that is accompanied with large displacements of part of the structure, characterized by an almost rigid body-motion. In these cases, the application of some invariants to the pictures (Minotti & et al. 2011), offers a suitable pre-requisite before the search of the internal displacement field.

One of the limits of these methods is therefore related with the high number of controlling points (finite element nodes) that increases computing time and affects the solution convergence.

RBF should in principle be able to reduce considerably the control points, choosing them in an optimal way, while describing the displacement field accurately.

Another important characteristic is the capability to increase the control points in the region where important gradients of displacement are present, so that local approximation can be reduced accordingly.

At the present time the methods proposed are reliable only if plane surfaces are analyzed, but some interesting solutions have been proposed to manage also non plane surfaces, through an independent identification of them, followed by the projection of the solution field, or through three dimensional computer vision techniques (Sutton & et al. 2009).

There are also limits on the strain field maximum values that can be identified; when the differences in displacements of the two frames are too high, it is necessary to apply the analysis on a closer sequence of pictures.

1.2 Presented method

In this study we focus only on the RBF morpher capability to represent the target deformed image using a proper points placement. A synthetic deformed image generated using FEM is used for the test so we known in advance the exact deforming field everywhere.

A small subset of FEM grid nodes is used to define RBF centers; FEM displacement is used at such point to fill the unknown value of the image matching problem.

The inverse method is used to define a conformal $R^2 \to R^2$ transformation, able to map the deformed points' distribution onto the original one. The main advantage of using the inverse transformation is the easiness of interpolation when looking-up on the overall target image. The value at each new pixel position is defined using its position as input in the interpolation field, the resulting inverse transformation retrieves a point in the original image. The color value at this position (that of course is not at a pixel position) is computed averaging the values of 4 neighbor pixels.

2 MATERIAL AND METHOD

2.1 FEM generated reference images

The image is generated using a simple geometry: a square (1m side) with a circular hole (0.2 m radius) is morphed by imposing a rigid motion (0.1 m) to the hole. Boundary conditions are imposed restraining the nodes on the external edges of the square; a constant displacement along the horizontal axis is imposed to all the nodes that belong to the hole.

Original image and target image are represented in Figure 1.

2.2 Radial Basis Function interpolation

Radial Basis Functions (RBF) are a very powerful tool created for the interpolation of scattered data; RBF are successfully used for a wide range of applications including mesh morphing (Biancolini 2011) and image manipulation (Arad & et al. 1994). They are able to interpolate everywhere within the space a function defined at discrete points giving the exact value at original points. The behaviour of the function between the points depends on the kind of radial function adopted that can be fully or compactly supported; some functions require a polynomial correction to guarantee compatibility for rigid modes.

A linear system of equations (of order equal to the number of source point introduced) needs to be solved for the coefficients calculation. Once the unknown coefficients are calculated, the data at an arbitrary point inside or outside the domain (interpolation/extrapolation) is expressed as the summation of the radial contribution of each source point (if the point falls inside the influence domain).

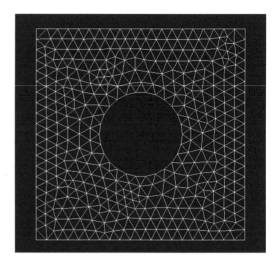

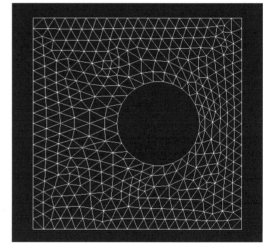

Figure 1. FEM generated images used for the test. Original mesh (top) and deformed mesh (bottom).

An interpolation function composed of a radial basis ϕ and a polynomial h is defined as follows:

$$s(\mathbf{x}) = \sum_{i=1}^{N} \gamma_i \phi(\|\mathbf{x} - \mathbf{x}_i\|) + h(\mathbf{x})$$

The degree of the polynomial has to be chosen depending on the kind of radial function adopted. A radial basis fit exists if the coefficients γ and the weights of the polynomial can be found such that the desired function values are obtained at source points and the polynomial terms gives 0 contributions at source points, that is:

$$s(\mathbf{x}_{k_i}) = g(\mathbf{x}_{k_i}) \quad 1 \le i \le N$$

$$0 = \sum_{i=1}^{N} \gamma_i q(\mathbf{x}_{k_i})$$

362

for all polynomials q with a degree less or equal to that of polynomial h. The minimal degree of polynomial h depends on the choice of the basis function. A unique interpolation exists if the basis function is a conditionally positive definite function. If the basis functions are conditionally positive definite of order $m <= 2$, a linear polynomial can be used:

$$h(\mathbf{x}) = \beta + \beta_1 x + \beta_3 y + \beta_4 z$$

The values for the coefficients γ of RBF and the coefficients β of the linear polynomial can be obtained by solving the system:

$$\begin{pmatrix} \mathbf{M} & \mathbf{P} \\ \mathbf{P}^\mathrm{T} & \mathbf{0} \end{pmatrix} \begin{pmatrix} \gamma \\ \beta \end{pmatrix} = \begin{pmatrix} \mathbf{g} \\ \mathbf{0} \end{pmatrix}$$

$$M_{ij} = \phi\left(\left\|\mathbf{x}_{k_i} - \mathbf{x}_{k_j}\right\|\right) \quad 1 \leq i \quad j \leq N$$

where $\mathbf{g}$ are the known values at the source points, $\mathbf{M}$ is the interpolation matrix defined calculating all the radial interactions between source points and $\mathbf{P}$ is a constraint matrix that arises to balance the polynomial contribution and contains a column of "1" and the x y positions of source points in the others two columns:

$$\mathbf{P} = \begin{pmatrix} 1 & x^0_{k_1} & y^0_{k_1} \\ 1 & x^0_{k_2} & y^0_{k_2} \\ \vdots & \vdots & \vdots \\ 1 & x^0_{k_N} & y^0_{k_N} \end{pmatrix}$$

Radial basis interpolation works for scalar fields. For the conformal transformation each component of the field is interpolated as follows:

$$\begin{cases} v_x = s_x(\mathbf{x}) = \sum_{i=1}^{N} \gamma_i^x \phi\left(\left\|\mathbf{x} - \mathbf{x}_{k_i}\right\|\right) + \beta_1^x + \beta_2^x x + \beta_3^x y \\ v_y = s_y(\mathbf{x}) = \sum_{i=1}^{N} \gamma_i^y \phi\left(\left\|\mathbf{x} - \mathbf{x}_{k_i}\right\|\right) + \beta_1^y + \beta_2^y x + \beta_3^y y \end{cases}$$

The radial basis method has several advantages that make it very attractive. The key point is that being a meshless method it's able to interpolate everywhere regardless the position of points used to generate the field. This means that the control points used for mapping of deformation can be selected everywhere, including special positions (i.e. constrained areas, or areas with displacement known in advance).

It is worthwhile to notice that the closed form interpolator can be differentiated analytically (the order depend on the class of continuity of the radial function adopted) and so the strain map can be recovered from the displacement field.

2.3 RBF setup

Presented study has been conducted using direct fit of RBF by means of a **MathCAD** worksheet (it comes

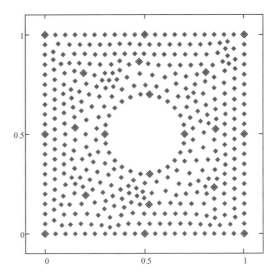

Figure 2. RBF reference points (20 in blue) are selected at FEM nodes grid positions (409 red).

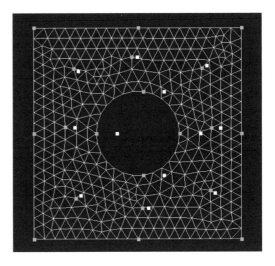

Figure 3. RBF reference points in original and target positions, overlap on the image to be deformed.

with **Lapack** solver for linear systems). This approach allows to evaluate the feasibility of the method very quickly (see Biancolini 2011).

In figure 2 the RBF points selection is shown: 20 of the 409 FEM grid points are used placing them on the boundary of the domain and in the bulk. The RBF points distribution is represented in Figure 3 where each point is plotted in its original and target position onto the original image.

The deformed image is obtained transforming the original one via the RBF mapping field. The position of each pixel of the new image is processed to retrieve the position onto the original image. The value is computed averaging the values at the four surrounding pixels at each location. As already explained in the

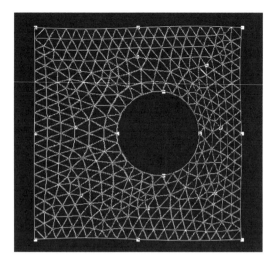

Figure 4. RBF deformed image represented with RBF points in the target position.

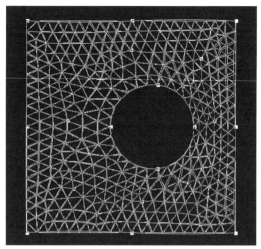

Figure 5. RBF deformed image compared with the target image.

introduction we have selected the inverse transformation (i.e. RBF centers defined in their target positions, and the vector field that map them in the original ones) because it allows to easily explore each pixel of the target image; neighbors interpolation is then implemented on the original image.

3 RESULTS

3.1 Spatial interpolation

The RBF field allows to compute the deformed image. The result is summarized in Figure 4 where the deformed image is represented together with RBF points at their target positions.

A direct comparison of deformed image as computed using RBF field and target image is exposed in Figure 5.

A good agreement is observed comparing the two images; nevertheless a quantitative evaluation of deformation is required in order to estimate the interpolation error of the RBF field. The FEM displacement at each nodal position of the FEM model used to generate the reference images is available.

3.2 FEM field reproduction

The inverse transformation field is available everywhere and so it can be used to calculate the inverse displacement at each nodal positions of the deformed FEM model.

Figure 6 presents a comparison between FEM nodes in deformed positions as computed by FEM solution (i.e. at exact target position in this numerical experiment) and as displaced by the RBF fields; the latter has been obtained computing the inverse displacement

value obtained by RBF at each FEM deformed position and then applying it (with reversed sign) to the original FEM mesh, so that the data are comparable.

The same quantities are used for a quantitative evaluation of the error; the inverse RBF displacement, changing the sign, can be compared to the displacement obtained from the FEM solution according to the equation:

$$err = \frac{|\delta_{RBF} - \delta_{FEM}|}{|\delta_{FEM}|} \cdot 100$$

Obtained error distribution is mapped onto the model surface using the surface plot represented in Figure 7. As expected, higher errors are observed in the constrained area where displacement are small.

About 75% of the point population have an error lower than 20% whilst a poor interpolation is achieved for the remaining ones in details:

- 42.8% below 5%
- 17.4% in the range 5%–10%
- 14.9% in the range 10%–20%
- 13.9% in the range 20%–50%
- 11.0% over 50%

4 DISCUSSION AND CONCLUSION

The use of RBF for deformation analysis by image processing has been explored in this study. The simplified experiment with RBF control points displacements known in advance allows to demonstrate that even a coarse point distribution (20 points in the presented results) allows to properly predict the target image, getting an error distribution (as verified by the FEM field used to generate the target) that is lower than 20% for about 75% of the nodal points of the FEM model.

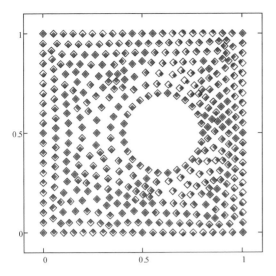

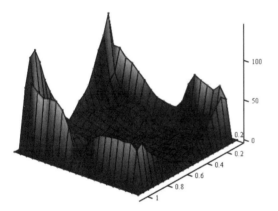

Figure 6. FEM nodes in deformed position as computed by FEM (open blue diamonds) and as interpolated using RBF field (red diamonds).

Figure 7. Error map: RBF displacement vs FEM displacement.

RBF method is meshless and extra points can be easily added in further steps to improve the results. The drawback is the cost of RBF using large dataset.

In order to use the proposed approach in an automated algorithm suitable for the measurement of the deformation field by image processing, a meaningful approach can be hierarchical; i.e. a first step is conducted using only a few RBF control points to get a first estimation of the field (in this case the number of unknown coefficients is twice the RBF centers number); the values of unknown coefficients are then calculated minimizing in some way the matching error (i.e. the difference between the target image and the deformed one); in further steps extra centers can be added using the field of the previous step to have proper guess values to be used in the minimization loop.

Among ongoing activities on this topic we remark: improving the field quality adding extra points (outside optimization loop) in constrained areas, the automation of image matching; the analysis of experimental images; the evaluation of the strain field using RBF differentiation and its comparison to the FEM one.

REFERENCES

Amodio D., Broggiato G.B., Salvini P., *Finite Strain Analysis by Image Processing: Smoothing Techniques*, Strain, Vol. 31, n. 3, 1995, pp. 151–157.

Arad, N., Dyn, N., Reisfeld, D., Yeshurun, Y. *Image warping by radial basis functions: application to facial expressions*, CVGIP: Graphical Models and Image Processing, 56 (1994), pp. 161–172

Besnard, G., Hild, F., Roux, S. *Finite-Element Displacement Fields Analysis; from Digital Images: Application to Portevin–Le Châtelier Bands*, Experimental Mechanics (2006) 46: 789–80.

Biancolini, M. E., *Mesh Morphing and Smoothing by Means of Radial Basis Functions (RBF): A Practical Example Using Fluent and RBF Morph* in Handbook of Research on Computational Science and Engineering: Theory and Practice, IGI Global, ISBN13: 9781613501160.

Flusser, J., Suk, T., Zitová, B. *Moments and Moment Invariants in Pattern Recognition*, John Wiley &Sons, (2009).

Minotti M., Marotta E., Salvini P., *Determinazione del campo di grandi spostamenti tramite l'elaborazione di immagini digitali*, Atti Convegno XL AIAS, 7–10 Settembre, Palermo, 2011.

Sutton M.A., Orteu J.J., Shreier H.W., *Image Correlation for Shape, Motion and Deformation Measurements*, ISBN 0387787461, Springer, 2009.

Sutton M.A., Turner J.L., Chao Y.J., Bruch A., Chae T.L., *Full field representation of discretely sampled surfaces deformation for displacements and strain analysis.* Experimental Mechanics, 31 (2): 168–177, 1991.

Torres, J., Menéndez, J.M. *A Practical algorithm to correct geometrical distorsion of image acquisition cameras*, International Conference on Image Processing, Singapore, 2004.

Computational Modelling of Objects Represented in Images – Di Giamberardino et al. (eds)
© *2012 Taylor & Francis Group, London, ISBN 978-0-415-62134-2*

Movement analysis based on virtual reality and 3D depth sensing camera for whole body rehabilitation

M. Spezialetti, D. Avola & G. Placidi
Department of Health Sciences, University of L'Aquila, L'Aquila, Italy

G. De Gasperis
Department of Electrical and Information Engineering, University of L'Aquila, L'Aquila, Italy

ABSTRACT: Movement analysis is a complex technique to measure and evaluate the movements of the human body. It can be adopted to support various activities, including rehabilitation. Over the last decade, many systems for capturing human movements have been developed. The most advanced motion capturing systems use image processing techniques through the analysis of live video streams collected from a set of video cameras. These systems are very expensive, need to be installed and used by professionals. Moreover, they are not intended for patient rehabilitation, at home. Rehabilitation can be facilitated and accelerated by using virtual reality and movement analysis assisting systems.

This paper describes a rehabilitation system for domestic whole body rehabilitation. It is composed by a virtual reality environment where spatial information of the human body are collected by a 3D depth sensing camera and inserted into the virtual world. The system is designed to perform data analysis by an off-line motion tracking algorithm. A complete set of rehabilitation exercises have been also implemented. A trial exercise is illustrated and discussed.

1 INTRODUCTION

Movement analysis of the human body is a growing field to measure movements and to evaluate their qualitative response to interactive stimuli. These techniques can be classified either as haptic-interfaces based or as vision based. In the first case, movements are computed by applying electro-mechanical devices on specific parts of the body being analyzed. In the second case, the same type of information is calculated by processing video streams used to capture human body movements. Vision based techniques allow more natural user experience. Moreover, they have many practical advantages compared to the haptic-interface devices, such as usability, portability, and generality.

The vision based methods are used to address different application domains, such as human-computer interaction (Moeslund et al. 2006), computer-based surveillance (Papalambrou et al. 2011), biomechanics or bioengineering (Chaczko et al. 2011). Often, vision based movement analysis systems are very expensive and their usage and set-up can result very complicated for non-professional users. In addition, they are not designed to perform rehabilitation activity in domestic environment.

In recent years, there have been developed several high-quality video dedicated capturing systems, such as *BTS Gemini* (BTS 2012) and *T-Series* (Vicon 2012). Movement analysis is performed using high-speed cameras and skin markers to track the changing positions and orientation of body segments (kinematic). Floor mounted force plates are used to measure the magnitudes and directions of the resulting forces exerted on the ground (dynamics), and surface electromyography (sEMG) electrodes are used to record neuromuscular activations (Winter 1990). These systems can need different technical requirements for specific application domains, such as for research, sport, clinical setting. For this reason customized data collection protocols and data reduction models are often implemented.

Rehabilitation is a repetitive and long duration therapy that is commonly divided in two categories: physical rehabilitation (PR) and functional rehabilitation (FR). The exercises for PR are mainly based on the force application to recover patient motion skills where muscles and joints are stressed. FR exercises are intended to recover dexterity and precision in movements, necessary for work and daily life. Each exercise must have different difficulty ranges defined by modifying some parameters. Traditional rehabilitation is often one-to-one (i.e. one therapist for one patient) or, in some case, many-to-one (i.e. many therapists for one patient): the costs are high and there is no way to monitor the activity the patient performs autonomously at home. Moreover, the verification is subjective and difficult: spatial and temporal data are missing. For these reasons, rehabilitation can be facilitated and accelerated if movement analysis and virtual reality based assisting systems are used (Ustinova et al. 2011, Spezialetti 2012). The advantages would be:

- the patient is tempted by an exercise that looks like a game;
- the patient can interact with a virtual environment, thus making possible to simulate a lot of stimulating situations;
- patient movements can be monitored and accurately analyzed;
- it is possible to calculate a series of numerical parameters (e.g. duration, repetition frequency, precision, score);
- exercise difficulty is objective and can be gradually modified;
- the patient can use the system at home and data can be transmitted and evaluated automatically (e.g. by a server-side application).

This paper describes a video based, virtual reality, rehabilitation system for domestic whole body physical rehabilitation. Different exercises designed like games can be loaded within a virtual reality environment with which a blob-man model, representing the real posture of the user, can interact in real time. The system is designed to analyse data by an off-line motion tracking algorithm. Our system is very simple to install, manage and use and it has been designed according to the current application's usability criteria (Jacko 2011) to improve installation and calibration tasks.

The paper is structured as follows. Section 2 shows the system architecture including the used technologies, the user interface and an exercise implementation. Section 3 summarizes the experimental results. Finally, section 4 concludes the paper.

2 SYSTEM ARCHITECTURE

The architecture of the developed system is based on three leading technologies suitably extended and connected. The first is the OptriCam™ 130 (Optrima™, 2012), a 3D depth sensing camera able to provide a depth map (i.e. 3D information) for the identification of different objects, and their distance from a viewpoint, within the video stream. The second is the iisu™ SDK (iisu™ SDK, 2012), a middleware that allows an easy communication process between the camera and a customised end-user application. The third is the Irrlicht Engine (Irrlicht, 2012), a cross-platform high performance real time 3D engine that allows the rendering of 3D data. Figure 1 summarizes the system architecture.

Our main contribute has been to develop the bridge-interface between iisu™ and Irrlicht Engine including the related movement processing algorithms. Moreover, different rehabilitation exercises have been also implemented.

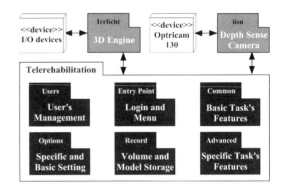

Figure 1. System architecture.

2.1 Supported technologies

The 3D depth sensing camera represents the central point of our system. The possibility of using stereo vision systems (Franchi et al. 2009) was considered, but the choice is limited by:

- the installation of a set of video cameras surrounding the scene can be very cumbersome for whole body applications and, often, impractical for domestic environment;
- the video cameras have to be calibrated;
- the software to manage the set of video cameras is computationally inefficient, thus resulting inappropriate for real time applications.

The last point is particularly important because the video-cameras management, four in the same time, and 3D scene reconstruction is completely demanded to the application software. On the contrary, by using the 3D depth sensing camera the same kind of information is furnished by the camera (i.e. by the embedded software working on the camera) which natively provides the depth map of the scene.

The OptriCam™ 130 is a time-of-flight (TOF) camera (Lange et al. 2001), that uses an infrared signal, modulated at 20 MHz, transmitted by 8 infrared LEDs to illuminate the scene. A 2D infrared electro-optical sensor, synchronized with the transmitters, collects the infrared map in four points per period of the modulating wave. In this way, from each pixel of the scene the phase difference between the received and transmitted modulated signal is calculated (i.e. the distance of the pixel from the camera is evaluated). This allows to construct the depth map for each frame collected by the camera (maximum temporal resolution: 1/30 s).

Another fundamental tool is the iisu™ SDK which provides a gesture-based interface to interact with a dedicated application. The tool uses a set of fixed primitives which can be suitably managed to build different sets of gestural commands. The iisu™ SDK data model, reported in Figure 2, is composed by different processing layers:

- **Source 3D**: allows to collect raw data from the device and to construct the depth map, the confidence map, the color image of the scene,

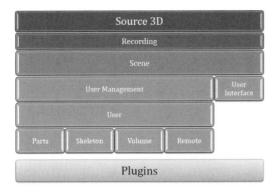

Figure 2. iisu™ SDK processing layers.

Figure 3. An exercise screenshot.

and other filtering methods to improve data quality;

– **Recording**: allows to save videos in a proprietary Softkinetc video format (SKV);

– **Scene**: transforms the depth map in a 3D coordinate system centered on the scene. Moreover, this layer implements calibration and scene setting operation, background removal, and the insertion/removal of cutting planes;

– **User interface**: allows to detect a series of natural human gestures to be used on third party developed applications;

– **User Management**: activates/manages the user inside the scene (up to 4 users at a time);

– **User**: detects the user and calculates some low-level parameters, such as body height, center of mass, direction of movement;

– **Volume**: calculates volumetric information of the user. In this layer, what is identified in the depth map as a user is transformed in a cloud of samples through sampling. By creating a set of spheres of uniform dimensions (the centroids) around these samples, it is possible to obtain a volumetric rendering of the user. Moreover, besides positions, other information are maintained about centroids: velocity (calculated starting from the position in the previous frame), temporal coherence, position in the 2D image;

– **Parts**: allows to detect some parts of the human body (head, hands, feet, pelvis, sternum) and to memorize their position, velocity and state (revealed/unrevealed);

– **Remote**: allows to track some human body parts and gestures useful for the user to manage a control input system;

– **Skeleton**: is used to associate a skeletonized model of the human body to the recognized parts of the user body. Moreover, it gives to the developers parameters and filters for tuning the fluency and naturalness of movements.

Though some of the previous layers can appear redundant (e.g. the head position is given in three of them), each layer is assigned to a specific task and

differences between them are referred to precision, abstraction level and computational complexity.

The last utilised tool is the Irrlicht Engine, an open source graphical engine which strengths are: multi-platform, OpenGL and Direct3D compliant, easily manageable and real time oriented.

As last remark, we highlight that all the extensions on the adopted technologies as well as the bridge-interfaces between them, and each rehabilitation exercise are implemented in C++.

2.2 The user interface

The user interface has been developed by using a general class (*Base*) that maintains common interaction features between different rehabilitation exercises including the events generated by the user interface. In a specific package, (*Entry Point*), a suitable class (*Menu*) is responsible for the visualization of the starting menu, the choice of the exercise, and the user identification. Another package, (*Advanced*), contains the extended classes to run specific exercises. The user (the patient) can run each exercise, set scene parameters (e.g. point of view, or scene illumination), and perform camera calibration (e.g. background removal, calibration of a fixed posture). The skilled user (the therapist) is also able to set some advanced options regarding: the model visualization and rendering, the exercise difficulty parameters, and the virtual 3D frame rate. Finally, the administrator user (the system manager) can assign and modify user privileges. An example of interactive interface during a rehabilitation exercise is shown in Figure 3.

2.3 A simple exercise

The implementation of rehabilitation exercises has to consider two main aspects: *real time interaction* (**rti**) and *off-line processing modality* (**opm**).

The real time interaction has to ensure the correct exercise execution: it is important the patient sees his 3D representation into the virtual world and interacts

Figure 4. Real pose (left) and the corresponding real time blob model represented on a balance board (right).

with it in real time. The patient has to control the exercise and some of its outcomes (such as: score, time of execution, and so on), but he has no need of using recognition/tracking functionalities at all. As shown in Figure 4, as user model we used a cloud of 150 uniform spheres obtained by iisu™ SDK volume layer. During the exercise execution, the spatial position of the centres of these spheres are memorized of each frame. We have 3 spatial coordinates per point, each represented by 5 bytes, corresponding to 5 characters. As an example, by considering 150 points per frame at 30 frames per second, a 10 min video occupies just 40.5 Mbytes. The system is designed to maintain the whole exercise in RAM memory to avoid unwanted jerky. When the exercise ends, data are stored into a file. These data are finally used by the *opm* that runs independently of the user.

The *opm* is important for accurate processing of patient movements inside the exercise scene. As shown in Figure 5, a skeletonized model of the human body segmented into 15 fixed points, is linked to the parts of the user body recognized by using the cloud of 150 spheres from each frame of the video stored into the file. The association is completely automatic and realized through a progressive particle filter (Chang et al. 2008). In particular, our algorithm computes the complete particle set of 150 spheres (for each frame) in order to estimate the position of the 15 body model parts. The system required an initialization: the first time it is used, the therapist adapts the model dimensions to the patient body. Through the therapist interface view, starting by a patient screen-shot collected by the camera, the therapist associates each part of the model to the corresponding part of the patient body, by using the mouse (in this phase, the therapist has to take care of excluding some parts from the model if the corresponding part of the patient body is missing, for example in case of limb amputation). Once the procedure ends, the system calculates the model dimensions and stores it in memory for successive usage. Each time an exercise starts, the patient locks his body parts to the positions assumed by the corresponding parts of the model represented the first time: in this way the exercise starts always in the same position and the filter initialization is automatically ensured.

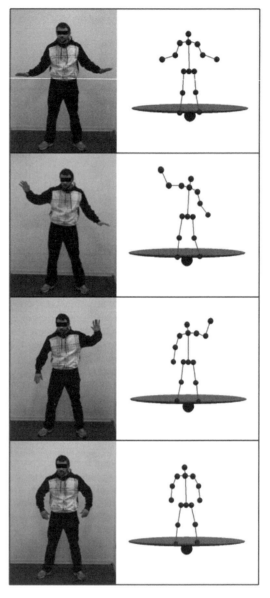

Figure 5. Real poses (left) and the corresponding off-line skeleton model represented on a balance board (right). Rows represent different times.

3 EXPERIMENTAL RESULTS

In order to test the developed system, we have implemented a set of trial exercises. In this section we focus on one of these to show the basic features of the system. The exercise consists in a balance task by simulating the user equilibrium over a board supported by a pivot. The effect of the user above the board is simulated by considering the spheres distribution and position with respect to the vertical axis passing through the user center of mass (the spheres have the same dimension and density). The user controls his balance over the board by bringing his center of gravity on the right or on the left. As the subject oscillates with respect to his center of gravity, his 3D representation is distributed on his feet proportionally.

The subject, while watching his cloud representation on the screen, has to keep his body balanced enough to avoid that the tilt angle of the board exceeds a threshold. The errors are recorded and for receiving the highest score, one would need to avoid losing the control of the board in a fixed temporal duration. During this time the task is made incrementally more difficult by increasing the board sensitivity (every 30s). At the end of the exercise, a file containing the user 3D representation is recorded, elaborated automatically in off-line, and the score is stored and shown to the user.

This simple exercise could be applied in neuro-rehabilitative treatment but has been also applied to perform brain activation measurements by using functional near infrared imaging (Bisconti et al. 2012). Figure 5 shows some screenshots of the exercise. In particular, the left part shows the user positions, collected by a secondary webcam, while the right part shows the results of the off-line processing, obtained through progressive particle filter to associate the skeletonized user model to the 3D cloud of spheres model.

We have observed that off-line processing (on different middle level hardware configuration) is very robust and efficient. In fact, we suppose it could be also used to track the human body in real time. The off-line modality allows a further file dimension reduction by a factor of 10 (we have to store just 15 objects for frame instead of 150) and to calculate a set of dynamic parameters for each body part (trajectory, velocity). Though the file contains less data then before, it carries all necessary information.

4 CONCLUSION

Motion analysis and virtual reality based assisting systems can facilitate and accelerate rehabilitation. Often, movement analysis this kind of systems are very expensive and their usage and set-up is oriented to professional users. In addition, they are not designed to perform rehabilitation activity in a domestic environment.

In this work, we discussed the design and implementation of a low-cost movement analysis system based on virtual reality and 3D depth sensing camera to be used for whole body rehabilitation for domestic usage. Some trial exercises have been implemented, both in real time and off-line modality, and an example has been also reported. Both real time and off-line modalities have been tested and some screenshots have been reported. The results demonstrated the good performance of the off-line calculation through the application of progressive particle filter. The file containing the calculated model can be efficiently forwarded to a server, stored into a database and accessed by the therapist for further calculations, evaluations and comparisons.

ACKNOWLEDGEMENTS

Special thanks go to the Softkinetic International SA/NV ("SoftKinetic") for providing us continuous technical support. Moreover, the Company also provided the DepthSense^{TM} 311 camera, for free, by which we can improve the proposed system.

REFERENCES

Bisconti, S. & Spezialetti, M. & Placidi, G. & Quaresima, V. 2012. Functional near-infrared frontal cortex imaging for virtual reality neuro-rehabilitation assessment. In *Proceedings of the 3th Edition of Computational Modeling of Objects Presented in Images: Fundamentals, Methods and Applications*. CompImage'12.

BTS, Bioengineering Technology and Systems 2012. http://www.btsbioengineering.com.

Chaczko, Z. & Kale, A. & Chiu, C. 2010. Intelligent health care a motion analysis system for health practitioners. In *Proceeding of 6th International Conference on Intelligent Sensors, Sensor Networks and Information Processin*, ISSNIP '10. IEEE Computer Society Publisher, 303–308, USA.

Chang, I.C. & Lin, S.Y. 2008. 3D human motion tracking using progressive particle filter. In *Proceedings of the 4th International Symposium on Advances in Visual Computing*, ISCV '08. Springer Berlin Heidelberg Publisher, 833–842, Las Vegas, NV, USA.

Franchi, D. & Maurizi, A. & Placidi, G. 2009. A numerical hand model for a virtual glove rehabilitation system. In *Proceeding of IEEE International Workshop on Medical Measurements and Applications*, MeMeA '09. IEEE Computer Society Publisher, 41–44, USA.

Gavrila, D.M. 1999. The visual analysis of human movement: a survey. *Computer Vision and Image Understanding*. Elsevier Science Inc. Publisher, 73(1): 82–98, New York, NY, USA.

iisuSDK^{TM}, Soft Kinetic International 2012. http://www.softkinetic.com/Solutions/iisuSDK.

Irrlicht Engine, IRR Licht 3D Engine 2012. http://irrlicht.sourceforge.net/2012/01/new-website/.

Jacko, J.A. (Ed.). 2011. Human-Computer Interaction. In *Proceeding of 14th International Conference in Human-Computer Interaction*, HCI '11. Springer Publisher, Lecture Notes in Computer Science (LNCS), 6791(2011):570, Orlando, FL, USA.

Keshner, E.A. 2004. Virtual reality and physical rehabilitation: a new toy or a new research and rehabilitation tool?. *International Journal of Neuroengineering and Rehabilitation*, JNER '04. BioMed Central Ltd. Publisher, 1(8): 1–4.

Lange, R. & Seitz, P. 2009. Solid-state time-of-flight range camera. *IEEE Journal of Quantum Electronics*. IEEE Computer Society Publisher, 37(3): 390–397, USA.

Moeslund, T.B. & Hilton, A. & Krüger, V. 2006. A survey of advances in vision-based human motion capture and analysis. In *Computer Vision and Image Understanding*. Elsevier Science Inc. Publisher, 104(2): 90–126, USA.

Optrima^{TM}, Soft Kinetic International 2012. http://www.softkinetic.com/Solutions/.

Papalambrou, A. & Soufrilas, P. & Voyiatzis, A.G. & Serpanos, D.N. 2011. A secure DTN-based smart camera surveillance system. In *Proceedings of the Workshop on Embedded Systems Security*, WESS '11. ACM Publisher, 3(1): 1–4, New York, USA.

Spezialetti, M. 2012. Development of applications supporting rehabilitation through iisu™ system – depth sensing camera. *Bachelor's Thesis, Supervisor: G. Placidi.* University of L'Aquila, L'Aquila, Italy.

Ustinova, K.I. & Leonard, W.A. & Cassavaugh, N.D. & Ingersoll C.D. 2011. Development of a 3D immersive videogame to improve arm-postural coordination in patients with TBI. *International Journal of Neuroengineering and Rehabilitation*, JNER '11. BioMed Central Ltd. Publisher, 8(61): 1–11.

Vicon, Vicon Motion Systems and Peak Performance Inc. 2012. http://www.vicon.com.

Winter, D.A. 2009. Biomechanics and motor control of human movement. *Book Series: Processing.* Wiley John and Sons (Eds.). John Wiley and Sons Inc. Publisher, 4 Edition: 384.

Computational Modelling of Objects Represented in Images – Di Giamberardino et al. (eds)
© 2012 Taylor & Francis Group, London, ISBN 978-0-415-62134-2

People detection and tracking using depth camera

Rui Silva
EXVA Technologies, Avepark – Zona Industrial de Grandra, Guimarães, Portugal

Duarte Duque
IPCA – Instituto Politécnico do Cávado e do Ave, Barcelos, Portugal

ABSTRACT: In this paper we present a method for real-time detection and tracking of people in video captured by a depth camera. For each object to be assessed, an ordered sequence of values that represents the distances between its center of mass to the boundary points is calculated. The recognition is based on the analysis of the total distance value between the above sequence and some pre-defined human poses, after apply the Dynamic Time Warping. This similarity approach showed robust results in people detection.

1 INTRODUCTION

Depth video cameras are starting to play a major role in videogames and novel Natural User Interface systems, which represent an interesting challenge to the computer vision field.

This kind of video sensors typically uses the signal reflected by an IR light to measure the total distance from the object to the camera. At some parts of the scene the distance can't be estimated and, therefore, the depth image will be severely affected by noise. These failures occur most frequently in surfaces that don't reflect the IR light, and the signal quality deteriorates as the distance increases.

Other related research studies try to minimize noise problems using techniques for smoothing [Lee], filling the image gaps [Edeler], or assuming distance restrictions in the acquired data. Despite this, it's an inevitable fact that the noise will interfere on the processing of the depth images, especially at the objects segmentation step. In order to perform the image segmentation and object detection in depth images, other works also use auxiliary data of color video cameras [Holz] or background subtraction techniques [Leens] to find moving objects. These methods present many limitations for a use in uncontrolled environments. For example, when the provided light for the color video camera is weak or its intensity varies along the time. Furthermore, if moving objects and persons are already present in the scene in the first acquisition of the background image, these techniques will have worse results.

In the next chapters, we present a new fast method to find and track people, or other kind of predefined objects, only using the data from a depth camera.

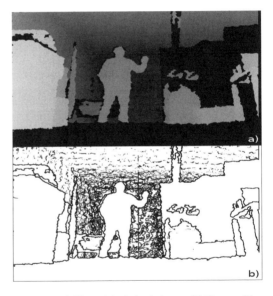

Figure 1. a) The original depth image. b) The resulting binary image with the main edges.

2 IMAGE SEGMENTATION

Our algorithm starts by making the image segmentation. This step is used to locate the boundary lines of the objects present in the depth image. For this purpose, we perform the edge detection based on the gradient values.

The gradient value of each pixel is obtained by convolving the Prewitt kernels with the original image, to calculate approximations of the first-order derivatives.

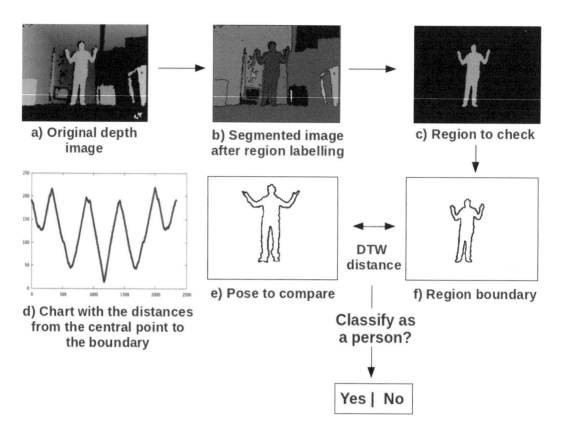

a) Original depth image

b) Segmented image after region labelling

c) Region to check

d) Chart with the distances from the central point to the boundary

e) Pose to compare

DTW distance

f) Region boundary

Classify as a person?

Yes | No

Figure 2. Scheme of the proposed approach.

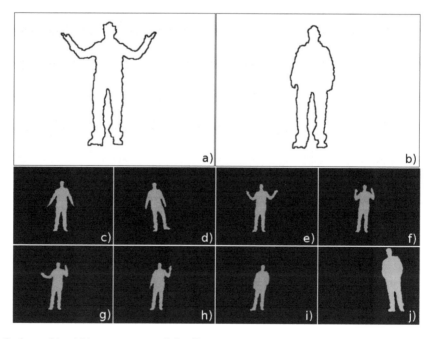

Figure 3. On the top, (a) and (b) represent two predefined human poses. Below, (c) to (j) are examples of correct classifications after calculate the DTW total distance with the above pose images.

374

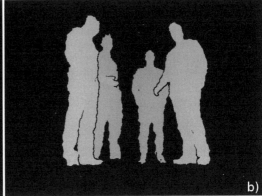

Figure 4. a) The original depth image. b) Detection of multiple persons with robustness to partial occlusion.

The segmentation is expressed in a binary image (Fig. 1.b) that contains just the main edges of the depth image, after applying a threshold to the gradient values.

3 DISTINCTION AND FILTERING OF OBJECTS

After the segmentation, we perform the labeling of connected components present in the binary edge image. A unique ID number is assigned to each connected region, contained within a closed edge. This allows us to distinguish the various objects.

We count the total number of pixels of each object, and measure the width and height of its bounding box.

As we're searching for people, in order to speed up the processing, we ignore all the regions that don't correspond to human body morphological characteristics.

Therefore, the object is deleted if:

- The total number of pixels is smaller than a predefined minimum;
- The total number of pixels is higher than a predefined maximum;
- The ratio width / height is higher than 0.8.

4 FEATURE EXTRACTION

To distinguish and compare the objects present in depth images, the shape of the object is one of the most expressive features.

We apply the Chain Code algorithm [Freeman] to read the position of the boundary points of the object in an orderly manner. The starting point for the Chain Code algorithm is defined by the boundary point located at the maximum value of the x-axis frequency histogram of the top half of the object.

Then, the values of the distance between the center of mass of the object and its boundary points, are used as the main feature for identify the object (Fig. 2.d).

This step was previously done for the poses that we use in the predefined set for the comparison process, and the arrays were stored in files for posterior use.

5 CLASSIFICATION OF OBJECTS

Finally, the descriptor generated in the previous step is compared with the predefined human body poses. To check the similarity of the object being assessed with each of the predefined pose, both arrays are processed with Dynamic Time Warping (DTW) algorithm [Myers].

If the total distance between the two arrays after the DTW is below a minimum preset value, the object is classified as a person (Fig. 2).

Depending on the pose to be compared, we have defined different threshold values. If we increase the value in these parameters, the number of poses accepted related to each predefined descriptor will also increase. This can be an advantage to classify different human movements with a low number of predefined poses, but also raises the number of false positives. To avoid this, when the boundary line it's not so expressive, which happens in poses that the legs and arms are close together, the value should be low.

6 RESULTS

In this section we present some experiments that we have made with the implementation of the proposed algorithm. In Figure 3 we can see that the application is capable of correctly classify different human poses. We can also verify that the shape of the accepted objects is related to the ones of the predefined set, but is not strictly equal to them. Figure 4b shows multiple persons detection, with good results even with partial occlusion.

The main problem when we perform the segmentation of depth images based on the edges, is that if two or more objects are at the same distance from the camera and touch each other at some point, no edge will be found at that zone. Consequently, the system will recognize them as being only one object.

In the future we will define a ground truth set. For this, we will do the manual segmentation of the objects in some depth images. So we can use those human-marked boundaries, to properly evaluate the results of our algorithm.

7 CONCLUSION

In this work we presented a new approach to automatically detect people with depth cameras. The proposed solution consists on an edge detection algorithm, an object filter based in morphologic characteristics, a feature extraction technique based on the shape, and comparing the degree of similarity with DTW.

The system seems to be sufficiently fast to detect and track multiple people, walking normally around the scene.

Despite the results obtained, the validation of this solution still lacks of an extensive test. Therefore, our main concern in the near future is to extend the experiments on different real-world scenarios.

REFERENCES

Edeler, T., Ohliger, K. Hußmann, S. & Mertins, A. 2010. Time-of-Flight Depth Image Denoising using Prior Noise Information. *ICSP*.

Freeman, H., 1961. On the encoding of arbitrary geometric configurations. *IRE Transactions on Electronic Computers*: 260–268.

Holz, D., Holzer, S., Rusu, R. B., & Behnke, S. 2011. Real-Time Plane Segmentation using RGB-D Cameras, *Proceedings of the Robocup Symposium*.

Myers, C. S., & Rabiner, L. R. 1981. A comparative study of several dynamic time-warping algorithms for connected word recognition. *The Bell System Technical Journal*: 1389–1409.

Lee, P.-J. 2010. Nongeometric Distortion Smoothing Approach for Depth Map Preprocessing. *BMSB*.

Leens, J., Piérard, S., Barnich, O., Van Droogenbroeck, M., & Wagner, J.M. 2009. Combining Color, Depth, and Motion for Video Segmentation. *ICVS09*: 104–113.

Selected pre-processing methods to STIM-T projections

A.C. Marques, D. Beaseley, L.C. Alves & R.C. da Silva
Instituto Tecnológico e Nuclear, Instituto Superior Técnico, Univ. Técnica de Lisboa, Sacavém, Portugal
Centro de Física Nuclear da Universidade de Lisboa, Lisboa, Portugal

ABSTRACT: Scanning Transmission Ion Microscopy (STIM) is a non-destructive microscopy that cross-sectionally probesthe energy losses of ions transmitted across very thin samples. By combining the generated 2D projection maps with sample rotation STIM lends itself to quantitative mass density analysis in 2- and 3-dimensions with micrometerresolution. A STIM system and results of first performance tests of pre-processing and reconstruction software ofa noncomplex sample are described. Selected pre-processing methods were applied to projection data prior to reconstruction, for evidencing its impact on the quality of reconstruction. In the process a new algorithm for removal of background noise and spikes arising from detector operation was developed. The principle is based on the determination of a mask limited by the object contours and setting outlyingpixels to zero. This approach is less prone to deleting very thin structures in the object that produced only a small but real energy signal. Furthermore it may be applied to 2D images obtained by other vision techniques.

1 INTRODUCTION

In STIM-T a 2-3 MeV proton beam is rastered across the surface of a sufficientlythin sample generating 2D energy loss maps in suitable detectors. These maps contain information on the integrated mass density distribution across the scanned surface. Three different beam-detector geometries can be employed, on-axis, off-axis and on-/off-axis, depending on whether the detector is: *i*) in the line-of-sight, directly exposed to the transmitted proton beam (on-axis); *ii*) away from it, by a small, forward, angle,collecting protons that were, either directly scattered by the sample in that direction (off-axis), or transmitted, collimated and then scattered in the detector direction by a thin mono-elemental film (on-/off-axis). The on-axis geometry provides true areal mass density maps but suffers from excessive count rates and high damage levels to the detectors, with the consequent degradation of performance and shortening of lifetimes. On the contrary, off-axis geometry avoids the count rate and detector damage problems but does not yield true areal mass density maps, as it is the measurement of scattered protons, and suffers from the interfering contributions of differing energy losses upon scattering by the sample atoms. The on-/off-axisgeometry offers a compromise solution: it allows near true areal mass density mapping, as it introduces only minor, equal weight interference, one which does not distort or alter the distribution, while still allowing use of detectors that are not capable of withstanding high count rates and radiation damage as the whole surface of the detector is used for detection rather than the parts in the direct line of the beam. In the present work data was acquired in the off-axis geometry, as it represents a common choice which is easier and quicker to implement, while able to cope with the restrictions imposed by detector performance and lifetime, when more rugged detectors are not available. By combining the 2D projection maps with sample rotation STIM lends itself to be used as a valuable tomographic technique (STIM-T) at sample sizes of a few tens of micrometres and sub-micron resolution, with many potentialapplications to the fields of biology, biomedicine, geology, environment, etc. Reproducibility in the measurements at such scaleputs strict requirements on focusing, beam stability, sample positioning and alignment with the vertical rotation axis, which has to be precise. Hence, technical developments on a new specific 3D setup have been implemented at the ITN nuclear microprobe. Despite the improvements, positioning and alignment problems with the rotation axis remain withsome degree of misalignment. Threetypes of solution were devised to mitigate these problems: *i*) assembly of a better positioning stage with 5+1 degrees of freedom,which is becoming ready to use, *ii*) development of a suitable alignment strategy, and *iii*) selection, implementation and test of pre-processing methods to correct forexperimental inaccuracies occurring during the projections acquisition. All steps prior to reconstruction, i.e. from raw data to the transformation into sinogram, are described. Special emphasis is given to the new algorithm for removal of noisy background around the sample in the projection. In this method the Canny edge detection algorithm proposed in 1986 by J. F. Canny is used to differentiate sample image from non-sample image domains, the latter being discarded or zeroed [1].Subsequent data reconstruction was performed with a modified version of the back filtered projection (BFP) based algorithm previously implemented by Oliver Steinbock [2].The morphology of atest sample was qualitatively analysedforsoftware testing purposes.

2 EXPERIMENTAL METHODS

The experiments were performed with a 2 MeV scanning proton microbeam at the nuclear microprobe, coupled to a 2.5 MV Van de Graaff accelerator [3]. A test sample consisting of a Formvar™ fiber with immobilized SiC grains ($\sim$30 µm in diameter) was analysed with a beam focussed to 3.3×3.9 µm^2 and rastered over a900 $\times$ 900 µm^2 synchronised with the detection and positional data collection. This provides for 2D mapping with parallel beam geometry.A Hamamatsu S1223-01 PIN diode with 15 keV energy resolution placed behindthe sample off-set by $\sim$20° from the beam directionwasused to measure the protons energy after transmissionthrough the sample. The projection set was acquired over an π angular range in steps of 10°. The sample rotation was controlled by a one-axis goniometer standing on top of a 2-axis translation table coupled to the vacuum chamber by a bellow. The sample was mounted on a needle clamped to the sample holder,which was screwed to the goniometer axis. The holder was specially designed for improving sample positioning and stationary while rotating but unfortunately, sample precession couldn't be avoided due to anunpredicted table tilt. The scan was computer controlled by a digital scanning system made by Oxford Microbeams™ and adjusted to follow the sample precession, constantly keeping itin the scan region. For each projection, the energy and the-beam position coordinates are saved for everyevent. Hence, the data can be reduced off-line into a single map obtained from statistical data treatment of the events occurring at each pixel. This can be directly performed with the OMDAQ computer code [4]. For converting channel number to energy a calibration sample was made from 6.3 µm thick Mylar™ foils cross mounted to provide four regions with different thicknesses: 6.3 µm, 12.6 µm 18.9 µm and 25.7 µm. The 2 MeV beam was set to scan the full area in order to produce a well-known calibration spectrum. The energy of each peak in the spectrum was estimated by means of Monte Carlo simulations SRIM software [5]Afterwards, the median energy per pixel for each 256 $\times$ 256 pixels projection was determined for reducing noise and preserving sample edges.

3 RESULTS AND DISCUSSION

Following the acquisition, the projection data was pre-processed to correct for noise and imperfect sample movement upon rotation. This pre-processing process is divided in sixmain steps: removal of noisy background and of spikes, nonlinear conversion of energy loss into mass, vertical and horizontal alignment,

correction of the rotational centre in the sinograms and removal of artefacts from reconstructed tomograms. Sincedetermining the correct overall composition of the sample was out of the scope of this study, asimple BFP-based algorithm waspreferred to satisfactorily reconstruct the off-axis STIM data.In fact the conversion of energy loss into areal mass density was merelyperformedfor reducing reconstruction artefacts.

3.1 Data pre-processing

With the off-axis geometry, the proton intensity at the detector was relatively high providing spectra with good statistics, although poorer resolution.This is understandable, since due to scattering from the sample the energy loss is not only a function of the mass density and composition (as is on-axis STIM), but also depends on the kinematic energy transfer. The effect of scattering from different sample atoms (e.g. H, C, O, Si, etc.) can occur in any place, from the front to the back of the sample, causing different energy loss contributions to the spectrum. The resulting effect is a spectral broadening, which will in turn decrease the contrast in imaging. However, in this case errors of less than 1.6% are introduced in the calculations of transmitted proton energies if data is treated as obtained in on-axis geometry. Therefore, no correction factor was applied and typical STIM on-axis pre-processing was used. Noise caused by the excessive irradiation of one particular spot of the detector surface, and by statistical variations in the collected datais removed. Non zero pixels outside the sample area are set to zero. In the literature three methods for zeroing those pixels are referred. They are listed below and each has at least onerecognized disadvantage, motivating the proposal of a fourth method that do not interfere with projection sample data.

1) Using a threshold for zeroing pixels with energy loss below a certain level [6, 7]. The threshold is often set to the energy resolution of the particle detector since statistical fluctuations are of that order of magnitude. The drawback is erosion of contoursin regions of the sample with very low energy losses.

2) Detecting the regions of the projection belonging to the sample by performing a 2D search from the largest energy value (certainlya part of the sample) to a contour with value equal to theenergy resolution [6]. The region outside the object is set to zero. This only works if the detector damage level is not high enough to mask the sample, otherwise a part of the sample would be also removed.

3) Retaining the measured projections while applying the method 1 to a copy set that will be used specifically for determining the centre of mass [7]. Then the original set of unthresholded projections will be shifted and backprojected. With this, the detector noise remains and still shows in the tomograms, its removal demanding applying a specific threshold implying that thin structures of the sample may be erased as well.

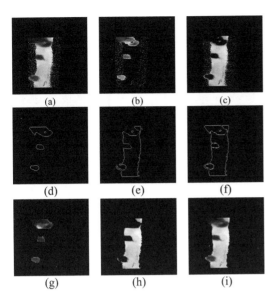

(a) (b) (c)

(d) (e) (f)

(g) (h) (i)

Figure 1. Median energyprojection map measured at 0°
before (a) and after thresholding for energy values below (b)
and above (c) 0.95 MeV. The corresponding edges images
obtained by the Canny method are shown from (d) to (f),
resulting (f) from the sum of (d) and (e). Sample data is readout from the measured file, line by line between edges, and
written in a zeros matrix of the same size of (a). The resulting
images are shown from (g) to (i) after spikes removal.

As with method 2, the method proposed here
employs contour identification to exclude – i.e. to
zero – the external region, not belonging to the sample.
It achieves this by means of the Canny method.The
major benefit is leaving the data inside the sample
contours unaffected. Evidently any unrealistic energy
values inside the sample area will not be removed.
However, in subsequent pre-processing steps it can certainly be set into more reasonable discrete values. For
illustrating the method, this is applied to the projection
takenat zero degrees and results shown in figure 1.

First, an energy threshold of ~ 0.95 MeV is applied
to each projection in order to define two distinct energy
ranges and thus, separate the signal coming from the
SiC grains and from the Formvar™ fiber by generating two extra images as shown in figure 1, from
(a)–(c). Note that the threshold parameter depends on
sample composition and hence, varies among projection sets taken from different samples. Indeed more
than one threshold may be needed for samples with
more than two compounds. Second, for obtaining the
edges image of the whole sample and from the grains
and Formvar™ fiber parts, the Canny algorithm was
applied. The resulting three edges images revealed to
have a significant number of gaps, compromising set-
tling a mask in the outer region of the sample, or sample
parts, and zeroing it. Filling the gaps by optimizing
the Canny method parameters, e.g. threshold, wasn't
enough. Therefore, before using the Canny algorithm
to the images showing parts of the sample, a median
filtered in two dimensions using a 4 by 2neighborhood

was applied. These two edges images were subsequently summed to produce the edges image of the
whole sample as illustrated in figure 1 from (d)-(f).
This image has few noticeable gaps and these are not
in consecutive lines, i.e. adjacent lines, which made
filling the gaps an easy task. This consists in counting the number of maxima existing in each line of a
given edges image, determining what their positions
are regarding the correspondent ones in the adjacent
lines (top and bottom lines), and in case a gap is
detected filling in the position vertically coinciding
with the top or bottom maximum – depending on their
relative positions as well. With this approach the outer
boundaries of the sample was established as the column indexes restricting the sample data to be readout
from the as measured projection and written in a new
zeroes matrix with its size. This newly created image
only gets a copy of the original data inside the determined sample contour as it was measured.Defining
three masks per projection as explained, allowed not
only producing projection images free of noise in the
outer region of the sample, but also permitted accessing to parts of the sample with similar composition. In
this case, the sample regions corresponding to SiC and
Formvar compounds were reasonably separated. Note
that because the density of the iron needle used to hold
the fiber is nearer to the SiC grains it also appears in
images of figure 1. However, the needle area is later
cropped from projections. The next pre-processing
step consisted in removing unrealistic energy values,
some probably still resulting from detector noise, from
the sample area in the projections set – i.e. dark pixels in bright regions and bright pixels in dark regions
are eliminated. This is done as explained in ref. 8,
i.e. by vertically and horizontally sweeping each projection looking for every pixel exceeding its pair of
neighbor pixels, or falling below the minimum of the
pair by 30%, and replacing them by the mean value of
the pair in analysis. The energy maps are next easily
converted into energy loss by subtracting the incident
beam energy to every pixel in the energy map. All
values without physical meaning, i.e. below zero and
above 2MeV (the incident beam energy) were set to
zero. The energy loss was converted into areal mass
density through the calculation of the Ziegler instantaneous stopping power for each element in the sample
[5]. The proton stopping power for the SiC and For-
mvar™ ($C_8H_{14}O_2$) compounds in the sample were
determined using the Bragg's additively rule [9]. Here
the elementary composition of the sample was firstly
assumed to be constant with only the density varying
throughout the sample and secondly, heterogeneous.
The first approach is common for the reasonsgivenin
ref. 7, howeversincethe hydrogen content in this sample is significant and may be a source of error not
negligible because of scattering, the heterogeneous
approach was also considered. In any case the inac-
curacy inthe tomogram reconstructed density values
is mainly due to experimental inaccuracies. However,
the aim was evaluating the mathematical consistency
of the sinogram and analyzing how it may affect the

sample morphology in the 3D image. This topic will be brought to discussion in the next section. Having performed the non-linear conversion of energy loss into areal density, the measured projection set is manipulated as in ref. 8to ensure that the middle of the sample is centered in each projection and that the center of mass motion in sinograms is adjusted on a sine curve as in ref. 10.This is possible because the position and shape of the sample should not change extensively in successive projections.

3.2 Data reconstruction

Back filtered projection is a simple method to infer the 3D structure of an object from a series of projections. The computational scheme relies on extracting the individual sinograms, $I_z(x, \theta)$, from the angular projections set (i.e. $P_\theta(x,z)$ set), where z, x and θ represent the vertical and horizontal cartesian coordinates in each projection and the rotation angle, respectively. The stack of slices normal to the rotation axis, i.e. of tomograms $T_z(x,y)$, are next computed via back projection of the filtered sinogram along parallel beams of angle θ and summation over all θ. Filtering the sinogram set with a Ram-Lak filter multiplied by a Hamming window in the frequency domain prior to the summation de-emphasizes high frequencies. The resulting slices are next stacked in a 3D array for visualization. This part of the tomographic procedure was done using the free "Tomography" software made by the Steinbock group and provided in their web site [2]. However, a few modifications have been introduced in order to reduce reconstruction artefacts. This was accomplished by introducing the tomogram pre-processing steps proposed by T. Andrea and listed below[8].

1) Negative density values are set to zero.
2) Tomogram is rescaled by making its total mass equal to the projections total mass.
3) Resulting tomogram is scaled to a factor indicating the number of pixels per micron.
4) Remove straight lines artefactsvia threshold application. All values below the density level at whichartefactscease are deleted.
5) Smooth tomogram in case the number of projections is below optimal. Noise is reduced but edges become compromised, i.e. blurred.

The tomograms reconstructed from the formvar fiber data set before and after applying the threshold step are shown in figure 2for emphasizingthe benefits of its use.

The resulting tomograms have not significant details, which is not surprising considering the low number of projections suffering from many experimental inaccuracies, including precession of the rotation axis. The reconstructed 3D probed volume of the test sample seen from one angle is shown in figure 3. The reconstructed surface of the formvarfibre is coloured in redclearly evidence one of holes corresponding toSiC grains immobilized in the fiber.

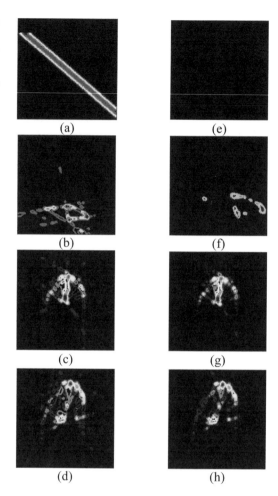

Figure 2. Tomograms before (a)-(d) and after (e)-(h) setting the threshold to 0.1285. The tomograms or slices from top to bottom correspond to the 1, 50, 150 and 170 heights.

Thesewere difficult to correctly reproduce, either from the projection set where they are isolated from the rest of the same, as illustrated in figure 1, or from the projection set where a homogeneous composition was assumed. In this case, the grains appear a bit more prominent and again, with the position in the fiber welldeterminedbut morphological poorly reconstructed. It was also tried to sum the projection sets isolating the grains and the fiber in order to obtain an heterogeneous set of the whole sample but the result was not significantly better because of artifacts being recurrently present. However, this last approach maybe promising and give better results with a larger set of projections acquired in the new setup being presently assembled and tested.

4 IMAGE BACKGROUND REMOVAL

Isolating objects from a background scene is a very important pre-processing step in many daily

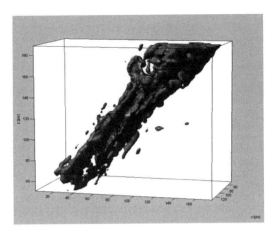

Figure 3. 3D image of the Formvarfibre. The wider part on top has as expected the corresponding hole of the immobilized SiC grain.

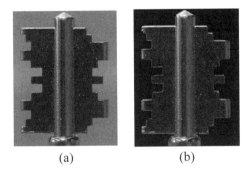

(a)   (b)

Figure 4. Image with light background before (a) and after (b) isolating the object.

life applications. Accurate and efficient background removal has been useful in several activities such as those related with object recognition, tracking persons, modelling and graphics. One of the common approaches is to use edge detection for tracing the boundary line between an object and the background for image segmentation. However, it is difficult getting the entirely boundary line free of gaps. Depending on the image complexity this can become a very challenging task. In this section, the efficiency of the simple method proposed in section 3.1 is tested with a different kind of image. In figure 4, the method was applied to a key photograph, which intensity grey scale ranges from 0 to 255. For obtaining a reasonable edges image of the key it was necessary to generate three edges images and add them. Hence, three thresholds were set: one covering the intensity range above 130, a second one covering the range above 106 and a third one defining the range below 190.

The main issues in this approach the number thresholds depends on the type of image as its value, which is not automatically determined. Another limitation concerns the criteria used to read out data from the original image in order to create a new one. The reading and writing limits rely on the outer most edges in each line,

thus if the sample has whole-like structures the background is not removed. In fact, if the noise is not too intense, typically it can be removed with the algorithm used for removing spikes from the tomography projection data. In order to overcome these limitations we are currently working on an extension of this image background removal method.

5 CONCLUSIONS

This preliminary study performed on a test sample shows for the first time the capability to perform 3D STIM-T experiments at ITN. Inefficiencies and inaccuracies in the experimental setup at the time of acquisition of the data here presented as in all data off-line analysis steps were clearly identified and are currently being improved. In fact, the STIM-T setup here described was abandoned and presently, a completely new sample stage is being installed. The sample rotation is now achieved through a motorized stage with a minimum angular displacement of 0.45° and positioning precision of less than 0.35°. This stage accommodates the sample holder, consisting of a 2 axis limited tilt (of ±17° and ±20°) precision goniometer, mounted on top of a XY-table with a ±5 mm range and micrometer precision. Finally, the sample itself is placed on a needle clamped to the head tip of the goniometer setup. An alignment procedure aiming at a stable, non-precessing rotation has been devised and implemented. Using a micro-TV camera coupled to the motorized stage control computer the sample is brought to close alignment with the (vertical) rotation axis by repeatedly acting on tilt axes of the goniometer head. Next the sample is rotated by the motorized stage and the XY displacements used until no lateral motion is perceived of its geometric rotation centre line. Even without this new setup the digital corrections devised for the pre-processing of projections have been demonstrated as a necessary, extremely valuable and powerful tool, capable of coping with most of the mishaps of a non-optimal setup. Energy calibration, removal of noise and spikes from the projection data as well as vertical and horizontal alignments worked fine. Only the rotation centre correction was not so successful because of the significant sample precession and few projections taken in this exploratory experiment. As a result tomograms are poor in revealing sample details and are marked by a considerable amount of star-like and straight line artifacts, which were reduced by pre-processing the tomogram but not completely avoided without losing sample data as well. Considering all these problems the 3D image reasonably reproduces the expected outer shape of the fibre and allows identifying the position of the SiC grains, although reconstruction is far from perfect. Grains size and the small number of projections in the π angular range, made it difficult to achieve a better reconstruction. For the purpose of isolating sample elements and separately calibrate them as areal density, the background removal algorithm was of great usefulness, making use of intermediate edges images, corresponding to different sample masses, and arriving to a final edges image, withreliable whole outer contour of the sample.

REFERENCES

John Canny, 1986. A computational approach to edge detection. Pattern Analysis and Machine Intelligence, IEE Transactions, PAMI-8(6):679–698.

Web site about Steinbock group software: http://www.chem.fsu.edu/steinbock/downloads.htm

L.C. Alves, M.B.H. Breese, E. Alves, A. Paúl, M.R. da Silva, M.F. da Silva and J.C. Soares 2006. Micron-scale analysis of SiC/SiCf composites using the new Lisbon nuclear microprobe. Nucl. Instr. and Meth. B: Beam Interactions with Materials and Atoms, 161-163: 334–338.

G.W. Grime, M. Dawson, 1995. Recent developments in data acquisition and processing on the Oxford scanning proton microprobe. Nucl. Instr. and Meth. B, 104:107.

J.F. Ziegler, J.P. Biersack, and U. Littmark, 1985. The stopping power and ranges of ions inmatter. Pergamon Press, New York.

A.Sakellariou, 2002. STIM and PIXE tomography – Three dimensional quantitative visualisation of micro-specimen density and composition. PhD thesis, School of Physics, The University of Melbourne.

T. Andrea, M. Rothermel, T. Butz, T. Reinert, 2009. The improved STIM tomography set-up at LIPSION: Three-dimensionalreconstruction of biological samples. Nucl. Instr. and Meth. B, 267: 2098–2102.

T. Andrea, 2008. Full- and limited angle ion micro-tomography (STIM and PIXE): Applications to Biological Samples. Diploma Thesis, University of Leipzig.

W.H. Bragg and R. Kleeman, 1905. On the α particles of Radium, and their loss of range in passing through various Atoms and Molecules. *Philosophical Magazine*, 10:318–340.

S. G. Azevedo, D. J. Schneberk, J. P. Fitch, H. E. Martz, 1990.Calculation of the rotational centers in computed tomography sonograms.IEEE Trans. Nucl. Sci., 37(4): 1525.

Computational Modelling of Objects Represented in Images – Di Giamberardino et al. (eds)
© 2012 Taylor & Francis Group, London, ISBN 978-0-415-62134-2

Affine invariant descriptors for 3D reconstruction of small archaeological objects

Maweheb Saidani & Faouzi Ghorbel
CRISTAL laboratory GRIFT research group, National School of Computer Sciences,
Manouba University, Tunisia

ABSTRACT: This paper addresses the problem of recovering 3D structure of small archeological objects under affine projection. Affine projection is an approximation to the perspective projection which works quite well when object size and depth are small compared with the distance between the camera and the object. A new technique for affine reconstruction from two affine images is developed, which consists in decomposing the problem into two subproblems: contour matching based on invariant descriptors and 3D recovering based on matching results and calibration parameters. First we compute affine invariants descriptors on extracted contours and then we estimate affine transformation that best align the matched contours: a scale factor α, a shift value l_0 and an affinity matrix A are computed. This step establishes point-to-point matching that preserves the circular order of contours points. Second we use matching results in addition with camera parameters, previously determined, to retrieve the whole 3D model. In order to validate the proposed method, some experiences on small archeological 3D objects in stereovision context are exposed.

1 INTRODUCTION

To date the techniques applied to the restitution of small archaeological objects are based on the exhausting calculation of 3D point clouds, which represent the outer surfaces of the objects. Laser scanning techniques are likely the best and most reliable for these purposes. However, the costs associated with these techniques prohibit their use apart from that most archeological founds are too fragile for repeated handling. Therefore, the alternative is the use of passive methods like stereovision. In stereovision pocess, matching is the most crucial step and there is already a large amount of work reported in the literature concerning stereo matching. Area-based methods (Trucco and Verri 1998) treat the problem without attempting to detect salient objects. These methods are sometimes called correlation-like methods because they use a rectangular window to parse the whole image. The principle drawback of this family of methods is the high computational cost and the restricted range of distortions. However, feature-based methods (Grimson 1985) aim at establishing point correspondences between two images. For that purpose, they extract some easily detectable features (e.g. intersection of lines, corners, etc.) from images and then use these points to compute the closest transformation based on a similarity metric. Unfortunately, the solution of this problem is far from trivial and usually relies on the assumption that the deformation is close to identity and that features provide a strong contextual evidence for matching landmark points. Another class of techniques used to match objects in different images is based on the analysis of their shapes. To apply these techniques, one must begin by extracting features from the objects shapes, such as a group of points, segments, boundaries, etc. In Computational Vision, this task is usually known as object segmentation. Frequently, following the segmentation of the objects from the input images, the matching between the extracted objects features are then accomplished. Some techniques begin by trying to align the objects, usually, this image alignment is refered to as image registration; to achieve this goal, here we propose to use a set of affine invariant descriptors applied on extracted contours from each image and normalized with affine arc length. When the similarity of the objects boundaries is quantified in the form of euclidean distance between their corresponding descriptors, the matching of points can be accomplished by the use of affine transformation parameters estimation, such transformation can be used in the context of stereovision as a good approximation of perspective projection when the camera is allowed to be at some distance from the scene.

In this paper, we adopt affine invariant descriptors to solve the contour correspondence problem and we make the following contributions:

- We formulate the correspondence problem as computing euclidean distance between affine invariant descriptors to accomplish the first step of matching, then, we estimate affine transformation parameters to get point-to-point correspondence with order preservation.

- We show that such matching algorithm have good results on stereoscopic correspondence and can be extended to match planar surfaces and facilitate 3D reconstruction of planar facets in calibrated stereo-images.

2 PROBLEM FORMULATION

In this section, we introduce notation and we formulate the problem. First, we deal with two different views of a 3D scene that we assume to be in affine conditions so we recall the affine camera model. Second, we review the matching algorithm of a pair of contours which have two principles steps: global and local matching. Otherwise, the use of parameterisation is a key step in matching process since it allows to have equal number of points in all considered contours, indeed we considerate affine arc length reparametrization (F. Ghorbel 1998) to normalize closed curve under affine transformation. Such normalization is based on B-spline functions which are robust relatively to multiple derivatives (F. Chaieb and F. Ghorbel 2003).

2.1 The affine camera model

The affine camera, introduced by Mundy and Zisserman (Mundy and Zisserman 1992) corresponds to a projective camera with it's optical center on the plane at infinity, in consequence all projections rays are parallel. It's projection matrix has only eight degree of freedom and can be expressed in the following form:

$$P_{aff} = \begin{pmatrix} P_{11} & P_{12} & P_{13} & P_{14} \\ P_{21} & P_{22} & P_{23} & P_{24} \\ 0 & 0 & 0 & P_{34.} \end{pmatrix} \quad (1)$$

The elements P31, P32 and P33 are equal to 0. The affine camera model can be decomposed in: (1) a 3D affine transformation between world and camera coordinate system; (2) parallel projection onto the image plane; and (3) a 2D affine transformation of the image. In homogenous coordinates we can express the affine camera as:

$$x = MX + t, \quad (2)$$

where $M = [M_{ij}]$ is a 2×3 matrix with elements $M_{ij} = Pij/P34$ and $t = (P14/P34, P24/P34)^T$ is a 2 vector giving the projection of the origin of the world coordinate frame (X = 0). A key property of the affine camera model is that it preserves parallelism. This approximation to the full perspective camera model works quite well when object size and depth is small compared with the distance between the camera and object.

2.2 Matching contours subscribing shapes: Global matching

Let's consider right and left projections of the same 3D object and let's γ_1 and γ_2 be two parametrizations of external closed contours which can be assumed to be related by an affine transformation and let $\triangle$ be the determinant of the matrix composed by the four complex Fourier coefficients of the coordinates. In (F. Ghorbel 1998), author proposed two sets of invariant descriptors I and J which are respectively given respectively by (eq. 3) and (eq. 4).

$$I_{k_1} = |\triangle_{k_1, k_0}|, I_{k2} = |\triangle_{k_2, k_0}|, I_k = \frac{\triangle_{k, k_0}^{k_1 - k_2} \triangle_{k_1, k_0}^{k_2 - k} \triangle_{k_2, k_0}^{k - k_1}}{|\triangle_{k_1, k_0}^{k_2 - k - \alpha}||\triangle_{k_2, k_0}^{k - k_1 - \beta}|} \quad (3)$$

for all $k - \{0, k_0, k_1, k_2\}$

$$J_{k_1} = |\triangle_{k_1, k_3}|, J_{k_2} = |\triangle_{k_2, k_3}|, J_k = \frac{\triangle_{k, k_3}^{k_1 - k_2} \triangle_{k_1, k_3}^{k_2 - k} \triangle_{k_2, k_3}^{k - k_1}}{|\triangle_{k_1, k_3}^{k_2 - k - \alpha}||\triangle_{k_2, k_3}^{k - k_1 - \beta}|} \quad (4)$$

for all $k - \{0, k_1, k_2, k_3\}$

Indeed matching closed contours leads to compute euclidean distance between their corresponding descriptors which is close for similar shapes and large for different ones due to the stability property (F. Chaker, F. Ghorbel, and M.T. Bannour 2007), (F. Chaker, F. Ghorbel, and M.T. Bannour 2009), (F.Chaker and F.Ghorbel 2010).

2.3 Matching points under contours: Local matching

Since we have previously established global matching between contours in stereo images. For each pair of contour which are said to be similar, we have to determine point-to-point correspondence; Formally, if we consider O_1 and O_2 two closed curves normalized by affine arc length, they are said to be related by an affine transformation if:

$$h(l) = \alpha A f(l + l_0) + b \quad (5)$$

where f and h are the reparametrization of O_1 and O_2 by the affine arc length.α is a scale factor and A is an element of $SL(2, R)$, b is an IR^2 vector of translation and l_0 is the shift value. So estimation of affine motion can be resumed on estimation of it's three parameters: affine matrix A, shift value l_0 and scale factor α if we consider a normalization under translation (M. Saidani 2011).

3 EXPERIMENTS AND RESULTS

Our interest is particularly focussed on small archeological 3D objects which are found on excavations and donated to museums. The range of objects includes coins, jewellery and other personal items, containers and tools. The types of material these objects are made from include bronze, iron, wood and worked bone. The objects date from prehistoric times to the 19th century. Storing 3D model of theses precious objects is necessaire but require laser scanning techniques in almost the time which are likely the best and most reliable

Figure 1.   Stereo pair of small archeological objects.

Table 1.   Euclidean distance between Affine descriptors of the left (column) and the right (line) contours subscribing shapes from stereo pair, for each contour the smallest distance is obtained between two views of the same object and this due to the stability of affine invariant descriptors.

	Mirror	Knife	Cup	Bottle
Mirror	**0.23**	1.6	0.76	0.45
Knife	1.6	**0.17**	0.66	0.98
Cup	0.76	0.66	**0.11**	1.03
Bottle	0.45	0.98	1.03	**0.37**

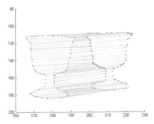

Figure 2.   Matching contours of "Cup" object extracted from right and left images.

Figure 3.   Matching results for all localized contours subscribing shapes in stereo pair.

for these purposes. However, the costs associated with these techniques prohibit their use. In our experimental study we consider a stereo pair of some archeological objects made in bronze (figure 1) and we proceed by the segmentation step for matching closed contours subscribing objects. If we consider the duality between closed contours and homogeneous regions for shape we can state that each closed contour represents the boundaries of the observed object surface. To obtain the whole 3D reconstruction, we will use generated closed contours up to some scale factor to fill the lack of informations about surface.

3.1   Stereo matching of contours

We consider typical stereo camera configuration used for capturing stereo images. The two cameras are mounted such that their optical axes are coplanar and aligned in parallel. The left and right cameras in the stereo system capture a pair of images simultaneously. In theses conditions we can assume that the perspective projection of a point according to it's coordinates in the left and the right image: (u, v) and (u_1, v_1) can be modelised by an affine transformation:

$$(u_1, v_1) = A_{aff}(u, v) + T, \qquad (6)$$

In our stereoscopic matching, we start by global matching of contours subscribing objects in two views using euclidean distance between their affine invariant descriptors (Table 1). We underline that this step is related to the segmentation procedure which produces a set of closed curves related to objets boundaries in the stereoscopic pair. Next step is matching points under closed contours which are said to be similar through the use of affine parameters estimation (figure 2). This step is done iteratively for all pairs of contours and final results are illustrated by (figure 3).

3.2   3D recovering of object's structure

There are two computational subproblems associated with 3D reconstruction from stereo images: feature correspondence and structure estimation. Since the first problem is explained above, we are going to focus on the second challenge. If the intrinsic and extrinsic parameters of the cameras are known via calibration procedure (Faugeras 1993), a 3D reconstruction in absolute coordinates is possible. However, the accuracy of the reconstructed structure is sensitive to the accuracy of calibration parameters. Moreover, any errors in solving the correspondence problem between two images also affect the accuracy of the reconstruction: matching errors induce to reconstruction errors. The problem of structure estimation in stereovision can be solved using a simple procedure known as triangulation (Faugeras 1993). In our experiments, a 3D reconstruction of matched points can be established for object profile. whereas, theses points are insufficents for surface estimation. For this goal, we proceed by generating iteratively closed curves up to some regular scale factor within the considered curves. The matching process can easily be applied to these new generated curves since reparametrization conserve the order of points. Such process produce dense point-to-point matching and leads to a first reconstruction of 3D points that represent the the whole observed planar suface (figure 4(a)), moroever, if we consider a simple geometry of object we can retrieve the 3D reconstruction of the whole volume (figure 4(b)).

4   CONCLUSIONS AND FUTURE WORKS

In this work, we have reviewed a stereo matching method based on matching contours by affine invariant descriptors and a dense point-to-point matching through the use of affine transformation estimation

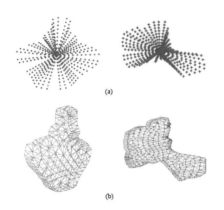

(a)

(b)

Figure 4.   (a) 3D contour reconstruction for planar surfaces,
(b) Final 3D models.

from reparameterized curves. Such matching have been exploited to do a 3D reconstruction of small archeological objects. Experimental study on real data has been accomplished and results have been exposed. In future works, we will apply our method to assist other conventional matching techniques in order to consolidate their results. Otherwise, we will use this method for estimating fundamental matrix of sensors in uncalibrated conditions. As another perspective, we intend to extend the method to reconstruction of piecewise planar models with the use of multiple views when the geometry of the object is complicated.

REFERENCES

Faugeras, O. (1993). *Three Dimensional Computer Vision: A Geometric Viewpoint* (2nd ed.). London: The MIT Press.
F. Chaieb and F. Ghorbel (2003). Un procd de reparame-trage p2-invariant de courbes partir de quelques longueurs quiprojectives. In *TAIMA*.
F. Chaker and F. Ghorbel (2010). Apparent motion estimation using planar contours and fouriers descriptors. In *VISAPP*.
F. Chaker, F. Ghorbel, and M.T. Bannour (2007). Con-tour retrieval and matching by affine invariant fourier descriptors. In *IAPR Conference on Machine Vision Applications*.
F. Chaker, F. Ghorbel, and M.T. Bannour (2009). Content-based shape retrievial using different affine shape descrip-tors. In *VISAPP*.
F. Ghorbel (1998). Towards a unitary formulation for invariant image description: application to image coding. In *Annales de Telecommunications*.
Grimson, W. E. L. (1985). Computational experiments with a feature based stereo algorithm. In *IEEE Trans. Pattern Analysis and Machine Vision*.
M. Saidani (2011). Shape matching by affine movement estimation for 3d reconstruction. In *QCAV*.
Mundy, J. and A. Zisserman (1992). Geometric invariance in computer vision. In *MIT Press,*.
Trucco, E. and A. Verri (1998). Introductory techniques for 3-d computer vision. In *IEE*.

Computational Modelling of Objects Represented in Images – Di Giamberardino et al. (eds)
© 2012 Taylor & Francis Group, London, ISBN 978-0-415-62134-2

Establishment of the mechanical properties of aluminium alloy of sandwich plates with metal foam core by using an image correlation procedure

H. Mata, R. Natal Jorge, M.P.L. Parente & A.A. Fernandes
IDMEC – Faculty of Engineering, University of Porto, Porto, Portugal

A.D. Santos
INEGI – Faculty of Engineering, University of Porto, Porto, Portugal

R.A.F. Valente
Department of Mechanical Engineering, University of Aveiro, Aveiro, Portugal

J. Xavier
CITAB, University of Trás-os-Montes and Alto Douro, Vila Real, Portugal

ABSTRACT: Cellular materials are very common in nature, being present on natural materials such as cork, wood, coral, bones, etc. This configuration has some advantages, since it combines high stiffness with a relatively low density [1]. A requirement for the application of new materials in modern industry is a thorough documentation of their properties and mechanical behaviors. In addition, the success of numerical simulation for analyses of structures and components rests upon the accuracy and efficiency of the applied material models and the parameters of those models [2]. The main goal of this work is to study, both experimentally and numerically, sandwich panels composed by a core of aluminum foam separating two sheets of aluminum, in order to assist in the design process of forming. DIC (Digital Image Correlation) is used to perform experimental characterization of sheets of aluminum from sandwich plates showing to be an efficient and reliable procedure to obtain both the flow curve as well as its anisotropic characteristics.

1 INTRODUCTION

The cellular materials are very common in nature, such as cork, wood, coral, bones, etc., and this configuration has some advantages, since it combines high stiffness with a relatively low density. However, due to difficulties in production and reproduction, only recently the use of cellular metallic materials began to be more common. These are obtained by manufacturing processes not yet fully controlled, and that is why its use is more limited, since these processes have been undergoing improvements and the quality and reproducibility of properties of the final product as well. This trend is due to the incentive industry by recognizing the great potential of application of this class of materials, combining high stiffness to low density, together with the capacity of energy absorption, vibration absorption and thermal insulation, leading to its use for various applications in the aerospace, and automotive industries, among others. Although the market for this type of material is already quite significant, it has been growing rapidly due to the improvement of manufacturing processes as well as calculation models [2–3].

The design of passive safety systems in transportations still have a great potential for development as a way to reduce deaths and injuries, which is also associated to the economic costs and social impacts associated with this problem. On the other hand, from an environmental standpoint, the use of advanced composite materials to this end can also represent an optimized level of energy efficiency. The impact energy absorption, with the use of a well-designed light-weight protection system, is directly related to the thermal efficiency and consumption of the engines, thus leading to a lower level of greenhouse gases sent to the atmosphere. It is within this framework that fits and makes sense the establishment of the mechanical properties of the aluminum sheets, which is associated to the sandwich plates with metal foam core.

This work describes a classical procedure to obtain the mechanical properties of the aluminum sheet outer layers using the traditional measure of displacement – force values, taking samples with three selected orientations: $0°$, $45°$, 90^v to rolling direction, with load cell and displacement transducer on the tensile test and using another different procedure called digital image correlation method using *ARAMIS* system in the same test.

DIC is an optical methodology for the measurement of complex materials and structures involving 3D

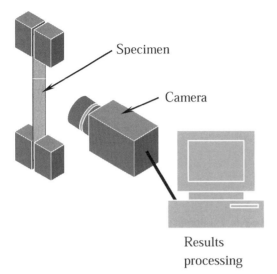

Figure 1. *Schematic illustration of the ARAMIS system.*

Aluminum Plates

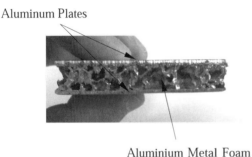

Aluminium Metal Foam Core

Figure 2. Structure of the composite.

Figure 3. Aluminum metal foam (Alporas foam $\bar{\rho} = 9\%$).

Figure 4. Aluminum sheets EN AW 5754.

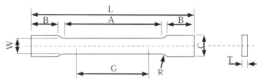

G=50, W=12.5, R=20, L=200, A=75, B=50, C≈20, T- Thickness [mm]

Figure 5. Specimens used on the tensile test.

2.1 Metal foam core

The foam is a closed cell foam Alporas (Figure 3), developed in Japan by Shinko Wire Company in the 90s, being an aluminum alloy AlCa1,5Ti1,5, with a density of 0,25 g/cm^3, has a relative density ($\bar{\rho}$) of about 9% and porous size about 4–6 mm [5].

2.2 Aluminum sheets

As previously mentioned, the aluminum sheets considered in the composites for the present work have 1 mm thickness, and are composed by an aluminum alloy (AlMg$_3$) EN AW 5754 none heat treatable, with a typical modulus of elasticity of about 70 GPa and a Poisson ratio about 0.33 [5].

3 EXPERIMENTAL SETUP

The characterization of the aluminum plate is done by performing tensile test. To this characterization three different directions of rolling 0°, 45° and 90° were being considering. To this end specimens were cut from the panel with the dimension shown in Figure 5.

Typically the aluminum alloys present a relative degree of anisotropy, and considering these three directions of rolling we can identify and measure the difference between those three directions.

In these work two methods to obtain the characterization were used:

- A standard mechanical system, with load cell and displacement transducer – global measurement;

deformations during loading [4]. The system provides a full-field, non-contact measurement of the material deformation based on digital image correlation (DIC). Figure 1 schematically represents the photomechanical set-up in which a high-resolution charged-coupled device (CCD) camera is coupled with a mechanical tensile test. The advantages of using this optical method are the simple procedure to obtain the stress-strain curve to compare with the results of standard system and principally a convenient method to determine the strain evolution across the whole region of interest, on the plane of the sheet, to obtain the r-values.

2 MATERIAL DISCRIPTION

The composite structure investigated in this work is a composite composed of three layers. The first out layer in aluminum sheet with 1 mm of thickness bonded a core of 8 mm metal foam, also in aluminum and another out layer also bonded in aluminum sheet with 1 mm thickness, Figure 2.

Figure 6.   Tensile test and *ARAMIS* system.

Table 1.   Number of specimens tested by direction.

0°	45°	90°	Total
1	1	1	3

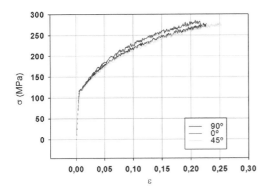

Figure 7.   True stress – strain curve of the 3 specimens.

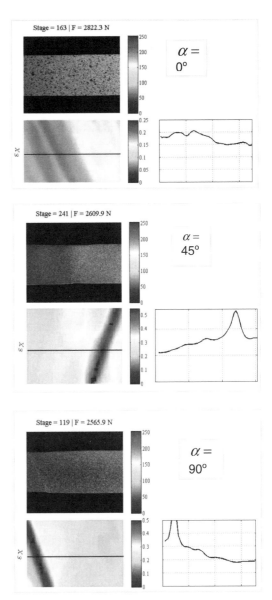

Figure 8.   Detail of the strain and necking obtained by Aramis system.

- A Digital Image Correlation method, providing local state of deformation across the surface of interest (Figure 6).

The Aramis® DIC-2D v6.02 software was used in this work [4]. The optical system was equipped with an 8-bit Baumer Optronic FWX20 camera (resolution of 1624 × 1236 pixels), coupled with an Opto Engineering Telecentric lens TC 23 06®. A speckle pattern was created on the central part of the specimens by using black-to-white spray. Images were recorded with an acquisition frequency of 1 Hz. Regarding the measuring parameters used in the Aramis system, a subset size and subset step of 15 × 15 pixels were selected, in a compromise between correlation and interpolation errors. The strain field was then numerically derivate from the displacements on a strain length base of 5 subsets.

4   RESULTS

These tests were performed; one for each direction and results are presented on the graph of Figure 7.

Observing the graph, the curves are similar in progress, but with slightly different depending on each direction, obtained using the traditional method of measure. Higher hardening for 0° when compared to 45° and for 90° the curve presents a serrated behavior typically of Al/Mg aluminum alloys at selected strain rate.

The curves of the graph on Figure 7 define the properties of the aluminum sheets.

Table 2. Lankford's r-value.

	Initial yield stress [MPa]	r-value
r_0	116.45	0.61
r_{45}	108.57	0.84
r_{90}	116.45	0.82

The plastic parameters of the model are the stress – strain curves resulted of the characterization of aluminum sheet and r-values to study the anisotropy [6].

Using the DIC system we obtain the local deformation and necking for any direction. Figure 8 shows the output of this optical system and it is visible the local map of strains obtained by the displacement measure on the two directions of the black points captured at each time by the camera.

It was also obtained for any direction and for any test the respective r-values as are shown in the Table 2.

In sheet metal forming applications we are generally concerned with plane stress conditions. Consider x, y to be the "rolling" and "cross" directions in the plane of the sheet; z is the thickness direction and considering x coincidente with direction 1, y coincident with direction 2 and z coincident with 3 (Figure 9), simplifying considering different strengths in different directions in the plane of the sheet, which is called planar anisotropy. $\sigma_{33} = 0, \sigma_{23} = 0$ and $\sigma_{31} = 0$ [6].

In a simple tensile test performed in the x-direction in the plane of the sheet, the flow rule for this potential defines the incremental strain ratios. Therefore, the ratio of width to thickness strain, often referred as Lankford's r-value, are

$$r_0 = \frac{d\varepsilon_{22}}{d\varepsilon_{33}} \qquad (1)$$

$$r_0 = \frac{d\varepsilon_{11}}{d\varepsilon_{33}} \qquad (2)$$

And for an angle of $45°$

$$r_{45} = \left(r_{45} + \frac{1}{2} \right) \left(1 + \frac{r_x}{r_y} \right) \qquad (3)$$

These r-values can be used directly to obtain the anisotropic yield stress ratios, the parameters used by the Hill criteria anisotropic model (Table 2) [7]. This r-values are also used to obtain some parameters for alternative anisotropic yield criteria, e.g. Barlat'91 [8], currently applied to modelling numerically the yield locus of aluminium sheets.

5 CONCLUSIONS

The exposed experimental work allows establishing the anisotropic mechanical properties of aluminum alloy of sandwich plates with metal foam by using image correlation procedure. Besides, it is possible to apply this characterization in models to the numerical simulation of complex loading cases, and try

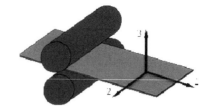

Figure 9. Coordinate system.

Table 3. Anisotropic yield stress ratios.

R_{ij}	
R_{11}	1
R_{22}	1.09
R_{33}	0.96
R_{12}	0.87

to numerically reproduce the real behaviour of this material in demanding applications.

Another conclusion from this characterization is the possibility to obtain anisotropic parameters which promote the study of the anisotropic effect on aluminum sheet as a resistive structure of the composite. The adopted DIC methodology has permitted an efficient and reliable determination of material properties (aluminum alloy sheet) namely the flow curves for different orientations to rolling direction as well as its anisotropic r-values.

ACKNOWLEDGEMENTS

The authors truly acknowledge the funding provided by Ministério da Educação e Ciência – Fundação para a Ciência e a Tecnologia (Portugal), by FEDER/FSE, under grant PTDC/EME-TME/098050/2008.

REFERENCES

M.F. Ashby et al., Metal foams: a design guide, Oxford, Butterworth-Heinemann, 2000.
A.G. Hanssen et al., Validation of constitutive models applicable to aluminium foams, International Journal of Mechanical Sciences 44 359–406, 2001.
H. Mata et al., Numerical modelling and experimental study of sandwich shells with metal foam cores, Key Engineering Materials 504–506, pp. 449–454, 2012.
J. Xavier et al., Stereovision measurements on evaluating the modulus of elasticity of wood by compression tests parallel to the grain, Construction and Building Materials 26, 207–215, 2012.
Technical Data Sheet, Gleich Aluminum, (2010).
Hill, R., A Theory of Yielding and Plasticity Flow of Anisotropy Metals, Proceedings of the Royal Society of London Series a-Mathematical and Physical Sciences, 193 (1948) pp. 281–297.
Abaqus, Inc. Abaqus Analysis User's Manual Version 6.10.
F. Barlat, D.J. Lege, J.C. Brem, A six-component yield function for anisotropic metals, International Journal of Plasticity 7 693–712, 1991.

Computational Modelling of Objects Represented in Images – Di Giamberardino et al. (eds)
© 2012 Taylor & Francis Group, London, ISBN 978-0-415-62134-2

Morphological characterisation of cellular materials by image analysis

A. Boschetto, F. Campana & V. Giordano
Dip. di Ingegneria Meccanica e Aerospaziale, Sapienza Università di Roma, Roma, Italy

D. Pilone
Dip. Ingegneria Chimica Materiali Ambiente, Sapienza Università di Roma, Roma, Italy

ABSTRACT: Physical and mechanical properties of metallic foams derive from their micro- and macro-structure. Unfortunately these foams have a very heterogeneous structure so that their properties may show a large scatter. In this paper image analysis has been used to determine size, morphology and distribution of cells in three aluminium alloy foams produced by means of different manufacturing processes. The study highlighted that difficulties deriving from cell cavity coalescence can be overcame by using the assisted watershed method. The morphometric analysis of cells showed that cell size, shape and distribution are strictly related to the manufacturing process. The proposed and analysed method appears to be a useful tool to correlate cell morphology with the manufacturing parameters and then to tailor foam properties to given applications.

1   INTRODUCTION

Metallic foams are cellular materials with a stochastic distribution of voids that can be regular and closed (foam) or complex and interconnected (sponge) (Wad01).

Many structural applications get benefit from the adoption of closed cell foams due to excellent stiffness to weight ratio and good shear and fracture strength (e.g. lightweight structures and sandwich cores). They have exceptional ability to absorb energy at almost uniform load allowing to design compact and light energy absorbers. Moreover they have an improved damping capacity that is larger than that of solid metals by up to a factor of 10 and by high flexural vibration frequencies. This makes them suitable for mechanical dumping and vibration control, in particular in the field of industrial tools. Open cell structures, thanks to high surface/volume ratio and to the possibility of controlling the pore size (Ash00), have potential interesting applications as support for catalytic converters for diesel engines, parts of heat exchangers, flow-through devices and electrodes for NiMH and NiCd batteries.

Without any doubt foam performances have a close relationship with their cell topology. Void density and pore size represent the most relevant variables affecting global structural response (Fus99). Shapes and spatial distribution of pores and struts are other important characteristics to understand local stress-strain distribution during loading, since the global response derives from the complexity of the meso-scale behaviour (Olu02, Zhu00 & Wen08). Systematic analysis and quantification of cell morphology may

improve not only the understanding of foam mechanical behaviour but also the evaluation of properties obtained with different manufacturing techniques (Lhu09), as anisotropies may be cause of response scattering.

Despite these preliminary remarks there are in literature only few attempts to link, by means of image analysis, cell size and distribution to either manufacturing parameters or macroscopic material properties. Generally speaking many X-ray tomography acquisitions have been done to better understand foam structure or its behaviour under different load conditions. For example in (Lie11) X-ray tomography has been applied to investigate the role of void morphological defects in fatigue behaviour, founding that it is less sensitive to foam missing walls than the quasi-static strength. In (Kol08) the effect of pressurised foaming on the pore size and on the distribution of solid fraction, as well as the effect of slow and fast cooling on the microstructure is investigated by means of analyses able to separate individual cells of a single void according to the watershed algorithm. A very recent paper (Lie11) introduced a model to predict the scatter of macroscopic properties in dependence of the network of struts. Starting from computed tomography data, statistical characteristics of the foam geometry were estimated. That paper highlighted that there are deviations between experimental and simulated values that could be explained by the simplifications used in the model. In fact the model has a more homogeneous structure than the real foam which shows large variations in the strut thickness.

A robust correlation between stochastic void distributions and Finite Element simulations of the foam mechanical response is one of the most debated aspects, often investigated through 3D reconstructions of foam specimens (Lhu09, Rod08). In addition some attempts to correlate void morphology and foam characteristics by a proper selection of shape indicators are also available. Considering that there are no standards available for characterising the structural morphometric parameters of foams by computed tomography, the approach used for bone morphometric parameters characterization can be adopted (Saa09). As far as the 2D analysis is concerned the main used parameters are mean cell size and cell size distribution, fraction of solid contained in the cell, connectivity between cells (connectivity is 0 when all the cells are separated and 1 when all the cell are connected) and cell orientation.

Considering the role played by the cell morphology on the final mechanical response, as in part already shown in previous research on aluminium foam characterisation (Cam08) and manufacturing (Qua11), this paper discusses how to evaluate relevant morphological components, by 2D image processing, for void shape measurements in metallic foams. In particular, after a general description of two type of aluminium foams here investigated (one obtained by an infiltration process and the other two by compact powder foaming), a general scheme of image analysis starting from 2D acquisitions is presented and discussed to highlight the critical aspects related to the data post-processing: (i) quality of the image (such as acquisition quality or calibration); (ii) type of segmentation (with and without a watershed algorithm to separate individual grains); (iii) morphological components to properly characterise cell shapes with different grades of regularity.

2 MATERIALS AND SPECIMEN PREPARATION

In this work three cellular metallic materials are investigated.

The first kind of material, hereafter referred to as AlSi7Mg3, was produced in our laboratory using an AlSi7Mg0.3 alloy (also referred with EN AC 42100). The specimen has a cylindrical shape, 28 mm in diameter and 50 mm in height. It was obtained by an infiltration process (also called replication) defined as a general three-step procedure (Deg02): (a) preparation of a leachable salt pattern; (b) infiltration of the pattern by pressure followed by solidification; (c) dissolution of the pattern. The process parameters were: plug speed of 32 mm/s, mould temperature of 500°C, melting temperature of 700°C, holding pressure of 25 bar, holding time of 30 s. In order to reduce the production costs, commercial sodium chloride, NaCl, was used as pattern material. Two graphite crucibles were introduced in a muffle; the first one was used to dry the salt, 9 g, the second one was used to melt the aluminium, about 40 g. The salt was sieved by means of a 3 mm mesh sieve in order to eliminate all the smaller

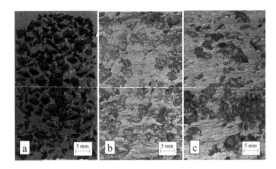

Figure 1.   Macrographs showing the structure of AlSi7Mg3 (a), AlSi7 #1 (b), AlSi7 #2 (c) specimens.

particles. Afterwards it was put in a mould cylindrical cavity, 28 mm in diameter and 100 mm in height. The molten aluminium was cast and a hydraulic cylinder was rapidly pushed forward to fill the mould and infiltrate the melt in the salt bed. Then the salt was dissolved in water.

The other two foams, hereafter referred to as AlSi7 #1 and AlSi7 #2, were supplied by IFAM in Bremen, Germany. They are obtained by compact powder technology from AlSi7 alloy, with nominal density 690 kg/m^3 and 1090 kg/m^3 respectively (Fra09).

Several samples, for every considered alloy, were sectioned minimising cell wall damage. They were physically sectioned by a metallographic precision cut-off machine. To this end a Minitom Struers (125 mm in diameter and 0.5 mm thick wheel, with bakelite bonded SiC abrasive) was used. A very slow cut was performed to preserve the integrity of the material. The surfaces were painted using a black dye in order to obtain a good contrast, essential for image analysis. The specimens were ground with a series of SiC papers and afterwards polished with 1 and 0.3 μm alumina.

Specimens' images have been captured by an Image Sensor Type CCD with 5.04 Megapixel resolution and f\2.8 Carl Zeiss optic lens. All the photos were taken in the same conditions putting the commercial digital camera on a tripod. Figure 1 shows the photos used henceforward for image analysis.

3 ELABORATION METHODS

Every image has been processed according to the procedure summarised in Figure 2 and detailed further down.

Calibration is the first step. It has the aim of adjusting perspective and radial distortions arising from aberrations in the sensor or from positioning errors (Rus06). It has been performed by an inverse methodology that fits the acquired image onto an experimental reference frame. For this reason the specimens have been microindented in nine known positions (space coordinates) by a microhardness tester. They have been adopted to find, by a least-square fitting, a pseudo-inverse solution of the transformation matrix, that

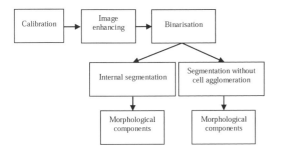

Figure 2. Image processing workflow.

relates the space coordinates to the image coordinates according to Faugeras intrinsic and extrinsic parameters (Fau93). By applying this procedure, as expected, no perspective error was appreciable and the maximum radial distortion was found at the maximum distance from the focus. The resulting sensitivity was $30.7\,\mu m$/pixel. After the cropping performed to include in the images only the specimen surfaces, every image has a resolution of 911×1625 pixel^2.

Image enhancing represents the second step. It consists of a set of operations that take place with the goal of improving the quality of pictorial information. Firstly, the original images have been enhanced by means of contrast equalisation and brightness correction. Then a median filter has been applied in order to avoid random error and impulse noise. This noise is caused by light reflection on the smooth surfaces of the faceted cells resulting from the infiltration process. The results have been converted to grayscale and then a conventional opening (neighbour coefficient equal to 1, three iterations) has been applied to remove very small features within the cavities, leaving untouched the microporosities on the cell walls.

The third step is the *binarisation*. It separates features (cell cavities) from background. The applied threshold value has been found by Kittler and Illingworth algorithm (Kit86).

After binarisation two different *segmentations* have been applied to analyse the agglomeration of overlapped cell. It is evident from Figure 1 that in many regions the analysed materials often present open cell structures, characterised by a lot of overlapping cavities. Since the overlapping area morphology characterises the material performance in many kind of applications (e.g. thermal, acoustic, fluidodinamic, . . .), it is necessary to identify and reconstruct the separated cells that form the agglomerated void. Besides, from a structural point of view, it is important to regard the agglomerate as a unique region because it is characterized by incomplete walls that do not take part to the deformation processes, while the interconnection walls of closed cells deform under loading. For these reasons an *internal segmentation* by assisted watershed has been applied to characterise agglomeration morphology of single open cells, while a *segmentation without cell agglomeration* has been run through automatic watershed to simply localise touching regions of single agglomerations. The assisted

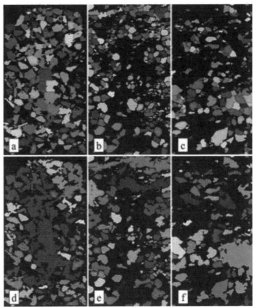

Figure 3. Labeled regions obtained with (a, b, c) and without (d, e, f) internal segmentation for the different investigated specimens: AlSi7Mg3 (a, d), AlSi7 #1 (b, e), AlSi7 #2 (c, f).

watershed algorithm creates the seed for the region segmentation by means of an user panel specifically built to allow the operator to directly see the original image overlapped to the catchment basins. It has been applied on gradient filtered images. The image including the segmented regions is calculated by subtracting the watershed boundaries to the original image. The automatic watershed adopts seeds that have been automatically reconstructed by means of morphological erosion referring to a structural element, defined as a disk shaped region having a radius of 10 pixel.

The last step of the applied procedure concerns with the computation of the *morphological components* according to the labelling results. Figure 3 shows the feature distribution resulting from the two segmentation algorithms and subsequent labelling.

Figures 3a, 3b, 3c are obtained by means of internal segmentation, Figures 3d, 3e, 3f by means of simple clustering made without cell agglomeration. The meaning of each figure may be understood by comparing them with the photos reported in Figure 1. The morphological indicators computed after labelling are discussed in the next section.

4 RESULTS AND DISCUSSION

The first relevant parameter useful to characterise foams represents the spatial distribution of cell density. It has been computed by partitioning the matrix data in terms of percentage of voids per unit area. The distribution has been evaluated according to the assisted watershed segmentation, that allowed to perform internal cell segmentation (Figure 4).

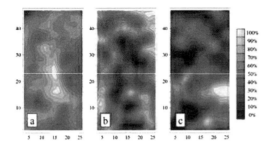

Figure 4. Contour plot showing void density distribution in AlSi7Mg3 (a), AlSi7 #1 (b), AlSi7 #2 (c) specimens.

Table 1. Cell intersection parameter, estimated density and measured density for each specimen.

	Cell Intersection Parameter [%]	Estimated density [kg/m³]	Measured density [kg/m³]
AlSi7Mg3	2.05	910	950
AlSi7 #1	3.57	1290	620 ± 110
AlSi7 #2	4.14	1470	860 ± 110

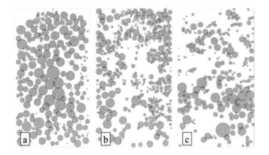

Figure 5. Spatial distribution of equivalent disks in AlSi7Mg3 (a), AlSi7 #1 (b), AlSi7 #2 (c) specimens.

This figure shows that the specimen obtained by means of the infiltration process (AlSi7Mg3) presents a void density quite uniform with some exceptions near the outer walls. This reflects the cell distribution imposed by the manufacturing process since the salt, used as *pattern*, has been chosen with a very tight granulometric distribution in order to adequately fill the mould. The specimens made by the compact powder process show a less uniform distribution with some very compact zones. In this case this may be attributed to the fact that bubble formation is not controlled by a *pattern* but by the gas evolution during foaming.

In Figure 5 an overview of the cell overlapping, or agglomeration, is given for each specimen according to the assisted watershed. It has been obtained by positioning cell equivalent disk areas at their centroids.

These images allow to understand the position of the agglomerations and their configuration. For example in the case of the infiltrated specimen (AlSi7Mg3) a large agglomeration is located at the specimen centre and it is mainly due to the overlapping of rather similar voids. On the other hand in the case of AlSi7 #1 agglomerations seem due to the overlapping of small voids surrounding a larger one. In the last specimen, AlSi7 #2, both kinds of morphologies may be recognised.

The *cell intersection parameter*, defined as the ratio between the total area of the intersections and the specimen area, is an interesting indicator since it gives information about the uniformity of void distribution. It is reported in Table 1. The lowest cell intersection parameter is obtained for AlSi7Mg3, which is the foam made by the infiltration process. Its value is 2.05% while the specimens made by compact powder technology are characterised by higher values: 3.57% for AlSi7 #1 and 4.14% for AlSi7 #2. This suggests that the infiltration foam has more homogeneous cell distribution with less agglomerations in comparison with the others.

Table 1 reports also the density of the three specimens estimated according to the acquired images. This *estimated density* has been found assuming that 3D voids are cubic, with a length computed by the equivalent square area of the 2D void that is obtained from the segmentation. For AlSi7Mg3 the estimated density is very close to the measured one (-4%), while for the other two acquisitions a relevant discrepancy is found due to a certain scatter between nominal and actual densities (maximum deviation from nominal values, measured on other similar specimens, of over 25%!) and to a possible estimation error related to the necessity of applying other formulas for the equivalent void volume used in the *estimated density* (in the case of spherical voids the estimated densities become respectively 970 and 1100 kg/m³).

By coupling the analysis of the *cell intersection parameter* with the density spatial distribution, important information about the mechanical response and thus about product development design, may be found. Low material density with low cell intersection parameter is desirable to gain improved energy absorption capability, since it represent closed cell structure.

From the frequency histogram of the *equivalent disk diameter* information about size distribution of voids can be obtained. Figure 6a shows the results concerning with internal segmentation. From the comparison among these three histograms the the cell size distributions' difference may be easily associated to manufacturing considerations. The histogram related to AlSi7Mg3 highlights that this foam presents a wide scattering that can be described by two distributions. The first one is characterised by a normal distribution (P value > 0.5) with a mean of about 2 mm. This value is close to the salt granulometry and confirms that this process may easily control the cell size. The second distribution is characterised by a high frequency peak at a diameter under 0.2 mm that reveals the presence of porosities caused by the fast cooling. Concerning the other two specimens, both obtained by compact powder technology, their histograms show that the void

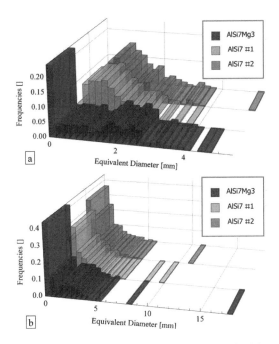

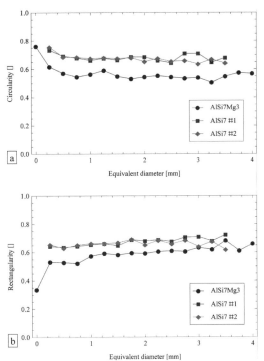

Figure 6. Equivalent diameter distributions obtained by means of internal segmentation (a) and simple clustering (b).

Figure 7. *Circularity* (a) and *rectangularity* (b) trends as a function of the *equivalent diameter.*

size distributions are quite similar and that they can be perfectly described by two Lognormal distributions (P value > 0.8) with a mean of about 1.25 mm.

Figure 6b shows the same kind of elaboration on the data obtained without *internal segmentation*. It gives results with some differences in comparison with those in Fig 6a. This may be extremely significant from the structural point of view. As already mentioned shape uniformity and closed cell structure are important for the structural response of foams, particularly for energy absorption, and hence the presence of extended agglomerated voids may significantly affect the response robustness. In the case of the investigated specimens, according to Figure 6b, it can be evinced that although the infiltration foam has the highest quantity of smallest voids it has also more large agglomerations than the other two. This is clearly shown by the bin related to the highest value of equivalent diameter, found for the AlSi7Mg3 foam at 19 mm.

For a technological characterisation of metallic foams it is of paramount importance to analyse also the regularity of the cells (circular, elliptical, jagged or with lobes). It can be associated to local stress concentration or response anisotropies. First of all cell shapes may be characterised by means of *circularity* that is the ratio of the equivalent disk perimeter to the void perimeter length. Circular shapes yield the maximum circularity value of 1 while this value decreases for complex shapes. A further important indicator is the *rectangularity* defined as the ratio of the object's area to the area of its minimum bounding box. Very small values of this parameter characterise slender objects.

In Figure 7a the *circularity* as a function of the equivalent diameter, for each specimen, is reported. The plot shows that its value is quite constant over the range 0.6–0.8, and the AlSi7Mg3 specimen has a lower *circularity* and, according to its *rectangularity* shown in. Figure 7b, a more slender void shape. Also in this case the difference among AlSi7Mg3, AlSi7 #1 and #2 are justified by their specific manufacturing processes, since the salt pattern adopted in the infiltration process determines the formation of jagged and slender cells.

5 CONCLUSIONS

In the study of cellular materials the image analysis is an important tool to better understand void characteristics. Nevertheless many aspects are difficult and not standardised yet. In this paper the problem of a correct investigation of the cell cavities has been studied by setting up dedicated procedures.

3D segmentation of cells in open-cell foams and, more generally, the coalescence of cell cavities creates many difficulties to a proper morphological analysis of void structures. Watershed algorithms have been applied to manage this problem. By assisted watershed the subdivision of overlapped cells is investigated in terms of *cell intersection parameter* and distribution of the *equivalent disk diameter*. They demonstrated good performance in highlighting the differences determined in the foam structure by the two adopted manufacturing processes (infiltration and compact powder technology). Moreover associating this analysis to the results related to automatic watershed, a clear comprehension of the mechanical response may be achieved, since open cell structures may be seen as structural defects if they are infrequent and not regularly distributed.

REFERENCES

Ashby, M.F., Evans, A.G., Fleck, N.A., Gibson, L.J., Hutchinson, J.W., Wadley, H.N.G. 2000. Metal Foams: A Design Guide, Butterworth-Heinemann publications.

Campana, F. & Pilone, D. 2008. Effect of wall microstructure and morphometric parameters on the crush behaviour of Al alloy foams. *Materials Science Engineering A* 479: 58–64.

Degischer, H.P. 2002. Material definition, processing, and Recycling. In Wiley-VCH Verlag GmbH & Co. KGaA (ed.) *Handbook of Cellular Metals: Production, Processing, Applications*.

Faugeras, O. 1993. Three-Dimensional Computer Vision: A Geometric Viewpoint, *The MIT Press, London, UK.*

Fraunhofer Institute for Manufacturing Technology and Applied Materials Research IFAM. 2009. *Annual Report 2008-09.* Bremen, Germany.

Fusheng, H., Zhengang Z. 1999. The mechanical behavior of foamed aluminum. *Journal of Materials Science* 34: 291–299.

Kittler, J., Illingworth, J. 1986. Minimum error thresholding. *Pattern Recognition* 19(1): 41–47.

Kolluri, M., Mukherjee, M., Garcia-Moreno, F., Banhart, J., Ramamurty, U. 2008. Fatigue of a laterally constrained closed cell aluminum foam, *Acta Materialia* 56 (5): 1114–1125.

Lhuissier, P., Fallet, A., Salvo, L., Brechet, Y. 2009. Quasistatic mechanical behaviour of stainless steel hollow sphere foam: macroscopic properties and damage mechanisms followed by X-ray tomography. *Mater. Letters* 63: 13–14.

Liebscher, A., Proppe, C., Redenbach, C., Schwarzer, D. 2011. Uncertainty quantification for metal foam structures by means of image analysis, *Probabilistic Engineering Mechanics*. In press.

Olurin, O.B., Arnold, M., Körner, C., Singer, R.F. 2002. The investigation of morphometric parameters of aluminium foams using micro-computed tomography. *Materials Science and Engineering* A328: 334–343.

Quadrini F., Boschetto A., Rovatti L., Santo L. 2011. Replication casting of open-cell AlSi7Mg0.3 foams. *Materials Letters* 65: 2558–2561.

Rodriguez-Perez, M.A., Solórzano, E., De Saja, J.A., García-Moreno, F. 2008. In DEStech Pub. *Porous Met. and Metall. Foams*. Pennsylvania, USA.

Russ J.C. 2006. The Image Processing Handbook, 5th ed., CRC Press Taylor & Francis Group, Boca Raton FL, USA.

Saadatfar, M., Garcia-Moreno, F., Hutzler, S., Sheppard, A.P., Knackstedt, M.A., Banhart, J., Weaire, D. 2009. Imaging of metallic foams using X-ray micro-CT. *Colloids and Surfaces A: Physicochemical and Engineering Aspects* 344 (20): 107–112.

Wadley, H.N.G. 2001. Cellular Metals Manufacturing: An Overview of Stochastic and Periodic Concepts. In J. Barnhart, M. Ashby, N. Fleck (eds). *Cellular Metals and Metal Foaming Technology,* Ver MIT, Bremen, Germany.

Wen-Yea Jang, Kraynik, A.M., Kyriakides, S. 2008. On the microstructure of open-cell foams and its effect on elastic properties. *Int. Journal of Solids and Structures* 45: 1845–1875.

Zhu, H.X. Hobdell, J.R. Windle, A.H. 2000. Effects of cell irregularity on the elastic properties of open-cell foams, *Acta Materialia* 48: 4893–4900.

Computational modelling and 3D reconstruction of highly porous plastic foams' spatial structure

I. Beverte & A. Lagzdins
Institute of Polymer Mechanics (University of Latvia), Riga, Latvia

ABSTRACT: A mathematical model and a PC code are developed for restoring the probability distribution functions for basic characteristics of structural elements (Polymeric struts) in highly porous (Porosity P>90%), open-cell plastic foams. Two characteristics are considered: a) Length and b) Spherical angle. Information, acquired from light microscope images, is used as input and verification data. A good correlation of the calculation results with experimental data, acquired from polyurethane foams compression test, is proved to exist.

1 INTRODUCTION

Symmetry of mechanical and physical properties of materials is connected to symmetry of their structure, e.g. free-rise plastic foams with regard to elasticity and strength properties are transotropic composite materials with the plane of isotropy perpendicular to the rise (RD) direction. Mathematical modelling of physical and mechanical properties of such materials requires an understanding of their intrinsic structure.

The available experimental studies on plastic foams' structure are dealing mainly with the cell's diameters' projections, measured from images in directions parallel and perpendicular to RD (Hilyard 1985 and Hawkins 2005). Histograms of cell's diameters' projections d_1^p, d_2^p and d_3^p determined with the help of light microscope or scanning electron microscope (LM and SEM), were presented (Hilyard 1985). A nonlinear relationship between moduli of elasticity E_1, E_2 and E_3 and extension degrees of the cells $K_1 = d_3^p/d_1^p$ and $K_2 = d_3^p/d_2^p$ was determined experimentally. Distribution of cells extension degrees in free-rise blocks of PUR and PVC foams in dependence of the place according to the height was investigated. It was found that in the middle of a PVC block the cells tend to be nearly spherical and have K_1, $K_2 \approx 2$ at the bottom and top of the block.

Both the experimental and theoretical methods comprise several drawbacks. The fact that projections are determined from images of different kind instead of the actual length of the struts is mainly neglected and the dimensions of projections of elements are assigned to the actual length. Therefore experiments were performed and a mathematical model and a computational code (*.xls) were developed for highly-porous PUR foams, permitting to determine distribution functions of polymeric struts': a) Length and b) Spatial orientation angles, using the images obtained by a penetrating light microscope (LM).

Table 1. Young's moduli of PUR foams in compression.

Characteristics	Number of a block				
	1	2	3	4	5
ρ_f, kg/m^3	33.1	37.4	54.6	81.0	75.0
E_1, MPa	4.2	5.6	9.4	18.8	22.0
ρ_f, kg/m^3	32.0	33.6	49.2	76.4	75.2
E_2, MPa	4.5	5.3	11.3	19.2	21.4
ρ_f, kg/m^3	33.2	38.5	54.4	80.3	73.9
E_3, MPa	10.5	7.5	20.0	26.5	31.5
E_3/E_1	2.4	1.3	2.1	1.4	1.4
E_3/E_2	2.3	1.4	1.7	1.3	1.4
Average	2.4	1.3	1.9	1.3	1.4
Place	1	4	2	5	3

2 EXPERIMENTAL

2.1 Mechanical testing

To estimate the mode of mechanical anisotropy, Young's moduli E_i, i,j = 1,2 and 3 in compression were determined, Table 1 for five industrially manufactured free-rise (RD parallel to OX_3) in a mould PUR foams' blocks.

Average densities of the blocks was $\rho_f = 33$, 37, 53, 79 and 75 kg/m^3. The outer crust of the blocks was cut off to retain homogeneous material with final dimensions $50 \times 50 \times 20$ cm. The specimens were rectangular prisms $100 \times 50 \times 50$ mm, strain rate $\dot{\varepsilon} = 10\%$/min. temperature T = +18°C. Four specimens, cut out from the bottom part of the blocks, were tested for each point. Moduli E_1, E_2 and E_3 were calculated at $\varepsilon_{ii} \leq 2\%$.

2.2 Investigation of structure

A column $20 \times 2.5 \times 2.5$ cm, height parallel to RD, was cut out from each block for investigation of structure. The column was situated in the geometrical centre of the blocks' plane X_1OX_2 to minimize possible influence of anisotropy due to contact with mould sides. Two cube specimens $2.5 \times 2.5 \times 2.5$ cm were cut out from each column. The 1-st cube was located on the bottom of the column/block, the 2nd in the centre of the column/block, difference in the coordinates X_3 of centres of cubes $\Delta X_3 = 8.75$ cm. A mark on the far left top corner of the blocks, columns and specimens was made to avoid unintended turns around OX_3.

A slice $0.5 \times 5 \times 25$ mm was cut from each side A, B and C ($X_3 = $ const., $X_2 = $ const. and $X_1 = $ const.) of the cube and attached to a glass slide. Light beams from two external sources were directed to the slice: 1) Parallel to the plane of slice and 2) In angle $\approx 30°$ with the vertical axis OX_3. $3 \times 2 = 6$ images, (*.jpg), $5 \times$, were taken by a digital camera DC, attached to the LM for each block as well as an image of a size of natural length 1 mm.

The focus of the LM was fixed on the upper struts of the surface. The depth of focus range h_f was defined as the range of distance of the image behind the lenses of the image-forming system (LM + DC) measured along the axis of the device throughout which the image had app. equal sharpness. To determine h_f and its display on images, several steel plates, each of thickness 0.05 mm were pressed firmly into a pile with an offset ≈ 0.5 mm from one another at one side and images of the stepped surface were taken. The approximate depth of focus range h_f was determined as sum of thickness of the plates, visible equally sharply in the images: $h_f \approx 0.35 \pm 0.05$ mm.

Elements on the images are projections of the a) foams structural elements struts, nods and faces on the image planes A, B or C and of b) angles formed by the struts with different axes. Projections of struts and angles were measured from paper-printed images of each slice, simultaneously comparing with the image files on PC LCD for a clearer presentation of structure.

2.3 Methodology

A local image area corresponding to an average character of structure was chosen for measuring the elements. Elements were measured in millimetres with a ruler, precision ± 0.5 mm. Scale was not taken into account at this stage. An on-screen measuring with the help of corresponding software was tried, but turned out to be less convenient. It is important for the observer to consider all elements one by one in the chosen local area of the image. Only obvious defects (E.g. local orientation of a group of struts differing from dominating orientation or untypically shaped strut), not corresponding with the average character of the elements may be omitted. A subjective tendency of the observer to: a) Shift to choosing longer projections and b) Not to notice/include very short projections was recognized and constantly corrected. A short projection could correspond to a) A short struts having angle $\alpha_1 \approx 0°$ with the image plane or to b) A long struts having $\alpha_2 \approx 90°$ with the image plane, Figure 1.

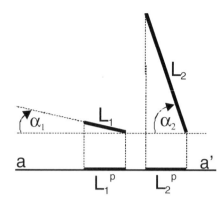

Figure 1. Projections L^p of the struts' length L on image plane aa'.

Since the foams Nr.4 and 5 have densities 1.5 – $2 \times$ higher than foams Nr.1, 2 and 3, their structural elements and projections visible on the images are smaller. To reduce potential subjective error of the observer, making choice of elements and measurements, the images of foams Nr.4 and 5 were printed in enlarged scale (141%) to maintain app. equal conditions for the observer. Scale was taken not yet taken into account.

A representative layer of foams was defined as a layer having thickness equal to the depth of focus range h_f. It was assumed that the characteristics and distribution of struts in the representative layer corresponds to the average character and distribution in foams' cubical specimens. Principles for including elements into the representative sample were formulated:

1) Only struts' projections visible with both ends entering the nods were considered for including in the representative sample. It was important for the observer to estimate properly how far reached the strut while entering the nod for each foams' density;

2) Only projections depicted sharply along their whole length or a certain part of the length were included in the representative sample. Elements with changing sharpness along length correspond to struts situated angularly to the image plane and having length reaching outside the representative layer. The number of such elements was divided by 2 for each group of length, since they belong to 2 representative foams' layers.

3) Blur elements were considered to be outside (Above or beyond) the representative layer and as such were not included into the sample. Results of measurements were depicted in Table 2.

Values of elements in the representative samples were sorted and arranged into $I = \text{ROUNDUP}(L_{max}^p)$ number of bins, where $\text{ROUNDUP}(L_{max}^p)$ is the value of biggest projection L_{max}^p in the sample rounded up to the nearest highest integer value. Bin width h = 1 unit,

$$h\,(i\text{-}1) < L^{pi}{}_K \leq hi,\ i = 1, 2, ..., I, \qquad (1)$$

Table 2. Experimental data of struts' length projections L^p (Scale: 1 unit = 1/55 mm).

Specimen	Image plane of slice	L^p_{Amax}, units	L^p_A of K^i_{max}	L^p_{BCmax}, units	L^p_{BC} of K^i_{max}
1-1	11A	11.00	1.5	–	–
	11B+11C	–	–	18.0	4.5
1-2	12A	11.50	2.5	–	–
	12B+12C	–	–	18.0	4.5
2-1	21A	13.00	2.5	–	–
	21B+21C	–	–	19.0	4.0
2-2	22A	12.00	2.5	–	–
	22B+22C	–	–	15.0	4.0
3-1	31A	7.00	1.5	–	–
	31B+31C	–	–	12.0	2.5
3-2	32A	8.00	2.5	–	–
	32B+32C	–	–	11.5	2.5
4-1	41A	6.03	1.5	–	–
	41B+41C	–	–	8.51	2.5
4-2	42A	8.16	2.5	–	–
	42B+42C	–	–	7.80	2.5
5-1	51A	6.03	1.5	–	–
	51B+51C	–	–	10.64	2.5
5-2	52A	6.38	2.5	–	–
	52B+52C	–	–	8.51	2.5

Table 3. Approximation results of experimental histograms.

Specimen	α^A	α^{BC}	$\Delta\alpha$	K1	$K1^{aver}$, place in range
1-1	0.15	0.06	0.09	2.50	2.45
1-2	0.12	0.05	0.07	2.40	(1)
2-1	0.13	0.09	0.04	1.44	1.28
2-2	0.09	0.08	0.01	1.13	(4)
3-1	0.31	0.16	0.15	1.94	1.85
3-2	0.23	0.13	0.1	1.77	(2)
4-1	0.31	0.22	0.09	1.41	1.24
4-2	0.27	0.25	0.02	1.08	(5)
5-1	0.29	0.18	0.11	1.61	1.53
5-2	0.29	0.20	0.09	1.45	(3)

Table 4. Anisotropy modes in dependence of mould's dimensions.

N	Proportion	Anisotropy	Spatial distribution ϕ	θ
1	$l_2 > l_1, l_3 >> l_1, l_2$	Orthotropy	$p_1(\phi)$	$p_2(\theta)$
2	$l_1 \approx l_2, l_3 >> l_1, l_2$	Transversal isotropy (Transotropy)	Uniform	$p(\theta)$
3	$l_1 \approx l_2 \approx l_3$	Isotropy	Uniform	Uniform

where k = 1, 2, ..., K^i – number of elements in the i-th bin. Values of projections (As central numbers of the bin) corresponding to the most populated bin with values K^i_{max} in it were determined. Histograms of the probability distribution $p(L^p) = K^i/N$ and projections spherical angles θ'_i were created. Histograms of struts' projections' L^p probability distribution for the cubes a) 1-1 with the most differing characteristics and b) 4-2 with the most equal characteristics in planes A and B, C. The histograms are asymmetric therefore a generalised exponential function was used for approximation: $F(x) = A\,x^q\exp(-\alpha x^b)$; were m, α and b are parameters determining sharpness of maximum and asymmetry. The sample of L^p values defined by $F(x)$ was determined calculating $F(x)$ in the points $x_i = h \times i$, i = 1, 2, ..., I and then the probability

$$p(x) = F(x) / \sum_{i=1}^{I} F(x_i), \quad \sum_{i=1}^{I} p(x_i) = 1 \qquad (2)$$

was calculated. Numerical analysis revealed that values A = q = 1, b = 1.5 and α varied in the limits $0.02 \leq \alpha \leq 0.35$ were appropriate to find the best fit between experimental and theoretical representative samples: p = p(x,α). The value of α, providing the least difference between L^p_{aver}, s and v of the experimental and theoretical samples was defined as the fitting one for approximating function p(x):

$$|L^p_{aver}{}^E - L^p_{aver}{}^T| \leq \varepsilon_1, \quad |s^E - s^T| \leq \varepsilon_2 \text{ and}$$
$$|v^E - v^T| \leq \varepsilon_1, \qquad (3)$$

where $\varepsilon_1, \varepsilon_2, and \varepsilon_3$- precisions. Numerical calculation results are presented in Table 3. Parameters α corresponding to probability distribution functions of struts'

projections in images parallel to planes A, B and C are denoted as α^A and α^{BC} and difference $\Delta\alpha = \alpha^A - \alpha^{BC}$. To characterise the anisotropy of different foams' blocks, anisotropy degree K_3 was implemented as $K1 = \alpha^A/\alpha^{BC}$, $K1 = 1.0$ for isotropic foams. Gradient of K1 in a block is characterised by $\Delta K1 = K1_1 - K1_2$.

3 THEORETICAL

3.1 Anisotropy modes

Let us consider *free-rise highly porous* (Porosity P>90%) foams produced in an open mould having dimensions l_1, l_2 and l_3 along axis OX_1, OX_2 and OX_3, correspondingly. Several cases of the foams structural anisotropy can arise in dependence of mutual proportions of mould's dimensions, Table 4.

Extension of cells in the foams' volume due to relatively big height of mould $l_{31} = l_3/l_1 >> 1$ and $l_{32} = l_3/l_2 >> 1$ has to be distinguished from local extension of cells in the outer crust of a block or due to contact with mould. The structure of foams' can be characterised in general by probability distribution functions of the struts': 1) length L and 2) spherical angles ϕ and θ, therefore mathematical modelling of foams' structure is performed in several stages.

3.2 Struts' length distribution in isotropic foams

Experimental investigations have revealed that struts length distribution is not uniform even in isotropic plastic foams. For isotropic foams having low porosity (P<30%), the distribution of radii R of un-interconnecting bubbles can be described well by the generalised exponential function (Berlin 1980). It is assumed that struts' length distribution in highly porous isotropic foams retains the same general character:

$$F(L, \alpha) = A\,L^q \exp(-\alpha L^P), \quad \sum_{i=1}^{l} F(L_i) = 1. \qquad (4)$$

The parameters of the distribution function of struts' length in isotropic foams are considered to depend solely on foams' formulation and other chemical conditions that determine different growth capacities of gaseous bubbles.

3.3 Struts' spatial distribution in isotropic foams

In isotropic foams the representative samples of the struts length projections on planes A, B and C are nearly similar. Then it can be assumed that struts are distributed spatially uniformly and a certain spatial angle $d\omega$ corresponds to each strut, $d\omega$ being measured by the corresponding square dS on the spherical surface, Figure 2. In the mathematical model the struts are considered to emerge from the centre O of a sphere, radius $L = 1$, the direction of the i-th strut being defined by spherical coordinates ϕ_i and θ_i. Then $d\omega = dS/L^2$. It is assumed that the spatially uniform distribution of struts' can be achieved by a uniform distribution of the equal elementary squares dS_0 corresponding to a single strut, on the surface of the sphere, centre of a square being intersection of the strut with sphere surface.

The array of spherical coordinates ϕ_i, θ_i corresponding to the quarter bands in the first octant of the sphere and ring bands on the upper segment is calculated for steps $d\phi_1 = d\theta = 2.5°$ (36 bands, $816 + 50$ elements) with a PC code "ANGLES", developed by the authors.

The values of length of 6628 struts of the General sample were distributed according to function in Equation 5 by the following method. A subsample of n=100 elements L (Length of a strut projection or the strut itself), having probability distribution $p = p(L,\alpha)$ was considered. The length of elements was defined in units, the bin width was 1 unit, possible number of bins L^{max}, frequency of a bin $n_i^1 = n\,p(L_i,\alpha)$. Values of the elements in a bin were represented by the upper limit value of a bin. Frequencies were rounded to the nearest integer value. The subsample was created distributing 100 elements in a matrix $[100 \times n1i^{max}]$, where n_i^1 is the last full bin, having at least one element. Distribution was performed by a PC code "MATRIX", (*.xls) developed by the authors.

The values of the chain were assigned repeatedly to the spatial positions of the array of 816 elements in the quarter bands of the first octant and 50 elements the upper segment. Since there is no reason for struts in isotropic foams to be directed predominantly in any

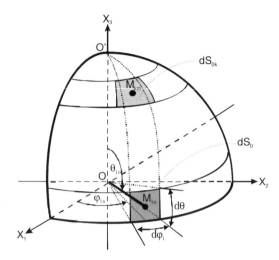

Figure 2. Elementary squares dS_{0k}, corresponding to a single strut OM_{km}.

direction, the angular distribution was assumed to be uniform: $p_1(\phi) = $ const. and $p_2(\theta) = $ const.

Then the General sample of 6628 elements with definite length distribution $p = p(L,\alpha)$ and spatially uniform angular distribution was defined. Due to symmetry the General sample was constructed by multiplying the elements in the array of the bands by 8 and elements in the upper segment by 2.

3.4 Transotropic foams 'struts' system

Coordinates X_1, X_2 and X_3 were calculated for the end point M of the n-th strut from the General sample of isotropic foams model, Figure 3:

$$X_{1n} = L_n \sin\theta_n \cos\varphi_n,$$
$$X_{2n} = L_n \sin\theta_n \sin\varphi_n \text{ and}$$
$$X_{3n} = \cos\theta_n,\; n=1, 2, ..., N. \qquad (5)$$

It was assumed that the foams' structural anisotropy, namely transotropy could be implemented into the mathematical model by linear coordinate transformations along axis OX_3, parallel to RD (Beverte 1997):

$$X_1' = X_1,\; X_2' = X_2,\; X_3' = K_3 X_3; \qquad (6)$$

where K_3 – a coefficient characterising the structural anisotropy of foams. After transformation $X_3' = K_3 X_3$ the n-th strut's length $L_n = OM$ became $L_n' = OM'$:

$$L_n' = L_n \sqrt{\sin^2\theta_n + (K_3 \cos\theta_n)^2} \qquad (7)$$

and its' spherical angle θ_n became θ_n', Figure 3:

$$\Theta_n' = \text{atan}\,(1/K_3 \tan\theta_n);\; \varphi_n' = \varphi_n = \text{const.} \qquad (8)$$

Thus 3 arrays characterising 6628 struts of transotropic foams' model were calculated: 1) Length L_n', 2) Angle Θ_n', and 3) Angle ϕ_n'.

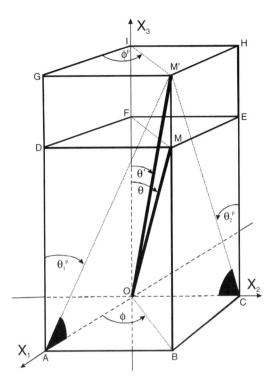

Figure 3. Modelling of transotropic foams' structure.

3.5 Projections of transotropic foams' struts

Elements visible on the LM images are projections of the foams' strut's length L' on the image plane.

Let us consider image planes parallel to $X_3 = \text{const.} = A$, $X_2 = \text{const.} = B$ and $X_1 = \text{const.} = C$. Projections of transotropic foams' strut's length L' on image planes B, C and A are:

$$L_2^P = CM' = [((L' \cos\theta')^{**}2 + L'\sin\theta'\cos\varphi)^{**}2)]^{1/2};$$
$$L_1^P = AM' = [((L' \cos\theta')^{**}2 + (L'\sin\theta'\sin\varphi)^{**}2)]^{1/2};$$
$$L_3^P = IM' = L'\sin\theta'. \qquad (9)$$

Projections of the strut's spherical angle θ' on planes B and C are:

$$\Theta_{1n}'^{P} = \text{atan} (\tan\theta'_n \sin\varphi_n);$$
$$\Theta_{2p}' = \text{atan} (\tan\theta'_n \cos\varphi_n) \qquad (10)$$

and projection of the angle ϕ_p on plane A is:

$$\varphi_p' = \varphi_n = \text{const.} \qquad (11)$$

Arrays of 1) Length projections L_1^P, L_1^P and L_3^P and 2) Projections of struts' spherical angles $\Theta_{1n}'^{p}$, $\Theta_{2n}'^{p}$

and ϕ_n' were calculated for 6628 struts of transo-tropic foams' model. When $K_3 = 1.0$, projections of isotropic foams' elements were determined. Nods were considered to be spherical in the presented mathematical model. A projection of a sphere's diameter equals to itself and can be determined from images.

3.6 Reconstruction of struts' system from LM images

The overall task for reconstruction of the struts' system is to: 1) Determine the distribution for struts' length projections on plane A from the corresponding LM images and then to 2) Find the corresponding extension degree K_3, that provides such distributions of struts length and orientation, that distribution of struts' projections length and angular orientation on planes B and C correspond to the experimentally determined ones from LM images B and C. If three images are taken, planes parallel to planes A, B and C of a transotropic foams' block, the general task is to reconstruct the 3-D structure of foams (The distribution of struts' length and spatial angles) with the help information gathered from images. A PC code PROJECTIONS was developed for numerical calculations.

3.7 Conclusions

A good correlation of the calculated results with experimental data was proved. The probability distribution functions, approximating the experimental data on strut's projections on image planes A, B and C are in a good correlation with the distribution functions derived from the mathematical model.

The proposed method and computational modelling could be used for reconstruction of 3-D structure of orthotropic highly-porous foams (Foaming in a narrow, high mould), too and for closed-cell foams comprising polymeric platelets instead of struts.

ACKNOWLEDGEMENT

The research was performed at *EU ERDF funded project 2010/0290/2DP/2.1.1.1.0/10/APIA/VIAA/053.*

REFERENCES

Hilyard, N.C. (ed.) 1982. Mechanics of Cellular Plastics. *London:* Applied Science Publishers LTD.
Hawkins, M.C., O'Toole, B. & Jackovich, D. 2005. Cell Morphology and Mechanical Properties of Rigid Polyurethane Foam. *J.Cell.Plast* 41 (May): 267–285.
Berlin, A.A. & Shutov, F.A. 1980. Chemistry and Technology of Gas-Filled Highpolymers. Moscow: Nauka (In Russian).
Beverte, I. 1997. Elastic Constants of Monotropic Plastic Foams. 1. Deformation Parallel to Rise Direction (Semi-axes Hypothesis). A mathematical Model. Mech. Compos. Mater. 33(6): 719–733.

Computational Modelling of Objects Represented in Images – Di Giamberardino et al. (eds)
© 2012 Taylor & Francis Group, London, ISBN 978-0-415-62134-2

Circularity analysis of nano-scale structures in porous silicon images

Sahadev Bera & Bhargab B. Bhattacharya
Advanced Computing and Microelectronics Unit, Indian Statistical Institute, Kolkata, India

Sarmishtha Ghoshal & Nil R. Bandyopadhyay
School of Materials Science and Engineering, Bengal Engineering and Science University, Howrah, India

ABSTRACT: Whilst regular structures like micro-test tubes and micro-beakers fabricated on Porous Silicon (PS) offer potential platforms for implementing various biosensors, controlling the uniformity of pores during electrochemical etching is a challenging problem. One important objective of such fabrication procedure is to ensure the circularity of pore boundaries. Thus to tune up and standardize the etching process, a fast image analysis technique is needed to evaluate and quantify the geometry of these nano-scale PS structures. In this paper, we present an automated approach to pore image analysis: given a top-view image of a PS chip captured by a Scanning Electron Microscope (SEM), the porous regions are segmented and each of the pore boundaries is approximated by a circle. We use a simple digital geometric technique to determine a best-fitting circle for each pore and compute the Hausdorff distance between them to estimate the quality of pore formation. Experimental results on several SEM images of PS structures have been reported to demonstrate the efficiency of the proposed method.

Keywords: circle fitting, digital geometry, nano-scale imaging, Porous Silicon, shape analysis.

1 INTRODUCTION

Porous Silicon (PS) based devices have recently emerged as a potential platform for exploring numerous applications to nano-biotechnology, e.g., medical diagnostics, in-vitro pathogen detection, gene identification, and DNA sequencing (Granitzer and Rumpf 2010; Stewart and Buriak 2000; Betty 2008; Ghoshal et al. 2010; Ghoshal et al. 2011). Because of its non-toxic nature and biodegradability, it is also highly suitable for implementing in-vivo biosensors and drug delivery modules (Anglin et al. 2008; Salonen et al. 2008). The intriguing property of visible light emission from electrochemically etched PS was observed long ago (Cullis and Canham 1991). PS chips with an average pore diameter ≤ 2 nm (Rouquerol et al. 1994) are called microporous and they admit photoluminescence (PL) at room temperature whereas, those with pore diameters >50 nm (Rouquerol et al. 1994), are called macroporous and they have applications to photonics, sensor technology and biomedicine (Lehmann 2003; Betty 2008; Saha et al. 2006; Lin et al. 1997; Reddy et al. 2001).

Many physical properties of PS, e.g., luminescence, refractive index, and heat conductivity are determined by the fraction of porosity. For biological applications, a uniform arrangement of porous structures called micro-test tubes or micro-beakers having diameters approximately $1 - 1.5 \,\mu$m are desirable for loading nanoparticles or drugs within the pores (Ghoshal et al. 2011). Techniques of creating uniform macroporous structures by controlling formation parameters have been reported in the literature (Vyatkin et al. 2002; Harraz et al. 2005). PS also provides a viable platform for observing surfaceenhanced Raman scattering (SERS), which is useful in detecting the presence of chemical and biological molecules (Chan et al. 2003; Jiao et al. 2010). Microbeakers on PS (<100 nm in width) with pore size ($>1.5 \,\mu$m in diameter) can be used as a SERS substrate for various bio-sensing applications. Ideally, these structures should be produced on the PS chip as a regular array of circular pores. However, because of the process uncertainty, pores often appear with deformed boundaries as observed from the captured SEM images (Ghoshal et al. 2011).

In this paper, we address the problem of estimating the circularity of the pore structures based on an image processing technique. In a PS chip, adjacent pores may merge to form a connected pattern of complex shape. Based on some geometric properties of a digital circle, we propose a segmentation technique to isolate each of the pores from the given SEM image. Once the boundary of a pore is extracted, a circle-fitting algorithm is deployed to determine a best fit followed by estimating the deviation from circularity by computing the Hausdorff distance (Rucklidge

1997) between the actual pore boundary and the fitted circle.

2 FORMULATION OF THE PROBLEM

In order to formulate the problem, we first observe certain characteristic properties of a PS chip image taken by a scanning electron microscope (SEM). A typical top-view SEM image of a PS chip is shown in Fig. 1, where the pore boundaries appear as nearly circular objects. As the top surface of a PS chip resembles a 3D terrain, some pores are focused and some are defocused in the captured image. The deep black objects of the image are in the focused plane and thus the corresponding pores appear with sharp boundaries in the image; the lighter black objects represent those pores, which are in the defocused planes and blurred. Thus to analyze the detailed structure of the pores, we first need to perform automatic segmentation to isolate each object from the image.

An ideal microporous structure should consist of an array of pores each having a circular contour. However, in real life, the contours seldom become perfectly circular. Further, during electrochemical etching process, two or more neighboring pores may often coalesce to form a composite porous structure. In the image, the contour of such a pore appears as a closed curve consisting of several nearly-circular arcs. For example, in Fig. 1, the objects labeled as 1, 2, 3 and 8 represent a single porous structure each, whereas, the objects labeled as $4-5$, $6-7$ and 9 indicate connected structures formed by overlapping of two or more circular pores. Sometimes, the overlap is peripheral as in the objects labeled as $4-5$, $6-7$ (Fig. 1), where the constituent seed pores are visibly perceptible. In such cases, the challenge lies in segmenting out each individual circular contour from the image of the connected object. This is necessary, because otherwise, an attempt to fit a single digital circle (Klette and Rosenfeld 2005) to the original contour of the connected pore would incur a huge amount of error. In some other cases when the overlap between neighboring pores are significantly large as in the object labeled 9 in Fig. 1, then segmentation may not be useful as the overall contour appears to be that of a nearly-circular pore. In such a case, however, fitting a single circle to the entire contour is likely to yield a fair approximation.

3 PROPOSED WORK

Given a SEM image, our algorithm consists of four steps: (i) segmentation of porous objects and extraction of their contours, (ii) isolation of individual circular components from the connected structures, if any, (iii) fitting a digital circle to the contour of each of the segmented pore images, and (iv) computation of Hausdorff distance between the actual contour and the fitted circle to estimate the goodness of circularity.

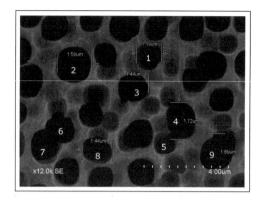

Figure 1. SEM image (1280×960) of microporous Si formed on p-type substrate with magnification $12000 \times$ (top-view).

The various steps of the algorithm are now described below.

3.1 Segmenting the porous objects from the input image

The input SEM image $\mathcal{A}$ of the PS chip is supplied as a gray-tone image, in which it is easy to identify three types of objects depending on the intensity value of the pixels. The deep black objects represent the focused pores, the medium black objects represent the defocused pores or small pores, and the gray background, which corresponds to the PS substrate. This is evident from the histogram (Gonzalez and Woods 2001) of the image (see Fig. 5(a)). Using these gray levels, we segment $\mathcal{A}$ and create two binary images one consisting of only the focused pores (deep black) $\mathcal{A}_1$ and another with only the defocused pores (medium black) $\mathcal{A}_2$ so that pores appear in black on a white background. Further analysis is performed on $\mathcal{A}_1$ and $\mathcal{A}_2$ separately.

Contour extraction: Once a binary image of the porous structure $\mathcal{A}_1$ is given, the next step is to extract a one-pixel thick contour for each of the nano-structures, which appear as a black object in $\mathcal{A}_1$. As discussed earlier, such a nano-structure may represent a single pore or a connected pattern of two or more pores. Let $\mathcal{A}_1$ contain m objects $b_1, b_2, \ldots b_m$ and n pores $a_1, a_2, \ldots a_n, m(\leq n)$.

Although there exist several algorithms for detecting boundary/edge of an object, for instance, those based on Canny edge detection (Chanda and Majumder 2009), Laplacian operator (Chanda and Majumder 2009), Sobel operator (Chanda and Majumder 2009), we here use a simple scan-type algorithm to output a single-pixel thick contour.

We first make a copy $\mathcal{B}_1$ of the image $\mathcal{A}_1$. Then the image $\mathcal{A}_1$ is scanned from left-to-right and top-to-bottom to check the *four neighbors* of each black pixel. If all the *four neighbors* of a black pixel are black in $\mathcal{A}_1$, then the corresponding pixel in $\mathcal{B}_1$ is replaced by a white pixel. Clearly, at the end of the scan, for

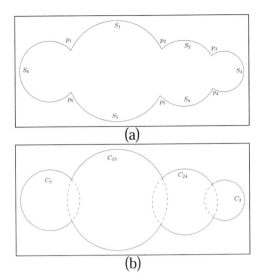

(a)

(b)

Figure 2. (a) A connected structure of 4 overlapping pores (b) isolation of pores.

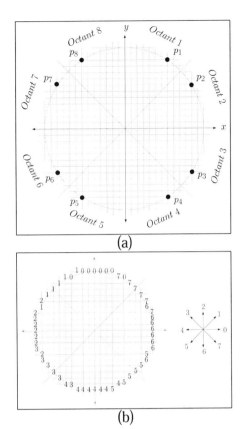

(a)

(b)

Figure 3. (a) Illustration of a special set of eight pixels (b) chain code of a digital circle.

each object b_i, $i = 1 \ldots m$ in $\mathcal{A}_1$, a one-pixel thick 8 *connected* closed curve c_i will be stored in $\mathcal{B}_1$. Let $C = c_1 \cup c_2 \cup \ldots \cup c_n$.

Isolating connected pores: Since some porous structure may appear as a connected pattern of overlapping pores, e.g., objects $4 - 5$, $6 - 7$ in Fig. 1, it is necessary to identify the underlying circular pores that form the pattern. Thus, after the one-pixel thick boundary of an object is extracted, we examine the curvature of the contour as follows. We assume that the closed contour consists of one or more circular arcs. In the former case, the structure will be approximated by a single fitting circle. Otherwise, the component arcs are identified and subsequently the constituent circular shapes are determined. In order to accomplish this, we check the chain code of the contour and make use of a property of a digital circle (Klette and Rosenfeld 2005).

Property of chain code of a digital circle: It is known that the chain code of a digital circle satisfies a simple differential property: for any two consecutive pixels along the circumference, the difference in chain code is either 0 or 1 or 7. Figure 3(b) shows the chain code of a digital circle. In other words, the angle between two consecutive pixel directions in the sense of 8-neighborhood is either 0 or $\pi/4$. However, if the boundary curve consists of two or more digital circles then at the turning points, the above differential property of the chain code will not hold. Fig. 2 shows an example of a closed curve consisting of six circular segments of four digital circles. Let $p_1, p_2, ..., p_6$ denote the six turning points and let $S_1, S_2, ..., S_6$ be the six segments. Since at a turning point (or joining point), two consecutive circular segments meet, there will be a sharp change in directional angle, which invalidates the differential property of the chain code.

Thus by checking the chain code along the boundary of the curve, we can easily identify the location of turning points and the arc segments. Ideally, the number of turning points should be an even; however, some of the turning points may disappear because of uneven electrochemical etching. If the number of turning points is 0, then it consists of a single pore. Otherwise, we need to identify the underlying pores in the connected pattern. For example, in Fig. 2, the total number of segments is six, but they belong to four circles, as segments S_1 and S_5 belong to the same circle. Similarly, S_2 and S_4 belong to the same circle. In order to identify these underlying circles, we use a technique of circular arc recognition based on a digital geometric method (Bera, Bhowmick, and Bhattacharya 2010). For each of the circular segments $S_1, S_2, ..., S_6$, let $C_1, C_2, ..., C_6$ be the corresponding fitted digital circle. The digital circles C_1 and C_5 will have nearly equal radii, and their centres are very close to each other; hence they can be treated as components of a single digital circle C_{15}. Similarly C_2 and C_4 are recognized as arcs of a single digital circle C_{24}. So, in this case, the connected structure can finally be decomposed into four digital circles C_6, C_{15}, C_{24}, C_3 (see Fig. 2). In the next subsection, we describe the procedure of fitting a circle to a digital arc in detail.

405

3.2 Fitting a circle to a closed digital curve

For each closed curve in C, we first identify the digital circular segments (arcs), and then compute a fitting circle for each of these segments. If there is no turning point, only one circle is computed, otherwise multiple fitting circles are computed and merged, whenever possible, as discussed earlier.

A geometric property of a digital circle: In real space $\mathbb{R}^2$, there exists a unique circle that passes through three fixed non-collinear points. Also, given any three points on a circle, the construction procedure based on perpendicular bisectors will produce the same circle. However, any three points chosen from the boundary of a circle in digital space $\mathbb{Z}^2$, may not yield the same circle in the euclidean space. Only in the special cases when the three points are chosen from a special subset $\mathbb{S}$ of grid points $\mathcal{G}$ of a digital circle, the reconstructed euclidean circle determined by the two perpendicular bisectors become unique. Such a special subset S can be constructed as follows: consider a pixel in the 1_{st} octant, and choose a pixel from each of the remaining seven octants by reflection. For example, let p_1 be a pixel in 1_{st} Octant and $p_2, p_3, p_4, p_5, p_6, p_7, p_8$ be the seven pixels obtained by reflection as shown in Fig. 3(a). Let $S = \{p_1, p_2, p_3, p_4, p_5, p_6, p_7, p_8\}$. Then S is a special subset of $\mathcal{G}$.

In our problem, since the boundary of the pore structures are seldom perfectly circular, the above method will not work. Hence, for a closed curve c_i in C, we extract the segments, and then from each segment $\mathcal{H}$ we choose three pixels p_1, p_2, p_3 at random. Using the coordinate values of the pixels p_1, p_2, p_3 the center and radius of the digital circle passing through these three pixels are computed from the intersection of perpendicular bisectors and by applying rounding-off techniques. The center and radius are stored in the two arrays Cn and $\mathcal{R}$ respectively. We repeat this procedure for a large number of triplet pixels chosen from the segment $\mathcal{H}$. Finally, the center and radius of the best fitting circle for the segment $\mathcal{H}$ are computed by taking the medians of the respective values stored in Cn and $\mathcal{R}$. If for two or more segments, the computed values of their radii and the locations of their centres are very close, we merge these segments to fit a single digital circle.

3.3 Computing Hausdorff distance

To estimate the deviation from circularity of the porous objects, we use Hausdorff distance (Rucklidge 1997) as a measure of error. The Hausdorff distance (HD) or Hausdorff metric measures how far two subsets of a metric space are from each other. Informally, two sets are close in the sense of Hausdorff distance if every point of either set is close to some point of the other set. Let X and Y be two non-empty subsets of a metric space (M, d). We define their Hausdorff distance $d_H(X, Y)$ by

$$d_H(X, Y) = max\{\sup_{x \in X} \inf_{y \in Y} d(x, y), \sup_{y \in Y} \inf_{x \in X} d(x, y)\}$$

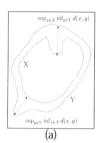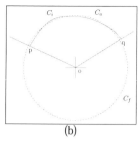

Figure 4. (a) Illustration of Hausdorff distance (b) computed segment in our experiment

where sup represents the supremum and inf the infimum. Figure 4(a) shows the Hausdorff distance between two sets. In our experiment, we also attempt to fit a circle to an open segment, as shown in Fig. 4(b), where C_i is a contour segment and C_f is the fitted digital circle. So, if we compute the Hausdorff distance between C_i and C_f, we will observe unexpected result as the input curve C_i is not closed. To tackle this problem, we consider the arc C_a between op and oq and compute the Hausdorff distance only between C_i and C_a to estimate the error of fitting.

The HD of the input digital curve and the fitted digital circle is computed to estimate the deviation from circularity. From the empirical evidence we conclude that if HD is within 30% of the radius of the fitted circle then the pore is fairly circular in shape.

3.4 Interpreting the structure of PS

The PS chip may consist of a large number of pores. We can classify them into three groups (best, good, bad) depending on their goodness of circularity and the requirement. For each pore we calculate the percentage error $(PE) = (HD \times 100)/r$, where r is the radius of fitted digital circle. Depending on the value of PE we classify the pores into three groups. Since our method also computes a median radius value of each pore, we can also estimate the volume of the pore, if the information about the height is known.

4 EXPERIMENTAL RESULTS

We have implemented the algorithm in C on the open-SUSE™ OS Release 11.0, HP xw4600 Workstation with Intel® Core™2 Duo, 3 GHz processor. We have performed the test on several SEM images of PS chips, computed the average diameter of each pore, and estimated their circularity from image analysis.

A step-wise demonstration of the proposed method on a sample gray-tone image is shown in Fig. 1. In this image, the diameters of some pores have been measured by the SEM along certain directions, which are marked in units of μm.

We first perform histogram thresholding to isolate the objects of the focused plane and produce a binary image containing nearly circular discs, each

406

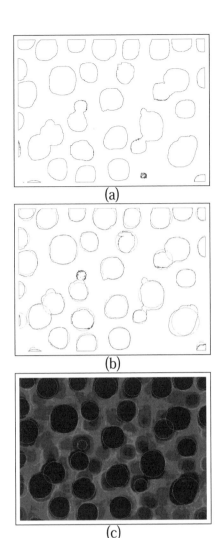

(a)

(b)

(c)

Figure 5. Snapshots of the experiment on the image shown in Fig. 1 (a) single-pixel contour extraction for each object (b) fitting a digital circle (c) superimposed fitted circles.

Table 1. Comparison of observed and computed features of the pores for the image shown in Fig. 1.

CN	NCP	Center	r	HD	PE	CD	OD
1	700	(734, 227)	70	21.10	30.14	1.16	1.19
2	776	(302, 275)	99	10.00	10.10	1.67	1.59
3	686	(638, 403)	84	14.56	17.33	1.39	1.44
4	787	(909, 581)	93	26.65	28.66	1.53	1.72
5	458	(829, 727)	46	14.14	30.74	0.76	–
6	545	(232, 637)	81	18.97	23.42	1.34	–
7	532	(141, 721)	86	13.00	15.12	1.42	–
8	810	(436, 751)	80	13.00	16.25	1.32	1.44
9	852	(765,1125)	104	25.94	24.94	1.72	1.69

121 pixels = 1 μm, '–' Not observed
CN – Circle Number, NCP – Number of Contour pixels
r – Radius (in no. of pixel)
HD – Hausdorff distance (in no. of pixels)
PE – Percentage Error = $(HD \times 100)/r$
CD – Computed diameter (μm)
OD – Observed diameter along a given direction by SEM (μm)

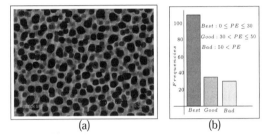

(a)   (b)

Figure 6. (a) Result for a PS image (b) circularity classification vs. frequency.

corresponding to a black porous structure of the input image. For each disc, a one-pixel contour is computed and a simple closed curve is obtained as in Fig. 5(a). Next, we analyze each curve by checking its chain code, and decompose it into segments, wherever applicable. A best fit digital circle is determined by computing the medians of the centres and radii arrays obtained by a repetitive experiment of randomly selecting triplet pixels from the segment for a large number of times. Figure 5(b) shows the image with fitted digital circles as obtained by our algorithm and Fig. 5(c) shows the superimposed circles on the actual images of the pores. We also report the location of the centre and diameter of the fitted circle and compute the PE for each fitted circle relative the actual pore. The measured and computed values of diameters are listed in the Table 1. It may be noted that the supplied values of

diameter correspond to a measurement along certain directions as shown in Fig. 1, whereas the computed one reflects the median value. This accounts for the observed differences in the values of diameter. As the pore labeled as 4 is elliptical in shape, the computed diameter differs significantly from the measured one. This is also reflected in the PE value. In our experiments, 121 pixel length is equal to 1 μm as per SEM characteristics.

This objects in the defocused plane can also be analyzed similarly. Fig. 6 shows the results of the experiment on another SEM image of a PS chip. In this image, we have analyzed a total 181 pores, and the fitting digital circles are computed. From the viewpoint of circularity, we classify them as "best" (if $PE \leq 30$) or "good" (if $30 < PE \leq 50$) or "bad" (if $PE > 50$). Thus out of the 181 pores, 112 pores are classified as best, 38 as good and 31 as bad.

5 CONCLUSION

In this paper, we have presented an automated technique for estimating the circularity of pores in PS

407

chips, purely based on image analysis. Our procedure is simple, fast, and relies on some basic principles of digital circles. We have run our experiments on several SEM images of PS chips, results of some of which are presented in this paper. The estimated diameter values are sufficiently accurate and conform nicely to the observed values. As the current practice of conducting manual evaluation of circularity in PS chips is a tedious and time-consuming process, the proposed automated procedure may be used as a powerful analysis tool. Improving the robustness and accuracy of the method may be further investigated as future research issues.

REFERENCES

Anglin, E., L. Cheng, W. Freeman, and M. Sailor (2008). Porous silicon in drug delivery devices and materials. *Advanced Drug Delivery Reviews 60*(11), 1266–1277.

Bera, S., P. Bhowmick, and B. Bhattacharya (2010). Detection of circular arcs in a digital image using chord and sagitta properties. *Proc. Eighth International Workshop on Graphics Recognition (GREC 2009),Lecture Notes in Computer Science (LNCS), Springer 6020*, 69–80.

Betty, C. (2008). Porous silicon: a resourceful material for nanotechnology. *Recent Pat Nanotechnol 2*(2), 128–36.

Chan, S., S. Kwon, T. Koo, L. Lee, and A. Berlin (2003). Surface-enhanced raman scattering of small molecules from silver coated silicon nanopores. *Advanced Materials 15*(19), 1595–1598.

Chanda, B. and D. Majumder (2009). Digital image processing and analysis. *PHI*. Cullis, A. and L. Canham (1991). Visible light emission due to quantum size effects in highly porous crystalline silicon. *Nature 353*, 335–338.

Ghoshal, S., A. Ansar, S. Raja, A. Jana, N. Bandyopadhyay, A. Dasgupta, and M. Ray (2011). Superparamagnetic iron oxide nanoparticle attachment on array of micro test tubes and microbeakers formed on p-type silicon substrate for biosensor applications. *Nano Express 6*(1), 540.

Ghoshal, S., D. Mitra, S. Roy, and D. Majumder (2010). Biosensors and biochips for nanomedical applications: a review. *Sensors and Transducers 113*(2), 1–17.

Gonzalez, R. and R. Woods (2001). Digital image processing. *Prentice Hall, second edition*.

Granitzer, P. and K. Rumpf (2010). Porous silicona versatile host material. *Materials 3*(2), 943– 998.

Harraz, F., K. Kamada, K. Kobayashi, T. Sakka, and Y. Ogata (2005). Random macropore formation in p-type silicon in hf-containing organic solutions. *J. Electrochem. Soc. 152*(4), C213–C220.

Jiao, Y., D. Koktysh, N. Phambu, and S. Weiss (2010). Dual-mode sensing platform based on colloidal gold functionalized porous silicon. *Appl. Phys. Lett. 97*(15), 153125–153127.

Klette, R. and A. Rosenfeld (2005). Digital geometry: Geometric methods for digital picture analysis. *Morgan Kaufmann Publishers*.

Lehmann, V. (2003). Trends in fabrication and applications of macroporous silicon. *Physica Status Solidi (a) 197*(1), 13–15.

Lin, V., K. Motesharei, K. Dancil, M. Sailor, and M. Ghadiri (1997). A porous siliconbased optical interferometric biosensor. *Science 278*(5339), 840–843.

Reddy, R., A. Chadha, and E. Bhattacharya (2001). Porous silicon based potentiometric triglyceride biosensor. *Biosensors and Bioelectronics 16*(45), 313–317.

Rouquerol, J., D. Avnir, C. Fairbridge, D. Everett, J. Haynes, N. Pernicore, J. Ramsey, K. Sing, and K. Unger (1994). Recommendations for the characterization of porous solids. *Pure and Applied Chemistry 66*, 1739–1758.

Rucklidge, W. (1997). Efficiently locating objects using the hausdorff distance. *International Journal of Computer Vision 24*(3), 251– 270.

Saha, H., S. Dey, C. Pramanik, J. Das, and T. Islam (2006). Porous silicon-based smart sensors. *Encyclopedia of Sensors, American Scientific Publishers 8*, 163–196.

Salonen, J., A. Kaukonen, J. Hirvonen, and V. Lehto (2008). Mesoporous silicon in drug delivery applications. *J. Pharmaceutical Sci. 97*(2), 632–653.

Stewart, M. and J. Buriak (2000). Chemical and biological applications of porous silicon technology. *Advanced Materials 12*(12), 859–869.

Vyatkin, A., V. Starkov, V. Tzeitlin, H. Presting, J. Konle, and U. Konig (2002). Random and ordered macropore formation in p-type silicon. *J. Electrochem. Soc. 149*(1), G70–G76.

Determination of geometrical properties of surfaces with grids virtually applied to digital images

S. Galli & V. Lux & E. Marotta & P. Salvini
Università di Roma "Tor Vergata", Roma, Italy

ABSTRACT: The objective of the present work is to detect the geometric features of surfaces through digital image comparison. The method does not require stereo image processing but is based on a single camera vision. The base of this work regards the displacements of a grid virtually applied on the surface. To this goal the real printed grid case is firstly discussed. The grid virtually attached to the pictures identifies a finite element mesh associated to the comparing images.

1 INTRODUCTION

Rigid body detection and surface out of plane measurements has been a subject extensively studied by many researchers [Pilet 2007]. In many cases the objective is directed to the recognition of objects, whatever is their spatial positioning and surface deformation. This is often accomplished by methods that make use of pattern of key points [Lowe 2004].

For the reconstruction of surface deformations many efforts have been done by researchers, in both 2D and 3D approach, many of these methods have been compared in [Sutton 2000]. The general achievements show that many benefits are gained if simultaneously using the entire area of the image acquired; this is to say not considering subsets separately. In this optics, some smoothing can be achieved by B-spline regularization [Cheng 2002] or by finite element formulation for the acquisition of the deformation field [Hild 2008] by DIC.

In the present paper the attention is devoted on the capability to extract the geometric shapes of surfaces, originally flat. This means that the interest is not only directed towards object recognition, but to the surface characteristics for the identification of its off-plane displacements. Within the paper, it is demonstrated that the use of an effective image, sufficiently variegated (such as a speckle image or any other non-uniform picture) is equivalent to a regularly meshed grid [Amodio 1995]. This equivalence is achieved by mapping the picture through a regular or non-regular mesh of quadrilateral sub-images (elements). These non-superimposed sub-images are managed as four node bilinear membrane elements, well known in finite element analysis. Each element of the grid maintains its peculiarity because it is characterized by a different color content and distribution (sub-image). This allows handling the surface identification even if part of the image is hidden to the camera.

2 METHOD DESCRIPTION

2.1 Frame of development

The geometric model adopted is the simple equivalent pinhole camera. According to this assumption, being f the distance between the pinhole and the image sensor, the following ratios can be written:

$$X' = -\frac{f}{Z}X; \qquad Y' = --\frac{f}{Z}Y; \qquad Z' = -f \qquad (1)$$

The negative sign is generally changed by considering the image projected between the viewing point and the object (Fig. 1).

In the pinhole assumption, the complex dioptric lens system is substituted by an ideal single lens, infinitesimally thin, thus the optical system agrees with the following assumptions [Sutton 2009]:

i) all parallel light rays are concentrated on the focus;
ii) no refraction is induced to all rays passing through the lens center;

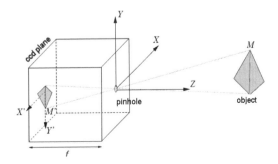

Figure 1. Scheme of pinhole camera view.

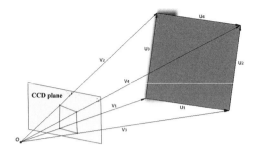

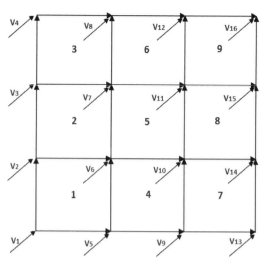

Figure 2. Vectors identifying the positioning of element nodes.

iii) all non-centered rays are deviated in correspondence of the middle plane.

Other assumed hypotheses, adopted when dealing with digital images are:

i) the optical axis is perfectly orthogonal to the sensor plane and centered on it;
ii) the sensor gauge is organized by a two perfectly orthogonal cell disposition.

2.2 Identification of grid images

In this section we assume that the image is simply constituted by a regular square grid. In the next section the association of a grid to a general image will be discussed.

Figure 2 shows the projection of a simple square on the CCD plane that is reversed, as usually. The geometry projected on the plane is given by the four vectors connecting the observation point to the square corners.

Once that the image is digitally acquired, each vector $\mathbf{v}$ associated to a point is known in its direction, while its magnitude remains unknown. Generalizing to the representation of a full grid, it is clear that all unknowns are represented by the moduli associated to the respective vectors. According to this logic, all vectors $\mathbf{U}$ are computed as differences of vectors $\mathbf{V}$ (Fig. 2). In the following, the vectors $\mathbf{V}$ will be addressed making use of respective unit vectors: $\mathbf{V} = \mathbf{m} \cdot \mathbf{v}$.

Making reference to the 3×3 grid represented in Figure 3, the first element gives the following equations:

$$\mathbf{U}_1 = \mathbf{V}_5 - \mathbf{V}_1 = m_5 \mathbf{v}_5 - m_1 \mathbf{v}_1$$
$$\mathbf{U}_2 = \mathbf{V}_6 - \mathbf{V}_5 = m_6 \mathbf{v}_6 - m_5 \mathbf{v}_5 \qquad (2)$$
$$\mathbf{U}_3 = \mathbf{V}_2 - \mathbf{V}_1 = m_2 \mathbf{v}_2 - m_1 \mathbf{v}_1$$
$$\mathbf{U}_4 = \mathbf{V}_6 - \mathbf{V}_2 = m_6 \mathbf{v}_6 - m_2 \mathbf{v}_2$$

Being the grid formed by equal squares, several conditions can be imposed to each of them - note that not all of them are independent - represented in Table 1.

Figure 3. Nomenclature of a 3×3 grid.

Table 1. Vector and scalar conditions for a square grid.

Geom. condition	Vector eq.	No. of scalar eq.s
$U_2 \parallel U_3$	$U_2 \times U_3 = 0$	3
$U_1 \parallel U_4$	$U_1 \times U_4 = 0$	3
$U_1 \perp U_2$	$U_1 \cdot U_2 = 0$	1
$U_2 \perp U_4$	$U_2 \cdot U_4 = 0$	1
$U_3 \perp U_4$	$U_3 \cdot U_4 = 0$	1
$U_1 \perp U_3$	$U_1 \cdot U_3 = 0$	1
$|U_1| = L$	$U_1 \cdot U_1 = L^2$	1
$|U_2| = L$	$U_2 \cdot U_2 = L^2$	1
$|U_3| = L$	$U_3 \cdot U_3 = L^2$	1
$|U_4| = L$	$U_4 \cdot U_4 = L^2$	1

The equations given in the previous table generates, using the modules as the unknowns, the following 14 equations (note that the first two are vector equations):

$$m_1 m_5 (\mathbf{v}_5 \times \mathbf{v}_1) - m_1 m_6 (\mathbf{v}_6 \times \mathbf{v}_1) - m_2 m_5 (\mathbf{v}_5 \times \mathbf{v}_2) - m_2 m_6 (\mathbf{v}_6 \times \mathbf{v}_2) = \mathbf{0}$$
$$m_1 m_2 (\mathbf{v}_1 \times \mathbf{v}_2) - m_1 m_6 (\mathbf{v}_1 \times \mathbf{v}_6) - m_2 m_5 (\mathbf{v}_5 \times \mathbf{v}_2) + m_5 m_6 (\mathbf{v}_5 \times \mathbf{v}_6) = \mathbf{0}$$
$$m_1 m_5 (\mathbf{v}_1 \cdot \mathbf{v}_2) - m_1 m_6 (\mathbf{v}_1 \cdot \mathbf{v}_6) - m_5 m_5 (\mathbf{v}_5 \cdot \mathbf{v}_5) + m_5 m_6 (\mathbf{v}_5 \cdot \mathbf{v}_6) = 0$$
$$m_1 m_5 (\mathbf{v}_1 \cdot \mathbf{v}_2) - m_1 m_6 (\mathbf{v}_1 \cdot \mathbf{v}_6) - m_5 m_5 (\mathbf{v}_5 \cdot \mathbf{v}_5) + m_5 m_6 (\mathbf{v}_5 \cdot \mathbf{v}_6) = 0$$
$$-m_2 m_6 (\mathbf{v}_2 \cdot \mathbf{v}_6) + m_2 m_5 (\mathbf{v}_2 \cdot \mathbf{v}_5) - m_5 m_6 (\mathbf{v}_5 \cdot \mathbf{v}_6) + m_6 m_6 (\mathbf{v}_6 \cdot \mathbf{v}_6) = 0$$
$$m_1 m_2 (\mathbf{v}_1 \cdot \mathbf{v}_2) - m_1 m_6 (\mathbf{v}_1 \cdot \mathbf{v}_6) - m_2 m_2 (\mathbf{v}_2 \cdot \mathbf{v}_2) + m_2 m_6 (\mathbf{v}_2 \cdot \mathbf{v}_6) = 0$$
$$m_1 m_1 (\mathbf{v}_1 \cdot \mathbf{v}_1) - m_1 m_2 (\mathbf{v}_1 \cdot \mathbf{v}_2) - m_1 m_5 (\mathbf{v}_1 \cdot \mathbf{v}_5) + m_2 m_5 (\mathbf{v}_2 \cdot \mathbf{v}_5) = 0$$
$$m_1^2 \sum_{i=1}^{3} v_{1(i)}^2 - 2 m_1 m_5 \sum_{i=1}^{3} \left(v_{1(i)} v_{5(i)} \right)^2 + m_5^2 \sum_{i=1}^{3} v_{5(i)}^2 = L^2 \qquad (3)$$
$$m_5^2 \sum_{i=1}^{3} v_{5(i)}^2 - 2 m_5 m_6 \sum_{i=1}^{3} \left(v_{5(i)} v_{6(i)} \right)^2 + m_6^2 \sum_{i=1}^{3} v_{6(i)}^2 = L^2$$
$$m_2^2 \sum_{i=1}^{3} v_{2(i)}^2 - 2 m_2 m_1 \sum_{i=1}^{3} \left(v_{2(i)} v_{1(i)} \right)^2 + m_1^2 \sum_{i=1}^{3} v_{1(i)}^2 = L^2$$
$$m_6^2 \sum_{i=1}^{3} v_{6(i)}^2 - 2 m_6 m_2 \sum_{i=1}^{3} \left(v_{6(i)} v_{2(i)} \right)^2 + m_2^2 \sum_{i=1}^{3} v_{2(i)}^2 = L^2$$

Table 2. List of unknowns produced by a single element.

Unknowns			
1	m_1^2	6	$m_2 m_6$
2	$m_1 m_2$	7	$m_2 m_5$
3	$m_1 m_6$	8	m_5^2
4	$m_1 m_5$	9	$m_5 m_6$
5	m_2^2	10	m_6^2

Table 3. Indexes for assemblage of elements given in Fig. 3.

Elem no.	Vertex	Submatr 1	Submatr 2
1	1 2 6 5	[1 2]	[5 6]
2	2 3 7 6	[2 3]	[6 7]
3	3 4 8 7	[3 4]	[7 8]
4	5 6 10 9	[5 6]	[9 10]
5	6 7 11 10	[6 7]	[10 11]
6	7 8 12 11	[7 8]	[11 12]
7	9 10 14 13	[9 10]	[13 14]
8	10 11 15 14	[10 11]	[14 15]
9	11 12 16 15	[11 12]	[15 16]

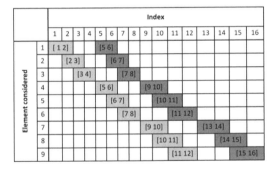

Figure 4. Layout of the assembled matrix.

All above equations give a non-linear (quadratic) system of equations where the unknowns are vector magnitudes. If one considers all possible combinations of products of unknowns as unknowns themselves, they turn to be 10 for a single element, with 14 equations each. As an example, for the element n. 1 of Figure 3, the ten unknowns are:

Previous table forms a system of equations that can be subdivided into two parts, so that the complete system represented in Figure 3 generates a full system as illustrated in the Table 3.

According to these matrix subdivisions, the assembled matrix assumes the forms given in Figure 4.

The full system of equations is therefore over-determined and the solution can be found solving all the quadratic unknowns involved. After the full solution, each vector magnitude is computed by the root square of the quadratic unknowns. Furthermore, the mixed product of the unknowns can be used to check the accuracy of the solution gained.

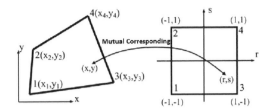

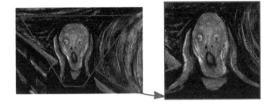

Figure 5. Physical/Natural coordinate systems.

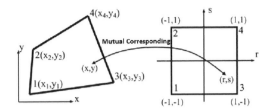

Figure 6. Change from physical to natural coordinates.

2.3 Virtual image embedded on a picture

If an image is present on a surface, this can be associated to a virtual grid. The point is to guarantee that the grid follows the changes of the image, due to movements of the surface that can be considered as a combination of rigid and deformable displacements.

This can be accomplished by considering the grid as a mesh of bilinear finite elements, whose movement guarantees the continuity of the surface.

Each element contains a part of the initially flat image; this information is maintained in a natural coordinate system as shown in Figure 5.

This is particularly suitable to compare elements that are initially irregular or become irregular due to large displacements on the image.

Because of this, each element sub-image is interpolated though a cubic spline approach. By this procedure, each element is always square-represented and keeps the same image content. In Figure 6 it is shown an example of how interpolation deforms the image.

3 RESULTS

The data here presented, apart the very next subsection, refers to all effective pictures taken with a focal distance equal to 29 mm, corresponding to a printed paper positioned at 1500 mm from the ideal lens center. All images have been obtained with an aperture equal to 1/8 to increase overall focus depth.

3.1 Test of the procedure on exact grids

The theoretical correctness of the procedure presented in the previous sections has first of all been investigated through the operation on a simulated grid (no pictures does exist) that has been deformed with one finite curvature or two finite curvatures. We can see that the reconstruction is perfectly accurate (no digital

Table 4. Noise effect on accuracy for various curvatures.

	No noise	Added noise
Plane grid	~1e-10%	4.06%
Single curvature	~1e-10 %	4.87%
Double curvature	23.13%	24.21%
D.curv. refined el.	13.61%	15.93%

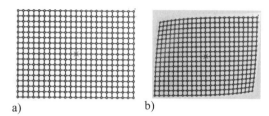

a)   b)

Figure 7. Digital pictures of the printed grid, a) before and b) after deformation.

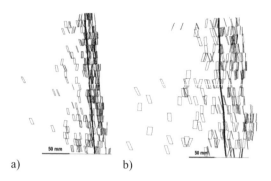

a)   b)

Figure 8. Comparison of overall result and square dispersion for; a) printed grid, b) virtual grid on speckle image.

Table 5. Angle identified when varying the point of view.

	Grid method	Speckle method	
Angel	results	16 × 16 el.	37 × 37 el.
5°	4.64°	3.96°	4.50°
10°	9.72°	8.23°	8.69°
15°	14.86°	9.64°	11.68°

error on pixel definitions is present since pixel positions are recorded with 12 digit precision), only if the local orientation of the grid causes a single curvature change ($R_c = 1430$ mm). But it is interesting to highlight that the application of a small noise (*1%* of the diagonal length of a single element) reduces considerably the precision if the results are very accurate, but does not appreciably changes the behavior if discrepancies are already encountered when data are not affected by noise.

From the above results, it is clear that the assumption of square grid to maintain its shape is very strong, difficult to obtain when double curvature are present. This means that the size of the elements of the grid should be taken as small as the curvature increases. As a matter of fact, the last row in Table 4 shows much better results in this case, as expected (double curvature keeping the same equivalent value than single curvature case).

3.2 Tests on pictures of printed grid

The grid is printed on a sheet of paper that is first photographed in a plane orthogonal to the focal axis, and the second picture considers the sheet deformed in various ways, such as shown in Figure 7.

Four cases are here presented; *i*) it concerns a 5-10-15° rigid rotation of the paper on a vertical axis as to generate a prospective view; *ii*) it regards a simple half-fold oriented as the vertical axis in the center, and folded at a corner; *iii*) the paper applied on a cylinder (diameter = 450 mm) with a vertical axis; *iv)* the paper applied on the same cylinder with the sheet base inclined of 30°.

Figure 8 a) shows the results obtained by the method described in §2.2 through top views. Two operating ways are compared: the black lines show the results when the over-determined solution is performed on all

squares at the same time, the blue lines consider the solution performed at each square, individually. The whole solution is much better than the second one, because pixel errors compensate each other.

The following Table 5 shows the errors introduced when varying sheet inclination. The results show that a rigid displacement on one of the grid axes can be managed by the method in the average by both images (grid or speckle above discussed).

The two cases of folds considered (*ii*) reveal a particular behavior of the algorithm. As a matter of fact, the mean square method tends to compensate the errors so that there is the tendency to keep flat in one direction. Referring to Figure 9 a) and b) the dispersion is illustrated when the fold is located in the center or in the corner, respectively. The maximum determined fold displacement respects the values imposed in the test within a *2%* error. Figure 10 b) results seem to be erroneously similar to the previous case, as the top views show. It is clear that the identification algorithm is affected by eventual curvatures non-aligned with the grid. In practice, since the algorithm tends to maintain the overall length at each element, it encounters some difficulties while managing elements that change all side-length due to curvatures imposed.

The results performed on cylindrical surfaces confirm the deficiencies previously indicated on double curvature grids. As a matter of fact the comparison between cases a) and b) in Figure 10 shows a much more evident dispersion of single computed squares (blue) when the sheet base is inclined toward horizon – relative full picture is visible in Figure 7 b).

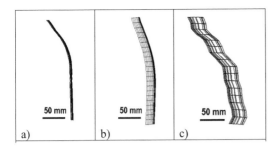

Figure 9. Top view of the profiles for case (*ii*): a) vertical fold b) corner fold, c) vertical fold for speckle image.

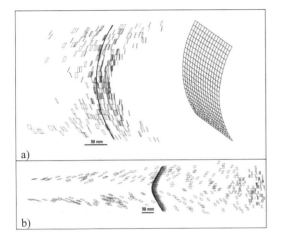

Figure 10. Dispersion of individually identified elements for: a) vertical cylinder, b) inclined cylinder.

The global results (black lines) of case a) are quite accurate (error in radius estimate lower than *1%*) while case b) evidently gives erroneous results.

3.3 *Tests on a printed speckle picture*

Indeed, the application of a virtual grid introduces some additional errors by respect to the printed grid. These errors amplify the deficiencies already evidenced before. The algorithm used to detect the displacement of the virtual grid by means of speckle deformed image is not discussed here, we only mention that it is based on a differential method whose convergence is achieved by means of minimum square errors. For example, one can compare the results given in Figures 11 and 12, a) and b), respectively. Case a) identifies the virtual grid imposed on the image to be taken as reference; case b) in Figure 11 shows good identification that can also be seen in Figure 8 b) through square dispersion. On the contrary, one can see that the corner correspondence on Figure 12 a) and b) is very poor, particularly at the left top corner where the deformations are the highest.

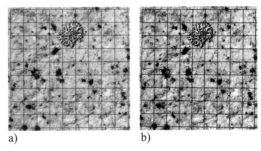

Figure 11. Virtual node locations before and after 5° inclination.

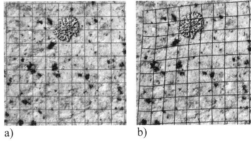

Figure 12. Virtual node locations before and after application on inclined cylinder

4 DISCUSSION AND CONCLUSIONS

The algorithm proposed here shows some difficulties when applied to structures that deforms with double curvatures. This can probably improves if the imposition on the length of the square corners does account of the effective length of them. In fact, when curvatures are present, the corner lengths cannot be anymore computed accurately by simple node distances. A better computation should accounts of effective surface distance through an iterative procedure, not implemented in the results here presented, but which represents an obvious future development.

From another point of view, the advantage of the method is that no regularization conditions are requested to the identifying surface, so that no additional conditions on surface deformed shape are a-priori imposed.

An idea of the solution accuracy can be reached from the dispersion of the results when computation regards each square element individually. When the dispersion is high the global results is correspondingly worse.

Use of speckle image instead of a printed grid is possible and a theoretical increment of information is available. However, the differential method, not discussed here, requires a limitation of the displacements introduced in the image to keep consistency. When the displacements are less than 50% of the virtual grid dimension the grid nodes moves correctly, for higher values accuracy problems become evident.

REFERENCES

Amodio D., Broggiato G.B., Salvini P., 1995, Finite Strain Analysis by Image Processing: Smoothing Techniques, *Strain, Vol. 31, n. 3, pp. 151–157.*

Cheng P., Sutton M.A., Schreier H.W., McNeill S.R., 2002, "Full-field Speckle Pattern Image Correlation with B-Spline Deformation Function, *Experimental Mechanics, 42(3), pp. 344–352.*

Hild F., Roux S., 2008, A Software for "Finite-element" Displacement Field Measurements by Digital Image Correlation, *Internal Report n. 269*, LMT-Cachan, UniverSud Paris.

Lowe D.G., 2004. Distinctive Image Features from Scale-Invariant Keypoints, *International Journal of Computer Vision, 60(2), pp. 91–110.*

Pilet J. Lepetit V., FUA P., 2007, Fast Non-Rigid Surface Detection, Registration and Realistic Augmentation, *International Journal of Computer Vision 76(2), 109–122.*

Sutton M.A., McNeill S.R., Helm I.D., Schreier H.Vr and Chao Y.J., 2000, "Photomeehanics," *Advances in 2D and 3D Computer Vision for Shape and Deformation Measurements, P. K Rastogi, ed., Topics in Applied Physics, Springer Verlag, 77, pp. 323–372.*

Sutton M., Orteu J.J., and Schreier H.W., 2009, *Image Correlation for Shape, Motion, and Deformation Measurements*, Springer, New York.

Computational Modelling of Objects Represented in Images – Di Giamberardino et al. (eds)
© 2012 Taylor & Francis Group, London, ISBN 978-0-415-62134-2

On the use of a differential method to analyze the displacement field by means of digital images

V. Lux, E. Marotta & P. Salvini
Università di Roma 'Tor Vergata', Roma, Italia

ABSTRACT: In the present paper a method to evaluate surface strains is discussed. The algorithm performs the analysis of the two comparing images, before and after the application of loads. The method is based on differential calculation of Jacobian matrix and the solution of an over-determined system of equations. The differential calculation is truncated at the first order and iteratively applied. In this work it is also presented a technique to increase the computation speed by a reduction of the number of system equations. Two different strategies are proposed: a partial grouping of pixels by equation averaging; the use of Hu's invariants applied to sub-images. The two methods are discussed and compared.

1 INTRODUCTION

One of the most interesting topic demanded to experimental mechanics is the capability to determinate the strain and the stress fields on specimens using non-invasive methods. The most promising methods involves photographic techniques, such as digital image correlation [Sutton 1983, Lagattu 2004, Sutton 2009, Cofaru 2010]. Usually the strain and motion analysis procedures make use of a sequence of pictures that follows the whole strain progress [Broggiato 2006]. In this paper a method that makes use only of two images is proposed; the first, taken with no applied load, the second one when all loads and deformations are settled.

There exist a 2-D digital Correlation technique as well as 3-D ones [Lu 1997]. Here we deal with 2-D technique which is based on the use of a single camera. However, most of the techniques are based on sub-image correlations. This means that the local correlation imposed on a sub-image does not have an influence on all other correlations performed on sub-images far away from the previous one. In this paper, according to former approaches [Cheng 2002, Broggiato 1995], we discuss a technique that solves the displacement fields as a whole, so that the continuity conditions is fulfilled in the processed image.

When consistent displacements are faced, the correlation techniques meet considerable difficulties to keep precision, as discussed in [Lagattu 2004].

The use of digital image processing to analyze the displacement field generally needs the solution of a very high number of equations; this can even attain the order of the number of pixels recorded into the image. Therefore, a technique directed to reduce the number of equations while increasing the efficiency is proposed.

2 METHOD

2.1 Differential method

The deformation field is calculated by the comparison of the original image and the deformed one. The approach used in this work is of a global kind [Broggiato 2006]. The problem consists in minimizing the error functional defined by the following formula:

$$E = \sqrt{\sum_i \left[I_d(x_i, y_i) - I_u(x_i, y_i) \right]^2} \qquad (1)$$

Where $I_u I_d$ represent the images before and after deformation. x_i, y_i are the *i-th* pixel coordinates into the images. The summation considers all pixels. Both images are grid-meshed through 4-node finite elements. To each element a sub-image is associated. At each sub-image (corresponding to a single finite element) the internal spatial distribution is based on the displacements of the four nodes bordering the element. The solution allows finding the node locations that make it possible to overlap the undeformed image onto the deformed one. Once the image is divided into sub-images, it is possible to write the formula (1) as:

$$E = \sqrt{\sum_j \left\| S_j(x) - S_{j0}(x_0) \right\|^2} \qquad (2)$$

Where S_j, S_{j0} are the *j-th* sub-image, while x and x_0 represent the nodal coordinates for the deformed state and the initial one, respectively. The dependence of x and x_0 is non-linear; for this reason it is not possible to directly find the solution, but an iterative procedure is needed. The algorithm consists in the linearization of the non-linear least squares problem. It is based on Taylor series expansion, truncated to its first order, of the sub-image when varying the generic nodal coordinate:

$$\mathbf{S}_j(X') = \mathbf{S}_0(X_0) + \sum_k \frac{\partial \mathbf{S}_0}{\partial \mathbf{x}_k}(x_k - x_{k0}) \qquad (3)$$

For the sake of clarity, the same is written in matrix formulation, evidencing the Jacobian matrix:

$$\begin{bmatrix} \frac{\partial \mathbf{S}_{1,0}}{\partial \mathbf{x}_1} & \frac{\partial \mathbf{S}_{1,0}}{\partial \mathbf{x}_2} & \cdots & \frac{\partial \mathbf{S}_{1,0}}{\partial \mathbf{x}_M} \\ \frac{\partial \mathbf{S}_{2,0}}{\partial \mathbf{x}_1} & \frac{\partial \mathbf{S}_{2,0}}{\partial \mathbf{x}_2} & \cdots & \frac{\partial \mathbf{S}_{2,0}}{\partial \mathbf{x}_M} \\ \vdots & \vdots & \cdots & \vdots \\ \frac{\partial \mathbf{S}_{N,0}}{\partial \mathbf{x}_1} & \frac{\partial \mathbf{S}_{N,0}}{\partial \mathbf{x}_2} & \cdots & \frac{\partial \mathbf{S}_{N,0}}{\partial \mathbf{x}_M} \end{bmatrix} \cdot \begin{Bmatrix} x_1 - x_1' \\ x_2 - x_2' \\ \vdots \\ x_M - x_M' \end{Bmatrix} = \begin{Bmatrix} \mathbf{S}_{1,0} - \mathbf{S}_{1,j}' \\ \mathbf{S}_{2,0} - \mathbf{S}_{2,j}' \\ \vdots \\ \mathbf{S}_{N,0} - \mathbf{S}_{N,j}' \end{Bmatrix} \quad (4)$$

or the equivalent $\mathbf{J} \cdot \Delta \mathbf{x} = \Delta \mathbf{s}$. Each term of the Jacobian matrix is composed by a number of elements that coincides with the number of pixels contained in the sub-image. Therefore, the matrix shows considerable dimensions. The computation is based on centered first derivate, evaluated by considering the pixels around each node. For this reason, the Jacobian matrix of the reference image is computed by respect to all possible node displacements. As mentioned before, the Jacobian matrix is composed of partial derivatives. Each matrix derivative is calculated by the four central points derivative formulation as:

$$\frac{\partial \mathbf{F}(x)}{\partial x} = \frac{1}{12h} \left[\mathbf{F}(x-2h) - 8\mathbf{F}(x-h) + 8\mathbf{F}(x+h) - \mathbf{F}(x+8h) \right] (5)$$

where h represents the discretization step. For example, the term $\mathbf{F}(x - 2h)$ represents the resulting sub-image when a bordering node is moved back of $2h$ pixels. The right choice of the parameter h is crucial. The solution requires the minimization of the error (2) through an iterative procedure. As a matter of fact the inversion of the Jacobian matrix is required (4).

$$\Delta \mathbf{x} = \mathbf{J}^{-1} \cdot \Delta \mathbf{s} \qquad (6)$$

The Jacobian matrix is a sensitivity matrix; however, its costly inversion must be performed just once. When the displacements of all nodes have been computed, it is easy to gain the internal strains, known by means of the nodal displacements of each element. This is made possible through the pre-multiplication of the vector of element nodal displacements by the matrix **B** (obtain by appropriate derivative of Q4 shape functions) [Zienkiewicz 1967] if first order approximation is accepted, or more sophisticated expressions if first order is not applicable.

3 CONVERGENCE ENHANCEMENT

3.1 *Methods introduced*

In this section, the optimal choice of the number of internal points, to get a quick convergence of the

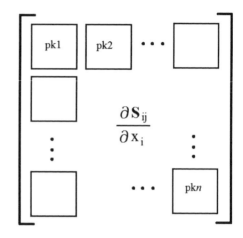

Figure 1. Grouping technique representation.

results, is discussed. A criterion to reduce the equations and to speed-up the solution process is introduced. As an example, for a high-definition 3000×4000 pixel image, meshed through almost 10000 elements having eight degrees of freedom each, a 240 billion of equations results. In this paper two possible techniques have been developed; the first considers possible grouping of pixels belonging to confined areas, the second one makes use of Hu's invariants to discriminate each sub-image.

3.2 *Method I: Grouping technique*

The technique of grouping is helpful to reduce the number of equations to solve. As cited before, the Jacobian presents a considerable dimension; in fact each term of the matrix is a partial derivative of the sub-image. The grouping method introduced in this paper, handles the single partial derivative of the sub-image. These are divided into sub-areas, and a sum of the values inside them is considered, as shown in Figure 1. By this way, the number of equations is reduced, and the iterative calculation speeds up. This manipulation reduces the dimension of each term of the Jacobian previously introduced, to a smaller matrix dimensioned by the number of divisions.

The size of the grouping is important, and must be chosen wisely. A considerable grouping amount is required for saving computational time, but it significantly reduces the data information, making it difficult to gain convergence in the iteration process.

3.3 *Method II: Hu's invariants*

Here we refer to moments as scalar quantities able to characterize a scalar field and possibly to point out some significant features. In statistics moments are widely used to describe the shape of a probability density function; in classic rigid-body mechanics to account of the mass distribution of a body, forming the inertia tensor. In mathematical terms, the moments are projections of a function onto a polynomial basis [Flusser 2009, Hu 1962]. General moments M_{pq} of an image $I(x,y)$, where p, q are non-negative integers and $r = p + q$ is called the order of the moment, are defined as

$$M_{pq} = \iint p_{pq}(x,y) \cdot I(x,y) \cdot dx\, dy \qquad (7)$$

Where $p_{00}(x, y)$, $p_{10}(x, y)$, ..., $p_{pq}(x,y)$, are polynomial basis functions defined in the domain. In this paper the image moments are used, consequently the function $I(x,y)$ is an image characterized by two coordinates x, y and a color value (e.g. RGB uses three values between 0 and 255 each).

A geometric moment of a discretized image is defined as a moment having the standard power basis $p_{pq}(x,y) = x^p y^q$. Therefore, it results:

$$m_{pq} = \sum_x \sum_y x^p \cdot y^q \cdot I(x,y) \qquad (8)$$

Moreover Hu's invariants are built up through a combination of geometrical moments that show the characteristics to not changing their values when some geometrical transformations are applied. Hu's invariants are originally seven, but other invariants could be computed. However, higher image invariants would increase considerably their magnitude, so that they cannot be managed together with the first seven.

The invariant moments are insensitive to translation, rotation and scaling transforms. In an index compact notation they are:

$\varphi_1 = m_{20} + m_{02}$

$\varphi_2 = (m_{20} - m_{02})^2 + 4m_{11}^2$

$\varphi_3 = (m_{30} - 3m_{12})^2 + (3m_{21} - m_{03})^2$

$\varphi_4 = (m_{30} + m_{12})^2 + (m_{21} - m_{03})^2$

$\varphi_5 = (m_{30} - 3m_{12})(m_{30} + m_{12})\left[(m_{30} + m_{12})^2 - 3(m_{21} + m_{03})^2\right] +$

$\qquad + (3m_{21} - m_{03})(m_{21} + m_{03})\left[3(m_{30} + m_{12})^2 - (m_{21} + m_{03})^2\right]$

$\varphi_6 = (m_{20} - m_{02})\left[(m_{30} + m_{12})^2 - (m_{21} + m_{03})^2\right] +$

$\qquad + 4m_{11}(m_{30} + m_{12})(m_{21} + m_{03})$

$\varphi_7 = (3m_{21} - m_{03})(m_{30} + m_{12})\left[(m_{30} + m_{12})^2 - 3(m_{21} + m_{03})^2\right] +$

$\qquad - (m_{30} + 3m_{12})(m_{21} + m_{03})\left[3(m_{30} + m_{12})^2 - (m_{21} + m_{03})^2\right]$

$$\qquad (9)$$

Only the first invariant moment has an intuitive meaning: polar moment of inertia.

In this section Hu's invariants are used to characterize each sub-image. Even using this technique, the solution requires the inversion the Jacobian matrix; in this case the Jacobian is not calculated through image differences, but differences on Hu's invariants. The image is meshed into sub-images by a grid; each sub-image is represented be its set of Hu's invariant moments. Hu's invariants are seven, but here, to

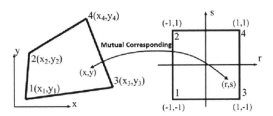

Figure 2. Isoparametric element representation.

increase the discriminant power, the computed invariants are doubled: they are computed on the sub-image itself and on its negative.

3.4 Natural coordinate system applied on elements

Both methods require comparing images to minimize the error. To this goal it is useful referring to the natural coordinate system used in the isoparametric element formulation [Zienkiewicz 1967].

Their characteristics are particularly suitable because, in the natural coordinate system, all the elements, as well as in the reference picture as in the deformed one, have the same shape and dimensions (a simple square as shown in Fig. 2). As a matter of fact, one of the difficulties encountered in both methods regards the change of the edges during displacement, now overtaken by using elements having the same domain shape, whatever is the image content. For the isoparametric four-node element, all internal points are mapped through natural coordinates r, s in the following manner:

$$\begin{cases} x(r,s) = \dfrac{1}{4}(1-r)(1-s)\,x_1 + \dfrac{1}{4}(1-r)(1+s)\,x_2 + \\[2mm] \qquad + \dfrac{1}{4}(1+r)(1-s)\,x_3 + \dfrac{1}{4}(1+r)(1+s)\,x_4 \\[3mm] y(r,s) = \dfrac{1}{4}(1-r)(1-s)\,y_1 + \dfrac{1}{4}(1-r)(1+s)\,y_2 + \\[2mm] \qquad + \dfrac{1}{4}(1+r)(1-s)\,y_3 + \dfrac{1}{4}(1+r)(1+s)\,y_4 \end{cases} \quad (10)$$

By this change of reference, an interpolation is required between the pixels in each element (only on sub-images of whole picture) and the values assumed in the $r - s$ coordinate system. According to the isoparametric formulation, the same interpolation used to locate any internal point is adopted to evaluate internal displacements. The use of $r - s$ coordinate system allows also to manage non-regular shaped elements.

4 RESULTS AND COMPARISON

4.1 Grouping technique vs Hu's invariant

In this section a comparison is proposed: i) full pixel computation; ii) grouping technique by varying packaging dimension; iii) Hu's invariant moments. All

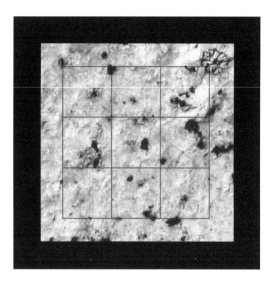

Figure 3. Reference undeformed image 500 × 500 pixel.

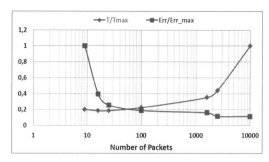

Figure 4. Cpu Time VS Error for the grouping technique.

Table 1. Grouping technique performance.

Number of packs	Iterations	Total time[s]	Jacobian Time [s]	Error [%]	Average Error [pixel]
4	Not convergence				
9	12	0,8886	0,0243	8,1729	0,2311
16	11	0,8166	0,0242	3,7483	0,1060
25	10	0,8252	0,0242	2,6902	0,0760
100	12	0,9963	0,0256	1,5607	0,0441
1600	13	1,5676	0,0531	1,4710	0,0416
2500	13	1,9551	0,0728	1,0258	0,0290
No enhancement	17	4,4014	0,2526	1,0848	0,0306

Table 2. Hu's invariant moments technique performance.

Number of invariants utilized	Iterations	Total time [s]	Jacobian Time [s]	Error [%]	Average Error [pixel]
14	236	20,7042	0,0367	10,0316	0,2837
12	236	20,7139	0,0382	10,0397	0,2840
10	258	22,6446	0,0384	22,3302	0,6316

techniques have been applied on the same reference image. This image shows the surface of a granite (Fig. 3). The use of this image is due to his particular distribution of color that is a natural sort of speckle. The image is divided in 3x3 elements and 16 nodes; the dimension of a single element is 100×100 pixels. The original image is digitally deformed by $\varepsilon_x = 0.02$ and $\varepsilon_y = 0.01$. The *i*) results (no enhancement) are obtained when the number of packets reaches 10000 (number of pixels in the sub-image).

Several tests are performed progressively decreasing the number of packets. The lowest number of packets considered is 9, while the maximum numbers is 10000, representing the solution without any enhancement technique. Both convergence time and final error of the displacements are compared. All computations are performed on an internal processor Intel®Core *i7-2600 K* with a clock set to 3.4 GHz.

The error is calculated through a ratio: in the numerator is the sum of the displacement differences between the identified final nodal position and the theoretical one (known due to strain imposed); in the denominator is the sum of the differences between the theoretical position and the initial one.

In Figure 4 errors and convergence times are normalized by respect to the highest values encountered. Note that this relative definition of error penalizes the lower displacements and this must be kept in mind when comparing different deformation magnitudes. As expected, the maximum errors are obtained with the minimum number of packets; the worst convergence time is obtained when no enhanced technique is used. Increasing the number of packets decreases the error convergence, but increases the calculation time and vice versa. It is possible to detect a crossing point of minimum error-time curves (Fig. 4, Tab. 1). It is interesting to highlight that the red curve identifies a well-defined knee, demonstrating that a strong grouping is possible while shortly affecting accuracy.

The results obtained with the method of Hu's invariant moments are shown in Table 2. This method does not show at the present time particularly good outcomes. The invariants allow to greatly reducing the number of equations of the system, but do not ensure acceptable results both in terms of computing time and precision of the solution. Even the use of non-speckle images helped to gain accurate results. However, at strain values of the order of some percent, the Hu's invariant method works acceptably. The increase of the convergence time is mainly due to the increment in the number of iterations required to converge.

Further tests have been performed using the grouping technique. In particular, some tests considered large strains applied, over 30%. It is interesting to observe that the grouping technique is able to manage also this amount of image differences, even when the non-enhanced technique is unable to reach convergence. Thus, the use of packets is not only able to speed up the solution time, but it is capable to organize information so that the convergence capability is stronger than before. As an example, in the case analyzed and proposed in Figure 5, presenting a strong deformation reaching 0.35 in both principal directions, calculated with 25 packages, the method returns a solution with an error close to 0.56% and 9.84 s of convergence time, whereas without the use of packages the solution does not converge at all.

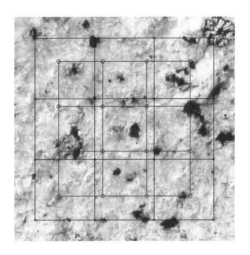

Figure 5.   Example large deformed image and grid solution obtained by grouping technique.

Table 3.   Grouping technique performance.

Number of packs	Iterations	Total time[s]	Jacobian Time [s]	Error [%]	Average Error [pixel]
25	26	39,7946	0,1358	7,2077	0,4189
No enhancement	410	926,9013	10,2175	6,7912	0,3947

To validate the grouping technique, another example is performed: the image used is not a speckle, but a generic image. In this case the deformations are not of the same magnitude of the preceding one, but they are set to $\varepsilon_x = 0.03$ and $\varepsilon_y = 0.01$; much more elements are used to mesh the image. In particular a grid of 10×10 elements, having 50×50 pixels each is used. The study was performed both with a number of packets equal to 25, and without any convergence enhancement technique.

This latter example shows once again the convenience of the use of packets, both in terms of accuracy and computational time. This convenience is stronger and stronger when increasing the number of sub-images managed. Incidentally, it is shown that the use of packets is profitable on non-speckled images.

5   CONCLUSION

In this work a differential method to determine the displacement field is presented. This method consists in the resolution of a great number of equations; for this reason the CPU time can be considerable. In the paper two different procedures to reduce the number of equations are presented. The first method consists on the grouping of the Jacobian matrix. Despite the reduction of the number of equations, the system is always over-determined, the solution converges with an error decreasing while increasing the number of

Figure 6.   Example deformed image and grid 10x10 solution obtained by grouping technique.

packets. This method is robust and reliable and allows convergence even when very high deformations occur.

A possible use is to apply this procedure as a preliminary calculation in order to approach the exact solution, then refining the results by omitting packets grouping. The second method uses Hu's invariants for the assembly of the Jacobian matrix. The number of the equations is reduces to the number of Hu's invariant moments. This second method designed to enhance the convergence is not as reliable as the previous one, as a matter of fact, even though the number of equations is lower than in the grouping method, the iterations required increase significantly. Furthermore, the increase in the total time is not accompanied by an accuracy rise.

REFERENCES

Broggiato G.B., Cortese L., Santuci G., 2006. Misura delle deformazioni a campo intero attraverso l'elaborazione di sequenze di immagini ad alta risoluzione, *XXXV Convegno Nazionale AIAS*.

Cofaru C., Philips W., Van Paepegem W. 2010. Improved Newton–Raphson digital image correlation method for full-field displacement and strain calculation, *Vol. 49, No. 33 Applied Optics*.

Flusser J., Suk T., Zitová B., 2009. Moments and Moment Invariants in Pattern Recognition, *John Wiley & Sons*.

Hu M.K., 1962. Visual Pattern Recognition by Moment Invariants, *IRE Trans. Info. Theory, vol. IT-8, pp.179–187*.

Lagattu F., Brillaud J., Lafarie-Frenot M.,2004. High strain gradient measurements by using digital image correlation technique, *Materials Characterization 53 (2004) 17– 28*.

Lu H., Zhang X., Knauss W. G., 1997. Uniaxial, shear, and poisson relaxation and their conversion to bulk relaxation: Studies on poly(methyl methacrylate). *Polymer Composites*, 18(2):211–222, April 1997.

Sutton M.A., Orteu J.J., Shreier H.W., 2009. Image Correlation for Shape, Motion and Deformation Measurements. Basic Concept , Theory and Applications, *New York, Springer*.

Zienkiewicz O.C., Taylor R.L. 1967, The finite element method, *McGraw-Hill*.

Computational Modelling of Objects Represented in Images – Di Giamberardino et al. (eds)
© 2012 Taylor & Francis Group, London, ISBN 978-0-415-62134-2

Image processing approach in Zn-Ti-Fe kinetic phase evaluation

V. Di Cocco & F. Iacoviello
Università di Cassino e del Lazio Meridionale, DICeM, Cassino (FR), Italy

D. Iacoviello
*Sapienza University of Rome, Dept. of Computer, Control and Management Engineering Antonio Ruberti,
Rome, Italy*

ABSTRACT: Hot-dip galvanizing is one of the most used methods to apply zinc-based coatings on steels, in order to provide sacrificial protection against corrosion over all the steel surface. In the last years new alloying elements on the Zn bath are studied to optimize properties oriented to new market requirements. In this work a new approach is proposed in order to quantify the main morphological properties of an ZnTiFe phase rich in Ti, present only in coatings produced by baths containing Ti. No quantitative studies have been yet carried out on intermetallic phases formation as consequence of Ti addition in the Zn bath. The innovative approach for ZnTi-based coatings has been applied to five different dipping times, thus allowing to evaluate the kinetic of formation.

1 INTRODUCTION

Hot-dip galvanizing is one of commercially most important protection method of steel surfaces. Zinc or zinc-based coatings provide a sacrificial protection against corrosion (in the galvanic protection zinc is less noble to steel at ambient condition), and play a role of a barrier to many aggressive environments, ASTM Handobook 1999, Marder 2000, Reumont et al. 2001, Sungh et al. 2008, Tzimas et al. 2001.

Galvanizing are still an important research field to optimize coatings in microstructure changes or many properties such as substrate adherence, corrosion behavior or simply external aspects, through the addition of alloying elements or different pretreatments. Another galvanizing field is in concrete constriction where studies by means of SEM (Scanning Electron Microscope) and EDX analysis (Energy Dispersive X-ray spectroscopy) showed that corrosion products formed on passivated hot dip galvanizing (HDG) are more stable in test electrolyte than on unpassivated surface. Furthermore studies also indicate that tests based on potential measurements can be developed to assess the performance of a passivator for zinc coated rebars exposed in concrete pore solution, Marder 2000.

Zinc-based coating layer formation is obtained by interdiffusion of zinc and iron atoms to generate a layer characterized by different chemical composition leading to different intermetallic phases according to the Zn-Fe diagram showed in Figure 1. From the iron-coating interface to external surface there are different intermetallic phases, which Zn content increases. Therefore the zinc coating is a multilayer system

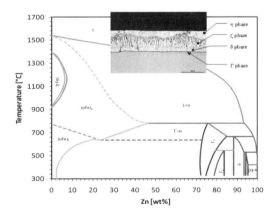

Figure 1. Zinc Coating phases diagram, Marder 2000.

meanly formed by four phases (Fig. 1), characterized by different thickness and mechanical properties. Outer layer is a ductile η phase with maximum Fe content up to 0.03%. The subsequent layer is named as ζ phase, which is isomorphous with a monoclinic unit cell and an atomic structure that contains a Fe atom and a Zn atom surrounded by 12 Zn atoms at the vertices of a slightly distorted icosahedron. The icosahedra link together to form chains and the linked chains pack together in a hexagonal array, Marder 2000. δ phase is a brittle one with a Fe content up to 11.5 wt%, with an hexagonal crystal structure. The last phase is a very thin face centered cubic layer named Γ phase and is characterized by a Fe content up to 29 wt%. Coating formation is governed by physical parameters (bath temperature, immersion time, pre-galvanizing surface temperature, etc.) and chemical parameters (bath and steel chemical compositions, flux chemical composition, etc.).

Processes are very important also on typologies of coatings; eg. in the galvanized steel strip, produced through a continuous hot-dip galvanizing process, the thickness of the adhered zinc film must be controlled by impinging a thin plane nitrogen gas jet, Yoon et al. 2009.

In the last years, there has been an increase in zinc coatings research, focusing both on coating procedures and mechanical behavior characterization, in order to optimize Zn layer thickness and mechanical performances, Tzimas et al. 2001. Three kind of research can be outlined to develop the optimizations, the first about material, when it is possible to choose it, the second is about pre-treatment, where it is possible to define chemical composition of flux or different temperature of pre-heat, and finally about chemical composition of the bath.

Presence of silicon in the material is very important to coatings formation and their properties. An excessive content in steels accumulates on the surface of substrate due to the limited solubility of silicon in the Γ layer. Due to Fe/Zn reaction that determines movement of the α-Fe/Γ interface towards the steel substrate, the α-Fe reach in silicon breaks and the particles enters in to λ layer through the Γ layer due to low solubility of silicon-rich α-Fe in the Γ layer. Then silicon-rich α-Fe dissolves in the λ layer and accelerates the growth of the λ layer to steel substrate, and the coating becomes loose, Balloy et al. 2007.

Some studies about pregalvanizing treatment show firstly that it is possible to replace the conventional industrial chloride flux used in galvanization by a vegetable oil like the linseed oil. Moreover it is also possible to use a mineral oil added with an acid function. Presence of mineral oil well protects the steel but its effects on galvanization are not so good. However addition of hydrochloric acid in the oil leads to improvements coated areas and adherence. Also natural fatty acid used in the flux operation leads to good galvanizations due to its light acidity, Balloy et al. 2007.

Other studies on the bath additions in the bath were carried out in order to evaluate the effects of alloying on the coating formation. For example in the last years the effect of strontium on the adhesive strength and corrosion resistance of hot-dip galvanized coating is studied. Presence of strontium improves both the properties through dendritic grain refinement (86% of grains refinements at 0.002 wt%), adhesion strength (best values at 33%) and corrosion resistance (at 30%), Vagge et al. 2009.

Another study shows that the influence of the SiO_2:Na_2O molar ratio of silicate solution on the properties of silicate coatings on electrochemical tests in comparison to HDG is a decrease of the corrosion current densities, and an increase of the polarization resistance and the total impedance values, enhancing the corrosion resistance, Yuan et al. 2010.

In order to painting to get a good adhesion, on the galvanized surface an organofunctional silane is deposited on hot-dip galvanized cold rolled steel, where γ-MPS treated samples. Low silane concentration gives a smooth appearance film and higher silane concentrations lead to more cracks in the silane film and ultimately detachment, Bexell et al. 2007.

When coated steel sheets are subjected to corrosive environments, their corrosion behaviors are affected, due not only to changes of the coating texture, but also of the microstructure. Basal plane texture coefficient would increase at lead content of zinc bath increasing, as well as would increase the texture coefficient of high angle pyramidal, low angle pyramidal and prism planes. Γ layer thickness would be increased with increasing the lead content of the zinc bath. Coatings that have a better corrosion resistance are characterized by greater basal texture coefficient and smaller Γ layer thickness , Asgari et al. 2008.

To prevent the penetration of the aggressive ion Cl^- in the outdoor exposition, a presence of oxid under coating is accepted. Moreover the galvanic performance of the coating improves by the presence of ZnO-rich inner alloy layers, as also evidenced when polarization studies were conducted with a different approach, Shibli et al. 2006.

Visual inspection puts into evidence the presence of these specific layers; image analysis techniques allow the automatic detection and quantification of the elements of interest in each specimen, considering also the evolution of these objects in time.

Longitudinal sections of the specimens were metallografically obtained and observed by means of an optical microscope . In the data the objects of interest are characterized by a bright and uniform gray level. A useful operation is the segmentation that is the partition of the image into region homogeneous with respect to a chosen property, for example the gray level. Different methods may be useful to segment the proposed images, but for the present application the classical Otsu method suits well, Otsu 1979, with an efficient and robust performance. Roughly speaking, the Otsu binarization method provides a threshold obtained by the minimization of the probabilistic variance for two normal distributions of the two classes, the background and the object, as will be re-called in the next section.

In this work an innovative bath containing Ti was used to generate Zinc based coating in five different dipping time (15, 60, 180, 360 and 900 s). Recent studies show bending damage of coating in presence of Ti, Di Cocco et al. 2012, but microstructural morphologies and thicknesses of phases are not yet well quantified. Each coatings microstructure was investigated in order to evaluate the presence and the morphological properties of a peculiar phase rich in titanium, typically present in ZnTi-based coatings. To this aim, image analysis techniques were considered for an automatic identification and quantification of the layers of interest.

Table 1.	Galvanized steel chemical composition (wt%).				
C	Si	Mn	P	S	Al
.090	.167	.540	.010	.004	.051

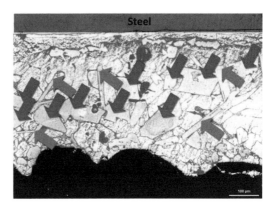

Figure 2. LOM image of 360 s dipping time coating section with presence of new kind phase in two morphologies: green arrows "slim" crystal, red arrows "fatty" crystal.

2 MATERIAL AND EXPERIMENTAL METHODS

For all the investigated dipping time, $80 \times 25 \times 3$ mm commercial carbon steel plates specimens with chemical composition showed in Table 1, were considered.

Prior to galvanizing, steels samples were degreased and rinsed with alcohol. Subsequently specimens were pickled in an aqueous solution 50% HCl at 25°C for 10 minutes, washed in fresh water, fluxed in an aqueous solution containing 280 g/l $ZnCl_2$ and 220 g/l NH_4Cl at laboratory temperature for 2 minutes and then dried for 10 minutes at 50°C. Then specimens were dipped for 1 minute considering an innovative galvanizing bath at 460 ± 2 °C, characterized by presence of 0.5% wt of titanium.

In order to identify the kinetics of phases grow for each investigated dipping time, longitudinal sections of the specimens were metallografically obtained and observed by means of an optical microscope (LOM).

Images are used to detect automatically the layers of interest; firstly the image is corrected by a γ-correction with $\gamma = 0.7$: this operation provides a brighter image. On this data a four levels segmentation is performed; a hierarchical Otsu segmentation is considered: firstly the image is binarized and then each region is furtherly binarized, and a four level segmentation is obtained. Each binarization is obtained by Otsu method: it yields the optimal threshold from the histogram of the image. The method assumes that the histogram of the data is bi-modal and therefore two classes of pixels are considered, the object and the background. The optimal threshold is obtained by minimizing the intra-class variance:

$$\sigma_\omega^2(t) = \omega_1(t)\sigma_1^2(t) + \omega_2(t)\sigma_2^2(t) \qquad (1)$$

where σ_i^2 are the variances of the classes and the weights ω_i represent the probabilities of the two classes separated by the threshold t.

3 RESULTS AND DISCUSSION

Hot dip galvanized coatings, carried out at different dipping time, are characterized by not negligible surface defects, due to not optimized physical parameters, still unknown for the innovative process with Ti in the bath. For this reasons Light Optical Microscopy (LOM) observation are carried out only in regions of coatings without defects or inhomogeneities using two pictures at high magnifications (200x) for each low dipping times coatings and low magnification for each

high dipping times coatings, due to different thickness of coatings.

From LOM analysis of metallographically prepared samples, the presence of a new phase is observed for all investigated dipping times. Figure 2 shows the presence of an inner δ phase and many crystal of a new phase, not present in any coating obtained from conventional baths, as shown in Figure 1. The new phase is characterized by regular boundary and two different morphologies: a "slim" crystal, probably due to nucleation of phase, and a "fatty" crystal morphology, due to high values of diffusion of Zn, Ti (from bath) and Fe (from steel).

It is a very brittle phase, often dispersed in ductile matrix formed by two phases: the first one characterized by compact morphology and the second one characterized by lamellar morphology "pearlite like".

Brittle phase crystals and double phases ductile matrix are present in an outer layer witch thickness depending on dipping time.

These kind of phases are present in all investigated cases and in particular the observations, carried out on the picture, highlight the following results (from two pictures of each dipping time):

- at 15 seconds of dipping time the presence of a developed δ phases and of a very thin outer layer is observed;
- at 60 seconds of dipping time the presence of a developed δ phases and of a growing outer layer is observed;
- at 180 seconds of dipping time the presence of a thin δ phases and of a thick outer layer are observed;
- at 360 seconds of dipping time an increase of δ phases thickness and an unchanged outer layer are observed;
- at 900 seconds of dipping time an increase of δ phases and outer layer thickness are observed.

In order to identify new brittle phase parameter the described image segmentation procedure is considered. In the segmentation procedure, one of the main difficulty is the identification of the best property that discriminates the objects of interest from the other elements and therefore apply the segmentation algorithm. For the considered data two information are useful: the gray level and the uniformity. More precisely, for the images caught up to 360 s the four levels segmentation with respect to the gray level allows a reliable identification, whereas in the images caught at 900 s what allows the enhancement of the layers is a segmentation with respect to the uniformity. This is due to different morphology of the brittle phase, which takes on a "fatty" morphology as a consequence of diffusion phenomena. The uniformity is evaluated as follows. Let z denotes the gray level, L the number of distinct gray levels and $p(z_i), i = 0,1, 2,..., L$-1 the corresponding histogram.

The Uniformity is given by:

$$\overline{U} = \sum_{i=0}^{L-1} p^2(z_i) \qquad (2)$$

It is maximum in the regions in which all gray levels are equal.

The four level segmentation (performed on the gray level for all the images except for the one caught at 900 s for which the uniformity was used) retrieves all the information of interest; in particular, the layers are the objects characterized by the higher level. By simple morphological operations they are enhanced and for each of them it is possible to define useful properties, for example the centroid and the area. In fact, what appear interesting is the distribution of the layers in the specimen and the percentage of the area in the brittle phase. The former is the y-axis of the centroid, whereas the latter is evaluated as the number of pixels constituting the object.

As an example of the segmentation procedure two images are shown, Figure 3 and Figure 4; they are specimens at 360 s and 900 s, respectively.

In Figure 3a and Figure 3b the original data (specimens at 360 s) and the four level segmentation are shown, whereas in Figure 3c the obtained layers are enhanced.

For the analysis of the specimens at 900 s, along with the original data of Figure 4a, the uniformity image is presented, Figure 4b; in Figure 4c the four levels segmentation obtained from Figure 4b is shown, whereas the layers are enhanced in Figure 4d.

Results of image analysis are further processed in order to identify micromechanisms of coating formation. Surfaces of coatings are expressed in terms of percentage referred to 900 s coating surface, to identify the kinetic showed in Figure 5.

Coatings sharply grow at low dipping times up to 180 s, and brittle phase decreases to the minimum value. In this conditions, kinetics are mainly driven by high diffusion of Zn and Ti atoms from the bath to the coating and by diffusion of Fe and Ti atoms from brittle phase to ductile matrix.

From 180 to 360 s we can see a not significant increase of whole coating thickness, due to a diffusion

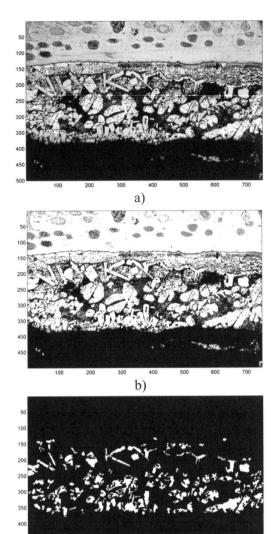

Figure 3. Analysis of specimens at 360 s; a) original data; b) four-levels segmentation; c) identified layers.

equilibrium between atoms from the bath to coating and from coating to brittle phase, which increases its importance.

From 360 s to 900 s both whole coatings and brittle phase increase.

Analysis of centroids positions, normalized with respect to 900 s conditions, (Figure 6) shows an increase up to 360 s.

At 360 and 900 s positions are roughly the same. Brittle phase takes whole external layer in "slim" dispersed morphology (at 360 s) and in "fatty" compact morphology (at 900 s). From 360 s, coating formation is driven by brittle phase which characterizes kinetics and properties.

424

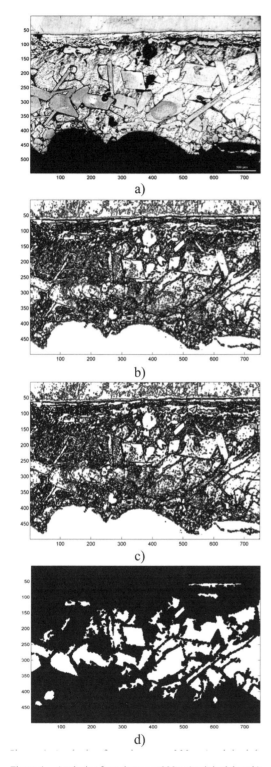

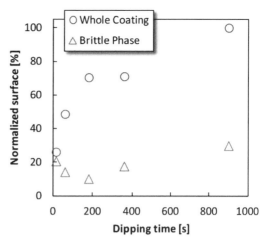

Figure 5. Normalized surface of whole coating (referred to 900 s coating) and brittle phase (referred whole coating at the same dipping time).

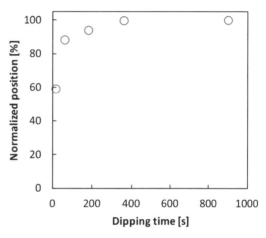

Figure 6. Brittle phase centroids position referred to 900 s coating.

At high dipping time, brittle phase is directly involved to diffusion phenomenon from the bath, due to its position, sometime covered by a thin layer generated by wettability of bath.

4 CONCLUSION

In this work a new approach is proposed in order to quantify the main morphological properties of a ZnTiFe phase rich in Ti, present only in coatings produced by baths containing Ti. The innovative approach has been applied to five different dipping time, thus allowing to evaluate the kinetic of formation.

The use of image analysis procedure may help in the automatic and objective evaluation of the coating formation. The main difficulties relies in the analysis of the specimens at 900 s; in this case what really defined

Figure 4. Analysis of specimens at 900s; a) original data; b) uniformity image; c) four-levels segmentation; d) identified layers.

the brittle phase was the uniformity in the gray level distribution.

Results of image analyses allow the evaluation of the kinetic of coating formation of an innovative hot dip galvanizing process in bath containing Ti. The following mechanisms can be summarized:

1) from 15 to 180 s the coatings are characterized by a sharply increase at the expense of the brittle phase; from 180 to 360 s the whole coatings are characterized by a plateau and from 360 up to 900 s we have an increase;

2) brittle phase decreases up to 180 s and increases from 180 up to 900 s;

3) at 360 and 900 s the position of the centroids shows an almost full complete extension in the outer ductile matrix.

REFERENCES

Asgari, H., Toroghinejad, M.R., Golozar, M.A. 2008. The role of texture and microstructure in optimizing the corrosion behavior of zinc Hot-Dip coated steel sheets, ISIJ International, Vol.48, No. 5, 628–633

ASTM Handbook Corrosion, Hot dip coatings 1999. 13.

Balloy, D., Dauphin, J.Y., Tissier, J.C. 2007. Study of the comportment of fatty acids and mineral oils on the surface of steel pieces during galvanization, Surface & Coatings Technology 202, 479–485.

Bexell, U., Grehlk ,T.M. 2007. A corrosion study of hot-dip galvanized steel sheet pre-treated with γ-mercaptopropyltrimethoxysilane, Surface & Coatings Technology 201, 4734–4742.

Di Cocco, V., Iacoviello, F., Natali, S., Zortea, L. 2012. Influence of 0.5%Ti and 1%Sn in intermetallic phases damage in Hot Dip Galvanizing Coating, Proceeding of IGF Workshop, Forni di Sopra (UD), Italy, ISBN 978-88-95940-37-3.

Marder, A.R. 2000. The metallurgy of zinc-coated steel, Progress in Materials Science 45, 191–271.

Otsu, N., 1979. A threshold selection method from gray-level histograms, IEEE Trans. Sys., Man., Cyber, 9, 62–66.

Reumont, G., Voct, J.B., Iost, A., Foct, J. 2001. The effects of an Fe-Zn intermetallic-containing coating on the stress corrosion cracking galvanized steel, Surface & Coatings Technology 139, 265–277.

Shibli, S.M.A., Manu, R. 2006. Development of zinc oxide-rich inner layers in hot dip zinc coating for bassier protection, Surface & Coatings Technology 201, 2358–2363.

Singh, D.D.N., Ghosh, R. 2008. Molybdenum-phosphorus compounds based passivator to control corrosion of hot dip galvanized coated rebars exposed in simulated concrete pore solution, Surface & Coatings Technology 202, 4687–4701.

Tzimas, E., Papadimitrou, G. 2001. Cracking mechanisms in high temperature hot-dip galvanized coatings, Surface & Coatings Technology 145, 176–188.

Vagge, S.T., Raja, V.S. 2009. Influence of strontium on electrochemical corrosion behaviour of hot-dip galvanized coating, Surface & Coatings Technology 203, 3092–3098.

Yoon, H.G., AHN, G.J., Kim, S.J., Chung, M.K. 2009. Aerodynamic investigation about the cause of check-mark stain on the galvanized steel surface, ISIJ International, Vol. 49, No. 11, pp. 1755–1761.

Yuan, M.R., Lu, J.T., Kong, G. 2010. Effect if SiO_2:NaO molar ratio of sodium silicate on the corrosion resistance of silicate conversion coatings, Surface & Coatings Technology 204, 1229–1235.

Computational Modelling of Objects Represented in Images – Di Giamberardino et al. (eds)
© 2012 Taylor & Francis Group, London, ISBN 978-0-415-62134-2

Quantitative characterization of ferritic ductile iron damaging micromechanisms: Fatigue loadings

V. Di Cocco, F. Iacoviello & A. Rossi
Università di Cassino e del Lazio Meridionale, DICeM, Cassino (FR), Italy

D. Iacoviello
Sapienza University of Rome, Dept. of Computer, Control and Management Engineering Antonio Ruberti, Rome, Italy

ABSTRACT: Due to the peculiar graphite elements shape, Ductile Cast Irons (DCIs) are able to combine the high castability of gray irons with the toughness of steels. This result is obtained by means of a chemical composition control and not by means of an extended, and expensive, annealing treatment of white irons (that is necessary to obtain malleable irons). Focusing ferritic DCIs, different damaging micromechanisms should be investigated considering both ferritic matrix and the role played by graphite nodules: according to the observed micromechanisms, it is evident that graphite nodules role can't be summarized with a simple "debonding", usually considered as the only effective in numerical simulations. In this work, damaging micromechanisms in fatigue loaded ferritic DCI specimens were investigated by means of Scanning Electron Microscope (SEM) during the fatigue test ("in situ" tests). Focusing graphite nodules, customized image processing procedures were optimized in order to quantify the evolution of damaging micromechanisms with the loading cycles number.

1 INTRODUCTION

1.1 Material

The combination of high castability and good mechanical properties (e.g., high damping capacity, good wear resistance and interesting tensile behaviour and toughness), along with low manufacturing costs and low volume shrinkage during solidification, allowed to all the ductile irons grades (DCIs) to be considered and applied in many applications, especially but not uniquely in automotive industry: gears, camshaft, crankshaft, trunk axles are only some applications (Toktaş et al. 2008).

Matrix controls these good mechanical properties (Ward 1962, Labrecque & Gagne 1998) and the main spheroidal cast iron grades could be classified as follows:

- ferritic DCIs are characterized by good ductility, with tensile strength values that are equivalent to low carbon steel;
- pearlitic DCIs show high strength values, good wear resistance and moderate ductility;
- ferritic-pearlitic grades properties are intermediate between ferritic and pearlitic ones;
- martensitic DCIs show very high strength, but low levels of toughness and ductility;
- bainitic grades are characterized by high hardness;
- austenitic DCIs show good corrosion resistance, good strength and dimensional stability at high temperature;

- austempered grades show a very high wear resistance and fatigue strength.

Considering these applications, fatigue loading could be considered as one of the most critical loading conditions to be considered: the analysis of the damaging micromechanisms could allow both an improvement of commercial grades (optimizing both chemical composition and microstructure) and the optimization of non destructive procedures to be performed on the operative components, in order to early identify the very first signs of fatigue damaging.

In the last decades, damaging micromechanisms analysis in DCI were mainly focused on static or quasi-static loading conditions (Berdin et al. 2001, Dong et al. 1993, Dong et al. 1997, Liu et al. 2002). The main damage micromechanism was often identified in voids growth corresponding to graphite nodules, cracks nucleation and growth corresponding to graphite nodules – matrix interface, with consequent micro-cracks coalescence generating a "final" macro crack: numerous studies provided analytical laws aiming to describe a single void growth, depending on the void geometries and matrix behavior. In the last years, other experimental activities allowed to define more precisely the damaging micromechanisms in DCIs (Iacoviello et al. 2008, Di Cocco et al. 2010), allowing to better characterize the role played by the graphite nodules: they are not to be merely considered as growing microvoids in a ductile matrix (ferrite), but they are characterized by peculiar damaging micromechanisms

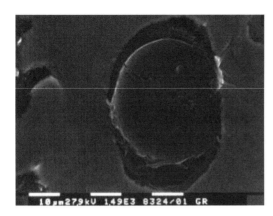

Figure 1. EN GJS350-22 DCI ($\sigma_{max} = 320$ MPa and $\sigma_{min} = 120$ MPa). Graphite nodule damaging morphologies.

with their own peculiar evolution (Fig. 1; damaged dark element is the cracked graphite nodule).

1.2 Image analysis applied to DCI damaging characterization

In order to identify and quantify the effects of damage, image analysis procedures may help. The first difficulty relies in the quality of the data; even if at a first sight the effects of damage appear evident, their automatic identification is quite difficult, due to the presence of scratches and dust. In the specimens the effects of damage have generally the shape of thin curvilinear rectangular; moreover there is also a darker kernel mainly positioned in the center of the specimen. In this work the interest will be focused on the former elements.

The elements of interest are generally not well separated one to each other and from the background. Therefore a segmentation procedure is advisable; the segmentation is the partition of the image in regions homogeneous with respect to a chosen property. For the data of this application the objects of interest are the darkest elements of the image; the direct segmentation yields a non-satisfactory result, due to many artifacts and to the confusion among the effects of damage. Therefore a suitable preprocessing is crucial.

Once the image is segmented (in the present application an eight levels segmentation was required) the damage effects are identified among the objects with the lowest grey level.

2 INVESTIGATED MATERIAL AND EXPERIMENTAL PROCEDURES

2.1 Investigated DCI and fatigue test procedure

In this work, the investigated fully ferritic DCI (EN GJS350-22) was characterized by the following chemical composition:

% C = 3.66
% Si = 2.72

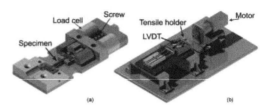

Figure 2. Tensile holder with microtensile specimen (a); fatigue testing machine (b).

% Mn = 0.18
% S = 0.013
% P = 0.021
% Cu = 0.022
% Cr = 0.028
% Mg = 0.043
% Sn = 0.010

This DCI is characterized by a very high nodularity of graphite elements (higher than 85%), with about 9-10% as graphite elements volume fraction.

Fatigue tests were performed considering microtensile specimens ($25 \times 2 \times 1$ mm). Specimens were ground and polished and fatigue loaded intermittently with a tensile holder and observed in situ using a SEM, considering 20 graphite elements. During fatigue tests, specimen deformation and applied load were measured by means of a Linear Variable Differential Transformer (LVDT) and two miniature load cells (10 kN each), respectively. Figures 2a and 2b show the tensile holder and the tensile test machine, respectively.

Fatigue tests were repeated three times and performed under load control conditions ($\sigma_{max} = 320$ MPa; $\sigma_{min} = 120$ MPa), with a loading frequency of 0.03 Hz. In order to perform scanning electron microscope (SEM) observations, fatigue loading was stopped every 1000 cycles (near final rupture, observations frequency is higher).

2.2 Image analysis procedure

As already noted, the image analysis procedure required a preprocessing. First the original data is filtered by a second order statistic filter; more precisely, the image is partitioned into square domains of chosen size and each element in the domain is replaced by the n-order element in the sorted set of neighbors specified by the nonzero elements in the same region. This filtering yields a more uniform image; but the obtained image was not suitable yet for the segmentation. It has been necessary to enhance the contrast by a contrast-limited adaptive histogram equalization, without enhancing the darker kernel damage.

The resulting image is segmented with respect to gray level. Some algorithms were tried, for example the discrete level set method (De Santis et al. 2007)) but for the present application the simple and effective Otsu method suits really well, (Otsu, 1977). It is based on the assumption that the histogram of the data is bi-modal

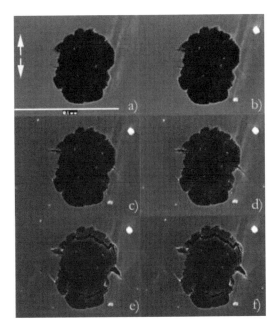

Figure 3. EN GJS350-22 DCI ($\sigma_{max} = 280$ MPa and $\sigma_{min} = 120$ MPa). SEM in situ surface analysis corresponding to the following loading cycles: (a) 1; (b) 14100; (c) 38300; (d) 40100; (e) 46000; (f) 48100 cycles (arrows show the loading direction).

and therefore two classes of pixels are considered, the object and the background. The optimal threshold is obtained by minimizing the intra-class variance:

$$\sigma_\omega^2(t) = \omega_1(t)\sigma_1^2(t) + \omega_2(t)\sigma_2^2(t)$$

where σ_i^2 are the variances of the classes and the weights ω_i represent the probabilities of the two classes separated by the threshold t.

An eight level segmentation was needed and obtained by a hierarchical sequential Otsu binarization. The damage is identified among the objects with the lowest level and higher eccentricity.

3 EXPERIMENTAL RESULTS AND DISCUSSION

3.1 Damaging morphologies in graphite nodules

Focusing the damage evolution in graphite nodules, after the first fatigue cycle no damage is observed (Fig. 3–5 a).

After more than 14000 load cycles, damage initiates both at graphite nodules – ferritic matrix interface

(mainly slip bands emission) and in graphite elements (Fig. 3 – 5, b). The main damaging micromechanism observed in graphite nodules is a sort of debonding of the graphite core obtained during solidification (both directly from the melt and during eutectic solidification) with respect to the graphite shell obtained

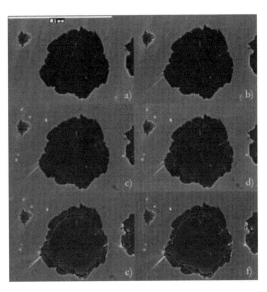

Figure 4. EN GJS350-22 DCI SEM damage analysis. Same loading conditions of Figure 3.

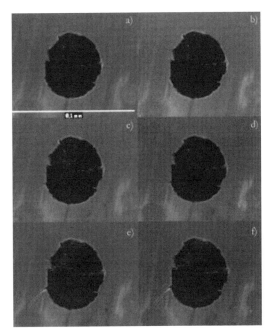

Figure 5. EN GJS350-22 DCI SEM damage analysis. Same loading conditions of Figure 3.

during cooling due to C solubility decrease in austenite grains. Microcracks in the nodule center are only seldom observed.

Unfortunately, all the observed nodules were far from the fracture surface. However, some observations near crack surface were performed (Fig. 6) and it is possible to confirm that the micromechanisms observed in graphite nodules (Figs. 3 – 5) increase

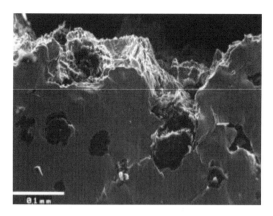

Figure 6. EN GJS350-22 DCI SEM damage analysis (near fracture surface).

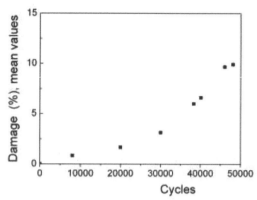

Figure 8. Damage evolution (%) mean values (six nodules) – fatigue cycles.

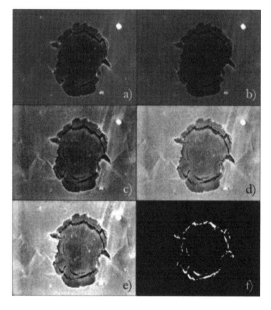

Figure 7. a) Original data; b) filtered image; c) contrast-limited adaptive histogram equalization; d) γ-correction; e) eight-levels segmented image; f) damage identification.

their importance up to the end of the fatigue test (final rupture).

3.2 Damaging morphologies quantitative characterization

In this subsection an example of the image analysis procedure is presented, along with the damage quantitative evaluation. The nodules of Figure 3 were considered; for the preprocessing phase the details of the analysis of the SEM image corresponding to 46000 loading cycles are presented (Fig. 7a). For the second order statistic filter a 4×4 domain is assumed, replacing each element of the domain with the element (of the same domain) with the lowest gray level, (Fig. 7b). The

contrast-limited adaptive histogram equalization operates on small regions of the image, and an 8×8 tile was assumed. Each tile's contrast is enhanced to obtain, as much as possible, a uniform histogram (see Fig. 7c). To avoid non desired enhancement of the dark kernel the contrast enhancement limit was kept low, equal to 0.016. It could be useful a γ-correction operation, with $\gamma = 0.4$ (Fig. 7d).

The described image preprocessing phase is crucial to enhance the damage effects that are of interest in this context and to reduce the other consequences of the experimental loading procedure. Once the image has been suitably preprocessed an eight levels segmentation is needed to identify the damage. The Otsu method was chosen for its effectiveness, even if, after the described preprocessing procedure, also the discrete level set method (De Santis et al. 2007) yielded satisfactory results. The eight levels segmentation by the Otsu method was obtained hierarchically (Fig. 7e); the objects with the lowest gray level and an eccentricity greater than 0.85 are identified as the damaged elements (Fig. 7f).

The same analysis was performed for all the images of Figure 3; a possible quantitative evaluation of the damage relies in the estimation of the percentage of the obtained area of Figure 7f, with respect to the initial total graphite nodule area.

Damage evolution in graphite nodules was investigated considering ten graphite nodules. Results dispersion is really low and damage (%) mean values evolution with fatigue loading cycles is shown in Figure 8.

The importance of these results is quite evident. In fact if the results will be confirmed by further investigations, according to this analysis it would be possible to quantify the fatigue damage evolution of a ferritic ductile iron by means of the analysis of the damage evolution of the graphite nodules. Considering that it is possible to perform the analysis of graphite nodules damage evolution also "in situ", e.g by means of portable digital microscopes, this investigation

430

technique could allow to assess the residual fatigue life of a component.

3.3 Conclusions

In this work, the fatigue damage evolution in a ferritic ductile cast iron has been investigated by means of scanning electron microscope observations of microtensile specimens, focusing the damage evolution in graphite nodules. According to the experimental results, it is possible to summarize that:

Image analysis may help in the identification and automatic evaluation of damage; different kind of damage requires difference processing and a significant image preprocessing was required.

The evolution of damaging in graphite nodules during fatigue loading is progressive with the loading cycles and, considering that it is possible to perform the analysis of graphite nodules damage evolution also "in situ", e.g by means of portable microscopes, this investigation technique could allow to assess the residual fatigue life of a component.

REFERENCES

Berdin, C. Dong, M.J. & Prioul, C. (2001). *Local approach of damage and fracture toughness for nodular cast iron.* Engineering Fracture Mechanics. 68: 1107–1117.

De Santis, A. & Iacoviello, D. (2007). *Discrete level set approach to image segmentation*, Signal, Image and Video Processing, Springer-Verlag London, 1(4): 303–320.

Di Cocco, V. Iacoviello, F. & Cavallini, M. (2010). *Damaging micromechanisms characterization of a ferritic ductile cast iron*, Engineering Fracture Mechanics. 77: 2016–2023.

Dong, M. J. Hu, Hu, G. K. Diboine, A. Mouline, D. & Prioul, C. (1993). *Damage modelling in nodular cast iron.* J. de Physique IV. 3: 643–648.

Dong, M. J. Prioul, C. & François, D. (1997). *Damage effect on the fracture toughness of nodular cast iron: Part I. Damage characterization and plastic flow stress modelling.* Metall. And Mater. Trans. A. 28A: 2245–2254.

Iacoviello, F. Di Cocco, V. Piacente, V. & Di Bartolomeo, O. (2008). *Damage micromechanisms in ferritic-pearlitic ductile cast irons.* Materials Science and Engineering A. 478(1–2) 181–186.

Labrecque, C. & Gagne, M. (1998). *Ductile iron: fifty years of continuous development.* Canadian Metallurgical Quarterly 37(5): 343–378.

Liu, J.L. Hao, X. Y. Li, G. L. & Liu, G. Sh. (2002). Microvoid evaluation of ferrite ductile iron under strain. Materials Letters. 56: 748–755.

Otsu, N. (1979). *A threshold selection method from gray-level histograms*, IEEE Trans. Sys., Man., Cyber, 9: 62–66.

Toktaş, G. Toktaş, A. & Tayanc, M. (2008). *Influence of matrix structure on the fatigue properties of an alloyed ductile iron.* Materials and Design 29: 1600–1608.

Ward, R.G. (1962). *An introduction to the physical chemistry of iron and steel making.* Arnold, London; St. Martin's Press, New York.

Interpretative 3D digital models in architectural surveying of historical buildings

M. Centofanti & S. Brusaporci

Architecture and Town Planning Department, L'Aquila University, Italy

ABSTRACT: Aim of the research is studying how 3D digital models can be used for historic architecture's representation and documentation. The question is what are the characteristics of a 3D model to be suitable as a restitution of an architectonical surveying, that is to be a correct scientific description of architectonical data. An historical architecture is made by spaces, surfaces, volumes, materials, constructive systems, etc. and it's the synthesis of modification and stratification processes occurred during centuries. In particular it's important the representation of the constructive elements. The realization of Architectonical Information Systems is useful. With the correlation between 3D digital components, historical documents and surveying data, 3D architectonical models can favour processes of knowledge, critical analysis and planning.

1 INTRODUCTION

Technologies and instruments for 3D architectonical surveying – first of all digital photo-grammetry and laser scanning – changed traditional procedures, favouring restitutions made by 3D models. In consequence the theme of architectonical data modeling and visualizing become a very important focus of interest. Object of this paper is how 3D digital models can favour historic architecture representation and documentation. This research has been developed by the authors within the Research of Relevant National Interest PRIN COFIN 2008 "Complex models for architectonic and urban heritage", principal investigator Mario Centofanti, financed by Italian Ministryfor University. Mario Centofanti wrote the 2nd and the 5th paragraph, Stefano Brusaporci the 3th and the 4th.

2 ARCHITECTONICAL SURVEYING AND 3D MODELLING

The problems deriving from using 3D digital models in architectonical surveying of historical buildings arise from two fundamental well-known considerations.

First of all an architecture is a complex organism, synthesis of spaces, surfaces, volumes, materials, illumination, etc., made by a technological constructive system; besides an historical building is the result of modification and stratification processes, reporting cultures and history of past ages.

Secondly architectonical surveying is a knowledge process finalized to the building's documentation, that is a scientific architectonical data description; the issue of an architectonical survey is influenced by the scale of representation – i.e. the model's levels of detail –, by the architectonical characteristics of the building (for example a ruin of classical age, a Renaissance palace,

Figure 1. The Palace of Margherita d'Austria in L'Aquila (before the earthquake of 2009).

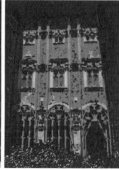

Figure 2. The photogrammetric point cloud of the court with seismic shoring structures.

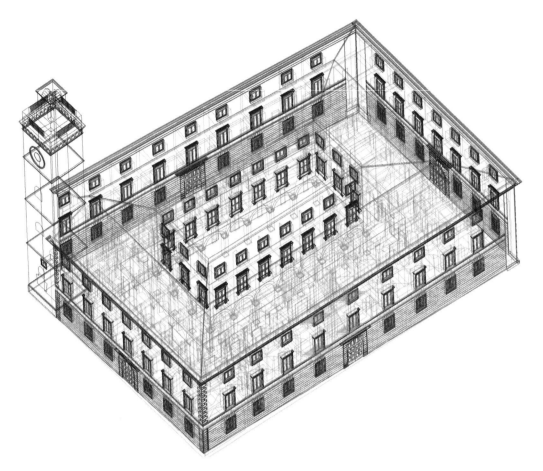

Figure 3. 3D model in wire frame visualization.

a baroque church, a rationalist building, etc.), by the aim of the surveying (for example a restoration).

Digital technologies enriched architectonical representation's modalities and processes. In particular architectonical surveying outcomes can be composed not only by bi-dimensional drawings but also by 3D models from which the most convenient renderings can be derived. These 3D models have to present suitable characteristics to be considered an architectonical surveying restitution (Docci & Maestri 2009. Docci & Chiavoni & Filippa 2011).

For example, an architectural 3D model for historical building restoration has to be capable of meeting multiple demands: it should allow a varied body of information to be collected and, at the same time, be processable and modifiable by the many professionals involved; it should be available for consultation by the competent bodies for approving the project, and browsable, queriable, measurable but not editable; it should ensure high quality, photorealistic viewing for communication of design choices. These are many aspects of the same digital media, and in order to meet these demands, they should have appropriate characteristics, not only metrical but also constitutive and qualitative. In this sense, the issue of defining

standards and formats for modeling is inescapable (Brusaporci 2011).

3 SIGNIFICANS AND SIGNIFICANCE

The articulation in the two sequential and synergic phases of modeling and rendering produces a postponement of the final restitution in regard to what happened in traditional drawing. From the same 3D model we have numerous representative models to describe an architecture: facades, plans, sections, axonometric or prospective views, etc. with different conditions of illumination, materials, context, etc.

About the 3D modeling the study points out that in architectonical surveying methods of reverse and 'direct' modeling are not in complete alternative but often are integrated, and it is necessary for several reasons: point clouds aren't a restitutive model, it's difficult have a laser scanning without gaps, point clouds of an entire building usually describe approximately architectonical particulars.

The reverse modeling can be referred to a process of synthesis because from discontinuous elements (point clouds) surfaces are interpolated; a 'direct' modeling

Figure 4. The rendering describes volumes, colors and the architectonical articulation of the facades.

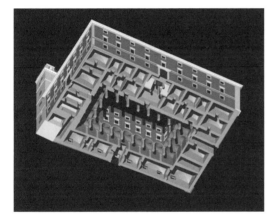

Figure 5. This view shows the palace's conformation: the internal yard, the porch, the entrance halls, the vaulted rooms.

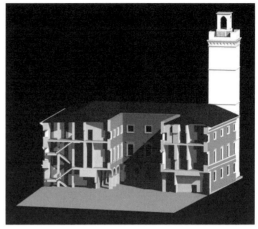

Figure 6. The cross section describes the system of hall - courtyard - stairs between the court and the entrance on Via delle Aquile.

Figure 7. Split model with the internal organization of the palace and the vaults of the groud floor.

method follows an analytic procedure, according to a preliminary conceptual division of the building in its components, in succession modeled and then assembled. Besides the choice of using MESH or NURBS surfaces is connected to critical considerations related to the model's documental aim and to the levels of detail that we want (Benedetti & Gaiani & Remondino 2010). Even if NURBS surfaces are tessellated during rendering, nevertheless can be useful for a correct geometrical description of the architectonic object and its particulars.

There is a wide series of tools available to surveyors for 3D restitutions. The most common software enable the extraction from point clouds of sections and orthogonal views of the clouds themselves, as well as the synthesis of MESH or NURBS surfaces or even models built on 3D primitives arranged using Boolean operations, and all of them have to be defined semantically in terms of their architectural significance (De Luca & *alii*. 2011).

A restitutive digital model for architectonical surveying requires the study of the constructive systems.

The necessity of a model to be made by not only building's surfaces but also constructive elements, underlines the problematic relation between 'representative' and 'represented' objects. This theme is also connected to the relationship between modeling and restitution scale: it's impossible realizing a model of the whole building, defined in all constructive components – the ontological problem of knowledge or copying the reality –. At the same time we cannot have a model too complex to be managed.

We linked models that, step by step, have more definition; it isn't only a problem of levels of details (LOD) – connected to the number of polygons used to modeling a surface – rather the number of components constituting the model, that is how much the 3D representation is mimetic or analogical of the building reality (Brusaporci 2010).

Further 3D models can be the support for architectonical information systems: this can be useful to enrich the model with other kinds of architectonical information (Trizio 2007; Trizio 2009). Indeed, about

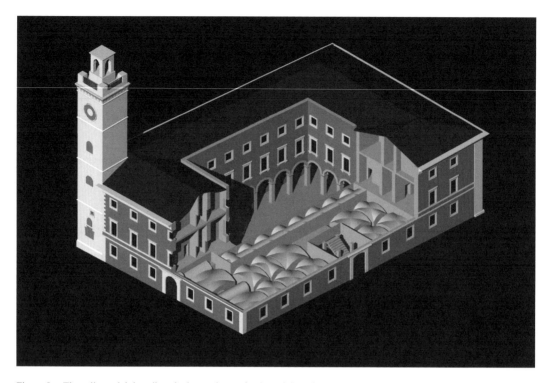

Figure 8. The split model describes the internal organization of the palace.

the knowledge process, the model favours the collection, correlation, systematization and analysis of the large and heterogeneous number of information deriving from architectonical surveying and historical-documental studies. These data can be related to the specific 3D component. According to its characteristics, the 3D model has metrical, geometrical and architectonical information, in addition about colors, materials, degradation, etc. The digital model, as three-dimensional interface, allows the surfing and the interrogation in an intuitive way, favouring historical building's analysis. Already during the Research of Relevant National Interest PRIN COFIN 2006 "Integrated software systems in architectural and urban heritage conservation, protection and exploitation", principal investigator Mario Centofanti, the research unit of L'Aquila University defined an information system for architecture called SIArch-3D, built by the integration of 3D models, also photorealistic, in GIS of ESRI Company (Centofanti & *alii* 2011; Brusaporci & Trizio 2010; Centofanti & *alii* 2008).

Reference can be made to the approach provided by Building Information Modeling software that ignores general basic objects - lines, splines, surfaces, volumes created through Boolean operations, etc. –referring to components such as vertical or horizontal closures, doors, windows, etc., and requires the definition of these elements in libraries, i.e. these components must be standardized (Mingucci & Garagnani 2011); the goal is to optimize the cost-time benefits of the planning process based on the industrialization, prefabrication and standardization of commercial components.

Figure 9. Model of the monumental stair case.

This aspect is difficult to reconcile with the specific needs of surveying historic architecture, where each element has its own historical and architectural value. Notwithstanding, BIM can offer interesting elements, in particular the main types of software on the market are Autodesk Revit, Archicad / Allplan, Bentley Systems, Digital Project. The first two, and the

Figure 10. Archival documents are linked to the respective architectonical components: 3D model of the existing main door and its first project (1830).

most popular, have parametric modeling limitations, while the latter two allow integration with complex 3D modeling.

4 CASE STUDY

The case studies is the town hall of L'Aquila, palace of Margherita d'Austria, an historical building characterized by important stratification and modification phenomena. The building has three levels, at the ground floor a porticoed quad, with a monumental staircase on the northern side, and it's connected to the squares that ring the palace by entrance-halls. The palace was built by Pico Fonticulano in the XV century, and widely modified after the earthquake of 1703 to realize the 'Collegi Giudiziari dei Tre Abruzzi'. The seismic event of 2009 damaged the building; at the same time gives the opportunity to study the constructive characteristics.

The palace has been surveyed with the integration of different methods and instruments (laser scanning, digital photogrammetry and direct measurements), and peculiarity is that shoring structures produced wide gaps in point clouds.

The documents sited in the Historical State Archives of L'Aquila have been studied and compared with the surveying data.

A 3D model has been built with levels of detail corresponding to the scale 1:50. This model favours critical-historical studies and the communication of the architectonical values, for example with spatial, metrological, proportional, constructive analysis. The model is also suitable to study and prefigure restoration projects (Brusaporci 2011).

5 CONCLUSIONS

3D model can be an important instrument for historical-critical analysis of buildings and it has an own documental value. The restitutive model of architectonical surveying is an element of knowledge and an autonomous text that can be analyzed and interpreted. In fact the model gives many useful information on the architectonical significant and at the same time it can be historicized and studied, not only about its relation with the building, but also as a document itself, expression of his historical and cultural context. Besides if the digital representation remove from the real architectonical object, however it aids the process of abstraction that is fundamental for interpretation and planning. (Centofanti 2010).

Therefore the digital model favours a simulation of the reality according to a knowledge role: it enrich our experience, indeed it supply more experience of what we can gather without the model's mediation (Maldonado 1992). In this way the research unit of L'Aquila University studied how 3D models for surveying can simulate geometries, spaces, materials, building's characteristics, degradation, modifications, etc. To sum up: the Architecture.

REFERENCES

Benedetti, B. & Gaiani, M. & Remondino F. 2010. *Modelli digitali 3D in archeologia: il caso di Pompei*. Pisa: Edizioni della Normale.

Brusaporci, S. 2011. Modelli 3D per il progetto di restauro. In Maria Linda Papa (ed.), *Il disegno delle trasformazioni*. Napoli: Clean.

Brusaporci, S. 2010. Sperimentazione di modelli tridimensionali nello studio dell'architettura storica. In Stefano Brusaporci (ed.), *Sistemi Informativi In-tegrati per la tutela la conservazione e la valoriz-zazione del patrimonio architettonico e urbano*. Roma: Gangemi.

Brusaporci, S. & Trizio, I. 2010. Dal rilevamento in-tegrato al SiArch-3D. In Stefano Brusaporci (ed.), *Sistemi Informativi Integrati per la tutela la conser-vazione e la valorizzazione del patrimonio architet-tonico e urbano*. Roma: Gangemi.

Centofanti, M. & Continenza, R. & Brusaporci, S. & Trizio, I. 2011. The Architectural Information System SIArch3D-Univaq for analysis and preservation of architectural heritage. *The international archives of the photogrammetry remote sensing and spatial information sciences* XXXVIII-5/W16.

Centofanti, M. 2010. Della natura del modello architettonico. In Stefano Brusaporci (ed.), *Sistemi Informativi Integrati per la tutela la conservazione e la valorizzazione del patrimonio architettonico e ur-bano*. Roma: Gangemi.

Centofanti, M. & Continenza, R. & Brusaporci, S. & Ruggieri, G. & Trizio, I. 2008. Il progetto del SIArch – Sistema Informativo per l'Architettura. *Disegnarecon* 2/2008.

De Luca, L. & Bussayarat, C. & Stefani, C. & Véron, P. & Florenzano, M. 2011. A semantic-based platform for the digital analysis of architectural heritage. *Computers & Graphics* vol. 35, n. 2, april 2011, p. 227–241.

Docci, M. & Maestri, D. 2009. *Manuale di rilevamento architettonico e urbano*. Roma: Laterza.

Docci, M. & Chiavoni, E. & Filippa, M. (ed.). 2011. *Metodologie integrate per il rilievo, il disegno, la modellazione dell'architettura e della città*. Roma: Gangemi.

Maldonado, T. 1992, *Reale e virtuale*. Milano: Feltrinelli.

Mingucci, R. & Garagnani, S. 2011. Strumenti digitali per la modellazione dell'architettura. *Disegnarecon* 7/2011.

Trizio, I. 2009. Indagini Stratigrafiche e Sistemi In-formativi Architettonici: il GIS della chiesa di S. Maria in Valle Porclaneta. *Arqueología de la Ar-quitectura* 6/2009: 91–111.

Trizio, I. 2007. GIS-technologies and Cultural Heritage: stocktaking, documentation and management. In *Rethinking Cultural Heritage. Experiences from Europe and Asia*. Dresden: Technische Universität Dresden.

Computational Modelling of Objects Represented in Images – Di Giamberardino et al. (eds)
© 2012 Taylor & Francis Group, London, ISBN 978-0-415-62134-2

Digital integrated models for the restoration and monitoring of the cultural heritage: The 3D Architectural Information System of the Church of Santa Maria del Carmine in Penne (Pescara, Italy)

R. Continenza
Architecture and Urban Planning Department, University of L'Aquila, L'Aquila, Italy

I. Trizio
Construction Technologies Institute, Italian National Research Council, L'Aquila, Italy

A. Rasetta
L'Aquila, Italy

ABSTRACT: The SIArch-Univaq is an Information System dedicated to the Architecture. It comes from a research made in the University of L'Aquila, Architecture and Urban Planning Department by the working group on the Architectural Survey and Graphic Representation. It was financed by the Italian Ministry of University under the PRIN 2006 measure. At present the SIArch-Univaq project is completed and tested on some different buildings classified as Historical Heritage, the software allows, at each building scale, to make a tri-dimensional model navigation asking for and consulting the large amount of informations stored in the database and, by mean of appropriate print layouts, having a very good support for the restoration or rehabilitation projects.

1 THE SIARCH – UNIVAQ: CREATION, DEVELOPMENT AND EVOLUTION OF A 3D GIS

It's crucial the importance that the availability, the accessibility and the logical organization of all data related to each element of the building architecture has in the process of training and knowledge management on the architectural heritage. The availability of a complete and accessible documentation of the data is, after all, a necessary methodological premise for the realization of any planning operation and design regarding protection, preservation and fruition.

On the other side, the nature itself of the data, related to the historical and architectural heritage, seems to conspire against a rational and efficient knowledge structure. The documentary and archival information, for example, are not easily integrated with those ones having a geometric-dimensional nature and even more with other ones related, for example, to the technological equipment and degradation processes.

That's why, many experts working on cultural heritage management, have was very interested in the integrated GIS database technologies. Numerous trials have been conducted since the early '90s, among them, the most significant one was, certainly, the "Risk Map" of the cultural heritage (see the website: www.uni.net/aec/riskmap/italian.htm or www.cartadelrischio.it), a project of the Superior Institute for the Preservation and Restoration (ISCR),

started in 1992 and finalized to the planning of interventions on the cultural heritage that was a trailblazer in many other experiments in various sectors. In order to prevent inhomogeneities in the orgtanization of databases, and so allowing the integration into a broader and complete framework, in 2007 the Region of Tuscany (Italy), has issued a set of rules (with the changes made to the text of the Law of the Tuscany Region – 3 January 2005 – L.R. n. 1), named: "Standards for the territorial government", introduced by L. R. n.41 of 27 July 2007 and L. R. 15 of the 20 March 2007, were enacted rules for the unification of GIS enabled profiling in the region territory. All local authorities was obliged, within its territories, have to adapt the planned initiatives to very specific application standards defined at the regional level.

Today, GIS technologies are widely used in many applications at local and urban level, from urban planning processes up to the management of the urban facilities and services, in all cases in which it is necessary to associate a map-based range of information, no matter what its size or the complexity.

The treatment of the eminently three-dimensional phenomena, in a GIS environment, such as architectural artifacts, appeared much more complex. In this field, our research group intended to study the ways of building a GIS database in which data could be combined with digital models of architecture made for this purpose (the SIArch-Univaq is the result of a research program that began in 2004 at the Department

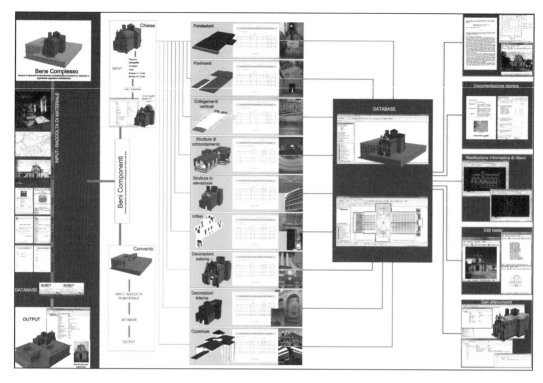

Figure 1.   The logical structure of the SIArch.

of Architecture and Urban Planning of the University of L'Aquila, Surveying of the Architecture Laboratory. The components of the research team are: Professor Mario Centofanti, Prof. Romolo Continenza, Prof. Gianfranco Ruggieri, Ing. Stefano Brusaporci, Ing. Ilaria Trizio).

Specific aim of this research was, and still it is, the construction of an information system directed to assist and to ensure a suitable cognitive support for all historical-architectural heritage conservation and restoration procedures as scheduling interventions maintenance and restoration planning.

The two main ways, on which the research has been moved, were, on the one hand, the construction of a model, at first two but later three-dimensional, driven to the detail level of a single building component and, on the other hand, the structuring of the database to be associated through a series of logical and functional relations that would allow an accurate representation of the phenomena to be described and to be documented.

In a first operational phase, the limited potentialities of the commercial software available have forced the creation of a rather complex series of agreements and cross-references between different two-dimensional descriptors, which engaged the research group for a long time allowing the close examination of issues relating to the structuring of the databases. The further evolutions of the software have, afterwards, allowed all the close examinations concerning the construction of a three-dimensional model that is suitable for the complete system structure.

2   APPLICATION TO THE CHURCH OF SANTA MARIA DEL CARMINE IN PENNE (PE)

The experimental application of the three-dimensional Architectural Information System (3D SIArch – Univaq) to the Church of Santa Maria del Carmine in Penne (PE), has been one of the test-bed of the functionality of the planning procedures. This test, which is one of the many similar proofs activated in order to test the procedure, benefited of a series of lucky coincidences that made possible to have two different sets of data collected in situ before and after the Abruzzo 2009 earthquake. Were indeed included in the procedure both the data collected during the various phases of the restoration site that affected the building between 2007 and 2008 and those detected during site repair of damage caused by the earthquake.

The overlap, the updating and the comparison of data entered into the database, topologically linked through the 3D-GIS to the three-dimensional model, allows an effective qualitative and quantitative reading of the phenomena that affected the building and an immediate and intuitive query model, conducted according to a user-friendly interface (Fig. 1).

3   THE DATASET AND THE PLANNING OF THE THREE-DIMENSIONAL MODEL

The methodology used for the design of the dataset regarding the SIArch of the church, follows, with

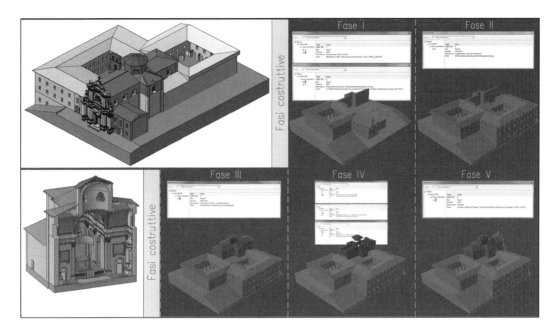

Figure 2.   Three-dimensional representation of the architectural complex in its construction phases.

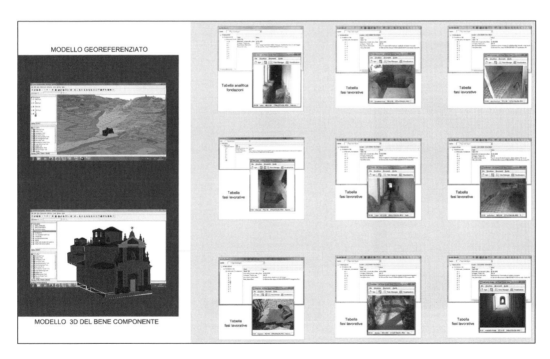

Figure 3.   The 3D-model of the church and queries concerning its elements.

some differences, the one used in other similar SiArch applications (Trizio 2010) where, at different levels of investigation (corresponded to a subsequent searches) are associated different datasets.

The levels of investigation used in this case are the same planned by ISCR to develop the "Risk Map". The levels started from the analysis of the "Architectural Complex", for which it is meant a set of elements functionally and formally connected in a self-organized architectural system, represented, in this case, by the convent and the adjoining church. For this level, the historical, metric, bibliographic, archival, graphic and photographic information related to the phases of the construction, to the condition of the property before

441

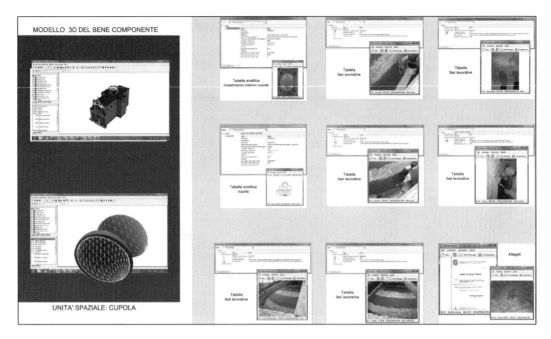

Figure 4. Results of the queries regarding the component element "dome".

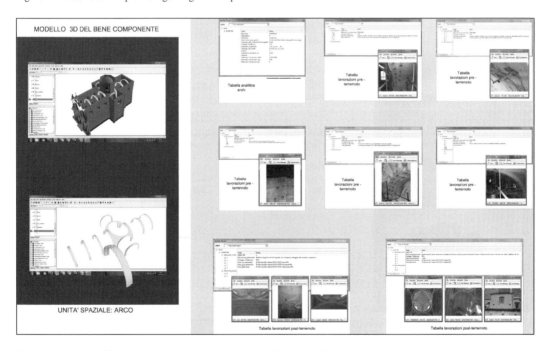

Figure 5. Results of the queries concerning the component element "arc".

the interventions and to the two restoration sites, inserted into the dataset, are alphanumeric, raster and vectorial (Fig. 2).

The subsequent level of investigation involved the analysis of the church as a "Building Composing an architectonic complex" (Fig. 3). For "Building Component", it is meant a set consisting of several "component elements", each of which fully identifiable within the whole. The architectural elements investigated in this level of detail, were: foundations, floors, vertical connections, elevation structures, horizontal ones, fixtures, interior decorations, outdoor decorations, roofing, plants a.s.o.. For each analyzed elements were included in the database the information described in the previous level (Fig. 4 and Fig. 5).

The three-dimensional model of the church and monastery, based on a survey (indirect, integrated with the direct one), was carried out in an appropriate modeling environment (Autodesk 3D Studio Max) and then imported through a procedure by now consolidated by the software, in the form ArcScene of the ArcGIS suite by ESRI, release 9.3.

Following the importation procedure, each three-dimensional element, has attributes common to any other feature class, is viewable through direct interrogation (Identify Results window), or through the attribute table, and it maintains, therefore, its dimensional and topological characteristics (georeferencing); it can also be modified in its graphical aspect and, finally, if it is equipped with a texture mapping, it preserves the visualization of the assigned mapping within the specific modeling environments.

4 CONCLUSIONS

The study – case was a very good trial of the 3D SIArch – Univaq as it allowed to test the procedure functionalities and to correct some that highlighted some critical points, allowing the integration of the queries, to different levels of scale, of the three-dimensional model with the needs posed by different required levels of use, such as: compilation of compared map – crackings, processing of damage maps caused by previous restoration interventions, etc., as well as the opportunity to have a wide and suitable working space to carry out further research activities.

REFERENCES

Benedetti, B. & Gaiani, M. & Remondino F. 2010. *Modelli digitali 3D in archeologia: il caso di Pompei*. Pisa: Edizioni della Normale.
Brusaporci, S. 2011. Modelli 3D per il progetto di restauro. In Maria Linda Papa (ed.), *Il disegno delle trasformazioni*. Napoli: Clean.

Brusaporci, S. 2010. Sperimentazione di modelli tridimensionali nello studio dell'architettura storica. In Stefano Brusaporci (ed.), *Sistemi Informativi In-tegrati per la tutela la conservazione e la valoriz-zazione del patrimonio architettonico e urbano*. Roma: Gangemi.
Brusaporci, S. & Trizio, I. 2010. Dal rilevamento in-tegrato al SiArch-3D. In Stefano Brusaporci (ed.), *Sistemi Informativi Integrati per la tutela la conser-vazione e la valorizzazione del patrimonio architet-tonico e urbano*. Roma: Gangemi.
Centofanti, M. & Continenza, R. & Brusaporci, S. & Trizio, I. 2011. The Architectural Information System SIArch3D-Univaq for analysis and preservation of architectural heritage. *The international archives of the photogrammetry remote sensing and spatial information sciences* XXXVIII-5/W16.
Centofanti, M. 2010. Della natura del modello architettonico. In Stefano Brusaporci (ed.), *Sistemi In-formativi Integrati per la tutela la conservazione e la valorizzazione del patrimonio architettonico e ur-bano*. Roma: Gangemi.
Centofanti, M. & Continenza, R. & Brusaporci, S. & Ruggieri, G. & Trizio, I. 2008. Il progetto del SIArch – Sistema Informativo per l'Architettura. *Disegnarecon* 2/2008.
Continenza, R. 2010. Il Progetto del sistema informativo architettonico SIArch-Univaq. In S. Brusaporci (eds), *Sistemi informativi integrati per la tutela, la conservazione e la valorizzazione del patrimonio architettonico e urbano*: 22–29. Roma: Gangemi.
De Luca, L. & Bussayarat, C. & Stefani, C. & Véron, P. & Florenzano, M. 2011. A semantic-based platform for the digital analysis of architectural heritage. *Computers & Graphics* vol. 35, n. 2, april 2011, p. 227–241.
Docci, M. & Chiavoni, E. & Filippa, M. (ed.). 2011. *Metodologie integrate per il rilievo, il disegno, la model-lazione dell'architettura e della città*. Roma: Gangemi.
Mingucci, R. & Garagnani, S. 2011. Strumenti digitali per la modellazione dell'architettura. *Disegnarecon* 7/2011.
Trizio, I. 2010. Il SIArch-Univaq della Villa Correr-Dolfin di Porcia (PN). Prospettive di un Gis 3D finalizzato alla catalogazione, al monitoraggio e alla salvaguardia del patrimonio storico e architettonico. In S. Brusaporci (eds), *Sistemi informativi integrati per la tutela, la conser-vazione e la valorizzazione del patrimonio architettonico e urbano*: 30–38. Roma: Gangemi.
Trizio, I. 2009. Indagini Stratigrafiche e Sistemi In-formativi Architettonici: il GIS della chiesa di S. Maria in Valle Porclaneta. *Arqueología de la Ar-quitectura* 6/2009: 91–111.

Computational Modelling of Objects Represented in Images – Di Giamberardino et al. (eds)
© 2012 Taylor & Francis Group, London, ISBN 978-0-415-62134-2

Cosmatesque pavement of Montecassino Abbey. History through geometric analysis

Michela Cigola
DART – Laboratory of Documentation, Analysis, Survey of Architecture and Territory
Department of Civil and Mechanical Engineering, Università di Cassino, Italy

ABSTRACT: The aim of this article is to show how images of various kinds, including photographs, surveying drawings, graphic rendering and all types of representation can support the analysis, study and documentation of historical floor surfaces in general, and in particular that in the abbatial basilica of Montecassino and other churches in its territory, once known as the "Land of Saint Benedict".

1 COSMATESQUE PAVEMENTS

The Cosmatesque school had its beginning in the early 12th century, and was chiefly active in Rome during the Romanesque period. The Cosmati (1) are a typical example of craftsmen whose artistic education and work were often hereditary, handed down in the family. Consequently, preferences for certain types of pattern or specific designs are characteristic of a particular group.

The Cosmati masters' work featured square or rectangular decorated panels set off by ribbons of mosaic wrapped around porphyry disks. The mosaic sections were always interspersed with white marble bands, essential in lending rhythm to the decorative scheme.

The can be no doubt that the main characteristic of a Cosmatesque pavement is the central strip that leads from the church entrance directly to the apse (Fig. 1).

The simplest type of central strip consists of a series of porphyry roundels joined by interlacing bands of mosaic and contrasting white marble in a guilloche pattern. This type became increasingly complicated, with later examples showing a highly complex design. Another common pattern for the central strip is the quincunx (Fig. 2), or in other words a square containing a central roundel surrounded by four other roundels, all connected by ribbons of mosaic and bands of marble. These two types of central strip were often combined to form extremely complex patterns.

After the central decorative motif was established, the entire floor surface was covered with rectangles in geometrical patterns that were usually repeated symmetrically around the longitudinal axis of the central design, with more attention devoted to the general effect of the surface as a whole than to the panel itself (Fig. 3).

2 DATING COSMATESQUE WORK THROUGH GEOMETRIC ANALYSIS

The Cosmati are a typical example of craftsmen whose artistic education and work were often hereditary, handed down in the family. Consequently, preferences

Figure 1.   Rome, Cosmatesque pavement in the Basilica of San Crisogono.

Figure 2.   Castel Sant'Elia, Cosmatesque pavement of the basilica.

Figure 3. Cosmatesque pavement in the Basilica of San Crisogono, panels with geometric patterns.

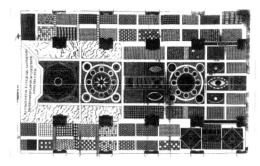

Figure 4. Montecassino, the pre-Cosmatesque pavement of the basilica, in an engraved from 1773.

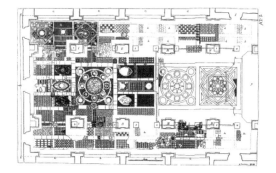

Figure 5. Drawing of the pre-Cosmatesque pavement of the basilica, Pantoni 1972 (6).

for certain types of pattern or specific designs are characteristic of a particular group. For this reason, they are studied by classifying them in families, and work can be grouped under the names of family members in cases where inscriptions indicating the craftsman's name survive, or attributed to the family group on the basis of stylistic characteristics and the choice of decorative patterns.

Dating and classifying Cosmatesque pavements invariably involves a number of problems, both because the work is almost always anonymous, and because of the lack of systematic studies of the topic (2). To attempt to establish a chronology and date Cosmatesque mosaics, it is necessary to bear three aspects in mind: the building, the work contained in it, and the pavement itself. For the first aspect, the building's architectural history must be retraced, while in the second stage, it is necessary to determine whether the church contains work signed by the Cosmati masters, and suppose that they were also responsible for the pavement As the third and final stage of dating, a rigorous stylistic and statistical analysis must be carried out of the geometric patterns and designs of the work in question, comparing it with other pavements that have already been dated and attributed.

Of these stages, that which is fundamental to dating is the systematic analysis of the geometric patterns and designs, which must be carried out in several interdependent steps.

The patterns appearing in Cosmatesque pavements can be divided into groups: patterns found in Roman pavements, patterns found in the pre-Cosmatesque pavements of the abbatial basilica of Montecassino and the other churches in its area (3 common patterns from the formative period of the Cosmati, i.e., the 12th and 13th centuries, and patterns associated with specific families, whose use in certain cases can be regarded as a sort of signature of the craftsman or family group.

A fairly complete examination of Cosmatesque pavements can be achieved on the basis of a comparative analysis of all these factors and by superimposing a series of variables, thus making it possible to assign a date and an attribution for the work.

3 THE MONTECASSINO PAVEMENT

One of the first examples of Cosmatesque work would appear to be the pavement of the abbatial basilica of Montecassino laid between 1066 and 1071.

Unfortunately, a full understanding of the Monecassino pavement is not possible, as it was covered by another marble inlay floor in 1720 (4) and is thus no longer visible. However, we can get an idea of what it was like from several drawings and the surviving parts which were relaid in some of the chapels of the monastery.

The first is an 18th century survey (5) (Fig. 4) executed with great care and attention to detail, representing all of the elements we find in Cosmatesque pavements, of which that at Montecassino is the first example.

The second drawing (Fig. 5) is the work of the monk and engineer Dom Don Angelo Pantoni, who during the bombed-out abbey's postwar reconstruction conducted important archeological investigations that led to the discovery of the pavement that had been believed lost. Much of it has been reconsigned to oblivion in an inaccessible air space beneath the new basilica, but several dozen panels had been removed at later periods to some of the monastery's chapels, including those of San Martino and the Santi Monaci (Fig. 6).

Figure 6. Remains of the pavement of the basilica of Montecassino in the chapels of the Santi Monaci and San Martino.

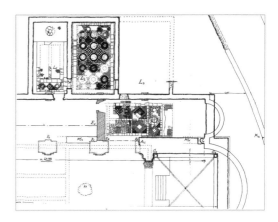

Figure 7. Remains of the pavement of San Vincenzo al Volturno (8).

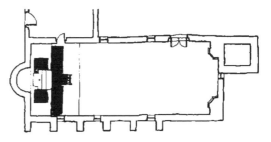

Figure 8. Plan of the church of Santa Maria Maggiore in Sant'Elia Fiumerapido, highlighting the areas where portions of the pavement remain (3).

From an analysis of this graphic evidence, the pavement appears to have featured a composition consisting of an orthogonal grid of light stone bands whose main purpose was to organize the rhythmic arrangement of the floor's compositional sequence, as well as to reflect the elements of the church's architecture and link them to the Cosmati work's panel layout (7).

The entire composition hinges on the large central strip, consisting of panels differing from those of the basilica's aisles both in size and in decorative scheme, which serves as a pathway towards the altar, with sections that grow increasingly complicated as they approach the presbytery. In the side aisles, the floor surface is filled in with rectangular panels that are clearly modeled on the columns of the medieval structure, rather than on the Baroque colonnade that replaced the 11th century architecture.

Of the twenty-one geometric patterns that can be identified with certainty, around half derive more or less directly from Roman and Late Antique examples (Plate I), and some of the latter were then commonly used in the Cosmatesque decorative language (Plate II); two patterns based on circular layouts appear only in this pavement (Plate III). Some of the patterns are also found in the pavements of other churches in the abbey's territory: that of the basilica of San Vincenzo al Volturno and that of the church of Santa Maria Maggiore in Sant'Elia Fiumerapido (Plate IV).

4 OTHER FLOOR MOSAICS IN THE MONTECASSINO AREA

4.1 *The abbey of San Vincenzo al Volturno*

Destroyed in the Second World War and later restored, the basilica of San Vincenzo al Volturno retains two sections of pavement that decorated both the basilica and several areas adjacent to it (Fig. 7). The first is located in the portion of the nave towards the counter-façade, and the remains indicate that the pavement was one of truly unique richness and complexity. In all probability, it can be dated to the first half of the 12th century, as the basilica of San Vincenzo al Volturno was solemnly reconsecrated in 1115 by Pope Paschal II, who had Cosmatesque mosaic pavements laid in many of the churches and basilicas he caused to be built or restored in Rome.

Though only a few vestiges remain of what was presumably an extremely elaborate Cosmatesque surface, we can still see a intricate motif centering around a large *rota*, of a size such that the motif probably occupied the entire width of the nave: before and after the *rota*, we can only guess that there were square panels marking the central strip from the entrance to the altar, but no memory remains of how they were decorated. The five geometric patterns that still fill the spaces

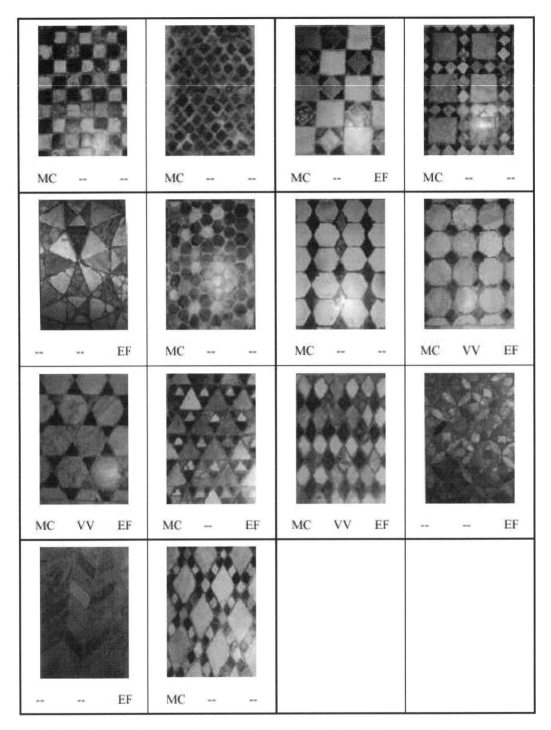

Plate I. Decorative geometric patterns of Roman or Late Antique origin found in the pavements of Montecassino, San Vincenzo al Volturno and Sant'Elia Fiumerapido.

between one *rota* and the other include squares enclosing checkerboard or star motifs, hexagons, octagons and 30 and 60 degree triangular shapes. All of these motifs will be found in the standard repertory used by the Cosmati in pavements dating from subsequent years (Plate II), while three are drawn directly from older examples (Plate I). All of the decorative geometric patterns are derived directly or with slight

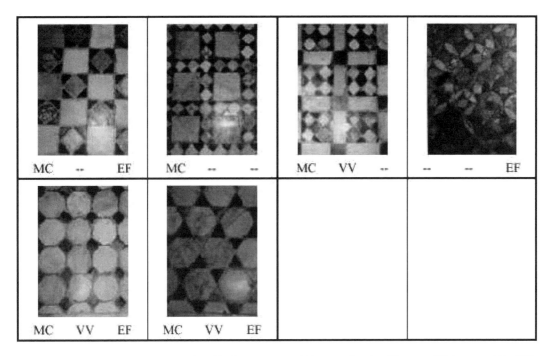

Plate II. Decorative geometric patterns found in the pavements of Montecassino, San Vincenzo al Volturno and Sant'Elia Fiumerapido which became part of the standard repertory of Cosmatesque pavements.

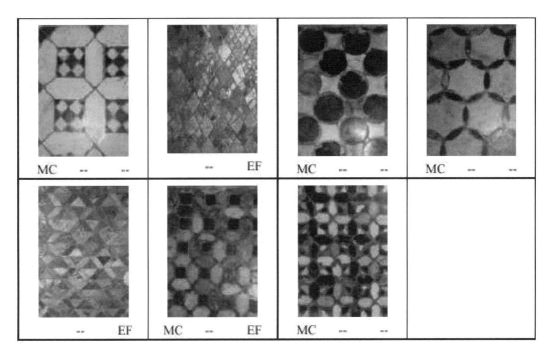

Plate III. Decorative geometric patterns found only in the pavements of Montecassino, San Vincenzo al Volturno and Sant'Elia Fiumerapido.

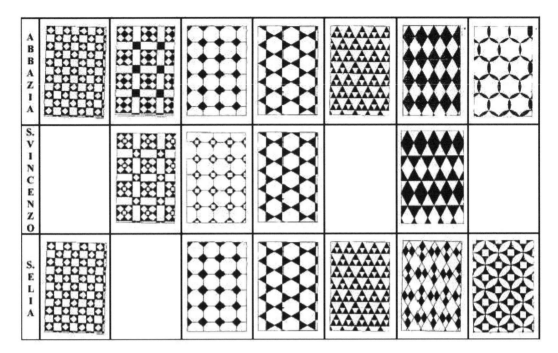

Plate IV. Decorative geometric patterns found in the pavements of Montecassino, San Vincenzo al Volturno and Sant'Elia Fiumerapido.

variations from those of the pavement of Montecassino (Plate IV).

4.2 The church of Santa Maria Maggiore in Sant'Elia Fiumerapido

The building, datable to the very end of the 12th century and restored after the war, is a rectangular hall church with the entrance on the long side.

The approximately 10 square meter section of pavement that is still visible (Fig. 8), postdating that in the abbey of Montecassino by around a century, is made up of rectangles of varying size arranged around the altar and bordered by white marble strips.

The distribution of the panels appears to have no direct relationship with the church's floor plan and, unlike most pavements of this type, puts no particular emphasis on the central strip.

For the most part, the geometric layouts are based on arrays of squares, rectangles, triangles, rhombuses and octagons, with only one of the patterns featuring circular elements.

Of the eleven geometric patterns that can be identified in the floor surface, almost all are part of the Roman and Late Antique decorative repertory (Plate I), while four were to become part of the decorations commonly used by the Cosmati masters throughout the entire period of their activity (Plate II). Another four are completely unknown, appearing for the first and only time in this pavement, and are thus to be ascribed to a more "pre-Cosmatesque" repertory (Plate III). Others are found among the patterns used for the pavement in the abbey of Montecassino (Plate IV),while some belong to the decorative repertory associated with the oldest family of Roman marble workers, that of *Magister Paulus.*

With the evidence available to us, it is risky to attempt to attribute the Sant'Elia pavement to the family of *Magister Paulus*, though the hypothesis is a fascinating one. We can only emphasize that the period in which these craftsmen were active (1108–1170) matches that in which the Sant'Elia pavement was executed, and that the family was active both in Rome and in the rest of Lazio.

5 CONCLUSIONS

A Cosmatesque pavement's period of construction and attribution can be identified with a good degree of approximation by analyzing its total surface and overall design, as well as the geometric layout of the individual panels of which pavements of this kind are made up.

A systematic study of the patterns typical of each family of craftsmen, in fact, is essential in an analysis of this kind, as each Cosmati family had certain particular patterns that would be included in the family's pavements as its hallmark.

A study of this kind, based on a thorough investigation of the stylistic and geometric features through photographs and/or drawings, is an example of how image analysis can contribute effectively to an understanding of historical architecture.

REFERENCES

[1] Cigola M. 1993. Mosaici pavimentali cosmateschi: Segni, disegni e simboli. *Palladio* Nuova serie anno VI n. 1: 101–110.

[2] Benevolo L. & Piazzesi A. & Mancini V. 1954. Una statistica sul repertorio geometrico dei Cosmati. *Quaderni dell'Istituto di Storia dell'Architettura* 5. Bonsignori: Roma. Glass D.F. 1980. Studies on Cosmatesque Pavements. *British Archeological reports Series* 82. Oxford.

[3] Cigola M. 2000. Pavimenti cosmateschi nel territorio cassinese. *Affreschi in Valcomino e nel Cassinate.* Cassino: Edizioni Università di Cassino: 231–247.

[4] Della Marra F. 1775.*Descrizione istorica del monasterio di Monte Cassino per uso e comodo dei forestieri,* 2ˆ edizione. Napoli 1775: 100.

[5] Gattola E. 1733.*Historia Abbatiae Casinensis.* Venezia: tav. VI.

[6] Pantoni A. 1972. Descrizione di Montecassino attraverso i secoli,. *Benedictina* XIX 2. Montecassino: 539–586.

[7] Cigola M. 1997. L'abbazia di Montecassino. Disegni di rilievo e di progetto per la conoscenza e per la memoria. *Disegnare. Idee, immagini* anno VIII n. 14: 43–52

[8] Pantoni A. 1980. Le chiese e gli edifici del Monastero di s. Vincenzo al Volturno. *Miscellanea Cassinese* 10. Montecassino: Tav. II

Computational Modelling of Objects Represented in Images – Di Giamberardino et al. (eds)
© 2012 Taylor & Francis Group, London, ISBN 978-0-415-62134-2

Laser scanner surveys vs low-cost surveys. A methodological comparison

C. Bianchini, F. Borgogni, A. Ippolito, Luca J. Senatore, E. Capiato,
C. Capocefalo & F. Cosentino
SDRA Department of History, Drawings and Restoration of Architecture,
Sapienza – University of Rome, Rome, Italy

ABSTRACT: The new developments in survey practice, until a few years ago represented only by laser scanner technologies, are now oriented towards low-cost solutions. New methodologies which make use of easily accessible instruments and software are beginning to spread in the domain of survey. The present study presents an analysis of the methodology called photomodelling. Based on the solid principles of photogrammetry and supported by new algorithms of multi image-matching, photomodelling makes possible to collect a great quantity of data by means of a simple camera. However, before the new methodology of survey can be considered a valid tool of analysis, it must undergo a series of verifications which will test its potentialities and possible limits. In the research presented here a comparison has been made between data obtained through photomodelling and those collected by suing two kinds of lasers: short range and long range.

1 INTRODUCTION

Architectural survey belongs with the most dynamic fields of research for the simple reason that it is directly linked to technological development. Precisely because of this close relation with the tools applied – be they software or hardware – we observe that survey is undergoing a significant process of revision from the traditional approach to one that is intimately associated with the enormous potentialities of digitizing.

Technological advancement in general and the great digital revolution in particular create innovative research instruments which easily find application in various disciplines of science. In the field of survey and representation, just like in other disciplines of science, the introduction of new informatics technologies has in a brief space of time revolutionized the way of conceiving and approaching the matter. While the fundaments of the discipline remain the same, for some time now we have had to take into account new instruments and new applications thereof.

In the field of survey alone the technological advancement has resulted in the development in two branches of research: the first one, the more tried and tested, deals mainly with data collecting with surveying instruments which can gather a great many detailed data in a shorter time. The other one, more specialized, studies and proposes new low cost solutions by means of which it is possible to obtain similar results using instrument easy to find at low prices.

In both above mentioned branches of research problems are analysed in their specificity and hence application of the results must needs be limited to the respective fields of study. The survey conceived in terms of a process of deep study with the aim of creating a correct representation of the object of study, involves an objective which often transcends the specific technique and the instrument applied in a process which has to be as extensive as possible.

In order to achieve the objective of study it is impossible to concentrate solely on a single specific technique. On the contrary, various techniques, methodologies and instruments have to be integrated to analyse the object in question. Thus, a confrontation between data obtainable with tried and tested instruments with the innovative low cost data acquisition technique seems to be inevitable, especially in photomodelling.

2 PHOTOMODELLING

Photogrammetry belongs with the traditional methodologies of instrumental architectural surveying. The recent technological revolution, especially inasmuch as it concerns digital photography and development of software, has greatly influenced this technique by introducing new photogrammetric instruments to substitute the analogue ones.

Undoubtedly the transition for the digital facilitated the access to a technique difficult to execute but - with the passage of time – opened up a way to apply fascinating alternative techniques. Worthy of mention in this context are new semi-automatic measuring instruments, including the 3D laser scanner. The wide distribution of these hardware instruments clearly indicates that they are appreciated by specialists.

Undoubtedly, the birth of photomodelling some time ago raised an interest in photogrammetry. They share their theoretical foundations, but the unique feature of photomodelling are the quick automatic procedures which can ensure data of quality and quantity like those obtainable through the laser scanner.

Digital photogrammetry provides the necessary information through manual application, following the principal vertexes and edges of the digital image. Photomodelling, on the other hand, identifies homologous points on the photograph disregarding the discontinuities but creating a cloud - of higher or lower density – which describes in the discrete way the surfaces analyzed.

In this case, just like with the laser scanner, one has to decide on the level of detail to be obtained. Then the instrument automatically provides the data with emphasis rather on their quantity than quality. This makes it impossible to determine directly the position of the selected edge that is only identifiable at the intersection of more planes present in the scene. This characteristic feature distinguishes photomodelling from digital photogrammetry, thus justifying the necessity to coin a new term for describing a different methodology that will represent a true and proper innovation in the domain of survey. Another advantageous feature of this technique is the low cost of indispensable instruments (the matter concerns mainly the cost of purchasing a camera) as well as the high velocity of data collecting. These characteristics, especially important for the work undertaken, are complemented by the aspect of an enormous potential. This concerns the possibility of extracting survey models even from photographs made in other historical periods. Understandably, the latter aspect is of extraordinary significance. It is enough to mention the possibility to effectuate reconstructions of historical objects or works of architecture almost completely in ruins by recreating them digitally with the new technique. In this manner survey data can be obtained unreachable through other technologies available on the market today.

The theoretic bases that distinguish this methodology are quite similar to the fundamental principles of photogrammetry. They only introduce algorithms of image-making and multi image-making which allow of automatic correlation of homologous points in more photograms. In this way it is possible to completely bypass the stage of manual restitution – the characteristic feature of digital photogrammetry – and thus achieve automatically the restitution of a great meanly single points. The study of the fundamental theoretical principles coupled with experimental field work made it possible to define a set of problems and potentialities of the methodology in question which, in turn, enabled the researchers to estimate the level of uncertainty of collected data. Special studies have been devoted to the relations between the colours and the geometry of the object, the influence of the lens aberration on the process of data collecting (Fig. 1) as well as to the problems inherent in measuring and calculating the uncertainty of results. Within the extensive panorama of the so called integrated survey - or

Figure 1. Capitals at the archaeological site of the town of Petra in Jordan. An example of three dimensional reconstruction effected with a camera after correcting the aberration of the lens. Images with and without photographic texture.

Figure 2. Virtual reconstruction by means of photomodelling of an item of ceramics found in the tombs of the archaeological site in Custumerium.

the survey typology which involves the application of more instruments to take full advantage of the single potentiality – it seems necessary to make a comparison between the tried and tested methodology (like the one which makes use of the laser scanner) and the new methodology of photomodelling, in order to verify the reliability and compatibility of the data acquired.

These ideas were used in solving problems concerning surveying archaeological material from the necropolis of the ancient town of Crustumerium situated a few kilometres away from Rome. The surveying of small and middle-sized objects was divided into two main stages. This procedure made it possible to compare the survey data obtained through photomodelling with those acquired with two different laser scanners: a short range one for small objects and a long range one for middle-sized artefacts. Among the small objects surveyed were ceramics discovered in one of the numerous tombs excavated in the past, now housed in the laboratories of Soprintendenza Archeologica di Roma, where the restoration works were completed (Fig. 2).

The finds were surveyed twice, each time with a different methodology: that of a short range triangulation laser and the one which applies the compact digital camera. As regards the middle-sized objects the different stages of excavation works in one of the tombs in the Crustumerium necropolis were documented and the data obtained with a digital camera were then compared with those gathered with a long range time-of-flight 3D laser scanner. The results obtained through the comparison of laser scanner survey and photomodelling have brought forth a number of problems of fundamental importance for the discipline of survey and representation. Questions concerning the level of data uncertainty (essential for the evaluation of any new methodology of surveying) as well as those regarding representation, pace of data acquisition and elaboration were all taken into account as elements valuable especially in archaeology.

3 LASER SCANNER SHORT RANGE VS PHOTOMODELLING

The idea was born out of the need to create three dimensional models representing finds through a virtual copy that would be useful for the purpose of study or simply for popularizing the discoveries. In order to make possible a comparative evaluation of the collected data, it was necessary to carry out a series of operations that would make the two sets of results compatible and hence comparable. For the purpose of the present study a comparison was made of the results obtained by comparing virtual models of a small oenochoe (wine jug), 130 x 181 mm. Taking into account the high level of data precision acquired with the triangulation 3D laser scanner (Minolta Vivid 9i), +/- 50μm at the distance of 0.6 m, the results of the comparison can be considered to provide the effective uncertainty of the technique of photomodelling, i.e. the difference between the real object and its virtual reconstruction.

The data collected with both survey techniques had the to be edited. In the case of triangulation 3D laser scanner this made possible to unify and to place correctly each shot. In case of photomodelling – it enables the researcher to obtain a point cloud and realize the surface of mesh interpolation. As far as the photographic model is concerned, which is completely beyond scale, it was necessary to find a factor that would make the model compatible with that obtained with the scanner. Given the complexity and significance of the operation it was decided to take into consideration the medium value of three different factors of scale obtained by comparing homologous measurements of the two models.

This procedure was found necessary because of the difficulties that appeared while searching manually for the homologous point in both models which – realized through different methodologies – presented a particularly different geometric definitions. When the 3D model constructed though the survey of the photographs was redimensioned (the dimensions of the model constructed through the survey of the photographs were adjusted), both models had to be aligned. This was done in to separate stages.

Table 1. Oinochoe n.16 di Crustumerium.

3D Laser Scanner data	
Number of photograms	13
Mesh model after editing	1.451.001 faces
Photography data	
Number of photograms	30
Mesh model after editing	216.520 faces
Calculation of deviation	
Maximum distance positive +10.400 mm	
Maximum distance negative	−10.391 mm
Average distance	−0.623 mm
Average distance positive	+1.029 mm
Average distance negative	−1.091 mm
Standard deviation	1.517 mm

In the first, purely manual, one, four point homologous between the models were found, thus defining a pre-alignment. The other one, realised with Geomagic Studio 10 software, is an automatic procedure able to sharpen the alignment, allowing a better superimposition of data. This automatic procedure makes it also possible to estimate the effective value of the result by calculating the so called average error, i.e. the medium value of the alignment. After the models have been purified, there followed the calculation of deviations, that is an analysis which carries out point evaluation on the whole surface by measuring the distance between the two models. This procedure seeks to distinguish a surface called Reference, a model of comparison, and a surface Test on which the actual evaluation of deviations will be carried out. Then a number of different methods to calculate the distance between surfaces exists. In this case it was decided to make the comparison showing the shortest distance between the reference surface and the test surface without limiting the direction. The results gained through this calculation are the following: Maximum Distance calculated between the two surfaces, distinguishing the positive and the negative on the basis of the direction of the normal to the test surface; Average Distance – calculating the medium value between positive values, negative values and the medium between the two; Standard Deviation which represents the medium from the absolute value of deviations without making distinction between the positive and the negative. The results are briefly presented in Table 1.

The feature that distinguishes the two models is the level of geometric detail of the models taken into consideration. While the data obtained through photography provide a geometric definition of the element surveyed, integrating everything through chromatic values of the texture, the triangulation 3D laser scanner applied in this study ensures a level of data acquisition thanks to which it is possible to evidence even the smallest deformations in the surface of the ceramic

Figure 3. Detail level, 3D model. Scanning data are characterized by numerous polygons which make possible a detailed interpretation of the geometries of the object analyzed. The data are further enriched by chromatic features provided by the acquisition capacities of the scanner.

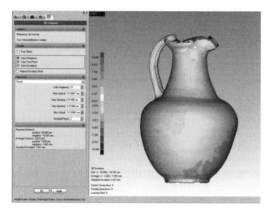

Figure 4. Calculating deviations makes it possible to quantify the shift between the two models. The picture shows some chromatic values of the so called deviation scale. The maximum values are attached to red and sharply blue shades of colour.

artefact. It is thus possible to register, for example, tiny fissures, classical stretching of the material, a clear evidence of the technique used in making the artefact, and even to detect the thickness of layer of the decorative pigment (Fig. 3).The deviation values presented in the table provide a valid tool of evaluation and comparison of the two data acquisition methodologies. A critical analysis of the data reveals that, although the medium values remain modest, there exist deviations which go beyond a millimetre to reach almost a centimetre of distance. To understand fully this problem it is necessary to refer to the graph that presents the calculation to notice the placement of the deviations (Fig. 4). In this case it is visible that the major part of the surface analyzed bears a chromatic variation which refers to deviation values between 0 and 0.5 mm to the progressively rise towards the zone where the collection of data is more and more difficult. The maximum value of the detachment does not correspond to a surface of the model but to the well specified and placed points where the absence of survey data made possible an ideal reconstruction of the part.

4 LASER SCANNER LONG RANGE VS PHOTOMODELLING

A survey of an archaeological object was selected to make a valid comparison between the long range laser scanner technology and photomodelling. In this case it was the documentation of an excavation campaign of a tomb discovered at the necropolis of the ancient town of Crustumerium.

The subject was researched precisely with the view to verify and test the limits and potentialities of the different technologies, even in restricted spaces particularly difficult to reach. The archaeological campaign of the particular tomb lasted five successive days, 60 hours of work including strictly archaeological procedures and those indispensable for documentation. It was possible to document with the tools provided by new technologies the most significant stages of the excavation, parallel to the classical graphic documentation carried out by the archaeologists. Before comparing the data obtained with the time-of-flight 3D laser scanner (Scanstation 2 of Leica) and those provided by photographs taken with a commercial compact camera of 12.0 megapixels, it was necessary to make a series of adjustments that would make the models obtained with the two different techniques of data acquisition compatible and hence – comparable. Taking advantage of the software and hardware at the disposal, it was possible to proceed with the comparison and obtain statistical values that could serve for further considerations. In total, 10 different stages of the excavation were analyzed. Each of them was surveyed with two methodologies and the results were compared by calculating deviations between the models.

An analysis of the values obtained by comparing models realized on the basis of the 3D long range laser scanner data and those derived from the photographs reveals some aspects on particular interest. The maximum and medium deviation values vary clearly in the different stages of excavation, reflecting the difficulties in data collecting during the surveying campaign. However, the values referring to medium deviation are included within the range of 3 and 9 mm. These data, though they represent a medium value in numerous measurements taken, offer a good point of departure for research on the reliability of three dimensional models constructed by means of photomodelling. Moreover, as has been illustrated in the preceding study, the undercuts and zones accessible with difficulty have brought out peaks of maximum deviation which influence the calculation of the medium, even though they are clearly described. For example, in the analysis of one of the ten stages of the excavation surveyed through the graph of deviations, even though the standard deviation being that of 0.6 cm, the maximum values concern specific zones where there number of surveyed points is insufficient. Despite all this, if we take into consideration only a part of the model, excluding the parts where the survey is incomplete, the comparison of the alignment operations with the deviation calculations yields the value of standard deviation on the level of 0.2 cm, i.e. three times lower than in the preceding case.

5 CONCLUSIONS

The comparison of methodologies has thrown a new light on the uncertainty level inherent in the technique of photomodelling as a survey instrument. Though it has not been possible to achieve a constant level of uncertainty for all the cases analyzed, various limits and potentialities have been diagnosed. Uncertainty and the detail level that can be achieved are directly connected to the intrinsic characteristics of the camera as well as to the external conditions of the object and its surroundings. The results of the study make it possible to conclude that at the moment the technique of photomodelling – verified and controlled by adequate procedures – can be considered a valid survey instrument for the realization of 2D and 3D graphic documentation.

REFERENCES

Bianchini, C., Borgogni, F., Capocefalo, C., Cappelletti, A., Cosentino, F., Ippolito, A. & Senatore. L. J. 2011. Il colore digitale per la caratterizzazione del rilievo archeologico. In *Colore e colorimetria contributi multidisciplinari vol. VII/A:* 481–488. Roma: Maggioli Editore.

Bianchini, C., Docci, M. & Ippolito, A. 2007. From plans to model: the unbuilt Vatican Basilica. In *From survey to the project:heritage & historical town centres:* 150–155. Firenze: Edifin.

Bianchini, C., Docci, M. & Ippolito, A. 2011. Contributi per una teoria del rilevamento architettonico. In *Disegnare Idee Immagini vol. 42:* 34–41. Roma: Gangemi Editore.

Bianchini, C. & Ippolito, A. 2008. Processo di elaborazione dei dati per la realizzazione del modello virtuale del plastico di Antonio da Sangallo. In *Actas del Congreso Internacional de Expresion Grafica Arquitectonica* : 73–82. Madrid: Instituto Juan de Herrera.

Bianchini, C. & Ippolito, A. 2009. Il Modello di Michelangelo per la Grande Cupola Vaticana: documentazione e prime analisi. In *Nuove ricerche sulla Gran Cupola del Tempio Vaticano:* 278–289. Roma: Edizioni PRE Progetti.

Bianchini, C. & Ippolito, A. 2010. Survey, modelling and analysis of vaulted structures and domes: towards a systematic approach. In *13° Congreso Internacional de Expresion Grafica Arquitectonica vol.1:* 95–100. Valencia: Universidad Politecnica de Valencia.

Borgogni, F. 2011. Rilievo integrato alla grande scala: il caso di Mérida. In *Metodologie integrate per il rilievo, il disegno, la modellazione dell'architettura e della città – PRIN 2007:* 41–46. Roma: Gangemi Editore.

Borgogni, F. & Ippolito, A. 2011a. I modelli 3D nei rilievi di architettura. In *Metodologie integrate per il rilievo, il disegno, la modellazione dell'architettura e della città:* 71–77. Roma: Gangemi Editore.

Borgogni, F. & Ippolito, A. 2011b. La costruzione dei modelli per il rilievo archeologico. In *Metodologie integrate per il rilievo, il disegno, la modellazione dell'architettura e della città:* 27–34. Roma: Gangemi Editore.

Borgogni, F., Ippolito, A. & Pizzo, A. 2010. Digital mediation from discrete model to archaeological model: the Janus Arch. In *FUSION of Cultures abstracts of the XXXVIII annual conference on Computer Applications and Quantitative Methods in Archaelogy: 317–320.* Granada: CAA

Capocefalo C. 2011. L'architettura illusoria e l'esperienza di Vallombrosa. In *Materia e Geometria:* 245–246- Firenze: Alinea Editrice.

Docci, M. & Ippolito, A. 2008. Il Ruolo del Disegno nel Progetto dell'Architettura Digitale. In *Rappresentazione e formazione tra ricerca e didattica – Studi e Ricerche sul Disegno dell'Architettura e dell'Ambiente vol. IV:* 195–216. Roma: Aracne Editrice.

Ippolito, A. 2005. Elaborazione tridimensionale e analisi del modello della cupola di Hagia Sophia. In *Metodologie innovative integrate per il rilevamento dell'architettura e dell'ambiente:* 42–47. Roma: Gangemi Editore.

Ippolito, A. 2007a. Dal reale al virtuale e ritorno: metodologie per la prototipazione. In *Informatica e fondamenti scientifici della rappresentazione – Strumenti del Dottorato di Ricerca in Scienze della Rappresentazione e del Rilievo:* 315–324. Roma: Gangemi Editore.

Ippolito, A. 2007b. Dalla nuvola di punti alla superficie. Analisi e problematiche. In *Metodi e tecniche integrate di rilevamento per la costruzione e fruizioni di modelli virtuali 3D dell'architettura e della città:* 32–43. Roma: Gangemi Editore.

Ippolito, A. 2007c. Studio proporzionale dell'ordine nel modello ligneo per il Nuovo San Pietro. *Disegnare Idee Immagini vol. 34:* 36–49. Roma: Gangemi Editore.

Ippolito, A. 2008. Percezione e mediazione digitale: il modello tridimensionale. In *Modelli di studio a scala locale della Carta del Rischio del Patrimonio Culturale ed Ambientale della Regione Siciliana:* 129–136.Palermo: Regione Siciliana Editore.

Ippolito, A. 2009a. Il Modello Digitale della Cupola di San Pietro. In *Nuove ricerche sulla Gran Cupola del Tempio Vaticano:* 206–231. Roma: Edizioni PRE Progetti.

Ippolito, A. 2009b. La modellazione delle superfici murarie del Tempio del Divo Claudio a Roma. In *Disegnare Idee Immagini vol. 38:* 76–85. Roma: Gangemi Editore.

Ippolito, A., 2011. Verso un approccio metodologico per il rilievo delle cupole. In *Le Cupole Murarie: Storia, Analisi, Intervento:* 343–360. Roma: Edizioni PRE Progetti.

Lambers, K., Eisenbeiss, H., Sauerbier, M., Kupferschmidt, D., Gaisecker, T., Sotoodeh, S., Hanusch, T., 2007. Combining photogrammetry and laser scanning for the recording and modelling of the Late Intermediate Period site of Pinchago Alto, Palpa, Peru. In *Journal of Archaeological Science 34(10):* 1702–1712.

Stumpfel, J., Tchou, C., Yun, N., Martinez, P., Hawkins, T., Jones, A., Emerson, B., Debevec, P., 2003. Digital reunification of the Parthenon and its sculptures. *Proceedings of Vast:* 41–50.

Computational Modelling of Objects Represented in Images – Di Giamberardino et al. (eds)
© 2012 Taylor & Francis Group, London, ISBN 978-0-415-62134-2

From survey to representation. Operation guidelines

C. Bianchini, F. Borgogni, A. Ippolito, Luca J. Senatore, E. Capiato,
C. Capocefalo & F. Cosentino
SDRA Department of History, Drawings and Restoration of Architecture,
Sapienza – University of Rome, Rome, Italy

ABSTRACT: Archaeological surveying undoubtedly belongs with the fields of research most elaborated in this domain of study. This is inevitably linked both to the scale of the problems tackled and – in more strictly 'physical' terms – to the fact that this operation is closely and indissolubly related to the continuous evolution of the archaeological excavation. The expectations related to archaeological surveying – to a much higher degree than to other types thereof – are much more complex and larger, ranging from a documentation of various materials found and/or of their traces to a diagnosing and differentiating the type of work to be done; from various types of finishing jobs to the documentation of different stages of destructive excavations. All this usually created ambiguous situations for the technicians in the stage of elaborating collected data and to those who did not participate in the work itself – in the stage of communication.

1 INTRODUCTION

The study and analysis of archaeological sites in the optics of the survey and representation have always lent themselves particularly well to development and control. In the past the field worker was often confronted by problems of various kinds rooted – on the one hand – in the instruments used in the stage of data acquisition, and linked to the necessity of communicating the registered data through various graphic models. In traditional surveying the tools guaranteed a high degree of metric reliability but collecting data involved enormous expenditure of money and time. This inevitably reflected back on the elaboration and communication that followed. In fact the amount of information that representation was demanded to transmit made elaborations almost impossible to use outside specialized applications to the point when the mutually unequivocal relation between the real object and its rendition through graphic models became ambiguous. With the passage of time, thanks to technological development, new instruments were invented which simplified the surveying operation through semi-automatic processes, at the same time securing a low level of uncertainty. Worthy special mention in this context is the 3D laser scanner. Thanks to its versatility it constitutes a new basis for data collecting. Consequently, it is the main source of problems involved in this field of research. Data acquired by laser scanner endow three-dimensional graphic models – static and/or dynamic – with a high level of precision which enable to clarify simplify and make the object of study immediately recognizable.

For a few years now the research conducted at the Department of History, Sketching and Restoration of Architecture has been focussed on establishing the so called 'operative protocol' that would define the methodology of surveying and representing of archaeological artefacts making full use of the above mentioned technologies. The point is to obtain results as rigorous as those achieved through traditional techniques and procedures.

The operative guidelines are dynamic and constantly evolving. They have been developed and verified during numerous surveying campaigns. They were conceived – and this point must be emphasised – not to make the process of surveying mechanical. Rather the idea behind it was to provide aid and a critical point of reference to the technician by optimizing the whole process. These reflections were born and developed also thanks to the collaboration of specialized archaeologists who made it possible to fully understand the aims and expectations connected with each surveying campaign. Among the numerous operations on local and international scale there are some surveying campaigns whose contribution to the definition of the guidelines was especially important.

On the territorial/urban scale there were the memorable surveying campaigns of some Roma theatres at important archaeological sites, such as the Roman Theatres at Petra and Jarash in Jordan, the Roman theatre and amphitheatre at the town of Mérida in Spain or the Roman theatre in Taormina (Fig. 1).

On the small scale surveys were conducted of the Tempio di Claudio and Janus Arch in Rome. Here, in many cases the surveying concerned objects of modest dimensions which required more detailed data for prototyping and therefore, for physically reproducing the object of interest (Figs 2–3). As already emphasized, in all these campaigns the synergy between the architects

Figure 1.   The Roman theatre and amphitheatre at the town of Mérida in Spain, Point Cloud.

Figure 3.   Arco di Giano in Rome, 3D texturized model.

Figure 2.   Arco di Giano in Rome, 3D Model.

and the archaeologists played an indispensable role. Indeed, it enabled them to integrate the correctness of acquisition, restitution restoration and communication characteristic of architects-surveyors with the intimate knowledge of the object of study – characteristic of archaeologists. These seem to be the optimal conditions to achieve the final aim of surveying.

2   OPERATING GUIDELINES

Just like with architectural surveying, or even more so with its archaeological variety, it is indispensable first to recognize and collect initial historical-iconographical documentation which will recreate a synthetic picture of the artefact in question. Each piece of information that helps to understand the object better is carefully studied in order to obtain a set of qualitative and quantitative data. An adequate project of the survey will be prepared on the basis of this general picture. Included therein will be a programme of successive works to be conducted, the choice of one or more surveying methods, of the time when the works will be conducted in situ or the time of restitution.

The first procedure seems to be absolutely necessary. It has to be done if the objectives are to be

achieved at a fixed period of time, reducing to the minimum the possibility of unexpected difficulties and obstacles. The operative guidelines presented in this study take this promise as the point of departure and is then articulated in two successive, interrelated paths: 1. 3D surveying, done by acquiring data by applying as best possible various methodologies and instruments , and 2. 3D survey carried out by restoring data through various representative models.

3   3D SURVEYING

The 3D surveying is the stage of collecting metric and historical-cultural data characteristic of the object under analysis. It is only the first step in the work and may seem a simple technical-mechanical operation. In point of fact, however, it of high theoretical value as it applies the correct methodology selected on the basis of the characteristic features of the studied artefact and the scale of the restoration of the final graphic model. This is only attainable when one is intimately familiar with all the instruments available. It is important to underline that this knowledge can by no means be limited to the ability to operate the apparatus. It also involves the knowledge of all the modalities and capacities of elaborating the collected data as well as to integrate various modalities in order to make the best use of the potentialities of each instrument used, short range laser scanners and of photomodelling for smaller artefacts. Each methodology seeks to provide a higher level of knowledge while constituting at the same time a better support at the stage of 2D and 3D restoration of the object under study. The tools in question are mainly the following:

– Long range laser scanner
– Short range laser scanner
– Topographic apparatus
– Digital photography

The instruments mentioned above are not all 'non-contact' ones, a feature undoubtedly preferable in the domain of archaeology because there is no danger of damaging the state of the analyzed artefacts. In view

Figure 4. The Roman theatres' point cloud at Jarash in Jordania.

of the fact that all the methodologies mentioned here can be mutually integrated, it has already become a consolidated and verified practice to link long range laser scanning with topographic measuring and a photographic campaign – in the case of particularly large artefacts; and to use the short range scanner coupled with photomodelling – when smaller object are analyzed.

In the former case the scanner has to be placed in more points in order to be able to survey the whole object. Thus it restores a series of partial point clouds which have to be re-connected. This operation can be carried out automatically by the operating software of the scanner which recognizes the targets preventively positioned near the artefact. The integration visibly more accurate ensures a better control over the 'registration' of partial point clouds and, consequently, lowers uncertainty levels. Procedures of this type were applied for surveying the Roman theatres at Petra and Jarash in Jordan (Fig. 4) and at the archaeological site at the town of Mérida in Spain. It is important to emphasize at this point that 3D laser scanners cannot yet do the so called process of discretization, i.e. cannot recognize the essential elements that describe an architectural element or – the more so – an archaeological element. Their function is limited to collecting a determined number of information per unit of surface area, which is a pure mechanical operation. Understandably then the information obtained with laser scanning has to be interpreted by an expert. Otherwise it will remain but a numerical model and as such must be considered a point of departure for further elaboration and not a final result. This is rooted in the fact that laser scanning provides such an enormous quantity of data that they may become a source of ambiguity unless elaborated by an expert armed with the knowledge of informatics and technical-theoretical competences which together enable him to interpret architecture correctly.

4 3D SURVEY

The next stage, that of 3D survey, is made up of operations which together make it possible to transform objective data of surveying into graphic elaborations by integrating different methodologies. It is at this stage in fact that the integration of data takes place

in the process of by superimposing and verifying the results obtained by different tools, making up for the shortcomings of each of them and optimizing the final elaboration. As has already been stressed, the elaboration of data and, consequently, the realization of graphic models must be carried out by experts who being able to operate specific software but – above all – having a certain degree of sensibility and awareness of the matter in question, can take correct 'interpretative choices'. The part that was done in advance in traditional survey, i.e. the discretization of the forms to be surveyed by choosing of significant points and geometries, is now transferred to the stage of restoration in the computer. The operative guidelines therefore, follow traditional preparations. They find in models of 'geometric' character a passage necessary to obtain the restoration of 'architectonic' type. Geometrical models discretize the forms in their elementary geometry and distinguish the proportions of the constituent parts thus being able to provide indications as to the relations and the unity of the measure. Architectonic models, on the other hand, provide a clear characteristics of the surfaces which ensures a direct mutually univocal relation between the real object and its representation. When details obtained through point clouds are unclear, which is often the case, recourse is made to photographic corrections or short range scanning.

Archaeological survey, moreover, makes necessary a multi-level communication that includes more types of information (from the state of degradation to the nature of the materials) and complements, the so called 'objective' survey with the survey of the 'critical' type, which takes into account specific and distinguishing aspects of the artefact. It is worth mentioning in this context the cognitive examination of the Tempio del Divo Claudio in Rome, a part of the Firb 2003 Project which aimed at inserting the data into computerized system of archive (Archaeological Informative System). The latter was divided into more levels in order to process various information according to different levels of deepening.

Critical survey, therefore, provides documentation which can serve more purposes, including those of monitoring and restoring the artefact. That is why a group of researches within the team conduct a series of experiments on the usefulness of reflectivity, a datum acquired together with spatial coordinates of each surveyed point by 3D laser scanner and linked to the quality of the surface bombarded by the beam of light. Such supplementary information made it possible in many cases to distinguish certain forms of degradation connected with humidity as well as to distinguish materials of different characteristic features (properties). However, due to the variations connected with various factors, the reliability of these information still await proper evaluation.

5 MODELS

Basically there are two types of models: the 'traditional', two dimensional ones, like plans, prospectuses and sections and three dimensional ones. The new technologies and techniques make it possible to obtain three dimensional models, static or dynamic, in a short period of time and with a high degree of reliability. Moreover, the 'neutral' representation of the artefact – two- or three dimensional – can now be integrated with the photo-realistic texture acquired directly by the tool used (3D laser scanner) or through a photographic campaign. The link between the picture and the surface is always controlled by semi-automatic processes, whose reliability is defined by the uncertainty level. The model obtained in this way, although highly realistic, does not have to simulate reality but simply to describe it by synthesizing numerous information contained in two dimensional models into a three dimensional unity.

The 3D texturized model allows of a full interpretation of the object and makes it possible to enlarge the perception of classical elaborations through interactive or immersive experiences able to simulate the navigation around and inside the object surveyed. Thanks to prototyping apparatus it is even possible to reproduce a physical copy of high communicative value.

With the above mentioned model as the point of departure and always in collaboration with archaeologists, it is possible to elaborate reconstructions of the artefact which show its initial state and help visualize them 'as they ought to have been'. They can then become the instrument indispensable for further study and learning.

Obviously, problems of representation scale must not be neglected with all models examined. The scale of representation should absolutely cohere with the final result to do away with communication problems. Therefore, each type of model is controlled and regulated by operative guidelines which determines various representation modalities and codifies their graphics and symbology, elements necessary for the correct communication of datum surveyed.

The final product of the process is then a virtualization of the real-life object under analysis represented through a series of models that describe the object in all its parts on pre-established scale and can collocate it virtually in a precise point of time, making it a scientific document of reference in the future.

6 CONCLUSIONS

The establishment of the operative guidelines for archaeological surveying seems to be indispensable to unify and regulate the procedures of data collecting, elaborating and restoration that will make the final result scientific in character, i.e. more objective and correct. The comparison with other experiences in this field, the experiments with different methodologies on large scale objects rendered it possible to establish a non rigid modus operandi which lends itself to adaptation in specific, concrete cases or to the exigencies of surveying, maintaining at the same the versatile character, of the survey.

REFERENCES

Bianchini, C., Borgogni, F., Capocefalo, C., Cappelletti, A., Cosentino, F., Ippolito, A. & Senatore. L. J. 2011. Il colore digitale per la caratterizzazione del rilievo archeologico. In *Colore e colorimetria contributi multidisciplinari vol. VII/A*: 481– 488. Roma: Maggioli Editore.

Bianchini, C., Docci, M. & Ippolito, A. 2007. From plans to model: the unbuilt Vatican Basilica. In *From survey to the project:heritage & historical town centres:*150–155. Firenze: Edifin.

Bianchini, C., Docci, M. & Ippolito, A. 2011. Contributi per una teoria del rilevamento architettonico. In *Disegnare Idee Immagini vol. 42*: 34–41. Roma: Gangemi Editore.

Bianchini, C. & Ippolito, A. 2008. Processo di elaborazione dei dati per la realizzazione del modello virtuale del plastico di Antonio da Sangallo. In *Actas del Congreso Internacional de Expresion Grafica Arquitectonica* : 73–82. Madrid: Instituto Juan de Herrera.

Bianchini, C. & Ippolito, A. 2009. Il Modello di Michelangelo per la Grande Cupola Vaticana: documentazione e prime analisi. In *Nuove ricerche sulla Gran Cupola del Tempio Vaticano:* 278–304. Roma: Edizioni PRE Progetti.

Bianchini, C. & Ippolito, A. 2010. Survey, modelling and analysis of vaulted structures and domes: towards a systematic approach. In *13° Congreso Internacional de Expresion Grafica Arquitectonica vol.1*: 95–100. Valencia: Universidad Politecnica de Valencia.

Borgogni, F. 2011. Rilievo integrato alla grande scala: il caso di Mérida. In *Metodologie integrate per il rilievo, il disegno, la modellazione dell'architettura e della città – PRIN 2007:* 41–46. Roma: Gangemi Editore.

Borgogni, F. & Ippolito, A. 2011a. I modelli 3D nei rilievi di architettura. In *Metodologie integrate per il rilievo, il disegno, la modellazione dell'architettura e della città:* 71–77. Roma: Gangemi Editore.

Borgogni, F. & Ippolito, A. 2011b. La costruzione dei modelli per il rilievo archeologico. In *Metodologie integrate per il rilievo, il disegno, la modellazione dell'architettura e della città:* 27–34. Roma: Gangemi Editore.

Borgogni, F., Ippolito, A. & Pizzo, A. 2010. Digital mediation from discrete model to archaeological model: the Janus Arch. In *FUSION of Cultures abstracts of the XXXVIII annual conference on Computer Applications and Quantitative Methods in Archaelogy: 317–320.* Granada: CAA

Capocefalo C. 2011. L'architettura illusoria e l'esperienza di Vallombrosa. In *Materia e Geometria*: 245–246- Firenze: Alinea Editrice.

Docci, M. & Ippolito, A. 2008. Il Ruolo del Disegno nel Progetto dell'Architettura Digitale. In *Rappresentazione e formazione tra ricerca e didattica – Studi e Ricerche sul Disegno dell'Architettura e dell'Ambiente vol. IV*: 195–216. Roma: Aracne Editrice.

Ippolito, A. 2005. Elaborazione tridimensionale e analisi del modello della cupola di Hagia Sophia. In *Metodologie innovative integrate per il rilevamento dell'architettura e dell'ambiente*: 42–47. Roma: Gangemi Editore.

Ippolito, A. 2007a. Dal reale al virtuale e ritorno: metodologie per la prototipazione. In *Informatica e fondamenti scientifici della rappresentazione – Strumenti del Dottorato di Ricerca in Scienze della Rappresentazione e del Rilievo*: 315-324. Roma: Gangemi Editore.

Ippolito, A. 2007b. Dalla nuvola di punti alla superficie. Analisi e problematiche. In *Metodi e tecniche integrate di rilevamento per la costruzione e fruizioni di modelli virtuali 3D dell'architettura e della città*: 32–43. Roma: Gangemi Editore.

Ippolito, A. 2007c. Studio proporzionale dell'ordine nel modello ligneo per il Nuovo San Pietro. *Disegnare Idee Immagini vol. 34*: 36–49. Roma: Gangemi Editore.

Ippolito, A. 2008. Percezione e mediazione digitale: il modello tridimensionale. In *Modelli di studio a scala locale della Carta del Rischio del Patrimonio Culturale ed Ambientale della Regione Siciliana:* 129–136.Palermo: Regione Siciliana Editore.

Ippolito, A. 2009a. Il Modello Digitale della Cupola di San Pietro. In *Nuove ricerche sulla Gran Cupola del Tempio Vaticano*: 206–231. Roma: Edizioni PRE Progetti.

Ippolito, A. 2009b. La modellazione delle superfici murarie del Tempio del Divo Claudio a Roma. In *Disegnare Idee Immagini vol. 38*: 76–85. Roma: Gangemi Editore.

Ippolito, A., 2011. Verso un approccio metodologico per il rilievo delle cupole. In *Le Cupole Murarie: Storia, Analisi, Intervento*: 343–360. Roma: Edizioni PRE Progetti.

Lambers, K., Eisenbeiss, H., Sauerbier, M., Kupferschmidt, D., Gaisecker, T., Sotoodeh, S., Hanusch, T., 2007. Combining photogrammetry and laser scanning for the recording and modelling of the Late Intermediate Period site of Pinchago Alto, Palpa, Peru. *In Journal of Archaeological Science 34(10)*: 1702–1712.

Stumpfel, J., Tchou, C., Yun, N., Martinez, P., Hawkins, T., Jones, A., Emerson, B., Debevec, P., 2003. Digital reunification of the Parthenon and its sculptures. *Proceedings of Vast*: 41–50.

Computational Modelling of Objects Represented in Images – Di Giamberardino et al. (eds)
© 2012 Taylor & Francis Group, London, ISBN 978-0-415-62134-2

Images for analyzing the architectural heritage.
Life drawings of the city of Venice

E.Chiavoni

Dipartimento di Storia, Disegno e Restauro dell'Architettura. Università Sapienza di Roma, Italy

ABSTRACT: For architects, analysing architectural heritage means performing all those operations that are useful for getting to know it, in order to gather together all the data and information needed for its promotion, regeneration or, quite simply, maintenance. These operations take place through drawing and surveying the urban context first of all, in order to understand the place, then the building as a whole, and lastly its most important details. It is precisely through drawing that a journey of subjective awareness is made through the space, and images are created which clarify not only the individual features but also the relationships that link them together. Drawing, therefore, not only represents a technique for illustrating, but also and especially a fundamental element for reading architectural heritage. The contrast between man's creation and the untamed environment, that is, the contraposition of built and natural environments in a context such as the city of Venice, provides much food for thought and inspires graphic experimentation. In this city, not only can the beauty of individual pieces of architecture be observed, but the images of the buildings themselves as they are reflected in the Lagoon can also be enjoyed. These images constantly change dimension, shape and even their colours, depending on how much light there is on a particular day and what time of day it is.It is this marriage between the vertical façades of the buildings and their horizontal reflections in the Lagoon that has been sought in freehand drawings from life, experimenting with different traditional drawing techniques such as pencil, coloured pencil, various coloured pens and watercolours.

1 INTRODUCTION

This article clarifies how interpreting the city through drawing can make a major scientific contribution to our knowledge of place.

Producing analytical drawings not only helps specialists understand the relationships and connections between architecture and its setting, but can also suggest ways of renewing and valorizing urban space.

Architecture is initially investigated in terms of what appears from outside and then, later, in its internals, its most intimate essence. The first approach always hinges on sensory perception, or in other words, on the relationship established between the observer and the city.

Each architect's direct experience in looking at urban places must be transmitted with clarity, though with an awareness that no representation will ever be able to transcribe all of reality.

Conveying the city's image activates processes of understanding and selection, and drawing, with its many representational tools and techniques, provides a subjective interpretation filtered by the draftsman's personality.

The conceptual process entails analyzing, selecting, interpreting, recording and communicating what we see, while the graphic process proceeds experimentally in seeking the techniques of drawing and

elaborating images whose potential is best suited to its aims. The gesture of drawing, whether it involves digital or traditional methods, continues to be an irreplaceable tool of critical research, and one which is fundamental in grasping and recording the complexity of urban spaces.

The drawing becomes a powerful theoretical and conceptual medium in which the autobiographical dimension, the sign as a form of communication, and the range of disciplinary models underlying it all merge together, transforming the drawing into an instrument of unheard-of potential. (quoted from Luca Molinari, Drawing is dead. Long live drawing, in Domus 956, March 2012)

2 EXPERIMENTATION

Drawing the city also means gaining a familiarity with its places, attempting a spatial interpretation of architecture, establishing a relationship between ourselves and the empty spaces, the green areas, investigating the light and the colours of the urban landscape.

This is what always happens, and it is also what the drawings included with this text are intended to convey: freehand drawings, as are needed to quickly transcribe the unstoppable flow of thought as it comes to grips with urban complexity, a universal language that uses a graphic alphabet made up of signs, signs that are apparently simple but that, joined together, are able to yield astonishing results.

Line drawing with pencils, coloured pencils, pens or markers, a job almost as painstaking as etching, that line after line defines the shapes, the differing directions of the signs that build up a dense mesh that identifies and personalizes the drawing's varied structures.

Patient, painstaking, but fast, because fixing an image on the sheet of paper must not take too long, as otherwise there is a risk of losing our concentration on the object we are analyzing; we must break down the most important aspects into their discrete parts, select and fix on paper the synthesis of signs that make the image of what we have before us recognizable. The time it takes to produce a drawing also takes its rhythm from the light, and in drawing from life it is important to get the shadows down immediately, before they change, as well as – in the case of a coloured drawing – the chromatic tones that change along with the changing light.

A drawing is never finished, it has a beginning but not an end, it could always be continued, almost indefinitely, until the paper itself wears out. At times, however, the expressiveness of certain drawings that are little more than sketches is far more intense than that which can be achieved with the richest and most elaborate representations. This study also addresses these aspects, and the different graphic productions are intended specifically to demonstrate the effectiveness of certain representations as compared with others, without, however, losing sight of our goal of

delineating the expression of a city as complex as Venice.

Why Venice? Not only is it a challenge to draw the elegant *palazzi*, the domes, the bell towers, the marble facades and brick houses; above all, it is disorienting to be faced with a double image of these places: one image lying on the vertical plane and another positioned horizontally, one seen straight on and one reflected on a moving surface, as unstable as that of the lagoon.

3 THE DRAWING

In approaching the study of a city, an architect cannot avoid analyzing its planimetric configuration, and Venice, in this regard, exhibits a shape that resembles a fish floating in the waters of the lagoon. To highlight this aspect, a plan of the city was drawn freehand with coloured pencils (Figure 1) to introduce the task of understanding and for use as a reference for all of the other representations. The various districts or *sestieri* – Cannaregio, Castello, Dorsoduro, Santa Croce,

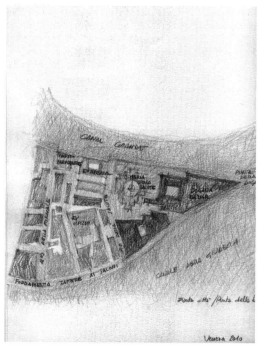

Figure 3. Plan of the city; detail of the Punta della Dogana area.

Figure 1. Freehand coloured pencil drawing of the plan of Venice.

Figure 2. Freehand black and white pencil drawing of the Venice lagoon, viewed from the gardens of the Biennale.

San Marco and San Polo – are shown in different colours and the lagoon in the background is translated into combinations of greens and blues.

This representation is a synthesis resulting from the study of a number of maps of the city.

A second drawing done from life and from a distance, specifically from the entrance to the gardens of the Biennale (Figure 2), shows the city's skyline and the direct relationship of its buildings, both monumental and private, with the lagoon. Though the drawing is extremely immediate, a quick pencil sketch, it achieves great clarity in conveying the harmony that links the built areas with the surrounding green, and the whole with the waters of the lagoon.

The next productions are drawings of a specific central area of Venice, the Punta della Dogana area where the historic customs house, after decades of abandonment, has been restored to public use as the city's new contemporary art center.

Designed by the architect Tadao Ando, this museum is of major cultural significance internationally because of its contribution to the spread of knowledge concerning the art of our time.

Another of the reasons that make this area particularly evocative is its close proximity to the basilica of Santa Maria della Salute, whose size and imposing dome make it one of Venice's chief architectural landmarks.

Figure 4.   Freehand drawing of Punta della Dogana with the dome of the church of Santa Maria della Salute.

Figure 5.   Drawing of the Punta della Dogana area by night.

Figure 6.   Freehand perspective drawing of Punta della Dogana, from life.

467

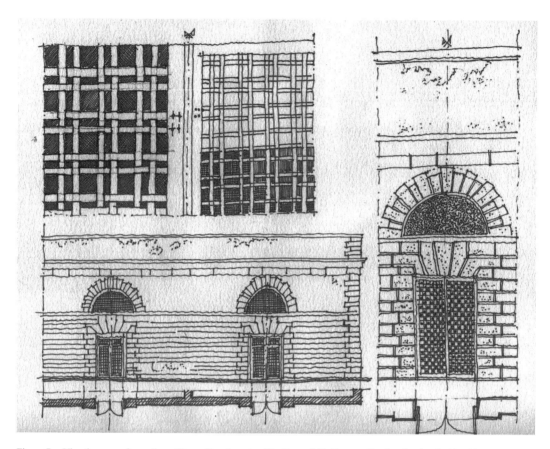

Figure 7. Visual survey of a portion of the wall and a gate of the Punta della Dogana. Freehand red and white felt pen drawing.

The first of these drawings is of the Punta della Dogana between the Grand Canal and the Giudecca Canal, and illustrates the buildings surrounding the Punta della Dogana museum in a detail of the city plan (Figure 3). It is a freehand line drawing executed with coloured pencils, enriched with the names of the buildings.

The next drawing, from life, is of the San Giorgio area, showing the Punta della Dogana and the Chiesa della Salute (Figure 4); it is a freehand line drawing, in pencil, and the composition emphasizes the relative proportions of the basilica and the customs house, with the horizontal spread of the museum building contrasting with the almost aspirational verticality of the church.

The architectural features of both buildings are also well depicted in this drawing: the splendid gateway to the customs house and the majestic, but at the same time elegant dome of the basilica rising on its massive drum. A few simple, rapid and almost uncertain pencil strokes hint at the water of the lagoon and its slight, continual movement. The right hand side of the drawing is penciled in very faintly to reflect how far away the buildings are.

Everything changes in a night view: the area is the same, but depicted from a more distant vantage point. The line drawing executed in colored pencil (Figure 5)

Figure 8. Freehand coloured pencil drawing of Venice.

is effective in showing the sharp contrast between the compact blue of the sky and the dark waters of the lagoon, its blue filtered by the reflections and glimmerings of artificial light spreading over the buildings and creating bright bands across the water. The outlines of the buildings are blurred, almost merging with the dense atmosphere and color of the night sky. The skyline is indistinct, and many of the towers and spires can barely be made out. The reflections of the buildings in the water are represented by dark, irregular patches of uncertain shape.

The work of interpreting urban spaces continues with drawings that approach their subject more closely,

Figure 9.   Freehand black and white pen drawing of Venice.

for a better view of the shapes, the dimensions, the proportions, passing from the general to the particular and bringing certain buildings or parts of them into sharper focus. A view of one of the facades of the customs house (Figure 6) drawn freehand in coloured pencil on-site is shown with a detail of the horizontal section to illustrate the inclined ground plan. The drawing highlights the building's shapes, geometries and masonry work, while the sky in the background also sets off the jutting watchtower at the end of the building that projects it, as if weightlessly, towards the lagoon.

The final interpretation shown is a true visual survey (Figure 7), several detailed line drawings done freehand with a red fine-tipped marker showing plan views, elevations and sections in rigorous proportion. The drawings represent a portion of the museum wall with its iron gate.

The drawings shown in the text are only a few of those produced for a study of the interpretation of the architectural heritage entitled *"Surveying and representation of color in architecture and its environment"* conducted in 2009 at the Ateneo Federato delle Scienze Umane, Arti e Ambiente. An analysis of drawing as an aid to understanding was carried out in several selected cities, Venice being one of the most significant.

REFERENCES

1-Chiavoni E. Matita e acquarello per catturare l'immagine urbana In: a cura di P.Albisinni, E.Chiavoni, L.De Carlo. *Verso un "disegno integrato", la tradizione del disegno nell'immagine digitale.* vol. 2°, p. 17-23, ROMA: Gangemi Editore 2010, ISBN/ISSN: 978-88-492-1971-5

2-Chiavoni E., Fabbri L.,Porfiri F., Tacchi G.L. Lettura dell'architettura contemporanea attraverso il disegno e la rappresentazione cromatica manuale.Il Museo Nazionale delle Arti del XXI secolo di Roma, anno 2011,ISBN 88-387-6042-x

3- Chiavoni E. The representation of colour and light in architecture through watercolours. In: *Colour & Light in Architecture* Università IUAV , Venezia, 11-12 novembre 2010, Verona: Knemesi, vol. 1°, p. 501–505, ISBN/ISSN: 978-88-96370-04-9

4-Chiavoni E. Matera, struttura, forma e colore, in *Disegnare, Idee, Immagini*, AnnoXXI , n.41 / 2010, pp 52-65, ISSN IT.1123-9247

5-Luca Molinari, Drawing is dead. Long live drawing, in *Domus* 956, March 2012, pag.68/73

6-*Acqua & Architettura. Rappresentazioni*, a cura di C. Mezzetti, M. Unali, edizioni Kappa, Roma 2011, ISBN 978-88-6514-078-9

Author index